Plant Systematics and Evolution Supplementum 3

A Biosystematic Study of the African and Madagascan Rubiaceae-Anthospermeae

Christian Puff

Springer-Verlag Wien New York

Dr. CHRISTIAN PUFF

Institut für Botanik der Universität
Wien, Austria

© 1986 by Springer-Verlag/Wien
Softcover reprint of the hardcover 1st edition 1986

With 126 Figures

Library of Congress Cataloging-in-Publication Data. Puff, Christian, 1949–   . A biosystematic study of the African and Madagascan Rubiaceae-Anthospermeae. (Plant
systematics and evolution. Supplementum; 3) Bibliography: p. Includes index. 1. Rubiaceae.
2. Botany – Africa. 3. Botany – Madagascar. I. Title. II. Series.
QK495.R85P84.   1986.   583′. 52.   86-20269.

ISSN 0172-6668
ISBN-13:978-3-7091-8853-8   e-ISBN-13:978-3-7091-8851-4
DOI: 10.1007/978-3-7091-8851-4

Foreword

Biosystematic studies on the *Rubiaceae* have a long tradition at the Institute of Botany in Vienna. Within this family the *Anthospermeae,* and especially its African and Madagascan members, are of particular interest because of several aspects in their evolution: 1) Perfection of anemophily within an otherwise nearly exclusively zoophilous family; 2) transitions from hermaphrodity to polygamy and finally dioecy; 3) differentiation from large and long-lived shrubs to short-lived herbs; 4) adaptive radiation from humid to seasonally dry, fire-exposed and xeric habitats. However, morphological diversity linked to sexual differentiation, modificatory plasticity, and eco-geographical polymorphism have for a long time hampered our understanding of the relationships among these African *Anthospermeae.*

Thus, it was imperative to put special emphasis on field observations and to carry out a variety of experiments with cultivated plants in addition to the analysis of an enormous herbarium material. The author, for this reason, carried out extensive field work, often under very adverse conditions, and covered most African countries from Ethiopia to Southern Africa and twice visited Madagascar. In this way a multitude of data was accumulated on the group in respect to germination and growth form, vegetative and reproductive morphology, anatomy and biology, embryology, karyology, crossing relationships, phytochemistry, distribution and ecology, etc.

The resulting biosystematic monograph of the African *Anthospermeae* covers 4 genera with 61 species. In regard to the breadth of its multidisciplinary approach and the depth of its documentation it ranks among the most comprehensive and advanced treatments available for African Angiosperm groups. It is hoped that it will not only serve as a model case for the evolutionary differentiation of such groups in tropical Africa, but also stimulate more interest in biosystematic and monographic studies on the flora of this continent.

Vienna, July 1986   F. EHRENDORFER

Contents

Abstract 1

A. Introduction 3

B. Materials and Methods 6

C. General Part 8

  1. Geographical Distribution and Habitats 8
  2. Habit and Growth Form Analyses 19
    2.1. Large to Medium-sized Shrubs 19
    2.2. Dwarf Shrubs 23
      2.2.1. Short-lived Shrubs ("Woody Herbs") 28
    2.3. Subshrubs 28
    2.4. Perennial Herbs 29
    2.5. Influence of External Factors on Habit and Growth Form . . 31
      2.5.1. Fire 31
      2.5.2. Browsing 33
  3. Stems 34
    3.1. Indumentum, Cortex, Cork 34
    3.2. Wood, Node and Petiole Anatomy 34
  4. Leaves 37
    4.1. Arrangement; Leafy Short Shoots 37
    4.2. Stipules 37
    4.3. Blades and Petioles 40
    4.4. Age and Persistence 42
    4.5. Anatomy 42
      4.5.1. Epidermis [incl. Trichomes and Stomata] 42
      4.5.2. Chlorenchyma 59
      4.5.3. Bundle Sheaths and Bundle Sheath Extensions . . . 63
  5. Inflorescences 65
  6. Flowers 72
    6.1. Pedicels 72
    6.2. Calyx 73
    6.3. Corolla 73
    6.4. Androecium 79
    6.5. Gynoecium 80
  7. Carpophores, Fruits and Seeds 81
    7.1. Carpophores 81
    7.2. Fruits and Seeds of *Anthospermum, Galopina* and *Nenax* . . 82
      7.2.1. Indumentum 82
      7.2.2. Persistent Calyx Lobes 83
      7.2.3. Fruits Dehiscing into Two Mericarps 84
      7.2.4. Indehiscent Fruits 89
      7.2.5. Fruit (Mericarp) Anatomy 89
      7.2.6. Seeds 94

7.3. Fruits and Seeds of *Carpacoce* 97
    7.3.1. Morphology. 97
    7.3.2. Anatomy. 98
    7.3.3. Seeds 103
8. Embryology 103
9. Karyology 107
10. Pollen 122
11. Phytochemical Data. 125
  11.1. Leaf Flavonoids (by R. D. WILSON). 125
  11.2. Iridoid Glycosides 130
12. Reproductive Biology 131
  12.1 Anemophily 131
  12.2. Sex Distributions and Sex Ratios 134
    12.2.1. Sex Forms of Individual Plants 134
    12.2.2. Sex Ratios and Sex Distributions in Populations . . 141
    12.2.3. Sex Distributions and Anemophily, and the Evolution
      of Dioecy 146
  12.3. Secondary Sex Characters and Sexual Dimorphism . . . 148
  12.4. Self-Compatibility, Auto- and Geitonogamy; Parthenocarpic
    Fruits and the Question of Apomixis 152
  12.5. Hybridization 156
  12.6. Diaspore Dispersal 161
  12.7. Germination Data 163
  12.8. Regeneration after Fire. 169

D. Systematic Part. 173

Tribe *Anthospermeae*, Subtribe *Anthosperminae* 173
  Key to the African and Madagascan Genera. 173
Anthospermum 174
  Regional Keys 178
  A. Key to Taxa Occurring in Tropical Africa 178
  B. Key to Taxa Occurring in the Flora Zambesiaca and the Flora of
  Angola Area 180
  C. Key to Taxa Occurring in the Flora of Southern Africa Area
  Excluding the SW Cape Floristic Region 181
  D. Key to Taxa Occurring in the SW Cape Floristic Region . . . 185
  E. Key to Taxa Occurring in Madagascar. 189
  The Taxa. 190
Nenax 393
  Key to Species and Subspecies 396
  The Taxa. 398
Galopina 427
  Key to Species 429
  The Species 430
Carpacoce. 446
  Key to Species and Subspecies 448
  The Taxa. 449
Taxa to be Excluded 465

E. Phylogenetic Relationships and Evolutionary Aspects 467

1. Relationships Within *Anthospermum* 467
  1.1. Introductory Remarks 467
  1.2. The *Anthospermum usambarense* Group. 475

Contents   IX

1.3. The *Anthospermum spathulatum* Group 479
1.4. The *Anthospermum whyteanum* Group 481
1.5. The *Anthospermum ternatum* Group 483
1.6. The *Anthospermum herbaceum* Group 485
1.7. The *Anthospermum pumilum* Group 488
1.8. Isolated SW Cape Taxa, Including General Comments on Taxa
Occurring in the SW Cape Floristic Region 492
1.9. Isolated Madagascan Taxa 497
1.10. Concluding Remarks 498
2. Affinities Between *Nenax* and *Anthospermum*, and Relationships
Within *Nenax* 501
3. Relationships Within *Galopina* and its Affinities to Other *Anthospermeae* 506
4. The Isolated Position of *Carpacoce* and Relationships Within the
Genus 508
5. Relationships Between *Anthosperminae* and Extra-African *Anthospermeae* 512
6. Thoughts on the Evolution of the *Anthospermeae* 513

Acknowledgements 520
References 521
Index of Taxa (in part D.) 533

Abstract

Detailed investigations of the genera *Anthospermum* (39 species; widely distributed in Africa S of the Sahara and Madagascar), *Nenax* (11 species; S Africa), *Galopina* (4 species; SE Africa) and *Carpacoce* (7 species; S Africa: SW Cape Floristic Region) are presented which are based on extensive field work, the observation of plants grown in the greenhouse and the study of fixed and dried material in the laboratory. These investigations include growth form analyses and notes on the influence of fire and browsing on habit and growth form; detailed data on stem and leaf, flower, fruit and seed morphology and anatomy and surface micromorphology; analyses of inflorescences; embryological and karyological data and pollen measurements; phytochemical information. Numerous data on the reproductive biology are presented: The secondary anemophily of the genera is described in detail; information on the sex forms of individual plants, and on the sex distributions and sex ratios within populations and species is given; the evolution of dioecy, and the secondary sex characters and the sex dimorphism of dioecious taxa is discussed. Evidence for the presence of self-compatibility and of auto- and geitonogamy of taxa, gathered in greenhouse experiments, is presented; the occurrence of parthenocarpic fruits is reported; no conclusive evidence for the occasionally presumed presence of apomicts within *Anthospermum* could be found. Experimental proof for the formation of viable hybrids between taxa, as well as a list of putative hybrids is given. Information on the dispersal of the diaspores, their morphological nature, and data on their germination are included. The differential regeneration behavior of taxa after a field fire is pointed out (plants killed by fire and regenerating from seed, vs. plants resprouting from the base).

In the Systematic Part (D.), keys to the African and Madagascan genera of the *Anthospermeae* and to the species (and subspecies) of each of the genera are given; for the large genus *Anthospermum*, five "regional" keys are produced to facilitate an easier identification of species and infraspecific taxa. Each genus is briefly circumscribed. For each species (subspecies, variety), a detailed description is included which takes into consideration the possible sexual dimorphism of a given taxon and growth form differences due to external factors (fire, browsing, etc.). Chromosome number, average pollen diameters, detailed habitat notes, a short description of the distribution, a distribution map, a complete listing of the specimens studied and, where appropriate, critical comments (on difficulties in the delimitation to other taxa, on variability, on relationships, etc.) are also given.

Anthospermum has the highest concentration of taxa in the SW Cape Floristic Region, followed by SE Africa and Madagascar (several new species endemic to the island are described); most of the tropical taxa are confined to montane areas and often have wide but (very) disjunct distributions.

Nenax occurs in the SW Cape Floristic Region, but several species also extend or are confined to the drier areas to the N and NE of that region.

Galopina is centered in SE Africa; only one, essentially afromontane species extends to the highlands of E Zimbabwe and neighbouring areas of Mozambique and to Malawi in the N and to the Cape Region in the SW.

Carpacoce is confined to the SW Cape Floristic Region; the genus comprises both species with a wide distribution in that area and species with very restricted or specialized distributions.

In Part E., the evolutionary relationships within each of the genera, the close affinities between *Anthospermum* and *Nenax*, the presumed alliances of *Galopina*, and the isolated position of the genus *Carpacoce* are discussed. In the speculations on the relationships between the genera of the *Anthosperminae*, the Macaronesian genus *Phyllis* is also included (which, in the previous chapters, had been largely ignored except for the occasional brief reference as it has already been studied in detail by another author). Finally, the relationships between the *Anthosperminae* and the extra-African subtribes *Coprosminae* and *Operculariinae* are also dealt with, and some ideas as to the evolution and the time-space development of the entire *Anthospermeae* are discussed.

A. Introduction

The family *Rubiaceae* is characterized by having predominantly zoophilous flowers. Like other large families of angiosperms (the *Compositae* or the *Rosaceae,* for example) it has, however, also produced a number of (secondarily) anemophilous representatives. In the *Rubiaceae,* it is the apparently natural tribe *Anthospermeae* and the monogeneric tribe *Theligoneae* which are comprised of exclusively wind-pollinated taxa (Puff 1982a)[1].

In connection with the shift from zoophily to anemophily the genera in question acquired a number of remarkable features which make a detailed investigation particularly rewarding (cf. Puff 1982a). The tribe *Anthospermeae*, widely distributed and centered in subtropical and tropical regions of the Old and New World (cf. map, Fig. 126), consists of approximately a dozen genera and comprises several hundred species. A detailed study covering all of these genera would be a quite impossible task and it was, therefore, decided to concentrate on the *Anthosperminae*. This subtribe, centered in Africa and Madagascar, is comprised of the genera *Anthospermum, Nenax, Galopina, Carpacoce* and of – as the "odd man out" – the Macaronesian genus *Phyllis*. The present study deals with all these genera except *Phyllis*. Frequently, however, reference to the latter, to other extra-African *Anthospermeae* and sometimes also to genera of the *Paederieae*, a tribe closely allied to the *Anthospermeae* (cf. Puff 1982a), will be made for comparative reasons.

While *Phyllis* is well known and has been the subject of recent investigations (Mendoza-Heuer 1972, 1977), the remainder of the *Anthosperminae* is not:

For S Africa, the most recent comprehensive treatment of the group is that of Sonder (1865) in Flora Capensis. Some definite improvements, primarily concerning taxa endemic to the SW Cape Floristic Region, were published by Salter (1937); these were also included in the Flora of the Cape Peninsula (Adamson & Salter 1950). Other than that, only few

---

[1] Note added September 1985: The Australian monogeneric tribe *Durringtonieae* (Henderson & Guymer 1985) also belongs here. While there is no doubt about the anemophily of *Durringtonia*, I am not fully convinced that it should be placed into a tribe of its own. More detailed investigations would be desireable.

additional publications are available, and frequently they are only comprised of descriptions of new taxa.

Southwest African (Namibian) taxa are dealt with in the Prodromus einer Flora von Südwestafrika (LAUNERT & ROESSLER 1966). *Anthospermum* is included in the second edition of the Flora of West Tropical Africa (HEPPER 1963), and VERDCOURT (1976) published a treatment of *Anthospermum* for the Flora of Tropical East Africa [No recent treatments for other Flora areas are available – cf. surveys by BAMPS (1976, 1981)]. All the above Flora treatments, however, have one principal shortcoming: the relevant genera were only studied for the region(s) concerned so that relationships to taxa outside the respective Flora areas remained undetected. For this reason, these accounts remain rather unsatisfactory – a fact that VERDCOURT (1976: 352), for example, was well aware of.

After some preliminary studies it transpired that an investigation of herbarium material alone would not yield satisfactory results. The need for the investigation and observation of living plants and populations, in particular for the study and understanding of reproduction biological features, became obvious.

Subsequently, a rigorous programme of field work was initiated which was carried out over a period of ten years (1976–1985). At first, it was planned to concentrate on S Africa because all genera of the *Anthosperminae* except for *Anthospermum* and *Phyllis* are endemic to the subcontinent and because the SW Cape Floristic Region, in which the *Rubiaceae* as a whole are poorly represented, has the by far highest concentration of taxa. It was not long, however, until interesting links between S African and tropical African taxa and obvious relationships between African mainland and Madagascan taxa prompted me to extend field studies to tropical Africa and Madagascar as well. In addition to numerous trips within S Africa, several expeditions were undertaken to Ethiopia, tropical East Africa (Kenya, Tanzania and Uganda), S-Central Africa (Malawi, Zambia, Zimbabwe) and Madagascar.

These expeditions not only yielded (1) valuable reproduction and population biological data, information on (2) the morphological variability within populations and of taxa as a whole, on (3) growth forms, on (4) the influence of environmental factors (burning and browsing in particular) and on (5) the ecology of the majority of taxa but also provided an opportunity to (6) obtain fixations for subsequent detailed morphological, anatomical and karyological work. Moreover, viable seeds of many taxa could be gathered for the subsequent cultivation of plants in the greenhouse. Thus the data presented in the following chapters stem to a large extent from field work, from field fixations and/or from plants grown from seeds which were collected in the wild.

Several specialists have been kind enough to carry out additional, detailed investigations on the plant material provided by me: Dr. E. ROBBRECHT (National Botanical Garden of Belgium, Meise) contributed palynological studies (ROBBRECHT 1982, 1985), Dr. J. KOEK-NOORMAN (University of Utrecht, Holland) dealt, in conjunction with me, with wood anatomical aspects (KOEK-NOORMAN & PUFF 1983), and Dr. R. D. WILSON (Chemistry Division, DSIR, Petone, New Zealand) investigated leaf flavonoids; his data are included in the present publication.

B. Material and Methods

The present investigations are based on the study of living material, plant parts fixed in the field (voucher specimens in herbarium WU) or in the greenhouse and herbarium specimens of the following herbaria and institutions: B, BM*, BOL*, BR*, BSB, COI, E, ETH*, FI, G, GH*, GOET, GRA, J*, K*, KMG, LISC, LISU, LMA*, LMU*, LY, M, MAL*, MO, NBG*, NH*, NHT*, NU*, P, POZ, PRE*, PRU*, PUC, S, SAAS, SAM*, SRGH*, STE*, TAN*, UPS*, US, W*, WAG, WIND*, WU and Z (abbreviations after Holmgren, Keuken & Schofield 1981; and asterisk denotes that the herbarium or institution was also visited).

Growth form analyses and data on sex distributions were largely obtained from plants raised from seed in the greenhouses of the Botanical Garden of the University of Vienna (H.B.V.) and from field observations. Cultivated plants were, furthermore, used for various reproduction biological experiments (pollen dispersal; crosses; proof of autogamy, etc.).

Plant parts for various morphological and anatomical investigations were fixed in FPA.

Plant material for microtome sectioning was embedded in wax after passing through the conventional series of media, and sections were stained either with safranin only or with both safranin and fast green.

For node anatomical studies, free-hand sections were made which were stained with phloroglucinol and HCl.

Both dried and fresh samples, the latter first fixed in the field or greenhouse in FPA or in 2.5% glutaraldehyde in cacodylate buffered sucrose (pH 7.4) and then critical-point dried, were studied under the SEM after coating with gold-palladium. SEM-graphs of sectioned leaves or fruits exclusively originated from samples sectioned with a sharp razor blade prior to critical-point drying. SEM studies were undertaken at both the EM Unit, University of the Witwatersrand, Johannesburg, and the Institute of Botany, University of Vienna, using a Joel JSM-T20 Scanning Microscope and an ISI 60 Microscope respectively.

Flower buds, shoot and root tips for chromosome counts were fixed in alcohol : acid acid (3 : 1). For most chromosome number determinations the acetocarmine squash technique was employed. In some cases, Feulgen preparations were made; for these, root tip fixations obtained from germinating seedlings were pretreated with 0.05% colchicine for half an hour at room temperature.

Average pollen diameters were obtained from measurements of fully developed pollen grains (three-nucleate and well stained with acetocarmine). The average size of 30 grains per plant was determined; the largest diameter of a grain was measured. Obviously deformed and "dwarf" or "giant" grains were ignored. Measurements were undertaken immediately after preparation since pollen grain diameters tend to increase their size in liquid with time.

Collections from the Flora of Southern Africa area are listed following the grid reference system as outlined by Edwards & Leistner (1971). Each dot on a distribution map represents any collection(s) made in a single 15′ × 15′ square; the precise location of any one collection within the square was ignored. Grid references of localities were largely extracted from Southern African Place Names (LEISTNER & MORRIS 1976); spelling of place names also follows this publication. In case of ambiguous spellings, both names are given. Farm localities (usually not included in LEISTNER & MORRIS 1976) have, as far as possible, been traced on official South African 1 : 50 000 or 1 : 25 000 maps; the spelling of Farm names has been adapted to that used on the official maps. Place names no longer in use are normally given in brackets following their modern equivalent; the Zoo-Historical Gazetteer (SKEAD 1973) has proved to be most valuable in establishing synonomies. Changes of Province borders in connection with the establishment of black "homelands" or "independent republics" (Transkei, Bophuthatswana, Qwaqwa, etc.) are ignored; localities from homelands are grouped with those Provinces under which they are listed in Southern African Place Names. In these independent homelands or states, place names have often been changed; synonyms are given as far as possible. Most of the information on various important collectors and their collecting localities and expeditions in S Africa was drawn from GUNN & CODD (1981). Additional information was, in some cases, obtained from HUTCHINSON (1946), JESSOP (1964; on SCHLECHTER's travels), McKAY (1943; on BURCHELL's trek), and DRÈGE (1843, 1847 a & b, 1848; localities of ZEYHER's collections; correlations of ECKLON & ZEYHER and DRÈGE collections).

Collections from outside S Africa are either arranged according to the geographical divisions of the respective Flora (Flora Zambesiaca, Flora of Tropical East Africa, etc.), or a simplified grid reference system is employed for the citations (see PUFF 1979 a for details). For the tracing of collecting localities the geographical and botanical gazetteers listed by HEPPER & BAMPS (1971) were consulted. Numerous grid references and details on collecting sites from Angola were kindly supplied by Prof. MENDES (LISC; personal communication). Important additional information on various collectors and their expeditions, finally, was obtained from papers published in the section "Histoire de l'exploration botanique de l'Afrique ..." of the "Comptes rendus de la IVe réunion plénière de l'A.E.T.F.A.T." (FERNANDES 1962).

A note on PUFF voucher specimens and collections: The collection numbers incorporate the collecting date — i.e., 780208-2/3 = February 8th, 1978; third plant of the second locality visited on that day.

Note that "S", "NW", etc. stands for *either* "South" *or* "southern", "North-West" *or* "north-western", etc.

Relevant literature published until mid 1984 is incorporated and discussed in the present publication. Not all more recent publications could be considered in detail, although reference is often made to articles published after the general deadline by means of footnotes.

C. General Part

1. Geographical Distribution and Habitats

Anthospermum is the most widely distributed genus, occurring over most of Africa S of the Sahara with the exception of the W African and Zaire basin lowland rain forest region (Guineo-Congolian Region) and the adjacent N areas ("Guinea-Congolia/Sudania Regional Transition Zone" and Sudanian Region sensu WHITE 1983 or Sudanian Domain of the Sudano-Zambezian Region sensu WHITE 1965, 1971). The most widely distributed and essentially afromontane species, *A. herbaceum*, also extends across the Red Sea to the mountains of the Yemen Arab Republic and adjacent parts of Saudi Arabia. Several species occur in and are endemic to Madagascar (map, Fig. 1 a).

The majority of tropical African taxa are essentially afromontane. They occur in all seven "Regional Mountain Systems" distinguished by WHITE (1978, map, Fig. 1); some Madagascan species (e.g., *A. emirnense*) occur in vegetation ± comparable and similar to montane vegetation on the African mainland. Some of the afromontane species are restricted to a few [e.g., *A. usambarense:* from the Imatong Mts. in the Sudan (cf. JACKSON 1956) S to the Nyika Plateau (Zambia/Malawi)] or to a single "Mountain System" (e.g., *A. vallicola:* Chimanimani Mts.). Others are characterized by wide, but markedly disjunct distribution ranges (e.g., *A. welwitschii:* from tropical E Africa to the Transvaal and also occurring in the highlands of Angola, or *A. pachyrrhizum,* occurring in the highlands of Ethiopia, and again roughly 1 500 km W on the Jebel Marra massif and a N satellite – cf. WICKENS 1976, QUÉZEL 1969). Often these species also occur in outliers of the Afromontane Region proper ("distant satellite populations of afromontane species" sensu WHITE 1978). In the W African Mountain System, there is only one species, *A. asperuloides,* which is restricted to the high mountains of the Cameroun and Fernando Po.

The Zambezian Region (sensu WHITE 1983; = Zambezian Domain of the Sudano-Zambezian Region sensu WHITE 1965, 1971) has only a few (but widely distributed) species (e.g., *A. ternatum:* from Angola to E Africa).

The highest concentration of *Anthospermum* species is found in S Africa. Some species occurring in the summer rain fall area (i.e., S Africa excluding the SW Cape Floristic Region) are widely distributed (e.g., *A. hispidulum:* from the Transvaal to the Transkei). Amongst these, both species with "closed" distribution ranges (i.e., without noteworthy disjunctions, e.g., *A. littoreum:* in the Tongaland-Pondoland Regional Mosaic of the Indian Ocean Coastal Belt sensu WERGER 1978 a, map,

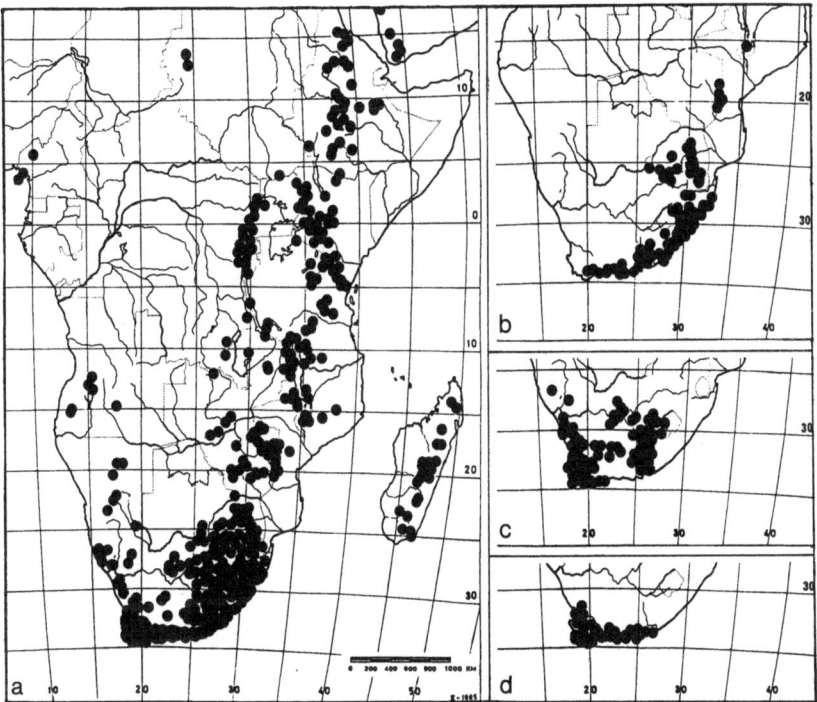

Fig. 1. Distribution of African and Madagascan *Rubiaceae-Anthospermeae. a Anthospermum, b Galopina, c Nenax, d Carpacoce*

Fig. 12, and WHITE 1983) and species with markedly disjunct distribution ranges (e.g., *A. monticola:* an example for a "Drakensberg-Sneeuberg species" — cf. HILLIARD 1978) occur. One species, *A. pumilum,* extends well beyond S Africa into the Zambezian Region (to S Tanzania). None of these taxa, however, reach the SW Cape Floristic Region. Other taxa have ± narrow distribution ranges (e.g., *A. streyi,* confined to sandstone areas on the Natal S Coast, or *A. basuticum,* endemic to the Drakensberg high plateau).

Numerous species are endemic to the SW Cape Floristic Region as delimited by GOLDBLATT (1978) and BOND & GOLDBLATT (1984). Some taxa are ± ubiquitous in the SW Cape (e.g., *A. aethiopicum*), but more

2*

commonly distribution ranges are less extensive (e.g., *A. bergianum*, restricted to the W and SW part of the Cape Region). A few taxa are endemic to very small areas (e.g., *A. ericifolium*).

A number of taxa radiate out from the SW Cape Region into the surrounding dry, karroid areas of the Karoo-Namib Region sensu WERGER (1978 b) (e.g., *A. spathulatum* subsp. *spathulatum:* to Namaqualand). Other taxa are confined to these dry, surrounding areas [e.g., *A. comptonii:* in the Western Cape/Karoo Domain sensu WERGER (1978 b), or *A. dregei* subsp. *dregei* in Namaqualand and the S Namib desert (i.e., Western Cape/Namaqualand Domain of the Karoo-Namib Region sensu WERGER 1978 b)].

Madagascar houses – in addition to species occurring in montane areas similar to those in mainland Africa (see above) – species which are confined to the Domaine du Centre (e.g., *A. palustre*) and the dry, western slopes of the Central Plateau ("Secteur des pentes occidentales" – "fôret basse sclérophylle" of HUMBERT 1965; e.g., *A. isaloense* – widely distributed, or *A. longisepalum* – narrowly endemic to the Isalo Massif). All Madagascan taxa are, however, confined to the East Malagasy Region sensu KOECHLIN (1972) and WHITE (1983) (= "Région du vent" or "Region orientale" – cf. PERRIER DE LA BÂTHIE 1936, HUMBERT 1955 and 1965, RAUH 1973).

Galopina, Nenax and *Carpacoce* are confined to S Africa (maps, Figs. 1 b–d) and their distribution ranges largely overlap with those of *Anthospermum.*

One of the four species of *Galopina, G. circaeoides,* is a very widely distributed, ± afromontane species which virtually "follows" afromontane forest and scrub to areas outside the Afromontane Region proper (i.e., it also occurs in "extensions" of afromontane vegetation; see WHITE 1978). Its range, which is disjunct in the North (Zomba Plateau/Malawi – highlands of E Zimbabwe – N Transvaal), extends well into the SW Cape Floristic Region; there it occurs in wooded kloofs which often contain at least some typical afromontane woody species. The other species of *Galopina* have more limited distributions and occur from the E Transvaal to the E Cape Province.

Nenax has, on the one hand, species which are endemic to the SW Cape Floristic Region (e.g., *N. acerosa*) or which are centered in the SW Cape and also extend into the hot, dry surrounding areas (e.g., *N. divaricata*). On the other hand, there are species which are largely confined to areas outside the SW Cape Floristic Region – i.e., to various parts of the Karoo-Namib Region sensu WERGER (1978 b) (e.g., *N. cinerea:* extending into the S Namib desert, or *N. arenicola:* centred in the W Coast

Fig. 2. Growth forms and habitats of *Anthospermum*. *a A. welwitschii* (PUFF 790211-2/3), population at the edge of afromontane forest on the summit of Piesangkop (S Africa, E Transvaal); the tallest plants are ca 3 m high. *b A. vallicola* (PUFF 790125-1/1), dominant in "heath zone" on the summit plateau of Mt. Inyangani (Zimbabwe); the plant in the foreground is ca 2 m tall

"Strandveld" sensu ACOCKS 1975). The mostly widely distributed species, *N. microphylla,* is centered in the E part of the Karoo Domain sensu WERGER (1978 b) and the adjacent "Kalahari-Highveld Regional Transition Zone" sensu WHITE (1983) and extends as far E as the central O.F.S. and the W, lower-lying parts of Lesotho.

Carpacoce is essentially a genus endemic to the SW Cape Floristic Region (as delimited by GOLDBLATT 1978 and BOND & GOLDBLATT 1984; map, Fig. 1 d). Only the widely distributed *C. vaginellata* barely extends beyond the E limits of the SW Cape Region. Except for the before mentioned species, the taxa have relatively narrow (e.g., *C. heteromorpha*) or very restricted distribution ranges (e.g., *C. gigantea,* known from a single locality only). Some taxa (e.g., *C. curvifolia*) show interesting disjunctions in their distribution ranges.

Habitats: Afromontane, shrubby species of *Anthospermum* frequently occur at the edge of forest (e.g., *A. welwitschii,* Fig. 2 a) or in scrub forest; such species may be (locally) dominant in the "ericaceous" zone or belt (e.g., *A. vallicola,* Fig. 2 b). *A. usambarense* is the only species that occasionally reaches the afroalpine belt sensu HEDBERG (1951); populations are known to occur up to an altitude of 4 100 m. A number of these species have distinct pioneer plant characteristics: they rapidly inhabit newly opened up habitats such as road cuttings, clearings, newly eroded slopes or young lava flows (e.g., *A. usambarense* on the Namlagira/Virunga Volcanoes: ORTH 1940, plate 43 b).

Numerous species are associated with rocky outcrops and slopes, rocky grassland and other rocky areas. They mostly grow in cracks of rocks and are often (cushion forming) dwarf shrubs. Examples are *A. dregei* subsp. *ecklonis, A. streyi* (Fig. 4 a) and *A. longisepalum,* occurring in sandstone in the W Cape Province, Natal and Madagascar respectively; *A. basuticum* and *A. pachyrrhizum* (Fig. 83 a) on basalt in the Drakensberg and Ethiopia respectively; *A. whyteanum,* frequently occurring on granite outcrops (see CHAPMAN 1962, plate 9); *A. ibityense* and *A. hispidulum* on

---

Fig. 3. Growth forms and habitats of *Anthospermum. a A. littoreum* (PUFF 791219-3/1), in coastal sand dunes near Qolora Mouth (S Africa, Transkei), also visible are *Ipomoea pes-caprae* subsp. *brasiliensis* (foreground, left), *Carpobrotus edulis* (background) and *Mimusops caffra* (trees in background); stems and branches mostly buried in sand; free, visible portions are ca 50 cm long. *b–c A. prostratum, b* plant (arrows: prostrate shoots) in fixed coastal sand dunes behind Agulhas Lighthouse (S Africa, SW Cape Prov.; PUFF 790912-3/1), *c* growth form: note short, erect lateral branches on the horizontal stems and adventitious roots arising at some nodes (PUFF 790911-2/1); the scale is 10 cm

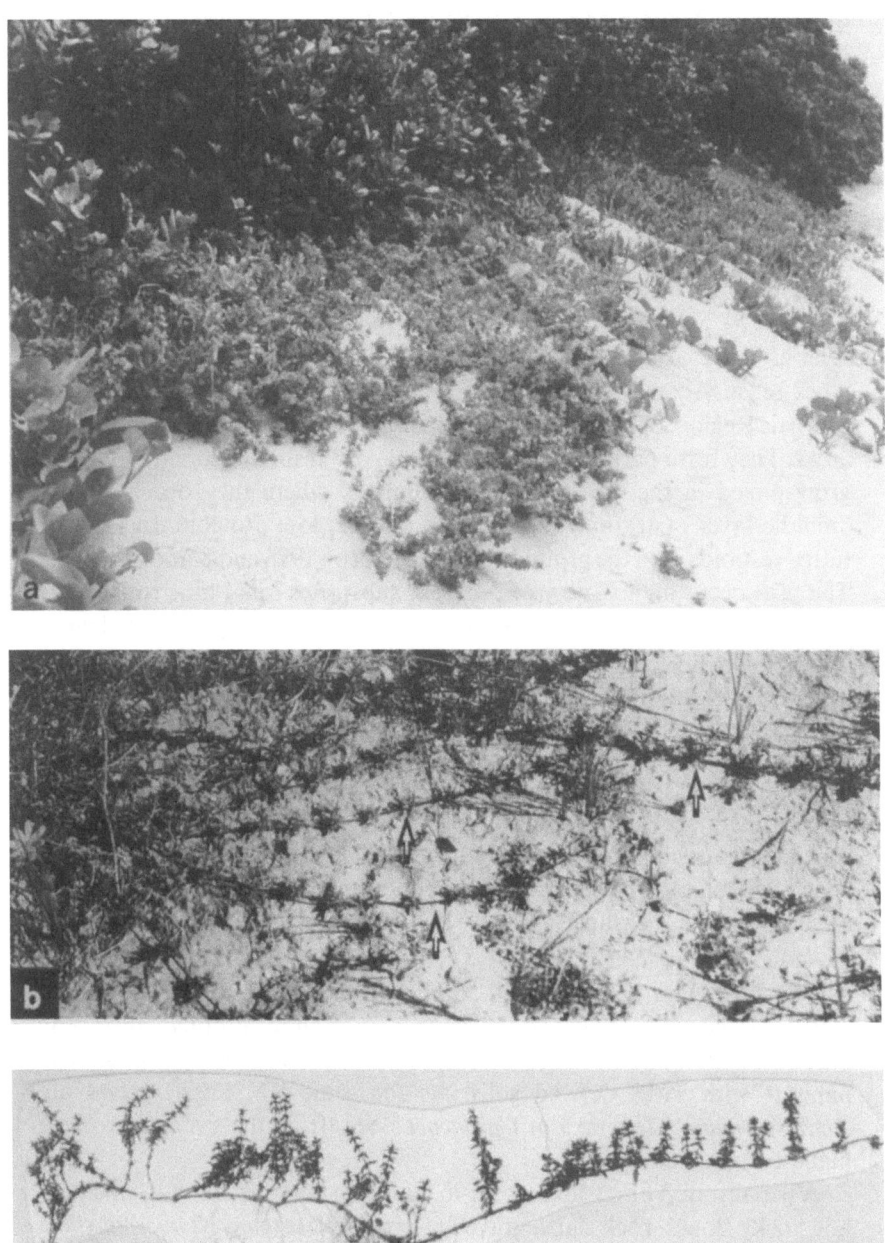

Fig. 3

quartzite in Madagascar and E South Africa respectively (the latter, however, also grows on sandstone, particularly in Natal).

A small number of taxa (e.g., *A. zimbabwense*) occur in pure grassland or grassy areas in woodlands (savannas).

Two (not closely allied) species, *A. littoreum* and *A. prostratum* (Figs. 3 a–c), grow in coastal sand dunes; the former often on the seaward-facing side of the first inhabited dunes.

SW Cape species of *Anthospermum* occur virtually from sea level to the highest peaks of the Cape mountains and are found in "Mountain Fynbos" (e.g., *A. spathulatum* subsp. *spathulatum*, Fig. 5), "Coastal Fynbos", "Coastal Renosterveld" and "Strandveld" – i.e., in all veld types sensu ACOCKS (1975) which are characteristic for the SW Cape Floristic Region. Species grow in rocky or gravelly terrain, or in sandy, flat areas. They form part of the "ericoid" and/or "restioid" communities or grow mixed in the "proteoid" communities, where they occur in the "middle layer" (stratum) made up of ericoid, low proteoid and large, tufted restioids or in the ground layer (cf. TAYLOR 1978 and KRUGER 1979). The taxa occur on a variety of different substrates (quartzitic soils, acid calcareous sandy soils, granite soils, heavy textured clayey soils or clayey loam; cf. LAMBRECHTS 1979), although the majority of SW Cape taxa appear to be confined mainly to – or at least show a clear preference for – Table Mountain Sandstone (TMS) or TMS derived soils (e.g., *A. bergianum*, Fig. 98). Certain species are ecologically euryotopic – e.g., *A. galioides,* a variable species, which is virtually omnipresent in the SW Cape; others are substrate-stable – e.g., *A. esterhuysenianum* (Fig. 4 b), which is apparently always associated with shale bands, or *A. aethiopicum,* which is a plant quite typical of fynbos occurring on relatively fertile, moderately moist ground.

Much of what has been said about habitats of SW Cape species of *Anthospermum* also applies to the genus *Carpacoce,* although at least a part of the distribution range of all species includes TMS areas or sandy habitats with TMS derived soils. As for some SW Cape species of *Anthospermum*, a few taxa of *Carpacoce* occur from the sea coast to the highest Cape mountains.

Virtually all *Nenax* taxa occur in ± dry to very dry habitats such as hot, rocky slopes, rock slabs, gravelly to rocky flats (e.g., *N. microphylla,* Fig. 6 a) or sandy areas. *N. cinerea,* nevertheless, often occurs along or in ephemeral rivers and water courses in hot and dry country, and some populations of *N. acerosa* subsp. *acerosa* were found in temporarily marshy areas and waterlogged soil (although the taxon typically occurs in dry, sandy flats).

Fig. 4. Growth forms and habitats of *Anthospermum. a A. streyi* (PUFF 790426-1/3), a cushion (ca 30 cm in diam.) growing amongst rocks at Inkonka Point, Oribi Gorge Nature Reserve (S Africa, Natal S Coast). *b A. esterhuysenianum* (PUFF 800923-3/2), forming a mat at base of restiad clumps on the summit of Sneeugat Peak, Witsenberge (S Africa, SW Cape Prov.); the portion shown is ca 20 cm

C. General Part

Galopina species primarily occur in forest edge vegetation, scrub, bush clumps, on road and stream banks or on the forest floor (afromontane vegetation; e.g., *G. circaeoides*) and also in more open, grassy areas (e.g., *G. crocyllioides*).

In all genera except for *Nenax,* there is a distinct wet → moist → dry gradient. Species confined to distinctly wet habitats, however, are few and

Fig. 5. *Anthospermum spathulatum* subsp. *spathulatum* (PUFF 791223-1/2). Population in dry mountain fynbos at the summit of Rooibergpas [S Africa, Cape Prov., S of Calitzdorp; see detailed study of that area by TAYLOR (1979)]; the tallest plants (arrows) are ca 2 m high; the only other taller shrub visible is *Colpoon compressum* (= *Osyris compressum*; Co)

are, in all three genera, herbaceous and never shrubby. In *Anthospermum,* for example, the Madagascan *A. palustre* grows in peat bogs, and the most widely distributed *Anthospermum* species, *A. herbaceum,* is most typically found – often shaded by scrub and other vegetation – along stream banks, edges of swamps and other wet places (although certain ecotypes have become established in less wet habitats; see p. 309). *Galopina circaeoides* (sharing with the former much of its distribution range) occurs in similar habitats, while the remaining species of *Galopina* occur in less wet, damp to ± dry and more open (sunnier) habitats. Similarly, the bulk of *Anthospermum* species occurs in relatively dry or ± damp habitats. In the SW Cape Region, *Carpacoce spermacocea* is characteristic for "hygrophilous fynbos" communities (sensu TAYLOR 1978) and occurs as undergrowth in scrub close to streams, at the edge of kloof forest or in

Fig. 6. Habitat and growth form of *Nenax microphylla*. *a* karroid vegetation between Hutchinson and Richmond (S Africa, Cape Prov.; PUFF 800103-3/1); the plants are confined to rocky outcrops (center); the majority of dwarf shrubs visible are *Compositae*; note several plants of *Euphorbia stellaespina*. *b* low, browsed plant from around Fauresmith (S Africa, O.F.S.; PUFF 790112-2/2); runner-like branches (thin arrows) produce new plantlets (thick arrows) where they root; the ericoid shrublet in the left foreground is *Sutera* sp.

vegetation of seepages or drainage lines. Three more species of *Carpacoce* (*C. gigantea, C. burchellii, C. curvifolia*) seem to be more common in fairly moist or damp habitats such as (temporarily) marshy areas, sheltered, shady places below cliffs or well drained, moist mountain slopes; the remaining taxa of *Carpacoce* occur in distinctly dry places.

Numerous taxa occur in habitats which are occasionally or regularly burnt (prior to the rainy season). The influence of fire on growth form and habit is discussed in the following chapter (2.5.1.); also see Regeneration after Fire (12.8.).

The altitudinal ranges of numerous taxa are remarkably wide. *A. herbaceum,* for example, is recorded from sea level to an altitude of over 3 000 m. The wide range of afromontane species without specific substrate requirements may be explained by the fact that the lower altitudinal limit of afromontane vegetation decreases with increasing distance from the equator, and that "outliers" of afromontane vegetation may occur at even lower altitudes. Also in many SW Cape taxa, wide altitudinal ranges are common, as identical substrates (e.g., TMS derived soils and sands!) may prevail from sea level to the highest peaks. Limited altitudinal ranges of some taxa, on the other hand, may be linked to a particular substrate (e.g., *A. basuticum* – largely confined to the basalts of the high Drakensberg).

Flowering periods: The majority of taxa of all four genera occurs in areas characterized by a seasonal climate with an often short rainy season and a long dry season. Plants tend to flower within a few weeks after the rains, although there are some exceptions. *A. bicorne,* for example, always seems to flower much later, i.e., only starts flowering some months after the end of the rainy season. Flowering periods tend to be rather long. In monitored populations of *A. pumilum* subsp. *pumilum* and *A. hispidulum,* for example, the (main) flowering period lasted for two to three months. In many taxa, however, populations contain odd specimens flowering out of season (so that flowering specimens of a taxon from all year round are often represented in herbaria), although there is a distinct main flowering season. Out of season flowering may extend to entire populations: in *A. spathulatum* subsp. *spathulatum* for example, populations were occasionally found which started flowering before the onset of the rains, while the typical peak of flowering is after the rains. Cultivated plants of species which exhibited a ± distinct flowering period in the field (e.g., *A. aethiopicum, A. ammannioides*) flowered either virtually all year round or several times a year in the greenhouse. Some *Carpacoce* species, in contrast to most other taxa/genera of the *Anthosperminae,* seem to flower rather sporadically (e.g., *C. scabra*) and may, as LEVYNS (1937) and SALTER (1937) pointed out, grow vegetatively for a number of years (e.g., no trace of flowering in three successive seasons in a population of *C. vaginellata* – SALTER l.c.).

2. Habit and Growth Form Analyses

The genus *Anthospermum* is comprised of large to medium-sized shrubs, dwarf shrubs, (few true) subshrubs, and perennial herbs. *Nenax* and *Carpacoce* species are primarily dwarf shrubs, and all *Galopina* species are perennial herbs.

Characteristic for almost all taxa is the presence of (often much-contracted) *short shoots*. Since short shoots do not contribute significantly to the branching patterns (although they are important in that they bear the bulk of the foliage of a plant), they are largely ignored in the following discussion of growth forms and in the growth form diagrams (Figs. 7, 11, and 12). It should be noted, however, that short shoots under certain circumstances (if, for example, long shoot apices are damaged by browsing animals) can elongate into long shoots.

Another fairly common phenomenon is the "unequal promotion" of lateral axes in the vegetative plant body. It occurs (sometimes more, sometimes less pronouned) across the whole growth form spectrum of the genera *Anthospermum*, *Nenax* and *Carpacoce* and is *not* confined to a particular category of growth form. "Unequal promotion" refers to situations where unequally strong lateral axes arise in the axils of an opposite pair of bracts (e.g., at a node); one lateral axis is much better developed than (promoted over) the other (cf., *N. microphylla*, Fig. 9). In several taxa, this trend is taken to the extreme, i.e., leads to the presence of only a single lateral element (long or short shoot) at a node. In such plants, the promoted lateral axes frequently (always?) follow a spiral. This can, for example, be seen in *N. cinerea* (Fig. 13 c), where lateral branches mostly arise singly at a node, but if they arise in pairs, one of the two branches is much weaker than the other. Similarly, "unequal promotion", although not as pronounced and regular as in the former example, can also be seen in *A. littoreum* (Fig. 7 b).

Formation of accessory shoots occurs in a number of taxa (e.g., fairly common in *A. herbaceum;* also in the inflorescence region: *A. paniculatum*, Fig. 29 b) but, as far as growth form and branching patterns are concerned, this is of minor importance and will, therefore, not be discussed in detail.

2.1. Large to Medium-sized Shrubs

Shrubby *Anthospermum* species attaining a height of one metre or more are considered in this category. In several species, individuals reach an average height of ca two meters, but odd plants may be up to three meters tall (e.g., *A. welwitschii*, Fig. 2 a). Taller plants (e.g., *A. usambarense*, with a recorded maximum height of 4 to 4.5 m: HEDBERG 1957: 176; VERDCOURT 1976: 331) are rare.

These larger shrubs have in common that they are few- or more often single-stemmed and grow ± erect (except *A. littoreum;* see below). The largest plants with a single main stem (diameters of up to 7 cm have been measured in the field) could, therefore, almost be called small trees.

Plants grown from seed and cultivated in the greenhouse developed very rapidly, attained a height of ca 45 to 75 cm within one year (e.g., *A. aethiopicum, A. ammannioides, A. usambarense*) and flowered for the first time after approximately one year to 18 months[1]. The young flowering plant of *A. aethiopicum* depicted in Fig. 7a shows the prominent main stem and the ± symmetrically arranged lateral branches. The longest (lowermost) lateral branches started to develop within the first six months, the next set of lateral branches during a second growth period (after ca 10 months) and the uppermost ones were the most recently produced. As growth continues, the lateral branches keep on elongating for a while and produce branches of a second (and higher) order. Eventually, however, the oldest (lowermost) lateral branches cease growth, die and break off, so that in older plants, the lower half to third of the main stem is ± bare of branches. Innovation, thus, is largely acrotonal.

These large, shrubby species invariably have inflorescences whose main axes remain vegetative ("prolificating" or auxotelic inflorescences; see 5.); inflorescence axes (e.g., main stem, branches) thus continue growth after flowering. Hence growth of these large shrubs is strictly monopodial, unless shoot apices are damaged or broken off (which apparently happens quite frequently in the natural habitat).

Due to damage of the main shoot apex at an early stage of development and subsequent vigorous growth of branches below, or due to the development of strong lateral branches near the base of the main stem, shrubs occasionally become *few-* or *several-*stemmed (e.g., *A.*

---

[1] Data on seasonal stem growth in length and diameter for plants of *A. galpinii* from a natural population are given by MOLL & al. (1982).

Fig. 7. Growth forms of *Anthospermum*. *a A. aethiopicum* (PUFF 790717-1/3), cultivated plant, 20 months old, flowering for the first time (for greater clarity the ternately arranged leaves and lateral branches are shown as opposite). *b A. littoreum* (PUFF 790424-1/1), cultivated plant, 18 months old, flowering for the first time; note unequal promotion of lateral branches arising at a node. *c A. pumilum* subsp. *pumilum* (PUFF 781203-2/1), development of unbranched flowering shoots following a veld fire; note dead, burnt off stumps (stems); compare with Fig. 8a. *d A. esterhuysenianum* (PUFF 800924-3/2), mat-forming flowering plant with prostrate, much-branched stems (compare with Fig. 4b); arrows point to axes having ceased growth. – Dead organs are indicated by interrupted lines; short shoots not drawn. The scale units are 10 cm. Further explanations in the text

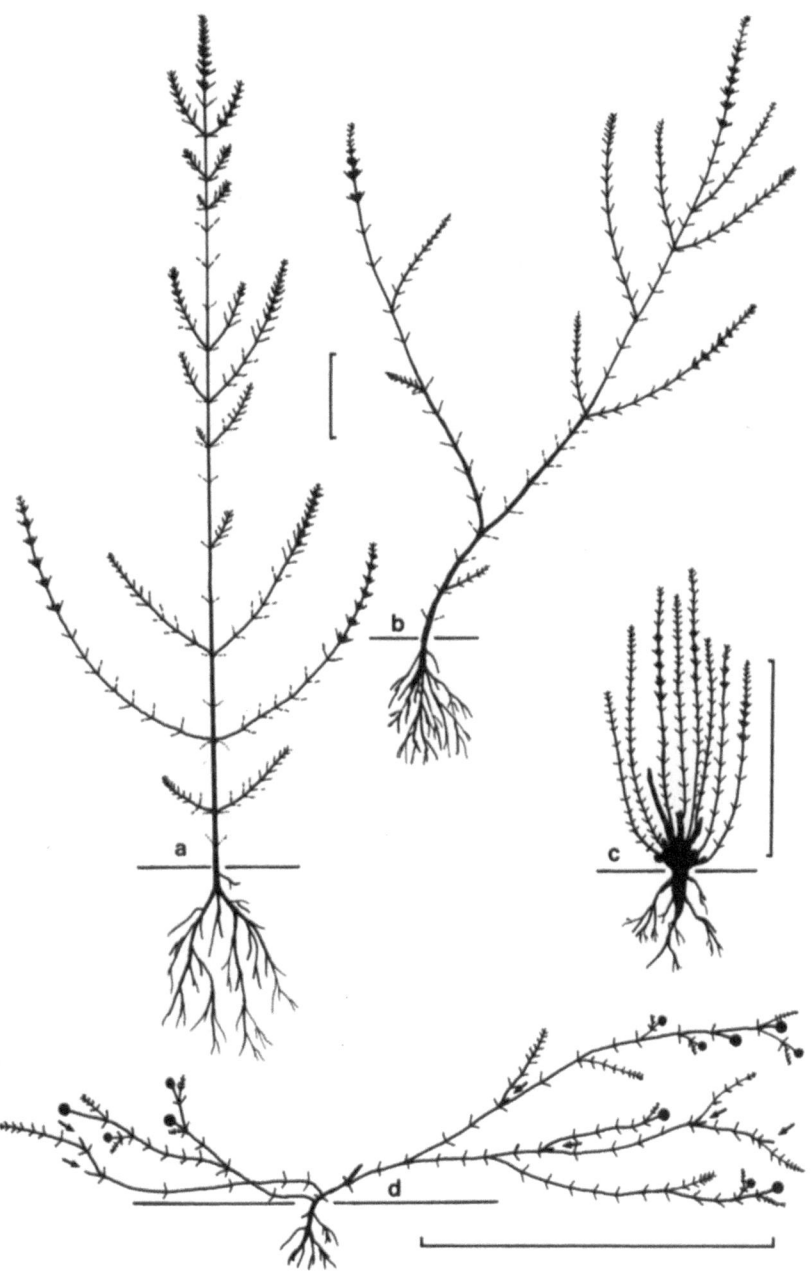

Fig. 7

Fig. 8. Growth forms and habitats of *Anthospermum*. *a–b A. pumilum* subsp.
pumilum, a plant in rocky grassland S of Cathcart (S Africa, E Cape Prov.; PUFF
790114-4/1); flowering shoots (ca 15 cm tall) produced after a veld fire. *b* old
individual; note woody, branched shoots, broadened woody base and new growth
near and at base (PUFF 780716-1/3). *c–d A. hispidulum* (PUFF 780716-1/2), note new
growth near and at base of stems. – The scale unit is 1 cm in *b* and *d* and 10 cm in *c*.
Further explanations in the text

spathulatum subsp. *uitenhagense,* Fig. 70 a). Frequently, only a single branch is produced at a node, which grows vigorously and pushes aside the main stem (a pseudo-sympodial-dichasial branching pattern may result: cf. Fig. 7 b, branching near the base, and Fig. 11 a). If such branches recur at other nodes in the lower part of the actual main stem, the shrub becomes several-stemmed. Obviously, in older individuals it is often impossible to determine which of the stems are the actual main axes and which are lateral branches.

Growth form differences of these large, shrubby species primarily concern the regularity and symmetry of branching, the number of lateral branches produced, the angles between main axes and lateral branches, the curvature of the branches, and the degree of branching of the lateral axes themselves. Plants of *A. spathulatum* subsp. *spathulatum,* for example, frequently exhibit a very symmetrical and regular branching pattern: branches are produced at virtually every node in the upper part of a plant, are – like the leaves – decussately arranged, and each branch of a pair is equally well developed; branches curve upward as shown in Fig. 7 a for *A. aethiopicum* but tend to be more regular; shrubs thus are ± cylindrical in appearance. In *A. vallicola* (Fig. 2 b), branches are held at an acute angle to the stems, and stems and branches are ± straight. As the lower branches are relatively long and the upper ones become increasingly shorter, the plants appear ± rounded in outline.

A. littoreum, a species of coastal sand dunes, can also be grouped with the large, shrubby species, although it differs in having procumbent to ascending stems and branches, which are normally, save for the terminal portions, covered in sand (cf. Fig. 3 a); axes covered in sand frequently root at the nodes. If grown in the greenhouse, the similarity to the typical erect shrubby species becomes obvious (Fig. 7 b). The stems, however, are too weak to stand upright by themselves and need artificial support.

2.2. Dwarf Shrubs

Dwarf shrubs, ca 30 to 50 cm tall, are, in general, many-stemmed. Their development largely corresponds to that of the large, shrubby species (see above), provided that they are characterized by proliferating inflorescence axes (cf., for example, Figs. 27 c–g; most dwarf shrubs in *Anthospermum* and *Nenax*). In species whose inflorescence main axes are terminated by flowers, growth is often sympodial-monochasial (see below).

Plants frequently are ± rounded bushes (cf. *A. streyi,* Fig. 4 a, or *A. pachyrrhizum,* Fig. 83 a), a habit which is caused by increased growth of basal lateral branches, particularly in younger plants (cf. *N. microphylla,*

Fig. 9). In older plants, acrotonal rather than continued basitonal innovation seems to play a bigger role; this results in rather dense, "closed" rounded bushes or cushions. Odd basal innovation shoots, nevertheless, are produced regularly (annually?) and these eventually replace older, woody shoots which have died off (cf. Figs. 8 b–d). Older

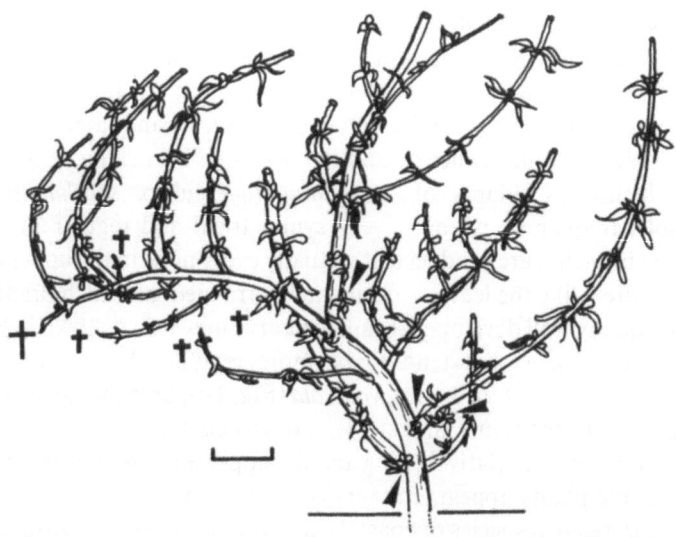

Fig. 9. *Nenax microphylla* (Puff 790710-1/1), base of cultivated plant, 2 years and 9 months old. The crosses mark shoots that have died off (larger cross: main axis). Leaves of both long and short shoots are mostly dead and shrivelled up, only leaves of innovation shoots (arrows) are green. Note unequal promotion of lateral shoots. – The scale unit is 1 cm. Further explanations in the text

plants often have thick, woody, not infrequently branched (tap) roots (cf. Figs. 8 b–d).

In dwarf shrubs in which the inflorescence main axes terminate in a flower [*Carpacoce* species, few *Nenax* and only one (?) *Anthospermum* species, i.e., *A. esterhuysenianum*], the continuation of growth is sympodial and the branching is almost always monochasial or anisotomous-dichasial (i.e., the dichasium branches are unequal; see "unequal promotion", beginning of this chapter). Due to the combination of both the unequal development of lateral axes in the vegetative region and sympodial branchings as a result of flowering (see, for example, *A. esterhyusenianum,* Fig. 7 d), the branching patterns may become highly complicated. The picture may be further obscured if the branching in the vegetative region becomes pseudo-sympodial-dichasial (see 2.1.). Such a

Fig. 10. Growth forms and habitats of *Nenax* and *Carpacoce*. *a N. cinera* (Puff 800102-4/1), growing in cracks of rocks near Whitehill Stn. (S Africa, SW Cape Prov.); note spine-tipped branches and irregular branching as a result of browsing; the arrows point to fruits; the plant shown is ca 20 cm tall. *b C. vaginellata*, part of plant (ca 10 cm) from Table Mt. above Cape Town (S Africa, SW Cape Prov.; Puff 791228-2/4). *c–d C. heteromorpha* (Puff 800101-3/1), in rich coastal fynbos in the Kogelberg Forest Reserve (S Africa, SW Cape Prov.); note curled, tightly packed leaves; the plant in *c* is ca 25 cm tall

branching pattern is characteristic of *N. coronata* and some *Carpacoce* species, namely *C. vaginellata* (Fig. 10 b) and *C. scabra* (Fig. 11 a). Detailed growth form and branching analyses of such plants, especially if they are older, are almost impossible to carry out.

Fig. 11. *Carpacoce scabra* subsp. *scabra* (PUFF 800914-6/1). *a* semischematic illustration of branching pattern. The starting point of shoots presumably developed in the current season are marked with thick arrows; thin arrows indicate possible continuation of growth in the following growing season; open circles stand for flowers or fruits; the positions of old flowers/fruits are marked with crosses. Note frequent sympodial-monochasial branching and trend to aniso-phylly of leaf pairs below flowers. *b* anisophyllous leaf pair below a flower. Only one leaf is well developed (= bract of lateral shoot); the leaf blade on the left is very poorly developed and hardly discernible from the triangular setae borne on the stipular sheath

N. microphylla occasionally forms dense clumps of intermingled plants. This comes about by vegetative reproduction: arching branches frequently root at the nodes where they touch the ground and tend to produce new plantlets (Fig. 6 b). These eventually become fully detached from the mother plant and thus give rise to clones which may be rather extensive, particularly if the plants are not permanently browsed by domestic stock (see also 2.5.2.).

Fig. 12. Growth form of *Anthospermum bergianum* (♀). *a* one year old plant, flowering for the first time. *b* flowering for the second time. *c* older plant (4 years old?); the inflorescences produced in the previous seasons remain clearly recognizable. – if inflorescence region. Further explanations in Fig. 7 and in the text

The growth form of *A. prostratum,* finally, also deserves special consideration: According to field observations supplemented by investigations of herbarium material, the upper parts of the main stems soon die off, and several lateral branches originating from near the base of the stems start to radiate out horizontally. They may reach about one meter in length, grow ± straight, branch occasionally, root at the nodes and give rise to short, erect, flowering shoots (Figs. 3 b–c). These erect shoots, bearing flowers only laterally, continue growth and again produce flowers in the following season. Eventually, however, they cease growth, and near their bases new prostrate long shoots are produced which, in turn, bear short, erect, flowering shoots. It was also observed that such cable-like long shoots can become detached from the "mother plant" and become independent new plants. This may result in the formation of ± large clones of criss-crossing and intermingled long shoot "cables". It appears that these prostrate long shoots never reach a great age (perhaps not more than two or three years?) and never become thick and distinctly woody; the growth form of *A. prostratum,* thus, approaches that of a subshrub (see 2.3.).

2.2.1. Short-lived Shrubs ("Woody Herbs")

In several taxa, shrub-like plants show a clear trend to and exhibit varying degrees of *shortlivedness*. In *A. bergianum,* for example, plants older than four years were never observed (annually produced ♀ in-florescence regions of plants with proliferating inflorescence axes which remain well discernible for several years allow a ± accurate age determination; cf. Figs. 12 a–c). Plants mostly show no trace of basal innovation shoots which could – like in dwarf shrubs reaching a greater age (e.g., *A. whyteanum*) – take over for old shoots that have died off. The same seems to hold true for *C. heteromorpha* (Fig. 10 c), although in that species the short internodes and the leafy short shoots at each node (cf. Fig. 10 d) make an exact age determination difficult if not impossible.

This shortlivedness is also very pronounced in *A. ternatum,* an erect, often single-stemmed, fast growing plant, which occasionally reaches a height of 100 (150) cm, and which may be quite woody at the base. Subsp. *randii* may even show a trend to a biennial habit, but this is not certain as no long-term observations could be made (also see, in this context, 2.5.1.).

2.3. Subshrubs

The ± intermediate nature of subshrubs [proper dwarf shrubs ↔ subshrubs ↔ perennial (pleiocorm) herbs] often makes it difficult to assign plants to one growth form or another. Growth form, furthermore, also seems to sometimes vary between populations of one taxon. True subshrubs occur in a few *Anthospermum* species and, perhaps, some populations of *Carpacoce spermacocea* (normally ± dwarf shrubby)[1].

---

[1] Plants that take on a subshrubby habit as a result of recurrent fires are not considered here; see 2.5.1.

A. paniculatum, for example, has shoots which terminate in a closed inflorescence (Fig. 29 b). Innovation takes place from below the inflorescence. The vegetative, woody body is never extensive and some plants are perhaps better classified as subshrubs rather than (proper) dwarf shrubs.

Also the growth form of *A. esterhuysenianum* approaches that of a subshrub. Plants have a quite woody tap root and short, woody (main) stems which (annually?) produce ± prostate innovation shoots near their base. These shoots apparently never reach a great age, remain relatively thin and become woody only in the basal portions. Their oldest parts may root at the nodes, become separated from the mother plants and continue growth independently. Together with the mother plants and younger plants produced from seed, they form dense mats which are rather characteristic of that species (Fig. 4 b). The basic growth form is ± comparable to that of *A. prostratum* described above.

In several taxa (e.g., *A. herbaceum, A. galioides*), gradual transitions from subshrubs (sometimes even ± proper dwarf shrubs) to pleiocorm herbs were observed.

2.4. Perennial Herbs

Perennial herbs are either represented as pleiocorm herbs [a few *Anthospermum* species, *Galopina* p. p., and (?) occasionally plants of *Carpacoce spermacocea* subsp. *spermacocea*] or rhizome herbs (*Galopina* p. p.;? sometimes plants of *C. spermacocea*). Forms ± transitional between pleiocorm and rhizome perennials also occur (*Anthospermum, Galopina*).

Herbaceous perennial species are characterized by an annual innovation of the entire aerial shoot system. In the pleiocorm herbs, innovation shoots arise from the short, perennating bases of the previous season's shoot generation; not infrequently the innovation shoots originate from accessory buds in the (hypo)cotyledonar region. Plants commonly develop a strong, woody, sometimes much-branched tap-root and numerous adventitious roots at the basal portions of the shoots.

Obviously, the relationship of pleiocorm herbs to subshrubs is close, the only difference being that the perennating shoot system is even shorter and further reduced. It, therefore, should not be surprising that within a given species, growth forms are not completely uniform. Particularly *A. herbaceum,* represented by a number of rather divergent ecotypes over its whole range of distribution (Figs. 80–82), is quite variable in this respect. Plants of *A. herbaceum* may also produce prostrate to ascending lateral shoots whose basal portions root at the nodes and are eventually covered in debris and soil, so that older individuals can become ± intermediate between proper pleiocorm and rhizome perennials.

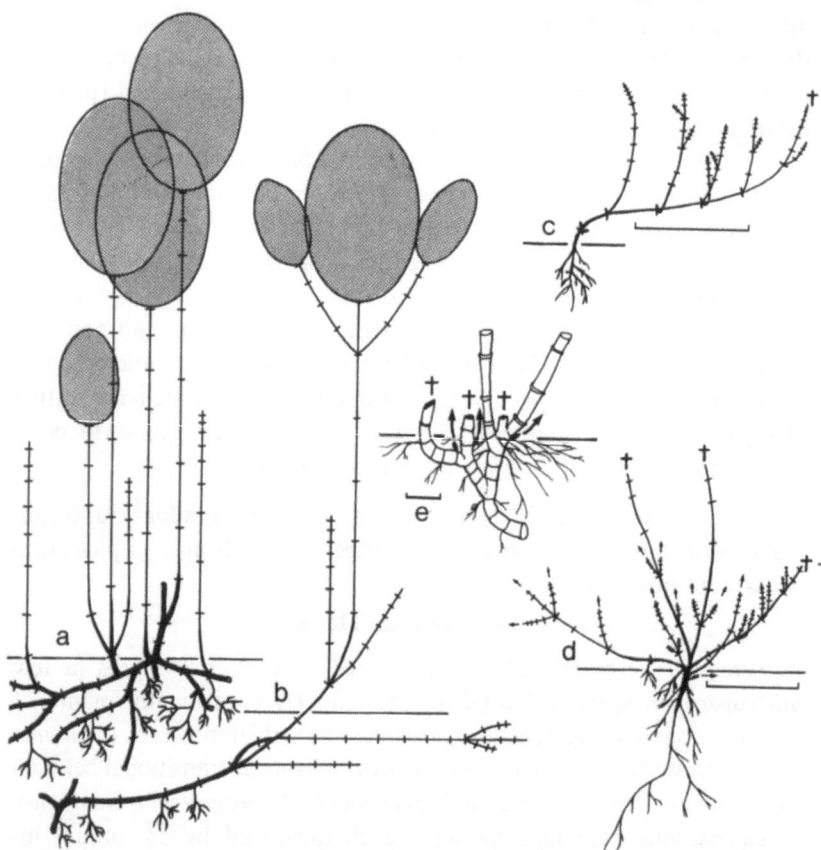

Fig. 13. Growth forms of *Galopina* and *Nenax*. *a* old plant of *G. circaeoides* (PUFF 791222-1/1) with extensive rhizome system and several flowering and vegetative aerial shoots (inflorescences shown as ellipses; aerial portions highly schematic). *b* *G. tomentosa* (PUFF 790415-2/1), some shoots continue to grow plagiotropically and will produce aerial shoots during the following growing season. *c* *N. cinerea* (PUFF 800102-4/2), cultivated plant, ca one year old; the main axis has ceased to grow (cross), lateral branches continue growth; note unequal promotion of lateral branches arising at a node. *d G. circaeoides* (PUFF 790909-6/1), cultivated plant, 18 months old; old flowering shoots have died off (crosses) and new shoots arise mainly from the base; some newly produced shoots are of hypocotyledonal origin; plagiotropic stems may root at nodes. *e G. crocyllioides* (PUFF 790416-1/1), part of rhizome; note innovation buds likely to continue growth (arrows); old, burnt off stems are marked with crosses. – Short shoots (*a–d*) not drawn. The scale units in *c* and *d* are 10 cm and 1 cm in *e*

Galopina circaeoides and *G. tomentosa* (Figs. 13a, b) are typical rhizome perennials. Adult plants have a ± extensive, plagiotropic, hypogeal rhizome system with adventitious roots and reduced scale leaves; the main (primary shoot) axis is often obscured and difficult to discern.

Based on observations of plants raised in the greenhouse, the development of *G. circaeoides* plants takes place as follows: The seedlings develop very rapidly and, within one year, develop a single, robust main shoot which is terminated by an inflorescence. In more robust plants, lateral shoots, originating from the lowermost part of the main stem (including cotyledonar shoots!), may also produce inflorescences later on in the season. Within one month after flowering, the fruits become mature and the inflorescence branches start to turn brown and dry up. At the same time, short shoots borne on the lower third of the stem start to elongate into long shoots. In the following growth period, these long shoots continue to grow and eventually produce inflorescences. Also, additional innovation shoots (mainly from accessory cotyledonar and hypo-cotyledonar buds) are produced which, on the one hand, grow plagio-tropically and remain hypogeal and, on the other hand, develop into prostrate to ascending aerial shoots. The latter soon develop adventitious roots at the basal nodes (Fig. 13 d) and eventually their basal parts also become part of the rhizome system. Up to an age of ca 3 years, the primary root system and the basal part of the primary shoots are still easily discernible, but in older plants the rhizome system becomes so extensive that primary shoots and roots are no longer distinct. The root system becomes strictly homorhizous; the rhizomes themselves become ± woody, and innovation shoots originate mainly from the slightly woodier, proximal portions of the plagiotropic shoots.

The growth form of *G. aspera* and *G. crocyllioides* approaches that of pleiocorm herbs. In these two species, rhizomes are short, little-branched, often ± erect and quite woody (cf. Fig. 13 e).

There are indications that the perennial herbs may be rather long-lived: plants of *Galopina* species, for example, having been cultivated for seven years, are still healthy and show no signs of dying off.

2.5. Influence of External Factors and Habit and Growth Form

2.5.1. Fire

Bush fires kill the aerial woody shoots of shrubby or dwarf shrubby species. In the growing season following a bush fire, several *Anthospermum* and a few *Nenax* and *Carpacoce* species are capable of regenerating from innovation buds situated at or near the base of the burnt stems ("sprouters" – see also 12.8.). Flowering aerial shoots produced after a fire are frequently unbranched and, furthermore, differ from older, woody branches of unburnt plants (of the same population) in internode length and leaf size and shape. If such aerial shoots are produced in larger numbers, fire-exposed plants become small, cylindrical to rounded bushes, hardly more than 20 cm tall (e.g., *A. pumilum* subsp. *pumilum,* Fig. 8 a and diagram, Fig. 7 c). Recurrent fires may cause the development of rather extensive, often almost disk-like woody bases ("xylopodia"), which can be up to 5 cm in diam. Such plants can be classified as suffrutices (the term *geoxylic* suffrutices should perhaps be avoided as it usually refers to the presence of massive, woody *subterranean* structures such as the "lignotubers" of *Leucospermum* sect. *Crassicaudex* − cf. ROURKE 1972; also see WHITE 1976)[1].

In taxa capable of resprouting after a fire, the suffruticose habit was never found to be obligatory[1]; plants again become many-stemmed (dwarf) shrubs if not exposed to fire for a number of years. As in other genera (e.g., *Otiophora* species, cf. PUFF 1981), both suffrutices (i.e., recently burnt plants) and dwarf shrubs (i.e., individuals in microhabitats sheltered from fire) are not uncommon side by side in a single population of one and the same taxon (e.g., *A. whyteanum*). In widely distributed taxa such as *A. pumilum* subsp. *pumilum,* growth form may subsequently also vary from region to region, depending on whether the plants grow in areas burnt regularly (for better grazing of domestic stock, for example; many parts of the Transvaal) or in regions not or only occasionally burnt (e.g., parts of the N Cape Prov.).

It is, furthermore, interesting to note that in taxa like the above-mentioned *A. pumilum* subsp. *pumilum,* the distinctly woody plants always produce − in addition to new growth in the upper (younger) parts of the plants − basal innovation shoots towards the end of the dry season or at the onset of the rainy season, no matter whether they have been burnt that year or not (see, for example, Fig. 8 b). While it would be tempting to conclude that this annual basal innovation is a genetic adaptation to burning, the fact that such basal innovations also occur regularly in shrubby taxa growing in habitats seldom or never exposed to fire (e.g., *A. hispidulum,* Figs. 8 c–d) presents a strong argument against this. It must also be stressed that such basal innovations are largely confined to several- to many-stemmed (dwarf) shrubs. Large, few- to single-stemmed shrubs usually do not have the faculty to regenerate via basal innovation shoots (buds) (they are normally killed by fire − see also 12.8.).

The process of basal innovation of many-stemmed dwarf shrubs is

---

[1] For examples of *obligatory* and *facultative* "geoxylic" suffrutices in African *Rubiaceae* see VOLLESEN (1981).

similar and comparable to the situation found in perennial herbs with pleiocorms or rhizomes, where aerial shoots die off annually by themselves. In this context in appears remarkable that all of the many-stemmed, basally innovating dwarf shrubs investigated wood anatomically are thought to be secondarily woody (KOEK-NOORMAN & PUFF 1983). The presence of basal innovation shoots in these taxa, which could be interpreted as a character protracted from ("passed on from") a herbaceous ancestral stock, may provide some additional support for the secondary nature of the woodiness in these taxa (this is a highly speculative assumption and can hardly be considered definite proof).

Misinterpretations of growth forms as a result of recurring fires should be avoided: *A. ternatum* subsp. *randii,* for example, is often recorded as being an annual. Plants often occur in fire-prone habitats, are killed by fire and regenerate from seed. Consequently, plants encountered in the field are mostly young flowering individuals in their first growing season. Careful investigations of populations, however, reveal that plants growing in microhabitats sheltered from fire can become at least a few years old. It is, therefore, rather misleading to call this taxon an annual.

2.5.2. Browsing

According to field observations, plants of the following taxa are browsed by sheep or goats: *Anthospermum dregei* subsp. *dregei, Nenax microphylla* and *N. cinerea* (frequently), *N. arenicola, A. comptonii, A. pumilum* subsp. *rigidum* and *A. dregei* subsp. *ecklonis* (occasionally), and *A. spathulatum* subsp. *spathulatum* and *A. monticola* (infrequently). Interestingly, the mentioned taxa are either confined to the drier parts of S Africa or browsed plants of more widely distributed taxa were only found in drier areas of their distribution ranges, while no trace of browsing was observed in taxa occurring in the wetter parts of S Africa or tropical Africa. This may suggest that these plants are not a favourite source of food and are only eaten in areas where animals have a limited choice of edible plants (e.g., in karroid and semi-desert areas).

Browsing results in a drastic change in habit and appearance of the plants. Like many other plants of the drier parts of S Africa and elsewhere (e.g., *Crocyllis anthospermoides,* cf. PUFF & MANTELL 1982 a), *N. microphylla* and, in particular, *N. cinerea* become distinctly spiny when browsed. In the latter (Fig. 10 a), all stems and major branches become spine-tipped. The plants remain low and the branching becomes very irregular (also see *N. arenicola,* Fig. 101 a); they often appear to be dead but some new long shoots (usually through elongation of short shoots) are produced on older stems and branches. Plants of *A. comptonii* may

become small, ± flat cushions, ca 20 cm in diameter when browsed (cf.
Fig. 96). The core of the cushion is made up of much-branched, short
woody stubs with only a few pairs of leaves on newly developed branches.
Continued browsing results in further branching of the shoots, and
browsing keeps the new branches short. *A. dregei* subsp. *ecklonis* behaves
similarly. Browsed plants of *A. spathulatum* subsp. *spathulatum* become ±
rounded, many-stemmed shrubs, hardly taller than 30 cm, with a dense
"network" of irregularly branched, intertwining twigs, while plants
normally are few- to single-stemmed, regularly (decussately) branched
and much taller (cf. Fig. 5).

3. Stems

3.1. Indumentum, Cortex, Cork

In all genera of the *Anthosperminae,* the young stems are usually
covered with papillae or hairs whose lengths range from ca 20 to 1 000 μm.
Except in *Galopina tomentosa,* where occasional multicellular hairs were
observed (in addition to unicellular ones), the papillae and hairs are
unicellular; their sculpturing is similar to that of the hairs and papillae on
leaves or fruits (compare Figs. 17 g and h–i).

Epidermal cells, cortex cells (primarily one to several subepidermal
layers) and some cells of the primary and secondary phloem (rays!)
are often filled with a dark brown to reddish substance (tannins?; see also
4.5.1.: Excretions). Stems, then, appear reddish to purplish-brown rather
than green (chloroplast-rich cortex cells).

Young stems may be ± 4-angular, but distinct collenchyma strands
forming angles are absent; older stems become ± round in section.

In older, woody stems (*Anthospermum, Nenax,* and *Carpacoce*
species), a thin layer of cork cells is formed by a phellogen originating in a
subepidermal layer; hairs are no longer present on older stems. Cork
layers usually peel off in papery strips or flakes. They are often reddish-
brown at first (cells with dark coloured contents!), but become greyish-
brown to dark grey (and thicker) on the oldest parts of the stem.
According to SOLEREDER (1899), cork formation in *Phyllis nobla* (*Anthosperminae*) takes place at the external limit of the "primary hard bast".

3.2. Wood, Node and Petiole Anatomy

The wood anatomy of *Anthospermum, Nenax* and *Carpacoce* species
(*Galopina* has no woody taxa) has been dealt with separately (KOEK-
NOORMAN & PUFF 1983). Interestingly, certain wood anatomical features
suggest that some of the *Anthospermum* species and all *Nenax* and

Carpacoce species investigated may exhibit "secondary woodiness" (i.e., woodiness which is interpreted as being phyletically in the process of increase, with the herbaceous condition representing the phyletic starting point). The remaining *Anthospermum* species investigated wood anatomically were found to possess "normal", typical rubiaceous wood.

The concept of secondary woodiness and the theory of paedomorphosis was created and, subsequently, strongly defended by CARLQUIST (1962, 1980 and literature cited therein). Although it is supported by several wood anatomists (see CARLQUIST 1980 for references), it has remained controversial and has been strongly opposed by some authors (see, for example, MABBERLEY 1974, 1982 and literature cited therein). In part E, an attempt is made to correlate wood anatomical findings with evidence gathered from other disciplines and to discuss the possibility of secondary woodiness and CARLQUIST's paedomorphosis theory as far as it concerns the *Anthosperminae*.

The n o d a l a n a t o m y of the genera of the *Anthosperminae* is more or less uniform. In a survey covering all *Galopina* and most *Anthospermum*, *Nenax* and *Carpacoce* species, nodes were always found to be u n i-l a c u n a r (one gap with one leaf trace; Figs. 14 a–d), as is the case for the majority of the *Rubiaceae* (cf. NEUBAUER 1981).

In most taxa, much contracted short shoots are borne in the axils of long shoot leaves. Due to the close proximity of the short shoot base and the subtending long shoot leaf, the vascular tissue portions which eventually form the vascular cylinder of the short shoot, may start to separate from the main vascular cylinder before the leaf trace branches out, or at least before the leaf trace gap closes again. There may thus be a "common gap" for leaf trace and vascular supply of the short shoot (Fig. 14 c). Unless carefully analysed, this may be mistaken for a 1–3 nodal pattern (one gap with three discrete traces). The occurrence of serially arranged accessory shoots may cause an even more complicated pattern. In the example presented in Fig. 14 k (*A. herbaceum*), the stipular sheath (with leaf traces lt) surrounds a central core consisting of the opened vascular cylinder of the main axis and, on either side, two pairs of interconnected vascular tissue portions which will eventually form the vascular cylinders of the lateral shoots ls_1 and the accessory shoots ls_2. At this stage, i.e., before discrete vascular cylinders are formed, it can already be seen that the leaf traces of the prophylls (i.e., the first leaf pair of the lateral shoot; ltp) are about to branch out.

There is some variation in the vascularization of the petioles, of the bases of the leaf blades and of the interpetiolar stipules. In taxa with distinctly petiolate leaves (i.e., all *Galopina* species, certain *Anthospermum*

species), the petioles are more often vascularized only by a single, fairly
large, ± semi-arc shaped bundle (*G. tomentosa*, Fig. 14e) rather than
three bundles (a larger median plus two smaller lateral ones). The latter
situation is met with more frequently in taxa with ± sessile leaves and
leaves with ± broad, cup-like stipular sheaths (some *Anthospermum*,
Nenax and *Carpacoce* species, e.g., *A. hispidulum*, Figs. 14f,i or *N.
microphylla*, Fig. 14j). The number of bundles in the petiole and the
lowermost part of the leaf blade may occasionally also be increased to five
(e.g., *A. herbaceum*, Fig. 14g).

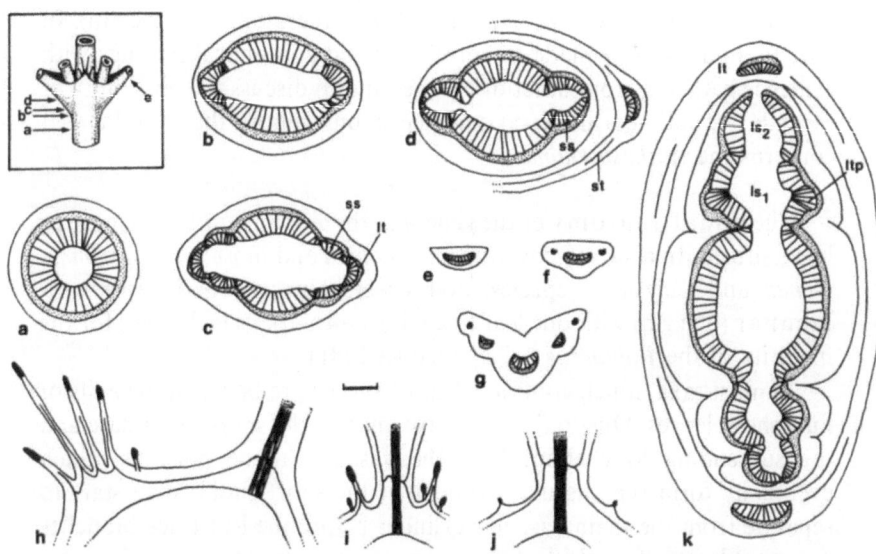

Fig. 14. Node and petiole anatomy of *Galopina*, *Anthospermum* and *Nenax a–e G.
tomentosa* (PUFF 790422-3/1); the insert shows where sections *a–e* were cut (lt leaf
trace, ss vascular tissue of short shoot, st vascular bundle supplying stipular
sheath). *f–g* petiole sections, *f A. hispidulum* (PUFF 790429-1/2), *g A. herbaceum*
(PUFF 780828-2/3). *h–j* stipules and petioles of *h G. tomentosa* (PUFF 790422-3/1), *i
A. hispidulum* (PUFF 790429-1/2) and *j N. microphylla* (PUFF 790112-9/1), note
colleters (black). *k* node section of *A. herbaceum* (PUFF 780208-2/3), explanations
in the text. – The scale unit is 1 mm for *a–e, h, k* and 0.5 mm for *f–g, i–j*

In all genera, the stipular sheaths are vascularized by ring-like bundles
which often run just below the upper edge of the sheaths. These bundles
invariable branch off laterally from the base of the prominent leaf trace
bundle (Figs. 14h–j); they may fork, and the ascending branches enter, as
small lateral bundles accompanying the larger median one, the petiole
and/or the leaf blade base (Figs. 14i–j). Large stipular "teeth" or fimbriae
are normally vascularized by bundles branching out from the ring-like

sheath bundle, while small stipular teeth or smaller, stalked colleters which occasionally also occur on the inside of the stipular sheaths are not supplied by vascular bundles (compare Figs. 14 h–j).

4. Leaves

4.1. Arrangement; Leafy Short Shoots

In *Galopina, Carpacoce* and in the majority of *Anthospermum* and *Nenax* species, the leaves are arranged decussately. In some *Anthospermum* species (e.g., *A. aethiopicum, A. galpinii*), in *N. hirta* and – very rarely – in *C. curvifolia,* leaves are arranged in whorls of three. *A. ternatum* (subsp. *randii*) may have leaves arranged decussately, in whorls of three or in whorls of four. In taxa with a whorled leaf arrangement, the nodes immediately above the cotyledons, nevertheless, still bear opposite leaves; it is only from the fourth to seventh node upwards that the switch to whorled leaf arrangement takes place (e.g., *A. galpinii, A. ammannioides;* field observations and observations on plants raised in the greenhouse!). In some taxa, the whorled leaf arrangement is obligatory (e.g., *A. whyteanum, A. aethiopicum*), while in others the leaf arrangement may differ between plant populations (e.g., leaves in whorls of three in most populations of *A. bergianum,* but decussate in some others), or even within populations (e.g., decussate or whorled in plants of *A. ternatum*).

Most taxa are characterized by the presence of leafy, often much-contracted short shoots which arise in the axils of long shoot foliage leaves. Size and shape of the short shoot leaves is often (but not always; see 4.3., below) similar or identical to that of the long shoot leaves so that, in the past, taxa have sometimes been described as having 12- or more-verticillate leaves. Pseudoverticillate leaf arrangement can be even more pronounced if leafy, *accessory* contracted short shoots are present as well (e.g., *A. streyi*).

4.2. Stipules

The interpetiolar stipules consist of distinct stipular sheaths which bear large, conspicuous to small, indistinct lobe- or bristle-like or fimbriate appendages.

The stipular sheaths (plus leaf blade bases) are either distinctly funnel-shaped (e.g., *C. vaginellata*) or cup-like. Cupulate sheaths may be ± broad and short (often not more than 0.5 mm long, e.g., *A. basuticum, A. ericifolium,* Figs. 15 b, d), broad and rather long (to ca 5 mm; some *Carpacoce* species, e.g., *C. curvifolia,* Fig. 15 g), or quite narrow (i.e., closely adpressed to the stems) and up to ca 2 mm long (numerous *Anthospermum* species) or much shorter (not more than 0.5 mm; most *Nenax* species, e.g., *N. cinerea,* Fig. 15 e).

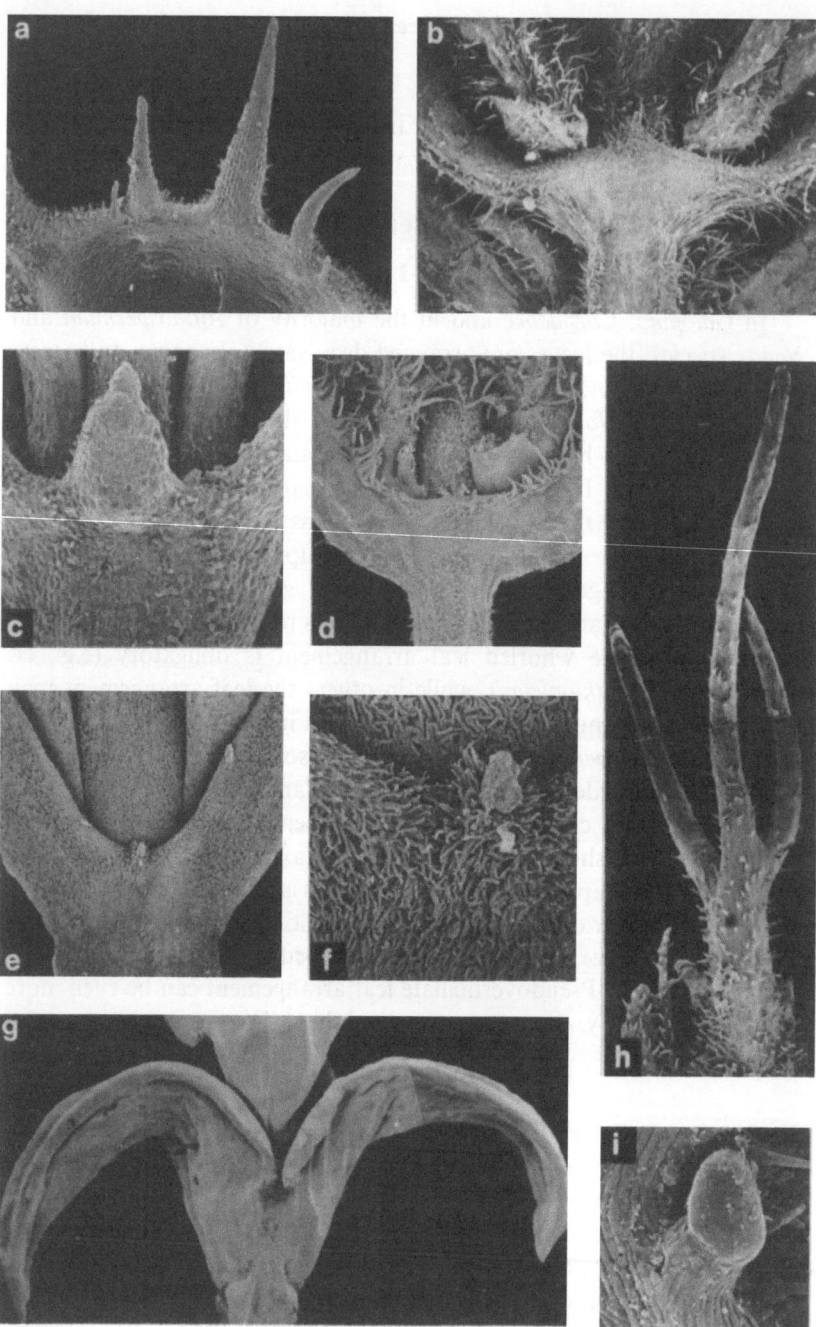

Fig. 15. SEMgraphs of stipules and stipular sheaths of *Anthospermum, Nenax, Galopina* and *Carpacoce. a A. welwitschii* (Puff 790210-2/1). *b A. basuticum* (Puff 790113-5/3). *c A. paniculatum* (Puff 790114-6/1); note triangular stipular lobe topped by conspicuous colleter. *d A. ericifolium* (Esterhuysen 28054). *e–f N. cinerea* (Puff 790711-1/1), *f* detail showing shrivelled colleter and characteristic indumentum. *g C. curvifolia* (Esterhuysen 27529). *h–i G. circaeoides* (Puff 780129-3/1), *i* stalked colleter on the rim of the stipular sheath. *b, d, g* herbarium material, all others critical point dried. *a–e, h* × 16; *f* × 70; *g* × 8; *i* × 93

Appendages on the stipular sheaths are sometimes small, ± triangular lobes (e.g., *A. paniculatum,* Fig. 15 c), but more commonly setae or fimbriae. In numerous *Anthospermum,* some *Galopina* species and in *Carpacoce spermacocea,* ± linear to linear-lanceolate setae or fimbriae occur in groups of 3 to 7 (sometimes more) on either side of the stipule (cf. Figs. 14 h, 15 a), but their number – even on individual plants – is often variable. The median seta is frequently longer and larger than the flanking (lateral) ones; the longest may be up to ca 5 (6) mm long (e.g., *A. herbaceum*). A solitary, large seta is less common, but does occur in some *Anthospermum* species (e.g., *A. ternatum;* to 2 mm long). In *Galopina circaeoides,* and to some extent in *G. crocyllioides,* the setae are fused basally for up to ca 2 mm (cf. Fig. 15 h); the former has the longest setae of all genera, sometimes reaching a length of over 10 mm. In several *Anthospermum,* most *Carpacoce* and in all *Nenax* species, the appendages of the stipular sheaths are much reduced and only present in the form of small, indistinct teeth-like structures (e.g., *A. ericifolium,* Fig. 15 d; *N. microphylla, N. cinerea,* Figs. 14 i, 15 e), or they are virtually absent (e.g., *C. curvifolia,* Fig. 15 g). There appears to be a distinct correlation between much reduced or ± absent stipular appendages and small, narrow stipular sheaths.

Stipular appendages are always colleter-tipped (cf., for example, Figs. 14 h–j, 15 f), although the colleters may only be clearly discernible on younger leaves. In addition, short-stalked colleters may occur on the rim or on the inside of the stipular sheath near the rim (*Galopina,* cf. Figs. 14 h, 15 i, and *Anthospermum* species, Figs. 14 i, 15 a). Colleters are a common occurrence on stipular structures of the *Rubiaceae* and are rather uniform anatomically (cf. SOLEREDER 1893, 1899, KRAUSE 1909). In the *Anthosperminae* (and in other *Rubiaceae*), the colleters consist of a central core of elongated cells surrounded by ± horizontally arranged, palisade-like cells (similar, for example, in *Alberta* and *Nematostylis,* cf. PUFF, ROBBRECHT & RANDRIANASOLO 1984, Figs. 14 B & F). While in numerous (tropical) woody *Rubiaceae,* the resinous excretion copiously produced by the colleters is thought to provide an effective protection against desiccation of young leaves and shoot apices, it is most doubtful that they have retained this function in the *Anthosperminae* (and, for that matter, in many other *Rubiaceae*).

The stipules are glabrous or, if an indumentum is present, it is confined to the outside of the stipules. Stipule indumentum corresponds to that on the stems and/or leaf blade bases and is, for this reason, not discussed separately.

4.3. Blades and Petioles

Galopina has the largest and broadest leaves of all the genera. The leaves are always distinctly petiolate; petiole lengths range from 0.5 to 14 (20) mm. The lanceolate, ovate to ± subcordate blades are up to 80 (105) mm long and to 30 (37) mm wide (Fig. 104).

Anthospermum, the largest genus, exhibits the greatest variation: Very common and most characteristic are leaves with subobsolete petioles and narrow (linear to linear-lanceolate) blades, which are usually not much longer than 20 mm and ca 2–3 mm wide (e.g., *A. streyi*, Fig. 4a); narrow, rather long blades (e.g., *A. vallicola*, blades up to 55 mm long) are uncommon. Some species have exceptionally small leaves, hardly ever exceeding 10 mm in length [e.g., *A. ericifolium*, or *A. esterhuysenianum*, Fig. 4b, with ± ovate blades only up to 6 (7.5) mm long and 3 (4) mm wide]. In addition, there are species with broader, frequently up to 6 mm wide, oblanceolate or elliptic blades, which can be up to ca 45 mm long (e.g., *A. welwitschii*); their petioles may be obsolete or up to ca 1 mm long. Even larger (up to 50 mm long or more) and broader (to 15 mm) leaves with more distinct petioles (to 6 mm) are less common (e.g., *A. ammannioides*).

In *Nenax*, narrow-leaved foliage – often strikingly similar to that of certain *Anthospermum* species – predominates. Leaves of *N. microphylla*, particularly those on older shoots, are extremely small (ca 2–4 mm long) and broadly (ob)ovate; those on younger shoots may be ± linear-lanceolate or narrowly oblong and to 7 mm long. Leaves on young shoots of *N. microphylla* may have short petioles (ca 0.5 mm long), otherwise the leaves of *Nenax* are epetiolate.

In *Carpacoce*, several taxa have narrow blades, which are similar to those of *Nenax*. Such leaves hardly reach 30 mm in length and are rarely more than 2 mm wide (e.g., *C. vaginellata*, Fig. 10b). In other taxa, blades are conspicuously recurved and much more rigid than in any of the other genera (e.g., *C. heteromorpha*, or *C. curvifolia*, Figs. 10c–d, 15g). The remaining taxa have broader (often ovate-lanceolate) and larger leaves; those of *C. gigantea*, being by far the largest in the genus, may be up to 70 (80) mm long.

Broad leaves have often ± flat or slightly recurved margins; they are usually ± membranaceous and less than 0.4 mm thick (all *Galopina*, some *Anthospermum* and a few *Carpacoce* species; e.g., *A. herbaceum*, *C. gigantea:* Table 2).

In addition, there are distinctly con- and revolutely rolled leaves ("echte kon- und revolute Rollblätter" of NAPP-ZINN 1973–1974). Revolutely rolled leaves – first mentioned for *A. hirtum* by KNOBLAUCH

(1896) – are widely distributed, especially in *Anthospermum*, while convolutely rolled leaves are relatively uncommon (e.g., in populations of *A. dregei* subsp. *dregei*, or in *N. cinerea*; Figs. 20 c, 24 a). Revolutely rolled leaves vary in the extent of the recurvature of the blades (compare, for example, Figs. 21 a–d), and, in some taxa, leaves may show two furrows if viewed from below – i.e., become "double-pocketed" (cf. CARLQUIST 1978) due to massive tissue extensions in the midrib region below (e.g., *A. longisepalum*, Fig. 21 d). Revolute blades are not infrequently ± asymmetrical (cf., for example, *A. monticola*, Fig. 22 a).

Leaves which are ± elliptic to round, ± triangular or semi-orbicular in section (i.e., needle-like or "ericoid" foliage; ericoid is used in a broadly descriptive sense only! Cf. Figs. 20 a–b, 23 a–b, 24 b–d, 25 e, g–h, 26 b–c, e, and 97 d for an overview of possible shapes and outlines) occur in species of all genera except *Galopina* and are often so similar that only certain anatomical details allow a distinction. Sometimes blades appear asymmetrical in section (cf., for example, *A. aethiopicum*, Fig. 23 a), although the degree of asymmetry may vary from plant to plant or even on a single plant.

Needle-like and ± round leaves may be slightly more than 1 mm thick (e.g., *N. divaricata, C. scabra* subsp. *scabra:* Table 2), but more often they have a thickness of less than 0.5 mm (e.g., *N. hirta:* Table 2).

It must be borne in mind that shrinkage upon drying of the leaves may result in shapes dissimilar to those described above (particularly in *Nenax* and *Anthospermum* species with "ericoid" leaves). Because of the large and thick-walled upper epidermal cells, the massive bundle sheath extensions in the midrib region and, in contrast, the rather loose, "shrinkable" mesophyll and the tender, thin lower epidermal cells, leaves that are ± elliptic in section when fresh, seem to be revolute and "double-pocketed" when dried. Due to shrinkage, dried leaves may also be thinner than fresh ones. Leaf thickness measurements presented in Table 2 only refer to fresh material or specimens preserved in FPA.

It should also be noted that in some taxa, the long shoot leaves are often distinctly shorter, broader and – in taxa with ericoid leaves – flatter than the short shoot leaves of one and the same plant (heterophylly; e.g., *C. scabra, A. galpinii*, or *N. acerosa* subsp. *macrocarpa:* compare Figs. 24 c and d). This may be particularly puzzling if newly produced long shoots (without short shoots) occur on one plant together with older long shoots which bear leaves on short shoots only (*N. acerosa* subsp. *macrocarpa!*). In *Galopina*, the short shoot leaves are often smaller than the long shoot leaves, but mostly have the same shape (especially in *G. circaeoides* and *G. tomentosa*).

4*

In dioecious taxa there may be considerable differences in leaf size, shape and cross-sectioned outline between ♂ and ♀ plants (see, in this context, 12.3.).

In a few *Carpacoce* and *Anthospermum* taxa, a slight trend to anisophylly can be observed. While anisophylly is certainly not very pronounced in the vegetative region (the leaves of *A. esterhuysenianum,* for example, are only very slightly anisophyllous), it can be more obvious in the inflorescence region (cf. anisophyllous bract pair of *C. scabra* subsp. *scabra,* Fig. 11 b; see 5. for details).

4.4. Age and Persistence

According to field studies and continued observations of cultivated plants, it appears that in shrubby taxa the leaves usually start dying off after one growth period (e.g., *A. ammannioides, A. aethiopicum, A. galpinii, A. spathulatum,* and others) and hardly become older (*N. cinerea?*). Dried, brown (long and short shoot) leaves, however, do not drop off readily but remain on the plants for several years (cf. *N. microphylla,* Fig. 9). Leaves, thus, are clearly *not* evergreen. This is particularly noteworthy as the highest concentration of taxa is found in the SW Cape Floristic Region, an area within mediterranean climatic conditions and characteristically *evergreen* sclerophyllous vegetation (cf. for example, KUMMEROV 1973).

4.5. Anatomy (Tables 1 and 2)

4.5.1. Epidermis

Size and Shape: In surface view, the epidermal cells typically appear hexa- to polygonal (with two sides often longer than the others), but there are also transitions to either almost round, rectangular or to more irregular shapes (cf. Figs. 16 a–d, 17 a–b). The anticlinal walls may be undulate. This is a fairly common phenomenon on the lower surface (e.g., *A. ternatum* subsp. *randii* or *G. circaeoides,* Figs. 18 c, 19 c), but on the upper surface distinctly wavy walls as in *A. esterhuysenianum* (Fig. 20 g) do not occur very often. It must be stressed that the cell shapes (as seen in surface view) are often variable and are, thus, of virtually no value as a distinguishing character between species or genera. In numerous species it is difficult if not impossible to make out the shapes of the epidermal cells because of the ± thick, continuous cuticles (see below). The average sizes of the upper epidermal cells in surface view are given in Table 1. Except for *Carpacoce* species, the lower epidermal cells are (considerably) smaller than the upper.

Table 1. Lengths and widths of upper epidermis cells (as seen in surface view) of some taxa of *Anthospermum, Galopina, Nenax,* and *Carpacoce* (averages; in μm)

Anthospermum aethiopicum	40–50 × 30–40
A. ammannioides	65–90 × 50–65
A. asperuloides	55–75 × 35–50
A. basuticum	40–50 × 30–40
A. bergianum	80–112 × 55–65
A. bicorne	56–80 × 50–95
A. comptonii	65–110
A. dregei subsp. *ecklonis*	80–105 × 40–55
A. esterhuysenianum	40–65 × 25–40
A. galioides subsp. *galioides*	40–65 × 35–50
A. galpinii	65–90 × 40–60
A. herbaceum	50–90 × 30–50
A. hispidulum	55–90 × 40–60
A. hirtum	50–95 × 30–70
A. isaloense	55–80 × 40–55
A. littoreum	55–75
A. longisepalum	40–55
A. monticola	40–80
A. pachyrrhizum	80–120 × 65–80
A. prostratum	55–80 × 40–65
A. pumilum subsp. *pumilum*	65–90 × 50–70
A. rosmarinus	70–95 × 55–70
A. streyi	50–100 × 32–50
A. ternatum subsp. *randii*	50–65
subsp. *ternatum*	135–160 × 70–100
A. vallicola	50–80 × 50
A. welwitschii	120–160 × 65–90
Galopina aspera	50–65 × 30–40
G. circaeoides	40–55 × 24–40
G. crocyllioides	50–65 × 35–50
Nenax divaricata	55–80 × 50–60
N. elsieae	80–95 × 40–55
N. microphylla	43–95 × 35–65
Carpacoce scabra subsp. *scabra*	50–90 × 32–45
C. spermacocea subsp. *orientalis*	55–110 × 36–44

The above survey of shapes and sizes refers to epidermal cells situated ± midway between the leaf margin and the midrib; the epidermal cells above and below the midrib are frequently, as depicted for *A. esterhuysenianum* (Fig. 20f), much elongated and ± rectangular.

The thickness of the upper and lower epidermal cells of a leaf differs considerably between *Anthospermum, Galopina* and *Nenax* (cf. Table 2). While in *Galopina* the upper epidermis is only 1.3 to 1.5 times as thick as the lower epidermis, this ratio is increased to ca 2.5 : 1 to 4 : 1 in *Nenax,* and

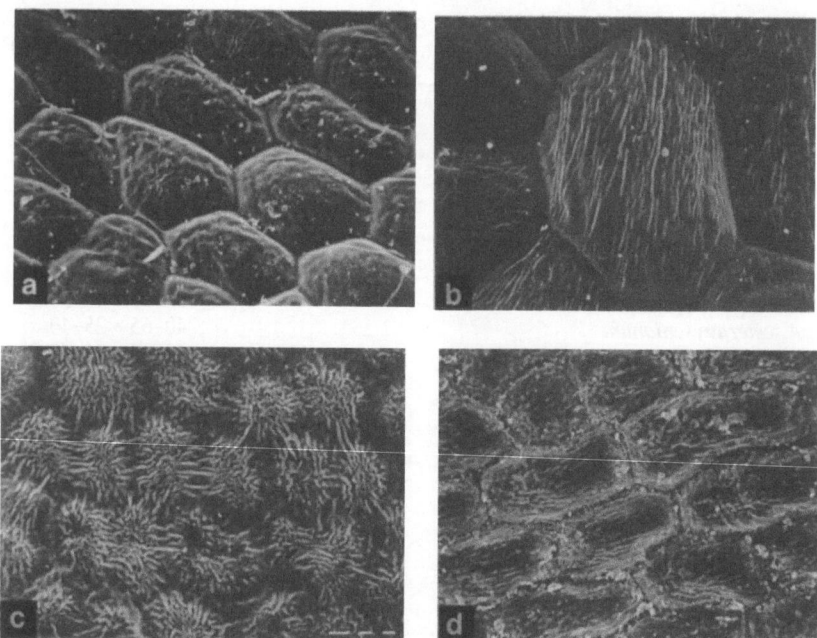

Fig. 16. SEMgraphs of leaf surfaces (upper epidermis) of *Anthospermum* and *Carpacoce*. *a A. herbaceum* (PUFF 781028-3/1). *b A. ternatum* subsp. *ternatum* (PUFF 781221-1/1). *c A. isaloense* (PUFF 800814-3/2). *d C. spermacocea* subsp. *orientalis* (PUFF 790909-5/2). – All critical point dried; all × 232

may be up to 9 : 1 in certain *Anthospermum* species. In *Carpacoce*, upper and lower epidermis cells are ± equally thick. In *Galopina*, the upper epidermal cells do not exceed a thickness of 50 µm, but in *Anthospermum* the cells may be more than 100 µm (e.g., *A. dregei* subsp. *dregei*, 100–160 µm), and in *Nenax* more than 200 µm thick. The lower epidermal cells may be less than 10 µm thick in *Anthospermum* and *Galopina*, but are commonly up to ca 30 µm thick or sometimes even thicker (certain *Anthospermum*, *Nenax* and *Galopina* species). The thickness of the upper and lower epidermal cells of *Carpacoce* ranges from 25 to ca 65 µm (consult Table 2).

Fig. 17. SEMgraphs of leaf surfaces and indumentum of *Anthospermum*, *Carpacoce* and *Nenax a A. longisepalum* (PUFF 800814-2/1). *b A. hispidulum* (PUFF 790225-1/2). *c C. scabra* subsp. *rupestris* (TAYLOR 6564). *d N. cinera* (PUFF 790711-1/1). *e A. hispidulum* (PUFF 780312-1/1), tip of revolute leaf from below. *f A. ericifolium* (ESTERHUYSEN 28054), leaf margin. *g N. hirta* subsp. *hirta* (PUFF 800910-2/1), stem hairs. *h–i* hair surfaces, *h N. hirta* subsp. *hirta* (fruit; PUFF 800910-2/1), *i A. galpinii* (fruit; PUFF 790303-2/1). – *a–c, e* critical point dried, all others herbarium material. *a–d, g* × 232; *e* × 35; *f* × 70; *h–i* × 1 628

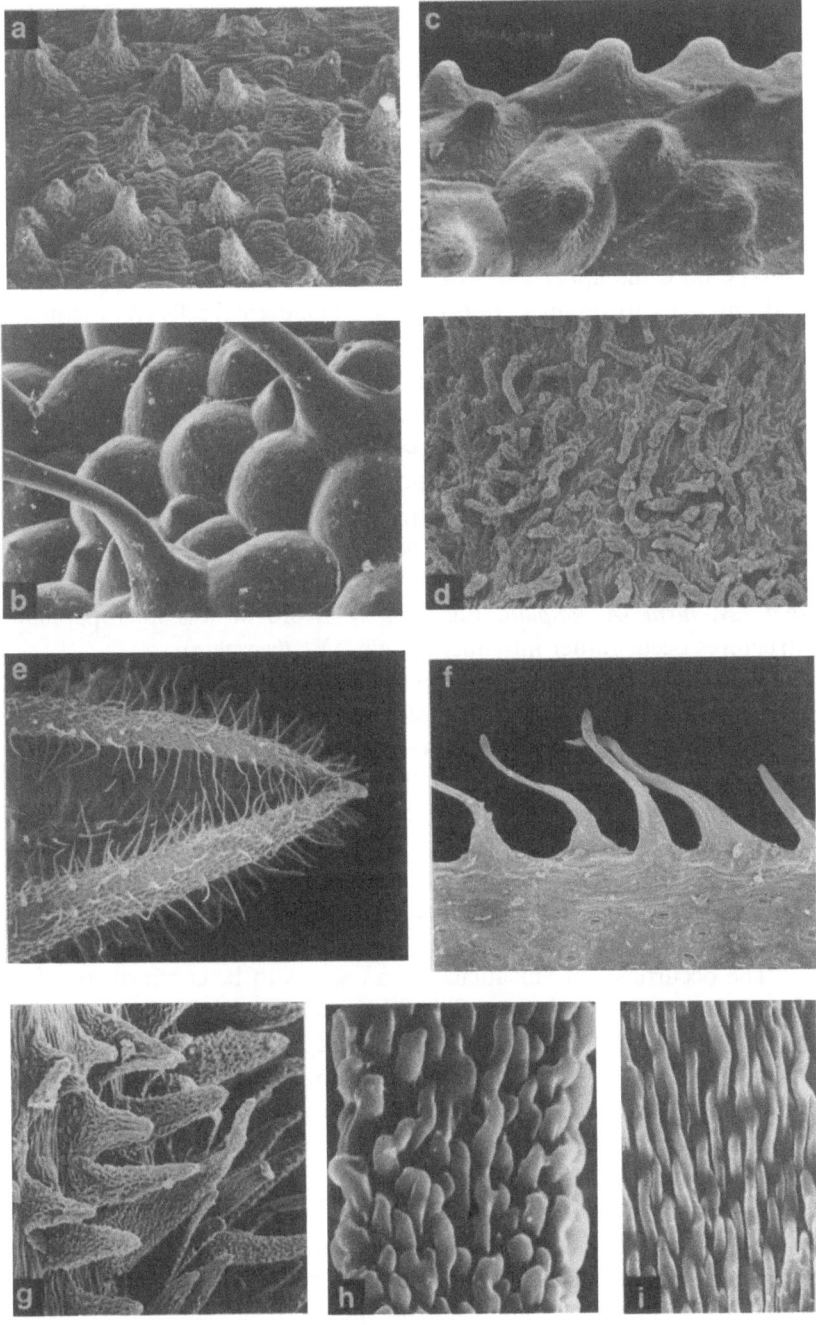

Fig. 17

In some xeromorphic species of *Anthospermum* and *Nenax*, the o u t e r w a l l s of the upper epidermal cells may be conspicuously thickened and cutinized. KNOBLAUCH (1896), for example, gives a thickness of 11.1 µm and 2.6–6.6 µm for the outer walls of the upper and the lower epidermal cells of *A. aethiopicum*. In general, the thickened outer walls of the upper epidermal cells are between 6 and 25 µm thick, with a peak at about 10–15 µm. The outer walls of the lower epidermal cells (usually not conspicuously thickened) are between 3 and 10 µm thick. In *Carpacoce* species, both the upper and lower epidermis may have thickened outer walls up to ca 10 µm thick. In *Galopina*, all epidermal cells are ± equally thick.

The superposed c u t i c l e s are approximately 1.5 to 4 µm thick. The cuticle surfaces may be ± smooth (e.g., *A. hispidulum*, Figs. 17 b, 18 b; *N. cinerea, young* leaf, Fig. 19 a; *G. circaeoides*, Fig. 19 c), but more commonly they exhibit some micro-sculpturing. Various kinds of striations predominate (compare Figs. 16–19). The striations on the epidermal cells s. str. are identical to those on the trichomes of the leaves (e.g., *A. longisepalum, A. hirtum*, Figs. 17 a and 18 f respectively), stems and fruits (e.g., *N. hirta, A. galpinii*, Figs. 17 g–i). More or less micropapillate structures occur rather infrequently (e.g., *N. divaricata*).

While cuticular sculpturings associated with stomata may, at least in some cases, be species-specific and of some systematic value (see p. 59), the surface characters of other epidermal cells appear to be too unspecific and variable to be of any systematic value. It also appears that the cuticular sculpturing may vary with the age of a leaf. In *N. cinerea*, for example, leaves from a young cultivated plant were found to be smooth and to lack sculpturing (Fig. 19 a); it is only in older plants that leaves develop their hairy covering and – with it – their characteristic cuticular sculpturing (Fig. 17 d).

The occurrence of epicuticular w a x seems to be confined to a few *Carpacoce* species. Rod- or ± flake-like structures or platelets appear to be primarily associated with stomata (e.g., *C. gigantea, C. scabra* subsp. *scabra*, Figs. 19 e, h). They are not always present on all leaves of a plant (disappearing with age?) or are entirely lacking in other plants of one and the same species (e.g., *C. scabra* subsp. *scabra*, compare Fig. 19 g and 19 h).

E x c r e t i o n s: In *Anthospermum, Nenax* and *Galopina* leaves, at least some epidermal cells are filled with a golden yellow to dark reddish-brown substance, which was first mentioned by SOLEREDER (1893) and KNOBLAUCH (1896) for *Anthospermum* species, *N. acerosa* and *N. hirta*. SOLEREDER (1893, 1899) referred to them as "resiniferous secretory cells" (more correctly to be called excretory cells – cf. NAPP-ZINN 1973–1974).

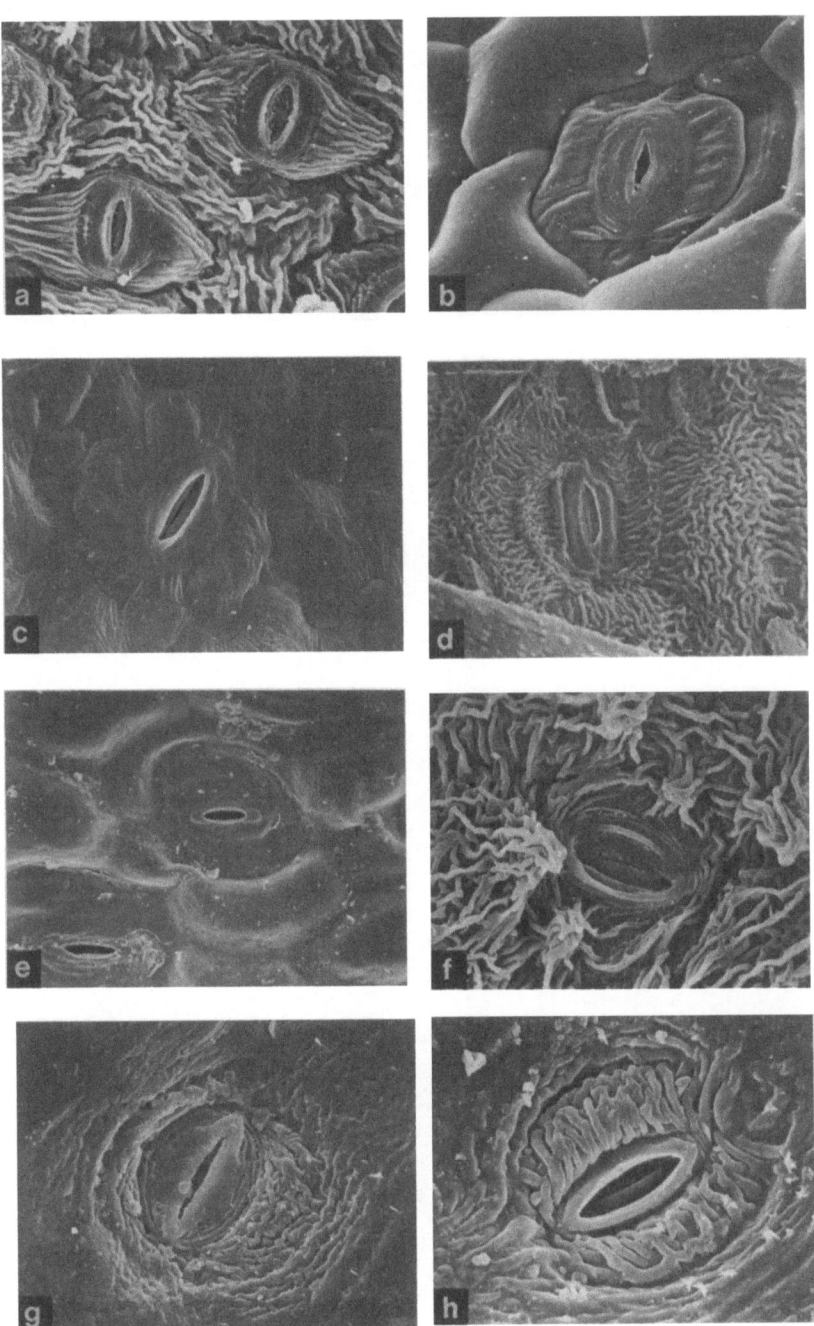

Fig. 18. SEMgraphs of stomata of *Anthospermum*. *a A. basuticum* (PUFF 790113-5/3). *b A. hispidulum* (PUFF 780312-1/1). *c A. ternatum* subsp. *randii* (PUFF 780215-3/1). *d A. rosmarinus* (KASSNER 2954). *e A. comptonii* (PUFF 790914-3/2). *f A. hirtum* (PUFF 800903-1/2). *g A. bicorne* (PUFF 800925-2/1). *h A. ericifolium* (ESTERHUYSEN 28054). *d, h* herbarium material; all others critical point dried; all × 700

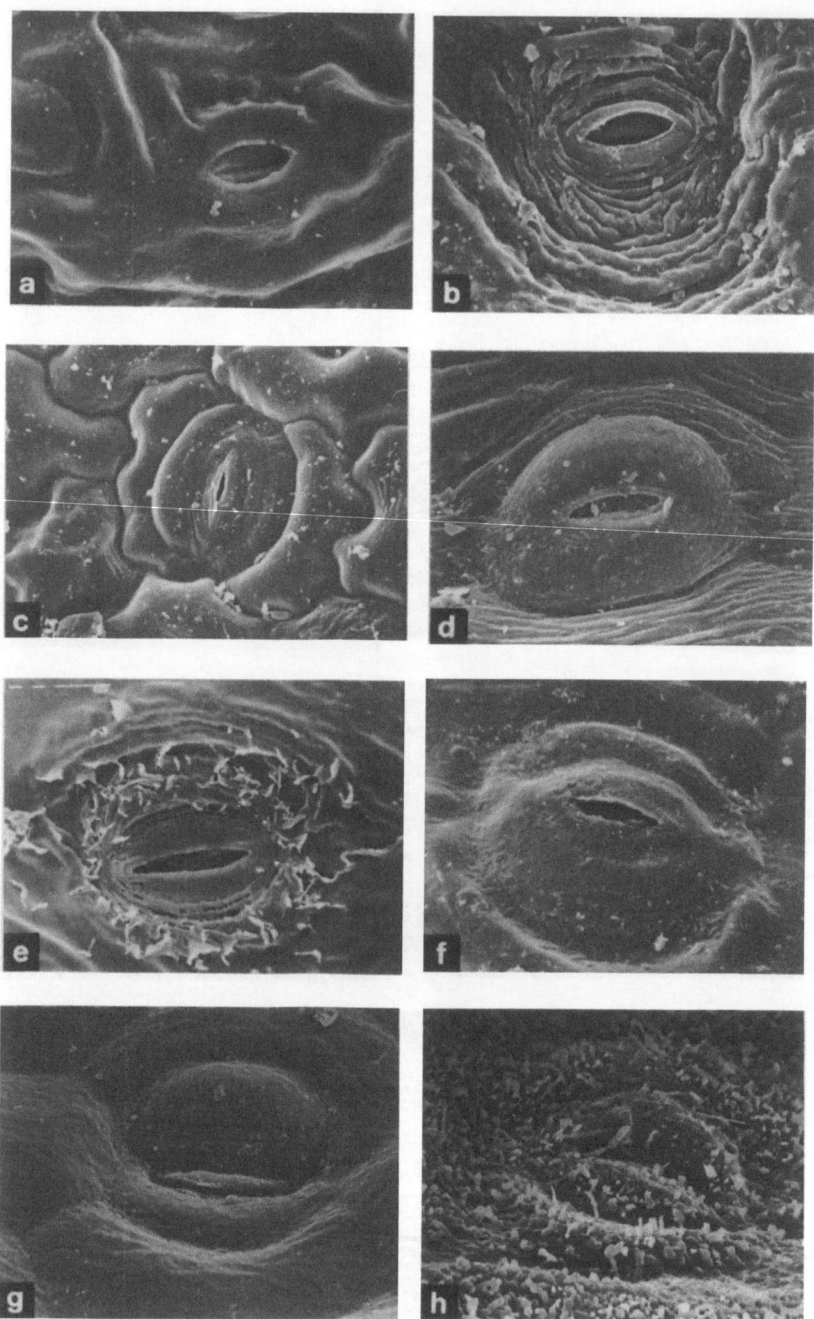

Fig. 19. SEMgraphs of stomata of *Nenax*, *Galopina*, and *Carpacoce*. *a N. cinerea*
(PUFF 800102-4/2), young leaf (hairs absent!) of cultivated plant. *b N. acerosa*
subsp. *acerosa* (PUFF 800920-3/1). *c G. circaeoides* (PUFF 781029-3/1). *d C.
spermacocea* subsp. *spermacocea* (PUFF 781228-3/1). *e C. gigantea* (TAYLOR 4241). *f
C. burchellii* (ESTERHUYSEN 20092). *g–h C. scabra* subsp. *scabra*, *g* PUFF 800914-6/1,
h ESTERHUYSEN 10961. – *b–d, g* critical point dried, all others herbarium material.
All × 700

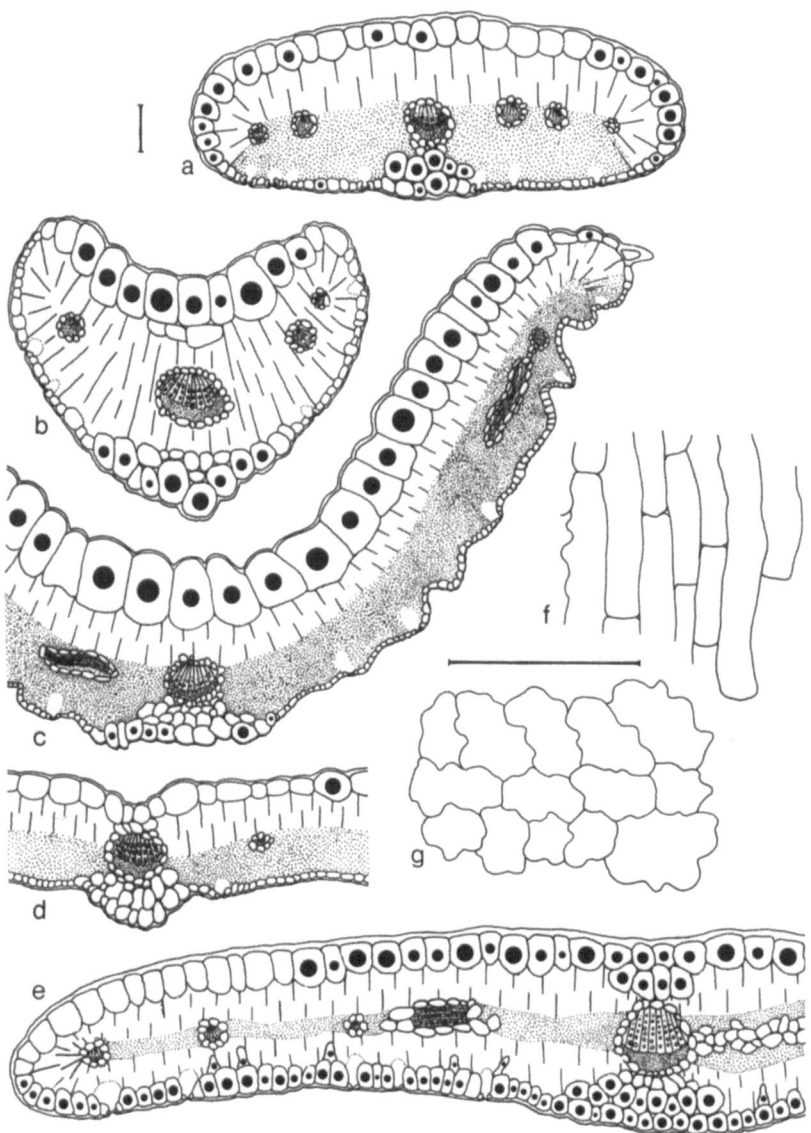

Fig. 20. Leaf sections and upper epidermis of *Anthospermum*. *a A. ibityense* (Puff 800730-1/2). *b A. bicorne* (Puff 800913-4/2). *c A. dregei* subsp. *dregei* (Puff 780810-2/1). *d A. pachyrrhizum* (Puff 810930-2/1). *e A. isaloense* (Puff 800814-3/2). *f–g A. esterhuysenianum* (Puff 800924-3/2), upper epidermis, *f* in the midrib region, *g* between midrib region and margin. – Semischematic camera lucida drawings; black dots: cells filled with tannin; chlorenchyma: palisade layers as straight lines, spongy mesophyll dotted. The scale unit is 0.1 mm

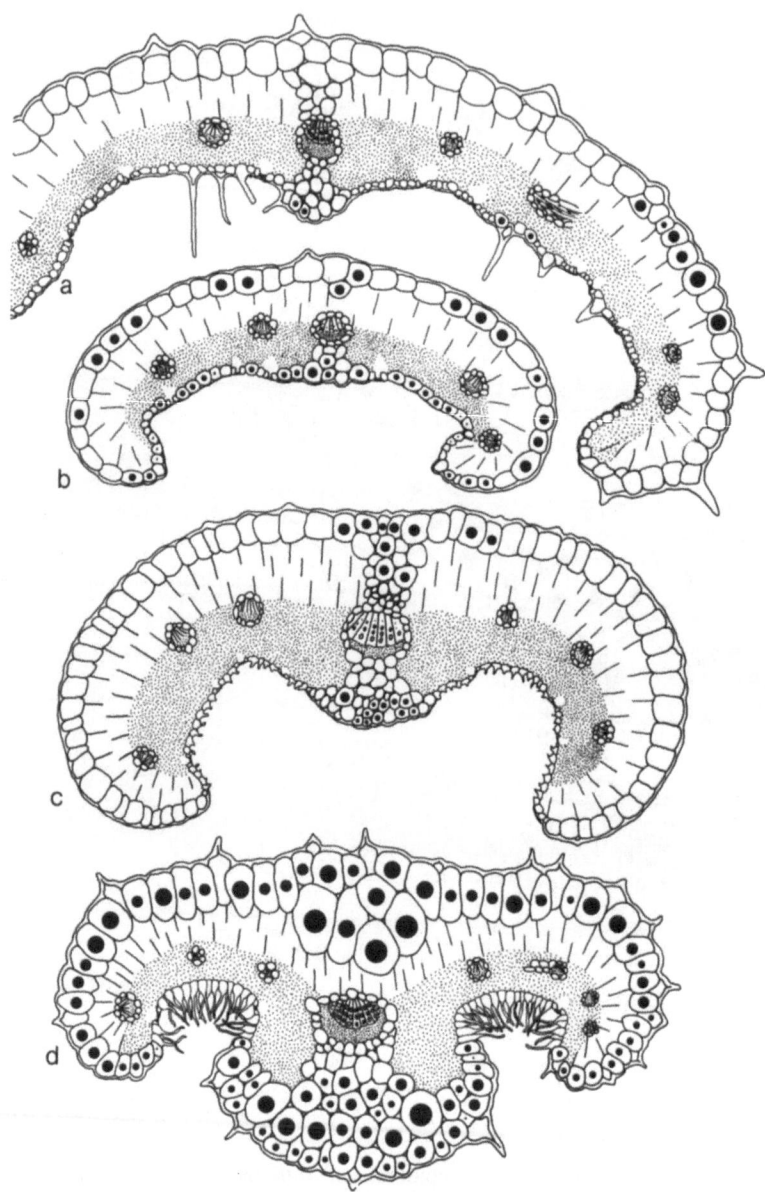

Fig. 21. Leaf sections of *Anthospermum. a A. hispidulum* (Puff 780312-1/1). *b A. zimbabwense* (Puff 790124-2/2). *c A. hirtum* (Puff 800903-1/2), note papillate lower epidermis. *d A. longisepalum* (Puff 800814-2/1). – See Fig. 20 for explanation and magnification

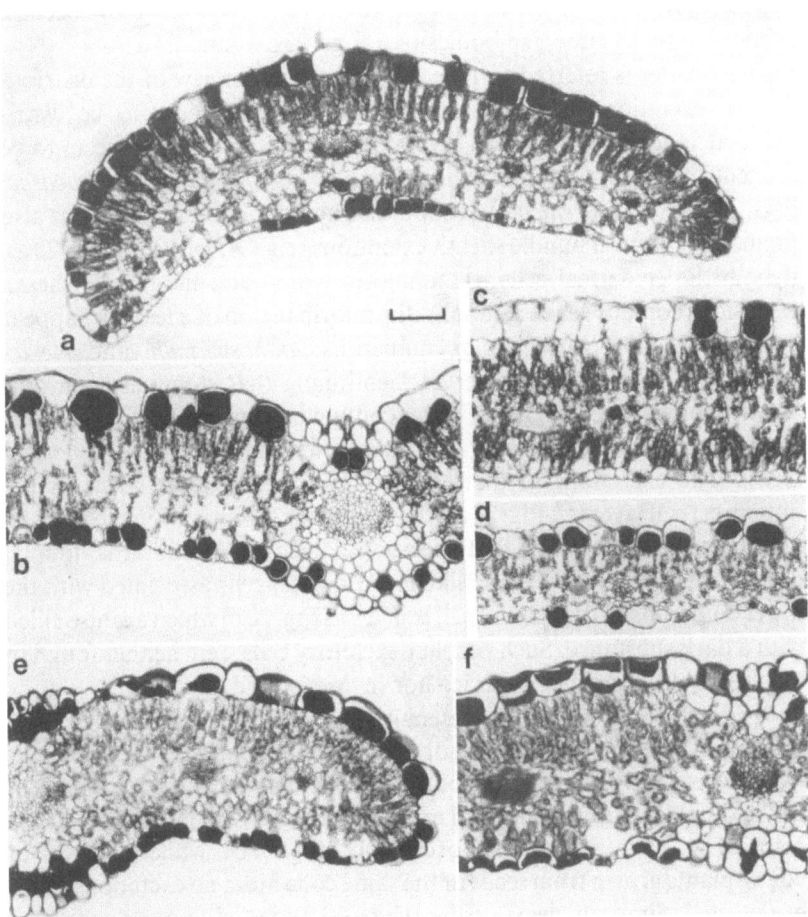

Fig. 22. Microtome sections of leaves of *Anthospermum* and *Galopina. a A. monticola* (Puff 780820-1/1), compare with Fig. 23c. *b–c A. welwitschii, b* Puff 781028-1/2, *c* Puff 781028-2/1, note mesophyll differences. *d G. crocyllioides* (Puff 790416-1/1). *e A. vallicola* (Puff 790125-1/1). *f A. usambarense* (Puff 780114-4/4). – Some epidermis and bundle sheath extension cells are filled with tannin (black); note idioblasts with raphids (*a, c–d*). The scale unit is 0.1 mm; *a–f* have the same magnification

Rousseau (1953) considers these excretions to be a mixture of essential (volatile) oils and phytosterines. Her conclusions, however, were based on the investigation of various members of the tribe *Rubieae*. The chemical nature of the excretions in the leaves of the *Anthosperminae* is as yet not known; perhaps they are tannins which are known to occur in leaves of the *Rubiaceae* (see, in this context, Haslam 1981). In stained microtome sections, the excretions contract to become dark, round bodies (cf. Figs. 22 a–f), and in critical point dried leaf sections they become cylindrical to ± cube-like bodies (see Figs. 23 a–d and 25 d).

The reader is referred to Figs. 20–25 for an overview of the distribution of excretions in the epidermis. They often appear to be distributed at random, although in several taxa the excretory cells seem to be concentrated in the midrib region (e.g., *N. acerosa* subsp. *macrocarpa*, Figs. 24 c–d, or *N. arenicola*). It should be noted that these excretions also frequently occur in bundle sheath extentions (e.g., *A. ibityense*, Fig. 20 a). If both the epidermal cells in the midrib region and the bundle sheath extensions contain dark excretions, the midrib region of a leaf may appear distinctly reddish brown (e.g., numerous *Anthospermum* and *Nenax* species). It, furthermore, deserves mentioning that stomatal apparatus and trichomes (papillae, hairs) never contain excretions (cf., for example, Figs. 21 a, c–d, *A. hispidulum*, *A. hirtum:* lower surface, and *A. longisepalum*). The excretions in the leaves of *N. cinerea* never occur in the outermost cell layer (cf. Fig. 24 a); the reason for this may lie in the unusual anatomy of the leaves of that species (see below). Attention is, finally, drawn to the peculiar, ± cone-like to cylindrical cells associated with the lower epidermis of the leaves of *A. isaloense* (Fig. 20 e) which are also filled with a dark substance. Such peculiar excretory cells were neither found in any other *Anthospermum* species nor in *Nenax* and *Galopina*.

The amount of excretions present in a leaf may vary. In *N. microphylla*, for example, newly developed short shoot leaves contained no excretions whatsoever (Fig. 24 e), whereas both the upper and the lower epidermis of leaves on older shoots contained excretions (Fig. 24 f); leaves from an old plant of *A. bicorne* contained excretions (Fig. 20 b), while in leaves of young plants grown from seeds of the same collection, no excretions could be detected. Similarly, leaves from the basal nodes of a young cultivated plant of *N. cinerea* contained no excretions, while – after several months – at least some epidermal cells of the basal leaves were filled with a dark substance. It thus seems that both the age of a *plant* and the age of a *leaf* may be responsible for at least some of the variation; it, however, remains unknown to what extent other factors (environmental conditions?) also play a role.

Such dark, "resiniferous" excretions are entirely lacking in *Carpacoce*.

SOLEREDER (1893) recorded "resiniferous, branched cells, often united in groups or series" in the *mesophyll* of *Phyllis nobla* leaves, but not in the leaves of any of the other *Anthospermeae* he investigated (*Nertera, Coprosma, Pomax*).

Similar dark excretions also occur in the stems (outside the vascular

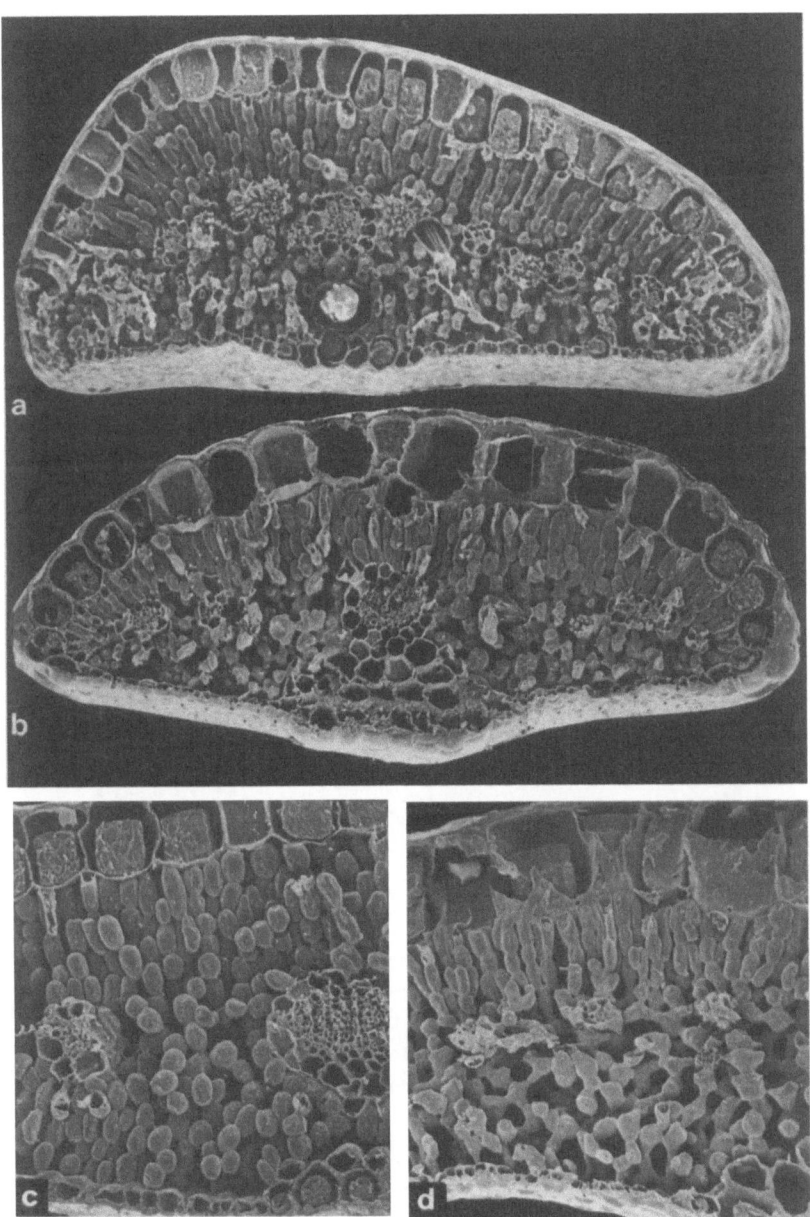

Fig. 23. SEMgraphs of critical point dried sectioned leaves of *Anthospermum. a A. aethiopicum* (Puff 790712-3/1). *b A. pumilum* subsp. *pumilum* (Puff 790122-2/1). *c A. monticola* (Puff 790113-3/1), compare with Fig. 22*a. d A. comptonii* (Puff 790914-3/2). — Note clumped contents of some upper epidermis and bundle sheath extension cells (tannins!), artificially opened idioblasts with raphids and differences in mesophyll structure. Further explanations in the text. *a–b* × 87; *c–d* × 124

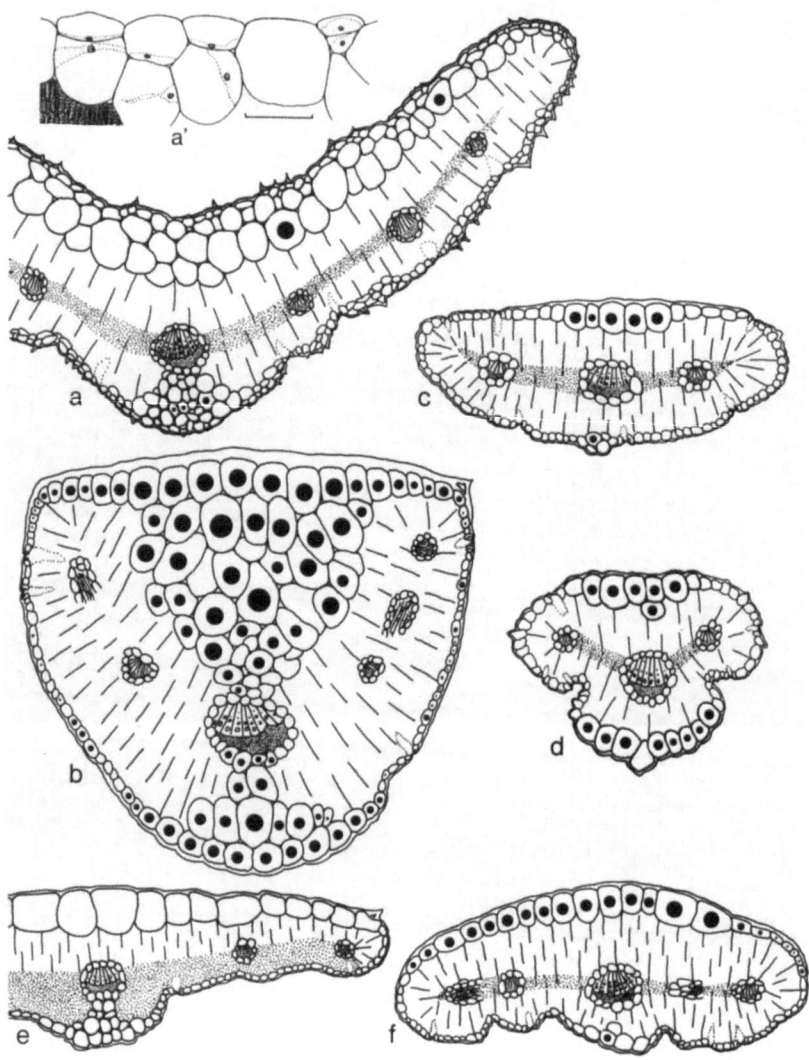

Fig. 24. Leaf sections of *Nenax. a–a′ N. cinerea* (PUFF 800102-4/2), *a′* developing multiple epidermis (young leaf from cultivated plant). *b N. coronata* (PUFF 800902-6/4). *c–d N. acerosa* subsp. *macrocarpa* (PUFF 800920-2/2), both from the same plant, *c* from long shoot, *d* from short shoot; note stomata on upper surface. *e–f N. microphylla* (PUFF 790710-1/10), *e* young leaf from newly developed long shoot, *f* leaf from short shoot. – See Fig. 20 for explanation and magnification

cylinder), the fruits (epidermis) and the testa cells of *Anthospermum*, *Nenax* and *Galopina* species.

The epidermis of *Nenax cinerea* and *Carpacoce hetero-morpha*: *N. cinerea* has a truly multilayered (multiple) epidermis (rather than a hypoderm; see NAPP-ZINN 1973–1974 for a discussion of these terms) (Figs. 24 a, 25 f). It was observed that in young leaves the epidermal cells of both surfaces undergo several periclinal divisions (cf. Fig. 24 a'), giving rise – on the outside – to small papillae which eventually cover the entire leaf surface (cf. Fig. 17 d). On the lower surface of the young leaf pictured in Fig. 19 a, no papillae are as yet developed, while in older leaves these papillae are so dense that it may even be difficult to detect the stomata (Fig. 15 e).

The origin of the two- to several-layered dermal tissue of *C. heteromorpha* (Fig. 26 e) may be identical to that of *N. cinerea*.

These two species are the only two taxa in the *Anthosperminae* that show a continuous several-layered dermal tissue.

According to VESQUE (cited by SOLEREDER 1893) a "hypoderm" is present in *Coprosma baueri* (subtribe *Comprosminae*). A reinvestigation of this species has shown that there is in fact a continuous layer of colourless cells between the actual epidermis and the mesophyll which is at least twice as thick as the epidermal layer; it is absent from the lower surface. Whether it is a hypoderm proper or a second epidermal layer could not be determined.

Trichomes (Papillae and Hairs): Papillae, with dome- or cone-shaped protuberances arising in the center of the cells (cf. Figs. 17 a, c), occur in selected species of all genera on the upper and the lower leaf surfaces, or sometimes only on the leaf margins. Such papillate protuberances are usually not more than 50 µm high, and sometimes even less than 10 µm. Their cuticle may be thicker and more conspicuously sculptured than that of the surrounding epidermal cells (cf., for example, *A. hirtum*, Fig. 18 f, papillae surrounding the stomata, or *N. cinerea*, Fig. 17 d).

Hairs, ranging from ca 100 to ca 1 000 µm in length, have the same basic structure as the papillae and always have a typical, ± "bulbous" base (e.g., *A. hispidulum*, Fig. 17 b). The hairs are unicellular in the genera *Anthospermum*, *Nenax* and *Carpacoce*. Odd hairs on the leaves of *Galopina* species (except for *G. crocyllioides*), however, may be two- to three-celled (especially on the midrib below); the majority of hairs on the leaves of *Galopina* species, nevertheless, is also unicellular. In *Nenax* and *Carpacoce*, the leaf surfaces bear no hairs; in a few species of these two genera (e.g., *C. heteromorpha, N. elsieae*), however, the leaf margins may have hairs ca 100 to 400 (500) μm long. In *Galopina* (except for *G. crocyllioides*) and certain *Anthospermum* species, hairs may be found on both surfaces (e.g., *A. hispidulum*, Figs. 17 e, 21 a), but these are often denser on the upper surface and on the midrib below than on the lower surface (e.g., *A. basuticum*). In several *Anthospermum* species, the hairs are confined to the leaf margins (e.g., *A. ericifolium*, Fig. 17 f). *A. longisepalum* is unusual and unique in having papillae on the upper surface and, on the lower side, hairs up to 100 μm long in the two furrows which bear the stomata (Fig. 21 d).

Stomata: The leaves of all four genera are hypostomatic with the exception of two *Nenax* species (*N. acerosa*, Figs. 24 c–d, *N. arenicola*) and several *Carpacoce* species (*C. vaginellata, C. scabra, C. gigantea, C. heteromorpha*, Figs. 25 e, 26 b–e); the latter have amphistomatic leaves.

In the *Rubiaceae*, amphistomatic leaves appear to be relatively rare and are – outside the *Anthosperminae* – only known to occur in one species of *Coprosma* (subtribe *Coprosminae*; FOWERAKER 1916), "*Gaillonia*"[1], *Crocyllis* and *Putoria* (*Paederieae*; QARER 1973, PUFF & MANTELL 1982 a and PUFF, unpublished), *Asperula* species (*Rubieae*; NETOLITZKY 1905), *Diodia* and *Spermacoce hispida* (*Spermacoceae*; HOLM 1907, MULLAN 1933), *Houstonia caerulea* (*Hedyotidae*; HOLM 1907), and *Nematostylis anthophylla* (= *Alberta loranthoides, Alberteae*; PUFF, ROBBRECHT & RANDRIANASOLO 1984).

The stomatal apparatus are of the typical "rubiaceous type" ("paracytic" sensu METCALFE & CHALK 1950; "para-twi-acyclic" in the morphological classification of stomatal complexes of PATEL 1979), and consist of guard cells which are surrounded by two parallel subsidiary cells (cf. Figs. 18 and 19).

The sizes of the guard cells range from ca 20×15 to 30×23 μm in *Galopina*, from ca 25×20 to 43×28 μm in *Nenax*, from ca 20×14 to

---

[1] See LÉONARD (1984) for comments on the "*Gaillonia* complex".

Fig. 25. SEMgraphs of critical point dried sectioned leaves of *Anthospermum*, *Carpacoce*, and *Nenax* a *A. bergianum* (PUFF 790718-1/1), vascular bundle and bundle sheath. b *A. welwitschii* (PUFF 790211-2/3). c *A. bergianum* (PUFF 790718-1/1), note absence of palisade cells; upper epidermis compressed. d *A. spathulatum* subsp. *spathulatum* (PUFF 790711-2/1). e *C. scabra* subsp. *scabra* (PUFF 790914-4/2), note stoma on upper surface; all vascular bundles are embedded in parenchymatous tissue. f *N. cinerea* (PUFF 790711-1/1), note small epidermis cells and subepidermal layers. g–h *N. divaricata* (PUFF 790712-2/1). – Further explanations in the text. a ×618; b–c ×124; e–h ×87

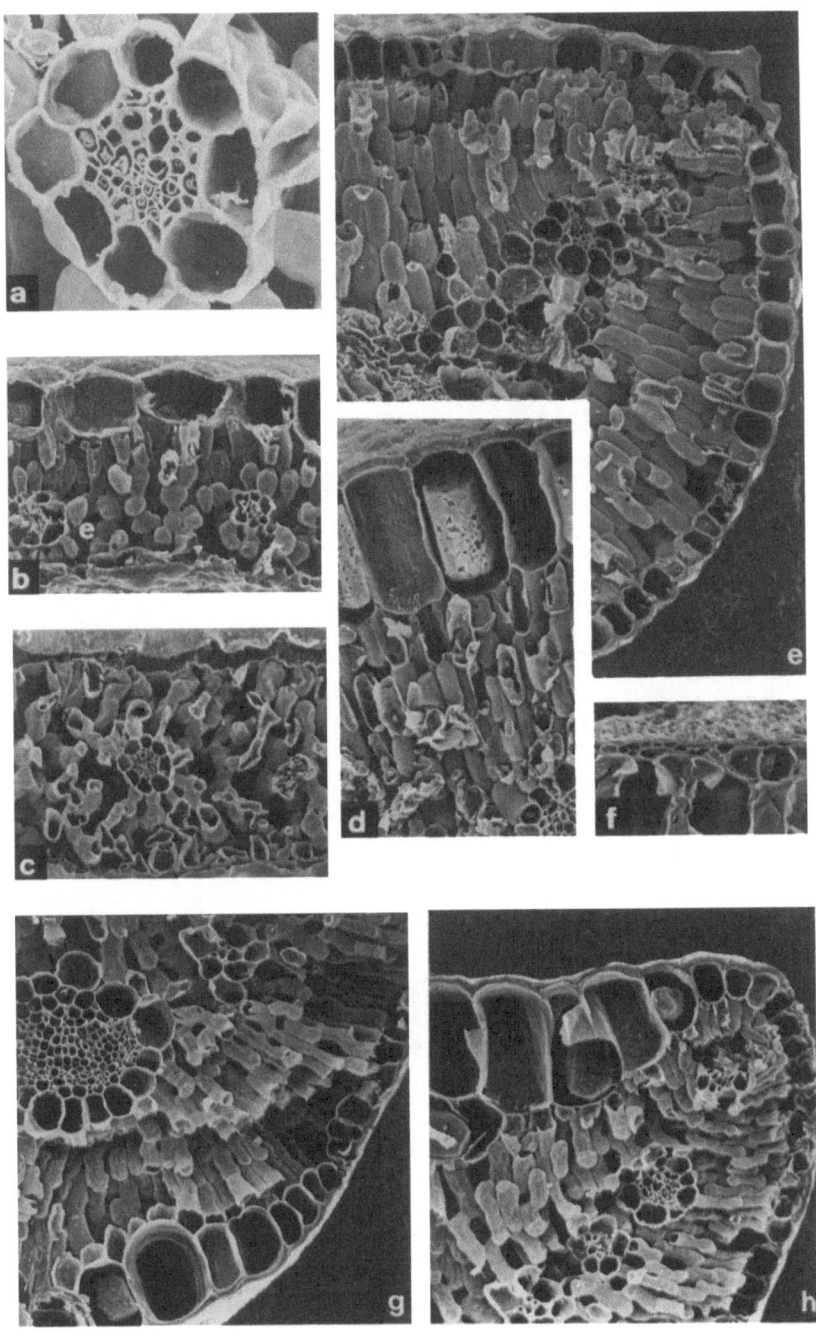

Fig. 25

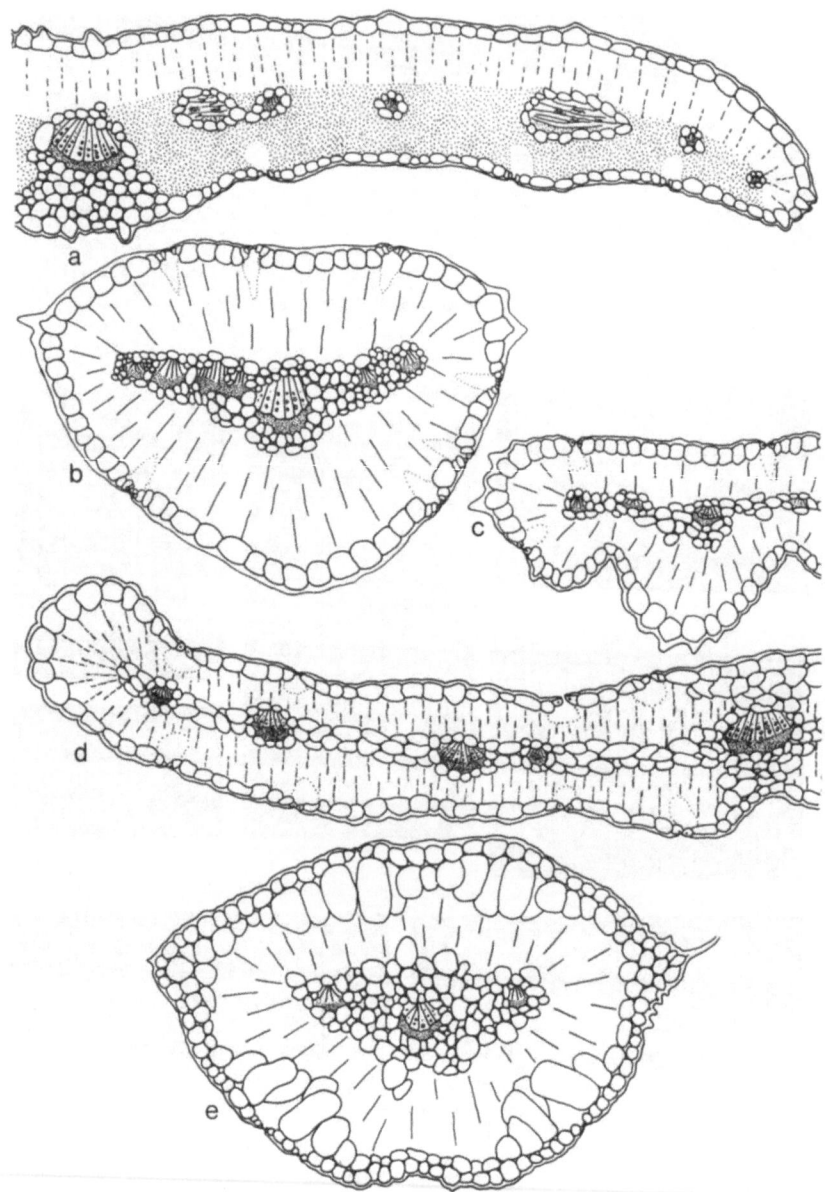

Fig. 26. Leaf sections of *Carpacoce*. *a C. spermacocea* subsp. *orientalis* (Puff 790909-5/2). *b C. vaginellata* (Puff 800913-4/3). *c C. scabra* subsp. *rupestris* (Taylor 6567). *d C. gigantea* (Taylor 4241). *e C. heteromorpha* (Puff 800917-4/1). – Note stomata on upper surface and continuous parenchymatous tissue between vascular bundles in *b–e*; interrupted straight lines indicate indistinct palisade cells (i.e., difficult to distinguish from spongy mesophyll). See Fig. 20 for further explanations and magnification

$42 \times 26\,\mu m$ in *Anthospermum*, and from ca 35×24 to $52 \times 35\,\mu m$ in *Carpacoce*.

In some *Anthospermum, Nenax* and *Carpacoce* species the stomata may be slightly, but not conspicuously sunk (e.g., *A. ericifolium, N. acerosa, C. scabra*, Figs. 18 h, 19 b, g). In a few *Carpacoce* species, they may be slightly raised (e.g., *C. burchellii*, Fig. 19 f). The stomatal apparatus are – at least partially – covered by the surrounding papillae or hairs in *A. hirtum, A. longisepalum* (Figs. 18 f, 21 d) and *N. cinerea*.

In numerous species of all genera except *Galopina*, the subsidiary cells (and occasionally also neighbouring cells) show conspicuous cuticular sculpturing. While, in a few taxa, this cuticular sculpturing is species-specific (e.g., the striations at right angles to the guard cells in *A. basuticum*, Fig. 18 a, or the massive cuticular ridges in *A. ericifolium*, Fig. 18 h), cuticular structures on the subsidiary cells cells are, in general, of little systematic value. The cuticular patterns may repeat themselves in different genera – the thick cuticular folds in *A. bicorne* (Fig. 18 g), for example, are also typical of numerous *Nenax* species (cf. *N. acerosa* subsp. *acerosa*, Fig. 19 b). In some taxa, the cuticular sculpturing of the subsidiary cells may vary from plant to plant; especially if the cuticular striations are normally weak (e.g., *A. hispidulum*, Fig. 18 b), they may be entirely absent from some subsidiary cells.

4.5.2. Chlorenchyma (Table 2)

In *Galopina* and in numerous *Anthospermum* species, the chlorenchyma is dorsiventral (i.e., consists of palisades and spongy mesophyll). The palisades are usually comprised of two or more layers of rather short, $\pm$ cylindrical cells which range from ca 20 to 45 (60) μm long (cf., for example, *A. aethiopicum, A. streyi, A. comptonii*, Figs. 23 a–b, d; 97 c). The spongy mesophyll, often marginally thicker than the palisade layer (consult Table 2), either consists of a network of branched cells (e.g., *A. comptonii*, Fig. 23 d) or of $\pm$ spherical cells (e.g., *A. welwitschii*, Fig. 25 b), which may sometimes be difficult to distinguish from the palisade cells. The spongy mesophyll may contain idioblasts filled with raphids (cf. Figs. 23 a, b) and/or (seldom) droplets of oil (e.g., occasionally in *A. galioides*).

Equifacial chlorenchyma, i.e., (i) a palisade parenchyma only, or (ii) an "isolateral" chlorenchyma with upper and lower palisades and a central spongy mesophyll, is typical for most *Nenax* and *Carpacoce* species (cf. Figs. 24 a–f, 26 b–e); it only occurs in few *Anthospermum* species (e.g., *A. bicorne, A. isaloense*, Figs. 20 b, e). In leaves with palisade parenchyma only, there are (4) 6 to 8, seldom more, rather tightly packed

Table 2. Leaf anatomical data of *Anthospermum*, *Nenax*, *Galopina* and *Carpacoce*. – L: leaf thickness (in µm) measured half-way between midrib and margin (without hairs). E: epidermis: thickness (in µm) of upper (Eu) and lower epidermis (El) [*: thickness of bulbous base of papillae only (*Anthospermum*) or of entire multiple epidermis (*Nenax*, *Carpacoce*)]; Eu : El: upper : lower epidermis

	L	Eu	El
Anthospermum aethiopicum (Puff 790712-3/1)	425–480	63–95	10–21
Knoblauch (1896)	–	60–77	–
A. *ammannioides* (Puff 790125-2/1)	200–240	35–50	15–25
A. *basuticum* (Puff 790113-5/3)	185–230	30–40	10–18
A. *bergianum* (Puff 800913-5/3)	310–370	55–105	31–45
(Puff 790718-1/1)	260–285	–	–
A. *bicorne* (Puff 800913-4/2)	450–500	65–145	20–55
A. *comptonii* (Puff 790914-3/1)	375–450	90–110	8–14
A. *dregei* subsp. *dregei* (Puff 780810-2/1)	310–375	100–160	12–25
subsp. *ecklonis* (Puff 790712-1/1)	255–325	55–95	10–17
A. *ericifolium* (Esterhuysen 35553)	200–225	50–75	31–43
A. *esterhuysenianum* (Puff 800924-3/2)	187–218	50–62	18–28
A. *galioides* subsp. *galioides* (790910-4/1)	220–280	28–45	10–18
(Puff 790910-3/2)	200–260	50–70	10–15
A. *galpinii* (Puff 790303-2/1)	210–262	31–50	12–18
A. *herbaceum* (Puff 780208-2/3)	175–189	25–38	12–25
(Puff 780409-1/1)	160–200	34–57	8–15
A. *hispidulum* (Puff 780312-1/1)	250–280	55–90	13–30
A. *hirtum* (Puff 800903-1/2)	280–375	60–85	9–15*
Knoblauch (1896)	–	–	–
A. *ibityense* (Puff 800730-1/2)	350–375	55–80	20–25
A. *isaloense* (Puff 800814-3/2)	310–375	80–105	45–75
A. *littoreum* (Puff 790116-1/1)	343–395	75–112	10–18
A. *longisepalum* (Puff 800814-2/1)	250–310	70–135	20–30*
A. *monticola* (Puff 780820-1/1)	230–280	37–50	15–25
(Puff 790113-3/1)	410–450	83–95	17–25
A. *pachyrrhizum* (Puff 810930-2/1)	220–250	40–75	20–32
A. *paniculatum* (Puff 790014-6/1)	230–265	35–50	12–22
A. *pumilum* subsp. *pumilum* (Puff 780415-1/2)	360–385	50–87	12–17
(Puff 790122-2/1)	200–250	60–90	15–23
A. *spathulatum* subsp. *spathulatum* (Puff 790910-6/1)	250–300	45–65	12–20
(Puff 790711-2/1)	350–425	80–100	14–22
subsp. *saxatile* (Puff 800924-3/1)	180–220	50–75	10–18
subsp. *uitenhagense* (Puff 790114-2/1)	400–450	32–45	10–20
A. *streyi* (Puff 790426-3/1)	370–425	85–105	8–16
A. *ternatum* subsp. *randii* (Puff 780215-3/1)	220–280	50–85	10–25
(Puff 790122-1/1)	160–200	45–60	10–17
subsp. *ternatum* (Puff 781221-1/1)	260–320	80–105	15–25
A. *usambarense* (Puff 780114-4/4)	280–375	43–68	17–50
A. *vallicola* (Puff 790125-1/1)	280–310	36–62	18–30
A. *welwitschii* (Puff 790211-2/3)	210–240	60–75	10–15
(Puff 791028-1/2)	280–310	56–75	12–18
(Puff 780211-2/1)	270–280	55–80	25–43

ratio (averages). E : M: epidermis : mesophyll ratio. Mesophyll: thickness of entire mesophyll (M), of palisade layer (Mp) and of spongy mesophyll (Msm) (in μm); Mp : Msm: palisade : spongy mesophyll ratio; M_1: palisades and spongy mesophyll not clearly distinguishable; M_2: upper palisades- spongy mesophyll- lower palisades present; M_3 : p: palisades only, s: spongy mesophyll only

Eu : El	E : M	M	Mp	Msm	Mp : Msm	M_1	M_2	M_3
4.5:1	1:3	235–320	105–160	106–160	1:1	–	–	–
–	–	–	–	–	–	×	–	–
2.1:1	1:2.3	130–160	70–90	60–80	1.1:1	–	–	–
2.1:1	1:3.2	130–180	60–80	60–100	1:1.1	–	–	–
2.1:1	1:2.3	190–220	–	–	–	×	–	–
–	–	–	–	–	–	–	–	s
2.8:1	1:1.7	220–270	–	–	–	–	–	p
9.1:1	1:2.6	260–300	75–110	150–188	1:1.9	–	–	–
7.5:1	1:1.3	175–220	75–95	100–125	1:1.3	–	–	–
5.6:1	1:2.4	190–230	70–95	115–140	1:1.5	–	–	–
1.7:1	1:1	93–105	–	–	–	×	–	–
2.5:1	1:1.6	106–143	–	–	–	×	–	–
2.6:1	1:3.8	165–215	80–85	85–120	1:1.2	–	–	–
4.8:1	1:2.2	140–185	60–85	80–100	1:1.2	–	–	–
2.9:1	1:2.7	140–185	60–75	80–110	1:1.4	–	–	–
1.7:1	1:2.5	112–125	50–62	62–75	1:1.2	–	–	–
3.9:1	1:2.2	115–140	45–65	70–80	1:1.4	–	–	–
3.4:1	1:2	155–220	80–95	75–125	1:1.1	–	–	–
6:1	1:2.8	200–280	90–150	90–130	1.1:1	–	–	–
–	–	–	–	–	–	–	–	s
3:1	1:2.9	250–280	100–140	135–155	1:1.2	–	–	–
1.5:1	1:1.4	175–250	145–210	30–60	3.9:1	–	×	–
6.4:1	1:2.5	250–280	95–125	156–218	1:1.7	–	–	–
4.1:1	1:1.2	125–180	60–90	60–95	1:1	–	–	–
2.2:1	1:2.9	168–200	75–93	95–110	1:1.2	–	–	–
4.1:1	1:2.8	300–340	–	–	–	×	–	–
2.2:1	1:2.2	170–190	60–95	100–110	1:1.3	–	–	–
2.5:1	1:3.2	180–200	80–95	100–110	1:1.2	–	–	–
4.8:1	1:3.5	260–300	125–145	125–170	1:1.1	–	–	–
3.9:1	1:1.5	125–150	–	–	–	×	–	–
3.4:1	1:3	180–250	80–110	90–140	1:1.2	×	–	p
5:1	1:2.6	250–310	–	–	–	–	–	p
4.1:1	1:1.5	100–125	55–65	45–70	1:1.1	–	–	–
2.6:1	1:7	350–400	–	–	–	–	–	p
7.7:1	1:2.5	235–300	110–140	125–160	1:1.2	–	–	–
3.9:1	1:2	155–187	90–100	65–93	1.2:1	–	–	–
3.9:1	1:1.6	90–125	45–60	45–65	1:1	–	–	–
4.6:1	1:1.6	160–190	65–80	95–115	1:1.5	–	–	–
2.4:1	1:3.4	215–315	90–135	125–180	1:1.3	–	–	–
2.1:1	1:3	200–235	90–120	110–125	1:1.2	–	–	–
5.1:1	1:1.8	140–160	70–80	70–80	1:1	×	–	–
4.3:1	1:2.6	200–220	–	–	–	–	×?	p
2.2:1	1:1.7	155–175	93–110	62–75	1.5:1	–	–	–

Table 2 *(continued)*

	L	Eu	El
A. whyteanum (PUFF 780215-3/1)	250–280	50–80	10–17
A. zimbabwense (PUFF 790124-2/2)	220–250	40–65	20–35
Galopina circaeoides (PUFF 780326-2/1)	70–110	16–25	9–22
G. tomentosa (PUFF 790304-3/1)	130–155	19–36	12–25
G. crocyllioides (PUFF 790416-1/1)	105–160	18–37	10–36
Nenax acerosa subsp. *acerosa* (PUFF 800907-3/1)	780–810	50–93	12–25
subsp. *macrocarpa* (PUFF 800920-2/2) long shoot	280–300	20–80	15–25
short shoot	400–450	25–80	19–25
N. arenicola (PUFF 790714-3/1)	350–400	42–75	21–25
N. cinerea (PUFF 800102-4/1)	300–375	75–155*	18–45*
N. coronata (PUFF 800902-6/4)	900–1000	55–130	20–40
N. divaricata (PUFF 790712-2/1)	990–1170	75–210	30–55
N. elsieae (PUFF 790913-6/1)	950–1100	70–210	25–45
N. hirta (PUFF 800910-2/1)	350–375	25–81	12–19
N. microphylla (PUFF 790710-1/1) young	180–210	50–115	15–30
old	250–310	35–85	12–35
Carpacoce curvifolia (ESTERHUYSEN 27381)	240–360	37–50	30–38
C. gigantea (TAYLOR 4241)	280–310	25–55	30–50
C. heteromorpha (PUFF 800917-4/1)	750–800	75–185*	75–185*
C. scabra subsp. *scabra* (PUFF 790914-4/2)	900–1010	50–65	45–65
subsp. *rupestris* (TAYLOR 6567)	350–450	30–50	30–60
C. spermacocea subsp. *orientalis* (PUFF 790909-5/4)	290–330	30–45	25–45
C. vaginellata (PUFF 800913-4/3)	700–750	50–65	50–70

palisade layers; their cells tend to be larger than in leaves with a normal, dorsiventral chlorenchyma, reaching lengths of 85 μm or (rarely) even close to 100 μm (see *C. scabra* subsp. *scabra* and *N. divaricata*, Figs. 25 e, g–h, for arrangement and shape of the palisade cells). In a few *Nenax* species, droplets of oil were observed in the equifacial chlorenchyma (e.g., occasionally in *N. acerosa* leaves).

Leaves with an equifacial mesophyll consisting of spongy mesophyll only, are confined to a few *Anthospermum* species (e.g., *A. bergianum*, Fig. 25 c).

The chlorenchyma structure may vary within taxa. In *A. monticola*, for example, leaves either have a distinctly dorsiventral chlorenchyma (Fig. 22 a), or all the chlorenchyma cells are of ± the same size and shape (Fig. 23 c). A comparison of *A. welwitschii* leaves showed that the leaves can either have a distinctly dorsiventral chlorenchyma (Fig. 22 b), a

Eu:El	E:M	M	Mp	Msm	Mp:Msm	M_1	M_2	M_3
5:1	1:2.5	170–220	65–95	95–125	1:1.4	–	–	–
3.7:1	1:1.3	155–175	65–75	80–100	1:1.3	–	–	–
1.3:1	1:1.5	50–60	20–30	25–30	1:1	–	–	–
1.5:1	1:1.9	80–95	37–50	35–45	1.1:1	–	–	–
1.4:1	1:1.6	65–90	35–60	25–30	1.7:1	–	–	–
3.8:1	1:4.7	400–440	–	–	–	–	–	p
2.5:1	1:2.9	180–230	–	–	–	–	×	–
2.4:1	1:2.7	160–250	–	–	–	–	×	–
2.5:1	1:3.6	280–310	–	–	–	–	–	p
3.6:1	1:1.2	155–190	–	–	–	–	–	p
3.1:1	1:2.3	250–320	–	–	–	–	–	p
3.3:1	1:2.5	450–475	–	–	–	–	–	p
4:1	1:2.7	435–510	–	–	–	–	–	p
3.5:1	1:2.3	140–175	–	–	–	–	–	p
3.7:1	1:1.3	100–170	50–75	50–105	1:1.2	–	–	–
2.6:1	1:2.5	165–250	–	–	–	–	×	–
1.2:1	1:2.8	155–280	–	–	–	×	–	–
1:1	1:2.1	140–190	–	–	–	×	–	p
1:1	1:1.9	190–310	–	–	–	–	–	p
1:1	1:7.7	400–475	–	–	–	–	–	p
1:1.1	1:2.2	140–200	–	–	–	–	–	p
1.1:1	1:3.3	200–275	95–210	105–155	1:1.2	×	–	–
1:1	1:6.3	350–410	–	–	–	–	–	p

chlorenchyma with rather similar palisade and spongy mesophyll cells (Fig. 25 b), or an upper and lower palisade layer and a central mesophyll (Fig. 22 c). Also remarkable are the differences observed in newly developed long shoot leaves and older short shoot leaves from one and the same plant (e.g., *N. microphylla*, Figs. 24 e–f).

Much variation in the chlorenchyma structure can be expected in taxa whose leaves have a rather loose palisade parenchyma and spongy mesophyll cells which are similar to the palisade cells; the chlorenchyma may, in some leaves, still be classified as dorsiventral and as homogenous (i.e., equifacial) in others.

4.5.3. Bundle Sheaths and Bundle Sheath Extensions

In all genera, both the midvein and the small lateral veins have distinct, parenchymatous bundle sheaths (see, for example, *A. bergianum*,

Fig. 25 a). The size of the bundle sheath cells ranges from ca 10 to 40 µm, but the cells occasionally reach a diameter of up to 50 to 70 µm around the median bundle.

In addition, the median bundles frequently have bundle sheath extensions (sensu Esau 1953, 1965; "BSE"). More commonly, BSE reach the lower epidermis only (e.g., *A. zimbabwense, N. cinerea, C. spermacocea* subsp. *orientalis,* Figs. 21 b, 24 a, 26 a); in a few taxa, they extend to the upper epidermis as well (e.g., *A. pachyrrhizum, A. hispidulum, A. hirtum, N. coronata, C. gigantea,* Figs. 20 d, 21 a, c; 24 b, 26 d). There may be some variation as to the extent of the BSE. In leaves of *A. herbaceum* from different populations, for example, it was observed that BSE, while always reaching the lower epidermis, do not always extend to the upper epidermis (see also *N. microphylla,* Figs. 24 e–f).

BSE, but not the actual bundle sheath cells, frequently contain a dark coloured substance (in all genera except for *Carpacoce*; see also 4.5.1.: Excretions).

BSE are thought to be derived from subprotodermal initials (cf. Napp-Zinn 1973–1974). Possibly, the subepidermal cells above and below the midvein which do not reach the actual bundle sheaths (occurring in some *Anthospermum* and *Nenax* species; e.g., *A. bicorne, N. divaricata,* Figs. 20 b, 25 g, h) are of the same origin.

Finally, attention is drawn to an unusual situation in most *Carpacoce* species: parenchymatous, colourless cells, indistinguishable from BSE cells, interconnect the median and the lateral bundles, so that all the bundles of a leaf appear to be embedded in a common parenchyma (see, for example, *C. gigantea* and *C. heteromorpha,* Figs. 26 d–e).

At the close of this chapter, mention should be made of Nontcheff's (1909) morphological and anatomical investigations of leaves of 52 species of *Cliffortia* (*Rosaceae*) which reveal numerous conspicuous parallelisms to the present study. *Cliffortia* not only has similar con- and revolutely rolled and round leaves/leaflets next to broad, flat leaves, but its leaves also exhibit anatomical conformities such as dorsiventral next to equifacial chlorenchyma, exceptionally large upper epidermal cells, and stomata on the upper and the lower surface in a few taxa. Significantly, *Cliffortia* has – apart from other conspicuous similarities – a distribution range similar to that of *Anthospermum* (although it is absent from Madagascar and poorly represented in tropical Africa; cf. Weimarck 1934); like the *Anthosperminae,* it has its highest species concentration in the SW Cape Floristic Region. "Ericoid" leaves of various unrelated Cape taxa (having evolved independently numerous times) have received the attention of many authors but, to my knowledge, no comprehensive, comparative morphological/anatomical study of leaves comparable to Nontcheff's investigations exists of genera with distributions extending beyond the SW Cape Region.

5. Inflorescences

Paniculate to thyrsic many-flowered inflorescences such as those depicted in Figs. 27 a–b (*Galopina, Anthospermum paniculatum; Phyllis*) can be considered the "basic type" of inflorescence in the *Anthosperminae*. They correspond to the "central type of rubiaceous inflorescence" which WEBERLING (1977) describes as being a terminal thyrsus or pleiothyrsus with an end flower[1].

With the exception of a few taxa, this "basic type" is variously modified. The following modifications and trends, which – in part – repeat themselves in the five genera of the *Anthosperminae*, can be recognized:

– the occurrence of "prolification": the inflorescence main axis does not produce a terminal flower and continues growth vegetatively (= auxotelic inflorescences sensu BRIGGS & JOHNSON 1979: 241) (cf. Figs. 27 c–h; in all genera except *Galopina*).

– the development of much-congested (partial) inflorescences – i.e., clustering of flowers (Fig. 27 c; *Anthospermum* in particular).

– increasing impoverishment of inflorescences (Figs. 27 b → e, f → g, h; in all genera). In this context another phenomenon is noteworthy: it was observed occasionally that in partial inflorescences (or cymes) which are comprised only of a terminal flower, a single lateral vegetative shoot arises in the bract pair below the end flower which continues growth and thus "overtops" the terminal flower (Figs. 27 k, 29 a, 7 d; in all genera except *Galopina*; also see below).

– unequal development, in the inflorescence region, of "lateral elements" (partial inflorescences, axes etc.) arising at a node (one promoted over the other, i.e., more-flowered, better developed than the other; Fig. 27 i; especially in *Carpacoce*; also in some *Anthospermum* and *Nenax* species).

Also noteworthy is the frequent occurrence of accessory shoots in the inflorescence region (Fig. 27 b; in *Anthospermum* and *Carpacoce* species in particular).

---

[1] Figures given as examples for this "central type" (WEBERLING 1977: Figs. 1 I, 5 I) much resemble panicles (such as Fig. 27 a in the present publication), and TROLL (1969: 359–370), in fact, describes many-flowered panicles for various *Rubiaceae*. On the other hand, these inflorescences could perhaps be termed pleiothyres (cf. TROLL 1964: 87). WEBERLING does not comment on this.

Not only in the *Anthosperminae* but also in numerous other *Rubiaceae*, the distinction between thyrses and panicles does not always seem to be clear-cut. See, in this context, BRIGGS & JOHNSON (1979: 245) for a discussion of the relationships between thyrses and panicles and gradations from panicles to "thyrsoids" in the *Myrtaceae*.

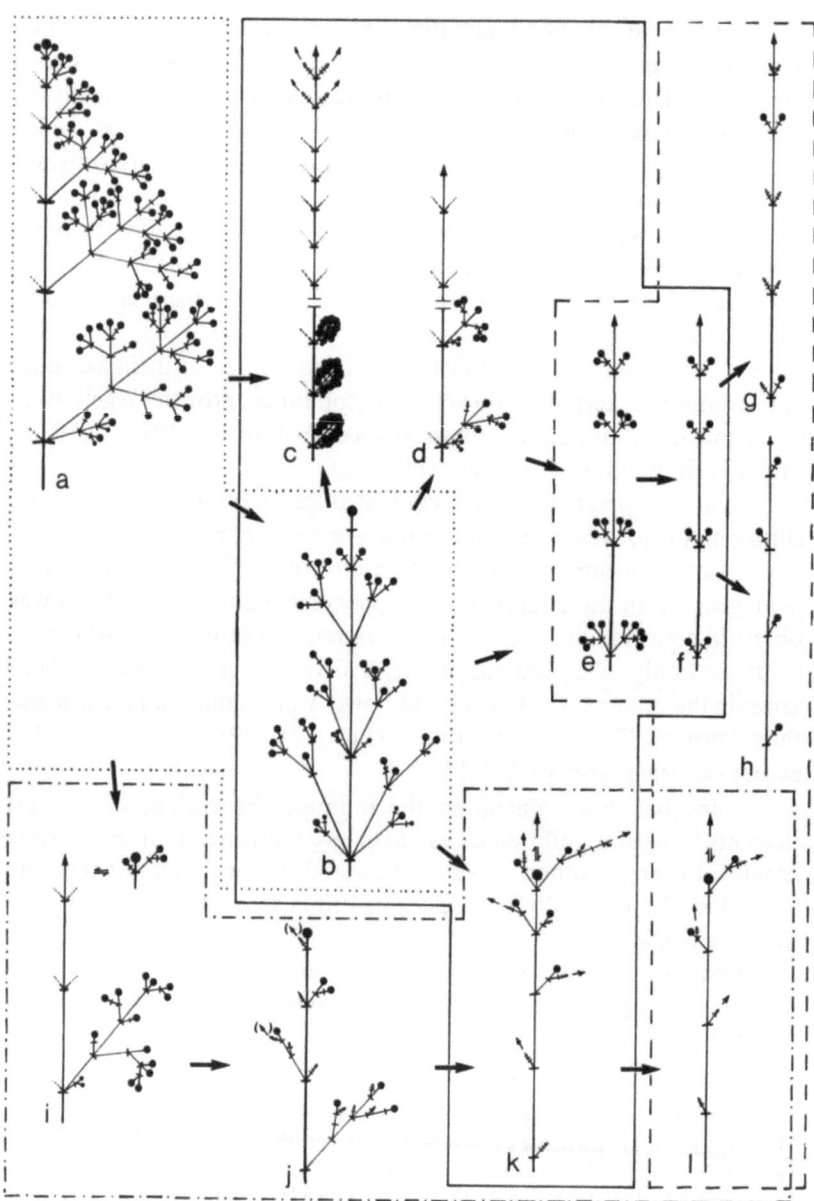

Fig. 27. Schematic representation of inflorescences of *Galopina, Anthospermum, Nenax* and *Carpacoce*. *a–b Galopina* (·····). *b–f, k Anthospermum* (———). *e–h, l Nenax* (– – – –). *i–l Carpacoce* (–·–·–·–). – Thick arrows indicate hypothetical progressions and relationships of inflorescence types; small arrows indicate continuation of growth of the respective axes. Further explanations in the text

Fig. 28. Inflorescences of *Galopina*. *a* *G. circaeoides* (PUFF 790911-8/1). *b* *G. tomentosa* (PUFF 790415-2/1). *c* *G. aspera* (PUFF 790221-3/1). *d* *G. crocyllioides* (PUFF 790416-1/1). – The scale unit is 10 cm

Galopina species show a reduction series from extensive, ± spheroidal to cylindrical, many-flowered bracteose inflorescences with long
internodes to fewer-flowered, more condensed, narrower inflorescences
(Figs. 28 a–d). The inflorescence of *G. crocyllioides* (Fig. 28 d) is often a
typical thyrse with branches only arising in the axils of the prophylls.

The inflorescences of *Phyllis* are similar to those of *G. circaeoides* and
G. tomentosa.

In *Anthospermum, A. paniculatum* is the only species with a typical
thyrse, which largely corresponds to that of *G. crocyllioides*, although it
often has accessory partial inflorescences (cf. Fig. 29 b). In the remaining
species, inflorescences with proliferating main axes (auxotelic inflorescences) are characteristic (although there are a few exceptions; see
below). Inflorescences are frondose or bracteose.

Common, particularly in ♀ plants, are condensed, ± cylindrical
inflorescence regions for which shortened main axis internodes, condensed partial inflorescences and greatly reduced peduncles and pedicels
are typical (e.g., *A. ammannioides, A. aethiopicum*, Figs. 54 b, c; 27 c).
There are indications that accessory partial inflorescences are also
involved in their formation. This, however, can hardly be proven directly,
as flowers are too densely clustered and spatial displacements and fusions
of axes, peduncles and pedicels impede exact analyses. The occurrence of
buds and flowers clustered together with mature fruits, however, indicates
the presence of accessory shoots (flowers on accessory partial inflorescences were always found to develop at a much later stage in, for
example, *A. paniculatum;* cf. Fig. 29 b).

Numerous species have cylindrical, leafy inflorescences which somewhat resemble those described above, but differ in being much fewer-
flowered and in having slightly longer main axis internodes. Such
inflorescences are slightly less conspicuous, but inflorescence regions are,
nevertheless, clearly distinguishable from the vegetative region. This, in
some species, may only apply to ♀ plants, while ♂ or ☿ + ♀ plants of the
same species may have very inconspicuous inflorescences (e.g., *A.
bergianum*, compare ♂ and ♀ inflorescences, Figs. 12 a–c and 98).

In a number of other species, the vegetative and inflorescence
internodes are of equal length and the bracts supporting the few- to one-
flowered partial inflorescences are indistinguishable from foliage leaves.
Such inflorescences, obviously, are very inconspicuous (e.g., *A. pumilum*
subsp. *pumilum*, Figs. 7 c, 8 a; *A. spathulatum* subsp. *spathulatum*,
Fig. 30 b).

In dioecious species, ♂ plants frequently have less conspicuous
(although often many-flowered) inflorescences, as the main axes inter-

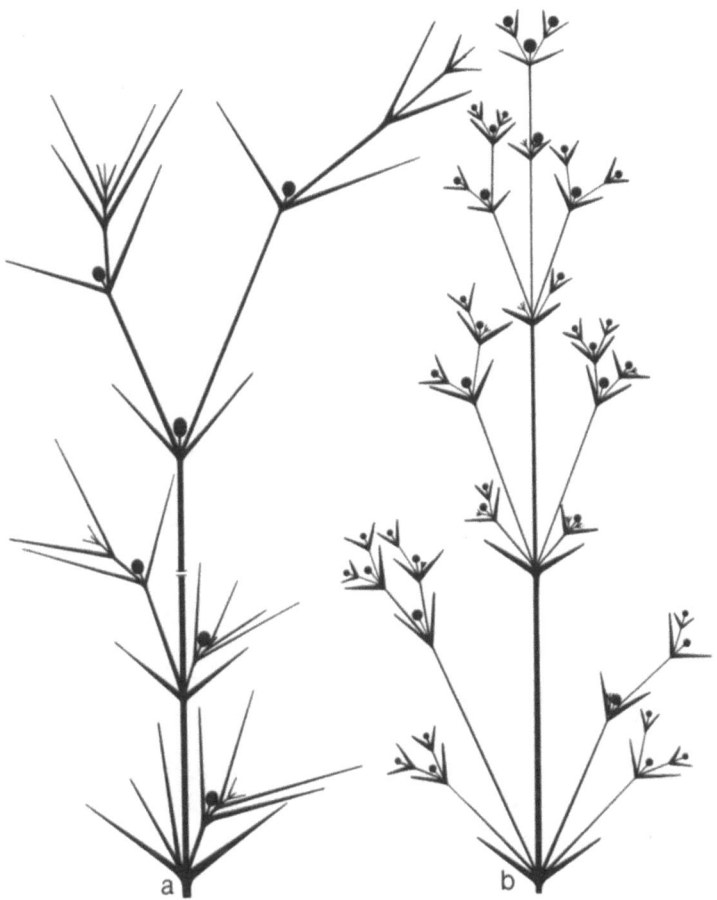

Fig. 29. Inflorescence diagrams of *Anthospermum. a A. esterhuysenianum,* note unequal promotion of lateral elements and very slight trend to anisophylly. *b A. paniculatum,* note presence of flowering accessory shoots at lower nodes. – Further explanations in the text

nodes are not as contracted as in the corresponding ♀ plants (e.g., *A. ammannioides, A. aethiopicum,* Figs. 54 a–d). Inflorescences of ♂ plants may, furthermore, be borne on branches of a higher order than those of ♀ plants (see also 12.3.).

In a few taxa, the partial inflorescences are several-flowered but only slightly contracted. Partial inflorescences, nevertheless, can appear dense and many-flowered due to the presence of numerous accessory shoots (e.g., *A. herbaceum,* Figs. 27 d, 30 a; the partial inflorescences often show a

Fig. 30. Inflorescences of ♀ plants of *Anthospermum*. *a A. herbaceum* (PUFF
791201-1/2), many-flowered axillary clusters. *b A. spathulatum* subsp. *spathulatum*
(PUFF 790711-2/1), few- to one-flowered partial inflorescences. – The portion
shown in *a* is ca 25 cm long and in *b* ca 12 cm long; st stigmas

trend to monochasial branching). It was also observed in these taxa that
branches in the inflorescence region may occasionally remain vegetative
and elongate considerably during a growth period. Eventually these
vegetative shoots may become longer than (i.e., "overtop") the entire
inflorescence and may give rise to another inflorescence – even in the same
growing season.

Unequal promotion of lateral elements arising at a node and
"overtopping" become obligatory in *A. esterhuysenianum,* which frequently also has a closed inflorescence system (i.e., a terminal flower
rather than a proliferating main axis; Fig. 29 a). There is either only a
single one-flowered partial inflorescence per node or, if two partial

inflorescences per node are present, one develops earlier and the flower has a much longer peduncle than the other. In both the vegetative and fertile regions, promoted lateral elements (vegetative shoots, partial inflorescences, etc.) follow a spiral (see also 2.). Bracts supporting the *promoted* lateral units may be marginally larger than those supporting the more reduced lateral units or than those in whose axils no lateral units arise. Vegetative shoots arising in the axils of bracts (prophylls) below the respective terminal flowers usually elongate and continue growth in the following growing season. Plants, therefore, exhibit complicated sympodial-monochasial or anisotomous-dichasial branching patterns (Fig. 7 d; see 2.).

The large genus *Coprosma* (subtribe *Coprosminae*) seems to exhibit a range of inflorescence types similar to that of *Anthospermum* (cf. OLIVER 1935) although − as far as it can be judged from the scanty data available − inflorescences never appear to be auxotelic and dense, ± cylindrical inflorescence zones such as in ♀ plants of, for example, *A. aethiopicum* and *A. ammannioides* (Figs. 54 b, c) seem to be lacking.

Nenax has the most reduced (impoverished) inflorescences within the subtribe (Figs. 27 e–h, l). Partial inflorescences are one-flowered or only occasionally three-(five-)flowered (Fig. 27 e; e.g., in *N. microphylla*). Inflorescence regions are ± conspicuous (due to slightly shortened internodes) only in ♀ plants of a few species (e.g., *N. hirta* or *N. acerosa* subsp. *acerosa*, Fig. 101 b). The inflorescences are frondose.

In species with a marked trend to unequal promotion of lateral elements (or in which only one lateral shoot is produced at a node, e.g., *N. cinerea*, Fig. 13 c), there is often only one flower per node (Fig. 27 h). In *N. arenicola*, inflorescence regions become even more inconspicuous: nodes with vegetative short shoots alternate with nodes bearing one (or occasionally two) one-flowered partial inflorescences (Figs. 27 g, 101 a). *N. divaricata* and *N. coronata*, in contrast to the other taxa of *Nenax*, very often (but not always) have inflorescence main axes with terminal flowers (closed inflorescences; Fig. 27 l). The inflorescence structure (and branching pattern) of *N. coronata* in particular is thus very similar to that of *A. esterhuysenianum* (see above).

Similar impoverished inflorescences occur in *Nertera* (subtribe *Coprosminae*; cf. WEBERLING 1977).

Carpacoce species have frondose to bracteose auxotelic or closed inflorescences. In all taxa, there is a (strong) trend to unequal promotion of lateral elements and a trend to monochasial partial inflorescences (Figs. 27 i–l).

In *C. spermacocea* subsp. *orientalis,* the basic structure of the ±

extensive inflorescences shows certain resemblances to those of *Galopina*, but instead of a terminal flower, the inflorescence main axis is often proliferating (Fig. 27 i). Partial inflorescences (at least those of a second order), furthermore, are often contracted and their flowers are frequently arranged in, or approach the arrangement of, a bostryx (or sometimes a cincinnus).

The remaining *Carpacoce* species have increasingly reduced (partial) inflorescences. Species with few- or only one-flowered partial inflorescences frequently exhibit – like *Anthospermum esterhuysenianum* and *Nenax* species (see above) – "overtopping", i.e., produce elongating vegetative shoots in the inflorescence region (Figs. 27 j–l; *C. scabra* subsp. *scabra*, Fig. 11 a).

C. vaginellata appears to be the only species in the genus in which inflorescences are *always* closed (Fig. 27 j).

In *C. spermacocea* and *C. scabra* in particular, bracts may be extremely anisophyllous, i.e., one blade of a pair of bracts is hardly developed (Fig. 11 b). This trend to anisophylly is conspicuously correlated with a, trend to monochasial partial inflorescences and/or unequal promotion of lateral elements at a node: the promoted lateral element always originates in the axil of the well developed bract (cf. *C. scabra* subsp. *scabra*, Fig. 11 a). This phenomenon has also been observed in various other *Rubiaceae* (cf. figures in PETIT 1964, WEBERLING 1977; *Theligonum:* cf. WUNDERLICH 1971; *Tricalysia:* ROBBRECHT 1979, Fig. 24; *Crocyllis:* PUFF & MANTELL 1982 a).

6. Flowers

Anthospermum, Galopina and *Carpacoce* have hermaphrodite, male or female flowers; *Nenax* has unisexual flowers only.

6.1. Pedicels

In *Anthospermum* and *Nenax,* the flowers are most commonly subsessile or very shortly pedicellate; pedicel lengths hardly exceed 3 mm. In dioecious species, the ♀ flowers are often ± sessile, while the ♂ flowers of the same species are (shortly) pedicellate [e.g., *A. ammannioides,* pedicels to 1.5 (2) mm in ♂].

In *Galopina,* the flowers have distinct peduncles and pedicels. Fruiting pedicels of *G. circaeoides* are filiform, slender and (7) 12–26 (37) mm long, those of other species are shorter and less slender (cf. Figs. 28 a–d).

In *Carpacoce,* the pedicels are either subobsolete (e.g., *C. vaginel-*

lata) or the fruiting pedicels may elongate to 7 or even 12 (20) mm as in *C. scabra* subsp. *scabra* and *C. spermacocea* subsp. *spermacocea*, respectively.

6.2. Calyx

The calyx is either entirely absent (*Galopina*, some *Anthospermum* and *Nenax* species) or consists of 4–5 (*Anthospermum, Nenax*) or 3–5 (6) (*Carpacoce*) lobes which crown the ovaries of ♂ and ♀ flowers. The calyx lobes often enlarge slightly after fertilization (see 7.2.2. for further details). Rudimentary ovaries of ♂ flowers may also be crowned by – or virtually only consist of – small, distinct to ± indistinct calyx lobes (compare ♂ and ♀ flowers of *A. vallicola, A. paniculatum*, Figs. 31 a–b and i, k, and *N. arenicola*, Figs. 33 d–e).

6.3. Corolla

In *Galopina*, the corollas are 4-merous, but odd flowers with 5-merous corollas also do occur occasionally.

In *Anthospermum* and *Nenax*, the majority of taxa has flowers with 4-merous corollas. In some, however, corollas are 4-5-merous, and in a few (e.g., *A. bergianum* or *N. acerosa*) 5-merous corollas are the rule (at least in ♂ flowers; ♀ flowers of the same species may, however, be 4-merous).

In *Carpacoce*, only *C. curvifolia* has 4-merous corollas; the remaining taxa have 5-6-merous or occasionally even 7-merous corollas (e.g., *C. heteromorpha*).

In all 4 genera, the corollas are creamy white, yellowish or yellowish-green to greenish. Corollas are often either streaked or tinged with red, reddish-brown, purplish-brown or purple (at least on the outside). In *Carpacoce* species in particular, the entire corolla may be dark brown to blackish (e.g., *C. spermacocea*), but this character is not consistent. It appears that dark coloured corollas or corollas with dark streaks are more evident in plants fully exposed to sunlight.

In all genera of the *Anthospermeae*, there are always distinct corolla size and shape differences between ♂̣/♂ and ♀ flowers. The ♀ corollas are, without exception, (considerably) smaller than the ♂̣ or ♂ corollas (or sometimes even absent: e.g., *A. paniculatum*). Corolla lobes of ♀ flowers are not only smaller and narrower but are also erect to ± spreading; they are never conspicuously recurved or curled back as in ♂̣ or ♂ flowers (cf., for example, *A. galioides* subsp. *galioides* and *A. paniculatum*, Figs. 31 g–h and i, k).

6*

Fig. 31. SEMgraphs of flowers and floral details of *Anthospermum. a–b A. vallicola* (Puff 790125-1/1), *a* rudimentary ovary and calyx of ♂ flower, *b* ♀ flower. *c A. galpinii* (Puff 790422-3/2), ♂ flower. *d A. ammannioides* (Puff 790125-2/1), ♂ flower. *e A. basuticum* (Puff 781126-1/7), ♀ flower. *f A. ternatum* subsp. *randii* (Puff 780215-3/1), ♀ flower. *g–h A. galioides* subsp. *galioides* (Puff 790910-4/1), *g* ♀, *h* ♀ flower. *i–k A. paniculatum* (Puff 790114-6/1), *i–j* ♂ flowers, *j* corolla removed to expose rudimentary stigmas, *k* ♀ flower. – All critical point dried; *a* ×23; *b–i, k* ×16; *j* ×46

Fig. 32. SEMgraphs of flowers and floral details of *Anthospermum herbaceum* (PUFF 780409-1/1). *a* ♀̂ flower. *b* ♂ flower. *c* ♂ flower, base of corolla tube opened to show rudimentary stigmas. *d–g* ♀ flowers, note increasingly smaller corollas; in *d* a rudimentary anther is clearly visible (detail: *e*), in *f–g* rudimentary anthers are absent. — All flowers taken from a single plant; all critical point dried; *c, e* × 23, all others × 16

Corolla tubes are either cylindrical, narrowly to broadly funnel-shaped or ± campanulate. In *Anthospermum*, tubes vary greatly in size and shape; in *Carpacoce*, they are cylindrical to narrowly funnel-shaped; in *Nenax*, they are ± cylindrical to broadly funnel-shaped and in *Galopina*, broadly funnel-shaped to campanulate tubes prevail.

⚥ and ♂ flowers: cylindrical tubes are hardly more than ca 0.5 mm in diameter; in *Anthospermum* species they are up to ca 3.7 mm long (e.g., *A. herbaceum*, cf. Figs. 32 a–b) and in *Carpacoce* species up to 5.5 mm long (e.g., *C. gigantea, C. heteromorpha*). Funnel-shaped tubes are hardly more than 1 mm long in *Anthospermum* and *Nenax* species (e.g., *A. ammannioides*, Fig. 31 d; *N. divaricata*, Fig. 33 g) but are up to twice as long in *Carpacoce* species (e.g., *C. spermacocea* subsp. *orientalis*, Fig. 34 b); campanulate tubes often do not exceed 0.5 mm in length (e.g., *G. circaeoides*, Fig. 33 i).

♀ flowers: in *Anthospermum*, *Galopina* and *Nenax* species, tubes are either narrow, ± cylindrical and often not longer than 0.5 mm (e.g., *A. galioides* subsp. *galioides*, *A. paniculatum*, Figs. 31 h, k; *N. elsieae*, Fig. 33 c) or they are subobsolete (e.g., *A. basuticum*, Fig. 31 e; *N. arenicola*, Fig. 33 d; *G. circaeoides*, Fig. 33 j). ♀ flowers with tubes up to 1.5 mm long occur only in some *Carpacoce* species (e.g., *C. spermacocea* subsp. *orientalis*, Fig. 34 c) and in those *Anthospermum* species, whose ⚥ and ♂ flowers have unusually long, cylindrical tubes (e.g., *A. ternatum* subsp. *randii*, Fig. 31 f). ♀ flowers with clearly discernible anther rudiments ("imperfect" ♀ as opposed to "perfect" ♀ without obvious anther rudiments; see 12.2. and Fig. 52) may have corollas ± intermediate in size between "perfect" ♀ and ⚥ or ♂ (e.g., *A. herbaceum;* compare Figs. 32 a–b and d–g).

The aestivation of the corolla lobes is valvate (cf. *N. arenicola*, Fig. 33 e; *C. spermacocea* subsp. *orientalis*, Fig. 34 a) as in most other tribes of the *Rubiaceae* (see survey in Wagenitz 1964). Lobes are linear, lanceolate to elliptic-lanceolate or ± triangular in ⚥ and ♂ flowers. In *Anthospermum*, *Nenax* and *Galopina*, the lobes of ⚥ and ♂ flowers are

Fig. 33. SEMgraphs of flowers and floral details of *Anthospermum*, *Nenax*, and *Galopina*. *a A. littoreum* (Puff 790414-1/1b), rudimentary ovaries of ♂ flowers; ♂ bud on the right. *b A. herbaceum* (Puff 781028-3/2), stigma. *c N. elsieae* (Puff 790913-6/1), ♀ flower. *d–f N. arenicola* (Puff 790714-2/1), *d* ♀ flower, *e* ♂ bud, *f* dorsal view of anther showing filament attachment. *g N. divaricata* (Puff 790712-2/1), ♂ flower; *h–k G. circaeoides* (Puff 781203-1/1), *h* ⚥ (or ♂?) bud, *i* ⚥ flower, *j* ♀ flower, *k* detail of petal. – All critical point dried; *a* × 23; *b* × 62; *c–e, g–j* × 16; *f* × 35; *k* × 121

Fig. 33

Fig. 34. SEMgraphs of flowers and floral details of *Carpacoce. a–e C. spermacocea* subsp. *orientalis* (Puff 790909-5/2), *a* ⚥ (or ♂?) bud, note apical appendages of petals, *b* ⚥ flower, upper part of stigma is broken off, *c* ♀ flower, *d* detail of petal of ⚥ flower showing apical appendage, *e* stigma surface, note pollen grains. *f C. vaginellata* (Puff 790912-1/1), ♀ flower. – All critical point dried; *e* × 62, all others

recurved or curled back (e.g., *A. paniculatum*, Fig. 31 i; *N. divaricata*, Fig. 33 g; *G. circaeoides*, Fig. 33 i); in *Carpacoce* the lobes are often less conspicuously recurved but are at least spreading (e.g., *C. spermacocea* subsp. *orientalis*, Fig. 34 b). In addition, the lobes (of ♀̂, ♂ and ♀ flowers) of *Carpacoce* are "hooded" or bear an apical appendage (e.g., *C. spermacocea* subsp. *orientalis*, *C. vaginellata*, Figs. 34 a–d and f), which may also bear some stiff bristles (e.g., *C. heteromorpha*). In *Anthospermum* and *Nenax*, lobes of ♀̂ and ♂ flowers range from ca 1.2 mm to 3 mm, and occasionally to 4 mm long and ca 0.5–2 mm wide; in *Galopina*, lobes are ca 1–1.7 (1.9) mm long and (0.3) 0.4–0.7 mm wide; in *Carpacoce*, they range from 2–5 (5.5) mm long and 0.5–2 mm wide. In ♀ flowers, the lobes are mostly linear or linear-lanceolate and are only 0.1–0.5 mm wide and ca 0.3–1 mm long (to 1.5 mm in *Carpacoce*).

Corollas are either glabrous (*Nenax*, *Carpacoce*, numerous *Anthospermum* species), papillate (*Anthospermum* species, *Galopina*) or hairy on the outside (*Anthospermum* species). Papillae (or hairs, as the case may be) are either confined to the outside of the corolla tube and the base of the lobes (e.g., papillae in *A. galpinii*, Fig. 31 c), or the lobes (e.g., hairs in *A. ternatum* subsp. *randii*, Fig. 31 f), occur only near the tip of the petals (e.g., papillae in *A. galioides* subsp. *galioides*, Fig. 31 h) or cover on the entire exterior surface (e.g., hairs in *A. ammannioides*, Fig. 31 d). There is often considerable variation within taxa as regards the occurrence of hairs and papillae on the corollas. In *A. herbaceum*, for example, the corollas may be entirely glabrous (cf. Fig. 32) or hairy, at least on the lobes; in *A. galpinii*, papillae may also occur over the entire corolla surface (compare with Fig. 31 c!). In *Galopina*, the outside of the corollas bears conspicuous, enlarged cells (ca 50–90 µm high) which give the surface a ± warty appearance (cf. *G. circaeoides*, Figs. 33 h, j–k); these are, however, not always present.

Bud shapes range from ± spheroidal to ellipsoidal to ovoidal (e.g., *A. galioides*, Fig. 31 g; *N. arenicola*, Fig. 33 e; *G. circaeoides*, Fig. 33 h; *C. spermacocea* subsp. *orientalis*, Fig. 34 a) to ± pyriform. In taxa with long-tubed ♀̂ or ♂ corollas (e.g., *A. herbaceum*, cf. Figs. 32 a–b) the limbs of the buds gradually expand into ovoidal to subcylindrical heads. Buds of ♀ flowers are usually ± spheroidal.

6.4. Androecium

The number of stamens equals the number of corolla lobes.

The filaments are attached to or near the base of the corolla tube (cf., for example, *A. herbaceum*, Fig. 32 c, right side). It is only in

Carpacoce species with unusually long-tubed corollas (e.g., *C. gigantea, C. heteromorpha*) that they arise around or just below the middle of the corolla tube. Filaments are filiform, slender, glabrous and (0.3) 0.4–1.2 (1.4) mm long in *Galopina*, ca 1–3.5 mm in *Anthospermum* and *Nenax*, and up to 6 mm in *Carpacoce*.

The anthers are yellowish to whitish (except for *Carpacoce* species which often have purplish to dark brown anthers), are ± ellipsoidal, ca 1–2.5 (3) mm long and 0.2–0.7 mm wide (occasionally smaller in *Galopina* and often somewhat larger in *Carpacoce*) and glabrous. They are dorsifixed, introrse (cf. *N. arenicola, N. divaricata,* Figs. 33 f–g) and dehisce longitudinally (cf. *A. ammannioides,* Fig. 31 d); thecae and pollen sacs are usually clearly discernible (cf. *C. spermacocea* subsp. *orientalis,* Fig. 34 b). The anthers of ⚥ flowers tend to be slightly smaller than those of ♂ flowers of the same plant/taxon (e.g., *A. herbaceum,* compare Figs. 32 a and b). Anthers with a high percentage of dead pollen (flowers ⚥ → functionally ♀) are smaller than those with viable pollen. Rudimentary, pollenless anthers of "imperfect" ♀ (see 12.2.) are usually less than 0.5 mm long (e.g., *A. herbaceum,* Figs. 32 d–f); "pure" or "perfect" ♀ have no anther rudiments.

6.5. Gynoecium

Flowers of *Anthospermum* (except for *A. ericifolium*), *Nenax* and *Galopina* species have two stigmas, while the majority of *Carpacoce* species has flowers with only one stigma. The common style is usually very short (ca 0.2–0.5 mm long) or virtually absent (*Galopina*); in some *Anthospermum* species, however, it may attain ca 5 mm in length (e.g., *A. ammannioides*). The stigmas are whitish, creamy-yellow, greyish or greenish-grey in *Galopina* species and in the majority of *Anthospermum* species. In most *Carpacoce, Nenax* and a few *Anthospermum* species (e.g., *A. paniculatum, A. hirtum*), the stigmas are reddish or purplish-red. Stigma colour does not always seem to be constant: in a number of *Anthospermum* species, for example, stigmas are either greyish-white or sometimes purplish-red; in *C. spermacocea* and *C. vaginellata,* flowers with greyish-white stigmas may occur next to those with the more common reddish stigmas (these data are largely based on observations of living plants as herbarium labels hardly ever contain the relevant information; the available data are incomplete and may, therefore, give a somewhat distorted picture of stigma colouration).

The stigmas, usually very conspicuous in ♀ plants (e.g., *A. herbaceum, A. spathulatum* subsp. *spathulatum,* Figs. 30 a–b), are up to ca 5 (6.3) mm long in *Galopina* species and up to 17 mm long in *Anthospermum, Nenax* and *Carpacoce* species. Stigma diameters range from 0.1–0.5 mm in

Anthospermum, Nenax and *Galopina*, and from 0.3–1.5 mm in *Carpacoce*. Stigmas of ♀ flowers are often thicker (and longer) than those of ♂ flowers of the same plant/taxon (e.g., *A. galioides* subsp. *galioides*, Fig. 31 g–h; *G. circaeoides*, Figs. 33 i–j), but this is not necessarily always the case (cf. *A. herbaceum*, Figs. 32 a and g). Invariably, the stigmas bear unicellular hairs which are ca 50–220 µm long (cf. *A. herbaceum*, Fig. 33 b, or *C. spermacocea* subsp. *orientalis*, Fig. 34 e). These hairs are distributed over the entire surface of the stigmas and are not confined to their inner portion, thus further increasing the receptive stigmatic surface; only in the lowermost part of the stigmas are these hairs sometimes absent on the outside (e.g., *A. galioides* subsp. *galioides*, Fig. 31 h, or *A. herbaceum*, Figs. 32 f–g).

The o v a r y is comprised of two fused, fertile carpels in *Anthospermum* (except for *A. bicorne* and *A. ericifolium*), *Nenax* and *Galopina*. In *Carpacoce* (except for *C. scabra*), one of the two carpels is infertile, variously reduced and modified (see also 7.3.). Except for *A. bicorne*, all taxa with only one fertile carpel/ovule also only have one stigma. Ovaries are small, from less than 1 mm to ca. 2 mm long in *Anthospermum*, *Galopina* and most *Nenax* species, and up to 4 mm long in *Carpacoce* and a few *Nenax* species. ♂ flowers frequently have clearly discernible, although often very small (0.6 mm long or less) rudimentary ovaries (e.g., *A. littoreum*, Fig. 33 a). Sometimes – particularly when the rudimentary ovaries of ♂ flowers are not quite so minute – there are also rudimentary stigmas which may be up to 0.8 (1.5) mm long (e.g., *A. paniculatum*, Fig. 31 j; *A. herbaceum*, Fig. 32 c).

Each locule (i.e., fertile carpel) contains a single, basally attached anatropous ovule (e.g., *A. zimbabwense*, Fig. 45 d; *C. scabra* subsp. *scabra*, Fig. 46); see 8. for further details.

For particulars concerning the indumentum of ovaries see 7.2.1.

7. Carpophores, Fruits and Seeds

7.1. Carpophores

The mericarps of all *Anthospermum* and some *Nenax* species are supported by carpophores. They are, however, absent in *Galopina* and *Carpacoce*, and in those *Nenax* species with indehiscent fruits. Morphologically, carpophores can be interpreted as modified pedicellar structures, i.e., as two unvascularized lateral outgrowths at the upper end of the pedicel providing support for the two carpels or mericarps. There is no evidence to support the assertion that these structures are homologous to pairs of bracteoles (cf. VERDCOURT 1976: 326). Their morphological structure is also not comparable to that of the carpophores of the *Apiaceae* [BAILLON (1880) may have placed the *Rubiaceae* next to the *Apiaceae* because of the presence of inferior overies and of carpophores!].

Carpophores often resemble tuning forks (cf. Fig. 37 e) or shallow, broad, U-shaped "headrests" (cf. Fig. 37 f). The two carpophore arms may be up to ca 2.5 mm long and may even reach up to ca ¾ the length of the mericarps in a few *Anthospermum* species (e.g., *A. spathulatum* subsp. *spathulatum*, Fig. 35 b, or *A. littoreum*, compare Figs. 37 e and 35 e). More often, however, they are shorter (1.5 to 1 mm long or less) and only support the lower third or lower quarter of the mericarps (e.g., *A. ternatum* subsp. *randii*, Figs. 35 c–d, *A. galioides*, Fig. 37 b or *A. asperuloides*, compare Figs. 35 i and 37 f). In the *Anthospermum* species with only a single fertile carpel (*A. bicorne*, *A. ericifolium*), the carpophores are distinctly asymmetrical (Fig. 36 b).

The hard, sclerified carpophores persist on the plants after the mericarps have been disseminated. They frequently bend slightly outwards as the fruits mature and may thus aid in and ease the dropping off of the mericarps (cf. *A. vallicola* and *N. divaricata*, Figs. 37 d and 38 g respectively).

Such carpophores, although absent from the other genera of the *Anthospermeae*, also occur in some genera of the allied tribe *Paederieae* (*Crocyllis*, "*Gaillonia* complex"[1]; PUFF & MANTELL 1982 a).

The fruits and seeds of *Anthospermum*, *Galopina* and *Nenax* and those of *Carpacoce* are dealt with separately because of striking morphological and anatomical differences.

7.2. Fruits and Seeds of *Anthospermum, Galopina* and *Nenax*

7.2.1. Indumentum

Fruits of *Anthospermum*, *Nenax* and *Galopina* are either covered with papillae, hairs or multicellular tuberculate (± wart-like) structures or are glabrous with a smooth to ± wrinkled surface.

Papillae and hairs on *Anthospermum* and *Nenax* fruits have the same structure as those covering the stems and leaves (see 3.1. and 4.5.1. for details). Papillae are roughly 30–60 (100) μm long, straight (e.g., *A. longisepalum*, Figs. 36 e–f, or *N. microphylla*, Figs. 41 e–f) or not infrequently curved. Larger papillae/short hairs (ca 80–300 μm long) are most common (e.g., *A. asperuloides*, Fig. 35 e, or *N. coronata*, Fig. 37 b). A dense cover of longer hairs (ca 350–800 μm long) is less common (e.g., *A. littoreum*, Fig. 35 e; curled hairs: *N. elsieae*, Figs. 38 e–f).

---

[1] See LÉONARD (1984) for comments!

Numerous taxa show some variation in the fruit indumentum. Sometimes the fruits of some populations in taxa with normally papillate fruits or fruits with rather short hairs may be entirely glabrous (e.g., *A. herbaceum, A. prostratum, N. divaricata,* or *N. microphylla* — compare Figs. 38 a–b and 41 e–f).

The fruits of *Galopina* are unique in having multicellular tuberculate or ± wart-like structures. They consist of two rows of short, pillar-like cells which are capped by a large, elongated cell (cf. *G. aspera,* Fig. 39 i); subepidermal cell layers may also be incorporated into these structures (cf. *G. circaeoides,* sections Figs. 39 k, m). In *G. crocyllioides,* the elongated capping cell is replaced by a unicellular hair (Fig. 39 j); these hairs are ca (300) 500–800 µm long. In fruits of *G. tomentosa,* such structures are sometimes absent altogether; here the fruit surfaces are often ± wrinkled (Figs. 39 d, h).

7.2.2. Persistent Calyx Lobes

The fruits of *Anthospermum* and *Nenax* are usually crowned by 4–5 persistent calyx lobes while the fruits of *Galopina* always lack calyx lobes.

In *Anthospermum,* the calyx lobes are commonly ca 0.5–1 (1.5) mm long, are either as long as they are wide or, more often, are longer than wide and narrowly triangular (e.g., *A. basuticum,* Fig. 35 f). Long, band-like calyx lobes reaching up to 3.7 mm in length occur in *A. longisepalum* (Fig. 36 e) and *A. ibityense.* In most taxa, the calyx lobes are of equal size or slightly subequal. Only in a few species are the calyx lobes distinctly unequal (two longer and two shorter lobes; e.g., *A. vallicola* or *A. pachyrrhizum,* Figs. 83 b–c), but this character is not always constant within a given species. In several taxa, the calyx lobes are indistinct and rather obscure and only up to ca 0.3 mm long (e.g., *A. streyi* and *A. galioides* subsp. *galioides,* Figs. 37 a–b). In these taxa in particular, there may be considerable variation: there may be no trace of calyx lobes on the fruits of one population, but obscure or small but distinct calyx lobes on the fruits of another. *A. ternatum* subsp. *randii* (Figs. 35 c–d) and *A. hirtum* serve as examples of species always without obvious calyx lobes. In *A. bicorne,* the reduced carpel, but not the mericarp containing the seed, bears two large calyx lobes (Figs. 36 a, c).

In *Nenax,* long and narrow calyx lobes are absent; the lobes are invariably (broadly) triangular, i.e., as long as they are wide, rounded or ± trapeziform; they are of ± equal size and are mostly less than 1 mm long. In *N. cinerea* and *N. microphylla* (Figs. 37 g, 38 a–b) the calyx lobes are often obscure or even absent, in a few other taxa they are rather massive and conspicuous (e.g., *N. elsieae,* Figs. 38 e–f).

7.2.3. Fruits Dehiscing into Two Mericarps

occur in all species of *Anthospermum* und *Galopina* and in several *Nenax* species.

The mericarps of *Galopina* are the smallest of all the genera; the mericarp lengths range from (1) 1.3–2.2 (2.6) mm and widths from (0.7) 1–1.4 (1.6) mm.

Anthospermum exhibits the greatest variability in mericarp size and shape. The smallest mericarps correspond to the lower size range in *Galopina* (e.g., *A. thymoides*, Fig. 37 c), the largest may be up to 3.5 (4.2) mm long and 1.5–2 mm wide (e.g., *A. paniculatum*, *A. bicorne* or *A. ericifolium*, Figs. 36 a, d). In the majority of taxa, however, the mericarps are usually ca 2–3 mm long and ca 0.8–1.5 mm wide.

The dehiscent fruits of *Nenax* tend to be larger than those of *Anthospermum*. Mericarp lengths range from ca 2–7.5 mm and widths from 1–8 mm. The fruits of *N. microphylla* and *N. cinerea* are ± inflated and round (*N. m.*, Fig. 39 a) or ± inflated and laterally compressed (*N. c.*, Fig. 10 a).

In dorsal or ventral view, the mericarps appear either elliptic, oblong to almost rectangular, ovate, obovate, round (Figs. 35 b, d–e, k; 37 b, h; 39 e–h) or ± heart-shaped (*N. cinerea*, Fig. 37 g).

There is much variation in the size of the commissures. In *G. tomentosa*, for example, the commissure is very small and often only ca half as long as the mericarp, since the mericarps circumscribe an angle of ca 75–90° (compare Figs. 39 c and h). In the other *Galopina* species, the commissures are larger, although the basic mericarp structure is the same (e.g., *G. aspera*, Fig. 39 f).

Mericarps with commissures as small as in *G. tomentosa* occur in only two *Anthospermum* species (*A. thymoides*, *A. palustre*, Figs. 37 c and 84 a–b). Otherwise, the commissures of *Anthospermum* and *Nenax* mericarps are larger, covering at least ¾ of the ventral surface; there is almost always a distinct and prominent to ± indistinct vertical "ridge" in the middle of

---

Fig. 35. SEM graphs of fruits and mericarps of *Anthospermum*. *a–b A. spathulatum* subsp. *spathulatum* (PUFF 791221-2/1), *a* fruit in oblique view, *b* mericarp, ventral side. *c–d A. ternatum* subsp. *randii* (PUFF 780215–3/1), *c* fruit in side view, *d* mericarp, ventral side. *e A. littoreum* (PUFF 790415-5/1), mericarp, ventral side. *f A. basuticum* (PUFF 790113-5/3), fruit, subtended by pair of bracts (prophylls), in side view. *g A. monticola* (PUFF 790908-1/1), persistent calyx. *h A. spathulatum* subsp. *uitenhagense* (PUFF 790114-2/1), mericarps/fruit as viewed from above. *i A. asperuloides* (MANN 1290), fruit in side view. *j–k A. herbaceum* (PUFF 810916-3/1), *j* fruit in side view, *k* mericarp, ventral side. – The arrows point to carpophores. *a–b, i* herbarium material, all others critical point dried; *g–h* × 23, all others × 16

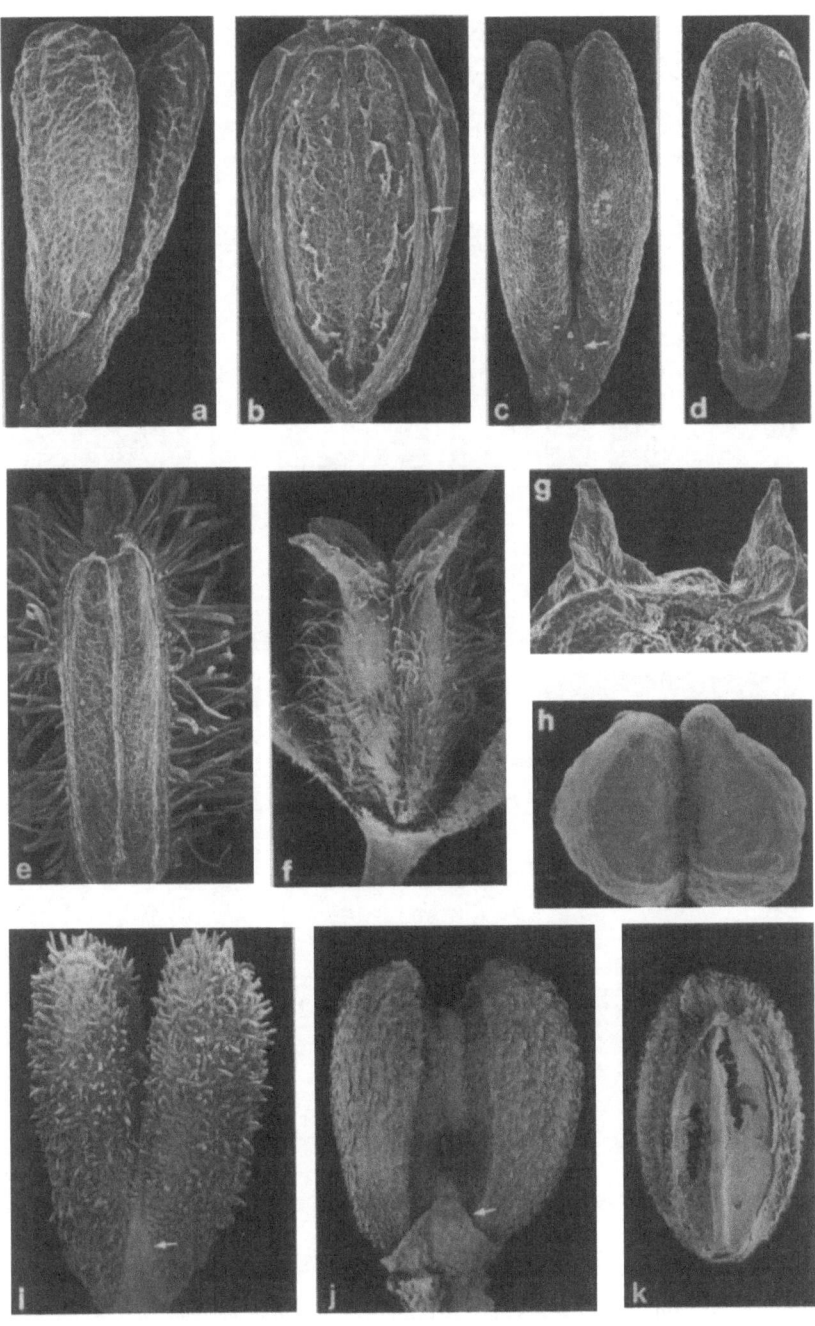

Fig. 35

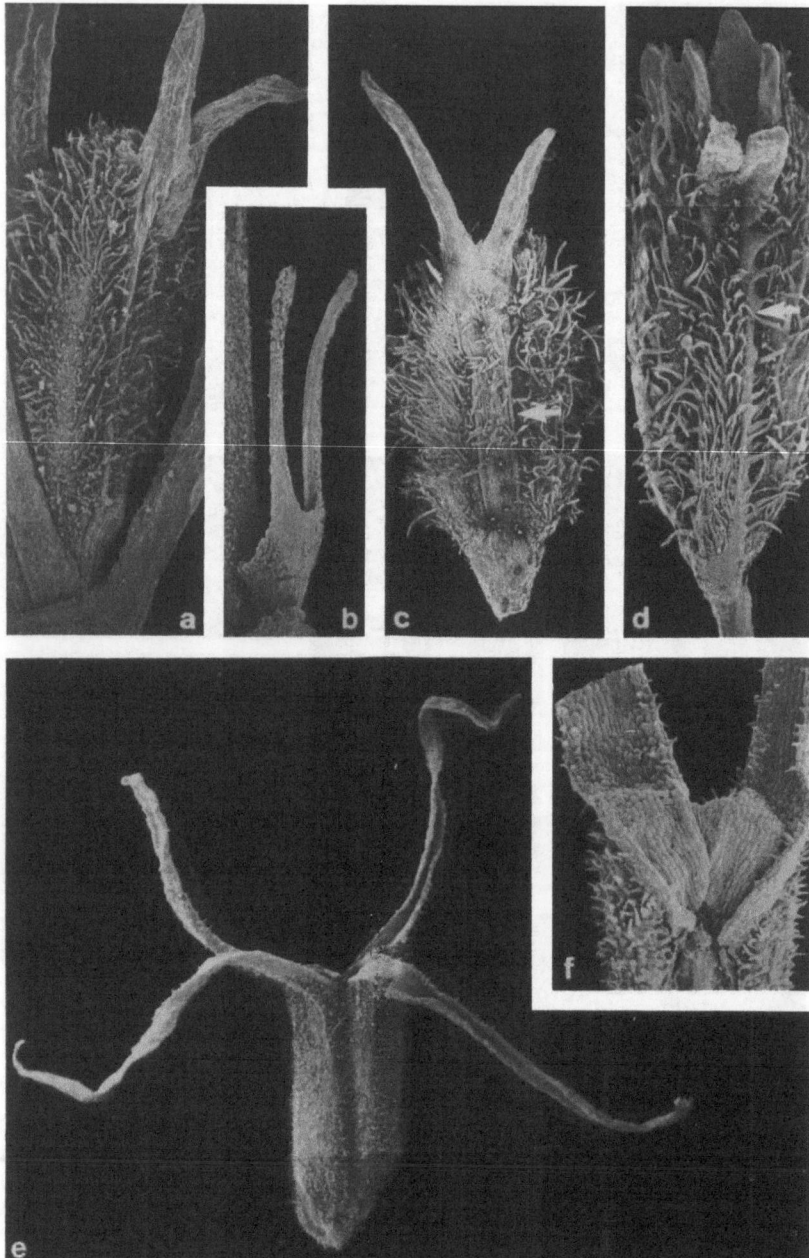

Fig. 36. SEMgraphs of fruits and carpophores of *Anthospermum. a–c A. bicorne* (Bolus 341), *a* fruit in side view, note reduced carpel with two enlarged calyx lobes on the right, *b* asymmetrical carpophore, *c* fruit with the reduced carpel (arrow) above and fertile carpel below. *d A. ericifolium* (Esterhuysen 28054), fruit with with reduced carpel (arrow) above and fertile carpel below. *e–f A. longisepalum* (Puff 800814-2/1), *e* fruit in side view, *f* upper part of mericarp with persistent calyx lobes. – All herbarium material; *a–e* × 16; *f* × 23

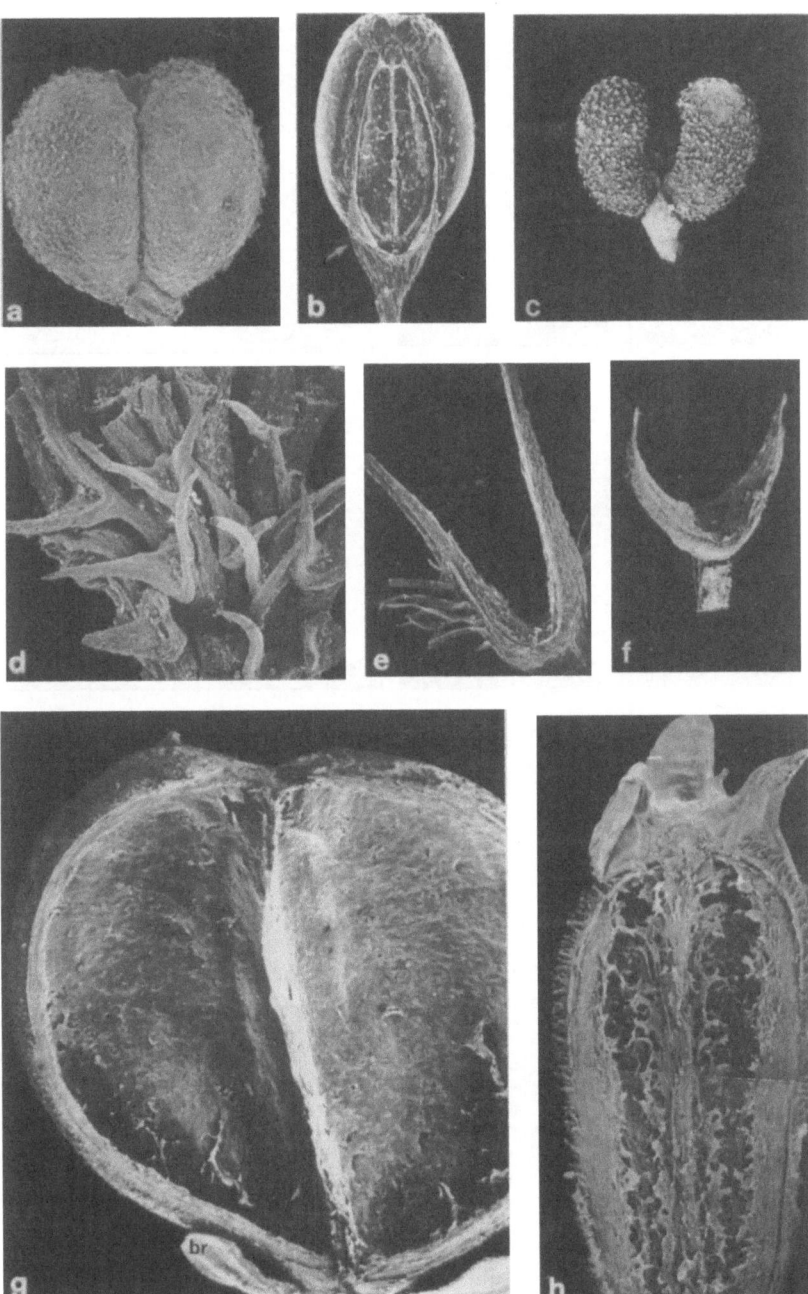

Fig. 37. SEMgraphs of fruits, mericarps and carpophores of *Anthospermum* and *Nenax. a A. streyi* (PUFF 790426-3/1), fruit in side view. *b A. galioides* subsp. *galioides* (PUFF 790913-2/1), mericarp (with attached carpophore: arrow), ventral side. *c A. thymoides* subsp. *thymoides* (PERRIER DE LA BÂTHIE 14426), fruit in side view (note small commissure!). *d–f* carpophores, *d A. vallicola* (PUFF 790125-1/1), *e A. littoreum* (PUFF 790415-5/1), *f A. asperuloides* (MANN 1290). *g N. cinerea* (PUFF 800102-4/1), mericarp, ventral side, subtended by a pair of bracts (br). *h N. coronata* (COMPTON 9649), mericarp, ventral side. − *a–b, e* critical point dried, all others herbarium material. All × 16

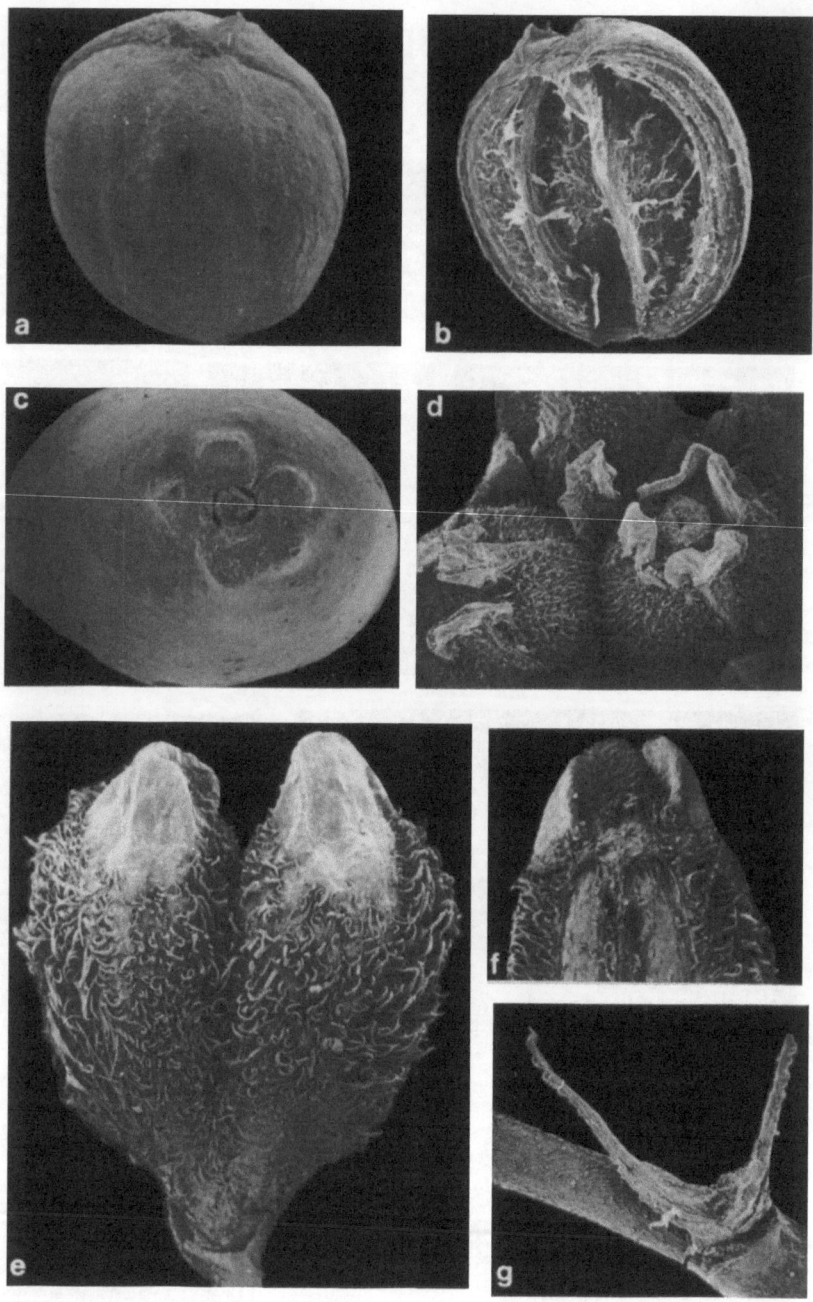

Fig. 38. SEMgraphs of fruits, mericarps and carpophores of *Nenax. a–b N. microphylla* (PUFF 790413-1/1), *a* fruit in side view showing the two mericarps, *b* mericarp, ventral side. *c N. arenicola* (PUFF 790714-2/1), indehiscent fruit as viewed from above; note calyx lobes. *d N. hirta* subsp. *hirta* (BOLUS 12764), fruits. *e–f N. elsieae* (ESTERHUYSEN 3656), *e* fruit in side view, *f* upper part of mericarp with persistent calyx lobes, ventral side. *g N. divaricata* (PUFF 790713-1/1), carpophore. – *a–c, g* critical point dried, all others herbarium material. All × 16

the commissure (e.g., *A. littoreum*, Fig. 35 e). The commissure may be ±
plane (e.g., *A. galioides* subsp. *galioides,* Fig. 37 b), or the tissue between
the central "ridge" and the sides of the mericarp may start to disintegrate
as the fruits mature (cf. *A. herbaceum*, Fig. 35 k) so that the ventral
mericarp surface appears to be hollowed out on either side of the "ridge"
(see, for example, *N. coronata,* Fig. 37 h). In this context, also see 7.2.5.

As the fruits mature, the sides of the mericarps occasionally start to
curl inward, a process which may continue after the mericarps have been
disseminated. Subsequently, the actual commissures may become obscured (e.g., *G. aspera,* compare Figs. 39 f and g).

7.2.4. Indehiscent Fruits

only occur in a few taxa of *Nenax* (e.g., *N. arenicola,* Fig. 38 c, section:
Fig. 41 c). The fruits are ellipsoidal, obovoidal to spheroidal and, except
for *N. hirta* subsp. *hirta,* are rather large, i.e., up to 8 mm long and 3.5 mm
wide.

7.2.5. Fruit (Mericarp) Anatomy

Mericarps of *Anthospermum, Galopina* and *Nenax* species always show
a clear differentiation of the pericarp into an exo- and an endocarp.

On the dorsal side of the mericarp, the exocarp is usually several- (ca
3–7-)layered (e.g., *A. bicorne,* Fig. 42 b) and consists of parenchymatous
cells. There may be local thickenings: in vertically ribbed mericarps, for
example, the exocarp is significantly thicker around and above the
vascular bundles (e.g., *A. paniculatum,* Fig. 41 a). In *G. circaeoides* (and
also in the other *Galopina* species?), the exocarp becomes somewhat
thicker where the tuberculate appendages are formed (Figs. 39 k, m).
There are always five, symmetrically arranged vascular bundles
embedded in the exocarp (see Figs. 40 a–c and 41 a) which, however, may
become crushed and difficult to see in fully mature fruits. Frequently, the
fruit epidermis cells (but apparently never the epidermal hairs) are filled
with dark excretions (e.g., *G. circaeoides,* Figs. 39 k, m). Occasionally, a
few raphid-containing idioblasts are present in the exocarp.

The endocarp invariably consists of sclerenchyma cells. It is
relatively thin and undifferentiated in *Galopina* species (e.g., *G. circaeoides,* Fig. 39 m), whereas in *Anthospermum* and *Nenax* species it is
usually more conspicuous. It may be massive and as thick as or thicker
than the exocarp on the dorsal side, but it is often thinner on the ventral
side (e.g., *A. bicorne, N. coronata,* Figs. 40 c and 42 a). Almost invariably,
the innermost endocarp layer consists of vertically arranged cells, while
the outer layers consist of horizontally orientated cells (e.g., *A. bicorne,*

7*

Fig. 42 b). On the ventral side, the median portions of the endocarps are usually raised and appear triangular to ± pointed in section so that the endocarps of paired mericarps frequently touch or almost touch each other (cf. *A. littoreum*, *N. coronata*, Figs. 40 a and c). This raised endocarp portion forms the vertical "ridge" running down the middle of the commissure (see 7.2.3., above). In numerous *Anthospermum* and *Nenax* taxa, the endocarps develop ± curved and sometimes rather massive lateral extensions ("arms"; cf. *A. palustre*, *A. bicorne*, *N. coronata*, Figs. 40 b–c and 42 a) which contribute considerably to the shape of the mericarps. Although distinct endocarp "arms" are not always present, it is conspicuous that the lateral endocarp portions (i.e., where the endocarp curves sharply from the dorsal to the ventral side of the mericarp) are always somewhat thickened (e.g., *A. littoreum*, Figs. 40 a and 41 b). These thickenings, as well as the proper arm-like lateral extensions, consist of horizontally arranged sclerenchyma cells (cf., *A. littoreum*, Fig. 41 b).

In *Anthospermum* and *Nenax*, the consistency and the longevity of the parenchymatous tissue on the ventral side of the mericarps are responsible for the appearance of the commissure. In ovaries or young fruits, a loose, spongy tissue consisting of large ± hexagonal cells (ca 60–100 μm, occasionally up to 200 μm in diameter) fills up the area between the two opposite mericarps. In *Nenax* species with indehiscent fruits, this tissue fills up the entire interior of the fruit (i.e., the area inside the endocarps; cf. *N. acerosa* subsp. *macrocarpa*, Fig. 41 g). Frequently, most cells of this tissue are filled with golden-brown excretions (appearing dark in stained microtome sections – cf. *A. littoreum*, Fig. 41 b, left side); some idioblasts with raphides may also be present. Upon maturation of the fruits, this loose tissue either (a) persists in its entity, (b) disintegrates locally or (c) disintegrates and breaks up to a large extent. If all or most of this tissue remains intact, the commissures of the mericarps are ± plane,

---

Fig. 39. SEMgraphs and microtome sections of fruits and mericarps of *Galopina*. *a G. circaeoides* (Puff 780326-2/1), fruit in side view. *b G. crocyllioides* (Puff 790416-1/1), fruit in side view. *c–d, h G. tomentosa* (Puff 790304-3/1), *c* fruit in side view, *d* detail, *h* mericarp, ventral side. *e–g, i G. aspera* (Puff 790211-3/1), *e* mericarp, dorsal side, *f–g* mericarps, ventral side, *i* mericarp surface. *j G. crocyllioides* (Puff 790416-1/1), mericarp surface. *k–m G. circaeoides* (Puff 781203-1/1), *k* whole mericarp, *l* detail of ventral side of mericarp, *m* detail showing part of radicle (upper left corner), endosperm, testa (filled with tannin), thin sclerenchymatic endocarp (right, thicker arrow), exocarp with vascular bundle (left) and idioblasts filled with raphids; note crushed integument cells below testa (left, thinner arrow) and fruit epidermis cells with dark contents. − *a–b, e–g* herbarium material; *c–d, h–j* critical point dried. *a–c, e–h* × 16; *d* × 35; *i–j* × 93; *k* × 49; *l* × 93; *m* × 149. Further explanations in the text

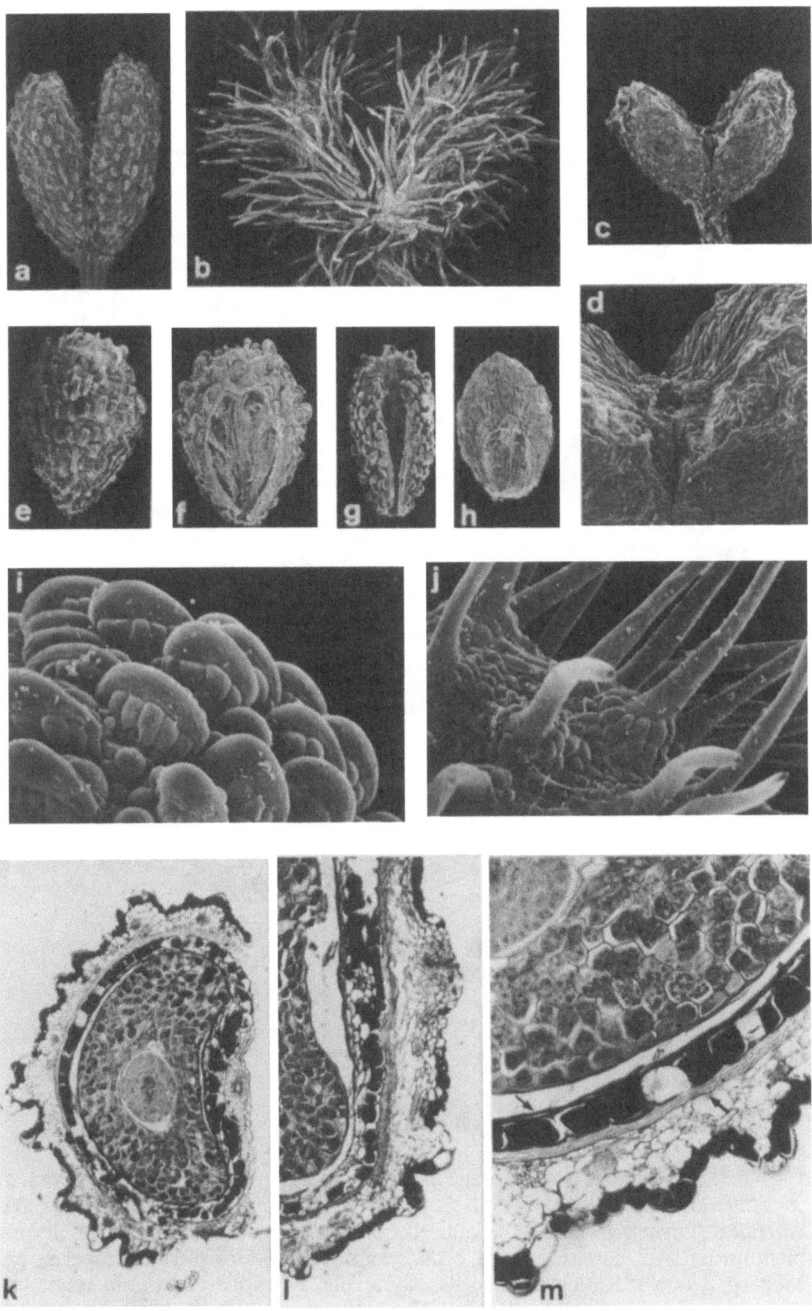

Fig. 39

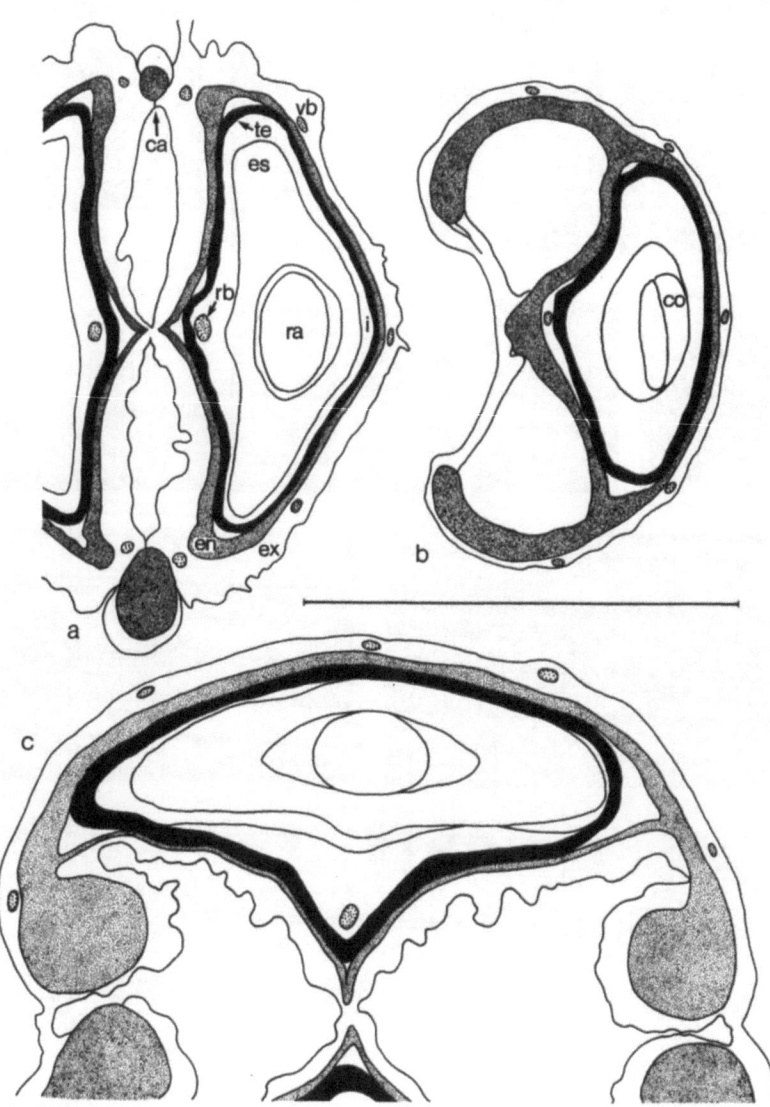

Fig. 40. Transverse sections of fruits or mericarps of *Anthospermum* and *Nenax*. *a A. littoreum* (PUFF 790415-5/1). *b A. palustre* (PERRIER DE LA BÂTHIE 3809). *c N. coronata* (PUFF 800902-6/4). – Semischematic camera lucida drawings; indumentum ignored; ca carpophore, co cotyledon, en endocarp, es endosperm, ex exocarp, i crushed integument cells, ra radicle, rb raphe bundle, te testa, vb vascular bundle. The scale unit is 1 mm

or a shallow groove is formed on either side of the median endocarp ridge (e.g., *A. littoreum* Figs. 35 e and 40 a). If only tissue cushions persist at the edge of the mericarps (as in *N. microphylla*, Fig. 41 f), the mericarps, in ventral view, show two lateral, ± broad ledges in addition to the median, vertical endocarp ridge (Figs. 38 b and 59 c). Fig. 37 h shows the disintegrated remains of this tissue covering the endocarp of *N. coronata*. As regards the deterioration of this tissue, there is obviously some variation within the taxa: In *A. paniculatum,* for example, the commissure is not always ± plane (Fig. 41 a), as the loose tissue to the left and to the right of the endocarp ridge may disappear almost entirely. *A. palustre* is unusual in that a strand of tissue connects the endocarp-"arms" and the median endocarp ridge so that each mericarp contains two large, air-filled pockets (Fig. 40 b). In *Nenax* species with indehiscent fruits, this loose tissue (cf. *N. acerosa* subsp. *macrocarpa*, Fig. 41 g) merely breaks up as the fruits increase in diameter; in mature fruits, there are only a few tissue shreds left (e.g., *N. arenicola,* Fig. 41 c).

In *Galopina*, the parenchymatous tissue on the ventral side of the mericarps does not differ markedly from the exocarp tissue on the dorsal side (except that no tuberculate structures are formed on the ventral surface). There may be a faint median, vertical ridge, but this is – in contrast to *Anthospermum* and *Nenax* – formed by the parenchymatous tissue and not by a raised, slerenchymatic endocarp portion (cf. *G. circaeoides,* Fig. 39 l). At maturity, the outermost ventral tissue layers may peel off, but at least a few inner layers persist to cover the endocarp.

In fruits of *Anthospermum* and *Galopina* and in *Nenax* species with dehiscent fruits, there is a distinct a b s c i s s i o n z o n e between the two mericarps. This is formed by the exocarp and consists of a narrow band of tissue made up of small, thin-walled cells (cf. *A. littoreum* and *N. microphylla,* Figs. 41 b and g respectively).

In *Nenax* species with indehiscent fruits, a continuous ring of sclerenchyma holds the two carpels together. In mature fruits, this sclerenchyma ring (representing the endocarp) is considerably thicker than (i.e., more than twice as thick as) the exocarp (cf. *N. arenicola,* Figs. 41 c–d) and is much more massive than in *Anthospermum* and *Nenax* species with dehiscent fruits. The sclerenchyma cells are arranged horizontally save for a ± distinct, additional zone of vertically arranged cells on the inside of the ring in the area between the two seed-containing endocarp cavities; horizontally and vertically arranged sclerenchyma layers are clearly discernible in Fig. 41 d.

In *A. bicorne* and *A. ericifolium,* only one of the two carpels is well developed and fertile. The second is much reduced and modified. As shown in Fig. 42 a for *A. bicorne,* there is only a band of undifferentiated, parenchymatous tissue instead of a second mericarp. Embedded in this tissue band are two larger vascular bundles (presumably supplying the two conspicuous calyx lobes, cf. Figs. 36 a, c), a smaller median and a few additional bundles. Anatomically, the fruit of *A. ericifolium* is similar.

7.2.6. Seeds

On maturation, the seed coats of *Anthospermum, Nenax* and *Galopina* seeds are reduced to an outer epidermis (i.e., an exotesta), although a few subepidermal, crushed integument layers (i.e., an endotesta) may remain discernible (cf. *A. bicorne,* Fig. 42 b or *G. circaeoides,* Fig. 39 m). This pattern of seed coat development is characteristic of most of the *Rubiaceae* (cf. NETOLITZKY 1926, WUNDERLICH 1971, CORNER 1976).

The testa cells are usually penta- to hexagonal in surface view (e.g., *G. circaeoides,* Fig. 42 c). They show no sculpturing and mostly have no thickenings at all or only very slightly thickened periclinal walls. They are filled with a brown substance (tannin?) which is already present in the integument epidermis of ovules with organized embryo sacs (see 8.); such dark substances are fairly common in the testa of rubiaceous seeds (see survey by WUNDERLICH 1971) and have also been recorded in the seeds of the allied tribe *Paederieae (Paederia scandens:* HASHMI & SIDDIQUI 1974). The testa cells are ca 10 to 25 (37) μm thick. In fully mature fruits, they are tightly adpressed to the endocarp so that it may be difficult to remove the testa from the sclerenchymatic endocarp. In microtome sections, the testa cells often appear to be attached to the endocarp rather than to the endosperm (presumably due to the fragile crushed integument layers between the actual testa and the endosperm; cf. *G. circaeoides,* Figs. 39 k, m). On the ventral side of the seed, several layers of integument

---

Fig. 41. Sectioned fruits of *Anthospermum* and *Nenax. a A. paniculatum* (PUFF 790114-6/1). *b A. littoreum* (PUFF 790415-5/1), abscission zone between two two mericarps; note vertical and horizontal layers of sclerenchymatic endocarp and cells filled with dark staining contents (tannin). *c–d N. arenicola* (PUFF 790714-2/1), *c* fully mature, indehiscent fruit, *d* detail showing thin exocarp and thick differentiated sclerenchymatic endocarp. *e–f N. microphylla* (PUFF 790112-1/1), *e* abscission zone between the two mericarps, *f* mericarp with small, ± flattened seed (note cotyledons in the endosperm cavity). *g N. acerosa* subsp. *macrocarpa* (PUFF 790910-4/2), young indehiscent fruit, oriented as in *c*; note large-celled parenchymatous tissue which will degenerate as the fruit matures (see *c*). *– b* microtome section, all others SEMgraphs of critical point dried material; *a, d–e, g* × 46; *b* × 68; *c, f* × 23

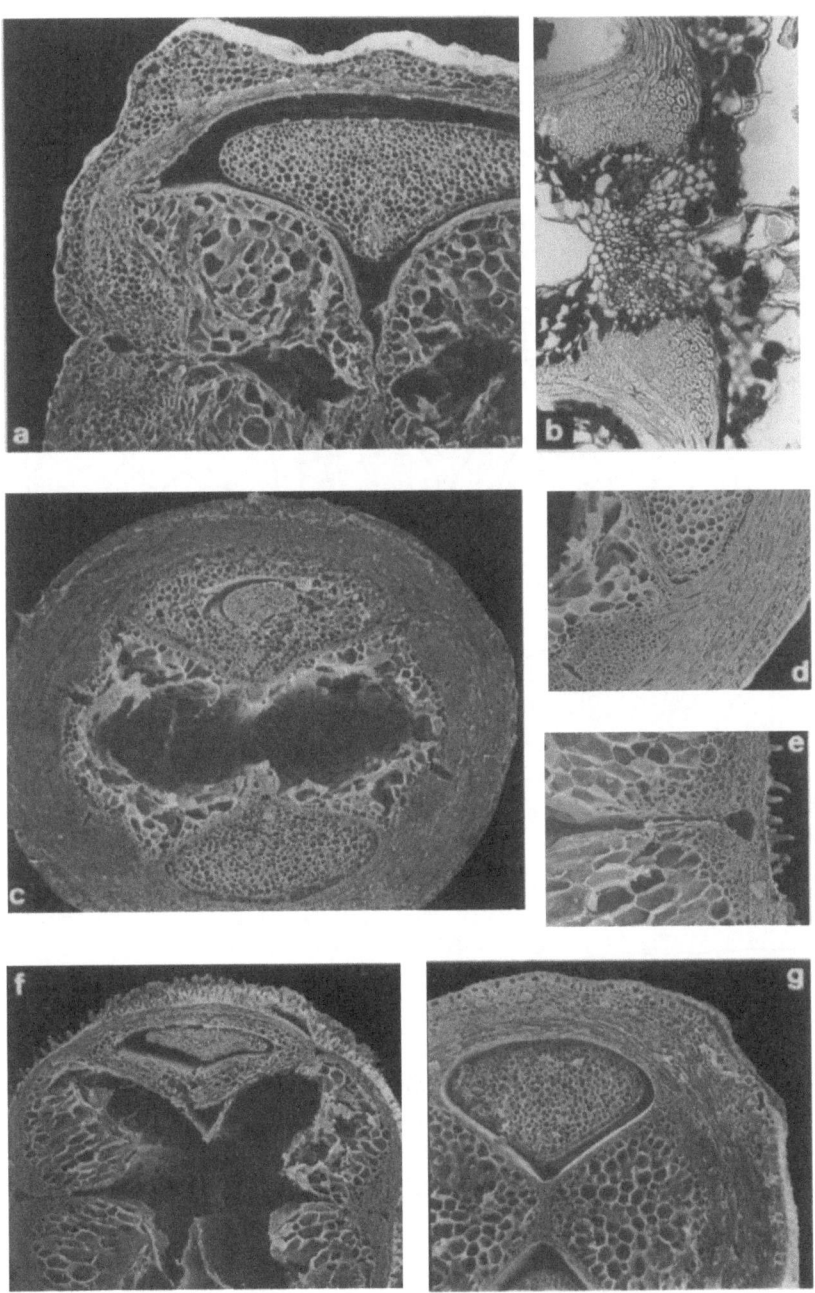

Fig. 41

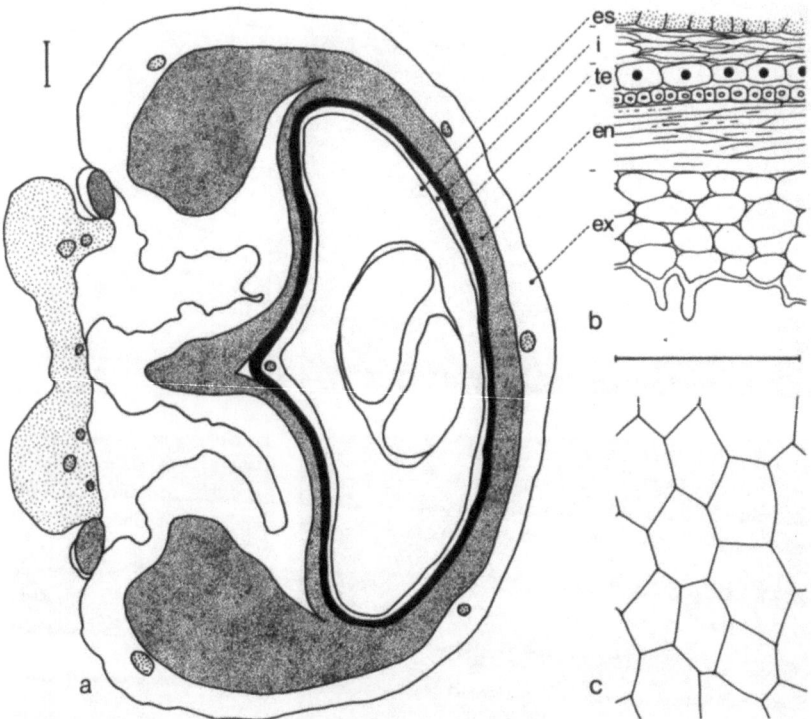

Fig. 42. Fruit and seed of *Anthospermum* and *Galopina*. *a–b A. bicorne* (PUFF 800915-5/1), *a* transverse section of fruit, note reduced carpel (left, dotted) *b* detail (testa cells filled with tannin: black dots; note differentiation of endocarp into a single layer of sclerenchyma cells running vertically and several layers running horizontally). *c G. circaeoides* (PUFF 790516-1/1), seed surface: testa cells. – See Fig. 40 for further explanations and magnification

cells remain intact, at least around the raphe bundle which usually remains clearly discernible (e.g., *A. littoreum*, *A. palustre*, *N. coronata* or *G. circaeoides*, Figs. 40 a–c and 39 k, l); in this area, the subepidermal layers as well as the outer epidermis cells of the seed may contain a dark substance.

The copious, "fleshy" endosperm consists of polygonal cells whose walls have no thickenings (e.g., *A. paniculatum*, *N. arenicola*, Figs. 41 a, c–d or *G. circaeoides*, Fig. 39 m). The embryo always lies in a distinct, central endosperm cavity (cf., *A. littoreum*, *N. coronata*, Figs. 40 a, c or *G. circaeoides*, Figs. 39 k, m).

7.3. Fruits and Seeds of *Carpacoce*

7.3.1. Morphology

Fruits of *Carpacoce* are always crowned by erect to spreading or curved persistent calyx lobes. They are large, leaf-like and ± (linear-) lanceolate (except for *C. curvifolia*, which has smaller, ± lanceolate to triangular lobes). In younger fruits, stipule-like fimbriate appendages may be discernible between the calyx lobes (e.g., *C. spermacocea* subsp. *orientalis*, Figs. 43 e and 44 h). The number of calyx lobes ranges from 3 (*C. gigantea*) to 5 and, occasionally, 6 (e.g., *C. vaginellata*). Calyx lobes may be distinctly unequal in some taxa (either only one lobe longer than the others as in *C. burchellii*, or more often two longer than the others as in *C. curvifolia* and *C. spermacocea;* Fig. 43 a). The longest lobes are ca 2–8 mm long (in *C. gigantea* they may be up to 12 mm long) and are not wider than 1.5 mm. Lobes are glabrous or subglabrous except for *C. heteromorpha* which has hairy lobe margins (Fig. 43 c).

Fruits are ellipsoidal or (narrowly) obovoidal, cylindrical to subglobose or ± turbinate. The fruits of *C. heteromorpha* (Fig. 43 c) are ca 1.5–3 mm in diameter, those of the other species are larger, up to 7.5 mm long and 2–3 mm wide (e.g., *C. vaginellata,* Fig. 43 d); fruits of *C. scabra* subsp. *scabra* may be slightly wider than long (Fig. 43 b).

Fruits of *C. scabra* are two-seeded, those of the other taxa are one-seeded (odd two-seeded fruits may, however, occasionally occur in some taxa – e.g., *C. vaginellata;* see below).

At maturity, fruits separate into exocarp and endocarp plus seed. The exocarp dehisces longitudinally into valves (cf. *C. scabra* subsp. *scabra,* Fig. 43 f; the number of valves equals the number of calyx lobes present).

The dispersal unit is the indehiscent, ± hard endocarp which encloses the seed (see 7.3.2. for details). The dispersal units of taxa with one-seeded fruits are ellipsoidal to almost cylindrical, or subglobose to ± pyriform. In the two-seeded fruits of *C. scabra,* the ventral side of the dispersal unit is plane to somewhat concave and the dorsal side convex (compare Figs. 43 f and j–k). The dispersal units are ca 1.2–3 mm long (slightly longer, up to 4–5 mm, in *C. spermacocea* and *C. vaginellata*), and ca 0.8–2 mm wide. Surfaces may be ± smooth at first, but the outermost layer(s) may soon flake off so that the dispersal units become rugose to ± muricate or somewhat ribbed (cf. Figs. 43 f–k and 44 f, i). In the one-seeded fruits, the dispersal units have a distinct vertical groove on the ventral side which, in *C. vaginellata,* widens at the base (Fig. 43 g). In *C. spermacocea* (Fig. 44 f) and *C. curvifolia,* the base of the dispersal unit is distinctly hollowed out below the vertical groove. This hollow is initially filled with a "plug" of spongy tissue which, as the fruit matures and dehisces, often and easily falls out. Such a spongy "tubercle" may sometimes also remain attached to the widened base of the dispersal unit of *C. vaginellata* (see 7.3.2., below). In *C. gigantea* and *C. heteromorpha* this tissue is absent from the base of the dispersal units (cf. Fig. 43 h–i).

7.3.2. Anatomy

The two-seeded fruits of *C. scabra*.

In Fig. 46 it can be seen that the exo- and endocarp is already differentiated at the flowering stage. The exocarp, in which the vascular bundles supplying the calyx lobes are embedded, is several-layered. Also the abscission zone at the base of the fruits (i.e., where the exocarp valves break off from the pedicels) is already preformed at the flowering stage. The relatively thin endocarp has a basal opening through which the funicle enters the locule. The funicles, which appear dark in unstained sections, pass into an also dark-coloured, solid central strand of conducting/transmitting tissue which extends into the stigmas. As the fruits mature, cell layers between exo- and endocarp (i.e., presumably the innermost, small-celled and thin-walled exocarp layers; cf. Fig. 46) start to disintegrate. The endocarp becomes more differentiated: on the ventral side of each dispersal unit, a narrow, vertically arranged zone of rather massive sclerenchymatic cells is formed (the ± smooth, median vertical area in Fig. 43 f). In the surrounding areas, the endocarp is made up of less distinctly sclerenchymatic cells and cells with only partially thickened walls. Some of these endocarp layers may themselves start to disintegrate so that the surface of the dispersal unit becomes rather uneven. Even at full maturity, the basal opening of the endocarp does not close (cf. Fig. 43 f); occasionally, dried up strands of tissue (remains of the conducting tissue) can be seen emerging from this opening.

Fig. 43. SEMgraphs of fruits and diaspores of *Carpacoce*. *a–d* fruits of *a C. spermacocea* subsp. *orientalis* (PUFF 790909-5/2), *b C. scabra* subsp. *scabra* (ESTERHUYSEN 10961), *c C. heteromorpha* (PUFF 800917-4/1), *d C. vaginellata* (SALTER 6432). *e C. spermacocea* subsp. *orientalis* (PUFF 790909-5/2), fruit as viewed from above; note stipule-like structures between calyx lobes and crown-like central outgrowth (compare with Fig. 44 h). *f C. scabra* subsp. *scabra* (ESTERHUYSEN 10961), diaspore, about to separate from the exocarp; note the two exocarp valves and the basal hollow in the diaspore. *g C. vaginellata* (PUFF 790912-1/1), diaspore, ventral side. *h–i C. heteromorpha* (PUFF 800917-4/1), diaspores, *h* side view, *i* ventral side. *j–k C. scabra* subsp. *scabra* (ESTERHUYSEN 10961), diapore, *j* side view, *k* dorsal side. – *b–d* herbarium material, all others critical point dried. *a–e* × 8; *f–k* × 16. Further explanations in the text

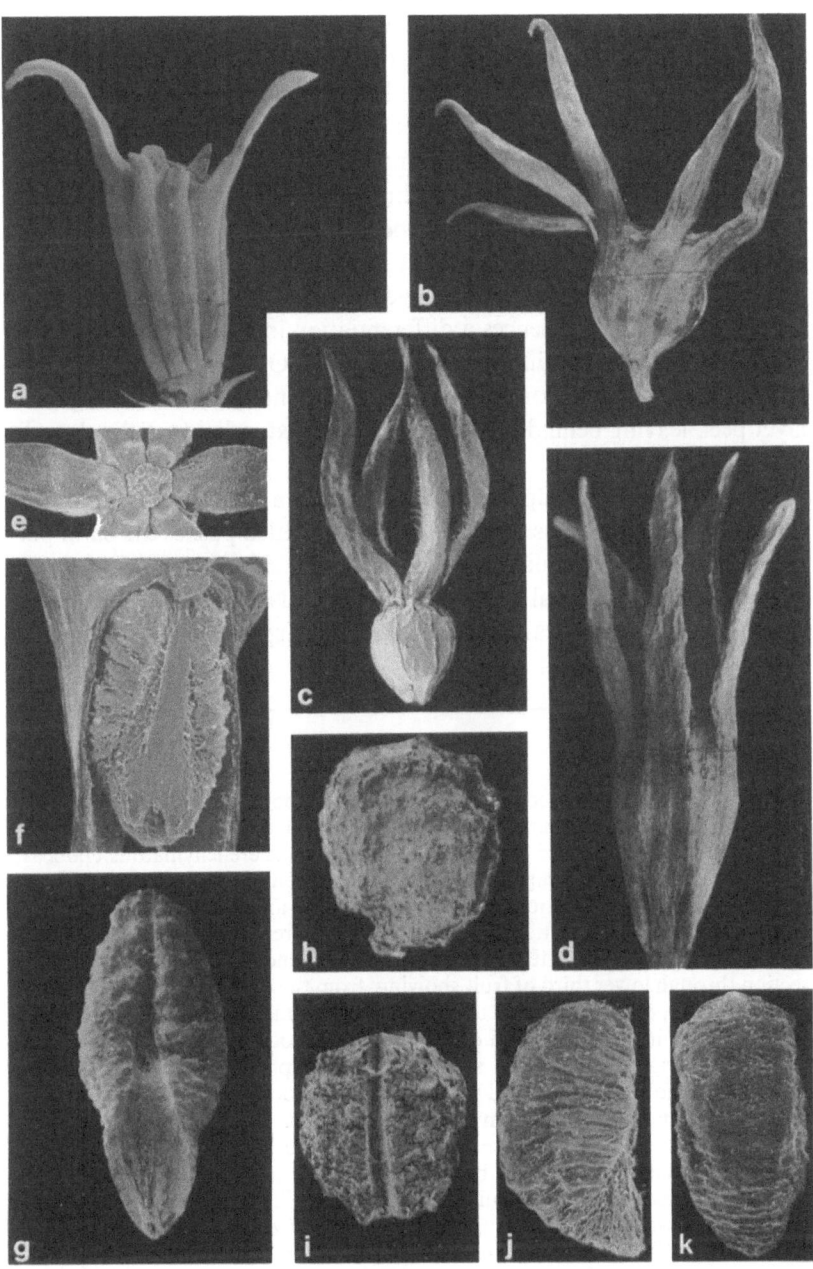

Fig. 43

One-seeded fruits.

In taxa with one-seeded fruits, one of the two carpels is much reduced
and modified to varying degrees. The fruits of *C. spermacocea* subsp.
orientalis may serve as an example (Figs. 44 a–h): as in two-seeded fruits,
the single seed is enclosed in a several-layered endocarp. The basal part of
the endocarp (enclosing the funicle: Figs. 44 a–c, f), however, becomes
very narrow and curved; it bears a small opening where the funicle enters
(Fig. 44 a; comparable with the opening in the endocarp of the two-seeded
C. scabra, Figs. 46 and 43 f). This opening may still be visible when the
fruits are fully mature (cf. Fig. 44 f; also in *C. vaginellata*, Fig. 43 g).
During maturation, the endocarp does not only separate from the
exocarp, but also experiences a differentiation into an outer layer and a
hard sclerenchymatous inner layer (Fig. 44 d). Outer layers may eventually come off in flakes, or only the outer tangential walls of a single cell
layer peel, leaving behind the partially thickened radial (anticlinal) walls
(Fig. 44 g).

The second carpel appears to be reduced to a "plug" or "cushion" of
spongy, parenchymatous tissue (Figs. 44 a–c) which pushes the basal part
of the endocarp containing the funicle to the side. A strand of tissue,
extending from the basal spongy tissue upwards to the apex of the fruit
and located in the vertical groove of the endocarp portion containing the

---

Fig. 44. SEMgraphs of sectioned fruits and diaspores of *Carpacoce. a–h C.
spermacocea* subsp. *orientalis* (PUFF 790909-5/2), *a–b* longitudinal section of fruit
showing lower part of seed attached to funicle, the sclerenchymatous endocarp
(thin arrow), the reduced carpel (★) forming a cushion of parenchymatous tissue,
the exocarp and the abscission zone between exocarp and pedicel (thick arrow); *c–
e* transverse sections, *c* section of reduced carpel (parenchymatous tissue; ★) and
funicle of the seed of the fertile carpel surrounded by endocarp (compare with *b*!), *d*
section through lower third of fruit showing, from the center outward, the radicle
of the embryo, endosperm with testa (testa cells hardly recognizable), endocarp
(differentiated into an inner and outer layer) and exocarp, *e* detail of *d* showing,
from left to right, part of the seed (in center: raphe bundle), the endocarp,
associated conductive tissue and remains of the reduced carpel (arrow), and the
exocarp; *f* diaspore, oriented as in *a*; note hollowed out basal portion, longitudinal
groove and parts of the endocarp coming off in flakes; *g* detail of endocarp, note
partially thickened walls of cells below the endocarp surface (left); *h* longitudinal
section of uppermost part of fruit showing stipule-like structures between calyx
lobes and crown-like central outgrowth. *i–j C. vaginellata* (PUFF 790912-1/1), *i*
diaspore, side view; *j* longitudinal section of fruit (oriented as in section *a*) showing
fertile carpel with endocarp (left; the seed has fallen out) and relatively large
reduced carpel (★). – All critical point dried. *a, h* ×35; *b–c, e* ×46; *d* ×23; *f, i–j*
×16; *g* ×93. Further explanations in the text

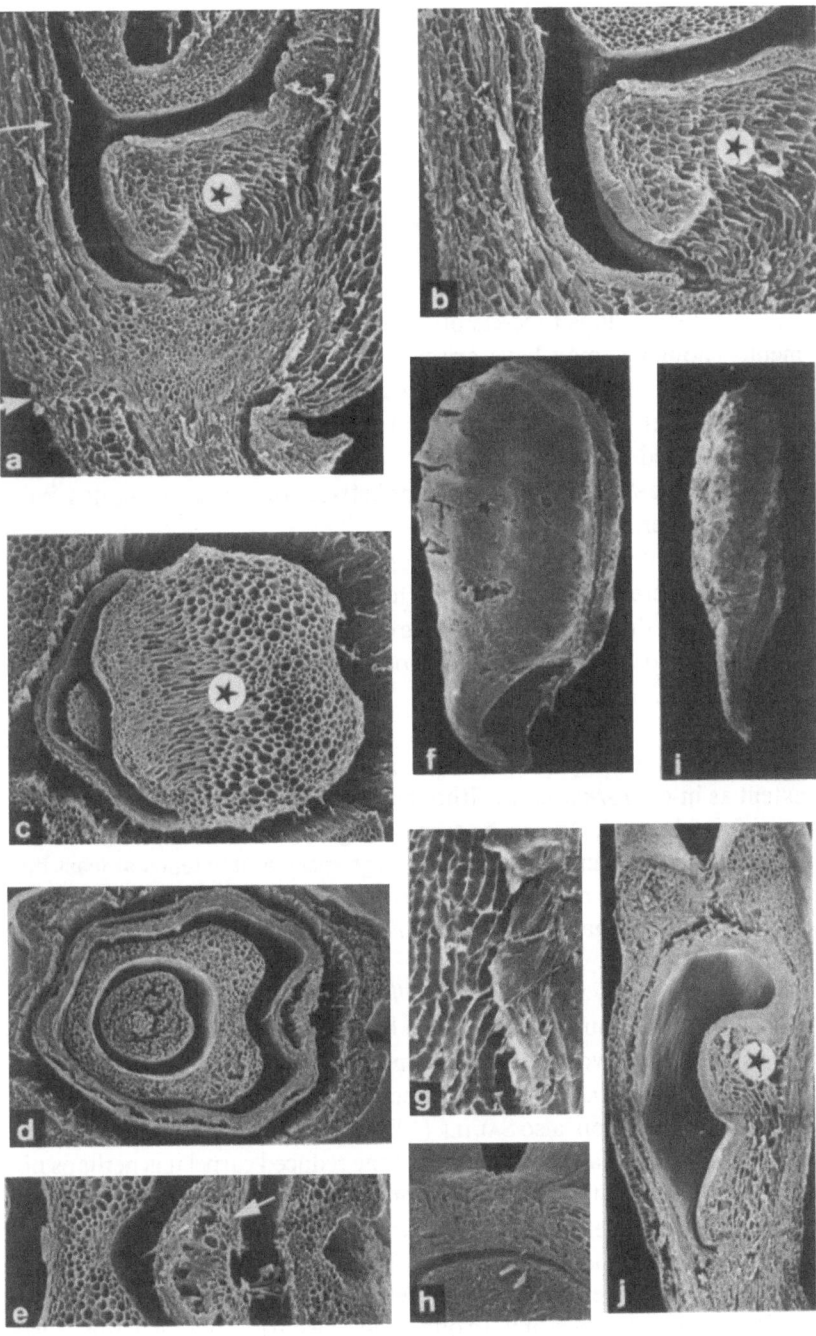

Fig. 44

seed (Figs. 44 d–e), is also likely to belong to the reduced carpel, although this groove most certainly also carries the conducting/transmitting tissue. The reduced carpel thus no longer shows a differentiation into an exo- and endocarp, although the remains of the infertile carpel and the fruit valves separate at maturity due to the deterioration of cell layers between them (Figs. 44 c–e). When the fruits dehisce, the spongy "plug" may remain in the hollowed out base of the endocarp, but more often it becomes detached from the endocarp and falls away. In her investigation of C. *spermacocea* fruits, Levyns (1937) came to the incorrect conclusion that "... the conspicuous tubercle of the fruit is a post fertilisation development. Though it may have arisen as a secondary structure from the rudiment of the second carpel, it would be incorrect to speak of it as a reduced carpel". It should be borne in mind, however, that no material of the two-seeded fruits of C. *scabra* was available to Levyns, which would have made the structural relationships between one- and two-seeded fruits obvious to her.

As the fruit matures, the small depression or cavity at the apex of the fruit (i.e., where the stigma was inserted) soon becomes filled with a dense mass of cells due to proliferating growth of tissue in the apical region (Figs. 43 e and 44 h). When the exocarp valves start to separate, this "crown" falls off easily.

In fruits of C. *vaginellata* the second carpel is not reduced to the same extent as in C. *spermacocea*, although the basic fruit structure of the two species is identical. As can be seen in Fig. 44 j, the spongy tissue of the reduced carpel occupies a relatively large space and extends at least half way up the endocarp containing the seed (also compare endocarp shapes, Figs. 44 f and i). Again, this spongy tissue falls off or is easily detached as the fruit dehisces.

It is noteworthy that in C. *vaginellata*, some plants may produce two-seeded fruits, although this seems to be a rather rare phenomenon [both plants with only two-seeded fruits and plants with one- and two-seeded fruits next to specimens with only one-seeded fruits were observed by me in a single population; also Salter (1937) made note of this phenomenon]. Considering the relatively large size of the reduced carpel it is perhaps not entirely surprising that such reversions to two-seeded fruits do occur in this species. The detection of odd two-seeded fruits in C. *heteromorpha*, however, is much more unexpected (e.g., in population Puff 800917-4/1). In that species, one of the two carpels is normally so reduced that it is difficult to detect a trace of it. In odd two-seeded fruits of C. *heteromorpha*, endocarps (plus seeds) are no longer ± subglobose, but their ventral surface is plane; in shape they are similar to the dispersal units of C. *scabra*.

7.3.3. Seeds

The seeds of *Carpacoce* have a very inconspicuous testa which may even be difficult to distinguish from endosperm cells (e.g., *C. spermacocea* subsp. *orientalis,* Figs. 44 d–e); no obvious subepidermal (endotesta) layers are visible except for a very small area surrounding the raphe bundle (cf. Fig. 44 e). The testa cells, in contrast to seeds of *Anthospermum, Nenax* and *Galopina*, do not contain a dark substance, they are polygonal in surface view, have no sculpturing and are only 10 µm or less thick.

The endosperm is copious. Endosperm cells are ± polygonal, their walls may be slightly thickened. The endosperm cavity containing the embryo is coated by a layer of distinct, small cells (cf. Fig. 44 d).

8. Embryology

The archesporium is unicellular in all genera of the *Anthosperminae* (cf. *G. circaeoides,* Fig. 45 b; *Phyllis,* FAGERLIND 1936) and in the presumably allied genus *Theligonum* (*Theligoneae;* KAPIL & RAO 1966). The nucellus is tenuinucellate and consists of a single, distinct layer of cells surrounding the archesporium (Fig. 45 b). The structure of the nucellus in the *Anthosperminae* thus corresponds to the "*Phyllis*-type" of nucellus in the *Rubiaceae* (cf. FAGERLIND 1937; = the general condition in the *Sympetalae*, MAHESHWARI 1950). The integument is massive and, except in *Carpacoce*, the outer integument epidermis cells are (much) larger than the other integument cells (compare *G. tomentosa, A. zimbabwense, Nenax cinerea*, Figs. 45 c–d, g, and *C. scabra* subsp. *scabra*, Fig. 46) and soon become filled with a dark substance (which, presumably, is the reason why any attempts to produce cleared preparations have failed).

In *A. aethiopicum, A. zimbabwense, G. circaeoides, G. tomentosa, N. acerosa, N. cinerea, C. scabra* subsp. *scabra* and *C. heteromorpha* the embryo sac development is similar to that of *Phyllis* (cf. FAGERLIND 1936). In the fully organized embryo sac, the three, relatively large antipodal cells are unequal in size, the median is often (considerably) larger than the others (cf. *C. heteromorpha* and *N. cinerea*, Figs. 45 a, f). Antipodal cells were always found to be uninucleate; divisions of nuclei as in *Phyllis* (4- and 8-nucleate antipodal cells, FAGERLIND 1936) could not be observed. The egg apparatus is characterized by relatively large synergids. The polar nuclei appear to fuse before the entry of the pollen tube, and the secondary embryo sac nucleus is often found suspended on plasma bands in the central portion of the embryo sac (cf. *N. cinerea*, Fig. 45 f).

The endosperm development is nuclear, as in most *Rubiaceae* (see survey by WUNDERLICH 1971). At later stages of development, the endosperm becomes cellular, it enlarges considerably and – as in numerous *Rubiaceae* – differentiates into two zones. Like in, for example, *Crocyllis* (*Paederieae*; PUFF & MANTELL 1982a, Fig. 16) the cells of the inner zone of the endosperm eventually degenerate so that the embryo comes to lie in a central cavity. The outer, compact zone consists of small, isodiametric cells, filled with storage materials (cf. *G. circaeoides*, Fig. 39m). As the seeds mature, the inner endosperm zone disappears completely, and the cavity (never entirely filled out by the mature embryo – cf. mericarp sections, Figs. 40a–c and 42a) becomes coated by a distinct layer of small endosperm cells (cf. for example, *C. spermacocea* subsp. *orientalis*, Figs. 44d–e). In the mature seed, the endosperm has enlarged so much that only the testa and perhaps a few crushed subepidermal layers are left of the integument (see 7.2.6. and 7.3.3.).

All *Anthosperminae* have distinct to ± indistinct obturators which appear ± finger-like in longitudinal section. In *N. cinerea*, *N. acerosa*, *G. circaeoides*, *G. tomentosa* and in *Phyllis nobla* they are quite large in young ovules (cf. Figs. 45b–c and FAGERLIND 1936, Fig. 1); their size varies in *Anthospermum* species (compare *A. zimbabwense* and *A. emirnense*, Figs. 45d and e). In *C. heteromorpha* and *C. scabra* subsp. *scabra* they are present but difficult to see because they are closely adpressed to the micropylar end of the ovule, leaving only a narrow gap (Figs. 45a and 46). There is little doubt that these obturators are outgrowths of the funicles. They clearly arise below the very base of the ovules, from where they grow downward (cf. *G. tomentosa*, Fig. 45c). They may also expand laterally to some extent (cf. *N. cinerea*, Fig. 45g). Obturator cells are isodiametric and hardly different in size and shape from integument cells (cf. *G. tomentosa* and *A. emirnense*, Figs. 45c and e) except for *Carpacoce*, whose obturators have ± cylindrical, elongated cells (cf. *C. scabra* subsp. *scabra*, Fig. 46). The obturators presumably cease growth before fertilization takes place.

Fig. 45. Ovules, embryo sacs and obturators of *Carpacoce*, *Galopina*, *Anthospermum*, and *Nenax*. *a C. heteromorpha* (PUFF 800101-3/1), ovule with fully organized embryo sac. *b G. circaeoides* (PUFF 790516-1/3), ovule with unicellular archesporium. *c G. tomentosa* (PUFF 790304-3/1), ovule with fully organized embryo sac. *d A. zimbabwense* (PUFF 790124-2/2), ovule with fully organized embryo sac; note small obturator (arrow); ovary wall differentiated into exo- and endocarp (en). *e A. emirnense* (PUFF 800729-1/1), obturator (finger-like projection on the right; bundle entering funicle on the left). *f–g N. cinerea* (PUFF 800901-1/10), *f* fully organized embryo sac with polar nuclei fused to form secondary embryo sac nucleus (2 synergids above, egg cell hidden; antipodal cells below), *g* view of raphe side of ovule showing lateral extension of obturator. – Note large integument epidermis in *c, d, g* and raphe bundle in *a, c–d, g*. Further explanations in the text.
Camera lucida drawings. The scale units are 0.1 mm except *e–f* (0.01 mm)

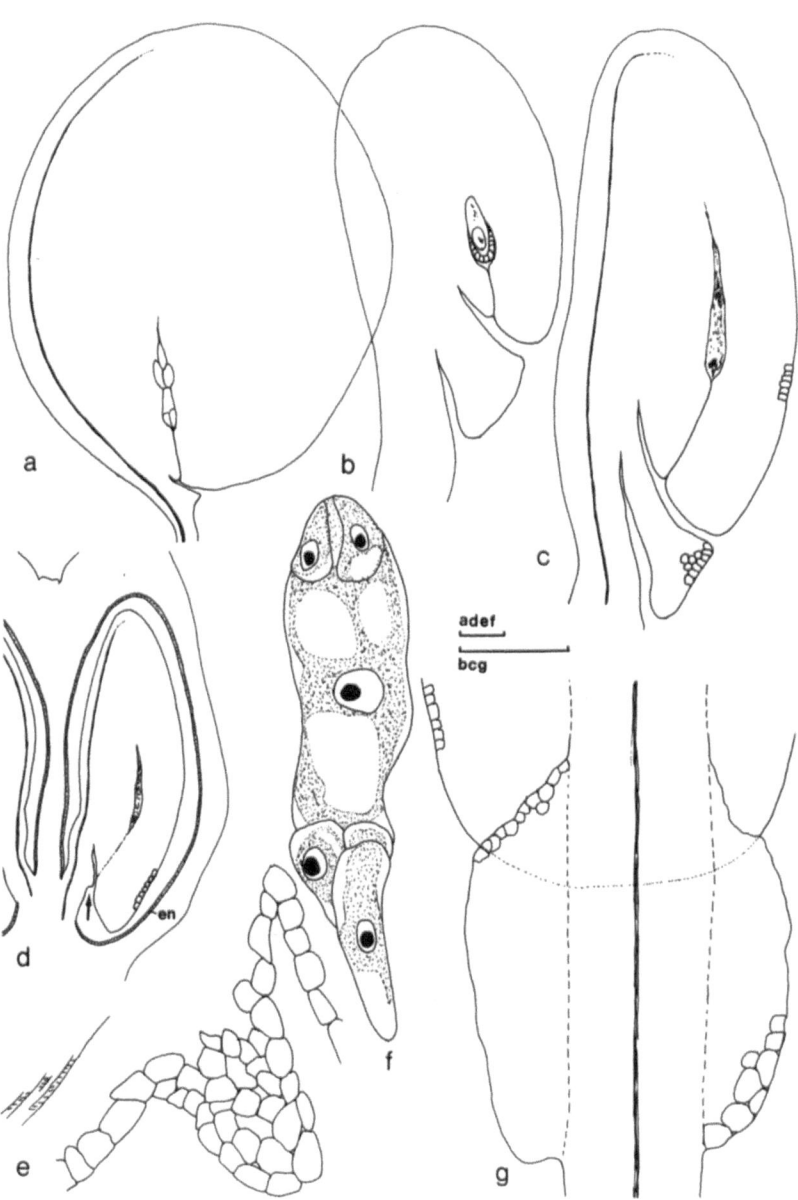

Fig. 45

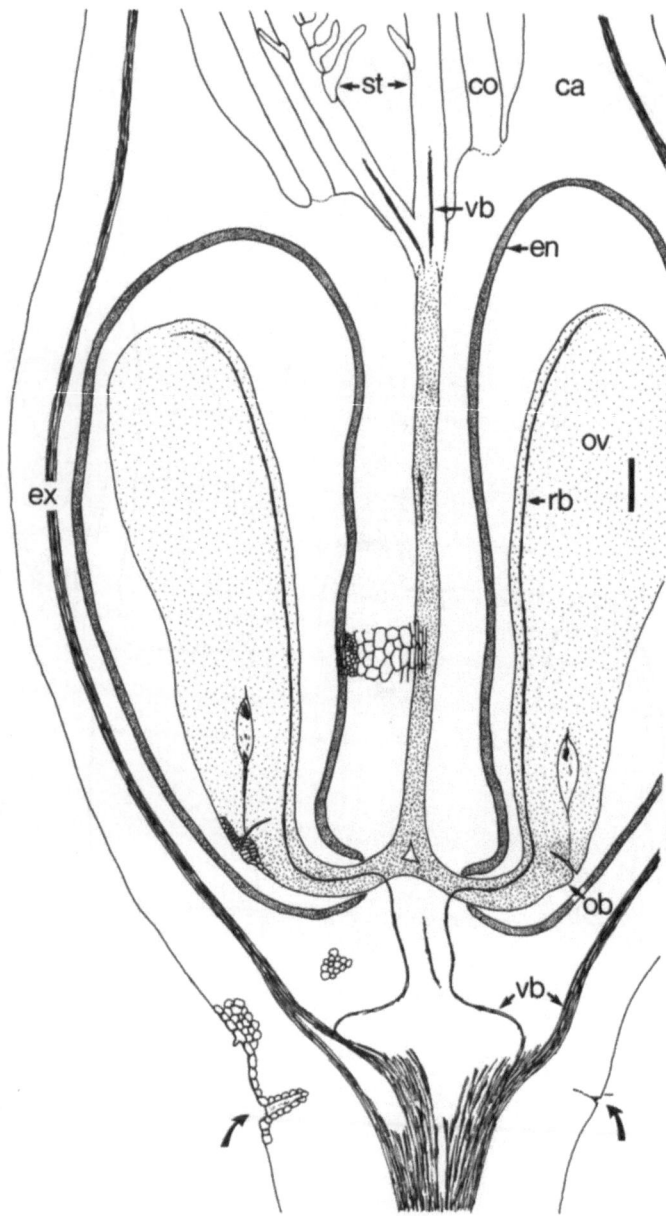

Fig. 46. Longitudinal section of ovary of *Carpacoce scabra* subsp. *scabra* (Puff 800914-6/1). – ca calyx, co corolla, en endocarp, ex exocarp, ob obturator, ov ovule, rb raphe bundle, st stigma, vb vascular bundle; the curved arrows point to abscission zone between exocarp and pedicel. Further explanations in the text. Camera lucida drawings. The scale unit is 0.1 mm

They probably are squashed during the maturation of the seed and, subsequently, can no longer be detected.

The vascularization of the ovules is simple: there is only one raphe bundle which remains unbranched and tapers out in the chalaza region or curves downward opposite the raphe for a short distance (cf. Figs. 45 a, c, and 46).

In all four genera, the ovary wall is differentiated into a distinct exo- and an (at first) thin endocarp (see also 7.2.5. and 7.3.2.). In *Anthospermum*, *Nenax* and *Carpacoce* in particular, this differentiation starts early. The two layers can already be distinguished when the embryo sac of the ovule start to become organized (cf. *A. zimbabwense*, Fig. 45 d and *C. scabra* subsp. *scabra*, Fig. 46).

9. Karyology

The basic chromosome number of the *Anthospermeae* is x = 11 (PUFF 1982 a). All taxa of *Galopina* and *Carpacoce* investigated are diploid (n = 11, 2 n = 22). The karyologically known *Anthospermum* species are also diploid except for *A. spathulatum*, which has di-, tetra- and hexaploids. *Nenax* has both di- and tetraploid taxa (Table 3).

Mitoses (acetocarmine squashes and Feulgen preparations): Metaphase chromosomes differ in size between (but not markedly within) genera. *Galopina* has smaller somatic chromosomes than the other genera of the *Anthosperminae* (compare Figs. 47 a and b)[1]. *Anthospermum* and *Nenax* chromosomes are similar; they may be smaller than those of *Carpacoce*. Chromosomes of the latter, however, could only be observed in acetocarmine squashes of shoot tips (3:1 alcohol-acidic acid field fixations) while mostly root tip mitoses were investigated in the other genera; sizes may, therefore, not be directly comparable. Somatic chromosomes, in general, are small, often only 1 µm long and hardly ever exceeding 2 µm in metaphase.

Root tip mitoses rarely showed irregularities. Only in *A. ibityense*, non-disjunction of chromosomes and "anaphase-bridges" were observed by KIEHN (1986). He also found that spontaneous polyploidizations (2x → 4x; sometimes 4x → 8x) in odd cells of seedling root tips (*Anthospermum* and *Nenax* spp.) appear to be quite common; their division, however, is regular.

---

[1] The small chromosome size is obviously correlated with a low DNA-content. Recent DNA-measurements of various *Anthospermeae* (*Anthospermum*, *Nenax*, *Galopina*, *Phyllis* and *Coprosma* spp.) showed that *G. circaeoides* has the lowest DNA-content of all investigated taxa of the tribe (KIEHN 1986).

Table 3. Voucher specimens

Taxon	n =	2n =	No. of individuals investigated
Anthospermum aethiopicum	11		2
		22	3
		22	3
	11		1
		22	3
	11		1
	11		2
A. ammannioides	11	22	2
	11		2
		22	1
A. basuticum		22	2
		22	1
A. bergianum		22	1
	11		2
	11		1
A. bicorne		22	3
		22	2
		22	1
A. comptonii		22	1
	11	22	2
A. dregei subsp. *dregei*		22	1
subsp. *ecklonis*	11		2
	11		1
	11		2
	11	22	2
A. emirnense		22	3
	11		2
	11		2
	11		1
A. esterhuysenianum		22	1
A. galioides subsp. *galioides*	11		1
	11		2
	11	22	2
	11		2
subsp. *reflexifolium*	11		2
	11	22	2

for chromosome counts

Locality	Coll. (Reference)
South Africa, Cape Prov., Kidd's Beach-Port Alfred road	Puff 790116-3/1 (WU)
−, −, ca 5–8 km W of Nieuwoudtville	Puff 790712-3/1 (WU)
−, −, Cedarberge, path to Sneeuberg	Puff 790717-1/3 (WU)
−, −, top of Montagu Pass, nr. Camfer Stn.	Puff 790909-4/1 (WU)
−, −, Baviaanskloofberge, Hankey − Willowmore rd.	Puff 791220-4/1 (WU)
−, −, Wuppertal, Tra-Tra River	Puff 800902-5/1 (WU)
−, −, Stanford − Papiesvlei rd., base of Perdeberg	Puff 800918-1/4 (WU)
Zimbabwe, Vukutu Farm W of Juliasdale	Puff 790124-1/1 (WU)
−, Rhodes Inyanga National Park, Circular Drive	Puff 790125-4/1 (WU)
−, Stapleford Forest Res., North Patrol	Puff 790127-1/2 (WU)
South Africa, OFS, Witsieshoek Gate	Puff 781126-1/8 (WU)
−, Cape Prov., top of Naudes Nek Pass	Puff 790113-5/3 (WU)
−, −, Cedarberge, Sneeuberg	Puff 790718-1/1 (WU)
−, −, top of Buffelshoekpas, Citrusdal − Ceres rd.	Puff 800903-2/3 (WU)
−, −, plateau of "Porterville Mts." (Skurweberge)	Puff 800914-1/2 (WU)
−, −, SW of Elim, on Viljoenshof road	Puff 800918-5/1 (WU)
−, −, Riviersonderend Mts., road to Jonaskop	Puff 800925-2/1 (WU)
−, −, top of Shaw's Mountain Pass	Puff 791224-1/1 (WU)
−, −, WSW of Barandas on road to De Rust	Puff 790902-2/2 (WU)
−, −, S side of Witteberge, above Farm "Fisantekraal"	Puff 790914-3/2 (WU)
Namibia (Southwest Africa), Farm "Kubub" S of Aus	Puff 780810-2/1 (WU)
South Africa, Cape Prov., Farm "Lokenburg", ca 34 km S of Nieuwoudtville	Puff 800901-1/2 (WU)
−, −, Clanwilliam − Calvinia rd., at Wuppertal turnoff	Puff 800902-4/1 (WU)
−, −, Leipoldtville − Elandsbaai rd., Kliphoutkop	Puff 800915-3/2 (WU)
−, −, base of Dasklip Pass, N of Porterville	Puff 800913-2/3 (WU)
Madagascar, Mt. Angavokely, E of Antananarivo	Puff 800726-2/1 (WU)
−, around Miadanandriana	Puff 850824-2/1 (WU)
−, Lac Tritriva, W of Antsirabe	Puff 800729-1/1 (WU)
−, Tampoketsa d'Ankazobe: Ambohitantely Reserve	Puff 850808-1/4 (WU)
South Africa, Cape Prov., Witsenberge, Sneeugat Peak	Puff 800924-3/2 (WU)
−, −, N of Riversdale	Puff 790911-1/1 (WU)
−, −, 16.5 km W of Skipskop, on Bredasorp road	Puff 790912-1/2 (WU)
−, −, Wildehondskloof, Montagu − Barrydale road	Puff 790913-2/1 (WU)
−, −, Cape of Good Hope Nature Reserve, above Platboom Bay	Puff 800907-2/1 (WU)
−, −, Tradouwspas, around Andries Uys bridge	Puff 790911-3/2 (WU)
−, −, Gysmanshoek Pass between Brandrivier and Heidelberg	

C. General Part

Table 3 (continued)

Taxon	n =	2n =	No. of individuals investi- gated
A. galpinii		22	3
	11		2
		22	2
	11	22	2
A. herbaceum	11		2
	11		1
		22	1
	11		1
	11	22	2
		22	1
	11		2
	11	22	2
		22	1
	11	22	2
	11		2
	11		1
	11	22	2
	11	22	2
		22	1
	11		1
	11		2
	11		1
		22	1
	11		1
A. hirtum		22	1
		22	2
		22	1

Locality	Coll. (Reference)
South Africa, Transkei, "Eagles Nest" above Port St. Johns	Puff 790415-4/1 (WU)
−, Natal, 13 km from Ngome on road to Vryheid	Puff 791214-1/1 (WU)
−, −, Nqutu−Qudeni rd., Sigqokwana ridge	Puff 790303-2/1 (WU)
−, −, Umtamvuna Nature Reserve	Puff 790422-3/2 (WU)
Ethiopia, ca 12 km N of Dessie, Lake Haik	Puff & al. 810926-2/3 (WU)
−, Harerge (Harar) Admin. Reg., between Kuni and Bedessa	Puff & al. 820921-2/3 (WU)
Kenya, Saiwa Swamp National Park, N of Kitale	Puff & Weber 780116-1/2 (WU)
−, Rift Valley Prov., ENE slope of Mt. Elgon	Lewis 5959 (K, MO) (Lewis 1966)
Tanzania, South Pare Mts., Chome Forest	Puff & Kibuwa 820901-1/4 (WU)
Uganda, Ruwenzori, Ibanda to Nyabitaba	Puff & Katende 830716-1/1 (WU)
Malawi, Mt. Mlanje, Chambe Plateau	Puff 780213-1/1 (WU)
−, Northern Region, Misuku Hills, Matipa Forest Res.	Puff 781221-2/2 (WU)
Zimbabwe, Salisbury−Mazoe rd., Christon Bank Nature Reserve	Puff 790123-2/1 (WU)
Swaziland, 10 km from Mhlambanyati on Malkerns rd.	Puff 790224-2/1 (WU)
South Africa, Transvaal, 24.8 m SE of Barberton	Lewis 6334 (K, MO) (Lewis 1966)
−, −, Soutpansberg, road to Bluegums Poort	Puff 791202-3/1 (WU)
−, −, ca 3–4 km from Josefsdal Border Post on road to Barberton	Puff 791212-2/1 (WU)
−, Natal, Nkandla Forest Reserve, ca 19 km S of Nkandla	Puff 790303-5/1 (WU)
−, −, South Coast, Marina Beach	Puff 790427-2/2 (WU)
−, −, Farm "Twinstreams" near Mtunzini	Puff 791217-3/1 (WU)
−, −, summit of Kamberg Mt.	Puff 800115-3/2 (WU)
−, OFS, Qwaqwa, path to Drakensberg Plateau (Mopeli Hut)	Puff 790217-1/4 (WU)
−, Cape Prov., Transkei, Butterworth−Qolora Mouth road, near Kei Mouth turnoff	Puff 791219-2/1 (WU)
−, −, Fort Cunyngham Forest Reserve, Dohne Peak	Puff 790115-5/8 (WU)
−, −, base of Buffelshoekpas, ca 15–17 km from Citrusdal	Puff 800903-1/2 (WU)
−, −, upper part of Dasklip Pass, N of Porterville	Puff 800913-3/1 (WU)
−, −, ca 3 km S of Elim, on Viljoenshof road	Puff 800918-4/1 (WU)

C. General Part

Table 3 (continued)

Taxon	n =	2n =	No. of individuals investi-gated
A. hispidulum	11	22	2
	11		2
		22	3
		22	1
A. ibityense		22	3
A. isaloense		ca 22	1
		22	1
A. littoreum	11		2
	11	22	2
	11	22	2
	11	22	2
A. longisepalum		22	1
A. madagascariense	11		2
A. monticola		18–22	1
		22	1
		22	2
	11		1
A. pachyrrhizum	11		2
	11		1
	11		2
A. paniculatum	11	22	2
		22	1
		22	1
	11	22	2
	11		1
A. perrieri	11	22	2
		22	2
A. prostratum		22	1
	11		1

Locality	Coll. (Reference)
Swaziland, Ngwenya Mtn.	Puff 790225-1/2 (WU)
South Africa, Transvaal, ca 24 km N of Bronkhorst-spruit, on Groblersdal road	Puff 790209-1/1 (WU)
−, −, Johannesburg, Northcliff	Balkwill sub Puff 790513-1/1 (WU)
−, Natal, Nomangci Hill, ca 12 km S of Nkandla	Puff 790303-3/1 (WU)
Madagascar, Mt. Ibity	Puff 800730-1/2 (WU)
−, Isalo Massif, ca 8–10 km W of Ranohira	Puff 800814-1/1 (WU)
−, Col de Tapias between Sakaraha and Ranohira	Puff 800814-3/2 (WU)
South Africa, Natal, Isipingo Beach	Lewis 6290 (K, MO) (Lewis 1966)
−, −, Richards Bay	Puff 790304-1/1 (WU)
−, −, South Coast, Marina Beach	Puff 790427-2/1 (WU)
−, Cape Prov., Gonubie Mouth, NE of East London	Puff 790116-1/1 (WU)
Madagascar, Isalo Massif, ca 8–10 km W of Ranohira	Puff 800814-2/1 (WU)
−, Tampoketsa d'Ankazobe: Mahatsinja – Ankazobe rd., 2 km NW of Kianagara turnoff	Puff 850807-3/1 (WU)
South Africa, OFS, Golden Gate National Park	Puff 780820-1/1 (WU)
−, −, Witsieshoek Gate	Puff 781126-1/5 (WU)
−, Cape Prov., Witteberge, Lundean's Nek – Belmore rd.	Puff 790113-3/1 (WU)
−, −, half way up Naudes Nek Pass, Rhodes side	Puff 790113-4/2 (WU)
Ethiopia, at Portuguese Bridge near Debre Libanos	Puff & al. 810930-2/1 (WU)
−, ca 2 km from Goha Tsion, edge of Blue Nile Gorge	Puff & al. 811011-2/1 (WU)
−, ca 16 km N of Gondar on Dabat road	Puff & al. 811005-4/1 (WU)
South Africa, Cape Prov., Hogsback – Seymour rd., near Cathcart turnoff	Puff 790114-6/1 (WU)
−, −, Fort Cunyngham Forest Reserve, Dohne Peak	Puff 790115-5/3 (WU)
−, −, E of Port Alfred, at Great Fish Point turnoff on Kidd's Beach road	Puff 790116-4/1 (WU)
−, −, Winterberge, 49 km from Adelaide on Tarkastad rd.	Puff 790117-6/1 (WU)
−, −, ca 7 km W of Karatara	Puff 790910-2/1 (WU)
Madagascar, 19 km N of Ambohimiadana, on rd. to Miadanandriana	Puff 850824-3/1 (WU)
−, Antananarivo – Antsirabe rd., around PK 36	Puff 850827-2/3 (WU)
South Africa, Cape Prov., ca 4–5 km N of Skipskop on Malgas – Swellendam road	Puff 790911-8/1 (WU)
−, −, inland from Cape Agulhas Lighthouse	Puff 790912-3/1 (WU)

Table 3 (continued)

Taxon	n =	2n =	No. of individuals investigated
A. pumilum subsp. *pumilum*	11		2
	11		1
		22	1
	11	22	2
		22	2
	11		1
subsp. *rigidum*	11		2
	11		1
A. spathulatum subsp. *spathulatum*	22		2
		22	2
	11		2
		66	3
	11		1
	11		1
	11		1
		22	2
		22	2
	20–22		2
subsp. *ecklonianum*		44	3
		44	2
subsp. *saxatile*		44	2
		44	3
subsp. *tulbaghense*	11		1
subsp. *uitenhagense*	11	22	2
	11		2
A. streyi	11	22	2
A. ternatum subsp. *randii*	11		2
	11	22	2
	11		1
		22	1
subsp. *ternatum*		ca 22	1
A. thymoides subsp. *thymoides*		22	3

Locality	Coll. (Reference)
South Africa, Cape Prov., ca 4–6 km outside Cathcart on Hogsback – Happy Valley road	PUFF 790114-4/1 (WU)
–, OFS, ca 10–12 km E of Parys, along Vaal River	PUFF 800109-1/1 (WU)
–, Natal, Royal Natal National Park, Sigubudu valley	PUFF 800106-1/1 (WU)
–, Transvaal, Swaershoekberge, Farm "Gelhout Kloof 195"	PUFF 781203-2/1 (WU)
Swaziland, Lebombo Mts., Blue Jay Ranch	PUFF 780408-1/3 (WU)
Zimbabwe, Rua (Ruwa) River, E of Salisbury Airport	PUFF 790122-2/1 (WU)
South Africa, Cape Prov., Valley of Desolation outside Graaff-Reinet	PUFF 790908-3/1 (WU)
–, –, Colesberg – Norvalspont rd., at km 19	PUFF 800104-1/1 (WU)
–, –, Giftberg plateau	PUFF 790713-1/3 (WU)
–, –, forestry rd. from top of Pakhuis Pass to Heuningvlei Forestry Stn.	PUFF 790716-1/2 (WU)
–, –, Farm "Lokenburg", ca 34 km S of Nieuwoudtville, Stinkfonteinberg	PUFF 790711-2/1 (WU)
–, –, nr. summit of Kamiesberg Pass	PUFF 790714-2/1 (WU)
–, –, Farm "Grootvlei", E foot of Skurweberge	PUFF 790718-2/1 (WU)
–, –, Karroopoort, Ceres – Calvinia road	PUFF 790719-3/1 (WU)
–, –, Tradouwspas	PUFF 790911-3/3 (WU)
–, –, Nuwekloof, Baviaanskloofberge, Hankey – Willowmore road	PUFF 791221-2/1 (WU)
–, –, Cape of Good Hope Nature Reserve, above Platboom Bay	PUFF 800907-2/2 (WU)
–, –, Slangkop near Darling	PUFF 800910-1/1 (WU)
–, –, Jonkershoek Forest Res., path to waterfalls	PUFF 791231-1/2 (WU)
–, –, Paarlberg, below "The Rocks"	PUFF 800904-2/1 (WU)
–, –, Riviersonderend Mts., ridges W of Jonaskop	PUFF 800921-1/1 (WU)
–, –, Witsenberge, Sneeugat Peak	PUFF 800924-3/1 (WU)
–, –, Paarlberg, below "The Rocks"	PUFF 800904-2/1 (WU)
–, –, Howiesonspoort SW of Grahamstown	PUFF 790117-1/1 (WU)
–, –, nr. Fraser's Camp, off Peddie – Grahamstown rd.	PUFF 791220-1/1 (WU)
–, Natal, Oribi Gorge Nature Reserve	PUFF 790426-1/3 (WU)
Malawi, Dedza Distr., Chongoni Forest Reserve	PUFF 780215-3/1 (WU)
–, Northern Region, Hora Mt.	PUFF 781230-1/1 (WU)
Zimbabwe, ca 20.5 km E of Headlands, on Rusape rd.	PUFF 790129-2/2 (WU)
–, Salisbury (Harare) – Mazoe rd., Christonbank Nature Reserve	PUFF 790123-1/3 (WU)
Malawi, Northern Region, Misuku Hills, from Kaseye Mission to Matipa Forest	PUFF 781221-1/1 (WU)
Madagascar, Ankaratra Massif, below Tsiafajavona	PUFF 850826-1/1 (WU)

C. General Part

Table 3 (continued)

Taxon	n =	2n =	No. of individuals investigated
A. usambarense	11	22	2
		22	
		22	3
	11		1
	11	22	3
		22	2
A. vallicola	11	22	3
	11		1
A. welwitschii	11	22	2
	11	22	2
	11		2
		22	2
	11	22	2
A. whyteanum	11	22	2
	11	22	2
		22	1
	11	22	2
	11	22	2
		22	1
A. zimbabwense		22	1
	11	22	2
	11		1
	11	22	2
Nenax acerosa subsp. acerosa	22		1
	22		2
subsp. macrocarpa	22		1
	ca 22		1
N. arenicola	22		2
		ca 44	1

Locality	Coll. (Reference)
Kenya, Mt. Elgon, 3 320 m	PUFF & WEBER 780114-4/1 (WU)
–, Mt. Kenya, 3 550 m	COE & KIRIKA 377 (UPS) (HEDBERG & HEDBERG 1977)
Tanzania, Mt. Meru, 3 450 m	BIE 66118 (UPS) (HEDBERG & HEDBERG 1977)
–, Kilimanjaro, Mandara to Horombo Hut	PUFF & KIBUWA 820903-2/1 (WU)
–, South Pare Mts., Chome Forest	PUFF & KIBUWA 820901-1/5 (WU)
Zambia, Nyika Plateau, nr. Zambian Resthouse	PUFF 781223-2/1 (WU)
Zimbabwe, summit of Mt. Inyangani	PUFF 790125-1/1 (WU)
–, Vumba Mts., summit of Castle Beacon	PUFF 790128-1/1 (WU)
South Africa, Transvaal, Strydpoortberge, Chuniespoort area, Donkerkloof	PUFF 790209-3/1 (WU)
–, –, The Downs area	PUFF 790210-2/1 (WU)
–, –, Westfalia Estate, Piesangkop nr. Duivelskloof	PUFF 790211-2/3 (WU)
Malawi, Mt. Mlanje, Lichenya Plateau	PUFF 780211-2/1 (WU)
–, Northern Region, Misuku Hills, Mughese Forest	PUFF 781219-1/4 (WU)
Zimbabwe, Vukutu Farm, W of Juliasdale	PUFF 790124-1/3 (WU)
–, Stapleford Forest Reserve, North Patrol	PUFF 790127-1/1 (WU)
–, Vumba Mts., summit of Castle Beacon	PUFF 790128-1/2 (WU)
Malawi, Mt. Mlanje, from Chambe Plateau to Chambe Saddle	PUFF 780211-2/2 (WU)
–, Northern Region, S. Vipya, Mt. Champhila	PUFF 781231-1/1 (WU)
Zambia, S end of Mafinga Mts., Zambian side	PUFF 781222-2/6 (WU)
Zimbabwe, Rhodes Inyanga National Park, Fort Nyangwe	PUFF 790124-2/2 (WU) PUFF 790124-4/1 (WU)
–, Stapleford Forest Reserve, North Patrol	PUFF 790127-2/1 (WU)
–, Vumba Mts., summit of Castle Beacon	PUFF 790128-1/3 (WU)
South Africa, Cape Prov., ca 1.5–2 km NE of Darling Bridge	PUFF 800920-3/1 (WU)
–, –, Cape of Good Hope Nature Reserve, Platboom rd.	PUFF 800907-3/1 (WU)
–, –, nr. Goudiniweg Stn., Worcester – Ceres rd.	PUFF 800920-2/2 (WU)
–, –, 3.5 km NW of The Fisheries (Gouritsmond)	PUFF 800927-2/1 (WU)
–, –, ca 8 km E of Hondeklipbaai	PUFF 790714-3/1 (WU)
–, –, ca 3 km SE of Graafwater – Lambert's Bay main road, on Leipoldtville rd.	PUFF 800915-2/1 (WU)

C. General Part

Table 3 (continued)

Taxon	n =	2 n =	No. of individuals investigated
N. cinerea		22	1
	11		2
		22	3
		22	1
N. coronata		22	1
N. divaricata		22	2
		22	1
	11		1
N. hirta subsp. hirta		22	2
		22	1
N. microphylla		22	2
		22	3
	11		1
	11	22	2
		22	2
	11		1
	11		1
N. namaquensis		22	1
Galopina aspera	11		1
	11	22	3
G. circaeoides		22	1
	11		2
	11	22	2
	11	22	4
	11		1
	11	22	2
		22	2
	11		1
G. crocyllioides	11	22	2

Locality	Coll. (Reference)
South Africa, Cape Prov., Akkerendam Nature Reserve N of Calvinia	Puff 790711-1/1 (WU)
	Puff 800831-5/1 (WU)
−, −, around Whitehill Stn. nr. Matjiesfontein	Puff 800102-4/1 (WU)
−, −, Farm "Blomfontein" between Sutherland and Fraserburg	Puff 800103-2/1 (WU)
−, −, W side of Pakhuispas, W of Leipoldt grave	Puff 800902-6/4 (WU)
−, −, "Boskop" area, ca 45–50 km S of Nieuwoudtville	Puff 790712-2/1 (WU)
−, −, Giftberg plateau	Puff 790713-1/1 (WU)
−, −, Farm "Sanddrif", E of Sederberg village	Puff 790717-3/3 (WU)
−, −, Slangkop near Dearling	Puff 800910-1/2 (WU)
−, −, NW base of Dassenberg, Mamreweg – Mamre rd.	Puff 800910-2/1 (WU)
−, −, 33 km from Bedford on Grahamstown rd., "Fish River Rand" area	Puff 790117-4/1 (WU)
−, −, Taandjies Nek area, W of Carnarvon, "Klipheuwels"	Puff 790710-1/1 (WU)
	Puff 800831-3/1 (WU)
−, −, Richmond – Hutchinson rd., ca 15 km W of R 2	Puff 800103-3/1 (WU)
−, OFS, Reddersburg – Smithfield rd., nr. Sonop turnoff	Puff 790413-1/1 (WU)
−, −, Wepener – Smithfield rd., ca 15 km SW of Vanstadensrus turnoff	Puff 800104-3/1 (WU)
−, −, ca 29 km S of Winburg, along N 1	Puff 800930-4/1 (WU)
−, Cape Prov., ca 7–8 km E of Kamieskroon	Puff 790714-1/1 (WU)
−, Transvaal, The Downs area	Puff 790210-5/1 (WU)
−, −, Westfalia Estate nr. Duivelskloof, Rakgwale Ridge	Puff 790211-3/1 (WU)
Malawi, Zomba Distr., Zomba Plateau	Puff 780208-2/9 (WU)
Zimbabwe, Vumba Mts., Bunga Forest	Puff 790128-2/1 (WU)
Swaziland, Farm "Stroma" nr. Mbabane on Mhlambanyati rd.	Puff 790224-3/1 (WU)
South Africa, Transvaal, Suikerbosrand Nature Reserve	Puff 790516-1/3 (WU)
−, −, ca 3–4 km from Josefsdal Border Post on road to Barberton	Puff 791212-2/2 (WU)
−, Natal, Nkandla For. Res., ca 16–19 km S of Nkandla	Puff 790303-5/2 (WU)
−, Cape Prov., Montagu Pass, nr. Keurrivier bridge	Puff 790909-6/1 (WU)
−, −, Prince Alfred Pass, Dieprivier Picnic Spot	Puff 791222-1/1 (WU)
−, Natal, Zuurberg, Ingeli Forest Reserve, Transkei border	Puff 790416-1/1 (WU)

Table 3 (continued)

Taxon	n =	2 n =	No. of individuals investigated
G. tomentosa	11		2
		22	3
		22	2
Carpacoce heteromorpha		22	2
	11		2
C. scabra subsp. scabra		ca 22	1
	11		2
C. spermacocea subsp. spermacocea		22	2
		22	2
		22	1
		22	2
subsp. orientalis	11		3
C. vaginellata		22	1
		22	2
	11		2

Meioses: Field fixations of flowers buds were used to investigate pollen mother cell (PMC) meioses (mostly acetocarmine squashes). PMCs in metaphase I and II differ in size within genera; PMCs of tetraploids tend to

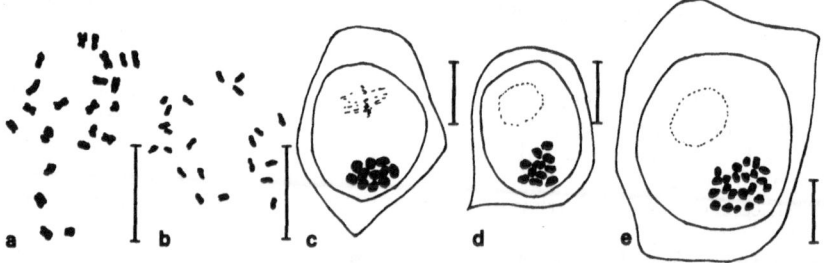

Fig. 47. Chromosomes of *Anthospermum*, *Galopina*, *Carpacoce*, and *Nenax*. *a–b* root tip mitoses, metaphase (Feulgen preparations), *a A. ammannioides* (2 n = 22; Puff 790124-1/1), *b G. circaeoides* (2 n = 22; Puff 790909-6/1). *c–e* PMC meioses, metaphase II (acetocarmine preparations), *c C. scabra* subsp. *scabra* (n = 11; Puff 800914-6/1), *d N. microphylla* (n = 11; Puff 800903-4/1), *e N. arenicola* (n = 22; Puff 790714-3/1). – The scale unit is 10 µm

Locality	Coll. (Reference)
South Africa, Natal, Palm Ridge Farm, ca 7 km N of Mtubatuba	Puff 790304-3/1 (WU)
−, −, Oribi Gorge Nature Reserve, Inkonka Point	Puff 790426-1/2 (WU)
−, Cape Prov., Transkei, above Port St. Johns	Puff 790415-2/1 (WU)
−, −, Kogelberg Forest Reserve	Puff 800101-3/1 (WU)
−, −, Vogelgat (Private) Nature Reserve, E of Hermanus	Puff 800917-4/1 (WU)
−, −, Ladismith Distr., Witteberg, S side	Puff 790914-4/2 (WU)
−, −, from summit of Pakhuispas to Heuning Vlei For. Stn.	Puff 800914-6/1 (WU)
−, −, Kogelberg Forest Reserve	Puff 800101-2/1 (WU)
−, −, Table Mt., top end of Nursey Gorge	Puff 800908-1/1 (WU)
−, −, Vogelgat (Private) Nature Reserve, E of Hermanus	Puff 800917-3/1 (WU)
	Puff 800917-6/4 (WU)
−, −, Montagu Pass	Puff 790909-5/2 (WU)
−, −, ca 16.5 km from Skipskop, on Bredasdorp road	Puff 790912-1/1 (WU)
−, −, Table Mt. plateau, between Nursery and Skeleton G.	Puff 800908-2/1 (WU)
−, −, from summit of Pakhuispas to Heuning Vlei For. Stn.	Puff 800914-6/2 (WU)

be larger than those of diploids. Inversely, metaphase chromosomes of tetraploids are smaller than those of diploids (*Anthospermum*, *Nenax*; compare Figs. 47 d and e). PMCs and chromosomes of *Carpacoce* are somewhat larger than those of *Galopina* and diploid *Anthospermum* and *Nenax* taxa (compare Figs. 47 c and d).

It is noteworthy that meioses occasionally do not take place synchronously in the anthers; prophases may be found next to fully developed pollen tetrads. In addition, meiotic irregularities (mainly clumping of chromosomes) and disturbed pollen development (deformed tetrads and pollen grains) were observed on numerous occasions (in *A. aethiopicum*, *A. galioides*, *A. prostratum*, *A. spathulatum* and *N. acerosa*, for example). Whether there is a direct connection between these disturbances and a disturbed anther development as such (♂ with fully developed anthers → ♂ with somewhat reduced but partially fertile anthers → functionally ♀ with sterile anther rudiments within an individual in, for example, *A. galioides*; see also 12.2.1.) could not be determined from the fixed material.

9*

10. Pollen

Pollen morphological studies of all genera of the *Anthosperminae* were carried out by ROBBRECHT (1982; 1985: additional observations on *Carpacoce* pollen). Pollen grains are 3-colporate and oblate, suboblate, oblate- or prolate-spheroidal, or subprolate. A rugulate exine is characteristic for all genera (see, for example, ROBBRECHT 1982, Fig. 7); only two *Carpacoce* species have an exine tending towards a tectum perforatum (ROBBRECHT 1982, 1985). Except for size differences (see below), the pollen of different species of a genus was found to be quite uniform. There were no significant differences in pollen morphology between species studied by ROBBRECHT (1982) and additional species investigated by me.

Average pollen diameters of *Anthospermum*, *Nenax*, *Galopina* and *Carpacoce*, correlated with ploidy level, are presented in Figs. 48 and 49. The pollen of *Carpacoce* is much larger than that of the remaining African genera and, for that matter, the largest within the tribe *Anthospermeae* (see ROBBRECHT 1982), although the genus, like the majority of species of the remaining *Anthosperminae*, is diploid.

Within species and sometimes also within populations of species there is often a conspicuously wide range of average pollen diameters. This phenomenon can (at least in part) be explained as follows:

(1) larger flowers (no matter whether ♂ or ♀̂) tend to have larger anthers and larger pollen than smaller ones. (2) ♂ flowers tend to have larger anthers and pollen than ♀̂ flowers on one and the same plant (cf. for example, flowers of *A. herbaceum*, Figs. 32 a–b) or from different plants of the same population. A ♂ individual of *A. hirtum* (population PUFF 800918-4/1), for example, had an average pollen diameter of 35 µm, a ♀̂ individual of the same population an average diameter of 32 µm. Similar differences in average pollen diameters were found between ♂ and odd ♀̂ plants in a single population of *A. bergianum*. In an individual of *A. herbaceum* (PUFF 780409-1/1) bearing ♂ + ♀̂ + ♀ flowers, the average pollen diameter from ♂ flowers was 28 µm, in contrast to a diameter of 25.2 µm from ♀̂ flowers. *C. spermacocea* subsp. *spermacocea*, having smaller flowers and anthers than subsp. *orientalis*, has, on average, smaller pollen than the latter (cf. Fig. 49).

This anther – pollen size correlation was observed in both non-dioecious and dioecious taxa of *Anthospermum*. In the diploid *A. galpinii*, for example, ♂ flowers with large anthers (ca 1.8 × 0.4 mm) were found to produce larger pollen than flowers with smaller anthers (ca 1.4 × 0.3 mm) taken from another plant of the same population (PUFF 790427-4/1). However, in *A. monticola*, another dioecious species with an unusually wide range of average pollen diameter, no such correlation could be

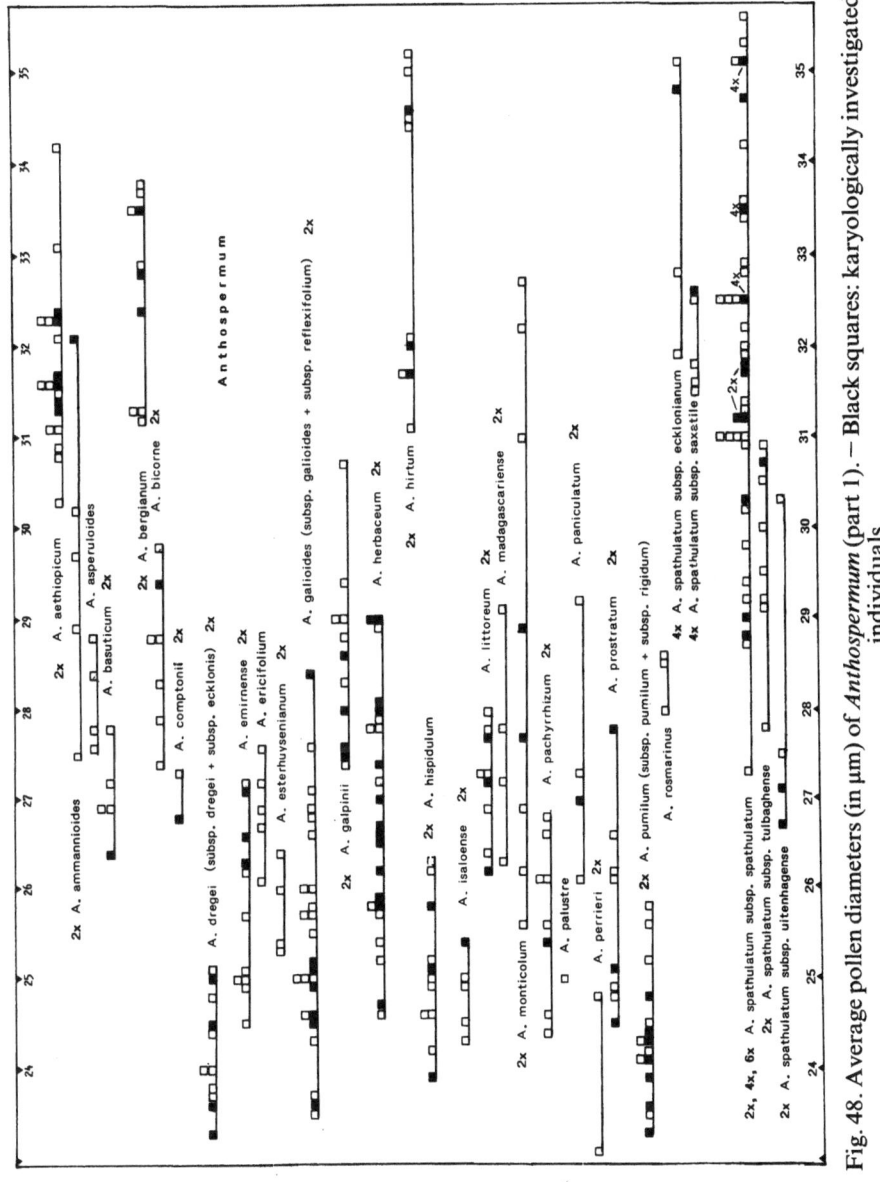

Fig. 48. Average pollen diameters (in μm) of *Anthospermum* (part 1). – Black squares: karyologically investigated individuals

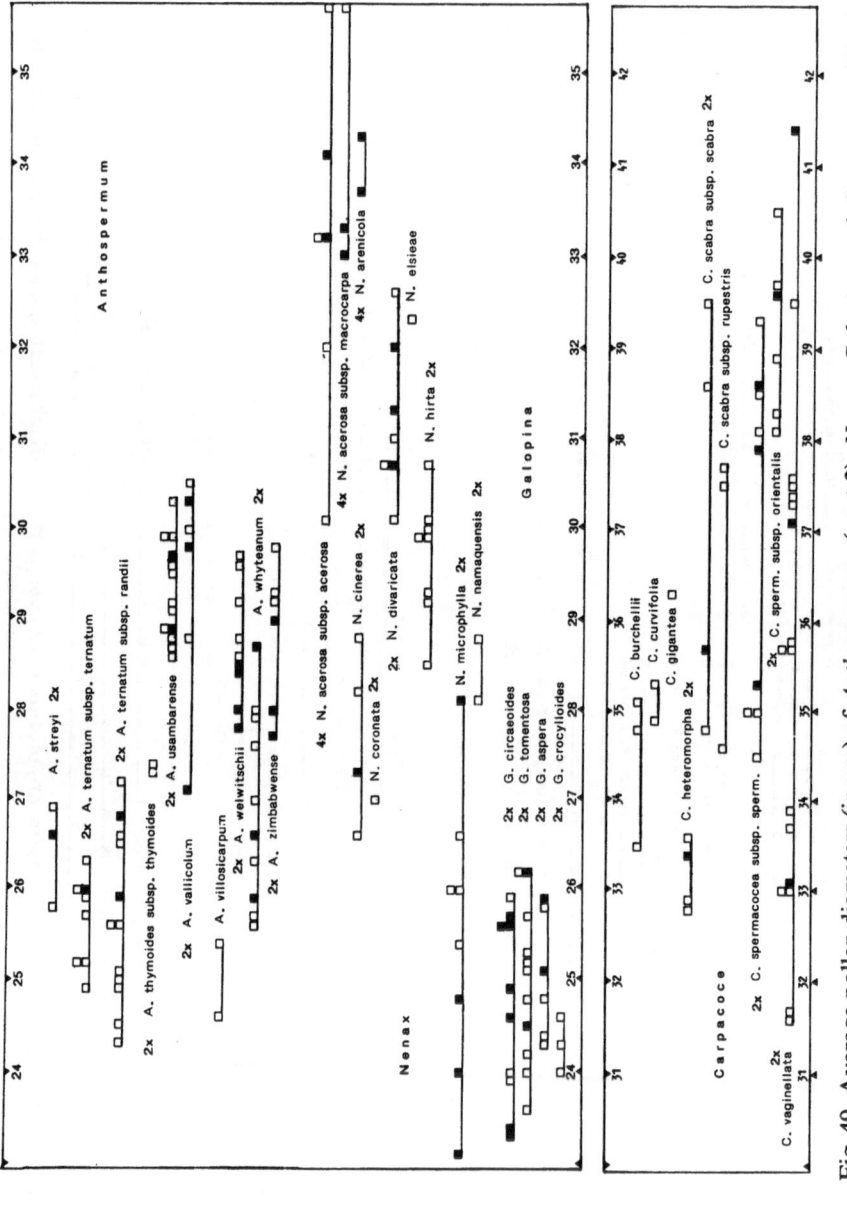

Fig. 49. Average pollen diameters (in μm) of *Anthospermum* (part 2), *Nenax*, *Galopina*, and *Carpacoce*. – Black sqaures: karyologically investigated individuals

detected. The patterns described above, therefore, do not always seem to be consistent and can, as yet, not be fully explained.

In a number of taxa, the percentage of dead pollen grains in an anther (flower) seems to be rather high (absolute figures are, however, not available). In some cases this may be due to the presence of "transitional" flower forms ($\male$ with well developed anthers → $\male$ with somewhat reduced but partially fertile anthers → functionally $\female$ with rudimentary, pollenless anthers on one plant) but in many other cases, there is no apparent reason for the high percentage of dead pollen grains (i.e., anthers seemingly well developed).

The wide range of average pollen diameters within certain taxa makes it difficult, if not impossible, to detect possible polyploids or changes of ploidy levels on the basis of pollen measurements. In *N. acerosa* subsp. *acerosa,* for example, it cannot be predicted if plants with small average pollen diameter (falling into the size range of the diploid *N. divaricata*) are diploid rather than tetraploid like karyologically investigated individuals with larger pollen (Fig. 49). It is, furthermore, impossible to draw a clear line of demarcation between di- and tetraploids of *A. spathulatum* subsp. *spathulatum*; size ranges of known tetraploids, for example, correspond to average diameters of the diploid *A. hirtum* (Fig. 48). On the other hand, it seems fairly safe to predict diploidy for karyologically unknown taxa like *A. asperuloides, A. ericifolium, A. rosmarinus* or *A. villosicarpum* or the karyologically unknown *Carpacoce* species (cf. Figs. 48 and 49).

11. Phytochemical Data

11.1. Leaf Flavonoids (by Dr. R. D. WILSON, Chemistry Div., DSIR, Lower Hutt, New Zealand)

Leaf samples of *Anthospermum, Nenax* and *Carpacoce* species and of *Galopina circaeoides* were investigated chromatographically. The methods employed largely correspond to those described in WILSON (1979, 1984).

Only flavonols were detected. They are represented by quercetin, kaempferol and myricetin as aglycones and glycosylated at the 3-position with glucose, rhamnose or arabinose and often also at the 7-position with glucose (see compilation, Table 4). The presence of myricetin is noteworthy, as its occurrence in leaves of the *Rubiaceae* is uncommon (cf. HEGNAUER 1973), while quercetin and kaempferol are widely distributed (cf. BATE-SMITH 1962).

Figs. 50 and 51 b depict in composite form the chromatograms obtained from 10 *Anthospermum*, 2 *Nenax* and 2 *Carpacoce* samples, and Fig. 51 a the chromatogram of *Galopina circaeoides*. In Table 4 the

Table 4. Flavonoid compounds of *Anthospermum, Nenax, Galopina* and *Carpacoce* species

Species	1 Qu	2 Qu-7-O-glu	3 Qu-3-O-arab	4 Qu-3-O-glu	5 Qu-3-O-glu-glu	6 Qu-3-O-glu-rham	7 Qu-3-O-glu-arab or -arab-glu	8 Qu-3-O-glu-glu-7-O-glu	9 Qu-3-O-glu-rham-7-O-glu	10 Qu-3-O-tetra-/triglycosides	11 Qu-3-O-glu-glu-rham or -glu-rham-glu	12 Ka	13 Ka-3-O-arab	14 Ka-3-O-glu	15 Ka-3-O-glu-glu	16 Ka-3-O-glu-arab or -arab-glu
Carpococe scabra subsp. *scabra* (PUFF 800914-6/1)		×		×			×			×						
Carpococe vaginellata (PUFF 800914-6/2)	×						×				×					
Galopina circaeoides (PUFF 790128-2/1)	×		×	×			×						×	×	×	×
Nenax cinerea (PUFF 800102-4/1)		×		×	×		×								×	
Nenax microphylla (PUFF 800930-4/1)	×	×		×	×	×	×	×	×						×	×
Anthospermum ? littoreum × *Anthospermum galpinii*[3] (PUFF 790424-1/1)	(×)			×		×?								×		
Anthospermum littoreum[4] (PUFF 790415-5/1)				×		×?								×		
Anthospermum galpinii[3] (PUFF 790426-1/1)	×			×		×?								×		
(PUFF 790415-4/1)	×			×		×?								×		
Anthospermum aethiopicum (PUFF 800924-4/1)	×	×[2]		×		×				×				×		
Anthospermum usambarense (PUFF 781223-2/1 A)	×			×		×				×		×		×		
Anthospermum emirnense (PUFF 800927-1/1)	×[5]			×		×				×[1]		×[5]		×		

Compound							
17 Ka-3-O-glu-rham	x	x	x	x			x
18 Ka-3-O-glu-glu-7-O-glu					x		
19 Ka-3-O-glu-rham-7-O-glu		x	x				
20 My	x^5		x	(x)			x
21 My-7-O-glu	x^2						x
22 My-3-O-arab			x				x
23 My-3-O-glu	x	x	x	x			x
24 My-3-O-glu-arab or -arab-glu							
25 My-3-O-glu-rham	x^1	x?	x?	x?			
26 My-3-O-glu-rham-7-O-glu	x^1	x					

Aglycones: Qu: quercetin; Ka: Kaempferol; My: myricetin. Sugars: glu: glucose; rham: rhamnose; arab: arabinose. Brackets: very weak.

[1] Qu- and/or My-3-O-glu-rham-7-O-glu; [2] Qu- and/or My-7-O-glu; [3] plus an unidentified Qu-3-O-diglycoside and My-3-O-diglycoside; [4] plus an unidentified Qu-3-O-diglycoside; [5] also found by PARIS & JACQUEMIN (1975).

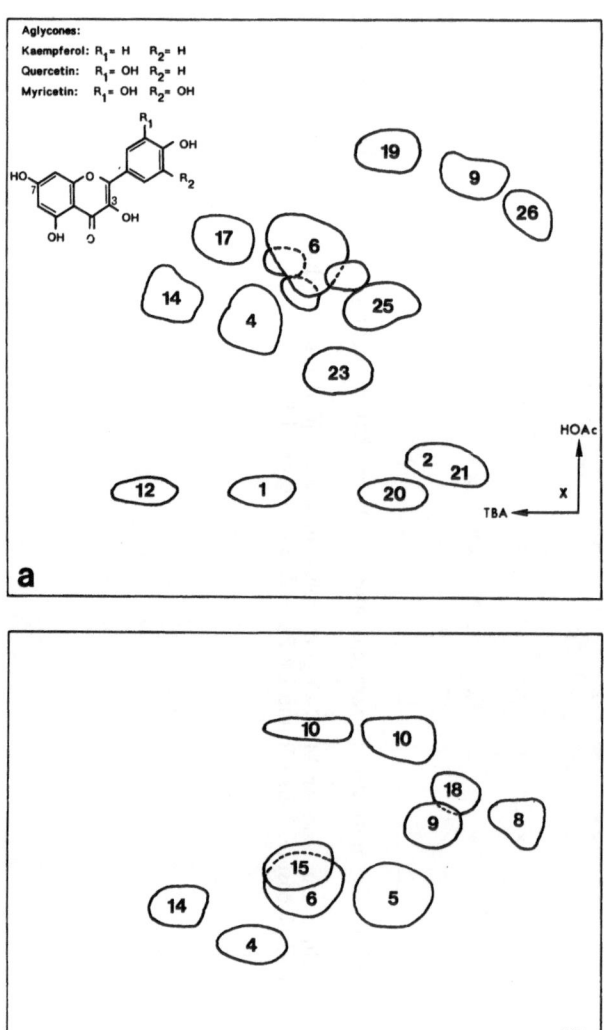

Fig. 50. Leaf flavonoids. Composite diagrams showing positions of spots on 2D-chromatograms. *a Anthospermum* (insert: chemical structures of aglycones). *b Nenax.* – See Table 4 for compounds corresponding to spot numbers

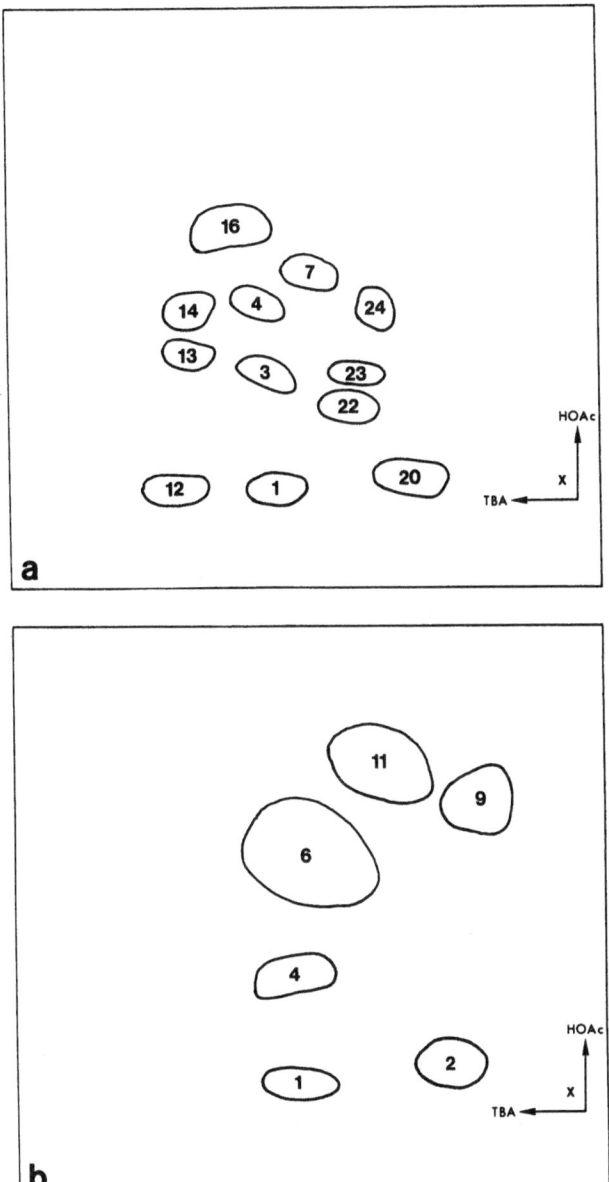

Fig. 51. Leaf flavonoids. Composite diagrams showing positions of spots on 2D-chromatograms. *a Galopina. b Carpacoce.* – See Table 4 for compounds corresponding to spot numbers

flavonols present are given for each sample/taxon separately. Two samples each of *A. littoreum, A. galpinii* (PUFF 790415-4/1) and the putative hybrid *A. littoreum × A. galpinii* (each sample from different plants which were grown from seeds of a single mother plant) were analysed; the compounds present were found to be identical.

Each species in the genera *Anthospermum, Nenax* and *Carpacoce* has its own distinct pattern. Hence flavonoid patterns should be useful in distinguishing different taxa and further chromatographic work may have potential.

The only attempt to identify and confirm the parentage of a putative hybrid by means of flavonoid patterns (? *A. littoreum × galpinii*) yielded inconclusive results, as the flavonoids in *A. littoreum* are all present in *A. galpinii* (see Table 4), and the expected patterns of a true hybrid would be (and are) the same as those of *A. galpinii*. Possibly the very weak presence of compounds 1 and 20 (quercetin, myricetin) in ? *A. littoreum × galpinii* may indicate hybridization, but this is rather speculative. Further attempts may be worthwile.

Genera apparently differ from each other in the presence or absence of certain compounds. Myricetin and its -3-O-glycosides, for example, were found only in some *Anthospermum* species and in *Galopina circaeoides*. *Carpacoce* contains quercetin, qu-7-O-glucoside and qu-3-O-glycosides only; *Galopina* appears to be the only genus in the subtribe which also has arabinose as a sugar. In general, however, there seem to be no *significant* differences between the patterns found in the four genera (except perhaps the absence/presence of myricetin?). Neither do the flavonoid patterns of these four genera of the *Anthosperminae* markedly differ from the genus *Coprosma* (subtribe *Coprosminae*; WILSON 1979, 1984).

11.2. Iridoid Glycosides

Asperulin (= asperuloside) and galiumglucoside have been detected in herbarium material of *Anthospermum aethiopicum* (KOOIMAN 1969). "Iridoids" are reported to occur in roots and branches of an unidentified Madagascan *Anthospermum* species and in *A. emirnense* by ANDRÉ & al. (1976). Asperulosidic glycosides are widely distributed in the *Rubiaceae* and are also known to occur in the genera *Phyllis* (subtribe *Anthosperminae*), *Coprosma* and *Nertera* (subtribe *Coprosminae*), and *Opercularia* and *Pomax* (subtribe *Operculariinae*) (see survey of KOOIMAN 1969). Paederoside (structurally closely related to asperulin, i.e., the acidic acid of the asperulin is replaced by the sulphur-containing thio-acidic acid; cf. HEGNAUER 1973) is believed to be responsible for the foetid odour of some

Carpacoce species. Plants of *C. spermacocea* subsp. *spermacocea* and subsp. *orientalis* always emit a strong malodour when the leaf or stem tissue is crushed. In *C. vaginellata,* plants of a few populations were found to produce a faint foetid smell (e.g., population PUFF 800914-6/2; collection TAYLOR 8592, herb. PRE, STE), while others were found to be odourless. The same can be said for *C. scabra* subsp. *scabra* (faint foetid smell in, for example, population PUFF 800914-6/1). Plants of *C. scabra* subsp. *rupestris* and *C. heteromorpha* were found to be odourless; no data are available for the remaining *Carpacoce* species. Plants of *Anthospermum, Nenax* and *Galopina,* according to my experience, are never foetid. For a survey of foetid odours in the remaining *Anthospermeae* and in other *Rubiaceae* see PUFF (1982 a).

12. Reproductive Biology

12.1. Anemophily

The following floral characters of the *Anthosperminae* are typical of an anemophilous syndrome:

The stigmas are long, exserted (e.g., *A. herbacum, A. spathulatum* subsp. *spathulatum,* Figs. 30 a–b, or *A. ammannioides,* Fig. 31 b) and their receptive surfaces are increased by hairs; they are dry. In the zoophilous *Rubiaceae,* the stigma surfaces are either non-papillate or bear unicellular papillae which are concentrated in distinct ridges, zones or heads; stigmas are dry or wet (see survey by HESLOP-HARRISON & SHIVANNA 1977). The stigma hairs in the *Anthosperminae* (conforming in structure with the common unicellular papillae of zoophilous *Rubiaceae*), however, cover the whole surface of the stigmas (perhaps a unique feature in the family?) and are not confined to their inner portion (e.g., *A. ammannioides,* Fig. 31 b). The receptive surface of the stigma is thereby significantly increased. The non-receptive stylar portion, on the other hand, is usually very small or hardly developed.

The anthers are well exposed and are often borne on ± long, slender filaments. They are relatively large and produce large quantities of pollen.

The pollen of the genera in question is dry and powdery. The grains, however, are not entirely smooth (cf. ROBBRECHT 1982) as in numerous other anemophiles. They are relatively small (average diameters ca 23–35 μm: *Anthospermum, Nenax, Galopina;* ca 31.5–41.5 μm: *Carpacoce;* cf. Figs. 48 and 49), conforming with the fact that pollen grains of anemophiles belong to the smaller size classes [although small grains are by no means confined to anemophilous taxa and occur, for example, also in entomophilous, herbaceous tribes such as the *Hedyotideae* (e.g., *Kohautia*: MANTELL 1985) or the *Rubieae*]. FAEGRI & AN DER PIJL (1971) list 20–30 (–60) µm as the "typical" diameter range for anemophiles. The pollen grains of the entomophilous tribe *Paederieae* are larger than those of the closely allied wind-pollinated tribes *Anthospermeae* (ROBBRECHT 1982) and *Theligoneae* (*Theligonum*; 20–29 × 27–33 µm: ERDTMAN 1952).

The large anthers and the relatively small size of the pollen grains imply that large quantities of pollen are produced. There are, however, no exact pollen counts to prove an increased pollen production or a changed (higher) pollen per ovule ratio in the *Anthospermeae* as compared to the entomophilous *Rubiaceae*.

As in other anemophilous plants, anthers open when warm and dry climatic conditions prevail (under natural conditions, flowers – and anthers – tend to start opening in the late morning; observations on several *Anthospermum* and *Nenax* species, and *Galopina circaeoides*). Anthers of newly opened flowers, placed under a dissecting microscope, will dehisce rapidly – almost explosively – within a few minutes due to the heat radiation from the light source. Similarly, mature buds open abruptly and the corolla lobes immediately curl backwards under these conditions (turgor mechanism).

Nothing is known about the transport distance and bouyancy behaviour of the pollen grains except that not a single pollen grain was found on simple pollen traps (slides coated with vaseline) placed at 2, 4, and 6 m away from isolated ♂ plants of *Anthospermum* species which were kept in a draughty hallway. Pollen was only found on the leaves below the (partial) inflorescences.

The *Anthospermeae*, like numerous other anemophiles, show a clear trend towards u n i s e x u a l flowers (and, subsequently, dioecy). In the *Anthosperminae*, the occurrence of dichogamy (protandry), considered to be a more important and widespread feature of anemophiles by BEACH & BAWA (1980), can hardly be interpreted as being a special adaptation to wind-pollination as protandry – often in combination with her-kogamy – is a widespread and common phenomenon in numerous (the majority of?) entomophilous *Rubiaceae*.

Also important is the a b s e n c e of any kind of o p t i c a l or c h e m i c a l a t t r a c t a n t. Perianths are small and insiginificant or even absent (♀ flowers) or larger, inconspicuously coloured and with recurved lobes (⚥ and ♂ flowers). In a few taxa, the relatively long, cylindrical to funnel-shaped corolla tubes (e.g., *A. herbaceum*, Figs. 32 a–b) of ⚥ or ♂ flowers are perhaps reminiscent of an ancestral entomophilous condition. Flowers lack any kind of odour; no nectar is produced and not even rudiments of disks or other nectar producing tissues are detectable.

A large number of the *Rubiaceae* has flowers with bicarpellate ovaries, and a sizeable number of entomophilous genera has only one ovule per locule. In a few genera, there are species in which one of the two carpels is reduced (e.g., *Otiophora:* ROBBRECHT & PUFF 1981). Seen in this context, the presence of only one or two ovules in the ovary of the genera in question cannot be considered a special adaptation (but a pre-adaptation) to wind-pollination. It is, however, noteworthy that several genera of the allied entomophilous tribe *Paederieae* have an increased number of carpels/ovules (up to 5; cf. PUFF 1982a).

The habitats of most taxa appear favourable for wind-pollination. Strong winds are characteristic for afromontane and afroalpine regions (cf. KILLICK 1963: Drakensberg; HEDBERG 1964) and, in fact, numerous taxa dominant in the ericaceous zone of the African mountains are anemophiles – e.g., *Phillipia* (*Ericaceae*), *Cliffortia* (*Rosaceae*) or *Stoebe* (*Asteraceae*), in addition to *Anthospermum*. Also in the SW Cape Floristic Region (where, judging from field experience, the percentage of anemophiles appears fairly high) and in the surrounding more arid regions strong winds are a frequent occurrence (cf. FUGGLE & ASHTON 1979; LEISTNER 1967: S Kalahari).

In angiospermous families with predominantly zoophilous flowers, the occurrence of some anemophilous taxa is a rather common situation. *Ranunculaceae, Rosaceae, Oleaceae* or *Asteraceae,* just to name a few, provide good examples for families in which small segments have progressed to "secondary" anemophily. Amongst these, flowers of numerous taxa exhibit features which are "intermediate" between "typical" zoophiles and "typical" anemophiles ("ambophiles" or "amphiphiles", GRONEMEYER 1967 and WOODHOUSE 1971, respectively). They occur next to "typical" zoo-(entomo)philes and "typical" anemophiles (e.g., *Acer:* HESSE 1979; see also literature cited therein) and may point to the "step-by-step" evolution of secondary anemophily.

In the predominantly zoo-(entomo)philous *Rubiaceae,* however, in which the genera of the *Anthospermeae* and the genus *Theligonum* (*Theligoneae;* WUNDERLICH 1971) are the only obvious "secondarily" anemophilous representatives[1], there are *no* taxa (genera/species) exhibiting such gradual transitions or a broad spectrum of intermediates between "pure" zoo-(entomo)philes and "pure" anemophiles. Taxa are *either* zoophilous *or* anemophilous. This – together with the wide (although now

---

[1] And also *Durringtonia* (*Durringtonieae*; HENDERSON & GUYMER 1985).

fragmented) ± subtropical to tropical and S hemisphere-centered distribution of the anemophilous taxa – may indicate that in the *Rubiaceae,* the
evolution of secondary anemophily occurred in the more "remote" past
rather than in the "recent" past. The overall similarity between the genera
may, furthermore, suggest that all anemophiles in the *Rubiaceae* may have
been derived from a common ancestral stock (i.e., the evolution of
anemophily is not likely to have recurred, independently, several times).

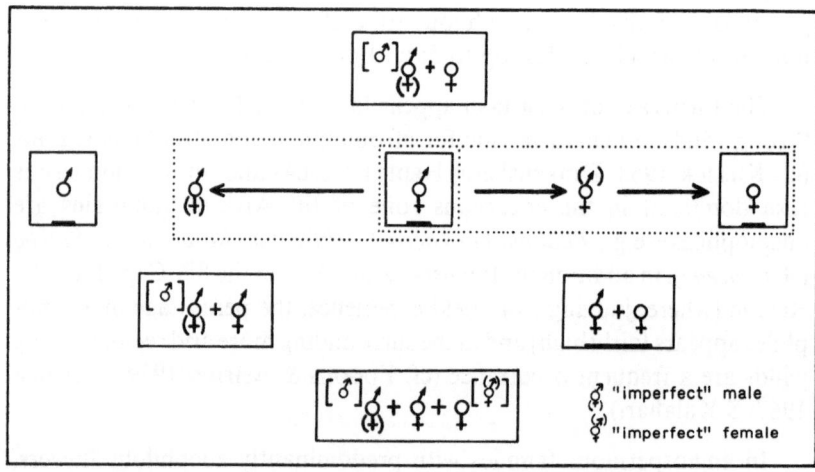

Fig. 52. Possible sex forms in individual plants of African and Madagascan
Anthospermeae. – A square or rectangle represents an individual; the most
common sex forms are underlined. The dotted rectangles stand for individuals in
which "transitions" (→) from one sex to another can be observed. Detailed
explanations in the text

12.2. Sex Distributions and Sex Ratios

12.2.1. Sex Forms of Individual Plants

The possible sex forms of individual plants are depicted in Fig. 52. The
following explanations and comments are necessary: To clarify the
relationships of and between sex forms and associated phenomena, the
terms "imperfect" male and "imperfect" female – as opposed to "pure"
male and "pure" female – are introduced (and mostly only used) in this
chapter. These terms are applied in a descriptive, but not functional, sense.

– "imperfect" male (⚲) refers to functionally male flowers having
small but still well discernible rudimentary ovaries and sometimes also
rudimentary stigmas, as opposed to "pure" males (♂) which lack obvious

ovary and stigma rudiments (calyx lobes may, however, be visible at the base of the corolla; they should not be confused with rudimentary ovaries). The distinction between small, well discernible ovary rudiments on the one hand and minute, hardly detectable ovary rudiments on the other, is not always entirely clear and this division is, therefore, somewhat arbitrary and subjective.

— "imperfect" female ($\overset{(\omega)}{\female}$) refers to functionally female flowers with well discernible but pollenless anther rudiments and corollas intermediate in size between those of $\overset{\wedge}{\female}$ flowers and "pure" female ($\female$) flowers without visible anther rudiments.

The most common sex forms in individual plants (Table 5) are $\overset{\wedge}{\female}$, $\overset{\wedge}{\female} + \female$, $\female$ and $\male$; $\overset{\wedge}{\female} + \overset{\wedge}{\male}$, $\overset{\wedge}{\male} + \overset{\wedge}{\female} + \female$, and $\overset{\wedge}{\male} + \female$ are less common.

Transitions from $\overset{\wedge}{\female}$ to $\overset{\wedge}{\male}$ (cf. Fig. 52; i.e., where the ovaries remain undeveloped and small and the stigmas do not elongate) can be quite common in some flowers of plants of, for example, *A. herbaceum*, *A. hirtum* or *A. hispidulum*. This may indicate the possible origin of andromonoecious ($\overset{\wedge}{\female} + \overset{\wedge}{\male}$) individuals. Similar transitions ($\overset{\wedge}{\female} \rightarrow \overset{\wedge}{\male}$) on gynomonoecious ($\overset{\wedge}{\female} + \female$) plants (again observed in *A. herbaceum*) may likewise indicate the origin of $\overset{\wedge}{\male} + \overset{\wedge}{\female} + \female$ plants and monoecious individuals ($\overset{\wedge}{\male} + \female$); the latter, however, are quite rare. On gynomonoecious ($\overset{\wedge}{\female} + \female$) individuals, furthermore, some flowers may be transitional "imperfect" females (i.e., $\overset{(\omega)}{\female}$; cf. Fig. 52. Also see 12.3.).

In addition to these "transitions" there are a number of other remarkable phenomena:

(1) Reversion of $\male$ flowers to $\overset{\wedge}{\female}$:
It was occasionally observed that on seemingly "pure" $\male$ plants (of dioecious taxa), a few odd flowers had reverted to $\overset{\wedge}{\female}$ forms. This seemed to be more common in those taxa where the $\male$ flowers had *relatively* conspicuous rudimentary ovaries and stigmas (e.g., *A. paniculatum*, Figs. 31 i–j or *A. littoreum*, Fig. 33 a). These reversals are not a general occurrence in a given species and there are interpopulational differences in their frequency; in general, however, they are the exception rather than the rule. For other cases of sex reservals (in non-*Rubiaceae*) see, for example, Freeman & al. (1984) or Lloyd & Bawa (1984) and also p.140.

(2) $\male$ plants with odd $\female$ flowers in dioecious taxa:
This unusual situation was observed in several cutivated plants of *A. aethiopicum*, *A. emirnense* and *A. littoreum* (but never in *all* plants of a population). There were often not more than two or three $\female$ flowers on an otherwise $\male$ inflorescence. This phenomenon does not occur in all populations or in all dioecious taxa, but how widespread it really is, is

Table 5. Sex forms in individual plants. – Data based on observations of cultivated refers to "imperfect" males or females respectively,

	♂	♂̸	♂̸+♀	♀	♂̸+♂
Anthospermum aethiopicum	×			×	
A. ammannioides	×			×	
A. asperuloides		×	×		
A. basuticum	×			×	
A. bergianum	×			×	
A. bicorne		×	×	×	
A. comptonii	×			×	
A. dregei	×	×	×	×	
A. emirnense	×			×	
A. ericifolium	×	×		×	×
A. esterhuysenianum	×	×	×	×	
A. galioides	×	×	×	×	×
A. galpinii	×			×	
A. herbaceum	×	×	×	×	×
A. hirtum	×	×	×	×	
A. hispidulum	×	×	×	×	×
A. ibityense		×		×	
A. isaloense	×			×	
A. littoreum	×			×	
A. longisepalum	×			×	
A. madagascariense	×			×	
A. monticola	×			×	
A. pachyrrhizum	×			×	
A palustre		×			
A. paniculatum	×			×	
A. perrieri			×	×	
A. prostratum	×			×	
A. pumilum	×	×	×	×	×
A. rosmarinus	×	×?	×?	×	×
A. spathulatum	×			×	
A. streyi		×		×	
A. ternatum	×	×	×	×	×
A. thymoides			×		
A. usambarense	×			×	
A. vallicola	×			×	
A. villosicarpum		×	×		
A. welwitschii	×			×	
A. whyteanum	×		×	×	
A. zimbabwense	×			×	
Nenax (all taxa)	×			×	
Galopina circaeoides		×	×	×	×
G. crocyllioides	×		×	×	×
G. aspera	×	×	×	×	

plants (C), of plants in the field (F), or of herbarium material (H). In Notes, ⚥(♂), ⚥(♀) and ♀ to perfect females (cf. Fig. 52 and p. 134)

♂+⚥+♀	♂+♀	Notes
		F, H; C: seldom ♂ + odd ♀ (Puff 791220-4/1)
		F, H; C: seldom ♂ + odd ♀; sex changes – see text
		H
		C, F, H
		H; F: occasionally ♂ + odd ⚥; Salter (1937): some plants ⚥ or ⚥ + ⚥(♀)
		F, H
		F, H
		H; F: occasionally transitions ⚥→⚥(♂) and ⚥→⚥(♀)→♀; trend to dioecy in subsp. *dregei*
		F, H; C: occasionally ♂ + odd ⚥, very seldom ♂ + odd ♀
		H: ♂ appear to be very rare
		H; F: occasionally also ⚥ + ♀→⚥(♂) + ♀?
		H; F: occasional transitions ⚥→⚥(♂) and also ⚥ + ♀→⚥(♂) + ♀? Trend to dioecy in subsp. *reflexifolium*
		F, H; C: seldom ♂ + odd ⚥ or ♂ + odd ♀
×	×	H; F: transitions ⚥→⚥(♀)→♀ (Fig. 32); also ⚥→⚥(♂)?
		H; F: transitions ⚥→⚥(♂) and ⚥→⚥(♀) ± common; also odd ⚥(♂) + ⚥ + ♀ plants?
		F, H; C: occasional transitions ⚥→⚥(♂); ♂ appear to be rare
		?; F: insufficient material; perhaps also ⚥ + ♀ and ⚥ + ♂?
		F, H
		H; C, F: sometimes ♂ + odd ⚥, very seldom ♂ + odd ♀
		?; F, H: insufficient material; perhaps also ♂ + ⚥?
		H, F
		H; F: seldom ♂ + odd ⚥
		H; F: seldom ♂ + odd ♀ or ♀ + odd ♂; very rarely ♂ + odd ⚥ + odd ♀
		?; H: insufficient material
		H; F: occasionally ♂ + odd ⚥; seldom ♂ + many ⚥
×		?; F, H: insufficient material
		F, H; Salter (1937): also ♂ + odd ⚥
		H; F: occasionally transitions ⚥→⚥(♂) and ⚥→⚥(♀)
		H: insufficient material
		H; C, F: seldom ♂ + odd ⚥
		F, H
		F, H: ♂ appear to be rare; often transitions ⚥→⚥(♀)
		H: insufficient material; also ⚥ and ♀?
		F, H; C: seldom ⚥→⚥(♀)→♀
		F, H
		H
		C, F, H: ♂ + odd ⚥ ± common in some areas
		F, H
		F, H
		H; C, F in part; C: very seldom ♂ + odd ⚥ (*N. microphylla*)
×		F, H; C: occasional transitions ⚥→⚥(♀)→♀, also ⚥ + ♀→⚥(♂) + ♀?
		H; trend to dioecy?
×	×	H; F: occasional transitions ⚥→⚥(♂)

Table 5 (continued)

	♂	⚥	⚥+♀	♀	⚥+♂
G. tomentosa	×	×	×	×	×
Carpacoce burchellii		×		×	
C. curvifolia		×		×	
C. gigantea		×		×	
C. heteromorpha		×			
C. scabra	×	×	×	×	×
C. spermacocea		×	×	×	×
C. vaginellata	×	×	×	×	×

difficult to judge (herbarium specimens do not give further indications). Long-term observations of populations would be needed.

This phenomenon of "leaky dioecism" also occurs in *Coprosma* (see, in this context, the controversy about the establishment of *C. pumila* on Macquarie Island: TAYLOR 1954; BAKER 1955; LLOYD & HORNING 1979; BAKER & COX 1984).

(3) Sex changes within a flowering season or from one season to the next in dioecious taxa and taxa with more complicated sex combinations:

Detailed observations were made on cultivated plants of *A. ammannioides* and *A. usambarense* (both essentially dioecious):

– *A. ammannioides* (PUFF 790124-1/1), plant C_{II}: December 30, 1980 – ♂ (flowering for the first time); soon (January 1981) also developing a few ⚥ flowers; March 15, 1981: plant bearing "pure" ♀ flowers. A similar pattern was observed in plant A_{XI}. In the remaining 11 cultivated plants of PUFF 790124-1/1 (5 ♂, 11 ♀) no irregularities were observed except for one functionally female (⚥) with "⚥" corollas and pollenless anthers.

– *A. usambarense* (PUFF 781223-2/1), plant A_I: February 2, 1981 – ♂ (flowering for the first time); after the first fertile ♂ flowers no more fertile anthers were produced and corollas became increasingly smaller; by March 15, 1981 the plant was pure ♀. The remaining 4 cultivated plants of PUFF 781223-2/1 showed no irregularities.

Based on herbarium studies it is suspected that odd plants of *A. aethiopicum* may show irregularities similar to those described above.

Numerous and detailed data referring to (3) are also available for *Galopina circaeoides* (it is easily cultivated, flowers quickly and regularly and the inflorescence structure allows for unproblematic investigations):

$\male + \male\female + \female$	$\male + \female$	Notes
	×	F, H
		?; H: insufficient material
		?; H: insufficient material
		?; H: insufficient material
		H; F: occasional transitions $\male\female \to \male$? Also $\female$?
		H; F: occasional transitions $\male\female \to \female \to \female$ or $\male\female \to \male$
		H; F: sometimes transitions $\male\female \to \male$?
		H; F: sometimes transitions $\male\female \to \male$; SALTER (1937): also $\male \leftarrow \male\female \to \female$

(a) *Sex change from year to year* in, for example, PUFF 790516-1/3, plant IV: inflorescences with many $\male\female$ and a few $\female$ flowers in 1980. In 1981, the same plant only had $\female$ flowers (but in that year, the shoots were not well developed and flowered later than usual; inflorescences were smaller and fewer-flowered; see also "Femalification" below).

(b) *Seasonal differences in sex distribution* within $\male\female + \female$ inflorescences of one and the same plant (observed on numerous plants): inflorescences bore $\male\female$ and $\female$ flowers in equal amounts in one year, but a shift to many $\male\female$ and few $\female$, or the reverse, few $\male\female$ and many $\female$ took place in the following year. The latter leads to the frequently observed

(c) *"Femalification"* of inflorescences and plants: On $\male\female + \female$ plants, inflorescences developed late in the season were, almost invariably, found to be purely $\female$. Also the latest developed flowers on an inflorescence were also almost all $\female$. From this follows that, on many-stemmed plants, individual flowering shoots may differ from one another in their sex distribution (this also occurs in some *Anthospermum* species!). Also, if plants do not grow too well (if, for example, kept in a too sunny place or under too dry conditions) and start flowering (markedly) later than other plants of the same population or, generally, appear to be in some kind of stress situation, there is a marked trend towards formation of purely $\female$ inflorescences[1]. It is, in this context, noteworthy that MELAMPY (1981), investigating sex-linked niche differentiation in $\male$/$\female$ *Thalictrum* species, came to the conclusion that $\female$ plants may be more stress tolerant than $\male$ ones.

This "femalification" is not confined to *Galopina* but was also observed in non-dioecious *Anthospermum* species (especially *A. herbaceum*) on numerous occasions.

---

[1] Also observed in cultivated plants of *Phyllis nobla*.

The observations described above imply an environment-dependent influence on the sex expression of gynodioecious (and di- or polyoecious) plants. This is a well known, documented phenomenon: BROCKMANN & BOCQUET (1978), for example, give a detailed account for *Silene vulgaris* subsp. *vulgaris,* and DOMMÉE & al. (1978) for *Thymus vulgaris.* SCHAFFNER (1935) reported sex reversals in varying degrees in the dioecious *Thalictrum dasycarpum* when plants were transplanted to a different habitat. He, furthermore, noted that the extent of reversal was in some cases decidedly different in different years. He concluded that "any sexual condition of an individual [without sex chromosomes – C. P.] is not determined by Mendelian sex genes but a physiological balance which is produced thru the interaction of the general hereditary potentialities of the cell on one hand and the environmental conditions on the other". VAN DAMME & VAN DELDEN (1982) documented the presence of two types of male sterility (functionally female flowers) in the gynodioecious *Plantago lanceolata* [♀ with reduced, pollenless anthers (= "imperfect" female in this account) and ♀ with hardly any anther rudiments and smaller corollas (= here called "pure" female)]. They presented evidence that these differences in expression of male sterility are cytoplasmatically determined ("plasmon types R and P") and, furthermore, considered the possibility that sex expression, at least to some extent, may be environment-dependent. A detailed genetical and cytological investigation of the situation in the *Anthospermeae* (including studies on correlations with environmental variables) would certainly be worthwhile.

(4) Precociousness of ♀ flowers:
In plants with ⚥ + ♀ flowers (and possibly also ♂ + ⚥ + ♀ plants), the ♀ flowers *always* open first on an inflorescence (*Galopina, Anthospermum* species). Detailed information is available for *G. circaeoides,* which is most suitable for observation:
Even if the terminal flower of an inflorescence should be ⚥ (not common!), it will not – as would be expected in a closed inflorescence – open first. The ⚥ flower apparently takes longer to develop and will only open one or two days (at the earliest) after the first lateral ♀ flowers have opened. Individual ⚥ + ♀ *inflorescences* as a whole are thus, at first, "*protogynous*"; an inflorescence may be ♀ for up to three days after the opening of the first flower. In contrast, however, individual ⚥ *flowers* (of *Anthospermum, Galopina* and *Carpacoce* in general) are always (strongly) *protandrous.* Similar findings were published for *Sanguisorba* sect. *Poteria* (NORDBORG 1967; heads with first flowering ♀ flowers above and strongly protandrous ⚥ flowers below). For certain *Apiaceae,* LOVETT DOUST (1980) reports protandry in individual flowers but "relatively

protogynous" umbellets and lower-order umbels which are protogynous with respect to higher-order (later) umbels; LOVETT DOUST interprets this apparent contradiction as "an avoidance of simultaneous competition for resources within the plant". Whether this interpretation can also be applied to *G. circaeoides* is open to discussion. It appears certain, however, that the initial female state may favour outcrossing in a particular inflorescence but that this does not altogether preclude geitonogamous pollination, as other inflorescences on the same plant may be further developed and have open ♀ or ⚥ flowers (see 12.4.).

12.2.2. Sex Ratios and Sex Distributions in Populations

Dioecious Taxa (Table 6).

Literature data (largely based on the investigation of non-rubiaceous plants) indicate that male-biased sex ratios appear to predominate in dioecious taxa. LLOYD & WEBB (1977) argue that (at least in some cases) this can be attributed to differential post-reproductive survival of the sexes associated with differences in reproductive effort (= DARWIN's, 1877, original hypothesis). LLOYD & WEBB, however, also do state that this is not necessarily always the case [see also, for example, MELAMPY & HOWE (1977) for a case of a sex ratio skewed in favour of ♀ (in *Triplaris, Polygonaceae*)]. In the dioecious *Anthospermum* species and *Nenax*, the sex counts of natural populations or of cultivated plants yield somewhat inconclusive results (Table 6). Ratios appear to vary from population to population; in some, males predominate, in others females. In *A. littoreum* (PUFF 790424-1/1), the ♀-biased ratio found in the field changed to a ♂-biased ratio in plants grown in the greenhouse from seeds taken from individuals of the same population (see Table 6). Taking the total of all populations of *Anthospermum*, the ratio ♂ : ♀ is 51% : 49%. As the number of flowering plants in numerous natural populations visited and of cultivated material was often small, the samples available may not always be representative. Taking this into consideration, the sex ratios presented here may often not be a true reflection of the actual situation.

Sex ratios for some *Coprosma* species (subtribe *Coprosminae*) are included in Table 7 for comparison. Similarly, as for *Anthospermum* and *Nenax*, populations do not always show a significant preponderance of males.

Non-dioecious Taxa (Table 7).

Within taxa of *Anthospermum*, *Carpococe* and *Galopina*, sex distributions were, in most cases, found to differ (markedly) from population to population. Percentages of ♂ and of ⚥ + ♀ individuals present, for example, may be reversed (cf. *A. galioides* subsp. *galioides*), and certain sex

Table 6. Sex distributions in populations of *Anthospermum* and *Nenax* (dioecious taxa). — C data from cultivated plants (grown from seed of one plant), F data based on field observations; number of plants observed in brackets

		♂	♀	
Anthospermum *aethiopicum*				
Puff 790717-1/3	C	69%	31%	(9/4)
Puff 791220-4/1	C	57%	43%	(17/13)
Puff 790712-3/1	C	61%	39%	(14/9)
Puff 800918-1/4	F	64%	36%	(7/4)
Puff 790909-4/1	F	38%	62%	(8/13)
Puff 791220-1/2	F	11%	89%	(1/8)
Total		**50%**	**50%**	
A. ammannioides				
Puff 790124-1/1	C	31%	69%	(5/11)
Puff 790125-4/1	F	40%	60%	(6/9)
Total		**35,5%**	**64,5%**	
A. bergianum				
Puff 800903-2/3	F	44%	56%	(14/18)
Puff 800914-1/2	F	43%	57%	(12/16)
Total		**43,5%**	**56,5%**	
A. emirnense				
Puff 800729-1/1	C	54%	46%	(7/6)
A. galpinii				
Puff 790415-1/1	C	54%	46%	(13/11)
Puff 790303-2/1	F	36%	64%	(10/18)
Total		**45%**	**55%**	
A. littoreum				
Puff 790424-1/1	C	70%	30%	(7/3)
Puff 790424-1/1	F	42%	58%	(7/10)
Total		**56%**	**44%**	
A. monticola				
Puff 781125-1/4	F	45%	55%	(27/33)
Puff 790113-3/1	F	58%	42%	(21/15)
Total		**51,5%**	**48,5%**	
A. paniculatum				
Puff 790114-6/1	F	59%	41%	(16/11)
A. pachyrrhizum				
Puff 811011-2/1	F	55%	45%	(21/17)
A. spathulatum subsp. *spathulatum*				
Puff 791221-2/1	C	75%	25%	(6/2)
Puff 790718-2/1	F	55%	45%	(23/19)
Total		**65%**	**35%**	
subsp. *ecklonianum*				
Puff 791231-1/1	C	64%	36%	(7/4)

Table 6 (continued)

		♂		
A. usambarense				
Puff 820901-1/5	F	45%	55%	(33/41)
Puff 820903-2/1	F	54%	46%	(43/37)
Total		**49,5%**	**50,5%**	
A. welwitschii				
Puff 791202-4/1	F	41%	59%	(16/23)
A. zimbabwense				
Puff 780125-1/3	F	61%	39%	(11/7)
Anthospermum				
Total		**51%**	**49%**	
Nenax *acerosa* subsp. *macrocarpa*				
Puff 800918-2/4	F	36%	64%	(8/14)
N. cinerea				
Puff 800102-4/1	F	37%	63%	(3/5)
N. divaricata				
Puff 790711-2/1	F	63%	37%	(12/7)
N. microphylla				
Puff 800104-3/1	F	42%	58%	(8/11)
Puff 800831-3/1	F	63%	37%	(12/7)
Total		**52,5%**	**47,5%**	
Nenax				
Total		**47%**	**53%**	

forms (especially ♂ and ♂ + ♀) are often missing altogether in some populations (cf. *A. herbaceum*). Lumping together the investigated non-dioecious *Anthospermum* species, it transpires that ♀̂ + ♀, ♀̂ and ♀ individuals are the most common sex forms in populations, while ♂, ♀̂ + ♂, ♂ + ♀̂ + ♀ and ♂ + ♀ individuals are rather rare. *Galopina* presents essentially the same picture. In *Carpacoce spermacocea*, there seems to be a shift from ♀ in favour of ♀̂ + ♀ or ♀̂, but the samples available are hardly representative.

It is cautioned against putting too much emphasis or value on the data presented in Table 7. "Spot-checks" of populations in the field may yield distorted data: As explained above (12.2.1.), shoots, for example, may be ♀̂ + ♀ when investigated, but the same plant may develop pure ♀ shoots later on in the season. Or shoots with open ♀̂ flowers may bear buds of ♀ flowers which are too small to detect in the field. Because of the very pronounced protandry it, furthermore, could not always be determined

Table 7. Sex distributions in populations of *Anthospermum, Carpacoce* and
of one plant), F data based on field observations; number of plants

		♂	♀
Anthospermum *dregei*			
subsp. *ecklonis*			
PUFF 800901-1/2	F	16%	19%
PUFF 800902-4/1	F	–	23%
Total		8%	21%
A. galioides subsp. *galioides*			
PUFF 790910-4/1	F	–	34%
PUFF 790911-1/1	F	–	53%
PUFF 790912-1/2	F	–	14%
PUFF 800903-2/1	F	–	50%
PUFF 800906-2/1	F	8%	8%
Total		1,5%	32%
A. hirtum			
PUFF 800903-1/2	F	–	–
A. pumilum subsp. *pumilum*			
PUFF 800105-1/1	F	–	90%
PUFF 800109-1/1	F	–	55%
Total			72,5%
subsp. *rigidum*			
PUFF 800104-1/1	F	27%	64%
A. herbaceum			
PUFF 790427-1/1	F	–	39%
PUFF 781125-2/1	F	6%	12%
PUFF 791201-1/2	F	–	–
PUFF 800105-2/3	F	–	8%
Total		1,5%	15%
Carpacoce *spermacocea*			
subsp. *spermacocea*			
PUFF 800908-1/1	F	–	14%
PUFF 800919-3/1	F	–	52%
Total			33%
Galopina *aspera*			
PUFF 790211-3/1	F	16%	5%
G. circaeoides			
PUFF 790909-6/1	C	–	10%
PUFF 790516-1/3	C	–	16%
PUFF 790114-7/1	F	–	44%
Total			23%
Coprosma *repens*		58,4%	–
C. grandifolia		52,2%	–
C. rhammoides		53,4%	–

---

[1] GORDON (1959).

[2] Plus 15 plants "sex not definite" (WILD & ZOTOV 1930); cited as *C. australis* in
GOODLEY (1964).

[3] Plus 5 plants "sex not expressed or mixed" (WILD & ZOTOV 1930); cited as
"♂ : ♀ = 69 : 34" in GOODLEY (1964).

Galopina (non-dioecious taxa). – C data from cultivated plants (grown from seeds observed in brackets. – Appendix: published data for *Coprosma*

♀+♀	♀	♀+♂	♂+♀+♀	♂+♀	
26%	39%	–	–	–	(5/6/8/12)
46%	31%	–	–	–	(3/6/4)
36%	**35%**				
57%	9%	–	–	–	(7/12/2)
27%	20%	–	–	–	(8/4/3)
58%	28%	–	–	–	(3/13/6)
37%	12%	–	–	–	(4/3/1)
34%	50%	–	–	–	(1/1/4/6)
42,5%	**24%**				
82%	18%	–	–	–	(37/8)
5%	5%	–	–	–	(17/1/1)
9%	36%	–	–	–	(23/4/15)
7%	**20,5%**				
–	9%	–	–	–	(3/7/1)
–	43%	18%	–	–	(19/21/9)
–	23%	41%	18%	–	(1/2/4/7/3)
24%	52%	–	16%	8%	(3/7/2/1)
42%	50%	–	–	–	(1/5/6)
16,5%	**42%**	**14,5%**	**8,5%**	**2%**	
72%	14%	–	–	–	(1/5/1)
18%	–	30%	–	–	(12/4/7)
45%	**7%**	**17%**			
53%	27%	–	–	–	(3/1/10/5)
30%	50%	–	10%	–	(1/3/5/1)
37%	21%	5%	16%	5%	(3/7/4/1/3/1)
22%	34%	–	–	–	(4/2/3)
30%	**35%**	**2%**	**8%**	**2%**	
–	41,6%	–	–	–	(652/465)[1]
–	47,8%	–	–	–	(47/43)[2]
13,6%	33%	–	–	–	(55/14/34)[3]

with certainty in the field, whether flowers were ♀ in a male state or ☿. These would have to be observed over a period of several days to see whether their stigmas would elongate; such observations, however, were only possible on a limited number of cultivated plants.

12.2.3. Sex Distributions and Anemophily, and the Evolution of Dioecy

The variability in sex expression and the often bewildering range of sex forms in certain taxa may be causally connected with the switch from zoophily to anemophily in the *Anthospermeae*. There may be a direct correlation between this (for the *Rubiaceae*) "new" mode of pollination (secondary anemophily – see 12.1.) and the trend to unisexual flowers (next to ☿) and, subsequently, the total separation of the sexes. Dioecy, as it is realized in numerous taxa of *Anthospermum* and all *Nenax* species, may be the most "ideal" situation for anemophiles, although this is not a generally accepted hypothesis. BAWA (1980: 33), for example, argues that "almost all monoecious and hermaphroditic wind-pollinated species are strongly dichogamous", and believes that "a certain degree of correlation between dioecy and wind pollination in the north-temperate regions" is misleading. Nevertheless, it is difficult to imagine how even strong dichogamy in, for example, *Anthospermum* could be advantageous if the long flowering period of a plant (allowing geitonogamous pollination!) is considered. At least for the *Anthospermum*, BAWA's argument (1980: 33) "certainly if plants are unisexual, their pollen will be trapped by conspecific rather than their own stigmas. But the same effect could be achieved by dichogamy ..." is hardly convincing.

In a summarizing article, BAWA (1980)[1] described the possible evolutionary pathways leading to dioecy: (1) direct evolution from hermaphrodites, (2) evolution via gynodioecy and (3) androdioecy, (4) via monoecy, and (5) from heterostyly. Various *Rubiaceae* are frequently used as examples for (5) [the ♀ are thought to be derived from the long style form, the ♂ from the short style form; BEACH & BAWA (1980) relate this to changes in pollination biology. See also BAWA & BEACH (1981) and literature cited therein]. *Anthospermum* has been reported to be heterostylous by both VERDCOURT (1958) and BIR BAHADUR (1968), but in the present investigations, heterostyly was not detected in any genus of the *Anthospermeae*. Thus possibility (5) cannot apply.

It is, in my opinion, not unlikely that the sometimes very pronounced protandry in ♀ flowers may give a clue as to the origin of ♂ flowers (see

---

[1] Also see BAWA (1984).

12.2.1. and Figs. 52 and 53: transitions ⚥̂ → ♂̂). If in a potentially ⚥̂ flower the stigmas no longer elongate and do not become receptive [i.e., the flower remains in its (initial) male state and the stigma (gynoecium) development is suppressed][1] the result is a ♂̂ or ♂ flower. Similarly, it may be speculated that ♀ flowers originate via transitions from ⚥̂ (⚥̂ → ⚥̂⁽ʷ⁾ → ♀; also see 12.2.1. and p. 148). It could thus be concluded that dioecy originates directly from ⚥̂ flowers. There are, however, strong arguments

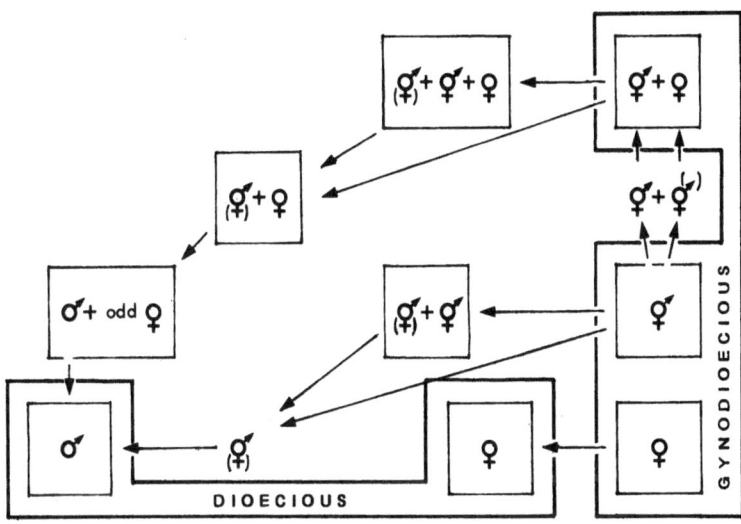

Fig. 53. Evolution of sex forms in African and Madagascan *Anthospermeae* (theoretical model). – Explanations in the text; also see Fig. 52

against this theory: No single species amongst the non-dioecious *Anthosperminae* is strictly ⚥̂, and no populations with only ⚥̂ individuals were detected in the field. There seems rather to be an underlying gynodioecious (⚥̂/♀) or ⚥̂/ ⚥̂+♀/♀ pattern in the non-dioecious taxa of *Anthospermum*, *Galopina* and *Carpacoce*. Thus it may be more likely that dioecy has evolved via ⚥̂/♀ or ⚥̂/ ⚥̂+♀/♀ as indicated in Fig. 53. In this theoretical model, the presence of ⚥̂+♀ individuals as well as ⚥̂ and ♀, and the occasional presence of (few) odd ♀ on ♂ individuals (see 12.2.1.) is incorporated, although it is realized that this is highly speculative.

The evolution of dioecy in the non-African subtribes *Coprosminae* and

---

[1] It is, in this context, noteworthy that ANDERSON (1973) put forward the theory that the initiation of *heterostyly* in the *Rubiaceae* may lie in a modification of the common protandry.

Operculariinae may have followed a different pathway. As reported earlier (Puff 1982 a) these genera have *protogynous* flowers. The tempting and easily imaginable theoretical derivation of ♂ from protandrous ⚥ is thus not applicable. Detailed information on genera of these subtribes are, however, lacking so that speculations of any kind would be inappropriate.

12.3. Secondary Sex Characters and Sexual Dimorphism

The occurrence of secondary sex characters, i.e., "differences between the sexes in structures other than the androecia and gynoecia" (definition of Lloyd & Webb 1977) is common in the *Anthosperminae*.

In most taxa, however, sexual dimorphism is confined to differences in petal (corolla) size between ⚥/♂ and ♀ flowers. The ♀ flowers are always (markedly) smaller than the ⚥/♂ (see 6.3. and Figs. 31–34), a condition also frequently encountered in other sexually dimorphic plants [see Baker (1948); Ponomarev & Demynova (1975) cited in Lloyd & Webb (1977); Kay & al. (1984), Dulberger & Horowitz (1984): also other differences – i.e., in calyx size, in corona structure, nectar production, etc. – in addition to corolla size differences in *Silene dioica* and *S. vulgaris*; Skottsberg (1944): Hawaiian *Rubiaceae*]. Whether – as, for example, in *Glechoma hederacea* – the larger corolla sizes of male-fertile flowers are physiologically associated with stamen development (i.e., related to the production of gibberellins by the developing stamens: Plack 1957, 1958) remains unknown. It can, however, frequently be observed in several taxa of *Anthospermum* (e.g., *A. dregei; A. herbaceum:* Fig. 32), *Galopina* (e.g., *G. circaeoides*) and *Carpacoce* (e.g., *C. vaginellata*) that in ⚥ flowers, the corollas decrease in size if the stamen development is disturbed and the anthers are smaller (although they still produce some pollen); functionally ♀ flowers with still relatively large, although pollenless anthers may have corollas ± intermediate in size between truly ⚥ and "pure" ♀ flowers without rudimentary anthers.

In several dioecious species of *Anthospermum*, sexual dimorphism extends beyond differences in floral characters. Most obvious and striking are perhaps the differences between the inflorescences of ♂ and ♀ plants. ♀ inflorescences tend to be more conspicuous due to the shortened internodes between the flowering clusters (partial inflorescences), and in several species, ♀ have rather compact, ± cylindrical inflorescence regions (e.g., *A. ammannioides, A. aethiopicum,* Figs. 54 b and c); the degree of shortening of the internodes, however, varies to some extent – even within populations. While it is assumed that the basic inflorescence structure of the ♂ and the corresponding ♀ is identical, it appears that the ♀ partial inflorescences (sessile flowering clusters) often have more flowers than

Fig. 54. Sexually dimorphic inflorescences in dioecious *Anthospermum* species. *a–b A. ammannioides* (PUFF 790125-4/1), *a* part of ♂ and *b* of ♀ plant. *c–d A. aethiopicum* (PUFF 790910-5/1), *c* part of ♀ and *d* of ♂ plant. – The scale unit is 10 cm. Further explanations in the text

those of the ♂[1]. In ♀, furthermore, the very closely spaced, clustered flowers no longer have distinct peduncles and pedicels; the latter may form a ± fused mass due to the close proximity of the flowers to each other (cf. *A. vallicola*, Fig. 37 d). In the corresponding ♂ partial inflorescences, in contrast, peduncles and pedicels are usually clearly discernible, although never very long.

---

[1] Interestingly, the situation appears to be reversed in *Coprosma* (subtribe *Coprosminae*): WILD & ZOTOV (1930) record "the large number of flowers in more or less dense inflorescences in the ♂" and "the fewer flowers and somewhat lax inflorescences in the ♀".

It appears that during one season ♂ plants of some dioecious *Anthospermum* species and certain *Nenax* species produce more flowers than ♀ plants (actual counts, however, have not been made because of the smallness and dense clustering of the flowers). This surmise agrees with published data, the majority of which record that there is an excess of ♂ flowers per inflorescence or ♂ inflorescences per plant (see survey by LLOYD & WEBB 1977; OPLER & BAWA 1978; HANCOCK & BRINGHURST 1980; BULLOCK & BAWA 1981; BARRETT & HELENURM 1981; BAWA, KEEGAN & VOSS 1982).

In some dioecious taxa of *Anthospermum*, ♂ and ♀ plants, furthermore, may exhibit differences in the degree of branching. ♀ shrubs tend to be less branched than the corresponding ♂ plants (e.g., *A. emirnense,* or *A. aethiopicum,* Figs. 54 c–d); the ♀ flowers are concentrated on a few inflorescence "cylinders" (cf. Fig. 54 c), while the ♂ flowers tend to be scattered over the whole shrub (cf. Fig. 54 d). Subsequently, the flowers or partial inflorescences are borne on axes of a different order: ♀ (partial) inflorescences may be confined to the main axis (stem) of the shrub and to the lateral branches of the first and second order, while those of ♂ are also borne on branches of a much higher (fourth or fifth) order.

The different degrees of branching may result in different shapes and habits of ♂ and ♀ plants. Often this is quite striking in *A. aethiopicum* (and probably also in *A. emirnense:* observed in population Lac Tri-triva/Madagascar): ♂ plants are rather "bushy" and have a fairly broad, rounded crown; the more sparsely branched ♀ plants appear much narrower and ± cylindrical. Similar differences were recorded in desert populations of *Simmondsia chinensis* (*Buxaceae*) by WALLACE & RUNDEL (1979); their findings that shrub *size* of ♂ and ♀ does not differ significantly also agrees with my observations on *Anthospermum*, which were carried out both in the field and on plants cultivated in the greenhouse under uniform conditions.

Finally, attention is drawn to differences in leaf size and shape between the sexes of dioecious taxa, a phenomenon which is not very common in the *Anthosperminae*. Also, it is not easily evaluated in the dioecious *Anthospermum* species and *Nenax* because the leaf blades are often rather narrow and ± needle-like, and measurement, which would have to be made in tenths of millimeters, are hardly possible in the field. A trend to slightly larger leaves on ♀ plants was occasionally observed in *Anthospermum emirnense, A. vallicola, A. welwitschii* and *A. spathulatum* subsp. *spathulatum.*

SALTER (1950) noted "leaves being largest in ♀" in (the non-dioecious!) *A. hirtum* and also mentioned the "smaller and less densely crowded" leaves in ♂ plants of *A. aethiopicum,* but ♂ and ♀ plants of that species cultivated in the greenhouses at HBV did not differ in leaf size and shape.

In *A. ammannioides,* however, leaf size and shape of ♂ and ♀ plants may differ quite markedly (cf. Figs. 54 a and b). It should be borne in mind, however, that the ♂ and ♀ branches depicted here represent extremes, and that leaf differences are not always so drastic and that there is, in fact, some overlap between leaves of ♂ and ♀ (compare Fig. 55). Leaf

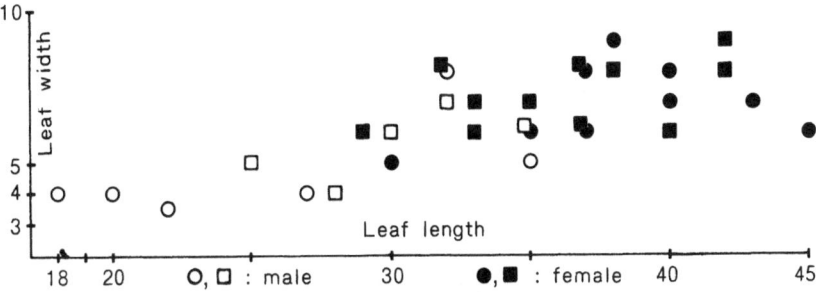

Fig. 55. *Anthospermum ammannioides,* leaf sizes of ♂ and ♀ individuals. – Round symbols: population Puff 790125-4/1, based on field data; squares: Puff 790124-1/1, measurements taken from plants grown from seed and cultivated under uniform conditions in the greenhouse. Each symbol represents an individual plant (average sizes of middle cauline leaves are given). – Measurements in mm. Further explanations in the text

measurements of ♂ and ♀ plants of *A. ammannioides* cultivated under uniform conditions in the greenhouse did not differ significantly from the random measurements taken in the field, although the variation and the extremes between ♂ and ♀ in the cultivated material did not seem to be as great (Fig. 55). This seems to indicate that the leaf variation in ♂ and ♀ is not or is only slightly influenced by the environment. This contrasts, for example, with the findings of WALLACE & RUNDEL (1979; *Simmondsia chinensis*), who recorded that the ♀ plants of desert populations had larger leaves than ♂ plants while plants from more mesic coastal environments did not exhibit any sex-related leaf dimorphism.

The production of larger, broader leaves in ♀ of dioecious species has also been noted in numerous other taxa, for example, *Silene alba* and *S. dioica* (= *Melandrium album* and *M. dioicum;* NIGTEVECHT 1966), *Mercurialis perennis* (MUKERJI 1936), just to name a few. Leaf dimorphism can, perhaps, also be interpreted as being an expression of the differences in reproductive strategies of ♂ and ♀ or, more precisely, their reproductive effort. LLOYD & WEBB (1977) speculate (p. 205) that "a greater reproductive effort by females is likely to be common, and that this explains a considerable number of the secondary sex characters ..."

In the survey of secondary sex characters by LLOYD & WEBB (1977) and in papers subsequently published (e.g., OPLER & BAWA 1978, BULLOCK & BAWA 1981), it was reported that ♂ plants often begin growth, prepare and begin to flower or mature earlier than the corresponding ♀ plants. Random field observations, in part, also confirmed that ♂ plants of some dioecious *Anthospermum* and *Nenax* species (e.g., *A. galpinii, A. zimbabwense, A. isaloense; N. cinerea*) do indeed start flowering earlier, but the reverse was also observed sometimes, and not uncommonly in different populations of the same species (e.g., *A. galpinii, A. aethiopicum, A. vallicola; N. microphylla*). ♂ and ♀ flowers of plants grown from seed and cultivated under uniform conditions in the greenhouse *always* virtually opened at the same time when they flowered for the first time (e.g., *A. ammannioides, A. emirnense, A. usambarense, A. littoreum, A. galpinii; N. microphylla, N. cinerea* and others).

Various studies indicate the presence of sex-linked niche differences and niche differentiation between the sexes (see survey by LLOYD & WEBB 1977, MEAGHER 1980, COX 1981, MELAMPY 1981), spatial segregation of sexes on elevational gradients (GRANT & MITTON 1979), or differential resource utilization by sexes of dioecious plants (FREEMAN, KLIKOFF & HARPER 1976). Little relevant information is available for the *Anthospermeae*. The one condition, however, that did become apparent during the course of this investigation was that plants growing under any kind of unfavourable condition or in "stress" conditions (i.e., often in somewhat drier habitats than the other plants of a population) were frequently ♀ – no matter whether they belonged to strictly dioecious species (e.g., *A. basuticum*) or to non-dioecious species (e.g., *A. herbaceum, G. circaeoides;* also see 12.2.1.). Whether this tendency has, in part, to do with differences in energy requirements for reproduction in ♂ and ♀ plants as suggested by PUTWAIN & HARPER (1972) is theoretically possible but there is at present no concrete evidence to support it.

12.4. Self-Compatibility, Auto- and Geitonogamy; Parthenocarpic Fruits and the Question of Apomixis

According to ANDERSON (1973), self-compatibility is "apparently widespread" in the (non-heteromorphic) *Rubiaceae*; little, concrete evidence, however, is given. MORTON (1972) states that "most of the African montane species I have grown in cultivation have proved to be self-compatible and in many cases self-pollinating". Amongst these,

Morton lists (in his Table 3) the genus *Anthospermum*, but he neither provides species names nor any particulars[1].

In view of these statements and observations, experiments on plants cultivated in the greenhouse were started. The most extensive experiments were carried out with plants of *Galopina circaeoides* (which, from a practical point of view, is easiest to work with) and set up as follows: Five plants (grown from seeds of one mother plant; Puff 790516-1/3) with ♂ + ♀ inflorescences were used for each set of experiments. Individual plants/pots were isolated from each other by placing them between separate closed double windows. There was no ventilation (virtually no air movement) between the double windows; the inner windows were only opened for daily investigations of the plants and occasional watering.

Experiments 1 and *2*: Cross pollination. Pollen from ♂ flowers of one plant were transferred by hand to marked ♀ flowers of another plant and vice versa.

Experiment 3: Only ♀ flowers were left on the inflorescence(s) of a plant; ♂ flower buds (easily recognized by their larger size) were removed so as to exclude the possibility of auto- or geitonogamy. → apomixis?

Experiment 4: Individual plants were left as they were; no air movement, no pollen vector. → autogamy?, apomixis?

Experiment 5: Pollen from ♂ flowers of a plant was transferred by hand to stigmas of marked ♀ flowers of the same plant. → geitonogamy?

Results:

1 & 2: All pollinated and marked ♀ flowers produced fruits with viable seeds except for a few flowers which were damaged during investigation.

3: No fruits were produced. Some ovaries enlarged slightly so as to resemble fruits but did not contain seeds (see Parthenocarpic Fruits, below). Thus apomixis is unlikely to occur.

4: Some but not all ♂ flowers produced fruits with viable seeds; no fruits were produced by ♀ flowers. As the possibility of wind-mediated pollen transport can be excluded, and considering the presumed absence of apomixis (see Experiment 3, above), autogamy must have occurred.

5: All pollinated, marked ♀ flowers set fruits with viable seeds – geitonogamy is successfull.

Results for *Anthospermum* species (obtained from experiments with *A. hispidulum*, Puff 790516-2/1, 790523-1/1, and *A. pumilum* subsp. *pumilum*, Puff 790520-1/1) were identical. In addition, it was observed that in a plant of *A. littoreum* (Puff 790424-1/1), an odd ♀ flower in a

---

[1] The species he studied was *A. asperuloides*. He had one potted plant in cultivation which set fruit (personal discussion with Prof. Morton, May 1984). As the species has ♂ or ♂ + ♀ plants, auto- or geitonogamy could have led to fruit set.

bagged cluster of ♂ flowers also developed a viable fruit – probably through geitonogamous pollination [*A. littoreum* is a dioecious species, and the occurrence of odd ♀ flowers on ♂ plants may provide a way for isolated ♂ plants to propagate! Also cf. "leaky dioecism", BAKER & COX (1984); 12.2.1.,page 135]. Similarly, in ♂ plants of other dioecious *Anthospermum* species in which some flowers reverted to ⚥, seed set was observed (auto- or geitonogamy?), but no detailed experimental evidence is available.

The results thus confirm the presence of self-compatibility and indicate that, at least as far as the studied taxa are concerned, allogamy is or can be supplemented by (facultative) auto- and/or geitonogamy. Protandry of individual ⚥ flowers may reduce selfing to some extent, but with respect to the ♂ and ♀ phases of a *whole* inflorescence or plant, it is not very effective – selfing through geitonogamous pollination is a strong possibility. In addition, it seems that – because of floral adaptations to anemophily – protandry in the *Anthospermeae* may be *relatively* less effective in the prevention of autogamy than in other *Rubiaceae*: although the stigmas only start to elongate and become receptive after the anthers have dehisced, the increased (hairy) stigma surfaces nonetheless reach down to the base of the corolla tube (the non-receptive stylar portion is very short or subobsolete!), thus bringing the receptive stigma surfaces and anthers into (relatively) close proximity to each other (much more so than in many zoophilous rubiaceous flowers with long styles which elevate the relatively small stigmatic surfaces well above the level of the anthers → dichogamy coupled with herkogamy!). The probability of some pollen remaining in the base of the corolla tube and coming into contact with the stigmatic surfaces seems quite high. It may also be that – in the absence of wind as pollen vector in certain habitats – at least some pollen grains (in ⚥ flowers) are dumped on to the immature stigmas (which are not "folded up" tightly and protected as in other *Rubiaceae* – cf. *Crocyllis*, PUFF & MANTELL 1982a, Fig. 15) and germinate after the stigmas have elongated and become receptive. Prerequisite for this to occur is, of course, that pollen remains viable for a few days. (See also Experiment 4, above.)

The proportion and extent of cross- and self-fertilization within a plant under natural conditions, and its variation from plant to plant and population to population remain uncertain. It is suspected, however, that for example in populations of *G. circaeoides* growing in sheltered forest undergrowth where wind is little effective as a pollen vector, auto- and geitonogamy may prevail, whereas in populations occurring in more exposed habitats (open grassland; grassland – scrub edge) there may be a marked shift to allogamy. A detailed and extensive study programme, however, has yet to be carried out.

Parthenocarpic Fruits.
Branchlets of cultivated ♀ plants of the dioecious species *A. aeth-iopicum, A. emirnense, A. littoreum, A. galpinii, N. microphylla* and *N. cinerea* were bagged before any flowers had opened. At anthesis selected unbagged ♀ flowers were hand-pollinated as controls. After fruit set, both the control and the experimental bagged branchlets were investigated. The ovaries of the bagged ♀ flowers had often increased in size – sometimes quite markedly so – but these "fruits" were always smaller than the "control" fruits. Both the fruits produced by the control flowers and by the isolated ♀ flowers were collected and then investigated anatomically (longitudinal and/or transverse sections). The "control" fruits mostly contained normally developed seeds and embryos whereas the "fruits" from bagged branchlets invariably contained shrivelled ovules and no fully developed seeds, although the pericarp structure was found to be near identical to that of the fertile fruits. Thus they could be termed pseudo- or parthenocarpic fruits. Such parthenocarpic fruits seem to occur rather frequently in species of *Anthospermum, Nenax* and *Galopina*[1]. Also seemingly dimorphic fruits – i.e., where one type of fruit is only half (to ¾) the size of the other – were frequently observed on otherwise normal plants. In such cases, mixtures of both fertile and parthenocarpic fruits are likely to be present; simple cross sections of mericarps are an easy way to check. According to my experience, parthenocarpic fruits may sometimes even be almost as large as fertile ones and may fall in the size range typical of fertile fruits of a given taxon as a whole.

The occasional occurrence of isolated, fruiting ♀ plants in the field (*Anthospermum, Nenax* species) has nourished speculations as to the apomictic nature of the fruits (e.g., SCHELPE, pers. comm.). Detailed information available for the dioecious *A. monticola* (PUFF 790908-1/1), however, suggests rather the presence of parthenocarpic fruits: The fruits of a single isolated individual of *A. monticola* (i.e., no other plants of the same species found in a radius of 100 m) were collected and approximately 200 mericarps were sown. Out of these, only 3 germinated. The other ungerminated mericarps were later investigated anatomically and it was found that although they were not noticeably smaller than those that had germinated only ovule rudiments were present. The 1.5% germination

---

[1] Seedless fruits, in general, appear to be relatively common. See GUSTAFSON (1942) for an extensive list of species [of non-*Rubiaceae*] with such fruits.

rate, nevertheless, seems interesting. It may be explained as follows: (1) any ♂ plants in the vicinity were too great a distance away to ensure a higher percentage of successful pollination (this would account for the few fertile fruits. The possibility that ♂ plants have disappeared, died or have been destroyed between flowering and fruiting of the ♀ plant seems too unlikely to be considered seriously); (2) the fertile fruits may be of *autogamous* origin – e.g., odd ⚥ flowers were produced on the ♀ plant [this is theoretically possible – cf. "leaky dioecism", BAKER & COX (1984); 12.2.1., page 135], or (3) the few fertile fruits *are* of apomictic origin. In that case it could be argued that apomixis is not yet fully established and is as yet not very successful.

At present, however, there is no clear indication for the occurrence of apomixis (several of the above observations, in fact, seem to provide evidence against it). But before detailed embryological work with special reference to the possibility of facultative apomixis is carried out, and before it has been fully documented whether perhaps the phenomenon of pseudogamy is a prerequisite for the development of apomictic fruits, its occurrence in (dioecious) species of *Anthospermum* and *Nenax* cannot altogether be discounted.

12.5. Hybridization

In the tribe *Anthospermeae*, the frequent occurrence of hybrids has been recorded among New Zealand species of *Coprosma* and has been suspected among Hawaiian species of the génus (OLIVER 1935). This is also confirmed by ALLEN (1961) who discusses "the prevalence of hybridism" among New Zealand *Coprosmas*. There have been some initial (successful) attempts to prove the hybrid nature of certain *Coprosma* plants by means of chemical analyses: TAYLOR (1964) analysed polyphenols quantitatively to study hybridization, and suggested that this method could be used successfully to study introgression and species relationships. WILSON (1979, 1984) used flavonoid profiles to detect hybrids.

In view of these findings and in view of the suspected occurrence of hybrids in the *Anthosperminae*, it was attempted (1) to produce artificial hybrids in the greenhouse, (2) to pay special attention to the occurrence of hybrids in the field, and (3) to use flavonoid patterns to detect hybrids.

(3) yielded inconclusive results (see 11.1.), but observations and data obtained from (1) and (2) produced valuable new information and evidence for the occurrence of hybridization in the *Anthosperminae*.

(1) Crossing experiments, carried out in the greenhouse: (a) Pollen from ⚥ + ♀ plants of *A. pumilum* subsp. *pumilum* (PUFF 790520-1/1) was transferred with a brush to newly opened ♀ flowers (*not* ⚥ flowers to avoid

the possibility of geitonogamy) of ♂ + ♀ plants of *A. hispidulum* (PUFF 790523-1/1). The ♀ flowers were then bagged to prevent any further exchange. (**b**) Pollen of ♂ *A. galpinii* (PUFF 790415-4/1) was transferred to stigmas of ♀ *A. welwitschii* (PUFF 781028-1/1).

In both cases, seed set was close to 100%. Germination of the *A. pumilum* × *A. hispidulum* seeds, however, was not good (less than 50%; but neither was it good in the seed material from which the parent plants were raised – see 12.7., Figs. 60 o–p and 61 i). Germination of *A. galpinii* × *A. welwitschii* seeds was better (ca 60%–70%). The hybrid plants grew well for several months but died (probably *not* due to reduced viability but due to cultivation problems) before reaching flowering stage. Since this happened while I was away on a collecting expedition, no vouchers or measurements of the hybrid plants are available; only the following previously made brief notes exist: 1. The hybrid plants had leaves intermediate in size and shape to the parent plants; 2. In the *A. pumilum* × *A. hispidulum* plants, the indumentum was as in *A. hispidulum* but less dense (*A. pumilum* has glabrous, *A. hispidulum* densely hairy leaves); 3. *A. galpinii* × *A. welwitschii* plants had a decussate leaf arrangement at the lower nodes but ternately arranged leaves above as in *A. galpinii* (*A. welwitschii* has strictly decussate leaves). The experiments, although not as extensive and as complete as originally planned, nevertheless, proved that (not even so closely allied) species are "crossable" and compatible.

Table 8. Odd hybrids between *Anthospermum* species. – * = occurrence of (putative) hybrids well documented (i.e., both parent species and odd hybrid plants observed in the field and dried specimens later studied in detail in the laboratory); brackets = (putative) hybrids presumably very uncommon; ? = occurrence of (putative) hybrids somewhat uncertain. – For further details see part D, Critical Remarks section of the respective species

(*A. aethiopicum* × *A. spathulatum* subsp. *spathulatum*)
A. basuticum × *A. monticola**– see Fig. 56
A. galpinii × *A. littoreum*
A. herbaceum × *A. hispidulum*
(*A. herbaceum* × *A. paniculatum*)* – see Fig. 57
(*A. herbaceum* × *A. pumilum* subsp. *pumilum*)
(*A. hirtum* × *A. spathulatum* subsp. *spathulatum*)?
(*A. ternatum* subsp. *randii* × *A. zimbabwense*)?
(*A. vallicola* × *A. zimbabwense*)*
(*A. whyteanum* × *A. ammannioides*)?
(*A. whyteanum* × *A. pumilum* subsp. *pumilum*)?
(*A. whyteanum* × *A. usambarense*)* – see Fig. 58
(*A. whyteanum* × *A. welwitschii*)*
A. whyteanum × *A. zimbabwense*

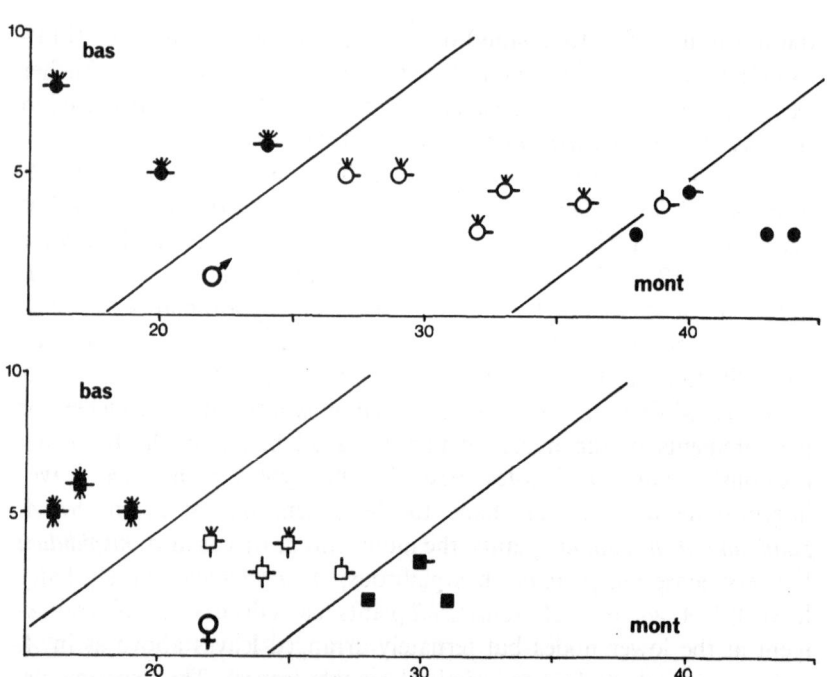

Fig. 56. *Anthospermum basuticum* (bas; Puff 781126-1/8), *A. monticola* (mont; Puff 781126-1/5) and hybrids (open symbols). O.F.S., Drakensberg, around terminus of Sentinel rd.; between "Witsieshoek Gate" and path to Mount-aux-Sources. – Horizontal axis: height of shrubs in cm; vertical axis: leaf blade length to width ratio (averages of a representative number of midstem leaves per plant; the highest ratios indicate ± terete blades due to strongly revolute margins). Leaf blade glabrous: ○, with a few hairs (margin, midrib): ♂, sparsely hairy on the upper surface: ♂, densely hairy: ♂; fruits/ovaries glabrous: ○, with few hairs: ♀, densely hairy: ♀; plant not densely leafy (leaves crowded near shoot tips only): ○, ± densely leafy: ○–, densely leafy: –○–. – Because of sexual dimorphism in some characters, ♂ and ♀ plants are plotted separately. "Pure" *A. monticola* occurs from ca 2 600–2 800 m, "pure" *A. basuticum* from ca 2 800–3 000 m, and hybrids from ca 2 700–2 900 m; exact altitude information, however, is not available for individual plants

(2) The detection of hybrids/hybrid populations in the field is not easy. One of the main difficulties is the variability of numerous taxa. In addition, there are often only few "good" and useable distinguishing characters between taxa. Leaf sizes and shapes and stipule structure of taxa often overlap. In numerous taxa, inflorescence morphology is basically too similar to allow a certain and definite separation. Floral characters – especially in species with complicated sex combinations – may be obscured by the presence of "transitional" flowers

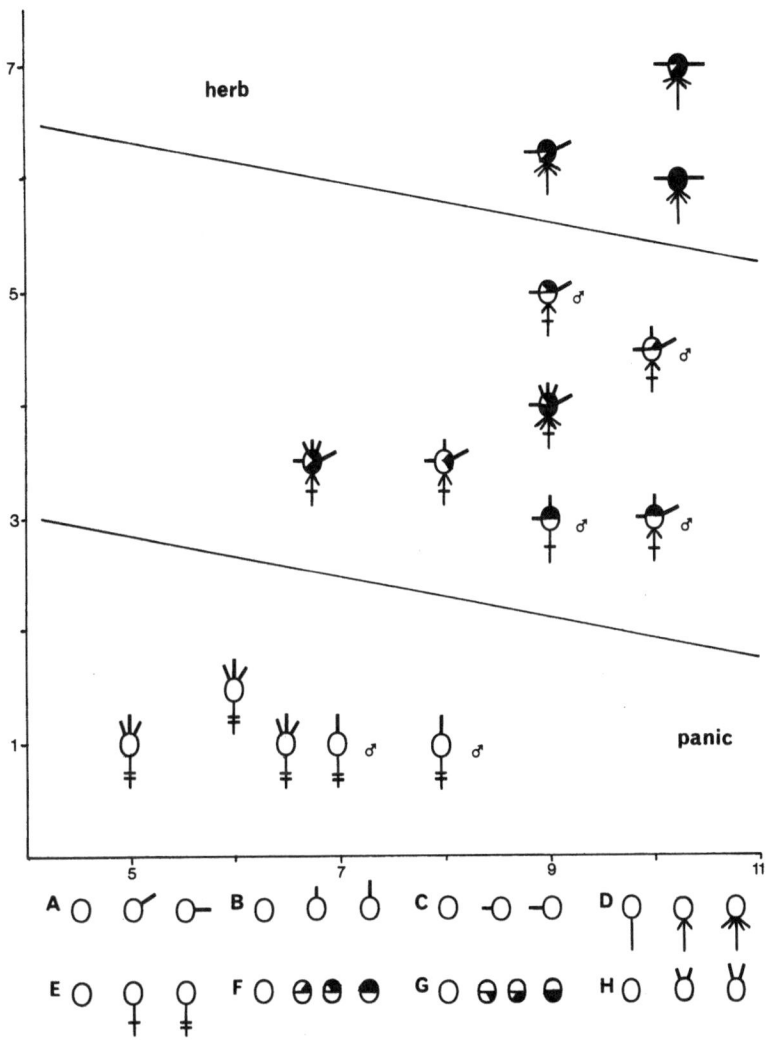

Fig. 57. *Anthospermum paniculatum* (panic; Puff 790115-5/3), *A. herbaceum* (herb; Puff 790115-5/5) and hybrids (Puff 790115-5/4). E Cape Prov., Fort Cunyngham Forest Res., Dohne Peak. – Horizontal axis: leaf length in mm; vertical axis: leaf width in mm. Symbols (left to right): A stems ± erect - ascending - prostrate. B inflorescence consisting of axillary flower clusters - partial inflorescences ± elongated - inflorescence a "typical" thyrsus (cf. Fig. 27 b). C leaf: petiole obsolete - 0.5–1 - 2–3 mm long. D leaf: stipular sheath with 1 - 3 - 5 setae. E short shoots ± elongated - ± contracted - much contracted. F hairs on upper leaf surface absent - very few - few - dense. G mericarp surface glabrous - sparsely - moderately - densely short-hairy/papillate. H mericarps without - with short, indistinct - distinct, ± massive calyx lobes. G, H: no data for ♂!

Fig. 58. *Anthospermum whyteanum, A. usambarense* and (putative) hybrid *A. whyteanum* × *A. usambarense* from the Nyika Plateau (Zambia/Malawi). *a A. whyteanum*, collected in cracks of rocks on a rocky outcrop, (dwarf) shrubs, ca 0.4–0.5 m (few to 1 m) tall (PUFF 781223-3/1); *b A. usambarense*, collected in forest edge scrub at base of rocky outcrop, shrubs to 2 m (few to 2.5 m) tall (PUFF 781223-3/2); *c* (putative) hybrid *A. whyteanum* × *A. usambarense*, collected in secondary scrub, shrubs to 1.2 m tall (FANSHAWE 9747, K). – *a–b* from around Chowe Rock (Malawi, near Zambian border), *c* "Nyika, Zambia" [no detailed locality information given]; photo montage: *A. whyteanum* and *A. usambarense* are added to sheet FANSHAWE 9747. – Further explanations in the text

($\vec{\male} \leftarrow \vec{\female} \rightarrow \female$; see 12.2.1.). Fruit characters may be unreliable due to the occasional presence of parthenocarpic fruits (see 12.4.). One can, thus, not always be certain whether an "atypical" specimen still belongs to one taxon (i.e., represents a geno- or/and phenotypic "extreme") or is of hybrid origin. More definite proof for the hybrid nature of individuals may often only be obtained from subsequent tedious laboratory investigations and measurements which must be exact to tenths of millimeters.

In *Carpacoce* and *Nenax* there is virtually no concrete evidence for the occurrence of hybrids. There are indications that a single known collection may represent an intergeneric hybrid between *Anthospermum* and *Nenax* (*A. pumilum* subsp. *pumilum* × *N. microphylla*; HANEKOM 606). As both taxa may occur in similar or the same habitats and as their distribution ranges overlap in part, this is not totally unrealistic.

There is little proof for hybrids amongst *Galopina* species. Different species very seldom occur in the same area or habitats. Possibly there are odd hybrids *G. circaeoides* × *G. tomentosa* – e.g., WELLS 3485 (see part D, *G. circaeoides*, Critical Remarks).

In *Anthospermum*, however, hybrids appear to be more common. Two categories of hybrids can be distinguished:

(a) Odd hybrids in localities where two species, by chance, happen to occur sympatrically and happen to flower at the same time. A (doubtlessly incomplete) survey is given in Table 8.

(b) Putative extensive hybrid populations. *A. hispidulum* and *A. pumilum* subsp. *pumilum* are likely to form such hybrid populations. The species are crossable and produce viable hybrids [see (1) Crossing Experiments, p. 156], but *definite* proof for the ocurrence of extensive hybrid populations (in, for example, parts of the Transvaal; see also part D, *A. hispidulum* Critical Remarks) is lacking. It is suspected that hybrid plants often look like "atypical", less densely hairy *A. hispidulum* plants. Further detailed field studies would be required. Also *A. usambarense* and *A. welwitschii* are thought to produce extensive hybrid populations (for details see part D, *A. usambarense*, Critical Remarks).

12.6. Diaspore Dispersal

Dispersal units are either the mericarps (*Anthospermum*, *Galopina* and *Nenax* p.p.), the entire fruits (*Nenax* p.p.) or the endocarps plus the enclosed seed (*Carpacoce*).

With few exceptions (see below), the diaspores seem to be rather unspecialized and to lack any obvious adaptations to a particular mode of dispersal. The relatively small weight (and size) of the diaspores (Table 9)

Table 9. Diaspore weights (in mg) of *Anthospermum, Galopina, Nenax, Carpacoce* and *Phyllis* species. – Average weights (n = 30). Diaspores: M = mericarp, F = entire fruit, E = endocarp plus seed

Anthospermum ibityense	(Puff 800730-1/2)	M	0.4
A. littoreum	(Puff 790415-5/1)	M	1.0
A. spathulatum			
subsp. *spathulatum*	(Puff 791221-2/1)	M	1.0
A. welwitschii	(Puff 781221-2/3)	M	0.3
Galopina circaeoides	(Puff 790909-6/1)	M	0.8
Nenax arenicola	(Puff 800915-2/1)	F	20.9
N. cinerea	(Puff 800102-4/2)	M	2.2
N. microphylla	(Puff 790710-1/1)	M	1.1
Carpacoce heteromorpha	(Puff 800917-4/1)	E	1.2
Phyllis nobla	(Bot. G. Zürich)	M	1.5

would indicate incidental dispersal by (strong) winds, although the diaspores are much heavier than typical "dust seeds" (ca 0.001 to 0.004 mg in *Orchidaceae, Pyrolaceae* and *Orobanchaceae*: van der Pjil 1982). Also granivorous birds, muddy feet or rain water transport may account for some dispersal. In my opinion, however, it must be strong winds that are responsible for the dispersal of the mericarps (and subsequently for the distribution pattern) of, for example, *A. dregei* subsp. *ecklonis*, a taxon which is always confined to isolated sandstone outcrops – often many dozens of kilometers apart from each other – in the W Cape Prov. (S Africa). The same is likely to hold true for most (all?) of the widely, but disjunctly distributed species (e.g., afromontane species: *A. usambarense*, etc.). In the case of the relatively heavier fruits of *Nenax* species (e.g., *N. arenicola*, Table 9), it is perhaps more likely that the diaspores are only rolled about by the wind rather than carried in the air.

   N. microphylla and *N. cinerea* are the only two species which show a ± obvious adaptation to wind dispersal. Their mericarps have a largely increased surface and appear ± hollowed out (cf. Figs. 37 g and 38 b); the lateral portions of the mericarps may function as wings. The two species are confined to (semi-)desert and karroid areas – i.e., regions in which the occurrence of wind-dispersed diaspores in numerous, taxonomically diverse taxa is well documented (see, for example, Leistner 1967: S Kalahari).

   It remains uncertain and somewhat doubtful if the exceptionally long, ± band-like calyx lobes of *A. longisepalum* (Fig. 36 e) and *A. ibityense* aid in the dispersal of the diaspores (epizoochory?). It was only observed in the field that groups of mericarps, held together by the entangled calyx lobes, lie on the ground next to the mother plant. The two enlarged calyx lobes of the sterile, modified carpel of *A. bicorne* (Figs. 36 a, c) most certainly do not serve a special purpose; they fall off as the fertile carpel/mericarp becomes detached from the carpophore.

It is possible that the unique "air pockets" of the mericarps of *A. palustre* (see section, Fig. 40 b) aid in their dispersal (nautohydrochory?; the species occurs in wet to moist habitats). The mericarps taken from herbarium specimens readily float in water but actual observations in the field are lacking. Also the diaspores of the coastal sand dune species *A. littoreum* may, to some extent, be dispersed by sea water currents. The long hairs covering the mericarps (Fig. 35 e) trap air so that the diaspores are not easily submerged (mericarps placed in a jar filled with sea water were still afloat after several weeks; they retained their viability and germination rates were not different from those of mericarps not exposed to sea water).

12.7. Germination Data

In *Anthospermum, Nenax* and *Galopina* germination takes place as follows: After the base of the mericarp has been forced open by the stretching hypocotyl and the elongating radicle (e.g., *A. ammannioides,* Fig. 59 a), growth of the seedling continues rapidly and the primary root and the hypocotyl double or triple in size within a day (e.g., *N. microphylla,* Fig. 59 c). Germination is epigeal. The hypocotyl is often curved into a loop at first and its base and its tip is at or below ground level. The hypocotyl eventually becomes erect and lifts the cotyledons above the ground. The cotyledons either free themselves from the mericarp as they emerge from the ground or they, sometimes, remain enclosed in the mericarp wall which forms as loose "hood". After a few days, however, the cotyledons escape to become phanerocotylar. Cotyledons are thin, green and foliage leaf-like (e.g., *A. ammannioides,* Fig. 59 b), i.e., they are "paracotyledons" in the terminology of DE VOGEL (1980); they show no trace of stipules (cf. Fig. 59 b), although (para)cotyledonary stipules do occur in other *Rubiaceae* (DUKE 1969) or "probably are the rule" in that family (DE VOGEL 1980). *Phyllis* (*Anthosperminae*), according to the survey of WUNDERLICH (1971), has "gland-like structures" in the axils of the cotyledons but – like for many other genera – detailed morphological-anatomical investigations which could prove the true stipular nature of these structures are lacking. Seedlings are of the "*Macaranga* Type" (DE VOGEL 1980), the most common seedling type in the dicotyledons; it is apparently characteristic for most (or all?) genera of the *Rubiaceae* (DE VOGEL 1980, and also DUKE 1969; tropical woody genera).

Germination experiments were carried out in petri dishes. Mericarps

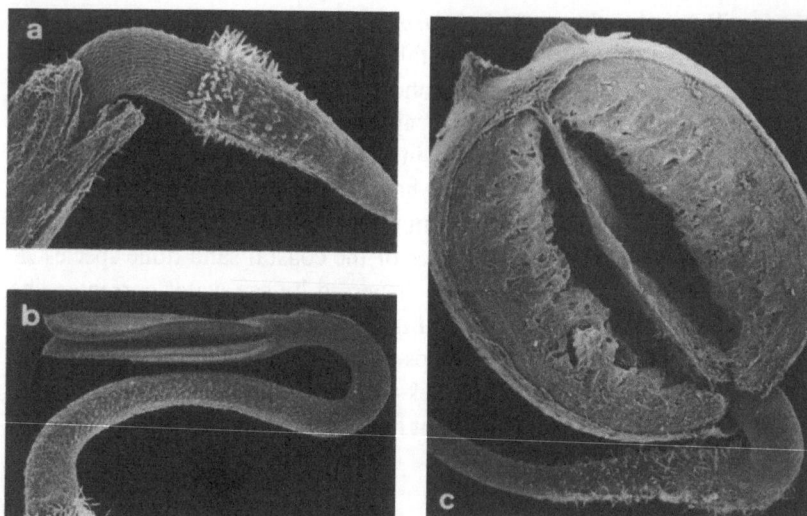

Fig. 59. SEMgraphs of germinating seeds and seedling of *Anthospermum* and *Nenax. a–b A. ammannioides* (PUFF 790124-1/1), *a* one day after start of germination, the pericarp splits open basally; *b* seedling, four days after start of germination, note absence of stipule-like structures between cotyledons. *c N. microphylla* (PUFF 790710-1/1), two days after start of germination. – All critical point dried. *a, c* ×35; *b* ×16

were placed on moist filter paper and kept under normal light conditions (day light). At the end of the germination experiments, mericarps that had not germinated were sectioned by hand to determine whether they contained (apparently) well developed seeds. Mericarps which were found to have originated from parthenocarpic fruits (i.e., mericarps without well developed seeds; see 12.4.) were excluded from the calculation of germination percentages. Depending on the availability of fruiting material, between (300) 200 and 50 (30) mericarps per plant, taxon or experiment were sown.

In many taxa, germination started within two to six days after sowing (cf. Figs. 60 and 61). In the following days there was usually a steep increase in germination; thereafter, germination leveled off and often ceased after approximately 8 to 15 days (e.g., *A. aethiopicum,* Fig. 60 i) or, in some taxa, even earlier. In more extreme cases, germination only lasted for three, or even only two days (e.g., *N. cinerea,* Fig. 60 a, or *A. spathulatum* subsp. *spathulatum,* Fig. 61 c). In these taxa, initial germination rates were often very high, and rates of more than 50% were not uncommon (e.g., 55% in the beforementioned examples); total germination capacity was often also high, reaching ca 70% to 95% (e.g., *A.*

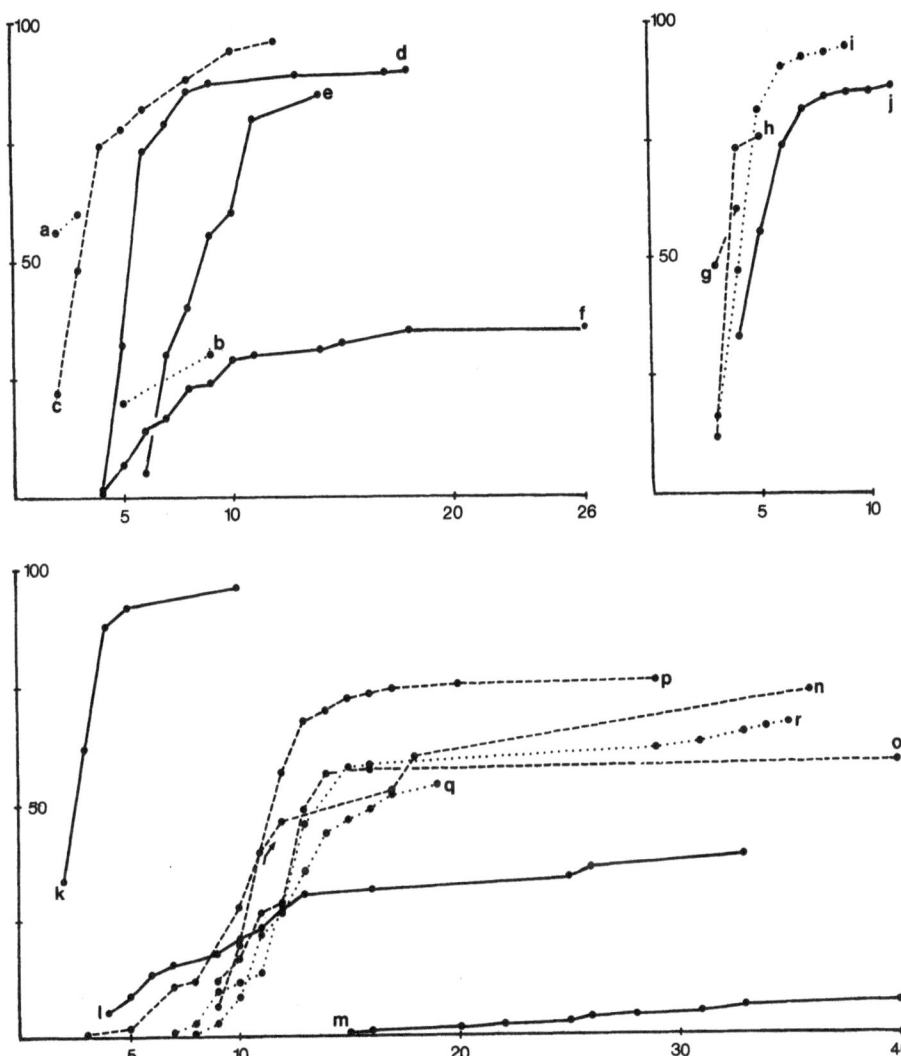

Fig. 60. Germination diagrams of *Nenax, Phyllis, Galopina*, and *Anthospermum. a
N. cinerea* (Puff 800102-4/2). *b N. microphylla* (Puff 790720-2/1B). *c P. nobla*
(Sunding s.n.). *d G. circaeoides* (Puff 790516-1/3). *e G. circaeoides* (Puff 790909-
6/1). *f G. tomentosa* (Puff 790422-3/1). *g–i A. aethiopicum, g–h* Puff 791220-4/1, *h*
sown ca one year later than *g, i* Puff 790716-3/1, *j A. aethiopicum*, data for fresh
seed converted from Levyns (1935). *k A. welwitschii* (Puff 780211-2/1). *l A.
usambarense* (Puff 781221-2/3). *m A. galpinii* (Puff 790415-4/1). *n–p A. hispidulum, n* Puff 790520-1/2, *o–p* Puff 790523-1/1, *p* sown ca one year later than *o. q–
r A. ibityense* (Puff 800730-1/2), *r* sown ca half a year later than *q. –* Horizontal
axis: days after sowing; vertical axis: germination in %. Mericarps taken from
individual plants

welwitschii, Fig. 60 k). *Phyllis nobla (Anthosperminae)* can be added to this group of taxa; its mericarps germinated rapidly, the germination percentage was high and the germination period short (Fig. 60 c).

In a second group of taxa, there was a more gradual and less steep increase after the start of germination. This was sometimes combined with a somewhat delayed start of germination (e.g., *G. tomentosa* or *A. galpinii,*

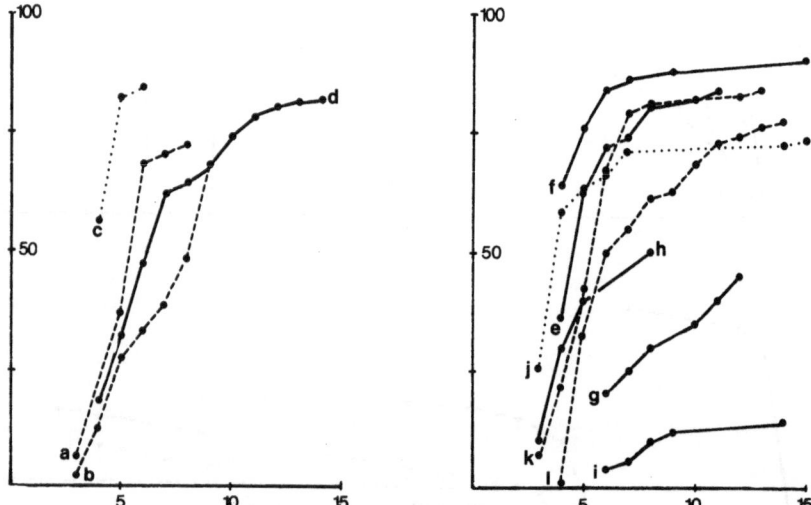

Fig. 61. Germination diagrams of *Anthospermum. a–d A. spathulatum* subsp. *spathulatum, a–b* PUFF 791231-1/2, *b* sown 16 months later than *a, c* PUFF 791221-2/1, *d* PUFF 790714-2/1A. *e–g A. emirnense, e–f* PUFF 800729-1/1, seeds from two different plants, *g* PUFF 800726-2/1. *h–i A. pumilum* subsp. *pumilum, h* PUFF 780408-1/3, *i* PUFF 790520-1/4. *j A. ammannioides* (PUFF 790124-1/1). *k–l A. littoreum, k* PUFF 790415-5/1, *l* PUFF 790424-1/1. – See Fig. 60 for explanations

Figs. 60 f and m). Germination sometimes continued up to 40 days after sowing (e.g., *A. hispidulum,* Fig. 60 o). Germination rates in this group of taxa tended to be lower with maxima between 75% and 35% (e.g., *A. ibityense* or *G. tomentosa,* Figs. 60 q and f), or even less (less than 10% in, for example, *A. galpinii,* Fig. 60 m).

Germination rates of fruiting material from different plants of a single population (*A. emirnense,* Figs. 61 e–f) were found to be similar, but those of mericarps from plants of different populations of a taxon where, in some cases, significantly different: While in *A. littoreum* (Figs. 61 k–l), for example, differences of germination totals were slight (less than 10%), germination in *A. emirnense* (Figs. 61 e–f and g) ranged from 45% to 90%. Similar extreme differences were found in, for example, *A. pumilum* subsp. *pumilum* (Figs. 61 h and i: 50%:15%). Within several taxa, the duration of the germination period was found to be markedly different (e.g., *A. spathulatum* subsp. *spathulatum,* Figs. 61 a–d: 3 to 11 days).

In order to document the change of germination capacity in relation to age, different batches of mericarps, originating from one plant, were sown at intervals ranging from ca half a year to 16 months. In *A. aethiopicum, A. hispidulum* and *A. ibityense* (Figs. 60 g–h, o–p and q–r), the mericarps sown later showed a *higher* (by ca 10%–18%) germination rate, whereas in *A. spathulatum* subsp. *spathulatum* (Figs. 61 a–b), mericarps sown later had a slightly lower (by ca 5%) germination rate. In all cases, the first batch of mericarps was sown within a year after harvesting. It is likely that mericarps remain viable for several years: LEVYNS (1935) germinated mericarps of *A. aethiopicum* from a single collection at yearly intervals over a period of five years. The average germination was between 80% and 90% in the first four years, and only dropped to 69% in the fifth year. In addition, she found that with increasing age, the germination period becomes more protracted (fresh mericarps: 12 days, Fig. 60 j; two and four years later: 18 and 39 days respectively). LEVYNS also noted that germination is, in *A. aethiopicum* at least, not stimulated by fire; equal numbers of seedlings occur in burnt and unburnt areas.

Age of the mericarps may be one of the factors to be considered when interpreting the germination data presented in Figs. 60 and 61. According to field observations, fruits may remain on the plants for (at least) 6 months to a year in woody taxa and rhizome perennials (fruits remaining attached to the old, dead and dried up aerial shoots); the actual age of the collected mericarps could, therefore, often not be established with certainty. It appears unlikely, however, that age alone is responsible for the protracted germination period and lower germination capacity of some of the taxa. Perhaps special, as yet unknown germination conditions are required by these taxa to bring about quicker and more successful germination. It is conspicuous that taxa with *rapid* germination and the highest germination percentages are often species with distinct p i o n e e r p l a n t properties which are found over a wide range of habitats and frequently also occur in disturbed areas such as road sides, clearings etc. *A. aethiopicum, A. emirnense, A. welwitschii* and *G. circaeoides* are examples of such species. Taxa with lower germination percentages and more protracted germination periods may be widely distributed but appear to occur in more "spezialized" habitats. It should be borne in mind, however, that the above observations are generalizations which do not always hold true. The widely distributed *A. pumilum* subsp. *pumilum,* mostly growing in cracks of rocks, for example, has a highly heterogenous germination behaviour (Figs. 61 h–i).

The germination behaviour of *N. cinerea* (a species occurring in dry to very dry localities) is similar to that of many desert plants: the mericarps mostly start to germinate almost immediately after a rain fall (taking advantage of the temporarily available moisture) but appear to exhaust their germinating capacity within a few days (two days in the experiments: Fig. 60 a).

It was also found that in *Carpacoce* species (*C. vaginellata, C. spermacocea* and *C. heteromorpha*) and *Nenax arenicola* and *N. acerosa,* diaspores failed to germinate although seeds were apparently viable (i.e., well developed). Neither attempts to germinate them in the dark or at higher temperatures and under strong artificial lighting (essential, for example, for the germination of many *Hedyotideae* – cf. CORBINEAU & CÔME 1980, 1982), or letting the seeds lie dormant for a year (post-ripening!) could bring about germination. These taxa obviously need very special, as yet unknown germination conditions. Since all of the mentioned "problem cases" are SW Cape endemics occurring primarily in acid, TMS-derived sands or sandy soils (except *N. arenicola*), the question arises whether there is a direct relation between changed germination behaviour and habitat (numerous SW Cape plants confined to TMS-derived substrates have mycorrhiza!).

Interestingly, the germination behaviour within the genus *Nenax* is heterogenous: no germination in *N. arenicola* and *N. acerosa* vs. rapid germination in *N. microphylla* and *N. cinerea* (Figs. 60 a–b). The two species whose seeds failed to germinate belong to the most specialized and "derived" within the genus: They are the only (documented) tetraploids within *Nenax* and are characterized by thick-walled, "stony" indehiscent fruits as dispersal units (vs. mericarps like in most other *Nenax* species; see 7.2.4). These fruits sit on ± flat and disk-like terminal pedicel portions (there are no carpophores!); at their base, i.e., where they are attached to the apical pedicel portions, the fruit wall is thin and hardly sclerified, thus virtually leaving a gap for the radicles to emerge (in fruits dehiscing into two mericarps, the radicles have to force the base of the mericarps apart). In order to test whether these thick-walled indehiscent fruits needed to rot before germination could take place, fruits of *N. arenicola* were kept submerged in water for several months, but the result was negative. This is not surprising as it is doubtful whether, under natural conditions, the dry habitats of that species could provide such permanently wet conditions. Due to the relatively greater weight of these fruits (in comparison to the light, thin-walled mericarps – cf. Table 9) the course of germination must be somewhat different from the pattern described at the beginning of this chapter: the seedling (i.e., its cotyledons) must emerge completely from the fruit at an early stage as it is unlikely that the hypocotyl, when it becomes erect during germination, would be able to lift the whole, heavy fruit above the ground. These indehiscent fruits, furthermore, are two-seeded; it is entirely uncertain how – and if – two seedlings (can) emerge from such a fruit. Further experiments and observations would be worthwile.

12.8. Regeneration After Fire

Plants occurring in fire-prone habitats are either killed by fire or survive and regenerate from a woody base ("sprouters"; see also 2.5.1.).

Plants killed by fire rapidly regenerate from seed ("seed regenerators")[1]. In studies dealing with veld burnings in fynbos and regeneration after fire (Table Mountain: ADAMSON 1935; Grahamstown: MARTIN 1966), it is well documented for the SW Cape endemic *Anthospermum aethiopicum* that it is (a) one of the first plants to regenerate from seed after a fire[2] and (b) that the species is one of the dominant and most abundant plants in the early phases of regeneration of fynbos (its capacity for quick germination and rapid growth is, furthermore, documented in 12.7. and 2.1.). In later stages, however, it gives way to other species due to its weak competitive vigour. *A. aethiopicum,* therefore, is yet another species exhibiting some ecological properties of pioneer plants. Various, essentially afromontane species of *Anthospermum* (e.g., *A. welwitschii* or *A. usambarense,* Fig. 62) are likely to have the same behavioural pattern.

Similar observations were also made on *Carpacoce spermacocea*: In the Ysternek Forest Reserve (E Cape Prov., S Africa), only a few old individuals were found in the dense hygrophilous fynbos community, which apparently had not been burnt for several years (PUFF 791222-2/1). This same locality was visited again nine months later (PUFF 800927-3/1). It had been burnt in the meantime and young plants of *C. spermacocea* were found in great abundance.

Sprouting species ("sprouters") grow rapidly after a fire and flower within the same season. On the summit of Pakhuispas (SW Cape Prov., S Africa), for example, plants of a population of *Carpacoce vaginellata,* which were found growing in dense, long unburnt mountain fynbos when

---

[1] Taxa belonging to this category are mostly (larger) single- to few-stemmed shrubs which do not produce innovation buds at or near the base of the stems.

[2] More detailed data are given by RICHARDSON & al. (1984): hot veld fire: August 23, 1980; first appearance of seedlings of *A. aethiopicum* after the fire: October 4, 1980.

Fig. 62. Regeneration from seed after fire. – Young plants of *Anthospermum usambarense* (ca 50–75 cm tall, mostly flowering for the first time) on slope burnt ca 18 months ago. *Phillipia* (vigorously resprouting from the base; note old, burnt stems) and bracken (*Pteridium aquilinum*, also resprouting) dominant in this community (Tanzania, South Pare Mts., Chome Forest; PUFF 820901-1/5)

the locality was visited for the first time (PUFF 790716-1/3), were burnt subsequently and had sprouted, grown vigorously and flowered when studied again 14 months later (PUFF 800914-6/1). *C. vaginellata,* however, is not an exceptional case. Numerous investigations of various fynbos communities have shown that sprouting species often account for the majority of plants in fynbos regenerating after fire (see, for example, VAN DER MERWE 1966 or TAYLOR 1972).

Table 10. Field observations on the fire survival of taxa of the *Anthosperminae*. In brackets a rating of the frequency of fires in habitats of a taxon is attempted: + + +: veld fires are likely to be a regular occurrence; + +: veld fires may occur occasionally; +: veld fires only occur under exceptional circumstances

A	B	C
Plants killed by fire; regeneration from seed	"Sprouters"	Indifferent: plants killed by fire or resprouting from the base after a fire

A

Anthospermum

aethiopicum (+ + +, + +; ADAMSON 1935, MARTIN 1966; RICHARDSON & al. 1984)
annamnioides (+ +, +)
emirnense (+, + +; occ. also B?)
galpinii (+ +; sometimes also B)
spathulatum (+, + +; rarely B)
ternatum (+ +, +; rarely B)
vallicola (+ +; occ. also B?)
welwitschii (+ +; sometimes also B)
usambarense (+ +, +)

Carpacoce

curvifolia (+, + +; occ. also B?)
spermacocea (+, + +)

B

Anthospermum

basuticum (+)
comptonii (+)
dregei subsp. ecklonis (+)
hispidulum (+)
pumilum (+ + +, + +, +)
whyteanum (+; occ. also A?)
zimbabwense (+ + +)

Carpacoce

burchellii (+ +; occ. also A?)
vaginellata (+ +)

Galopina

crocyllioides (+ +, + + +)

Nenax

microphylla (+; occ. also A?)
acerosa subsp. acerosa (+, + +)

C

Anthospermum

bergianum (+ +; more commonly A?)
galioides (+, + +)
herbaceum (+, + +; B: RICHARDSON & al. 1984)
hirtum (+, + +)
isaloense (+ + +; more commonly A?)
paniculatum (+ +; more commonly B?)

Galopina

aspera (+ +; more commonly B?)

"Sprouters" in the genus *Anthospermum* which occur outside the SW Cape Floristic Region (e.g., *A. pumilum* subsp. *pumilum* and *A. hispidulum*, cf. Figs. 7 c, 8 a and 8 c–d respectively) behave identically. For *A. herbaceum*, for example, it was documented that plants resprouting after a fire had attained a height of 75 cm within 4 months of the fire (RICHARDSON & al. 1984: graph, Fig. 5, as *A. lanceolatum*).

In Table 10 an attempt is made to summarize the present knowledge of fire survival properties in taxa of the *Anthosperminae*. It must be stressed, however, that the information presented here is far from complete. Data are often based on observations of few or even only a single population, and for numerous taxa no reliable field data are available. Some amendments may be necessary once additional information becomes available. Especially the category "sprouters" may require further special attention. Resprouting after fire may not always be obligatory; it may be that "sprouters" under certain circumstances and in some habitats, behave as "seed regenerators". As sprouting species have innovation buds at or close to (but not below) the soil surface, it might be expected that exceptionally hot fires (MARTIN 1966, for example, records a momentary rise in surface temperature to above 500 °C during a fire) – or, under certain circumstances, even average fires? – will kill the perennating organs. It can also be imagined that "out of season" fires (i.e., occurring in dry spells in the rainy season rather than in the dry season before the rains) may kill "sprouters" (which, in the *Anthosperminae* in general, do not need a fire to trigger the production of innovation shoots towards the onset of the rainy season).

D. Systematic Part

Tribe *Anthospermeae* CHAM. & SCHLECHTEND. ex Dc., Prod. **4**: 343, 578 (1830); HOOK. f. in BENTH. & HOOK. f., Gen. Pl. **2**(1): 26 (1873); K. SCHUM. in ENGLER & PRANTL, Nat. Pflanzenfam. **IV, 4:** 127 (1891); emend. PUFF in Bot. J. Linn. Soc. **84**: 370 (1982).

= *Anthospermeae* CHAM. & SCHLECHTEND. (pro "sectio") in Linnaea **3**: 309 (1828).

Characteristics: see PUFF (1982 a).

Subtribe *Anthosperminae* BENTH., Fl. Austral.: 401, 429 (1867); emend. PUFF in Bot. J. Linn. Soc. **84**: 370 (1982).

Characteristics: see PUFF (1982 a).

Key to the African and Madagascan Genera[1]

1. Mature fruits separating into exocarp-valves and endocarp plus seed; flowers ⚥, ♀ or (less commonly) ♂, mostly with only 1 fertile carpel and 1 stigma; corollas 4–7-merous, lobes distinctly hooded; calyx lobes mostly large, leaf-like. S Africa, SW Cape Floristic Region only .... **4. *Carpacoce*** (p. 446)

1*. Fruits either separating into 2 mericarps or indehiscent; flowers ♂, ⚥, ♀, sexes variously distributed, with 2 fertile carpels[2] and 2 stigmas; corollas 4–5-merous, lobes not hooded; calyx lobes mostly small or subobsolete .... 2.

2. Perennial herbs; leaves strictly decussate, blades relatively large and thin, distinctly petiolate; inflorescences terminal, paniculate to thyrso-paniculate; fruits without calyx lobes, separating into 2 mericarps, never supported by carpophores; SE Africa ....**3. *Galopina*** (p. 427)

2*. Large shrubs, dwarf shrubs, short-lived shrubs, subshrubs or perennial herbs; leaves decussate or in whorls of 3 or 4, blades mostly rather small and narrow, frequently ericoid, often without distinct petioles; inflorescences mostly variously congested, frequently much

---

[1] Excluding the Macaronesian genus *Phyllis*.

[2] Except for two *Anthospermum* species.

reduced and inconspicuous, in dioecious taxa often dimorphic (more conspicuous in ♀ than in ♂); fruits often with small calyx lobes, separating into two mericarps and supported by carpophores or indehiscent and without carpophores .... 3.

3. Growth form and leaf size, shape and arrangement variable; flowers ♂, ⚥, ♀, sexes variously distributed; ♂, ⚥ with long, cylindrical to short, broadly funnel-shaped to ± campanulate tubes; ⚥, ♀ mostly with greyish- to whitish-(green) stigmas; fruits always dehiscing into 2 mericarps, always supported by carpophores; widely distributed in Africa and Madagascar .... **1. Anthospermum**

3*. Many-stemmed, often intricately branched, dioecious dwarf shrubs; leaves frequently ericoid, mostly decussate; inflorescences always much reduced, partial inflorescences 1–3-flowered; ♂ flowers never with long, cylindrical tubes; ♀ often with reddish-purplish stigmas; fruits dehiscent or indehiscent, sometimes distinctly inflated, not always supported by carpophores; S Africa, centered in the SW and W Cape Prov. .... **2. Nenax** (p. 393)

Anthospermum L., Sp. Pl.: 1058 (1753), Gen. Pl., ed. 5: 479 (1754); SONDER in HARVEY & SONDER, Fl. Cap. **3**: 26 (1865); HOOK. f. in BENTH. & HOOK. f., Gen. Pl. **2** (1): 140 (1873); K. SCHUM. in ENGLER & PRANTL, Nat. Pflanzenfam. **IV, 4**: 129 (1891); DYER, Gen. S. Afr. Flow. Pl. **1**: 622 (1975); VERDC. in Fl. Trop. E. Afr., *Rubiaceae* **1**: 324 (1976); PUFF in Fl. S. Afr. **31**, 1 (2): 8 (1986).
Type species: *A. aethiopicum* L., Sp. Pl.: 1058 (1753).

− *Ambraria* HEISTER [Syst. Plant.: 11 (1748)] ex FABRICIUS, Enum. Meth. Pl. (Hort. Med. Helmstad.), ed. 2: 435 (1763) [based on *A. aethiopicum*]; non *Ambraria* CRUSE [see *Nenax*].

Dioecious (♂, ♀) or non-dioecious (⚥, ⚥ + ♀, ♀, or occasionally ♂ + ⚥ + ♀, ⚥ + ♂, ♀ + ♂ or ♂) shrubs, dwarf shrubs, short-lived shrubs subshrubs, or perennial herbs. Leaves decussate or occasionally in whorls of 3 (rarely 4), often in seemingly larger numbers at the nodes due to much-contracted, leafy short shoots ("pseudo-verticillate"), blades ± broad and large to ± ericoid and small, shortly petiolate to sessile, with ± cup-shaped stipular sheaths with one to many setae or fimbriae on either side. Inflorescences frequently leafy and rather inconspicuous, made up of mostly subsessile, many- to very few-flowered cymes, in dioecious taxa often sexually dimorphic (♀ inflorescences contracted, ± cylinder-like; e.g., Fig. 54). Flowers mostly subsessile, subtended by a pair of leafy bracts, ♂, ⚥ or ♀, 4–5-merous; ♂, ⚥: corolla tubes cylindrical, broadly funnel-shaped to ± campanulate, ± long to rather short, lobes ± lanceolate, recurved, anthers yellowish to whitish, exserted, dangling on long slender filiform filaments; ♀: corollas much smaller, tubes cylindrical, lobes mostly erect, linear to ± lanceolate; ⚥, ♀: style 0 or very short; stigmas 2 (only in *A. ericifolium* 1), long exserted, hairy, greyish- to whitish-(green), seldom purplish-red; ovary bicarpellate and biovulate (only in *A. ericifolium* and *A. bicorne* 1 carpel reduced), crowned by 4–5 large, conspicuous to small, indistinct calyx lobes, or calyx lobes ± lacking. Fruits crowned by the persistent calyx lobes, supported by ± U-shaped carpophores, dehiscing into two mericarps.

Chromosome Number (Table 3): $x = 11$; $n = 11$, $2n = 22$ (seldom $n = 22, 33, 2n = 44, 66$).

Average Pollen Diameter (Fig. 48 and 49): 23.1–35.6 μm.

Distribution (Maps, Fig. 1 a and 63): Widely distributed in Africa S of the Sahara; the highest concentration of taxa is found in S Africa. One species (*A. herbaceum*) extends into S Arabia (Yemen Arab Republic and adjacent parts of Saudi Arabia); several taxa are endemic to Madagascar.

Critical Remarks: From a taxonomist's point of view, the species of *Anthospermum*, in general, are "difficult". Especially those unfamiliar with the genus and without field knowledge of the plants could have problems in identifying and distinguishing certain species. Some of the reasons for this are:

– The considerable environment-induced morphological variability in numerous taxa: individuals exposed to fire vs. plants sheltered from fire (e.g., *A. pumilum* subsp. *pumilum*, Figs. 8 a and b); browsed plants vs. not browsed individuals (e.g., *A. comptonii*, Fig. 96). See C.2.5. for details.

– More widely distributed taxa often contain a number of "Forms"[1] which may be very distinct locally. The are, however, often connected with other "Forms" by a series of morphologically intermediate populations so that it is not possible to clearly define and delimit them or to recognize them taxonomically. The more prominent of these are pointed out in the text and sometimes documented photographically (e.g., *A. herbaceum*, Figs. 80–82; *A. galioides*, Figs. 92–94); some are also included in the keys.

– The occurrence of characters which are constant in some but

---

[1] In part D and E the term is very often used in the sense of ecotypes or geographical races. Its application, therefore, is ± comparable to that of "variant" in Babcock's (1947) *Crepis* monograph. It is not intended to denote a specific taxonomic category.

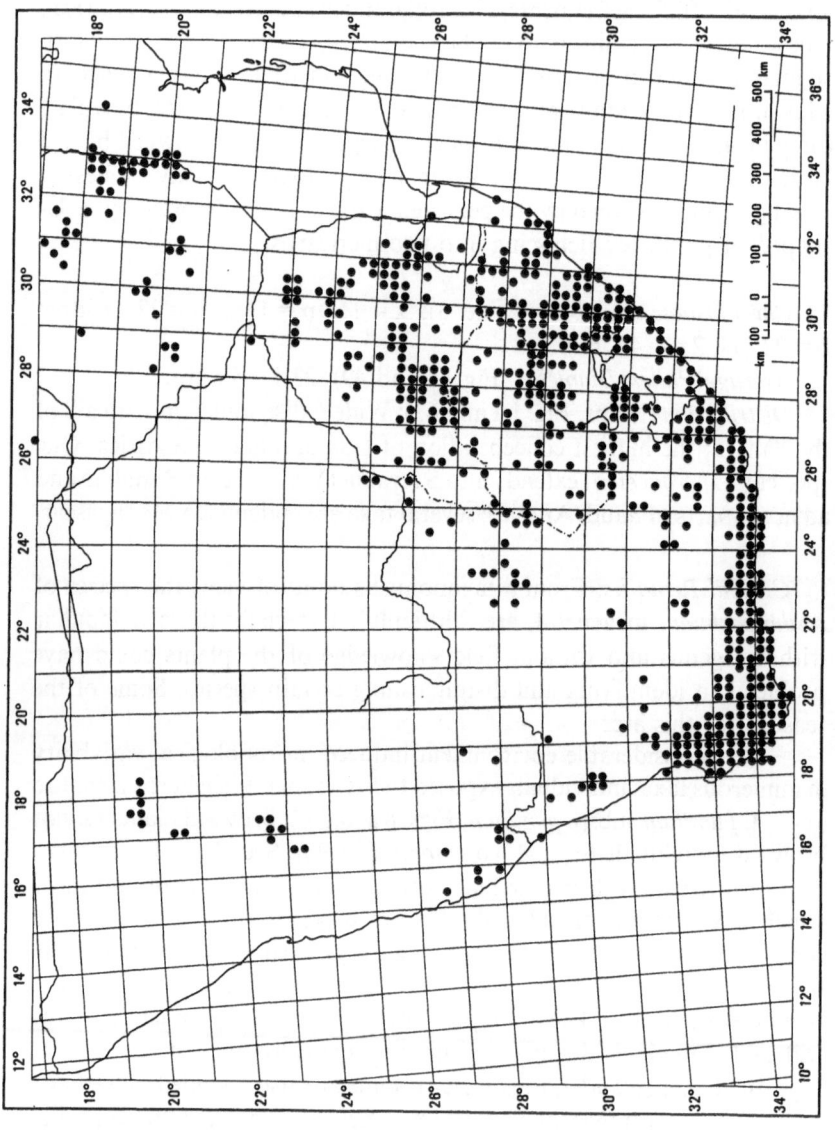

Fig. 63. Distribution of *Anthospermum* in S Africa (all taxa)

variable in other taxa. Notably the arrangement of the leaves – e.g., decussate *or* in whorls of 3 within a species (and sometimes even within a single population), or either strictly decussate or strictly in whorls of 3 in given taxa (see C.4.1. for further details). For this reason (and also in view of the variable sex distributions within taxa; see below) populations should be studied with care in the field and representative samples should be collected.

   – The occasional occurrence of flowers "transitional" between ☿ and pure ♀ (with corollas intermediate in size between ☿ and ♀!); the occasional reversal of ♂ flowers to ☿ in some essentially dioecious taxa; the marked protandry of ☿ flowers which may makes it difficult to decide in a dried specimen whether a flower is ♂ or ☿; the variability of sex distributions in populations of non-dioecious taxa. See C.12.2. for details.

   – The sex dimorphism in dioecious taxa which may extend beyond corolla size and shape differences to dimorphic inflorescences (see, for example, Fig. 54!) and, occasionally, to slight leaf size and shape differences (see C.12.3. for details). It is, therefore, recommended that both ♂ and ♀ are collected whenever possible.

Although there are a number of "recognizable", presumably allied species groups within *Anthospermum*, I decided against a formal further subdivision of the genus into subgenera and sections for the following reasons: None of these species groups are so well-defined in their character states that an unambiguous separation from another group is possible. The characters, in general, are too variable within each of these groups to allow a certain distinction. Various "trends" repeat themselves in different groups (for example, the trend to dioecy in essentially non-dioecious species groups at populational level). Thus, in my opinion, a further subdivision would only result in additional confusion, as especially someone who is not very familiar with the genus would probably experience considerable difficulties in placing a given taxon into the correct subgenus or section.

   In the following systematic treatment it was, nevertheless, attempted to group together presumably allied species.

   For convenience, 5 ± regional keys are provided which cover the respective Flora areas (Flora of West Tropical Africa; Flora du Zaire, Rwanda et Burundi . . . Flore de Madagascar et des Comores). Taxa which extend from one area to another or which are more widely distributed appear in each of the relevant keys, whereby regional "peculiarities" are taken into account (i.e., leaf size and shape data for one and the same taxon are not necessarily the same in different areas).

   Because of the lack of other suitable and usable characters in herbarium material, it is often necessary to use calyx lobe, corolla and fruit sizes or hair lengths as differential characters. In this context it should be noted that measurements exact to tenths of millimeters are essential for correct identifications.

Regional Keys

A. Key to Taxa Occurring in Tropical Africa

(Flora of Tropical East Africa and Flora of Central Africa area; W Africa, Sudan and Ethiopia)

1. Leaf surfaces ± densely covered with spreading hairs ca 0.2–0.6 (0.9) mm long; leaves mostly in whorls of 3; mericarps hairy............. 8.

1*. Leaves glabrous or with hairs on the margins and/or the midrib below or on the lower surface only, rarely with scattered, very short hairs on the upper surface; leaves decussate or in whorls of 3; mericarps glabrous, papillate or hairy...................................... 2.

2. Often scrambling, straggling or trailing perennial herbs with some-times ± woody rootstocks; fruits with ± broad, conspicuous longitudinal grooves between the mericarps (Fig. 37 j) (variable and widely distributed in E tropical Africa)... **25. A. herbaceum** (p. 300)

2*. Neither growth form nor fruits as above .... 3.

3. Erect, mostly single-stemmed shrubs to 3 (4.5) m tall; plants usually dioecious .... 4.

3*. Many- to several-(few-)stemmed, rounded dwarf shrubs, subshrubs, short-lived shrubs or "woody herbs" .... 5.

4. Leaves decussate, (7) 10–35 × (1) 1.5–3.5 mm, oblanceolate to ± linear-lanceolate, tough and ± leathery and with conspicuous but small upper epidermis cells; mericarps with calyx lobes to 1 (1.2) mm long; to 2 400 (2 700) m [Kenya (K 3, K 6), Tanzania (T 2, T 7) and Zaire (Haut-Katanga); distribution disjunct]....................................
....**1. A. welwitschii**[1] (p. 190)

4*. Leaves in whorls of 3 or, less often, decussate, (2) 3–10 (15) × 0.5–

---

[1] *A. welwitschii* and *A. usambarense* can form viable hybrids; putative hybrids will either key out to one or the other species. Extensive hybrid populations are believed to exist in Tanzania (T 7). See *A. welwitschii* and *A. usambarense*, Critical Remarks, for details.

1.5 (2) mm, narrowly obovate to ± linear, ± thin, not with conspicuous upper epidermis cells; mericarps with calyx lobes to 0.6 mm long; to 4 100 (4 500) m (widespread in the tropical E African mountains) .... **2. A. usambarense**[1] (p. 196)

5. Mericarps with 2 unequal calyx lobes to 1.3 (1.7) mm long; mostly ± rounded dwarf shrubs or subshrubs (Ethiopia and Sudan) ....
.... **26. A. pachyrrhizum** (p. 321)

5*. Mericarps not crowned by distinct calyx lobes or calyx lobes minute
.... 6.

6. Subshrubs with numerous short, unbranched stems to ca 15 cm long from an often ± massive, disk-like woody base; mericarps to 2 mm long, covered with scattered, ± hook-like hairs [only one collection from Tanzania (T 7)] ..... **29 (a). A. pumilum subsp. pumilum** (p. 330)

6*. Erect subshrubs, short-lived shrubs or "woody herbs" with several to few longer stems; mericarps 1.9–3 mm long, glabrous or covered with whitish spreading hairs .... 7.

7. Leaves ± linear and sometimes ± needle-like due to strongly revolute margins, (4.5) 6–25 (30) × (0.5) 0.8–1.2 (1.5) mm; stipular sheaths with 1 hairy, subulate seta [Tanzania (T 4, T 7, T 8) and Zaire (Haut-Katanga)] .... **19 (a). A. ternatum subsp. ternatum** (p. 283)

7*. Leaves (narrowly) oblanceolate to ± linear-lanceolate, 5–10 (15) × 0.8–1.5 (2) mm; stipular sheaths with mostly 3–5 glabrous setae (W Africa) .... **22. A. asperuloides** (p. 294)

8. Leaves (20) 30–50 × 3–7 (10) mm, linear-lanceolate to linear-oblong [Tanzania (T 4, T 7) and Zaire] .... **20. A. rosmarinus** (p. 289)

8*. Leaves (2) 3–13 (15) × 0.8–3 (5) mm, oblanceolate, narrowly elliptic-lanceolate to ± linear-lanceolate[2] .... 9.

9. Leaves 10–13 × 1–3 mm, narrowly elliptic-lanceolate; mericarps 2.4–2.7 × 1.2–1.3 mm, obovate, not crowned by distinct calyx lobes [Kenya (K 1) and S Ethiopia] .... **21. A. villosicarpum** (p. 292)

9*. Leaves (2) 3–11 (15) × 0.8–3 (5) mm, oblanceolate to ± linear-lanceolate; mericarps (1.6) 1.8–2.2 × 0.7–1 mm, ± oblong, mostly with 2 calyx lobes (0.2) 0.3–0.7 mm long [Tanzania (T 8)] ....
.... **16. A. whyteanum** (p. 268)

---

[1] See footnote on previous page.

[2] The sterile collection Burtt 1303 (K) from Tanzania, Kondoa Distr. (T 5) will key out here. See 19 (b). *A. ternatum* subsp. *randii*, Critical Remarks.

B. Key to Taxa Occurring in the Flora Zambesiaca and the
Flora of Angola Area

(Zambia, Malawi, Mozambique, Zimbabwe; Angola)

1. Erect, mostly single-stemmed shrubs ca 1–3 (4.5) m tall.............. 2.
1*. Several- to many-stemmed dwarf shrubs, subshrubs, "woody herbs"
    or perennial herbs .... 6.

2. Leaves decussate .... 3.
2*. Leaves in whorls of 3.. 4.

3. Leaves 10–35 × (1) 1.5–3.5 mm, strictly decussate [Angola; Zambia
    (W, E) and Malawi (N, S)].........................**1. A. welwitschii** (p. 190)
3*. Leaves (2) 3–10 (15) × 0.5–1.5 (2) mm, occasionally decussate, more
    commonly in whorls of 3; plants mostly densely leafy above,
    "ericoid" in appearance [Zambia (E) and Malawi (N)] `................
    ....**2. A. usambarense** (p. 196)

4. Leaves 0.5–2 (2.5) mm wide, mostly glabrous............................. 5.
4*. Leaves (2) 2.5–10 (15) mm wide, surfaces very shortly hairy
    [Zimbabwe (E) and Mozambique (MS)] ..... **5. A. ammannioides** (p. 209)

5. Leaves (2) 3–10 (15) mm long; stipular sheaths with 3–7 (8) setae;
    calyx lobes ca 0.2–0.6 mm long [Zambia (E) and Malawi (N)]........
    ....**2. A. usambarense** (p. 196)
5*. Leaves (10) 15–35 (55) mm long; stipular sheaths mostly with 1 seta;
    calyx lobes (0.5) 0.8–1.8 (2.1) mm long [Zimbabwe (E) and
    Mozambique (MS)].....................................**6. A. vallicola** (p. 212)

6. Plants dioecious; leaves always in whorls of 3, glabrous, to 8 (12) mm
    long; ♂ corolla tubes (± broadly) funnel-shaped, 0.5–1 mm long;
    mericarps 1.4–1.7 mm long [Zimbabwe (E) and Mozambique (MS)]
    ....**3. A. zimbabwense** (p. 201)
6*. Plants not dioecious; leaves decussate or in whorls of 3 (4), if in whorls
    of 3 either variously hairy or glabrous and to 25 (30) mm long; ♂ and ♀
    corolla tubes cylindrical to (narrowly) funnel-shaped, to 3.7 mm long;
    mericarps to 3 mm long .... 7.

7. Leaves on both surfaces densely covered with spreading hairs 0.2–
    0.5 (0.9) mm long; mostly distinct dwarf shrubs; mericarps usually
    crowned by calyx lobes ca (0.2) 0.3–0.7 mm long [widely distributed

from Malawi (N) and Zambia (E) S to Zimbabwe]........................
....**16. A. whyteanum** (p. 268)

7*. Leaves glabrous or variously hairy (but not as above); perennial herbs or subshrubs; mericarps not crowned by distinct calyx lobes or calyx lobes minute, to 0.3 mm long.. 8.

8. Leaves in whorls of 3, decussate or, occasionally in whorls of 4; stipular sheaths with 1 hairy, subulate seta ca (0.7) 1–3 (3.7) mm long; mericarps 2–3 × 0.8–1.2 mm .... 9.

8*. Leaves strictly decussate; stipular sheaths with 1 shorter, glabrous seta or 5 (–7) setae; mericarps smaller and broader.................. 10.

9. Leaves ± linear and sometimes ± needle-like due to strongly revolute margins, (4.5) 6–25 (30) × (0.5) 0.8–1.2 (1.5) mm; fruits mostly hairy, at least near the apex [Angola; Zambia (N, E), Malawi (N, S), Mozambique (N) and Zimbabwe (N)]................................
.....................................**19 (a). A. ternatum subsp. ternatum** (p. 283)

9*. Leaves ovate- or oblong-lanceolate to ± linear-lanceolate, (10) 12–45 (55) × 1.5–5 (8) mm; fruits glabrous or hairy (widely distributed)
....**19 (b). A. ternatum subsp. randii** (p. 285)

10. Often scrambling, straggling or trailing perennial herbs; leaves 5–55 × (1) 2–25 mm, ovate to lanceolate; ♂ and ♀ corolla tubes cylindrical, (1.5) 2–3.7 mm long; fruits with ± broad, conspicuous longitudinal grooves between the mericarps (Fig. 35 j) [widely distributed from Zambia (N) and Malawi (N) to Zimbabwe and Mozambique (MS)]................................ **25. A. herbaceum** (p. 300)

10*. Mostly subshrubs with numerous short, unbranched stems; leaves (4) 6–12 (22) × (0.5) 0.8–1.5 (2) mm, linear-lanceolate; ♂ and ♀ corolla tubes narrowly funnel-shaped, (0.5) 0.7–1.4 (1.7) mm long; fruits not as above [Angola; Zambia (S) and Zimbabwe (C, E, W)] ....
.....................................**29 (a). A. pumilum subsp. pumilum** (p. 330)

C. Key to Taxa Occurring in the Flora of Southern Africa Area Excluding the SW Cape Floristic Region

1. Flowers in distinct terminal thyrsic or thyrso-paniculate inflorescences (from Transkei and E Cape Prov. to the SW Cape).
....**15. A. paniculatum** (p. 264)

1*. Flowers in clusters of many to few at nodes, or ♀ flowers sometimes in ± condensed, cylindrical inflorescence zones............................. 2.

2. Plants of coastal sand dunes; at least lower parts of stems usually buried in sand; mericarps densely covered with spreading hairs to 0.8 mm long.. **10. A. littoreum** (p. 224)

2*. Habitat and habit not as above; mericarps glabrous or variously hairy, but not as above... 3.

3. Often scrambling, straggling or trailing perennial herbs; ♀ and ♂ flowers with cylindrical corolla tubes to 3.7 mm long......................
    .... **25. A. herbaceum** (p. 300)

3*. Shrubs or subshrubs, erect or mat- or cushion-forming perennial herbs; ♀ and ♂ flowers with much shorter and broader (funnel-shaped) corolla tubes or, if tubes long and cylindrical, habit not as above... 4.

4. Leaves densely covered with spreading hairs ca (0.1) 0.3–0.5 mm long
    .... 5.

4*. Leaves glabrous, papillate, or hairs on the margins and/or the midvein only .... 6.

5. Leaves   2.5–5 (6) × 0.8–1.2 (1.8) mm;   mericarps   2–2.5 × 1.5 mm; dioecious, cushion- or mat-forming dwarf shrubs of high altitudes (high Drakensberg areas of the Cape Prov., Natal, Lesotho and the O.F.S.)...**14. A. basuticum** (p. 262)

5*. Leaves 5–12 (20) × 1–3.5 (5) mm; mericarps 1.7–2 × 0.7–1 mm; plants not dioecious, not occurring above 2 000 (2 300) m (from the Transvaal and Swaziland to the Transkei)**17. A. hispidulum** (p. 274)

6. Low, ± cushion-forming dwarf shrubs, rooting at the nodes, either occurring at high altitudes (above 2 400 m; Lesotho) or much-browsed...**13. A. monticola** (p. 257)

6*. Growth form different, plants not rooting at the nodes, confined to lower altitudes or, if at high altitudes, perennial herbs.............. 7.

7. Rounded dwarf shrubs, low subshrubs or perennial herbs; mericarps roundish in side view; plants not dioecious................................. 8.

7*. Single- to several-stemmed, mostly erect shrubs to 3 m tall; mericarps elongated in side view; plants dioecious .... 14.

8. Leaves linear to narrowly ovate-lanceolate, to 1.5 (2) mm wide; mericarps glabrous or with ± hooked or curled hairs................ 9.

8*. Leaves broader, often ovate or ovate-lanceolate, to 6 (10) mm wide; mericarps glabrous, ± tuberculate or papillate, but never with curled or hooked hairs... 11.

9. Leaves mostly distinctly needle-like, to 0.8 (1) mm wide; often mat- or cushion-forming, very densely leafy dwarf shrubs (Natal S Coast only) .... **30. *A. streyi*** (p. 347)

9*. Leaves to 1.5 (2) mm wide, sometimes linear but not distinctly needle-like; subshrubs with numerous unbranched stems or ± robust dwarf shrubs .... 10.

10. Mostly subshrubs with numerous unbranched stems; leaves (4) 6–12 (22) × (0.5) 0.8–1.5 (2) mm, linear to narrowly lanceolate; ♀ and ♂ flowers with corolla tubes to 1.4 (1.7) mm long, lobes to 1.9 (2.5) mm long (from Transvaal to E and NE Cape) ....
.... **29 (a). *A. pumilum* subsp. *pumilum*** (p. 330)

10*. Quite robust dwarf shrubs; leaves 3–6 (8) × 1–1.5 mm, ovate-lanceolate, often tough and thickish; ♀ and ♂ flowers larger, corolla tubes to 1.5 (1.9) mm long, lobes to 3 (3.2) mm long (O.F.S. and Cape Prov. excl. the SW Cape) **29 (b). *A. pumilum* subsp. *rigidum*** (p. 342)

11. Diffusely and divaricately branched robust dwarf shrubs with tough, thickish, ovate to oblong-lanceolate leaves to 4 (4.5) mm wide; fruits glabrous, without ± broad and conspicuous longitudinal grooves between the mericarps (Namaqualand and S Namibia) ....
.... **33 (a). *A. dregei* subsp. *dregei*** (p. 369)

11*. Low cushion- or mat-forming perennial herbs or ± erect and tufted subshrubs; fruits mostly tuberculate and with ± broad and conspicuous longitudinal grooves between the mericarps (cf. Fig. 35 j)
.... 12.

12. Leaves ± succulent in nature; stipular sheaths often with only 1 seta; near the sea (Natal and Transkei) ....
.... **25. *A. herbaceum*** ("Salt Spray Zone Form") (p. 300, 311)

12*. Leaves not succulent; stipular sheaths with several setae; growing at high altitudes or in (burnt) grassland .... 13.

13. Low, matted or cushion-forming perennial herbs; leaves small; partial inflorescences few- to one-flowered ....**25. *A. herbaceum*** ("High Altitude" or "Trampled Grassland Form") (p. 300, 309)

13*. ± Erect, several- to many-stemmed subshrubs with ± massive, almost rosette-like ± woody bases ....
.... **25. *A. herbaceum*** ("Burnt Grassland Form") (p. 300, 309)

14. Leaves in whorls of 3 .... 15.

14*. Leaves decussate .... 16.

15. Leaves to 9 (13) mm long, strictly in whorls of 3; mericarps glabrous or papillate; ♀ plants mostly with conspicuous, condensed cylindrical inflorescence zones (SW Cape species, barely extending into the E Cape Prov.) .... **11. *A. aethiopicum*** (p. 227)

15*. Leaves to 20 (27) mm long, sometimes also decussate; mericarps densely covered with straight hairs ca 0.1–0.2 mm long; ♀ plants mostly not with very conspicuous, condensed inflorescences (Natal to E Cape Prov.) .... **9. *A. galpinii*** (p. 220)

16. Leaves (0.5) 0.8–1.5 (2) mm wide, narrowly oblanceolate to ± linear; mericarps densely covered with straight hairs ca 0.1–0.2 mm long .. ..**9. *A. galpinii*** (p. 220)

16*. Leaves wider or, if linear (-lanceolate), mericarps not shortly hairy .... 17.

17. Leaves (7) 10–35 × (1) 1.5–3.5 mm, oblanceolate to elliptic; stipular sheaths with 3–5 (–7 or 8) setae; ♀ plants mostly with conspicuous, condensed cylindrical inflorescence zones; usually large, single-stemmed shrubs (Transvaal; and Cape: Mafeking?)........................ ....**1. *A. welwitschii*** (p. 190)

17*. Leaves (2) 3–12 × 0.5–2 mm; stipular sheaths mostly with a single median seta; ♀ plants usually not with conspicuous cylindrical inflorescence zones; mostly small, several- to many-stemmed shrubs .... 18.

18. Leaves (ob)ovate, oblong or (ob)ovate-lanceolate, (0.7) 1–2 mm wide; fruits with conspicuous calyx lobes 0.3–0.6 (0.8) mm long (Natal, Lesotho, O.F.S. and interior of the Cape Prov. to the Nieuweveld Mts.)..**13. *A. monticola*** (p. 257)

18*. Leaves lanceolate to linear-lanceolate, 0.5–1.5 mm wide; fruits with indistinct or smaller calyx lobes... 19.

19. Mericarps narrowly obovate to oblong, not distinctly ribbed on the dorsal side, often seemingly truncate above (E Cape Prov.; Humansdorf Distr. to the Groot Kei).. ....**12 (b). *A. spathulatum* subsp. *uitenhagense*** (p. 245)

19*. Mericarps broader, mostly with 3 ± distinct ribs on the dorsal side and with calyx lobes to 0.4 mm long (SW Cape taxon, extending into the E Cape Prov.) ....**12 (a). *A. spathulatum* subsp. *spathulatum*** (p. 234)

D. Key to Taxa Occurring in the SW Cape Floristic Region[1]

1. Leaves in whorls of 3.. 2.

1*. Leaves decussate .... 3.

2. Tall shrubs with small, ± needle-like glabrous leaves to 1.2 (2) mm wide; corolla lobes 4 (widely distributed) .... **11. A. aethiopicum*** (p. 227)

2*. Small short-lived shrubs or subshrubs; leaves mostly ± imbricate, to 3 (3.5) mm wide, at least the margins ciliate; corolla lobes 5 (SW Cape s.str.) .... **37. A. bergianum*** (p. 384)

3. Fruits with only 1 fertile and well developed carpel/mericarp (SW Cape s.str.)... 4.

3*. Fruits with two fertile carpels/mericarps.................................... 5.

4. Reduced carpel with 2 calyx lobes ca 1.4–2 mm long, fertile carpel with 3 much smaller, indistinct calyx lobes; leaves needle-like, to 0.5 mm wide ....**39. A. bicorne*** (p. 392)

4*. Calyx lobes of fertile and reduced carpel subequal, ca 0.2–0.6 mm long; leaves linear-lanceolate, to 1.2 (1.4) mm wide........................
....**38. A. ericifolium*** (p. 389)

5. Flowers in distinct terminal thyrsic or thyrso-paniculate inflorescences (from the George Distr. to the E Cape Prov. and the Transkei)..**15. A. paniculatum** (p. 264)

5*. Flowers in clusters of many to few at nodes, or (seldom) solitary and terminal; ♀ flowers sometimes in ± condensed, cylindrical inflorescence zones... 6.

6. Plants prostrate (but not mat-forming), with long, trailing stems radiating out from a common base, rooting at the nodes and bearing short, ascending to ± erect lateral branches; dioecious; only in coastal dune sand (from the W Coast to the Port Elizabeth Distr.)
.... **32. A. prostratum*** (p. 364)

---

[1] The geographical delimitation of the Cape Floristic Region follows BOND & GOLDBLATT (1984).

Taxa endemic to the SW Cape or nearly so (i.e., sometimes extending to the E Cape Prov., etc.) are marked with as asterisk.

"SW Cape *s. str.*" refers to the area (roughly) from the Clanwilliam Distr. S to the Cape Peninsula and to the Caledon and Bredasdorp Distr.

6*. Growth form different; if plants prostrate, stems not rooting at the nodes, not with short ascending to erect lateral branches and not dioecious .... 7.

7. Leaves, stipules and younger stems (or at least leaf margins near base) ± densely covered with whitish spreading hairs ca 0.2–1 (1.5) mm long; corollas 5-merous.. 8.

7*. Leaves and stems glabrous, very shortly hairy or papillate; if leaves with hairs to 0.4 mm long, hairs confined to the margins and corollas 4-merous.. 10.

8. Leaves 4–6 (7.5) × (1.4) 2–3 (4) mm, ovate to lanceolate; flowers solitary and terminal on branches, often overtopped by short branches arising from below; plants usually mat-forming (confined to the higher parts of the SW Cape mountains)............................
................................**35. A. esterhuysenianum var. hirsutum*** (p. 379)

8*. Leaves larger, (4) 5–25 (30) × (0.5) 0.8–3 (4.5) mm, ovate-lanceolate to linear-lanceolate; flowers in axillary clusters; plants not mat-forming.. 9.

9. Plants mostly dioecious (♂ plants occasionally with odd ♀ flowers), densely leafy; leaves mostly ± imbricate, blades (4) 5–12 (15) × (1.2) 1.5–3 (3.5) mm; ♀ plants with ± condensed, cylindrical inflorescence zones.................................**37. A. bergianum*** (p. 384)

9*. Plants not dioecious, not densely leafy; blades larger, (6) 8–25 × (0.5) 0.8–3 (4.5) mm; ♀ inflorescences inconspicuous ....
....**36. A. hirtum*** (p. 380)

10. Plants dioecious; ♀ plants sometimes with ± conspicuous cylindrical inflorescence zones.. 11.

10*. Plants not strictly dioecious; sex distributions variable (☿, ☿ + ♀, ♀, etc.); ♀ plants never with conspicuous, ± condensed cylindrical inflorescence zones.. 16.

11. Tall single-stemmed shrubs to many-stemmed dwarf shrubs; ♂ corolla tubes short, to 1.2 mm long, (broadly) funnel-shaped to ± campanulate; mericarps narrow and elongated in side view, often ± triangular in section and 3-ribbed (very variable and very widely distributed).. 12.

11*. Subshrubs or low dwarf shrubs; ♂ corolla tubes ± cylindrical to narrowly funnel-shaped, to 1.7 (2) mm long; mericarps broad and roundish in side view, ± semi-terete in section, not ribbed...... 15.

12. Mericarps (1.5) 1.7–2.3 × 0.6–0.8 (1) mm, ± oblong, glabrous or shortly hairy, not distinctly ribbed dorsally; ♂ corolla lobes ca 1.2–2 (2.2) mm long; leaves (2.2) 2.5–4.5 (5.2) × 0.5–1 (1.3) mm (SW Cape s.str.) ....**12 (e). A. spathulatum** subsp. *tulbaghense** (p. 255)

12*. Mericarps larger, 2.2–4.2 × 1–2.2 mm, (broadly) obovate to ± oblong, often 3-ribbed dorsally, usually glabrous: ♂ corolla lobes (1.5) 1.9–3.4 mm long; leaves (3) 3.5–18 (25) × 0.5–2.5 (3) mm .... 13.

13. Leaves 8–18 (25) × 0.5–1.8 (2.2) mm, linear-(ob)lanceolate to linear, fine and ± needle-like; mostly single-stemmed (± narrowly) cylindrical, densely leafy shrubs ca 1–2 mm tall, in habit and appearance resembling *A. aethiopicum;* ♀ plants with rather conspicuous, condensed inflorescence zones .... **12 (d). A. spathulatum** subsp. *ecklonianum** (p. 251)

13* Leaves smaller and relatively broader, (3) 3.5–12 × 0.5–2.5 (3) mm; growth form not as above, plants often smaller, single- to many-stemmed shrubs of dwarf shrubs; ♀ plants with less conspicuous condensed inflorescence zones .... 14.

14. Several- to many-stemmed low cushion- or ± mat-forming dwarf shrubs with rather tough, closely spaced leaves; partial inflorescences 3- to 1-flowered; mericarps mostly with conspicuous calyx lobes to 1.5 (1.9) mm long (confined to the highest parts of the SW Cape mountains) ....**12 (c). A. spathulatum** subsp. *saxatile** (p. 248)

14*. Rather tall single-stemmed shrubs or small, often several-stemmed dwarf shrubs; partial inflorescences ± many- to few-flowered; mericarps with calyx lobes to 0.6 (0.9) mm long (variable and widely distributed)........**12 (a). A. spathulatum** subsp. *spathulatum** (p. 234)

15. Leaf blades distinctly recurved (at least near tip); ♂ corolla lobes 1.4–2.4 (2.9) mm long; mericarps ca 2–2.4 mm long (± widely distributed, especially in the mountains)... ....**31 (b). A. galioides** subsp. *reflexifolium** (p. 360)

15*. Leaf blades straight (but margins revolute); ♂ corolla lobes 2.7–3.2 mm long; mericarps 2.7–3 mm long (only in the interior, often in dry, karroid valleys)..............................**34. A. comptonii*** (p. 373)

16. Plants distinctly woody and dwarf shrubby, erect and ± cylindrical to ± rounded .... 17.

16*. (Short-lived) perennial herbs or subshrubs; if ± dwarf shrubs, plants prostrate and mat-forming.. 20.

17. Leaf blades distinctly recurved (at least near tip) (± widely distributed, especially in the mountains) ....
..............................**31 (b). A. galioides** subsp. **reflexifolium*** (p. 360)
17*. Leaf blades not distinctly recurved... 18.

18. Fruits papillate (or occasionally shortly hairy); blades narrowly lanceolate to linear (due to strongly revolute margins), ca 5–10 × (0.5) 0.8–1.5 mm, margins mostly glabrous (SW Cape s.str.)....
. **31 (a). A. galioides** subsp. **galioides*** ("Papillatum Form") (p. 353)
18*. Fruits mostly glabrous; blades broader, oblong or obovate-lanceolate (less commonly linear-lanceolate), 3–9 (12) × 1–2.5 (3) mm, margins often very shortly hairy............................ 19.

19. Several-stemmed, ± erect dwarf shrubs; mericarps 2–2.6 (2.8) mm long; ♂ and ♀ corollas glabrous, lobes (1.7) 2.2–3 (3.2) mm long (only in the Gouritz R. valley)..
....**29 (b). A. pumilum** subsp. **rigidum** (p. 342)
19*. Often rounded, low dwarf shrubs; mericarps (1.5) 1.8–2.2 mm long; ♂ and ♀ corollas mostly a little hairy outside, lobes 1.5–2.5 (3) mm long (only in the W part of the SW Cape s.str., confined to TMS areas)
....**33 (b). A. dregei** subsp. **ecklonis*** (p. 370)

20. Straggling, trailing or ± ascending, sometimes rather short-lived perennial herbs; leaves mostly ± lanceolate, to 30 mm long; stipular sheaths with 3–5 setae; ♂ and ♀ corolla tubes cylindrical, to 3.7 mm long; fruits often ± tuberculate and with ± broad and conspicuous longitudinal grooves between the mericarps (Fig. 35 j) (± afromontane species; from the Heidelberg Distr. E-wards) ....
....**25. A. herbaceum** (p. 300)
20*. Growth form different; leaves smaller, to ca 10 mm long; stipular sheaths mostly with 1 seta; ♂ and ♀ corolla tubes short, ± cylindrical, to 1.8 (2) mm long; fruits not as above..................................... 21.

21. Plants often ± mat-forming, stems prostrate to ± ascending; leaves relatively broad, to 3 (4) mm wide, ovate, oblong to ± lanceolate..
.... 22.
21*. Plants not mat-forming, stems erect to ± ascending; leaves narrower, to 2 (3) mm wide, linear-lanceolate to ovate-lanceolate............ 23.

22. Flowers solitary and terminal on branches, often overtopped by lateral shoots arising from below; corollas 5-merous (only in the higher parts of the SW Cape mountains)..
.............**35 (a). A. esterhuysenianum** var. **esterhuysenianum*** (p. 378)

22*. Flowers in axillary clusters; corollas 4-merous (not occurring in the
SW Cape mountains) ....
..... **31 (a). A. galioides** subsp. **galioides*** ("Prostrate Form") (p. 353)

23. Leaf blades distinctly recurved (at least near tip) ....
.... **31 (b). A. galioides** subsp. **reflexifolium*** (p. 360)
23*. Leaf blades straight, spreading to ascending (but margins
occasionally revolute) ... **31 (a). A. galioides** subsp. **galioides*** (p. 353)

E. Key to Taxa Occurring in Madagascar

1. Dioecious, mostly single-stemmed, erect shrubs ca (0.5) 0.75–2 m tall
.... 2.
1*. Several- to many-stemmed dwarf shrubs ca 7–40 (50) cm tall,
subshrubs or perennial herbs with ± woody rootstocks .... 4.

2. Leaves linear, ± needle-like due to strongly revolute margins, 0.3–
0.7 (1) mm wide .... **7. A. madagascariense** (p. 215)
2*. Leaves oblanceolate to linear-lanceolate, not ± needle-like, 1.2–
4 (5) mm wide .... 3.

3. Mericarps not crowned by distinct calyx lobes or calyx lobes minute,
ca 0.1–0.2 mm long; ♀ plants often with ± condensed, cylindrical inflorescence zones; leaves ± thin, upper surface not shiny and without large, ± conspicuous epidermis cells .... **4. A. emirnense** (p. 206)
3*. Mericarps with 2 calyx lobes ca 1–1.7 mm long; ♀ inflorescences not as
above; leaves rather coriaceous, upper surface usually shiny and often
with large, ± conspicuous epidermis cells ..... **8. A. isaloense** (p. 218)

4. Ovaries and fruits crowned by 4 (very rarely 5) narrow, ± band-like
calyx lobes ca 1.2–3.7 mm long .... 5.
4*. Ovaries and fruits not crowned by distinct calyx lobes .... 6.

5. Leaves and fruits glabrous; calyx lobes 1.2–1.7 (2) mm long ....
.... **23. A. ibityense** (p. 298)
5*. Leaves and fruits densely papillate or very shortly hairy; calyx lobes
ca 2.2–3.7 mm long .... **24. A. longisepalum** (p. 299)

6. Stems and leaves hairy .... 7.
6*. Stems scabrid and leaves glabrous .... 8.

7. Leaves (6.5) 8–12 (15) × (1) 1.5–2.5 (3) mm, with hairs ca 0.3–0.7 mm long; mericarps 1.7–2.4 mm long, ± oblong in side view, covered with whitish, ± spreading hairs to ca 0.2 mm long, commissure almost as long as mericarp..**18. *A. perrieri*** (p. 280)

7*. Leaves ca 4–6 × 1–2 mm, with hairs ca 0.1–0.3 mm long; mericarps 1.4–1.7 mm long, ± kidney-shaped in side view (cf. Fig. 37 c), covered with often curved and flattened very short hairs less than 0.1 mm long, commissure very short ....
............................**28 (b). *A. thymoides* subsp. *antsirabense*** (p. 328)

8. Mericarps ± kidney-shaped in side view, covered with often curved and flatted very short hairs less than 0.1 mm long (Fig. 37 c) ....
............................**28 (a). *A. thymoides* subsp. *thymoides*** (p. 327)

8*. Mericarps oblong to ± elliptic in side view, divergent, glabrous (Fig. 84 a)... **27. *A. palustre*** (p. 324)

1. ***Anthospermum welwitschii*** HIERN, Cat. Afr. Pl. Welw. **2**: 500 (1898); BRENAN in Mem. N.Y. Bot. Gard. **8**: 455 (1954); Verdc. in Fl. Trop. E. Afr., *Rubiaceae* **1**: 330 (1976); PUFF in Fl. S. Afr. **31**, 1 (2): 12 (1986). Type: Angola, Huila, Panda forests, near Eme, WELWITSCH 5335 (BM, holo.!; G, K, iso.!)[1].

= *A. cliffortioides* K. SCHUM. in Bot. Jahrb. **30**: 416 (1901). Type: Tanzania, Usafua, Beya-Berg (= Mbeya Mt.), GOETZE 1082 (B, holo. †; BM, BR, E, B, K, iso.!).

= *A. uwembae* GILLI in Ann. Naturhist. Mus. Wien **77**: 18, fig. 1 (1973). Type: Tanzania, Njombe Distr., Uwemba, GILLI 399 (W, holo.!) [see also *A. usambarense,* Critical Remarks].

− *A. ammannioides* sensu auctt. Afr. austr., non S. MOORE.

Dioecious or ± dioecious (seldom odd ♀ flowers on ♂ or ♀ plants), single- or (seldom) several-stemmed, ± erect shrubs, mostly much-branched above, ca (0.5) 1–3 m tall. Stems to 70 mm in diameter near the base; branches mostly ± regular and paired, often ascending; younger parts shiny reddish-brown to greyish, glabrous or ± densely covered with short whitish hairs or papillae to 0.3 mm long, less often with spreading hairs to 0.6 mm long; older parts brownish-grey to blackish, glabrescent,

---

[1] According to VERDCOURT (1976) the holotype should be in herbarium LISU. The specimen, however, was not amongst the material sent to me on loan.

epidermis peeling; internodes somewhat longer to shorter than the leaves; plants often ± densely leafy above. Leaves decussate, mostly pseudo-verticillate; blades (7) 10–35 × (1) 1.5–3.5 mm (on new growth occasionally to ca 45 × 6 mm), (narrowly) oblanceolate, elliptic to ± linear-lanceolate, narrowed to the base, glabrous or with a few short hairs or papillae to 0.2 mm long near the base of the flat to somewhat revolute margins and occasionally also on the midrib below, older leaves often tough and ± leathery and with conspicuous but small upper epidermis cells; apices acute to acuminate or ± mucronate; petioles subobsolete or to ca 1 mm long; stipular sheaths glabrous or shortly hairy, cup-shaped, 1–2.5 mm long, with 3–5 (–7,8) setae, the median ca (0.5) 0.7–4.5 (5.6) mm long. Inflorescences ± extensive, dimorphic, in ♀ often quite contracted, dense, ± cylindrical inflorescence zones. Flowers subsessile (in ♂ occasionally with peduncles and/or pedicels to 0.7 mm long), in clusters of many (very many: ♀) at the nodes; ♂, ♀ (rarely ⚥). Corolla 4-merous, greenish-yellow to pale yellow, occasionally purplish tinged on the outside, glabrous or with odd hairs < 0.1 to 0.2 mm long. ♂: tube (0.5) 0.7–1.2 mm long, funnel-shaped, lobes (1.2) 1.5–2.2 (2.7) × (0.4) 0.6–0.9 (1.1) mm, ± oblong-lanceolate, recurved; stamens 4, filaments exserted for (0.6) 0.9–1.5 (1.9) mm, anthers 1–2 × 0.3–0.7 mm; rudimentary ovary small, often crowned by 4 calyx lobes ca 0.1–0.4 × 0.2–0.3 mm, sometimes also with rudimentary stigmas to ca 0.9 mm long (flowers occasionally transitional ♂ → ⚥; sometimes also "true" ⚥ and transitions ⚥ → ♀ with at least partly reduced anthers and corollas smaller than in ♂). ♀: tube 0.2–0.5 mm long, ± cylindrical, lobes 0.2–0.7 × 0.1–0.2 mm, ± linear-lanceolate, ± erect; style ± 0–1 mm long; stigmas 2, 3–7.5 (10) mm long, greyish to whitish(-green); ovary ca 0.6–1 × 0.3–0.6 mm, mostly glabrous, crowned by 4 sometimes unequal (2 longer than the others) calyx lobes. Fruits mostly reddish-brown and shiny, supported by U-shaped carpophores to ca half as long as the fruits; each mericarp 1.5–2.7 × 0.7–1.2 mm, oblong, elliptic to ± obovate, dorsal side ± shallowly convex, glabrous or (seldom) ± sparsely covered with whitish, spreading hairs to 0.2 mm long, ventral side plane to shallowly concave, with a ± prominent median vertical ridge, commissure large; mericarps crowned by 2 ± triangular calyx lobes ca (0.3) 0.5–1 (1.2) × 0.2–0.5 mm, one occasionally longer than the other. – Figs. 2a, 15a.

Chromosome Number (Table 3): n = 11, 2n = 22.
Average Pollen Diameter (Fig. 49): 27.8–29.7 µm.
Habitat (Fig. 2a): Typically at the edge of afromontane forest or scrub; occasionally in grassland, in *Protea* veld or *Tarchonanthus* scrub, in streamside vegetation or on rocky slopes. Sometimes in disturbed sites

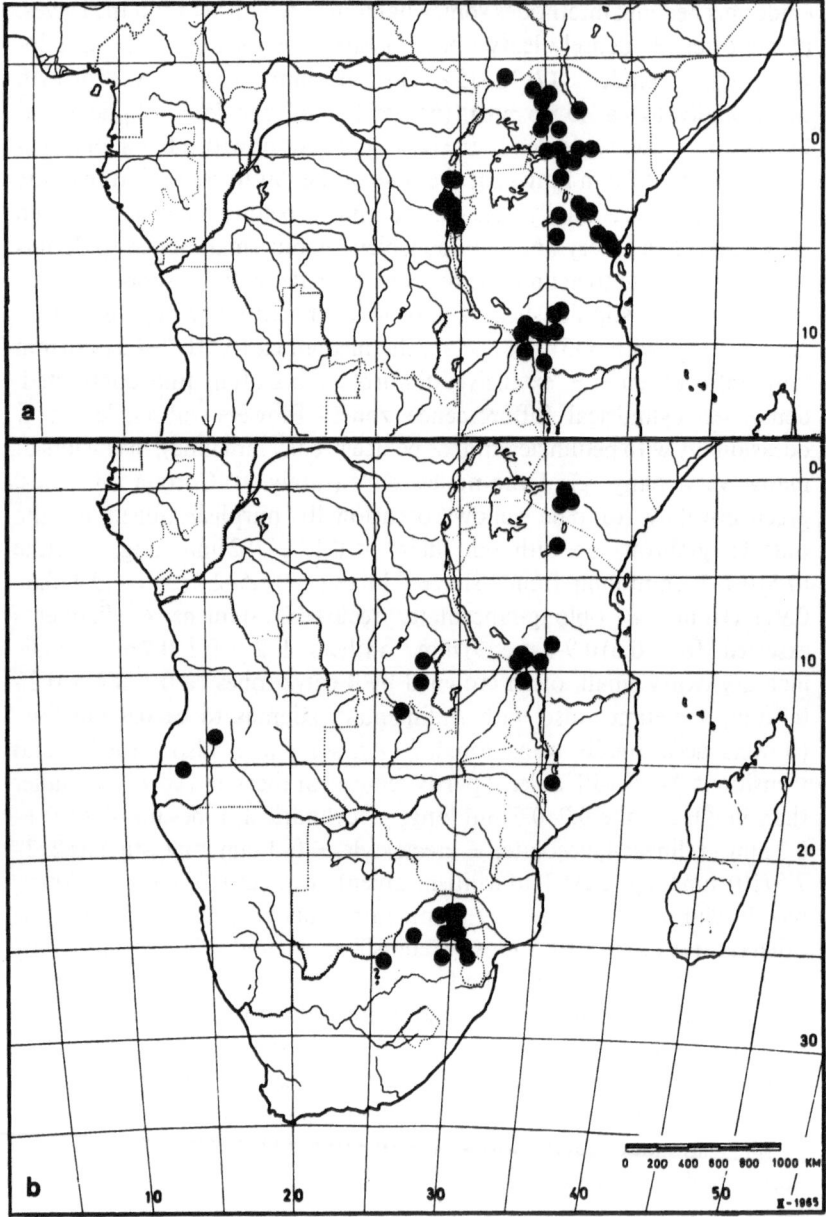

Fig. 64. Distribution of *Anthospermum. a A. usambarense, b A. welwitschii*

(roadsides, wattle plantations, etc.). Growing in humus-rich sandy loam, in ± dry sandy ground or in rough lava soils.—Ca (1 000) 1 300–2 400 (2 700) m.

Distribution (Maps, Figs. 64 b, 66 d): Disjunct. Kenya, Tanzania, Malawi, Zaire (Shaba) and adjacent parts of Zambia, Angola, S Africa (Transvaal; and Cape: Mafeking?). Also see Critical Remarks, below.

*Critical Re*marks: *A. welwitschii* has a remarkably fragmented distribution range (unrivalled by any other species of *Anthospermum*!). It occurs in Kenya S of the equator and in adjacent N Tanzania, and again in S Tanzania, but it is absent from the high E African mountains (Mt. Kenya, Mt. Elgon, Ruwenzori, Kilimanjaro, etc.). It is only found again in the extreme N and S of Malawi (Misuku Hills to the Nyika Plateau; Mt. Mlanje) but not in the mountains in between. Interestingly, it – unlike numerous other ± afromontane taxa – also "misses out" the E highlands of Zimbabwe [where it seems to be "replaced" by the related *A. ammannioides* (5.)] to recur again in the Transvaal.

Considering its wide distribution, *A. welwitschii* shows relatively little variation. The variation is found primarily in the fruit indumentum (hairy mericarps occur throughout its range but are less common than glabrous ones), the size of the persistent calyx lobes, the stem indumentum (longish hairs are uncommon), leaf size differences and sex distributions. S African material appears to be quite strictly dioecious, while in the other areas ♂ or ♀ plants with odd ⚥ flowers do occur occasionally, and individual plants sometimes even show transitions ♂ → ♀ via ⚥ (similar to the situation described for plants of *A. ammannioides* grown in the greenhouse. See C.12.2.2., p. 138). Odd plants with ternately arranged leaves do occur in S Tanzania but they are extremely rare; leaf arrangement is almost exclusively decussate.

The larger leaves with their tougher, sometimes ± leathery consistency, their ± conspicuous upper epidermis cells and their decussate arrangement, and the fruits with the longer calyx lobes generally distinguish *A. welwitschii* from the closely allied *A. usambarense* (2., below). *A. welwitschii,* moreover, does not extend to such high altitudes as *A. usambarense,* although the two can sometimes occur in the ± same habitats.

The problems encountered in the delimitation of *A. welwitschii* and *A. usambarense* in S Tanzania (T 7) and N Malawi are discussed in the Critical Remarks section of the latter. "*A. uwembae*" forms (e.g., GILLI 399 and MILNE-REDHEAD & TAYLOR 10775 from a nearby locality, and PAGET-WILKES 906 in MO (not in K!)][1] are not "typical" *A. welwitschii* but

---

[1] The sheet in K is *A. usambarense*!

may be of hybrid origin (*A. welwitschii* × *A. usambarense*; also back-crosses with "typical" *A. welwitschii*?). The few collections from Zambia (W: near Solwezi) and adjacent Zaire (Kundelungu Plateau) also differ somewhat from "typical" material, but a hybrid origin for these collections seems improbable, as no other taxa of *Anthospermum* are known from these localities.

Numerous collections of *A. welwitschii* only consist of inflorescences (especially of the more conspicuous and extensive ♀ ones!). As the leaves in the inflorescence region tend to be smaller than those in the vegetative region, and as inflorescence leaf sizes of *A. welwitschii* approach the sizes of leaves in the vegetative region of *A. usambarense,* such herbarium specimens may, subsequently, be difficult to separate from *A. usambarense* (also with often ± condensed ♀ inflorescences). It is, therefore, recommended that vegetative shoots be collected in addition to flowering and/or fruiting material.

In S Africa, the collection from Mafeking (KASSNER 1541, November 1907; plotted with a question mark on maps, Figs. 64 b, 66 d) is typical *A. welwitschii* but seems to lie considerably outside the species' distribution range. From my field observations it would seem that the country around Mafeking (where I did not find any *Anthospermum* species) is too dry and does not provide suitable habitats for *A. welwitschii*. On his travel to Broken Hill (= Kabwe/Zambia) KASSNER, however, did pass through Mafeking (cf. "My Journey from Rhodesia to Egypt", KASSNER 1911, map, Fig. 1), and the collecting date of the *A. welwitschii* collection corresponds to his itinerary. This, nevertheless, leaves the possibility open that a label mix-up occurred and that the specimen in question may have been from an earlier (1897?) KASSNER trip to the E Transvaal.

Collections

Angola. – **Benguela:** Chicala, Chicangala, around Cuima R., GOSSWEILER 12375 (BM, LISC). – **Huila:** Panda forests nr. Eme, WELWITSCH 5335 (BM, G, K); Humpata, Buraco de Bimbe, MENDES 3818, 3318 a (LISC); – , Bimbe, Chela escarpment, TORRE 8587 (LISC).

Zaire. – **Haut Katanga (Shaba):** Kundelungu Plateau, Luando R., 10 km SW of Mt. Kibwe wa Sanga, LISOWSKI, MALAISSE & SYMOENS 11517 (BR); – , Luanza R., 6 km NNW of Katshupa, MALAISSE 4376 (BR); – , 60 km from S end of plateau, SCHMITZ 4926 (BR).

Kenya. – **K 3**: Nakuru D., Menengai Crater, KOKWARO & MATHENGE 3394 (BR, K); Naivasha D., Lake Naivasha, behind Fisherman's Camp, GILBERT 4709 (ETH); – , Mt. Longonot, DÜMMER 5105 (GH, US), s.n. (K); – , near Njorowa Gorge, Hobley's Volcano, VERDCOURT 3201 (BR, K, PRE). – **K 3/6**: between Oljoro-o-Nyon R. and Lake Naivasha, crossing Mau escarpment, MEARNS 651 (US), 669 (BM, US). – **K 6**: between the Oljoro-o-Nyon and Narok R., MEARNS 375 (US); Masai D., Siyabei Gorge near Narok, BALLY 6593 (K), GLOVER & SAMUEL 3199 (K).

Tanzania. – **T 2**: Masai D., Ngorongoro Crater, BALLY 2427 (G, K). – **T 7**: Iringa D., Mufindi area, Nyamalala, PAGET-WILKES 906 (MO; atypical – approach. *A. usambarense*); Mbeya D., Kawiro R., nr. Marupindi's, also N side Mporoto (= Poroto) Mts., ST. CLAIR-THOMPSON 1141 (K); – , ("Usafua"), Mbeya Mt., GOETZE 1082 (BM, BR, E, G, K), KERFOOT 1685 (K), MILNE-REDHEAD & TAYLOR 10226, 10228 (K), PROCTER 1188 (K); Njombe D., ca 8 km S of Njombe, MILNE-REDHEAD & TAYLOR 10775 (B, BR, K, LISC; ± atypical); – , near Uwemba, GILLI 399 (W; ± atypical).

Zambia. – **W**: Solwezi, Luamisamba R., MUTIMUSHI 3119 (K, SRGH). – **E**: Nyika Plateau, FANSHAWE 7385 (K).

Malawi. – **N**: Misuku Hills, Mugesse (= Mughese) Forest, PUFF 781219-1/4 (WU); – , Willindi Forest Res., PUFF 781220-1/9 (WU). – **S**: Mt. Mulanje (= Mlanje), BRASS 16495 (BR, K, MO, SRGH, US), 16515 (K, MO, SRGH), BRUMMITT 11277 (K, SRGH), CHAPMAN 570 (BR, K, MAL, MO), HILLIARD & BURTT 6173 (E, LMU, NU), NEWMAN & WHITMORE 613 (BM, BR, SRGH, WAG), PUFF 780211-2/1 (W, WU), RICHARDS 16638 (K, LISC, SRGH), WILD 6185 (BR, K), 6219, 6220 (K, SRGH).

South Africa. Transvaal – **2229** (Waterpoort): Soutpansberg, ca 8–11 km W of Louis Trichardt-Wyllie's Poort rd., on track to Farm Bluegum Poort (**-DD**), BOTHA 863 (PRE), PUFF 791202-4/1 (WU). – **2230** (Messina): nr. Entabeni (**-CC** or **-CD**?), HUTCHINSON & GILLETT 4294 (BM, K); Venda, Tate Vondo For. Res. (**-CD**), HEMM 497 (PRE), PUFF 791201-1/3 (WU). – **2328** (Baltimore): Pietersburg D., Farm Leipzig, Blaauwberg (**-BB**), TSCHEUSCHNER s.n. sub TVL Museum no. 29528 (K, PRE). – **2329** (Pietersburg): Blaauwberg plateau (2328-**BB** to 2329-**AA**), CODD & DYER 9068 (K, NH, PRE, SRGH), ESTERHUYSEN 21452 (BOL, PRE), SCHOLES s.n. sub PUFF 781218-2/1 (W, WU); Louis Trichardt (**-BB**), BREYER s.n. sub TVL Museum no. 19437 (PRE); nr. Houtbosdorp, on rd. to Mooketsi (**-DD**), PUFF 791209-2/1 (WU). – **2330** (Tzaneen): Soutpansberg D., Farm Rustfontein, ca 14 km E of Louis Trichardt (**-AA**), SCHLIEBEN 7174 (B, BR, G, K, M, , PRE, US); Westfalia Estate nr. Duiwelskloof, Piesangkop (**-CA**), PUFF 790211-2/3 (BR, J, W, WU), SCHEEPERS 851 (B, BM, K, M, PRE, SRGH); Wolkberg, New Agatha Forest Res. (**-CC**), MULLER & SCHEEPERS 112 (LISU, PRE). – **2427** (Thabazimbi): Kransberg, off Bakkers pass (**-BC**), PUFF 781202-1/2 (W, WU); Waterberg D., Rankin's pass (**-DB**), ACOCKS 23578 (BR, K, M, PRE). – **2429** (Zebediela): Pyramid Estate nr. Potgietersrust (**-AA**), GALPIN 8989 (K, PRE); Strydpoortberge, Chuniespoort area, Donkerkloof (**-BA**), PUFF 790209-3/1 (W, WU), VAHRMEIJER 2428 (K, MO, PRE). – **2430** (Pilgrim's Rest): The Downs, Mamotswiri Peak area, Farm Haffenden Heights (**-AA**), PUFF 791208-1/2 (WU); Shilouvane (= Shiluvane Mission) (**-AB**), JUNOD 1055 (G); Letaba D., top of Mt. Lebojana (**-AD**), VENTER 1168 (PRE); Marieskop (**-DB**), VAN DER SCHIJFF 4987, 6128, 6185 (PRE); Ohrigstad Dam Nature Res. (**-DC**), EDWARDS 4055 (K × 2, PRE, SRGH), JACOBSEN 2649 (PRE), PUFF 781028-1/1 (J, W, WU). – **2529** (Witbank): Middelburg Distr., Farm Doornkop 273 JS, above Klein Olifants R. (**-CB**), FOURIE 652 (PRE); nr. Middelburg (**-CD**), KASSNER 435 (BR). – **2530** (Lydenburg): Farm Zwagershoek (= Kwaggashoek), nr. [SW of] Lydenburg (**-AB**), OBERMEYER 172 (PRE × 2); around Sabie (**-BB**), HUMBERT 10844 bis (P). – **2531** (Komatipoort): Barberton (**-CC**), THORNCROFT 1077 (BM, S). – Not traced: Steenkamp Berge, Middelburg (Distr.?), RUDATIS 2512 (STE). – Additional collections: BOLUS 11084 (BOL); BREMEKAMP & SCHWEICKERDT 429 (PRE); CODD 3965 (PRE), 8332 (K, M, PRE); DAHLSTRAND 1849 (PRE); GERSTNER 5733, 5987 (PRE); HEMM 497 (J); JEPPE 13 (J); MOSS & ROGERS 22038 (PRE); PUFF 790210-2/1 (J, W, WU), 790701-1/3, 790702-1/2 (BR, J, W, WU), 791201-3/1, 791202-2/1 (WU); REHMANN 6033 (BM, K);

THERON 3386, 3410 (PRE). **Cape — 2525** (Mafeking): Mafeking (-**DC**), KASSNER 1541 (E; see Critical Remarks).

2. Anthospermum usambarense K. SCHUM. [in Bot. Jahrb. **17**: 165 (1893) & in Abhandl. Preuss. Akad. Wiss. **1894**: 69 (1894): in observ.] in Bot. Jahrb. **28**: 112 (1899); ORTH in WALTER, Vegetationsbilder **25** (8): t. 43 a & b (1940); ROBYNS, Fl. Spermatoph. Parc Nat. Albert **2**: 372, t. 37, fig. 15 (1947); BRENAN in Mem. N.Y. Bot. Gard. **8**: 455 (1954); HEDBERG, Afroalp. Vasc. Pl.: 176, 328 (1957); DALE & GREENWAY, Kenya Trees & Shrubs: 425 (1961); VERDC. in Fl. Trop. E. Afr., *Rubiaceae* 1: 331 (1976).

Syntypes: Tanzania, Usambara [Mts.], HOLST 420 (B†); −, nr. Kwai, EICK s.n. (B†); neo- (= topo-) type: −, −, Matondwe Hill, head of Kwai valley, DRUMMOND & HEMSLEY 1351 (B!, BR!, K!, S!).

= *A. leuconeuron* K. SCHUM. in Bot. Jahrb. **30**: 416 (1902).
Type: Tanzania, Usafua, Beya-Berg (= Mbeya Mt.), GOETZE 1081 (B, holo. †; BM, BR, E, G, K, iso.!) [see Critical Remarks].

= *A. prittwitzii* K. SCHUM. & K. KRAUSE in Bot. Jahrb. **39**: 570 (1907).
Type: Tanzania, "beim Lager Kidoko, VON PRITTWITZ & GAFFRON, 28. August 1901" (B, holo. †; E, iso.![1]) [see Critical Remarks].

= *A. keilii* K. KRAUSE in Bot. Jahrb. **48**: 431 (1912).
Type: Tanzania, "Zentralafrikanisches Zwischenseenland", bei Usumbura, 2000 m, KEIL 18 (B, holo †).

= *A. aberdaricum* K. Krause in Notizbl. Bot. Gart. Berlin **10**: 609 (1929).
Syntypes: Kenya, Aberdares, Kinangop summit, FRIES & FRIES 2574 (S, lecto.!, selected here; B†; BR!, K!, UPS!); −, −, Sattima summit, FRIES & FRIES 2465 (B†; K!, UPS!).

Dioecious, single-stemmed, mostly much-branched, erect shrubs (0.5) 0.75–3 (4.5) m tall. Stems to ca 35 (–?) mm in diameter near the base, branches mostly ± regular, in threes (less commonly paired), often ascending to ± erect; younger parts reddish-brown to greyish, usually ± densely covered with whitish spreading hairs < 0.1 to 0.7 (1) mm long; older parts brownish-grey to dark grey, glabrescent, epidermis peeling; internodes mostly as long as or somewhat shorter than the leaves; plants ± densely leafy. Leaves in whorls of 3 or (less often) decussate, pseudo-

---

[1] The label reads "*A. prittwitzii* K. SCHUM. & K. KRAUSE, Kidoko. 1901. No. 57₁, leg. v. PRITTWITZ". There can be little doubt that this collection is to be considered an isotype.

verticillate; blades (2) 3–10 (15) × 0.5–1.5 (2) mm (on new growth occasionally to 20 × 3 mm), narrowly obovate, oblanceolate, oblong-elliptic, lanceolate to ± linear, narrowed to the base, glabrous or with a few papillae to 0.1 mm long on the margins and/or midrib below; margins often (strongly) revolute; apices acute to ± acuminate or ± mucronate; petioles subobsolete; stipular sheaths shortly hairy, ± broadly cup-shaped, (0.7) 1–2 mm long, with 3–7 (8) setae, the median ca 1–4.1 mm long. Inflorescences ± extensive, dimorphic, on ♀ often quite contracted, dense, cylindrical inflorescence zones. Flowers subsessile, in large numbers (in ♀ often more than in ♂) clustered at the nodes; ♂, ♀ (very rarely also odd ⚥). Corolla 4-merous (very seldom 5-merous), greenish to creamy yellow, often dark purplish-red or -brown tinged outside, glabrous or with odd hairs to 0.3 mm long. ♂: tube 0.7–1.6 mm long, funnel-shaped, lobes 1.4–2.4 (2.7) × 0.6–1 (1.4) mm, ± oblong-lanceolate, recurved; stamens 4, filaments exserted for (0.6) 1–1.7 mm, anthers (0.7) 1–1.7 × 0.3–0.7 mm; rudimentary ovary small and mostly with 4 well discernible calyx lobes ca 0.2–0.5 × 0.2–0.3 mm, occasionally also rudimentary stigmas present. ♀: tube 0.3–0.7 mm long, ± cylindrical, lobes 0.2–0.7 × 0.1–0.3 mm, ± linear-lanceolate, erect to ± spreading: style ± 0 to 0.5 mm long; stigmas 2, 3–6.1 mm long, greyish; ovary ca 0.7–0.8 × 0.5–0.6 mm, mostly glabrous, crowned by 4 subequal or 2 longer and 2 shorter calyx lobes. Fruits reddish-brown, shiny or greyish-brown, supported by short, broad U-shaped carpophores; each mericarp 1.2–2.2 (2.4) × 0.7–1 mm, ± obovate, oblong to elliptic, dorsal side ± shallowly convex, glabrous or (seldom) ± densely covered with whitish spreading hairs to 0.2 mm long, ventral side plane to shallowly concave, with a ± prominent median vertical ridge, commissure large; mericarps crowned by 2 ± triangular calyx lobes ca 0.2–0.6 × 0.3–0.4 mm, one often larger than the other. – Fig. 62.

Chromosome Number (Table 3): n = 11, 2 n = 22.

Average Pollen Diameter (Fig. 49): 28.6–30.3 µm.

Habitat (cf. ORTH 1940: pl. 43 a & b): A typical component of the ericaceous belt or heath zone of the higher tropical E African mountains and occasionally also extending into the alpine belt. Also found at the edge of montane forest or bush clumps, in the grassland/forest border and occasionally in upland (rocky) grassland, in streamside thicket or on rocky outcrops. Growing in sandy to gravelly soil and sometimes also on ± fresh lava and in ± boggy ground. – Ca 1 700–4 100 (4 500) m.

Distribution (Map, Fig. 64 a): Centered in tropical E Africa. Occurring from the S Sudan (Imatong Mts.) S to the mountains of NE Tanzania and, in the W, from the Zaire-Uganda-Rwanda border area S to the N end of

Lake Tanganyika (Burundi). Also in S Tanzania (from the mountains around the N end of Lake Malawi to the Songea Distr.) and on the Nyika Plateau (Malawi/Zambia border).

Critical Remarks: Typical forms of *A. usambarense* are rather tall, quite regularly branched shrubs with ± closely spaced, small leaves (hardly reaching 10 mm in length) arranged in whorls of three. Habit (the density of the foliage in particular), leaf size, shape and arrangement, however, do vary to some extent. Leaf sizes and shapes sometimes vary even on a single plant: new growth produced after the formation of ♀ inflorescences, for example, may have much larger leaves than the same shoot below the inflorescence region. Ternate leaf arrangement, although characteristic for the majority of specimens, is not found throughout. In some populations individual plants may have both opposite and ternate leaves, in others leaf arrangement may be strictly decussate. Not all populations are always strictly dioecious – in a few, odd ♂ plants with some ⚥ flowers were observed. As in numerous other taxa of *Anthospermum*, the fruits vary from (typically) glabrous to hairy.

Because of this variability it may – particularly in S Tanzania (T 7) and in the adjacent parts of Malawi (especially the Misuku Hills) – become difficult to separate *A. usambarense* from the more widely and disjunctly distributed *A. welwitschii* (1., above). Several specimens[1] approach *A. welwitschii* in having larger and broader leaves than "typical" *A. usambarense* and decussate leaf arrangement or both decussately and ternately arranged leaves on one plant. There seem to be no reliable and usable characters to allow an entirely certain identification of these collections except for the absolute midstem leaf size differences. It is realized that this is a rather "technical" and unsatisfactory way to separate the two taxa, but no better solution can be offered at present.

As both "typical" *A. usambarense* and "typical" *A. welwitschii* also occur in the same above-mentioned area (e.g., "typical" *A. usambarense* is common on the Nyika Plateau; the type of *A. cliffortioides* [= *A. welwitschii*] is from Mbeya Mt., T 7) and as greenhouse experiments prove that the species can be crossed, I suspect that "problematic" specimens such as those described above may be collections originating from (extensive?) hybrid swarms. I, however, do not have definite proof for this – simply because there do not seem to be sufficiently reliable and usable characters for a detailed analysis. The types of *A. leuconeuron, A. prittwitzii* and *A. uwembae* may, in addition to the specimens listed in the

---

[1] For example SCHLIEBEN 637, 851 and PERDUE & KIBUWA 11039 from T 7 and PUFF 781221-2/3 from N Malawi (listed as "atypical" in Collections).

previous footnote, represent such putative hybrid collections[1]. The types of the first two more closely approach *A. usambarense,* while the type of *A. uwembae* more closely resembles *A. welwitschii.*

Other odd, unusually large-leaved forms (but with mostly ternately arranged leaves) occur on Mt. Kilimanjaro (e.g., EVANS & ERENS 1043, WETTSTEIN & WETTSTEIN s.n.; both gathered in forest – i.e., at lower altitudes than "typical" *A. usambarense* which is common higher up the mountain), on the Usambaras (GEILINGER 1464; very much like *A. welwitschii*) and in Rwanda (BECQUET 749). In all of these localities, however, hybridization between *A. usambarense* and *A. welwitschii* can be excluded with some certainty as a possible reason for the unusual morphological features of these collections.

Plants from the E highlands of Zimbabwe which, in the past, have often been identified as *A. usambarense* are now split off as *A. zimbabwense* (3., below; see there for further details).

A. usambarense is not only allied to both *A. welwitschii* and *A. zimbabwense* but also shows affinities to *A. vallicola* (6.) and *A. ammannioides* (5.) [like the former endemic to E Zimbabwe (and Gorongosa Mt.)] and to *A. emirnense* (4.) and *A. madagascariense* (7.) (both Madagascan endemics). These species appear to form a rather well defined group of essentially (afro)montane, tall shrubby (except *A. zimbabwense*), dioecious taxa.

There are strong indications that *A. usambarense* occasionally hybridizes with *A. whyteanum* (16.) in both N Malawi and S Tanzania (i.e., in the N-most part of the distribution range of the latter species). See *A. whyteanum* for details and cf. Fig. 58.

Collections

Sudan. – Equatoria Prov., Torit D., Imatong Mts., Mt. Kinyeti ("Kineti", "Kinelti"), CHIPP 74 (K), JACKSON 934 (BM), THOMAS 1855 (BM, K), TOTHILL 13511 (K); –, –, Dumuso, JACKSON 1516 (K, ± atypical).

Uganda. – U 1: Karamoja D., Mt. Morongole, Dale U. 259 (K), THOMAS 3285 (K); –, Moroto Mt., EGGELING 2902 (K), KATENDE & LYE 418 (MO), WILSON 505 (K); –, Mt. Debasien (= Kadam), EGGELING 2710 (K), THOMAS 2934 (K). – U 3: Mbale D., above Bulambuli, TOTHILL 2373 (K); – Uganda side of Mt. Elgon, DALE U. 48 (K), DÜMMER 3348 (BM, BOL, K), OSMASTON 3991 (K), SYNGE 900 (BM, S, SRGH), 1911 (BM), THOMAS 2676 (K), TOTHILL 2454 (K).

Kenya. – K 1: Northern Frontier D., Nyiru Mt., KERFOOT 1999 (K). – K 2: Turkana D., Moruassigar (= Murua Nysigar), NEWBOULD 7192 (K). – K 2/3:

---

[1] SCHUMANN (1902) noted that GOETZE's type specimens of *A. leuconeuron* and *A. cliffortioides* (= "typical" *A. welwitschii*) (both from Mbeya Mt.) were growing side by side, but GOETZE apparently did not collect "typical" *A. usambarense* there [although it is known from Mbeya Mt.: DAVIES M 25 (K)]. Thus there is no direct and concrete evidence or clue as to the hybrid nature of *A. leuconeuron.*

Cherangani Hills, POWLES 21 (K), THULIN & TIDIGS 251 (K, S), TWEEDIE 4228 (K). – **K 3**: Trans-Nzoia D., Mt. Elgon, DALE 3206 (K, MO), LUGARD & LUGARD 403 (K), PUFF & WEBER 780114-4/1, 780115-1/3 (WU), TAYLOR 3438 (BM, S); Nandi (forest; escarpment), JOHNSTONE s.n. (K), SCOTT ELLIOT 6881 (BM, K); Timboroa, IRWIN 290 (K), TWEEDIE 3662 (K); SW Mau Forest, WHITALL 241 (K); Naivasha D., Kipiperi Hills, LAWTON 1690 (K). – **K 4**: Kiambu D., Kibata, S Kinangop Forest Res., KERFOOT 1450 (K); Nanyuki D., Mt. Kenya, Sirimon Track, GILLETT 16258 (EA, FI, G, GH, K, M, US, WAG). – **K 5**: Kisumu-Londiani D., Tinderet Forest, KWIN 106 (K). – **K 6**: Narok D., Enesambulai (= Nasompolai) valley, GREENWAY & KANURI 14535 (K, PRE). – Additional collections: BATTISCOMBE 837, 1167, 1319 (K); DALE 896 (BR); FRIES & FRIES 614 (BR, K, MO, S), 1487 (BR, K, S), 1569 (S), 2465 (K), 2574 (BR, K, S); GARDNER 1113, 1318 (K); GLOVER, GWYNNE, SAMUEL & TUCKER 2119 (K, PRE); GILLETT 16889 (BR); HEDBERG 1630 (K, S); KOKWARO 2346 (K); MCLAUGHLIN B2818 (G, K); NAPIER 1253 (K); POLHILL 229 (BR, K, PRE); SCHELPE 2609 (BM, BR), 2645 (BM); SYNGE 1840 (BM); G. TAYLOR 1421 (BM, S), 1534 (BM, G, S), 1538 (BM); Rev. TAYLOR s.n. (BM); WHYTE s.n. (K × 3); ZOGG, KRAMER & GASSNER 269/5 (WAG).

Tanzania. – **T 2**: Masai D., Ngorongoro Conservation Area, Empakaai Crater, FRAME 166 (BR, MO), RAYNAL 19222 (P); Mbulu D., Oldeani Volcano, BURTT 4229 (K), GEILINGER 3657 (K); –, Hanang Mt., BURTT 2300, 4043 (K), GEILINGER 3396 (K), GREENWAY 7663 (K, PRE); Arusha D., Mt. Meru, HEDBERG 2279, 2288 (K, S); –, Ngurdoto National Park, Tululusie (= Tulusia) Hill, GREENWAY & KANURI 12319 (BR, K, PRE); Moshi D., Mt. Kilimanjaro, from Mandara Hut to Horombo Hut, PUFF & KIBUWA 820903-2/1 (WU). – **T 3**: Pare D., South Pare Mts., Chome Forest, PUFF & KIBUWA 820901-1/5 (WU); Lushoto D., W Usambara Mts., Mkuzi, NE of Lushoto, DRUMMOND & HEMSLEY 2049 (B, BR, K, LISC, LMA, S); –, Nedelemai (= Ndelema, Derema), MTUI 26 (K). – **T 7**: Iringa D., Dabaga Highlands, Kibengu, POLHILL & PAULO 1518 (B, BR, K, LISC, PRE, SRGH); –, E Mufindi, GREENWAY 3494 (K, PRE); Mbeya D., Poroto Mts., RICHARDS 9728 (BR, K), 9819 (K); Rungwe D., Rungwe Mt., RICHARDS 14296 (K); Njombe D., (U)Kinga Mts., Kingika (= Kinjika Diude) Mt., GOETZE 949 (BM, BR, K), 949 a (E), 949 b (BR, G, P); –, Lupembe area, upper Ruhudje R., SCHLIEBEN 637 (B, BM, BR, G, K, M, S; atypical), 851 (BR, G; atypical). – **T 8**: Songea D., Matengo Hills, Luiri Kitesi (= Liwiri-Kiteza), MILNE-REDHEAD & TAYLOR 10509 (B, BR, K, LISC, SRGH), 10510 (B, BR, K, LISC). – Additional collections: BALLY 855 (FI); BRENAN & GREENWAY 8193 (K); BUCHWALD 25 (BM, BR × 2, COI, K); BURTT 2342, 4075 (K); CARMICHAEL 1277 (FI, K); COTTON 23 (K); CUISINIER s.n. (G); DAVIES E.35, M.25 (K); DRUMMOND & HEMSLEY 1351 (B, BR, K, S), 2729 (B, BR, K, LISC, S); GEILINGER 1464 (K; very atypical), 4522, 5061 (K); GILLMAN 976 (K); GOETZE 1081 (BM, BR, E, G, K); GREENWAY 3756 (K), 4613, 8390 (K, PRE); GROTE 3093, 3094, 3095 (K); HEDBERG 1282 (K, S); HEPPER & FIELD 5350 (K); KORITSCHONER 856, 975 (K); LUCHMAN 19 (K); MAGOGO 2221 (WU); NGONYANI 25 (BR, K); PAGET-WILKES 906 (K); PEDRO 3764 (LMA); PERDUE & KIBUWA 11039 (BR, K; atypical); PETER 8826, 8931, 12088, 16547, 46754, 55392, 55405, 55434, 55532 (B, K), 46653, 55653, 55665 (B); POLE EVANS & ERENS 1043 (BR, E, K, P, PRE; atypical); PROCTER 1671, 1811 (K); RICHARDS 23040, 24044 (K); GILBERT ROGERS 576 (BR, K, S); SCHLIEBEN 4837 (B, BM, BR, G, LISC, M, P, S); SEMSEI 2114 (K); SHABANI 1041 (K); ST. CLAIR-THOMPSON 965 (K); STEELE 72 (K); STOLZ 1339 (G, K, M, NU, S, W, WAG), 2009 (BM, BOL, BR, G, GH × 2, K, P, PRE), s.n. (BR); VOLKENS 834 (E, G, K), 834 a (PRE); von PRITTWITZ 57_1 (E); WETTSTEIN & WETTSTEIN s.n. (M × 2, one atypical); WILLIAMS 517 (BR, K, PRE); ZUMER Z 17 (WAG).

Zaire. – **Lacs Edouard et Kivu:** Virunga Mts., Nyamuragira Volcano, Stauffer 217 (K, WAG), 239 (K, P, PRE); Kivu Terr., Mt. Muhi, Hendrickx 5276 (BR), Kinet 7 (BR, J, US). – Additional collections: Burtt 3143 (G, K), 3182 (K); Christiaensen 630 (BR, J); de Witte 1582 (BR); Germain 3764 (BR); Ghesquiere 5147 (BR); Hoier 2bis (BR); Humbert 8159 (B, BR, P); Lebrun 4893 (BR), 7902, 8755 (BR, K); Léonard 329, 340 (BR, K); Linder 2098 (GH); Louis 5067 (BR, K); Scaetta 1622 (BR, K).

Rwanda. – Préf. Ruhengeri, Mt. Mgahinga, Hendrickx 5481 (BR); Rugege Forest, between Mt. Muzima and Bigugu, Auquier 2752 (BR, K, WAG); Préf. Cyangugu, Nyungwe Forest, Bouxin 525 (BR), Deuse 249 (BR, K); Rutova, Butare-Cyangugu rd., km 64, Auquier 2778 (BR, K, M, MO, WAG). – Additional collections: Auquier 3996 (BR, K, MO); Becquet 749 (BR, J, US; atypical); Christiaensen 1590, 1594 (BR, J); Deuse 1168 (BR); Hendrickx 7684 (BR, J); Michel 5146 (BR, J, K); Pierlot 1577 (BR); Symoens 5416 (BR); van der Veken 10915 (BR), 10998 (BR, WAG).

Burundi. – Kayanza Terr., Rwegura, Lewalle 2731 (MO), 3661, 3173 (BR, K); Ngozi Terr., N'Dora-Kayanza rd., Marlier 1493 (BR, J); Bujumbura Terr., Gakara-Karonge rd., Lewalle 6721 (BR, G); Mwaro Terr., Kisozi, Mt. Mugero, Lewalle 5523 (K; atypical).

Zambia. – E: Nyika Plateau, Fanshawe 9764 (K, SRGH), Webster 2 (SRGH); –, nr. Zambian Resthouse, Puff 781223-2/1 (BR, J, WU).

Malawi. – N: Nyika Plateau (various localities), Brass 17265 (BM, BR, GH, K, MO, PRE, SRGH, US), 17268 (BR, GH, K, MO, PRE, SRGH, US), 17318 (BR, K, MO, PRE, SRGH, US); Brummitt 10700 (K, SRGH × 2); Brummitt & Synge WC 27 (K, MAL, SRGH); Chapman 734 (BM, K, MAL, SRGH), 751 (MAL, SRGH); Pawek 11206 (K, MO, PRE, SRGH, WAG); Phillips 1030, 1155 (K), Puff 781223-3/2, 781224-2/1, 781225-2/1 (WU); Misuku Hills, Matipa Forest Res., Puff 781221-2/3 (WU; atypical).

3. *Anthospermum zimbabwense* Puff, spec. nova, *A. usambarensi* affine sed praecipue habitu, mericarpiis lobis calycis conspicuis destitutis et setis stipularum plerumque solitariis brevioribusque differt.

Type: Zimbabwe, Rhodes Inyanga National Park, grassland at the base of Mt. Inyangani, ca 2 150 m, Puff 790125-3/1 (WU, holo.; BM, SRGH, iso.).

Dioecious, erect, several- to many-stemmed subshrubs or dwarf shrubs ca 0.2–0.6 (1) m tall with often quite thick, ± disk-like woody bases to ca 30 mm in diameter (especially in plants regularly exposed to fires). Stems ± unbranched to much-branched, branches often more numerous in ♂ than ♀, ascending to ± erect, ca 1.5–3 mm in diameter, ± terete; younger parts greenish-grey or reddish-brown, ± densely covered with whitish-grey spreading hairs ca 0.1–0.3 mm long; older parts reddish-brown-grey, glabrescent, epidermis sometimes peeling; internodes normally longer than or as long as the leaves. Leaves in whorls of 3, pseudo-verticillate; blades (3) 4–8 (12) × 0.5–1.5 (3) mm, linear to linear-lanceolate or narrowly oblanceolate, glabrous; margins mostly revolute;

14*

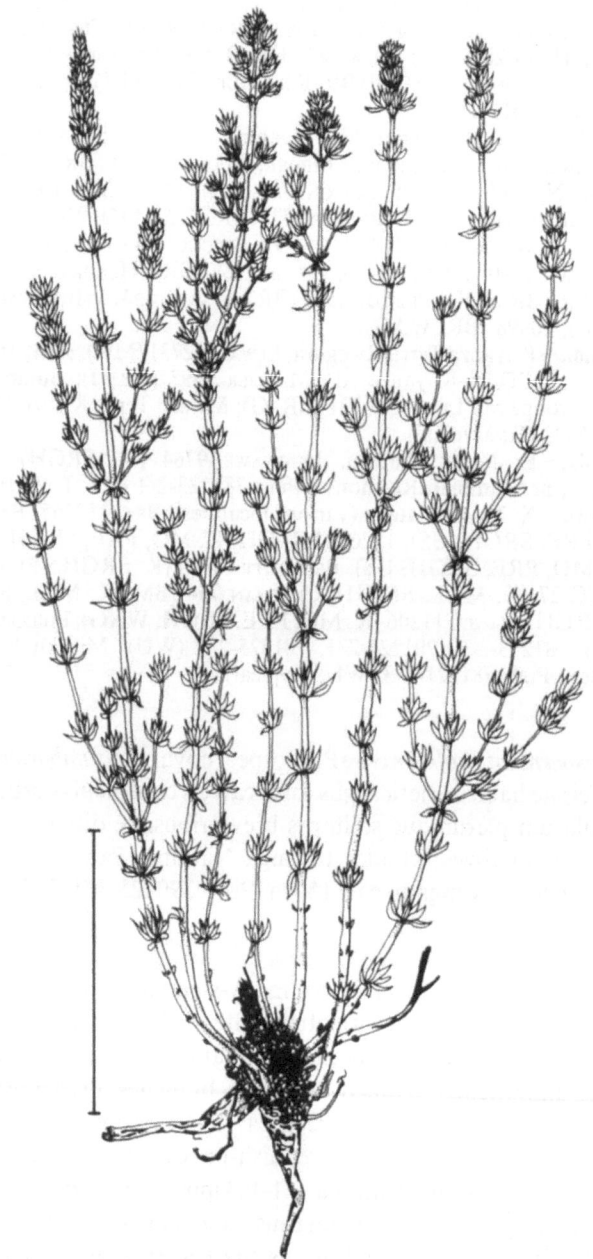

Fig. 65. *Anthospermum zimbabwense*, habit (PUFF 790125-3/1). – The scale unit is
10 cm

apices acuminate and frequently recurved; petioles subobsolete; stipular sheaths densely covered with whitish spreading hairs 0.1–0.3 mm long, (broadly) cup-shaped, ca 0.8–1.4 mm long, with a median, narrow seta ca 0.7–1.2 mm long, often flanked by an additional shorter seta on either side. Inflorescences in ♂ plants ± inconspicuous, made up of ± sessile, ± widely spaced several-flowered axillary clusters; in ♀ plants more conspicuous, ± compact, cylindrical inflorescence regions, made up of ± sessile, quite closely spaced many-flowered axillary clusters. Flowers ♂, ♀; corolla 4-merous, greenish-yellow to yellow, in ♂ usually a little hairy outside. ♂: tube 0.5–1 mm long, (± broadly) funnel-shaped, lobes 1.3– 2 × 0.5–0.8 mm, lanceolate, recurved; stamens 4, filaments 1–1.4 mm long, anthers (0.8) 1–1.3 × 0.2–0.5 mm; rudimentary ovary minute, hardly discernible. ♀: tube 0.2–0.4 mm long, ± cylindrical, lobes 0.3–0.5 × 0.1– 0.2 mm, ± linear-lanceolate, erect; style ± 0, stigmas 2, 3.4–9.1 mm long, greyish; ovary 0.7–0.8 × 0.6 mm, ± glabrous to a little hairy, crowned by 4 indistinct calyx lobes or calyx lobes ± lacking. Fruits reddish-brown, supported by small U- to V-shaped carpophores to ca half as long as the fruits; each mericarp ca 1.4–1.7 × 0.7–0.9 mm, ± obovate, dorsal side convex, ± glabrous or with a few whitish hairs ca 0.1 mm or less long, ventral side slightly concave and with a median vertical ridge; commissure relatively large; mericarps crowned by 2 indistinct, ± triangular calyx lobes hardly longer than 0.1 mm or calyx lobes ± lacking. – Figs. 21 b, 65.

Chromosome Number (Table 3): n = 11, 2 n = 22.

Average Pollen Diameter (Fig. 49): 27.7–29.8 μm.

Habitat: Typically found in afromontane grassland, often growing in association with bracken (*Pteridium*); sometimes in dense growth surrounding forest patches and in scrub along streams. – Ca 1 650–2 150 (2 500) m.

Distribution (Map, Fig. 66 c): Zimbabwe (E) and Mozambique (MS). Occuring in the E highlands of Zimbabwe (from Inyanga Distr. S to Melsetter Distr.) and doubtlessly also common in the neighbouring areas of Mozambique, although so far only collected in two localities.

Critical Remarks: *A. zimbabwense,* which usually occurs in fire-prone habitats, resprouts after a fire. Plants burnt regularly and ± frequently (every other season or even annually) develop quite massive woody bases from which the new aerial flowering shoots arise. In plants growing in habitats less frequently swept by fire or in places protected from fire (rather the exception in *A. zimbabwense!*), stems may grow for several seasons and become relatively tall, woody and branched. As in *Otiophora inyangana,* whose habitat and distribution range is ± identical to that of

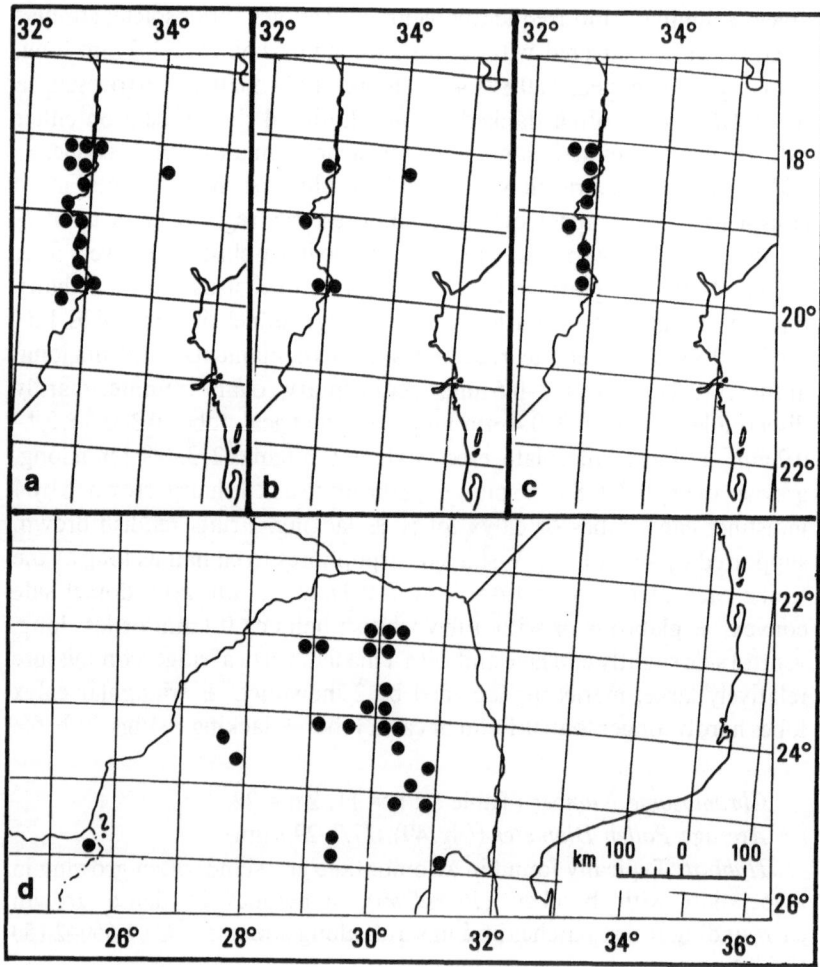

Fig. 66. Distribution of *Anthospermum*. *a A. ammannioides, b A. vallicola, c A. zimbabwense, d A. welwitschii* (S Africa only)

A. zimbabwense (cf. PUFF 1981), fire appears to be one of the main factors responsible for the morphological variability of the species.

Specimens of *A. zimbabwense* have previously often been identified as *A. usambarense* (2., above). In my opinion, the two species are doubtlessly closely allied, but *A. zimbabwense* deserves species rank for the following reasons: It differs markedly from *A. usambarense* in habit and growth form. Regeneration after fire is by resprouting (see above); plants thus normally are subshrubby. *A. usambarense*, on the other hand, is killed by fire, i.e., is a typical "seed regenerator" (cf. Fig. 62); it is a large (mostly single-stemmed) shrub to 3 (4.5) m tall[1]. *A. zimbabwense,* furthermore, differs from *A. usambarense* in various other morphological features: it has smaller flowers, frequently smaller leaves which are *always* ternately arranged, stipular sheaths with only 1 (3) smaller seta(e), and fruits which never have distinct calyx lobes as in *A. usambarense.*

A. usambarense is centered in tropical E Africa and does not extend further S than the Nyika Plateau (Zambia/Malawi border) – i.e., roughly 800 km N of the E highlands of Zimbabwe. Both taxa may have originated from a common ancestral stock, and *A. zimbabwense* may, in isolation, have evolved in E Zimbabwe. According to WILD (1964: 132) the "Inyangani Subcentre" sensu WEIMARCK (1941) [which ± corresponds to the distribution area of *A. zimbabwense* (and to that of the endemic *A. ammannioides* and *A. vallicola,* 5. and 6., below)] may have been isolated for more than one to two million years.

There are some indications that *A. zimbabwense* occasionally forms hybrids with *A. vallicola* and *A. whyteanum* (16.) (see the Critical Remarks section of the two species for details and also cf. Table 11). The occurrence of hybrids between *A. zimbabwense* and *A. ternatum* subsp. *randii* (19 b.) is also suspected but these appear to be very uncommon.

Collections

Zimbabwe. [E:] – **1832** (Umtali): nr. Inyanga Down village (**-BA**), NORLINDH & WEIMARCK 4643 (BM, K, MO, PRE, SRGH); Inyanga D., Litte Connemara, "World's View" (**-BB**), PUFF 790125-6/3 (J, WU), SEAGRIEF C.A.H. 3105 (K, SRGH), WEST 4464 (SRGH); Rhodes Inyanga National Park, Circular Drive on slopes of Inyangani (**-BD**), SEAGRIEF C.A.H. 1048 (BR, PRE, SRGH); Stapleford Forest Reserve (**-DB**), HUMBERT 15709 (P), PUFF 790127-2/1 (J, WU). – **1932** (Melsetter): Vumba Mts., Castle Beacon (**-BA**), PUFF 790128-1/3 (WU); Umtali D., Banti Forest (**-DB**), EXELL, MENDONÇA & WILD 174 (BM, LISC, SRGH); Melstetter D., Mt. Peni (**-DD**), SWYNNERTON 6100 (BM). – Additional collections: BAMPS, SYMOENS & VANDEN BERGHEN 26 (BR, WAG); CHASE 1295 (BM, K), 4342 (BM, K, LISC, SRGH); DALE SKF 748 (SRGH); DEHN s.n. (M); GARLEY 511 (BR, K, SRGH); GILLILAND 311 (BM, K); NORLINDH & WEIMARCK 4857 (BR, COI, K, M, MO, SRGH, WAG); PUFF 790124-2/2, 4/1, 790126-1/2 (J, WU), 790125-1/3 (WU), 790125-3/1 (BM, SRGH, WU); WHELLAN & DAVIES 971 (K, SRGH); WEST 4421 (LISC, SRGH), 4427 (K, LISC).

Mozambique. [MS:] – **1832** (Umtali): from Vila Manica towards Penhalonga (**-DD**), PEDRO & PEDRÓGÃO 6931 (LMA). – **1932** (Melsetter): Tsetserra (**-BD**), EXELL, MENDONÇA & WILD 293 (BM, LISC, SRGH).

---

[1] *Young* plants of *A. usambarense* (individuals flowering for the first time, i.e., ca one year old) may, however, superficially closely resemble *A. zimbabwense* (field obs.!).

4. Anthospermum emirnense BAKER in J. Bot. (London) **20**: 139 (1882).
Type: Central Madagascar, PARKER s.n. (K, holo.!).

Dioecious, erect, mostly single-stemmed shrubs ca (0.5) 0.75–2 m tall.
Stems ca 3 cm in diameter near the base, much-branched above, branches
ascending, ca (2) 3–4.5 mm in diameter; younger parts ± terete, often
reddish-purlish tinged, covered with papillae < 0.1 mm or whitish
spreading hairs to 0.3 mm long, older parts reddish-brown to dark grey,
glabrescent, epidermis peeling off; internodes often longer than the leaves.
Leaves decussate or, less frequently, also in whorls of 3 (sometimes on one
and the same plant), pseudo-verticillate; blades most commonly ca 10–
20 × 1.2–2.5 mm (in ♂ often somewhat smaller than in ♀), narrowly
oblanceolate to ± linear-lanceolate and with ± reflexed margins [on new
growth not uncommonly to 30 (40) × 4 (6) mm, oblanceolate to oblong-
lanceolate, narrowed to the base and often with flat margins][1], ±
glabrous, shortly hairy near the base on the lower surface and/or the
midrib below or with short, forwardly directed, prickle-like hairs on the
margins; apices ± acute; petioles subobsolete or, only on the largest
leaves, to 1–2 mm long; stipular sheaths hairy, broadly cup-shaped, ca
0.9–1.7 mm long, with (1) 3 (less commonly more) hairy, ± filiform setae,
the longest ca (0.7) 1–2 mm. Inflorescences ± extensive, dimorphic, in ♀
often quite contracted, ± cylindrical inflorescence zones. Flowers
subsessile, in large numbers (in ♀ often more than in ♂) clustered at the
nodes; ♂, ♀ (very rarely also odd ☿). Corolla 4-merous (very rarely also 5-
merous), yellowish(-green), glabrous. ♂: tube 0.5–0.7 (1) mm long,
broadly funnel-shaped, lobes 1.4–2 × 0.7–0.8 mm, lanceolate, recurved;
stamens 4, filaments 1–1.7 mm long, anthers 0.9–1.2 × 0.3–0.4 mm;
rudimentary ovary (very) small but well discernible. ♀: tube 0.3–0.7 mm
long, ± cylindrical to narrowly funnel-shaped, lobes 0.2–0.4 × 0.2–
0.3 mm, ± lanceolate, ± erect; style ± 0; stigmas 2, (2.4) 3–5.6 mm long
(shorter in odd ☿); ovary ca 0.7–0.9 × 0.4–0.6 mm, glabrous to scabrid-
hairy, crowned by 4 (very) inconspicuous calyx lobes. Fruits reddish
(-brown) supported by narrow, U-shaped carpophores to ca half as long as
the fruits; each mericarp 1.3–1.9 × 0.7–0.9 mm, ± oblong to obovate,
dorsal side convex, glabrous to covered with whitish hairs or papillae to
0.1 mm long, ventral side plane to ± concave and with a median vertical
ridge; commissure large; mericarps crowned by 2 indistinct, ± triangular
calyx lobes ca 0.1–0.2 mm long or calyx lobes ± lacking.

---

[1] Herbarium specimens rather often only consist of (parts of) inflorescences
(especially of the more conspicuous ♀ plants!) with *foliage leaf-like bracts*
subtending the flower clusters. These are considerably smaller (ca 5–10 × 0.5–
1 mm) than true foliage leaves.

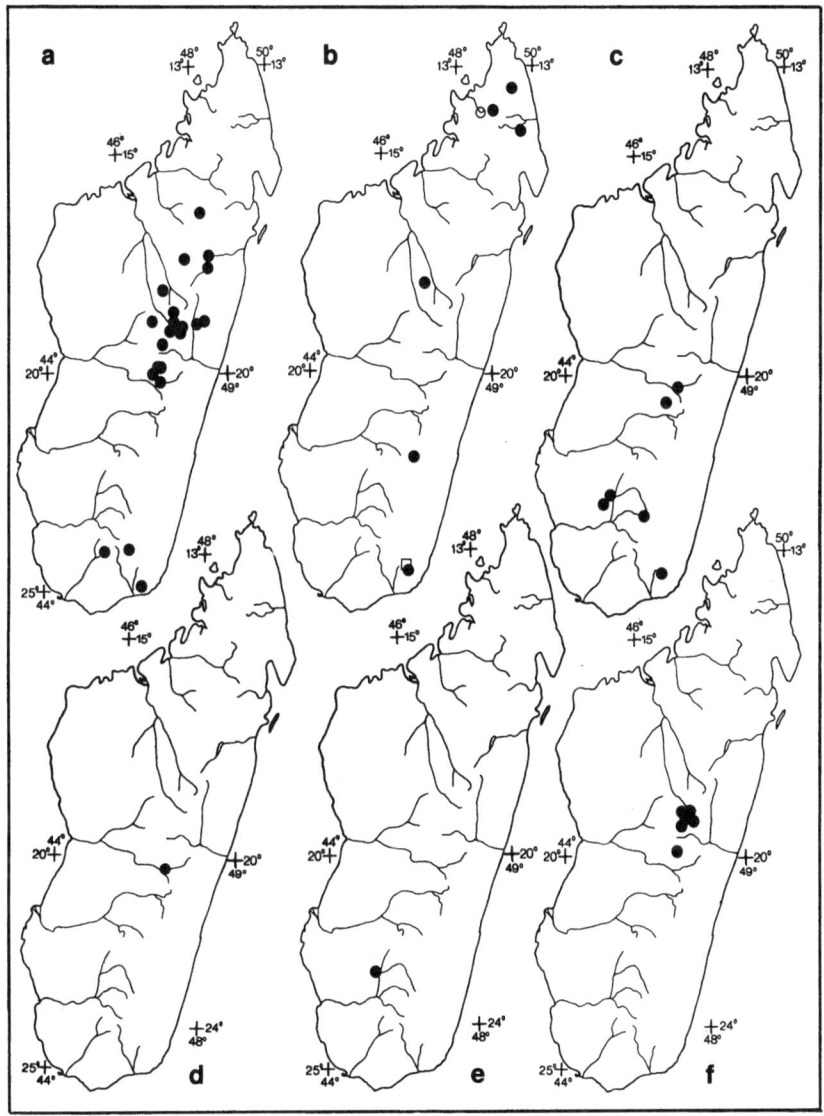

Fig. 67. Distribution of *Anthospermum* in Madagascar. *a A. emirnense, b A. madagascariense* (square: "var. *australe*"; open circle: "var. *humile*"), *c A. isaloense, d A. ibityense. e A. longisepalum, f A. perrieri*

Chromosome Number (Table 3): n = 11, 2 n = 22.

Average Pollen Diameter (Fig. 48): 24.5–27.2 μm.

Habitat: Growing at the edge of wet (montane) forests and in clearings, in dry rocky areas and not uncommonly in ± disturbed sites (road sides, edge of pine plantations or cultivated land). Sometimes also found, together with *Phillippia, Helichrysum,* etc., as undergrowth in remnants of dry sclerophyllous forest ("Tapia"; *Uapaca bojeri*– *Chlaenaceae* forest). Usually occurring over gravelly or red, lateritic soil. – Ca (?200–) 800–1 700 m.

Distribution (Map, Fig. 67 a): Madagascar. From the Tampoketsa d'Analamapitso (= Analamaitso Plateau; Bemarivo R.) and Lake Alaotra S to NW of Ft. Dauphin, but apparently absent from the Andringitra massif (± disjunct!).

Critical Remarks: *A. emirnense* is a very variable species. Variation is mainly found in size, shape and arrangement of the leaves (decussate and/or in whorls of 3) and internodes lengths. According to field observations in central Madagascar, even plants of a single population may show considerable variation in the above mentioned characters.

As in a number of other dioecious species, ♂ plants tend to have smaller leaves than ♀ plants (see C.12.3.), and the leaves on the main axis of a plant and on individuals flowering for the first time may be significantly larger than those on lateral branches or on older plants respectively. All this contributes to the variability of *A. emirnense*.

Narrow- and small-leaved forms of *A. emirnense* may, in habit, closely approach *A. madagascariense* (7.) but are easily distinguished by floral and fruit characters (see p. 217 for details).

A. emirnense shows a close affinity to the afromontane species group around *A. usambarense* (*A. welwitschii, A. ammannioides,* etc.) from the African mainland. It agrees with the E Zimbabwean *A. ammannioides* (5., below) in more characters than with any other species of that group, but these similarities should perhaps not be read as evidence for a direct, close relationship of these two particular taxa. I am quite convinced that this entire *A. usambarense* group is a natural one originally derived from a common ancestral stock (see E.1.2.).

Collections

Madagascar. – **1648**: Analamapitso (= Analamaitso), Bemarivo R., PERRIER DE LA BÂTHIE 3619 (?3819) (P). – **1747**: Mahobo (?Mahabe), SCOTT ELLIOT 1931 (E). – **1748**: L. Alaotra, Herb. Jard. Bot. Tana 3924 (P); SE of –, Mt. Ankaroka, HUBERT & COURS 17541 (P). – **1846/1947**: Imerina, around Marmarivo (= Miarinariva?), RUSILON 6, 109 (G). – **1847**: Tampoketsa d'Ankazobe, Ambohitately Reserve, PUFF, IGERSHEIM & RAJEMISOA 850808-1/4 (TAN, WU); Ambohimanga, DECARY 660 (P); Tananarive (= Antananarivo) and surroundings, herb. BLACKBURN s.n. (K), BOJER "VI.7" (W), DECARY 6770 (P), GOUDOT s.n. (G), HUMBERT 11049 (P), HUMBERT & PERRIER DE LA BÂTHIE 2198 (P); –, Mt. Angavokely, BERNARDI 11057, 11057 bis (G, K), BOSSER 13709 (TAN), PUFF 800726-2/1 (WU). – **1848**: Moramanga, AFZELIUS s.n. (S × 2, sub PRE 57576, 57577), MELLER s.n. (K), SCHLIEBEN 8081 (B, BM, BR, G, K, M, TAN); – D.,

Analamazaotra forest, Viguier & Humbert 1027 (B, P), SF 240 (TAN). – **1946**: Lake Andraikiba, Forsyth-Major s.n. (G); L. Tritriva, Puff 800729-1/1A, -1/1B (WU). – **1947**: around Arivonimamo, Bosser 1066 (TAN); around Miadanandriana, Puff, Igersheim & Rajemisoa 850824-2/1 (TAN, WU); Ankaratra region, Rousson s.n. (P); Andramasina, Descoings 3060 (TAN), Puff, Igersheim & Rajemisoa 850825-2/2 (TAN, WU); around Antsirabe, Humbert & Swingle 4609, 4652 (P), Perrier de la Bâthie 3939, 4021 (P). – **2047**: from Col des Tapias to S end of Mt. Ibity, Puff 800730-2/1 (WU). – **?2145**: Mt. Belambana, between Mangoky and Mananura basin, Perrier de la Bâthie 3919 (P). – **2445**: around Ampandrandava, between Beliky and Tsivory, Seyrig 577 (P). – **2446**: Mandrare basin, Sakamalio R. valley (tributary of Manambolo R.), Humbert 13391 (P). – **2546**: around Fort Dauphin, Mt. Oniva (Taviala), N of Ranopitso (= Ranopiso), Humbert 5853 (P). – Additional collections: Baron 653 (BM, K), 3950 (BM, E, K), 1752, 2039, 2122, 3322, 5192 (?5132), 5210, 5565 (all K); Benoist 929, 1627 (TAN); Bojer s.n. (G, K, W); Bosser 1121, 5194 (TAN); Descoings 2792, 2936, 3150, 3182 (TAN); Goudot s.n. (G × 4); Herb. Jard. Bot. Tana 1087, 1118 (TAN); Herb. Stat. Agric. Lac Alaotra 1832 (TAN); Hildebrandt 3513 (W), 3815 (BM, G, K, M, P, US, WU); Hilsenberg & Bojer s.n. (BM); Hodgkin & Stansfield 222, 273, 285 (K); Humblot 653 (K, P); Le Myrne & Vitus s.n. (P); Parker s.n. (K); Perrier de la Bâthie 3896 (P); Puff, Igersheim & Rajemisoa 850825-1/1, 850826-2/3, 850827-1/1, -2/4 (TAN, WU); Rakotozafy 9, 25 (TAN); Rusilon 19, 24 (G); Shufeldt 96 (US).

5. Anthospermum ammannioides[1] S. Moore in J. Linn. Soc. (Bot.) **40**: 102 (1911).

Type: Zimbabwe (Rhodesia): "Gazaland", Melsetter, Swynnerton 2156 (BM, holo.!; K, iso.!).

Dioecious or ± dioecious (♂ or ♀ plants occasionally with odd ♀ flowers), erect, single-stemmed shrubs ca 1–3 m tall. Stems to 40 mm in diameter near the base, much-branched above, branches ascending to ± erect, ca 4–7 mm in diameter; younger parts reddish-brown to greyish, densely covered with greyish-white hairs ca 0.2–0.7 mm long; older parts greyish-brown to greyish, glabrescent, epidermis peeling; upper internodes as long as or, more commonly, shorter than the leaves; branches usually without conspicuous leafy short shoots. Leaves in whorls of 3; blades 18–45 (65) × (2) 3–10 (15) mm (in ♂ often smaller than in ♀), narrowly (ob)lanceolate to oblong-lanceolate, narrowed to the base, both surfaces sparsely to ± densely covered with witish spreading hairs c. 0.1– 0.3 mm long, but lower surfaces often more densely hairy, particularly on the midvein, and upper surfaces sometimes ± glabrous; margins usually ± flat; apices acute to ± acuminate; petioles subobsolete or ca 2–6 mm

[1] The species epithet is based on the genus name *Ammannia* L. (*Lythraceae*). Moore's original spelling ("*ammanioides*") is, therefore, incorrect.

long on the largest leaves; stipular sheaths covered with hairs 0.3–0.7 mm long, broadly cup-shaped, ca 0.5–1.5 mm long, with a median, hairy seta ca (0.7) 1.5–4 (6.5) mm long, sometimes with an additional small seta on either side. Inflorescences in ♂ plants ± inconspicuous, made up of ± sessile, ± widely spaced several- to many-flowered axillary clusters; in ♀ plants compact, ± cylindrical inflorescence zones ca 20–45 (70) mm long and ca 15–20 mm in diameter, made up of ± sessile, closely spaced very many-flowered axillary clusters. Flowers subsessile (♀) or mostly with distinct pedicels ca 0.7–1.5 (2) mm long (♂); ♂, ♀ (sometimes also odd ⚥). Corolla 4-merous (very rarely 5-merous), greenish-yellow, sometimes buds purplish tinged, at least in ♂ covered with whitish spreading hairs on the outside. ♂: tube (0.5) 0.7–1 mm long, (± narrowly) funnel-shaped, lobes 1.5–2.4 × 0.6–1 mm, lanceolate, recurved; stamens 4, filaments 1.2–1.9 mm long, anthers 1.2–1.7 × 0.3–0.5 mm; rudimentary ovary minute, hardly discernible. ♀: tube 0.3–0.5 (0.8) mm long, ± cylindrical, lobes 0.2–0.5 × 0.1–0.2 mm, ovate, ± erect; style ca (0.5) 2–5 (6.5) mm long, stigmas 2, (4.8) 6–14 mm long, whitish-grey to greyish-green (much shorter in odd ⚥); ovary 0.7–1 × 0.6–0.8 mm, glabrous, crowned by 4 indistinct calyx lobes or calyx lobes ± lacking. Fruits reddish brown, supported by ± narrow, U-shaped carpophores up to ca half as long as the fruits; each mericarp (1.5) 1.8–2.2 (2.4) × 0.8–1 mm, ± oblong to obovate, dorsal side convex, glabrous, ventral side plane to ± concave and with a prominent median vertical ridge, commissure large; mericarps crowned by 2 indistinct, ± triangular calyx lobes ca 0.1–0.2 mm long or calyx lobes ± lacking. – Fig. 31 d, 54 a–b.

Chromosome Number (Table 3): n = 11, 2n = 22 (Fig. 47 a).

Average Pollen Diameter (Fig. 48): 27.5–32.1 μm.

Habitat: At the edge of afromontane forest, in scrub of stream banks or in bush clumps amongst rocks. Often associated with *Hypericum revolutum, Aloe arborescens, Myrica microbracteata, Rapanea melano-phloeos, Maesa lanceolata, Myrsine africana, Chrysanthemoides monilifera* subsp. *septentrionalis,* and occasionally also growing together with (the less common) *A. vallicola* (6., below). Sometimes also found in ± disturbed sites (old plantations, road banks, etc.); seldom in grassland (also see Critical Remarks). Recorded as occurring over granite and quartzite, in dolerite and in red, clayey soils. – Ca (1 100) 1 400–2 300 (2 500) m.

Distribution (Map, Fig. 66 a): Zimbabwe (E) and Mozambique (MS). In the E highlands of Zimbabwe (from the Inyanga Distr. S to the Melsetter Distr.) and in the neighbouring areas of Mozambique; also on Gorongosa Mt. (Gorongosa National Park).

Critical Remarks: *A. ammannioides* is somewhat variable in leaf size, shape and hairiness. The size and shape of the leaves, to some extent, depend on the sex of the plants (see C.12.3.: leaf length and width differences in ♂ and ♀ plants of a population; Figs. 54 a–b and 55). Leaves are never entirely glabrous but in some forms, hairs are only found on the lower surface (on the midvein in particular).

In ♀, the old inflorescences remain on the plant as compact, ± cylindrical, black-grey structures. The inflorescence main axes may either continue to grow vegetatively after flowering or cease growth so that the old ♀ inflorescences are found either lower down on the branches or in a terminal position respectively (cf. Fig. 54 b).

Plants may sometimes occur in habitats which are occasionally (e.g., bush clumps amongst rocks) or ± frequently and regularly (e.g., grassland) exposed to fires. They are killed by fire, but *A. ammannioides* is a good "seed regenerator". Its seeds germinate well and quickly (see C.12.7. and Fig. 61 j) and, according to observations on plants raised from seed in the greenhouse, the first flowering takes place within one year after seedsowing.

A. ammannioides is perhaps most closely allied to *A. welwitschii* (1., above), a widely and disjunctly distributed species which, however, is not known from E Zimbabwe and the neighbouring parts of Mozambique. In my opinion, *A. ammannioides* is a "good" species which is readily distinguished from *A. welwitschii* by its larger and broader, hairy and *ternately* arranged leaves, its fruits without distinct persistent calyx lobes, and its more compact, ± broadly cylindrical ♀ inflorescence zones.

Collections

Zimbabwe. [E:] – **1832** (Umtali): Inyanga village (**-BA**), Fries, Norlindh & Weimarck 2598 (BM, K, M, PRE, SRGH); Inyanga D., Little Connemara, "World's View" (**-BB**), Puff 790125-6/1 (WU), Seagrief C.A.H. 3104 (SRGH); West 4468 (SRGH); Vukutu Farm, W of Juliasdale (**-BC**), Puff 790124-1/1 (WU); Rhodes Inyanga National Park, Circular Drive above Pungwe Falls (**-BD**), Puff 790126-2/1 (WU); Stapleford Forest Reserve, "North Patrol" (**-DB**), Barrett 69/56 (B, K, PRE, SRGH), Puff 790127-1/2 (WU); Penhalonga (**-DC**), Galpin 9227 (PRE), Matineau 394 (SRGH). – **1932** (Melsetter): Vumba Mts., Castle Beacon (**-BA**), Puff 790128-1/5 (WU); Maritz Nek, N of Cashel (**-BD**), Cleghorn 1051 (SRGH); Lionhills Forest Reserve, S of Melsetter (**-DD**), Goldsmith 30/68 (K, LISC, MO, PRE, SRGH). – **1933** (Vila Pery): Chimanimani Mts., Longgulley (**-CC**), Noel 2029 (SRGH). – **2032** (Chipinga): Melsetter D., Kasipiti (**-BA**), Loveridge 1579 (BR, K, SRGH). – Additional collections: Bamps, Symoens & van den Berghen 147 (BR, SRGH); Brain 9383 (K, SRGH); Burrows 196 (SRGH); Chase 340 (BM, K, NU, SRGH), 5653, 5654 (BM, K, PRE, SRGH); Coates Palgrave s.n. sub SRGH 70603 (SRGH); Corner s.n. (E); Fries, Norlindh & Weimarck 3788 (BR, MO, SRGH); Gilliland 260, 325 (BM, K, PRE, SRGH); Goodier & Phipps 66 (SRGH); Greatrex s.n. sub SRGH 14925 (SRGH); Henkel sub Eyles 2610 (SRGH); Humbert 15556, 15717, 15791 (P); Loveless s.n. sub SRGH 144405 (SRGH); McGregor 78/37 (SRGH); Peter 30870 (B, K); Phipps 1177 (BR, K, PRE, SRGH); Puff 790124-3/1, 790125-2/1, -4/1 (WU); Rattray 953 (K, SRGH); Rushworth 772 (BR, SRGH); Simon 859 (K, PRE, SRGH); Schelpe 565 (BOL, LISC, SRGH); Swynnerton 2156, 2157 (BM, K); Taylor 3222 (NBG); vanden Berghen 75 (BR); West 4172, 4599 (SRGH).

    Mozambique. [MS:] – **1833** (Vila Gouveia): Báruè, serra de Choa, 16 km from Vila Gouveia (**-AA**), Torre & Correia 18685 (LISC, LMU). – **1834** (Villa Paiva de Andrada): Gorongosa (National Park) (**-AC**), Macêdo 1731 (LMA), Simão 1096 (LISC, LMA), Tinley 1837 (SRGH). – **1932** (Melsetter): from Vila Manica to Vumba (**-BB**), Pedro & Pedrógão 6756, 6761 (LMA), serra de Vumba, Mendonça 2544 (LISC); serra Zuira, Tsetserra (**-BD**), Torre & Correia 15557, 15699 (LISC), Torre & Pereira 12742 (LISC); Manica, Rotanda (**-DB**), Torre & Correia 13147 (LISC), Barbosa 1615 (LISC). – **1933** (Vila Pery): Chimanimani Mts., between Skeleton Pass and the Plateau (**-CC**), Grosvenor 361 (K, LISC, SRGH).

6. *Anthospermum vallicola* S. Moore in J. Linn. Soc. (Bot.) **40**: 103 (1911). Type: Zimbabwe (Rhodesia): "Gazaland", Chimanimani Mts., 7 000 ft., Swynnerton 2155 (BM, holo.!).

    Dioecious, erect, often single-stemmed shrubs ca 1–3 m tall with ± rounded crowns. Stems to 50 mm in diameter near the base, much-branched above, branches ascending, ca 2–5 mm in diameter, ± terete; younger parts reddish-brown to dark grey, sparsely to densely covered with very short whitish hairs; older parts often dark grey to blackish, glabrescent, epidermis peeling; internodes much shorter than the leaves. Leaves conspicuously crowded near the tips of the branches, in whorls of 3, pseudo-verticillate; blades (10) 15–35 (55) × (0.5) 1–2 (2.5) mm, linear to linear-lanceolate, glabrous, upper epidermis shiny, with large, conspicuous epidermis cells; margins revolute to ± flat; apices acuminate and sometimes ± curved; petioles subobsolete; stipular sheaths hairy, broadly cup-shaped, ca 1–2 (2.5) mm long, with a median, narrowly triangular seta ca (0.3) 0.7–1 (1.5) mm long, sometimes with an additional shorter seta on either side. Inflorescences in ♂ plants ± inconspicuous, made up of ± sessile, ± widely spaced several- to many-flowered axillary clusters; in ♀ plants compact, ± cylindrical inflorescence regions, made up of ± sessile, closely spaced (very) many-flowered axillary clusters. Flowers subsessile (♀) or mostly with distinct pedicels and peduncles to ca 1 mm long (♂); ♂, ♀. Corolla 4-merous, greenish-yellow to pale yellow, sometimes reddish-tinged outside, glabrous. ♂: tube 0.6–0.8 (1) mm long, (broadly) funnel-shaped, lobes 1.7–2.4 (2.8) × 0.7–1 mm, lanceolate, recurved; stamens 4, filaments ca 1.4–1.7 (2) mm long, anthers (1.2) 1.5–1.7 (2) × 0.3–0.5 mm; rudimentary ovary minute, sometimes hardly discernible from the pedicels but always with 4 ± linear to linear-lanceolate calyx lobes ca 0.1–0.4 mm long; occasionally also rudimentary stigmas present. ♀: tube 0.7–1.4 mm long, ± cylindrical, lobes 0.4–0.8 × 0.1–0.2 mm, linear to linear-lanceolate, ± erect to spreading or ± recurved; style ca 0.5–1.5 mm long, stigmas 2, 5–9.5 (14) mm long, thickish (to ca. 0.5 mm in diameter), greyish to greyish-green; ovary ca 1–1.5 × 0.8–1.3 mm, glabrous, shiny, crowned by 4 long calyx lobes. Fruits reddish brown, shiny, supported by ± broad and massive U- or V-shaped carpophores to ca half as long as the fruits; each mericarp (1.8) 2.2–3 (3.5) × 1.3–1.7 (1.8) mm, oblong, dorsal side shallowly convex, glabrous, ventral side ± plane and with a median vertical ridge; commissure large; mericarps crowned by 2 narrowly triangular to ± deltoid calyx lobes ca (0.5) 0.8–1.8 (2.1) mm long, often unequal in size, one sometimes only half as long as the other. – Figs. 2 b, 31 a–b, 37 d.

Chromosome Number (Table 3): n = 11, 2 n = 22.

Average Pollen Diameter (Fig. 49): 27.1–30.5 μm.

Habitat (Fig. 2 b): In mixed sclerophyll scrub (for a detailed description see PHIPPS & GOODIER 1962: 308 and GOODIER & PHIPPS 1962: 8; Chimanimani Mts.) and also in "heath" communities dominated by *Phillipia*. Often in craggy areas, amongst rocks or boulders (primarily dolerite and quartzite), or in glens, where it usually forms large populations. – Ca 1 700–2 600 m.

Distribution (Map, Fig. 66 b): Zimbabwe (E) and Mozambique (MS). In the E highlands of Zimbabwe from Mt. Inyangani S to the Chimanimani Mts. (in the latter also on the Mozambique side), and on Gorongosa Mt. (Gorongosa National Park, Mozambique).

Critical Remarks: A very distinct species that cannot be mistaken for any other taxon occurring in S Central Africa. It is probably allied to *A. welwitschii* (1.) and *A. madagascariense* (7., below) (both belonging to the *A. usambarense* group), although this affinity does not appear to be a very close one. The large shrubby habit in combination with dioecy and the fruits with conspicuous persistent calyx lobes, and the afromontane habit appear to indicate such a relationship. With *A. madagascariense* it, furthermore, shares the long narrow, ternately arranged leaves.

In the Chimanimani Mts., *A. vallicola* appears to hybridize occasionally with *A. zimbabwense* (3.) which also occurs there although it seems to be less common than the former. See Table 11 for a comparison of selected characters of *A. vallicola, A. zimbabwense* and putative hybrid collections (e.g., FINLAY 20, SRGH; GROSVENOR 196, LISC, SRGH; WILD 4578, LISC, MO, SRGH; all ♂, all collected between 2 300 and 2 450 m).

Table 11. Comparison of selected caracters of *A. vallicola*, *A. zimbabwense* and putative hybrids

	A. vallicola	A. zimbabwense	putative hybrids
Height (cm)	100–300	20–60 (100)	45–60
Internodes	much shorter than the leaves	longer than or as long as the leaves	as long as or slightly longer than the leaves
Leaves			
size (mm)	(10) 15–35 (55) × (0.5) 1–2 (2.5)	(3) 4–8 (12) × 0.5–1.5 (3)	7–12 × 0.5–1
upper surface shiny, epidermis cells large	+	−	+
♂ flowers			
bud	long, obovate	short, broadly elliptic and abruptly constricted below	intermediate
rudimentary ovary: calyx lobes	+ (conspicuous, to 0.4 mm)	± (ca 0.1 mm) or −	± , indistinct

Collections

Zimbabwe. [E:] – **1832** (Umtali): summit of Mt. Inyangani (**-BD**), PUFF 790125-1/1 (BR, J, W, WU). – **1932** (Melsetter): Vumba Mts., Castle Beacon (**-BA**), PUFF 790128-1/1 (BR, J, W, WU); Melsetter D., Mt. Peni summit (**-DD**), CHASE 6415 (B, K, LISC, PRE, SRGH), GOLDSMITH 22/68 (BR, K, LISC, MO, SRGH). – **1933** (Vila Pery): Chimanimani Mts., Bundi Plain (**-CC**), DRUMMOND 9106, 9106A (K, SRGH). – Additional collections: BURROWS 483 (SRGH); CHASE 6037 (BM, BR, COI, K, LISC, PRE, SRGH); DRUMMOND 8921 (K, SRGH); GILLILAND 901a (BM, K); GOODIER & PHIPPS 25 (K, LMA, SRGH); NOEL 2120 (SRGH), 2355 (K, SRGH); NORLINDH & WEIMARCK 4993 (BM, BR, K, M, MO, PRE, SRGH); RUSHWORTH 952 (K, SRGH); SWYNNERTON 2155 (BM); WILD 2850 (K, SRGH).

Mozambique. [MS:] – **1834** (Vila Paiva de Andrada): Gorongosa (National Park) Mt. (**-AC**), MACÊDO 1974 (LMA), TINLEY 1838 (K, SRGH), 1968 (SRGH). – **1933** (Vila Pery): Manica, Chimanimani Mts., DUTTON 100 (LMA); – , E of Point 71 (**-CC**), HALL 436 (SRGH). – Imprecise locality: "Manica e Sofala", PEDRO 2304 (LISC).

7. *Anthospermum madagascariense* HOMOLLE ex PUFF, spec. nova, *A. emirnensi* simile sed foliis linearibus plus minusve aciformibusque et fructibus lobis calycis conspicuis coronatis differt.

Type: Madagascar, Andringitra Massif, brousse éricoide vers 2 600 m, February 1929, PERRIER DE LA BÂTHIE 14465 ♂ et ♀ (P, holo.)[1].

   (– *A. madagascariense* HOMOLLE var. *australe* HOMOLLE
    – var. *humile* HOMOLLE
    – var. *laxum* HOMOLLE
    – var. *macrocarpum* HOMOLLE, nom. nuda)

Dioecious, erect, mostly single-stemmed shrubs ca (0.5) 0.75–2 m tall. Stems much-branched, branches ascending to erect, ca 2–4 mm in diameter, terete; younger parts reddish-brown, densely (less commonly ± sparsely) covered with whitish spreading hairs ca 0.1–0.2 (0.4) mm long, older parts brownish-grey, glabrescent, epidermis peeling; internodes (much) shorter than the leaves, plants densely leafy above. Leaves in whorls of 3 or occasionally decussate (sometimes on one and the same plant), pseudo-verticillate; blades (5) 7–18 (25) × 0.3–0.7 (1) mm, linear, ± needle-like due to strongly revolute margins, glabrous; apices acute to acuminate; petioles obsolete; stipular sheaths hairy (less commonly subglabrous), cup-shaped, ca 0.7–1.5 mm long, with (5) 3 (1) setae, the longest ca (0.7) 1–2.4 mm. Inflorescences ± inconspicuous, made up of ± sessile, several- to few-flowered axillary clusters. Flowers ♂, ♀; corolla 4-

---

[1] Another sheet, PERRIER DE LA BÂTHIE 14465 ♂ (P), must be considered as a paratype.

merous, yellowish, glabrous[1]. ♂: tube (0.5) 0.7–1.4 mm long, (narrowly) funnel-shaped, lobes 2–2.6 (3.2) × 0.8–1 mm, lanceolate, recurved; stamens 4, filaments ca 1.4–2 mm long; anthers 1.5–2.2 × 0.3–0.5 mm; rudimentary ovary very small but well discernible and crowned by small calyx lobes. ♀: tube 0.2–0.5 (0.8) mm long, cylindrical, lobes 0.3–0.7 × 0.1–0.2 mm, linear-lanceolate, ± erect; style 0.5–1 mm long; stigmas 2, 4.4–8 mm long, greyish; ovary ca 0.7–1.5 × 0.3–0.9 mm, glabrous[2], crowned by 4 calyx lobes. Fruits reddish(-purplish)-brown, supported by small, U-shaped carpophores; each mericarp ca 2.4–3.7 × 1.4–2 mm, ovate to ± oblong, dorsal side convex, glabrous[2], ventral side ± plane and with a median vertical ridge; commissure large; mericarps crowned by 2 linear-lanceolate calyx lobes ca 0.3–0.7 (1.7) × 0.2–0.3 (0.4) mm, one often longer than the other.

Chromosome Number (Table 3): 2 n = 22.

Average Pollen Diameter (Fig. 48): 26.3–29.1 µm.

Habitat: Mostly occurring in ericaceous scrub and growing in rocky areas or on scree slopes amongst rocks (gneiss, quartzite). Often associated with *Phillipia*. – Ca 1 450–2 600 m².

Distribution (Map, Fig. 67 b): Madagascar. Mostly confined to the highest mountains of the island and occurring from Anjanaharibe Mt. and the Tsaratanana in the N to the Andohahela massif in the S but not recorded from the Ankaratra and apparently rare in central Madagascar.

Critical Remarks: *A. madagascariense* is a quite variable species. Mme. A. M. HOMOLLE has, on herbarium sheets, indicated her intention to divide the species into the varieties "*laxum*", "*macrocarpum*", "*australe*" and "*humile*" (next to the "typical" *A. madagascariense*).

"Var. *laxum*" refers to plants from the Andringitra massif with the longest leaves [to 18 (25) mm long] which do not appear as crowded as in other plants because of their decussate (rather than ternate) arrangement. Since it (1) is not possible to distinguish them from "typical" *A. madagascariense* by any other characters and since (2) leaf size and "crowdedness" of the leaves are variable and "extremes" are connected with each other by a series of intermediate forms, I prefer not to recognize this variety.

"Var. *macrocarpum*", in my opinion, only refers to a collection with fully mature fruits.

Both "var. *australe*" and "var. *humile*", each represented by a single

---

[1] But see "*A. andringitrense*", Critical Remarks, below!

[2] But see "var. *australe*" and "var. *humile*", Critical Remarks, below!

collection, agree with *A. madagascariense* in all characters except for the ovaries and fruits. "Var. *australe*" (Beampingaratra massif, Maloto valley, fôret sur latérite de gneiss; 800–1 500 m, HUMBERT 6304; GH, P) has fruits which are densely covered with whitish spreading hairs ca 0.3–0.5 mm long and crowned by 4 long, linear-lanceolate calyx lobes. "Var. *humile*" (Manangarivo massif, 1 400 m, PERRIER DE LA BÂTHIE 3807; P) has fruits which are covered with short whitish hairs or papillae 0.1 mm or less long and crowned by 4 broad calyx lobes which are ± united in twos. These two specimens (not in Collections) are provisionally grouped with *A. madagascariense* (also included in map, Fig. 67 b; specifically marked) and are, at present, not formally recognized. More material would be needed in order to draw definite conclusions.

Yet another collection from the Andringitra massif (gneiss, 2 000 m, PERRIER DE LA BÂTHIE 3492; P; not mapped; not in Collections), provisionally named "*A. andringitrense*" by Mme HOMOLLE, must be mentioned in connection with *A. madagascariense*. It agrees with the latter in general habit, leaf size and shape, stipules, and has fruits which are identical to those of "var. *humile*" (see above). Flower size and shape fall in the range of *A. madagascariense* but the corollas are ± hairy on the outside, and the only collection available has ⚥ + ♀ flowers (whereas all known *A. madagascariense* plants are strictly unisexual). The internodes, furthermore, are as long as or even longer than the (decussately) arranged leaves; foliage therefore, is not as "dense" as in *A. madagascariense*. More and better material, however, would be needed. "*A. andringitrense*" may turn out to be a good and distinct new species.

A. madagascariense is clearly allied to *A. emirnense* (4.) and, consequently, to the African (afromontane) – Madagascan *A. usambarense* group. *A. madagascariense* normally occurs at higher altitudes than *A. emirnense* and in more typically montane habitats comparable to those in afromontane regions of mainland Africa. The distribution ranges of the two species hardly show any overlap (compare maps, Fig. 67 a and b). *A. madagascariense* is distinguished from *A. emirnense* not only by its narrower, more crowded leaves but also by the absence of the rather conspicuous, ± congested ♀ inflorescences; the mericarps, moreover, differ in bearing distinct persistent calyx lobes and the ♂ flowers differ in having narrowly funnel-shaped corolla tubes and longer lobes (differences in the corollas are also reflected in quite different bud shapes: long, obovate, abruptly constricted below in *A. madagascariense*; short, broadly elliptic in *A. emirnense*). In both species, the leaf arrangement is variable (decussate and/or in whorls of 3), although in *A. madagascariense* ternate leaves are more common, while in *A. emirnense* more plants have

15*

decussately arranged leaves. As in some other *Anthospermum* species (e.g., *A. ternatum*), leaf arrangement may vary within a single plant, but on a single individual it is always the lower nodes that bear decussate leaves whereas ternate leaf arrangement is confined to the upper (younger) parts of the plant.

Collections

Madagascar. – **1349**: Mt. Anjanaharibe, Herb. Stat. Agric. Lac Alaotra 3847 (TAN). – **1449**: Tsaratanana massif, PERRIER DE LA BÂTHIE 16124 (P); Marojejy massif, HUMBERT 22760 (P), HUMBERT & COURS 23847 (P). – **1747**: Tampoketsa d'Ankazobe, Mahatsinja-Ankazobe rd., ca 2 km NW of Kiangara turnoff, PUFF, IGERSHEIM & RAJEMISOA 850807-3/1 (TAN, WU). – **2246**: Andringitra massif, HEIM s.n. (P), HUMBERT 3801 (G, P), PERRIER DE LA BÂTHIE 3940, 3941, 13611, 14428 (all P), 14465 (P × 2). – **2446**: Andohahela massif, HUMBERT 6185 bis, 13655 (P). – Imprecise localities: between Tamatave and Antananarivo, MELLER s.n. (K); "NW Madagascar", BARON 5210 (K, P).

8. *Anthospermum isaloense* HOMOLLE ex PUFF, spec. nova, foliis laterioribus (sub)coriaceisque ab *A. madagascariensi*, inflorescentiis paucifloris et fructibus lobis calycis conspicuis coronatis ab *A. emirnensi* praeclare differt.
Type: Madagascar, Col des Tapias, between Ranohira and Sakaraha, ca 1 050 m, PUFF 800814-3/2 (WU, holo.; BR, K, P, iso.).

Dioecious, erect, single- or sometimes few-stemmed shrubs ca 75–150 cm tall. Stems to 30 mm in diameter near the base, mostly much-branched above; branches ascending, often ± regular and paired; younger parts ± 4-angled, scabrid, older parts terete, reddish-brown to dark grey, glabrescent, epidermis peeling; internodes shorter than or as long as the leaves. Leaves decussate, pseudo-verticillate; blades ascending or ± erect, 10–25 (35) × (1.5) 2–4 (5) mm, linear-lanceolate to oblanceolate, rather coriaceous, glabrous, upper surface usually shiny and epidermis cells often large, ± conspicuous; margins flat to slightly revolute; apics ± acute to mucronate; petioles subobsolete; stipular sheaths glabrous, cup-shaped, ca 0.7–1.2 (1.4) mm long, with a median, gland-tipped seta ca 0.3–0.6 (0.8) mm long. Inflorescences inconspicuous, made up of ± sessile, ca 6- to 2-flowered axillary clusters. Flowers ♂, ♀; corolla 4-merous, greenish-yellow to pale yellowish, sometimes with a few hairs on the outside. ♂: tube 0.7–0.9 mm long, broadly funnel-shaped, lobes 1.7–1.9 × 0.5–0.8 mm, lanceolate, recurved; stamens 4, filaments 1–1.4 mm long, anthers 1.2–1.5 × 0.2–0.4 mm; rudimentary ovary very small. ♀: tube 0.4–0.8 mm long, ± cylindrical, lobes 0.2–0.7 × 0.1–0.2 mm, ± linear-lanceolate, erect; style ca 0.3–0.7 mm long; stigmas 2, 2.9–5.8 mm long, greyish; ovary ca 1.5×0.8–0.9 mm, glabrous, crowned by 4 sometimes unequal calyx lobes. Fruits reddish-brown, supported by small, $\pm$ U-shaped carpophores; each mericarp 1.8–2.4×0.8–1.1 mm, $\pm$ oblong to obovate, dorsal side $\pm$ convex, glabrous or $\pm$ scabrid, ventral side $\pm$ plane and with a median vertical ridge; commissure large; mericarps crowned by 2 $\pm$ linear to narrowly triangular calyx lobes ca 1–1.7×0.2–$0.4\,(0.7)$ mm, one sometimes slightly longer than the other.

Chromosome Number (Table 3): $2\,n = 22$.

Average Pollen Diameter (Fig. 48): 24.3–$25.4\,\mu$m.

Habitat: Typically found as undergrowth in (remnants of) *Uapaca bojeri* – *Chlaenaceae*(– *Asteropeia*) forest ("fôret basse sclérophylle" of HUMBERT 1965; see RAUH 1973, Fig. 48 below, for a photograph of $\pm$ the type locality), but sometimes also (at high altitudes) in ericaceous scrub. In gravelly or sandy areas or between rocks; primarily in sandstone-derived soils but also over quartzite, gneiss or lateritic clay. – Ca $(400)\,600$–$1\,800\,(2\,000)$ m.

Distribution (Map, Fig. 67 c): Madagascar. More or less confined to the S part of the W slopes of the Centre Domaine of the E Malagasy Region (sensu HUMBERT 1965), i.e., from Mt. Ibity and the Itremo SW to the Isalo massif and S to the mountains around the Mananara R. basin (= NW of Ft. Dauphin).

Critical Remarks: *A. isaloense* is a "good" new species which appears to occupy a somewhat isolated position amongst the Madagascan *Anthospermum* species. It may be ($\pm$ distantly?) related to the African (afromontane) – Madagascan group of species centered around *A. usambarense* which is represented in Madagascar by *A. madagascariense* (7., above) and *A. emirnense* (4.). The former is easily distinguished by its much narrower (linear) leaves; *A. emirnense* differs primarily in having more-flowered, congested ♀ inflorescences and fruits without distinct persistent calyx lobes.

A. isaloense is, apart from deviating wood anatomical features (cf. KOEK-NOORMAN & PUFF 1983), also distinguished by a leaf anatomical character which is apparently unique in the genus: the leaves have peculiar $\pm$ cone-like cells filled with dark contents which are associated with the lower epidermis (Fig. 20 e).

Collections

Madagascar. – **2046**: 10 km W of [Massif de l']Itremo, BOSSER 9929 (TAN); Ambatofinandrahana, DECARY 14981 (P), RAUH 817 (TAN). – **2047**: Mt. Ibity, PERRIER DE LA BÂTHIE 3506, 3920 (P). – **2245**: Ranohira, RAUH 10377 (TAN); Isalo massif, HUMBERT 2819 (G), PERRIER DE LA BÂTHIE 3889, 16682 (P); – , W of

Ranohira, HUMBERT & SWINGLE 5000 (P), PUFF 800814-1/1 (WU); –, Col de Tapias, between Ranohira and Sakaraha, PUFF 800814-3/2 (BR, K, P, WU). – **2346:** Mt. Vohipolaka, N of Betroka, HUMBERT 11723 (P). – **2446:** Mananara basin, S of Imonty, between Andohahela massif and Elakelaka, HUMBERT 14069 (P). – Not traced: Ampaninona, BASSE s.n. (P).

9. *Anthospermum galpinii* SCHLECHTER in J. Bot. (London) **35**: 342 (1897); PUFF in Fl. S. Afr. **31**, 1 (2): 13 (1986).
Type: South Africa, Cape Prov. (Transkei), West Gate above Port St. Johns, ca 1 100 ft., December 1896, GALPIN 3472 (BOL, holo.!; GRA, K, PRE, iso.!).

Dioecious, mostly single-stemmed, erect and ± cylindrical to roundish shrubs ca 0.6–2 (2.5) m tall (in fire-exposed habitats sometimes not more than 25 cm tall and several-stemmed). Stems to ca 5 cm in diameter near the base, much-branched above, branches mostly ± erect to ascending, often in threes; younger parts ± angled, reddish to yellowish-grey, densely covered with whitish papillae to 0.1 mm long; older parts terete, yellowish-grey, greyish-brown to dark grey, glabrescent, epidermis often peeling; upper internodes shorter than to as long as the leaves. Leaves in whorls of 3 or (seldom) decussate, pseudo-verticillate; blades (6) 10–20 (27) × (0.5) 0.8–1.5 (2.5) mm, ± narrowly oblanceolate, linear-lanceolate to ± linear, narrowed to the base, glabrous or sometimes lower half of midrib below shortly hairy; apices acute to ± acuminate; margins often strongly revolute; petioles subobsolete; stipular sheaths mostly densely hairy, cup-shaped, ca 2 × 1 mm, with a median seta ca (0.4) 0.6–1.3 mm long, often with an additional smaller seta on either side. Inflorescences inconspicuous (♂) to ± conspicuous and somewhat contracted (♀), made up of ± sessile, several- to few-(3-)flowered axillary clusters. Flowers subsessile (♀) or occasionally with peduncles to 2 mm long (♂), ♂, ♀. Corolla 4-merous, yellow to greenish-yellow, with short white hairs on the outside or glabrous (especially ♀). ♂: tube 0.4–1 mm long, (broadly) funnel-shaped to subcampanulate, lobes 1.4–2.2 × 0.7–1 mm, ± lanceolate to ovate-lanceolate, recurved; stamens 4, filaments ca 1.3–2 mm long, anthers 1.3–2 × 0.3–0.5 mm; rudimentary ovary minute, shortly hairy and crowned by 4 small calyx lobes. ♀: tube 0.5–0.8 mm long, cylindrical, lobes 0.2–0.5 × 0.1–0.2 mm, ± triangular to lanceolate, erect; style 0 or ca 0.2–0.4 mm long; stigmas 2, (2.5) 3–5 (7) mm long, greenish-grey to greyish-white; ovary ca 1 × 0.8 mm, densely covered with short hairs and crowned by 4 small calyx lobes. Fruits reddish-brown to greyish, supported by narrow U- or fork-shaped carpophores ca 0.8–1.4 mm long; each mericarp ca 1.8–3 × 0.7–1.2 mm, elliptic, oblong to ± obovate, dorsal side convex, densely covered with whitish hairs or papillae ca 0.1–0.2 mm long, ventral side ± plane to slightly concave and with a prominent median vertical ridge, commissure almost as large as the mericarp surface; mericarps crowned by 2 rounded to ± triangular calyx lobes ca 0.1–0.3 mm long or calyx lobes ± lacking. – Fig. 31 c.

Chromosome Number (Table 3): n = 11, 2 n = 22.

Average Pollen Diameter (Fig. 48): 27.4–30.7 μm.

Habitat: On rocky ridges, on rock outcrops or at the edge of gorges or kloofs; often in grassland – forest borders, or in forest edge or riverbank scrub. Not uncommonly associated with *Passerina, Euryops, Stoebe* and *Psoralea* (e.g., Table Mt., Pietermaritzburg Distr.; cf. KILLICK 1959: 30). Rather frequently in fire-prone habitats. Sometimes also growing in disturbed sites (road sides, along railroad lines, etc.). Mainly over (confined to?) sandstone. – Ca 100–1 900 m.

Distribution (Map, Fig. 68 b): S Africa. From Natal (Zululand and N Natal) through the Transkei (Wild Coast) to the E Cape Prov.

Critical Remarks: From the Wild Coast of the Transkei to the Natal Coast area and Zululand, *A. galpinii* is well distinguished and not easily confused with any other species occurring in that area, but at the "fringes" of its distribution range both in the E Cape Prov. and in the Natal Midlands and N Natal, the separation of *A. galpinii* from other species can become rather problematic.

In the E Cape Prov., *A. galpinii* may be difficult to separate from the rather closely allied *A. aethiopicum* (11.). Its leaves are commonly, like those of *A. aethiopicum*, ternately arranged, and the leaf sizes and shapes of the two species may overlap to some extent; both species are ± large shrubs. For this reason it may, in some cases, not be possible to identify ♂ specimens (the size and shape of the ♂ flowers of the two species scarcely differ!) with 100% certainty. ♀ flowering specimens or fruiting plants are usually easier to distinguish as *A. galpinii* has shortly hairy fruits and *A. aethiopicum* typically has glabrous fruits; ♀ individuals of the latter, moreover, have more congested, distinctly cylindrical inflorescence zones (cf. Fig. 54 c). There are, however, exceptions: odd specimens of *A. aethiopicum* may have papillate or shortly hairy fruits and these then may rather closely resemble those of *A. galpinii*. SCHARF 1696 (PRE) from Groendal Wilderness Reserve (Uitenhage Distr.), for example, has shortly hairy fruits much like *A. galpinii*, while other collections from the same area (e.g., SCHARF 1444) are more "typical" *A. aethiopicum*. On the other hand, FLANAGAN 1216 (♀; provisionally placed with *A. galpinii*) has papillate, *A. aethiopicum*-like fruits but the less dense and contracted

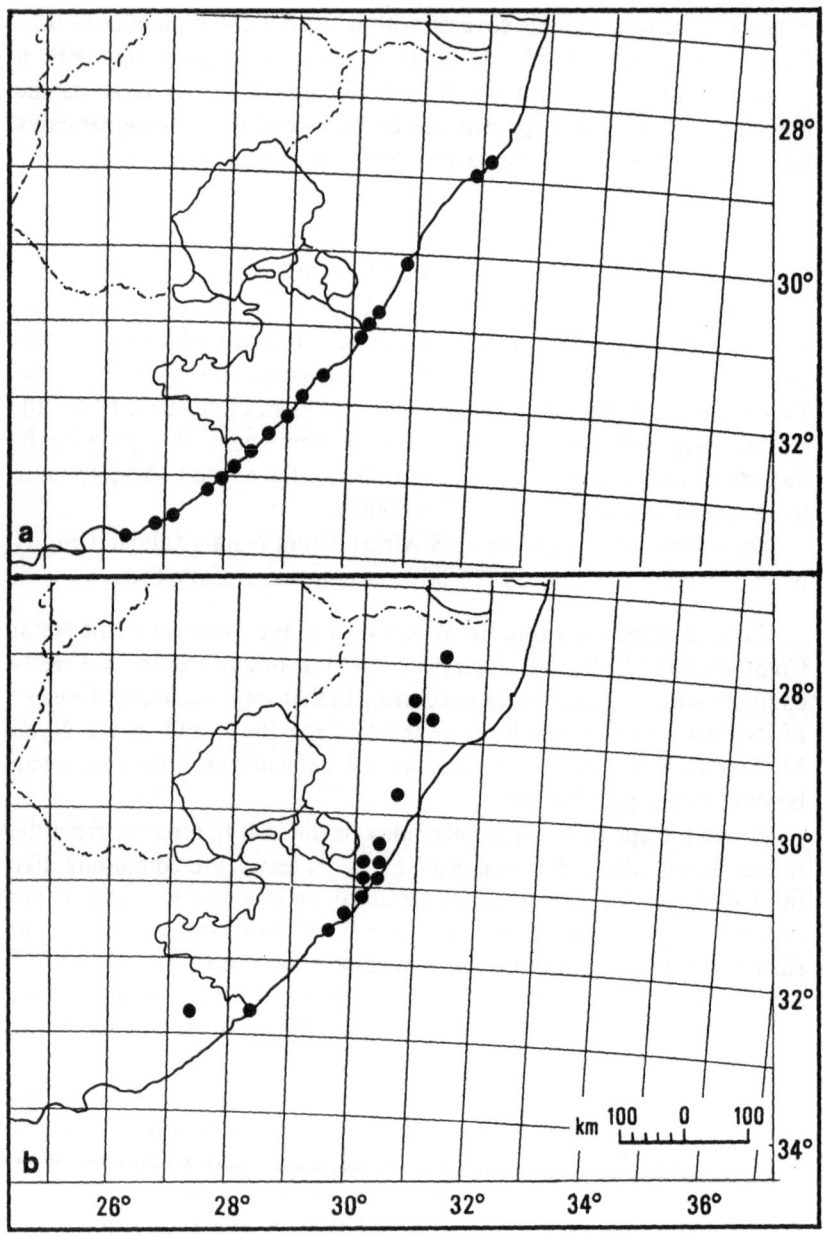

Fig. 68. Distribution of *Anthospermum. a A. littoreum, b A. galpinii*

inflorescences typical for *A. galpinii*. Some other E Cape Prov. collections of *Anthospermum* with decussate leaf arrangement, although agreeing with *A. galpinii* in their shortly hairy fruits and similar leaf sizes and shapes, seem to be very close to *A. spathulatum* (12.) or ± intermediate between *A. galpinii* and *A. spathulatum*[1]. The species' exact distribution and delimitation in the E Cape Prov., thus, remains somewhat uncertain and is not fully understood. It would certainly be worthwhile carrying out additional field work in this area, where both SW Cape taxa and taxa confined to the summer rainfall area of S Africa meet, and where some other taxa (e.g., *A. spathulatum*) also become "difficult".

In Natal, it is the delimitation between *A. galpinii* and *A. monticola* (13.) that may become rather problematic. Typically, the two species are easily separable (*A. galpinii* with ternate, larger and narrower leaves and shortly hairy fruits with minute calyx lobes; *A. monticola* with decussate, smaller and broader leaves and glabrous fruits bearing larger, conspicuous calyx lobes; cf. Fig. 35 g). In the Natal Midlands and N Natal, however, they seem to approach each other more closely in some characters. Natal collections of *A. monticola* tend to have larger and narrower leaves than elsewhere (cf. Fig. 72); leaves may thus rather closely match those of *A. galpinii*. *A. galpinii* plants from that area more often show decussate rather than the typical ternate leaf arrangement, and leaves tend to become somewhat broader. A definite and unambiguous placement to either *A. galpinii* or *A. monticola* of ♂ specimens (i.e., of individuals lacking the quite reliable ovary and fruit characters) may not be possible[2].

A number of collections from the former "Weenen County" (Estcourt Distr.) are also difficult to place. They appear to be somewhat intermediate between *A. galpinii* and *A. monticola* and also show some characteristics of *A. pumilum* subsp. *pumilum* (29.). Habit and some leaf characters of these collections agree with *A. galpinii*, but leaf arrangement is decussate as in *A. monticola* and *A. pumilum*; the fruits are shortly hairy but are relatively broad and have rather conspicuous calyx lobes which resemble those of *A. monticola*[3].

---

[1] Cathcart D., Toise R. (3227-DA), Acocks 9209 (K, PRE); Amatole R., Cata Forest Res. (3227-CA), Story 3191 (K, PRE); Stockenstrom D., Readsdale (3226-BC), van Gadow 491 (GRA). Not mapped.

[2] E.g., Vryheid D., Hlobane (2730-DB), Johnstone 343 (E, NU), or Zwaartkop (= ? Swartkop, Pietermaritzburg D., 2930-CB), Wylie sub Medley Wood 10141 (BOL, K, NH). Not mapped.

[3] E.g., between Peniston and Haviland Sidings (2829-DD), Acocks 10494 (NH); Estcourt (2929-BB), Acocks 10507 (NH, PRE); −, above Dalton Bridge, Bushmans R., Wright, West & Acocks 6 (BOL, NH, PRE). Not mapped.

The exposure to veld fires (populations growing on small rocky outcrops surrounded by grassveld, etc.) may contribute considerably to the morphological variability of *A. galpinii*. Although the plants – like other (large) shrubby species of *Anthospermum* – are more often killed by fires (cf. C.12.8.), they occasionally behave as "sprouters" and ± regularly burnt individuals remain low and become several-stemmed. Their flowering shoots may look very atypical in that the leaves are rather broad, large and ± thinnish.

As regards the presumed alliance between *A. galpinii* and *A. littoreum* (10., below) and the occurrence of odd hybrids between the two species see *A. littoreum*, Critical Remarks.

Collections

South Africa. Natal – **2731** (Louwsburg): 13 km from Ngome tow. Vryheid, nr. Ngome Forest Reserve (-CD), PUFF 791214-1/1 (WU). – **2830** (Dundee): Nqutu-Qudeni, ca 8 km S of Babanango turnoff, Sigqokwana ridge (-BD), PUFF 790303-2/1 (BR, J, WU); Nkandla D., The Heights, Qudeni (-DB), EDWARDS 2236 (NU, PRE). – **2831** (Nkandla): Nkandla Forest Reserve (-CA), LAWN 490 (NH), PUFF 790303-4/1 (BR, J, WU), VENTER 3396 (PRE). – **2930** (Pietermaritzburg): Table Mt. (-DA), KILLICK 603 (NU × 2, PRE). – **3030** (Port Shepstone): Dumisa Stn. (-AD), RUDATIS 1112 (BM, G, K, S, W; STE as 836), 1113 (BM, G, K, PRE, S, W; STE as 837); Oribi Flats (-CA), THODE 5047 (GRA, STE); Oribi Gorge Nature Reserve, Inkonka Point (-CB), DAVIDSON 1618 (J), KERFOOT 7270 (J), PUFF 790426-1/1 (WU); Pt. Edward-Izingolweni rd., Beacon Hill (-CC), PUFF 790423-2/3 (WU); just S of Margate airport (-CD), PUFF 790427-4/1 (WU). – **3130** (Port Edward): Umtamvuna Nature Reserve (-AA), PUFF 790422-3/2 (WU), STREY 9860 (K, NH, PRE, SRGH). – Imprecise locality ("Natal", "Port Natal", "Zululand"): GERRARD (& MCKEN) 1360 (BM, K, S, W). – Additional collections: FORBES & MUNDAY 454 (J); GERSTNER 4375 (PRE); GLEN 374 (J); MOLL 5497 (NH, PRE); NICHOLSON 1087 (PRE), 392, s.n. sub NH 63673 (NH); PUFF 791214-2/1 (WU); STREY 5857 (K, M, NH, NU, PRE), 6847 (E, K, M, NH, NU, PRE), 7051 (K, NH, NU, PRE), 7580 (B, BR, K, M, NH, NU, S). **Cape** – **3129** (Port St. Johns): Kambati (-BD), STREY 8682 (K, NH, PRE, SRGH); around Port St. Johns (-DA), GALPIN 3472 (BOL, GRA, K, PRE), 3473 (BOL, PRE), LEVYNS 6909 (BOL), MOSS 2408 (BM, J), PUFF 790415-4/1 (BR, J, WU). – **3227** (Stutterheim): Dohne Peak (-CB), PUFF 790115-5/6 (J, WU). – **3228** (Butterworth): nr. Kei River mouth (-CB), FLANAGAN 1216 (BOL, PRE; approaching *A. aethiopicum*).

10. *Anthospermum littoreum* L. BOLUS in Ann. Bolus Herb. **2**: 96 (1917); PUFF in Fl. S. Afr. **31**, 1 (2): 14 (1986).
Syntypes: South Africa, Cape Prov., East London, SIM 1483 (BOL!, PRE!), RATTRAY 38 (BOL, lecto.!, selected here; US!; GRA!, NH, NU, WU-photo!), – , – , Kentani Distr., nr. Black Rock Cove, PEGLER 2139 (BOL!, PRE!, US!, W!); – , – (Transkei), Port St. Johns, GALPIN 2850 (BOL!, GRA!, K!, PRE!).

= *A. ambiguum* GREVES in J. Bot. (London) **63**: 203 (1925).
Type: South Africa, Cape Prov. (Transkei), Port St. Johns, Moss 2456
(BM, holo.!; J, iso.!, K, NH, NU, PRE, WU – photo!).

Dioecious or ± dioecious (♂ plants occasionally with odd ♀ flowers),
± procumbent, somewhat straggling or ± erect shrubs; at least older
parts usually deeply buried in sand. Stems ca 0.5–3 m long, ca 3–8 mm in
diameter near the base and ca 1.5–3 mm in the midstem region, buried
parts often rooting at the nodes; stems mostly much-branched, branches
spreading to ± ascending, lower branches often buried in sand; younger
parts greyish, greenish-grey or greyish-brown, densely covered with
whitish spreading hairs 0.1–4 mm long; older parts greyish-brown,
glabrescent, epidermis often peeling; upper internodes mostly as long as or
somewhat shorter than the leaves. Leaves decussate, pseudo-verticillate;
blades (6) 8–15 (20) × (2) 2.5–5 (7) mm, obovate, oblanceolate, oblong to
± elliptic, narrowed to the base, ± thick and fleshy in nature, glabrous or
lowermost part of midrib below hairy, often ± discolourous; apices acute
to ± mucronate; margins flat to ± revolute (occasionally strongly
revolute in dry state); petioles subobsolete or to ca 1.5 mm long; stipular
sheaths hairy, cup-shaped, ca 1–1.5 mm long, with a median seta ca 0.4–
0.8 (1.5) mm long, often with an additional smaller seta on either side.
Inflorescences inconspicuous (♂) to ± conspicuous and somewhat
contracted (♀), made up of ± sessile, several-flowered (to occasionally ±
many-flowered: ♀) axillary clusters. Flowers subsessile or sometimes
shortly pedicellate (♀), ♂, ♀ (occasionally also odd ♀). Corolla 4-merous,
creamy yellow, yellow to greenish-yellow, hairy on the outside at least near
the base (♂) or glabrous (♀). ♂ (♀): tube (0.3) 0.4–0.6 (0.8) mm long,
broadly funnel-shaped to subcampanulate, lobes 2–2.4 × 0.6–0.8 (1) mm,
oblong-lanceolate, recurved; stamens 4, filaments 1.8–2 mm long, anthers
1.7–2 × 0.3–0.5 mm; rudimentary ovary minute, densely hairy. ♀: tube ca
0.3–0.7 (1) mm long, cylindrical, lobes 0.2–0.5 (0.7) × 0.2 mm, ± lanceolate to ovate, erect; style to 1 mm long; stigmas 2, (3) 4–6.7 (10) mm
long, greyish; ovary ca 0.8–1.2 mm long, densely hairy, crowned by 4
minute calyx lobes. Fruits greyish-brown or greyish, supported by
narrow, U- or fork-shaped carpophores ca (0.7) 1.1–1.5 (1.8) mm long;
each mericarp (1.7) 2–3 (3.5) × (0.7) 1–1.4 (1.8) mm, oblong to ± elliptic,
dorsal side shallowly convex, densely covered with whitish spreading hairs
ca (0.2) 0.4–0.8 mm long, ventral side ± plane or somewhat concave and
with a prominent median vertical ridge, commissure almost as large as the
mericarp surface; mericarps crowned by 2 ± triangular calyx lobes ca 0.1–
0.4 × 0.1–0.2 mm, usually hidden amongst the mericarp hairs. – Figs. 3 a,
33 a, 35 e, 40 a.

Chromosome Number (Table 3): n = 11, 2 n = 22.

Average Pollen Diameter (Fig. 48): 26.2–28 µm.

Habitat (Fig. 3 a): Confined to coastal areas and mostly growing in dune sand; found on the seaward-facing side of the first dunes, in stabilized dunes amongst scrub or in grassy patches, occasionally at the edge of dune forest. Often associated with *Aloe thraskii, Mimusops caffra, Dracaena hookeriana, Strelitzia nicolai, Eugenia capensis* subsp. *capensis, Passerina rigida, Scaevola plumieri* (= *S. thunbergii*), *Carpobrotus* and *Ipomoea pes-caprae* subsp. *brasiliensis.* – Ca 0–30 m.

Distribution (Map, Fig. 68 a): S Africa. Along the Indian Ocean coast from Natal (Zululand) through the Transkei to the E Cape Prov. (Alexandria Distr.).

Critical Remarks: *A. littoreum* is an unmistakable species, well distinguished by its habitat, habit and its long-hairy fruits. It is believed to be closely allied *A. galpinii* (9., above) which, over most of its distribution range, occurs in the same area, but always further inland. *A. galpinii* differs, in addition to growth form (erect shrubs), in having longer and narrower leaves mostly arranged in *whorls of 3* and fruits which are densely covered with *short* hairs, but it is similar in general flower and fruit structure and in being dioecious.

A. littoreum and *A. galpinii* may, in some areas, grow in the same vicinity (e.g., in the Natal – Transkei border area where coastal sand dunes and sandstone outcrops often occur close together) and are believed occasionally to form hybrids. The collections PUFF 790424-1/1 (from around the Umtamvuna bridge) almost certainly originate from a population consisting of *A. littoreum* and of hybrids *A. littoreum* × *A. galpinii*. Putative hybrids are characterized by having fruits with hairs intermediate in length between those of *A. littoreum* and *A. galpinii* and (decussately arranged) leaves approaching *A. galpinii* in width (i.e., narrower leaves than 'pure' *A. littoreum*).

Collections

South Africa. Natal – **2831** (Nkandla): Mtunzini (**-DD**), LAWN 1496 (NH). – **2832** (Mtubatuba): Richards Bay area (**-CC**), PUFF 790304-1/1 (J, WU), VENTER 6441 (PRE), WARD 5804 (PRE). – **3030** (Port Shepstone): Isipingo Beach (**-BB**), LEWIS 6290 (K, MO, US); Marina Beach, "Black Rocks" (**-CD**), PUFF 790427-2/1 (WU). – **3130** (Port Edward): Port Edward (**-AA**), RUMP s.n. (E, MO, NH sub 21776, NU); between Glenmore Beach and Portobello Beach (**-AB**), PUFF 790430-1/1 (WU). – Additional collections: FORBES 10 (NH), s.n. (sub NU 3591, sub STE 15199); FORBES & OBERMEYER 57 (NH, PRE); VAN DER BIJL s.n. sub BOL 17059 (BOL); VAN RENSBURG 134 (NU); VENTER 1836 (NH, PRE); WARD 280, 4235 (PRE). **Cape** – **3129** (Port St. Johns); around Tshani (**-CC**), PUFF 790414-1/1, 1/1 b (WU); Port St. Johns and surroundings (**-DA**), ALLSOPP & BRUCKNER 130

(NU); GALPIN 2850 (BOL, GRA, K, PRE); MOSS 2414, 2456 (BM, J); MUIR 4690 (NH); PUFF 790415-5/1 (BR, J, WU); SCHÖNLAND 3854 (GRA), 4035 (GRA, PRE). – 3227 (Stutterheim): Bulugha forest nr. East London (-DD), VON GADOW 32 (GRA). – 3228 (Butterworth): Elliotdale D., The Haven (-BB), GORDON-GRAY 1357 (GRA, NH, NU); Mazeppa Bay (-BC), THERON 1213 (BOL, PRE); at Trennery's (Qolora Mouth area) (-CB), PUFF 791219-3/1 (WU); Cintsa West (-CC), STREY 11204 (E, K, MO, NH, NU, PRE, SRGH, WAG), 11205 (NH, PRE). – 3326 (Grahamstown): Alexandria (-CB), OSBORNE 122 (GRA, PRE); nr. Port Alfred (-DB), BURCHELL 3821 (K), TYSON s.n. sub herb. MARLOTH 8586, s.n. sub PRE 17645 (PRE). – 3327 (Peddie): East London (-BB), RATTRAY 38 (BOL × 3, GRA, US); Kleinemonde (-CA), BRINK 60 (GRA). – Additional collections: ACOCKS 10994 (K, PRE), 15779 (PRE); BRITTEN 2088 (K, PRE × 3), 5195 (GRA, K, PRE × 3); COMPTON 13130 (NBG); DOWNING 7 (K, PRE); FLANAGAN 455 (NBG, PRE, SAM); GALPIN 5664 (GRA, K, PRE), 10443 (PRE); PAGE s.n. (K); PEGLER 2139 (BOL, PRE, US, W); PUFF 790116-1/1, -5/2 (WU); RYCROFT 1882 (NBG); SALISBURY 144 (SAM); SALTER 379/43 (BM); SIM 1483 (BOL, PRE); TYSON s.n. sub TVL Museum no. 17103 (PRE).

11. *Anthospermum aethiopicum* L., Sp. Pl.: 1058 (1753), ed. 2: 1511 (1763); CRUSE in Linnaea 7: 132 (1832); PUFF in Fl. S. Afr. **31**, 1 (2): 14 (1986). Syntypes: [South Africa, Cape Prov. ("Habitat in Aethiopiae"; "Caput bonae spei")] herb. CLIFFORD [in herb. BANKS] (BM, lecto.!, selected here); no collector given (LINN 1233.1!)[1].

= *A. aromaticum* SALISB., Prod.: 59 (1796), nom. illeg. alt.

= *A. ambrosiacum* MOENCH, Suppl. Meth. Pl.: 269 (1802), nom. illeg. alt.

= *A. aethiopicum* L. *var. ternifolium* CRUSE, Rub. Cap.: 11 (1825), in Linnaea **6**: 9 (1831); SONDER in HARVEY & SONDER, Fl. Cap. **3**: 28 (1865).
Type: none given[2].

[Frutex africanus ambram spirans, PLUKENET, Almagesti Mantissa Botanici: 159, t. 183, Fig. 1 (1700)].

Dioecious, erect, single-stemmed shrubs ca (0.5) 0.75–2 m tall. Stems to ca 20 mm in diameter near the base, few- to much-branched above, branches often regular and ternate, curving upwards to ± erect, in ♂

---

[1] The lectotype specimen is pictured and described in LINNAEUS, Hortus Cliffortianus: 455, t. 27 (1737).

[2] Presumably any of ECKLON's *A. aethiopicum* collections from the Table Mt. (e.g., ECKLON 24; GOET!, PRE!, S!). In 'Plantae Ecklonianae. *Rubiaceae*' CRUSE (1832: 132) lists under '*A. aethiopicum* L. em. CRUSE Msct.' (= *A. aethiopicum* var. *ternifolium* in his 1831 publication) ECKLON's collections from the "2te u. 3te Höhe des Tafelberges etc.".

frequently more numerous than in ♀; younger parts ± terete, reddish-purplish tinged, reddish-brown to greyish-brown, densely covered with whitish papillae to ca 0.1 mm long or, less commonly, glabrous; older parts greyish-brown to dark grey, becoming glabrescent, epidermis peeling; internodes shorter to slightly longer than the leaves, on new growth occasionally much longer. Leaves in whorls of 3, pseudo-verticillate; blades 3–9 (13) × (0.3) 0.5–1.2 (2) mm (in ♂ often smaller than in ♀), narrowly lanceolate or oblanceolate to linear, glabrous or occasionally with some very short hairs near the very base; apices acute to ± mucronate; margins mostly (strongly) revolute; stipular sheaths shortly hairy or, less commonly, subglabrous, (broadly) cup-shaped, 0.9–1.2 mm long, with a triangular to subulate median seta ca 0.5–1.5 (2.2) mm long, sometimes with 1 (2) additional minute setae on either side. Inflorescences extensive, dimorphic, in ♀ very conspicuous, much contracted, ± cylindrical inflorescence zones from a few to ca 17 cm long. Flowers subsessile, in large numbers (in ♀ often more than in ♂) clustered at the nodes; ♂, ♀. Corolla 4-merous, creamy yellow, greenish-yellow or pale yellow, glabrous or with a few whitish papillae to 0.1 mm long on the corolla tube and/or the margins of the lobes. ♂: tube (0.2) 0.4–0.9 (1.2) mm, (broadly) funnel-shaped, lobes (1.2) 1.6–2.5 (2.9) × 0.6–1.2 (1.4) mm, oblong-lanceolate, recurved; stamens 4, filaments exserted for (0.6) 0.9–1.7 mm, anthers 1–1.8 × 0.3–0.7 mm; rudimentary ovary small but well discernible, crowned by minute calyx lobes. ♀: tube 0.1–0.4 (0.6) mm long, cylindrical, lobes 0.1–0.5 (0.7) × 0.1–0.2 (0.4) mm, lanceolate to linear, ± erect; style ± 0–0.7 mm long; stigmas 2, 4–6.3 mm long; whitish- to greenish-grey; ovary ca 0.6–0.9 × 0.5–0.7 mm, glabrous or papillate, crowned by 4 indistinct small calyx lobes. Fruits reddish-brown to dark grey-brown, supported by U-shaped carpophores ca ⅓ (½) as long as the mericarps; each mericarp (1.4) 1.7–2.5 (2.7) × 0.8–1 (1.2) mm, oblong, elliptic to narrowly obovate, dorsal side (± shallowly) convex, glabrous and often with large, conspicuous epidermis cells or, less commonly, covered with whitish papillae less than 0.1 mm long, ventral side plane to shallowly concave and with a prominent median vertical ridge, commissure large; each mericarp crowned by two indistinct rounded to ± triangular calyx lobes ca 0.1–0.2 mm long or calyx lobes ± lacking. – Fig. 7 a, 54 c–d.

Chromosome Number (Table 3): n = 11, 2n = 22.
Average Pollen Diameter (Fig. 48): 30.3–34.2 μm.
Habitat: Found in all principal vegetation types of the SW Cape Floristic Region [Coastal Fynbos, Mountain Fynbos (including Arid Fynbos) and Coastal Renoster(bos)veld (cf. TAYLOR 1978, KRUGER 1979)]

except for the Strandveld (Veld Type No. 34 of Acocks 1975). Often growing in damp or moist to ± wet sites (along streams, valley bottoms, well drained slopes, wet depressions or ditches), especially in the drier parts of the SW Cape Region. Perhaps most abundant in areas where a moderately high rainfall is experienced and in relatively fertile soils (clayey soils, clay and gravel or clay sand), but also in TMS derived sand or sandy peat and over Witteberg quartzite. Levyns (1935) states that "many farmers consider it a valuable plant indicator. Soils on which it grows abundantly are regarded as being suitable for the cultivation of deciduous fruits". In coastal areas (from the Riversdale Distr. E to the Bathurst Distr.), it also occurs in (fixed) sand dunes. It may, furthermore, form extensive populations in disturbed sites (road sides, clearings, at the edge of orchards, etc.). – Ca 10–1 300 (1 600) m.

Distribution (Map, Fig. 69 a): S Africa. Widely distributed in the SW Cape Prov. In the E, it extends beyond the limits of the SW Cape Floristic Region to the Albany, Bathurst and Peddie Distr.; in the N, it extends to the Calvinia Distr. (Nieuwoudtville area).

Critical Remarks: A. aethiopicum and *A. spathulatum* (12. below) are the only two tall shrubby species of *Anthospermum* occurring in the SW Cape Floristic Region. The two species were often confused in the past (cf. *A. spathulatum*, introductory comments, and also p. 252 and Table 13) but *A. aethiopicum* is easily separated from *A. spathulatum* – even vegetatively – by its strictly ternately arranged leaves. It is, furthermore, characterized by its pronounced sexual dimorphism (see C.12.3. and Fig. 54 c–d) and the conspicuous ♀ plants with their characteristic contracted, dense cylindrical inflorescence zones. *A. aethiopicum* is also quite distinct in its ecological requirements (favouring rather fertile soils and damp to moist habitats, or habitats with a moderately high rainfall!).

As compared to *A. spathulatum*, *A. aethiopicum* shows relatively little variation. Variation primarily occurs in the size of the ♂ flowers, the mericarp surface and the leaves (leaf size and shape differences, however, appear to be largely sex- and age-dependent). Unlike several other SW Cape taxa of the *Anthosperminae*, no distinct ecotypes can be distinguished.

Although *A. aethiopicum* and the subspecies of *A. spathulatum* (subsp. *spathulatum* in particular) may grow together in the same area, I have on no occasion observed that populations of the two intermingle. Populations of *A. aethiopicum* and *A. spathulatum* subsp. *spathulatum* were always found to differ in their micro-habitats if found growing side by side. *A. spathulatum* subsp. *spathulatum* is usually found in drier and less fertile conditions than *A. aethiopicum* (compare, however, *A. spathulatum* subsp. *ecklonianum*, Critical Remarks).

It is possible that *A. aethiopicum* occasionally does form hybrids with *A. spathulatum* subsp. *spathulatum,* although in areas where I have seen both taxa growing next to each other, I have never been able to find any definite proof for the presence of hybrids.

In the E fringes of its distribution range, the distinction of *A. aethiopicum* from the doubtlessly closely allied *A. galpinii* can become problematic. See *A. galpinii* (9.), Critical Remarks, for details.

Collections

South Africa. C a p e − **3119** (Calvinia): ca 5–8 km W of Nieuwoudtville, on rd. to Vanrhynsdorp (**-AC**), PUFF 790712-3/1 (WU). − **3218** (Clanwilliam): plateau of Piketberg, Farm Vergesig (**-DC**), PUFF 800914-4/1 (WU). − **3219** (Wuppertal): SE of Uitkyk Pass (**-AA**), ACOCK(s) 3068 (S); Wuppertal, along Tra-tra R. (**-AC**), PUFF 800902-5/1 (WU). − **3225** (Somerset East): Somerset East D., Bosberg, along Visrivier (**-DA**), VAN DER WALT ('P.T.v.d.W.') 299 (PRE). − **3318** (Cape Town): Boundary Farms Baarhuis and Zonquasfontein, N of Darling (**-AB**), BOUCHER 3289 (PRE, STE); Kirstenbosch (**-CD**), BOLUS 18309, 18310 (BOL), ESTERHUYSEN 11778 (BOL), 26752 (BOL, MO), FORBES 176 (NH), GUTHRIE s.n. sub BOL 18308 (BOL), PUFF 791227-1/1 (WU); Paarlberg (**-DB**), DRÈGE s.n. (E, G × 3, K, P, W; partly mixed with *A. spathulatum*); Cape Flats, ca 30 km E of Cape Town (**-DC?**), LAVRANOS 3757 (MO); Stellenbosch (**-DD**), WORSDELL s.n. (K). − **3319** (Worcester): base of Witsenberg, path to Sneeugat (**-AA**), PUFF 800924-4/1 (WU); Gydo Pass (**-AB**), WALL 292 ('562/56'?) (S); Wabooms R., foot of Waaihoek (**-AD?**), ESTERHUYSEN 8943 (BOL); Orchard Siding (**-BC**), ROGERS 16597 (K; *A. spathulatum* in other herbaria); Du Toit's Kloof (**-CA**), BARKER 5982 (NBG); Worcester D., Brandwagt (= Brandwag) (**-CB**), VAN BREDA 362 (K, PRE); French Hoek (**-CC**), PHILLIPS 1159 (SAM), 8359 (PRE); ca 30 km from Worcester on rd. to Viliersdorf (**-CD**), PUFF 800102-2/1 (WU); Worcester- De Doorn rd., nr. De Wet Station (**-DA**), PUFF 800102-3/1 (WU); Robertson D., Bosmans Kloof Pass at McGregor (**-DD**), ESTERHUYSEN 4479 (BOL). − **3320** (Montagu): Witteberg, around [S of] Matjiesfontein (**-BC**), HUMBERT 9796 (P, PRE); ca 9 km E of Montagu on rd. to Barrydale (**-CC**), LEVYNS 473 (BOL, STE); between Duiwelsbos and Koloniesbos streams, N of Swellendam (**-CD**), PUFF 800919-4/2 (WU); Tradouwspas area (**-DC**), PUFF 790911-3/1, 790913-4/3 (WU); ca 11 km N of Heidelberg, S of Duiwenkoksrivier Drift, on rd. to Gysmanshoek Pass (**-DD**), PUFF 791223-4/1 (WU). − **3321** (Ladismith): Seven Weeks Poort (**-AD**), PHILLIPS 1446 (SAM); 14 km W of Swartberg Pass summit, on rd. to 'The Hell' (**-BD**), PUFF 840908-3/2 (WU); Gysmanshoek Pass between Brandrivier and Heidelberg (**-CC**), PUFF 791223-3/2 (WU). − **3322** (Oudtshoorn): Swartbergpas (**-AC**), PUFF 790915- 3/1 (WU), WALL s.n. (S); Robinson Pass (**-CC**), PUFF 791222-3/3 (WU); Montagu Pass (**-CD**), DREYER s.n. sub STE 18663 (STE), PUFF 790909-4/1, -7/1 (WU); Buffelsrivier, W of Kammanassie (**-DB**), THOMPSON 1387 (PRE, STE); ca 9–10 km W of Karatara, Farm Kleinbegin (**-DD**), PUFF 790910-1/1 (WU). − **3323** (Willowmore): ca 8 km S of Willowmore (**-AD**), THERON 1375 (K, PRE); Uniondale (**-CA**), PATERSON 3034 (GRA); top of Prince Alfred Pass (**-CC**), PUFF 800927-4/2 (WU), RYCROFT 2315 (NBG); Louterwater (**-DC**), DAHLSTRAND 1706 (PRE × 2). − **3324** (Steytlerville): Joubertina-Humansdorp rd., nr. Kompanjiesdrift Stn. (**-CC**), PUFF 800928-5/1 (WU); −, nr. Assegaaibos Stn. (**-CD**), LYNES 1697 a, 1697 b, 1763 (BM), PUFF 800928-6/1 (WU); Hankey-Willowmore rd.,

Baviaanskloofberge, above Farm Kleinplaas (**-DA**), Puff 791220-4/1 (K, WU);
Elandsberg, just opposite (S of) Cockscomb Mt. (**-DB**), Puff 840921-6/3 (WU);
Kareedouw- Humansdorp rd., ca 2 km SE of Two Streams Station (**-DC**), Puff
800928-7/1 (WU). – **3325** (Port Elizabeth): plateau of Suurberg, nr. Farm
Zuurberg Manor (**-BC**), Puff 800929-1/1˙(WU); Suurbergpas, N descent, between
Farms Aansvilla and Viewlands (**-BD**), Puff 800929-4/1 (WU); Uitenhage D.,
Groendal Wild. Res., Zunga catchment basin (**-CA**), Scharf 1444 (K, PRE), 1696
(PRE; ± appr. *A. galpinii*); Vanstadens Mts., NW end of watershed (**-CC**),
Thompson 1854 (PRE, STE); W of Port Elizabeth, tow. Witteklip Canyon (**-CD**),
Rodin 1048 (BOL, K, MO, PRE, US); Port Elizabeth, Humewood (**-DC**),
Paterson 779 (GRA). – **3326** (Grahamstown): (mts. near) Soutar's Post (**-AA**),
Burchell 3471 (K); Sidbury (**-AC**), Bayliss 91 (PRE); Howiesons Poort area
(**-AD**), MacOwan 161 (BM, K, S), 16233 (W), Puff 790117-1/2 (WU), Schönland
603 (GRA); Albany D., above Blaaukrantz (= Bloukrans) Stn. (**-BC**), Sidey 993
(S); nr. 'Frasers Camp' (ca 38 km E of Grahamstown) (**-BD**), Puff 791220-1/2
(WU); ca 11–12 km E of Alexandria/Cookhouse crossroad on Grahamstown-
Port Elizabeth rd. (**-CA**), Puff 791220-2/1 (WU); ca 19 km from Kenton-on-Sea
on rd. to Salem (**-DA**), Puff 790116-6/1 (J, WU); Kowie (= Port Alfred) (**-DB**),
Britten 2112 (PRE), 5012 (GRA). – **3327** (Peddie): Kidd's Beach- Port Alfred rd.,
ca 2 km beyond Keiskamma R. bridge (**-AB**), Puff 790116-3/1 (WU). – **3418**
(Simonstown): Hout Bay (**-AB**), Gillett 431 (STE); Cape of Good Hope Nat.
Res., Platboom rd. (**-AD**), Puff 800907-4/1 (WU); Sir Lowrys Pass (**-BB**),
Meebold 12167 (M); Kogelberg Res., Paardeberg (**-BD**), Grobler 6100 (STE; *A.
spathulatum* in PRE). – **3419** (Caledon): Houw Hoek (**-AA**), Bayliss 957 (PRE),
Puff 800916-1/1 (WU); half way between Kleinmond and Botrivier (**-AC**), Puff
800102-1/1 (WU); Vogelgat (Private) Nature Res., E of Hermanus (**-AD**), Puff
800917-1/1 (WU); Genadendal Mts. (**-BA?**), Galpin 3991 (GRA), 4102 (GRA,
PRE); Stanford- Papiesvlei rd., Perdeberg above Farm Flouhoogte (**-BC**), Puff
800918-1/4 (WU); Caledon D., Rooi Els (**-BD**), Leipoldt 4158 (BOL); Bredasdorp
D., ca 12 km inland from Gans Baai and Danger Point (**-CB**), Stokoe s.n. sub SAM
59525 (SAM); SW Koueberge, ca 4.5 km NNW of Nuwepos Farm house on rd. to
Papiesvlei (**-DA**), Puff 800918-2/1 (WU). – **3420** (Bredasdorp): for ca 3.5 km from
Swellendam on Heidelberg rd. (**-AB, -BA**), Puff 790911-4/1 (WU). – **3421**
(Riversdale): ca 16 km from Riversdale on rd. to Heidelberg (**-AA**), Puff 790911-
2/1 (WU); Riversdale, 'Kattestaart' (**-AB**), Marloth 7455 (PRE); [Farm] Boplaat
(Bovenplaats) nr. Albertinia (**-BA**), Muir 841 (PRE), 1955 (PRE × 3), 16747 (J),
s.n. sub BOL 25127, 25127 A (BOL); Ystervarkfontein- Albertinia rd., ca 5–7 km
inland from Farm Driefontein (**-BC**), Puff 790910-5/1 (WU). – **3422** (Mossel
Bay): Mossel Bay (**-AA**), Lindeberg s.n. (S); ca 24 km W of George (**-AB?**), Lynes
1660 (BM); Sedgefield (**-BB**), Bayliss 6865 (K, MO, S, WAG). – **3423** (Knysna):
Knysna (**-AA**), Rogers 22868 (PRE); Plettenberg Bay (**-AB**), Bowie s.n. (BM),
Rogers 27101 (K). – **3424** (Humansdorp): Zoet-Rug, Robhoek (**-AA**), Fourcade
1443 (BOL, K, STE); Clarkson (**-AB**), Thode A 854 (GH, K, NH); Kareedouw-
Humansdorp rd., around Dieprivier bridge (**-BA**), Puff 800928-8/1 (WU);
Humansdorp (**-BB**), Rogers 2990 (GRA). – Imprecise localities ("CBS") or no
locality given: Banks s.n. (W); Bauer '94' (W); Bergius, Mund(t) & Maire s.n. (G);
Brossard s.n. (P); coll. Burman [herb. Delessert] '81' and s.n. (G × 4); Carmichael
221 (BM); Gillett s.n. sub STE 30969 (STE); Gueinzius s.n. (WU); Harvey (E, K),
s.n. (K; PRE sub TVL Museum no. 10602); herb. Jacquin s.n. (W); Jordaan s.n.
sub STE 30968 (STE); MacWilliam s.n. (G); Moricand & Moricand s.n. (G);
Mund(t) & Maire s.n. (G); Oldenburg 654 (BM); Osbeck s.n. (S); Pappe s.n. (K);
herb. Portenschlager s.n. (W); Scott Elliot s.n. (E); Sickmann s.n. (G); Sieber

232   D. Systematic Part

[Flora Cap.] 6 (BM, W; partly mixed with *A. spathulatum*); Sonnerat s.n. (P); Sparrman s.n. (S; mixed with *A. spathulatum*); Thorn 187, 558 (K); Thunberg s.n. (S × 5); herb. Ventenat s.n. (G; mixed with *A. galioides*); Verreaux s.n. (G × 2); Wallich s.n. (BM, G); Wawra 208 (W); Wright s.n. (GH). – Additional collections: Archibald 5213, 5228, 5230 (PRE); Bayliss 4707 (GH, MO), 6782 (BR, K, MO, WAG), 7567 (BR × 2, G, MO × 2, US, WAG; M, GH as 7625?), 8676 (GH); Bolus 25124 (BOL); Bonomi s.n. (LY); Boucher 538 (K, PRE, STE), 2308 (PRE, STE; mixed with *A. spathulatum*); Britten 1535 (PRE; mixed with *A. spathulatum* subsp. *uitenhagense*), 2509 (PRE); Burchell 262 (GH, K), 864 (K); Coppejans 1369 (BR); Dahlstrand 261 (GRA, J, PRE), 1980 (J), 2227 (PRE, STE), 3219 (PRE); Dix 38 (BOL), 40 (BOL, GRA); Drège 7661 (MO); Ecklon 12 (S), 24 (GOET, PRE, S), 24b (M), 25 (E, GOET, S), 25b (K, M, MO, PRE, W), 28b (MO), 85 (a?), 189 (S), 246 (K), 1267 (MO), s.n. (S × 3; partly mixed with *A. galiloides*); Ecklon & Zeyher 2307α (BOL, GOET, M × 2, MO, S, SAM, W), 2307β (W; probably label mix up); 'Fauna & Flora' s.n. sub STE 30966 (STE); Forward s.n. sub GRA A1612 (GRA); Fourcade 410, 5662 (BOL), 1505 (BOL × 2, GRA, PRE, STE); Fries, Norlindh & Weimarck 199 (BM, K), 227 (LISC, M, PRE), 1136 (PRE); Galpin 218 (PRE); Gane (Rogers?) 33 (J); Garside 29 (K); Gillett 655, 1446 (STE); Gueinzius s.n. (S); Hafström s.n. (S); Hugo 578b, 2639 (STE); Humbert 9395 (P), 9530 (P, PRE); Johnson 261 (NBG); Jordaan 204 (STE); Keet 599 (GRA, STE; mixed with *A. spathulatum*), 759 (PRE, STE), 857 (STE; mixed with *A. spathulatum*); Krauss s.n. [1206, 1207] (G, W); Kruger 1409, 1529 (STE), 1516 (PRE, STE); Kuntze s.n. (K); Levyns 271, 3619, 10333 (BOL); Liebenberg 183 (BOL); Long 276 (K), 277 (K, PRE); Lynes 1036, 1037, 1065, 1624 (BM); MacOwan 1882 (BM, G, GH, K, SAM), 3050 (GRA); Marloth 7509 (PRE); Meebold 82, 213 (M); Montgomery 424 (STE); Moss 8246 (BM, J); Muir 5436 (PRE); 'Nature Cons.' 89 (PRE); Olivier 1148 (NBG); Peter 50217 (B, K), 50579, 41296 (B); Phillips 1752 (SAM); 'Pica Survey' 679 (PRE); Pillans 7432, 7891 (BOL), 10131 (BR, MO); Prior s.n. (K × 2); Puff 790116-5/1, 790117-3/1, 790717-1/3, 791221-3/3, 791226-5/1, 791230-3/1, -4/1, 800101-6/1, 800921-3/2, 800926-1/1 (WU); Purcell s.n. sub SAM 90324–90330 (SAM); Rennie & Rennie 103 (BOL); Rogers 17745 (J), 22870 (PRE), 27231 (B, J), 28993 (SAM); Salter 9324A, 9324B (BM); Scharf 1581, 1893 (K, PRE), 1878, 1892 (PRE); Schlechter 5676 (BR, COI, GH, GRA, NBG, PRE, STE, WU); Schönland s.n. (GRA); Selling 15 (S); Stokoe s.n. sub SAM 59524 (SAM); Story 2817, 2818, 3173 (PRE); Thorncroft sub Rogers 27327 (MO); Tyson 673 (PRE, SAM); Wall 292, s.n. (S); Walter & Walter s.n. (B); Wasserfall 530 (NBG); White 5514 (PRE); Wilms 3181a (G, WU); Worsdell s.n. (K); Zeyher s.n. sub SAM 16051 (SAM; mixed with *A. spathulatum*).

12. *Anthospermum spathulatum* Sprengel, Syst. Veg. 1: 399 (1825); Cruse, Rub. Cap.: 13 (1825), in Linnaea 7: 133 (1832); Puff in Fl. S. Afr. 31, 1 (2): 16 (1986).
   Type: "C.B.S.", no collector given[1].

---

[1] Cruse (1825, 1831) claims having seen and studied "Sprengel's" specimens of *A. spathulatum* (collected by an unknown collector but certainly not by Sprengel) and stated (1831) that specimens determined by Sprengel are in herbarium B. These have probably been destroyed, but a specimen in G, closely matching Cruse's description, is labelled '*A. spath.* Sprg.' and is clearly marked as

The type material of the most widely distributed and most common Cape species of *Anthospermum*, i.e., *A. aethiopicum* (11.), *A. spathulatum* and *A. galioides* (= *A. ciliare*, 31.) (and also that of their synonyms), are often made up of miserable, small fragments. One of the reasons for this is that collections of DRÈGE or of "commercial" collectors like ECKLON and/or ZEYHER were divided up into numerous duplicates. This, on the one hand, resulted in smaller and smaller specimens and, on the other hand, gave rise to numerous mix-ups. Apparently after the distribution of the duplicates, specimens were sometimes divided up even further – the best example of this is perhaps the rather complete collection of Cape *Anthospermeae* specimen-fragments (including many types) in GANDOGER's herbarium (LY). It consists of bits and pieces which SONDER had removed from his herbarium [see GANDOGER (1913) and also NORDENSTAM (1980)].

As numerous Cape species are similar in their leaf characters (size and shape, arrangement and stipules) and as the type material frequently only consists of small twigs bearing either only ♂, ♀ or only ♀ flowers or only fruits, and because no information whatsoever is given on growth form and size of plants, highly confusing situations arise, and the typification of taxa sometimes poses major problems. Considering the miserable state of the specimens and the lack of information, it is not too surprising that KUNTZE (1898), for example, reached that conclusion that the low, dwarf-shrubby or hardly woody *A. galioides* (= *A. ciliare*) is to be included in the tall, shrubby *A. aethiopicum*!

To date, the species *A. aethiopicum* and *A. spathulatum* have consistently been confused, and even SALTER, who carefully studied the species occurring in the Cape Peninsula (1937, 1950), failed to see the differences between the two species. *A. spathulatum* has, as "opposite-leaved" varieties such as var. *oppositifolium*, var. *ecklonianum*, and var. *montanum*, been included in *A. aethiopicum* (which, however, is characterized and easily recognized by its *always* ternately arranged leaves).

CRUSE at first (1825) created *A. aethiopicum* var. *oppositifolium* (next to "var. *ternifolium*"), stating that the former approaches *A. spathulatum* SPRENGEL. He, however, retained the latter as a discrete species. By 1831, he changed his mind and considered *A. spathulatum* to be a synonym of his *A. aethiopicum* var. *oppositifolium*. A year later, he changed his mind again. He described *A. aethiopicum* as having strictly 3-whorled leaves and excluded *A. aethiopicum* var. *oppositifolium* to include it as a synonym under *A. spathulatum* which he again recognized as a distinct decussate-leaved species. At the same time (1832) he recombined var. *ecklonianum* (considered a variety of *A. aethiopicum* in 1831) to include it under *A. spathulatum*.

CRUSE's 1832 publication, in which – in my opinion – satisfactorily resolved the complex situation, has, unfortunately, been either overlooked or ignored. SONDER, in Flora Capensis (1865), followed CRUSE's publication from 1831 and paid no attention to the article published in the following year. In addition, SONDER described the new species *A. tricostatum*. He considered this species to be allied to "*A. aethiopicum* var. *β*" [= *A. aethiopicum* var. *oppositifolium* in Flora Capensis 3: 28). The type specimens of *A. tricostatum* SONDER are *female* fruiting plants, while the type specimen(s?) of *A. spathulatum*, according to the data given by CRUSE (1825), are *male*. This may have been the reason why it did not occur to SONDER that his *A. tricostatum* was conspecific with CRUSE's "*A. aethiopicum* var. *oppositi-*

---

being from the Berlin herbarium. It can, with little doubt, be considered to be an isotype of *A. spathulatum*.

16*

folium" [i.e., *A. spathulatum* (subsp. *spathulatum*)]. SALTER (1937, 1950), in turn, must have followed SONDER's treatment and probably did not see the type specimens. He merely stated that (in the Cape Peninsula) "var. *oppositifolium* CRUSE is scarcely worth distinction as a variety" (1937) and "var. *oppositifolium* CRUSE only differs in having all leaves constantly opposite and is almost as common as the typical form" (1950).

A. *spathulatum* is perhaps the most complicated and most variable of the *Anthospermum* species centered in the SW Cape Floristic Region. It is also the most widely distributed Cape species of the *Anthosperminae* extending beyond the limits of the SW Cape Region to the N and E and into the karroid and drier interior of the Cape Prov. (cf. Fig. 69 b–d). It occurs from sea level to the highest mountain tops and over a wide array of substrates. A. *spathulatum* is the only *Anthospermum* species known, to date, to comprise di-, tetra- and hexaploids.

I have subdivided *A. spathulatum* into five subspecies. Of these, the widely distributed subsp. *spathulatum* is difficult and variable and contains a number of ± distinct ecotypes, while the delimitation of some in the smaller, less widely distributed and often ecologically quite well defined subspecies is on the whole more clear-cut. See the Critical Remarks sections of the respective subspecies for further details.

12 (a). Subsp. **spathulatum**; PUFF in Fl. S. Afr. **31**, 1 (2): 16 (1986).

= *A. aethiopicum* L. var. *oppositifolium* CRUSE, Rub. Cap.: 11 (1825), in Linnaea 6: 10 (1831); SONDER in HARVEY & SONDER, Fl. Cap. **3**: 28 (1865); SALTER in J. S. Afr. Bot. **3**: 109 (1937), in ADAMSON & SALTER, Fl. Cape Penins.: 732 (1950).
Type: none given[1].

= *A. aethiopicum* L. var. *montanum* SONDER in HARVEY & SONDER, Fl. Cap. **3**: 28 (1865).
Type: none given[2].

---

[1] CRUSE (1825) did not cite a type specimen for "*A. aethiopicum* var. *oppositifolium*" but noted in 1831 that several ECKLON specimens are "var. *oppositifolium*".
None of these specimens is in herbarium B (†?) nor is there an ECKLON collection in any other herbarium which bears a handwritten identification by CRUSE. It, nevertheless, appears to be safe to select ECKLON 5 from Table Mountain (S!) as a lecto-(neo-?)type.
[2] Sonder considered *A. aethiopicum* var. *montanum* to be identical to *A. aethiopicum* L. ζ. *alpinum* ECKLON & ZEYHER [Enum. Pl. Afr. Austr.: 366 (1836) (nomen non valide publ.; sine descr.)]. ECKON & ZEYHER 2307 ζ ['in vertice montis "Kasteelsberg" (altit. V), in montibus prope "Simonstown" (Cap), et in montibus "Hottentottshollandsberge" supra "Palmietrivier" (Stellenbosch)'] can, therefore, be considered the type (SAM, holo.!; FI, G, LY, M, MO, P, PRE, iso.!).

= *A. tricostatum* SONDER in HARVEY & SONDER, Fl. Cap. **3**: 28 (1865). Syntypes: South Africa, Cape Prov., Rietvallei, ECKLON & ZEYHER s.n. (S!); −,−, between Driekoppen and Bloodrivier, DRÈGE 9550 (S, lecto.!, selected here).

Dioecious, single- to several-stemmed, ± erect, cylindrical to rounded shrubs or dwarf shrubs ca 0.3–1.5 (2) m tall. Stems to ca 10 mm in diameter near the base and ca 1–4 mm in the midstem region, much to sparsely branched above or from the base upwards, branches often regular, paired, ascending to ± divaricate or ± irregular and intricate (browsed individuals); younger parts light grey, greyish-brown or occasionally reddish-brown, ± densely covered with whitish papillae to 0.1 mm long; older parts often with long fissures, brownish-grey to dark grey, becoming glabrescent, epidermis often peeling; internode lengths very variable, shorter to (much) longer than the leaves. Leaves strictly decussate, pseudo-verticillate; blades (3) 4–12 × 0.5–2.5 (3) mm, obovate, oblanceolate, lanceolate to linear-lanceolate, narrowed to the base, glabrous, upper surface often ± shiny and with conspicuous but small epidermis cells; apices acute to ± mucronate; margins revolute to ± flat; stipular sheaths mostly shortly hairy, cup-shaped, 0.5–1 mm long, with a median seta ca 0.3–1 mm long. Inflorescences inconspicuous to sometimes ± conspicuous and somewhat contracted (♀, especially when in fruit), made up of ± sessile, ± many- to few- (3- to 1-)flowered axillary clusters. Flowers subsessile or sometimes with pedicels ca 0.2–1 (1.4) mm long (♀, especially when in fruit); ♂, ♀. Corolla 4-merous, yellow to greenish-yellow or pale whitish-yellow, occasionally purplish tinged on the outside, glabrous. ♂: tube 0.5–1.2 mm long, (broadly) funnel-shaped, lobes 2–3.4 × 0.7–1.2 (1.5) mm, oblong-lanceolate, recurved; stamens 4, filaments exserted for (0.7) 1–1.7 mm, anthers 1.2–2.5 × 0.3–0.7 mm; rudimentary ovary minute but well discernible and often crowned by 4 calyx lobes ca 0.1–0.3 × 0.2–0.3 mm. ♀: tube 0.1–0.4 mm long, cylindrical, lobes 0.3–0.8 × 0.1–0.3 mm, ± linear-lanceolate, erect to ± spreading; style ± 0–0.5 mm long; stigmas 2, (1.5) 2–7.6 mm long, whitish, greyish or greenish-grey; ovary ca 0.9–1.5 × 0.5–0.8 mm, glabrous, crowned by 4 calyx lobes. Fruits shiny reddish-brown to dull greyish-brown, supported by often ± large U-shaped carpophores ca half (or more than half) as long as the fruits; each mericarp 2.4–4.2 × 1.2–2.2 mm, (broadly) obovate to ± elliptic, dorsal side shallowly convex, distinctly to rather obscurely 3-ribbed (hence mericarps often ± triangular in section), glabrous or, very rarely, shortly hairy, ventral side plane to shallowly concave and with a ± prominent median vertical ridge, commissure large; mericarps crowned by 2 distinct to obscure ± triangular calyx lobes 0.3–

0.6 (0.9) × 0.5–0.7 mm or, less often, calyx lobes ± lacking. – Figs. 5, 30 b, 35 a–b.

Chromosome Number (Table 3): n = 11, 22; 2 n = 22, 44, 66.

Average Pollen Diameter (Fig. 48): 27.3–35.6 μm.

Habitat (Fig. 5): In all Veld Types of the SW Cape Floristic Region (S portion of the Strandveld, Coastal Renosterveld, Coastal and Mountain Fynbos sensu ACOCKS 1975, KRUGER 1979). Also in "Karoo and Karroid" and "False Karoo" Veld Types (ACOCKS 1975) adjoining the Fynbos Biome, notably in the Mountain Renosterveld (Veld Type No. 43), Karroid Broken Veld (No. 26), Western Mountain Karoo (No. 28) and in the Valley Bushveld (No. 23c: Fish River Scrub). Occurring on rocky slopes and ridges or on outcrops and cliffs, in gravelly areas or in sandy flats and sand dunes. Mostly in (relatively) dry habitats. Growing over a variety of substrates (TMS, granite, Witteberg quartzite and limestone) in sandy to clayey and clayey-gravelly soils or in stony ground. – Ca 0–1 500 m (–1 830 m in the Namaqualand Distr. and in the interior of the Cape Prov.).

Distribution (Map, Fig. 69 b): S Africa. Widely distributed in the SW Cape Prov. and extending N to the Namaqualand Distr. [Kamiesberg area (also cf. ADAMSON 1938) and the Groot Pellaberg near the Orange R.] and E to the Kingwilliam's Town and Keiskammahoek Distr.; also extending slightly into the drier, interior parts of the Cape Prov.

Critical Remarks: Subsp. *spathulatum* is very variable and may be somewhat heterogenous. The most variable characters are discussed below:

(i) *Growth Form:* Perhaps the majority of subsp. *spathulatum* populations consists of almost exclusively single-stemmed and rather tall (ca 75 cm and more) individuals (Fig. 5). Plants may, however, occasionally also start branching from the base with the basal branches often becoming as strong as the main stem, resulting in the formation of rather 'bushy' and ± rounded many-stemmed shrubs (in this it, by the way, differs markedly from *A. aethiopicum* (11., above)]. This capacity to produce basal shoots also appears to be the reason why plants of subsp. *spathulatum* can resprout after fires (*A. aethiopicum,* in contrast, is a strict seed regenerator).

Especially in the drier parts of its distribution range, plants are occasionally browsed by sheep and goats. Browsed individuals usually become rather many-stemmed, intricately branched low rounded bushes, seldom reaching 75 cm in height; they tend to be very sparsely leafy and the clustered partial inflorescences are often few- or only 1-flowered.

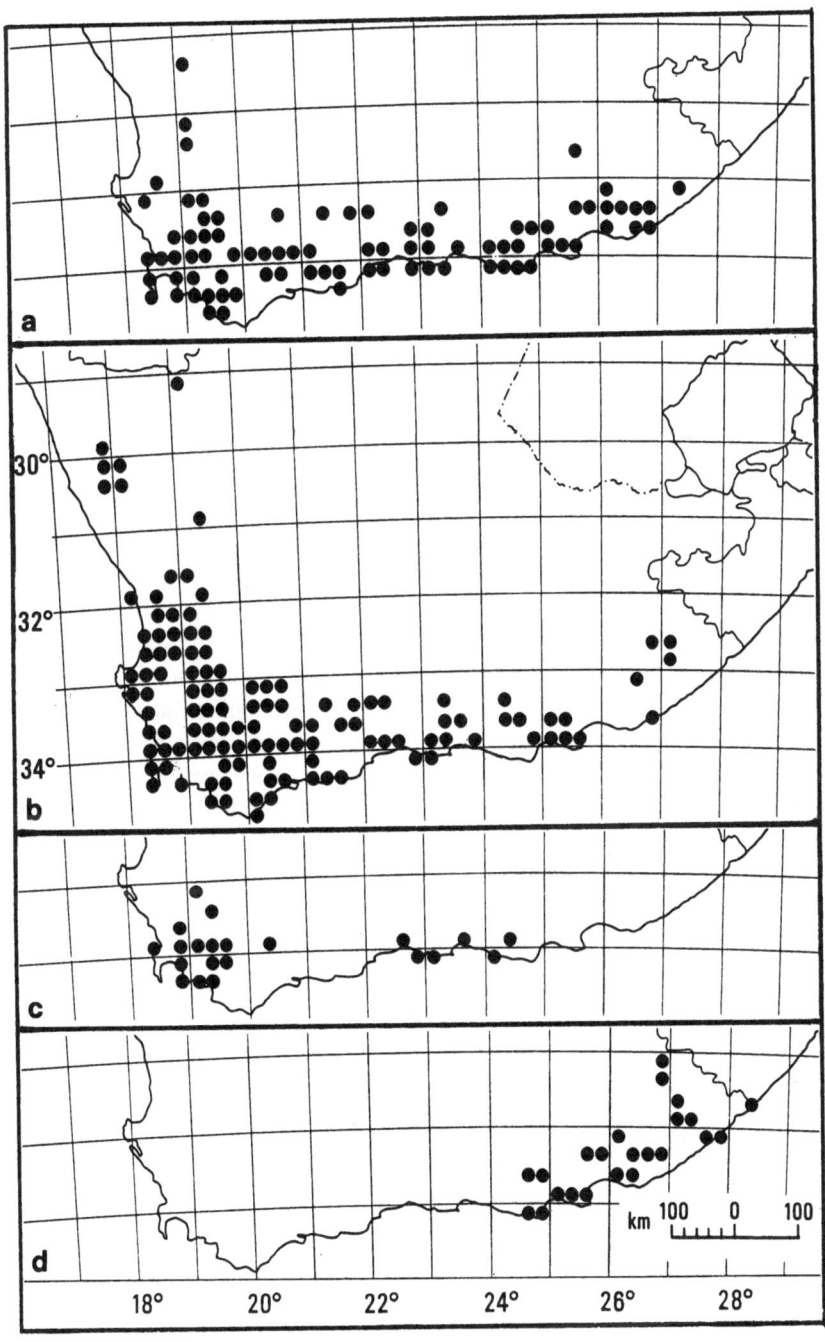

Fig. 69. Distribution of *Anthospermum. a A. aethiopicum, b–d A. spathulatum, b* subsp. *spathulatum, c* subsp. *ecklonianum, d* subsp. *uitenhagense*

(ii) *Female Inflorescences:* Next to rather conspicuous and somewhat contracted inflorescence regions made up of ± many-flowered axillary clusters, there are ♀ plants with indistinct, diffuse inflorescence zones with few- and even 1-flowered partial inflorescences (e.g., Fig. 30 b). Numerous intermediates occur between these two extremes.

(iii) *Fruits:* Variable in shape from being rather large and broad (especially in the upper half) to smaller and narrower; the three dorsal mericarp ribs range from being very conspicuous to ± indistinct. Plants with rather few-flowered partial inflorescences often have fruits with distinct (although never very long) pedicels, while the fruits of plants with several- to many-flowered partial inflorescences usually are subsessile.

The mericarps are normally glabrous, but shortly hairy fruits do occur, even though only rather infrequently (in a few out of several hundred collections).

(iv) *Leaf Size and Shape:* Extremely variable. Unusually broad [to 2.5 (3) mm] and, on the other extreme, almost linear leaves (only ca 0.5 mm wide) are, however, somewhat less common.

(v) *Chromosome Number:* Subsp. *spathulatum* contains di-, tetra- and hexaploid plants. The position of the polyploids within the subspecies and their origin, however, poses a number of unsolved (and unsolvable?) questions. While tetraploidy in the high mountain subsp. *saxatile* and in subsp. *ecklonianum* appears to be constant (this seems to be further supported by the — although admittedly few — large average pollen diameters of karyologically unknown individuals) and diploidy in subsp. *tulbaghense* and in subsp. *uitenhagense* is certain, the situation is not fully understood in subsp. *spathulatum*.

The only documented hexaploid (from the Kamiesbergpas E of Kamieskroon; PUFF 790714-2/1) does not in any way differ morphologically from other (diploid) collections from NW localities; it is a glabrous-fruited form and not one of the more unusual collections from that area with shortly hairy fruits (e.g., PEARSON 6334). Unfortunately, no pollen diameters are available for this hexaploid collection, as the counts originated from germinated seeds of fruiting material.

Also tetraploids such as those from the Giftberg Plateau (PUFF 790713-1/3) — with (large!) average pollen diameters of 34.7 and 35.1 μm (cf. Fig. 48) — are morphologically inseparable from known diploids.

The polyploids are perhaps merely spontaneous autopolyploids, although — in view of the prevailing dioecy in *A. spathulatum* — this explanation is not very satisfactory and not fully convincing.

In spite of this variability, it is possible to distinguish some "Forms" (mostly in the sense of morpho- and/or ecotypes) within subsp. *spathu-*

latum which, however, do not appear to be distinct enough to receive formal taxonomic recognition as there are too many intermediate "connecting" Forms between them. They are discussed here with a view to indicating "trends" to certain character combinations.

(a) *"Typical" subsp. spathulatum* (Fig. 5): Single- to few-stemmed, rather tall to ± medium-sized shrubs with normally regular, ± ascending branches, obovate to ± lanceolate leaves, somewhat contracted ♀ inflorescences and many- to several-flowered partial inflorescences; the fruits are usually distinctly ribbed on the dorsal side and mostly sessile. Widely distributed and ecologically eurytopic.

(b) *"Latifolium" or "Broad-leaved Form"*: Rather robust plants with leaves often considerably broader than in the "typical" Form; partial inflorescences few- to even only 1-flowered, ♀ inflorescences not contracted and inconspicuous (cf. Fig. 30 b); the fruits are usually larger than in the "typical" Form and are relatively broad, conspicuously ribbed on the dorsal side and mostly distinctly pedicellate. This Form seems to be concentrated in the interior parts of the Cape Prov. [from Farm Lokenburg (S Calvinia Distr.) to Karoo Poort, Witteberg and Seven Weeks Poort] but is entirely absent from the SW Cape s.str.

(c) *"Dune Sand Form"*: Rather small bushy, several- to many-stemmed (dwarf) shrubs with rather fine and often narrow, ± linear leaves[1]. The shiny, indistinctly ribbed and subsessile fruits tend to be smaller than in the other Forms; the inflorescence zones of ♀ plants are rather inconspicuous and the partial inflorescences are usually few-flowered. These Forms were observed and recorded in coastal areas from the Cape Peninsula to as far E as the Port Elizabeth Distr. They, however, appear to be entirely absent from the W Coast of the Cape Prov.

A dune sand population discovered near Cape Agulhas Lighthouse (PUFF 790912-3/2) differed quite markedly from the above described "Dune Sand Form" in having rather broader and quite succulent leaves, in having mostly only 1-flowered partial inflorescences and a low, ± spreading habit. No other collections matching this material could be found.

The most essential characters of these Forms are summarized in Table 12. It, however, needs to the stressed again that numerous collections do not exactly match any of the above Forms as they are ± intermediate in one or several of their characters.

---

[1] Small branches (especially of ♂ plants) may be confused with and may quite closely resemble specimens of subsp. *ecklonianum,* although the leaves of the "Dune Sand Form" tend to be shorter than those of subsp. *ecklonianum.* If herbarium sheets contain no information on growth form and habitat (as it is, unfortunately, often the case), a distinction between the two may become difficult.

Table 12. *Anthospermum spathulatum*. Comparison of selected distinction

	Growth form and habit			Leaf size (mm)
	A	B	C	
subsp. *spathulatum*				
a "typical" Form	+/−	+/ ± /−	−/ ±	(3)4–10 × 0.5–1.8 (2.3)
b "Latifolium Form"	+/−	+	−	4–12 × (1)1.5–2.5(3)
c "Dune Sand Form"	−	−/+	± /+	5–10 × 0.5–0.8
subsp. *uitenhagense*	−	−	±	(3)3.5–8(10) × 0.5–1.2
subsp. *saxatile*	−	−	+	(3.5)5–8(12) × 1–2
subsp. *ecklonianum*	+	+	+	8–18(25) × 0.5–1.8(2.2)
subsp. *tulbaghense*	−	−	±	(2.2)2.5–4.5(5.2) × 0.5–1 (1.3)

A: +: taller than 1 m, −: less than 1 m tall. B: +: single-stemmed, erect, −: several- to many-stemmed. C: +: plants densely leafy, −: not densely leafy.

Collections of subsp. *spathulatum* from the E-most localities [Amatole Range, Hogsback Mt., Pirie (3226-D and 3227-C); e.g., Story 3660, Rattray 231, Sim 1328] are somewhat atypical in appearance and in their fruit characters (Rattray 231, for example, has fruits with rather short calyx lobes and the mericarps are seemingly truncate above). These collections, however, clearly differ from *A. spathulatum* subsp. *uitenhagense* which occurs in the same general area (although in different habitats) (compare maps, Fig. 69 b and d). In some respects these specimens appear to approach *A. galpinii* (9.)(see also comments, p. 223).

Plants of *A. spathulatum* subsp. *spathulatum* may be confused with *Nenax* (especially with *N. acerosa* subsp. *macrocarpa* and *N. divaricata*) if not studied carefully. Many-stemmed, ± rounded, low shrubby forms of subsp. *spathulatum* in particular, may be rather similar to taxa of *Nenax* in appearance. It is primarily ♂ specimens that can cause confusion: the ♂ flowers of various *Nenax* species to some extent overlap in size and shape with those of subsp. *spathulatum*. Small ♂ branches — as they are often represented on herbarium sheets — have, therefore, frequently been incorrectly identified. With some experience, however, even vegetative small fragments can be identified with certainty: the *Nenax* species likely to be confused with *A. spathulatum* always have (at least a few) hairs on the basal parts of the leaf margins; their blades are not narrowed to the base.

characters of the subspecies and the "Forms" of subsp. *spathulatum*

Inflorescence		Fruits			Chromo-some no.	Average pollen diam. (μm)
D	E	mericarp size (mm)	F	G		
+/ ± /−	± /+	2.4–3.6 × 1.4–1.8	+/ ±	−/ ±	2x, 4x, 6x	27.3–35.6
−	−	(2.7)3–4.2 × 1.5–2.2	+	+/ ±	2x	29.8–31.8
−	−/ ±	2.5–3.2 × 1.2–1.7	±	−	2x	30.9–32.6
±	± /+	2.2–3 × 1–1.3	±	−	2x	26.7–30.3
−	−	2.4–3 (3.4) × 1.2–1.6	± /+	−	4x	31.5–32.6
+/ ±	+/ ±	2.5–3.1 × 1–1.5	−/ ±	−	4x	31.9–35.1
±	+	(1.5)1.7–2.3 × 0.6–0.8(1)	−/(±)	−	2x	27.8–30.9

D: +: ♀ plants with distinct, ± cylindrical inflorescence zones, −: ♀ plants with inconspicuous inflorescences. E: +: axillary flower clusters many-flowered, −: few- to 1-flowered. F: +: mericarps ribbed dorsally, −: not ribbed. G: +: pedicellate, −: (sub)sessile.

In contrast, the leaf blades of subsp. *spathulatum* are always glabrous and distinctly narrowed to the base. The habitually similar *N. acerosa* subsp. *macrocarpa,* moreover, has often quite conspicuously dimorphic leaves (the long shoot leaves rather short and broad; the short shoot leaves in their axils longer and more needle-like). This never occurs in subsp. *spathulatum.*

Collections

South Africa. Cape − **2917** (Springbok): Little Namaqualand, Misklip (= Farm Mesklip?) (**-DD**), ESTERHUYSEN 5840 (BOL, PRE). − **2919** (Pofadder): Groot Pellaberg (**-AA**), VAN JAARSVELD & PATTERSON 6774 (NBG). − **3017** (Honde-klipbaai): Farm Dassiesfontein nr. Kamieskroon (**-BB**), VAN BREDA 4211 (K, PRE); Rietkloof Mt. (**-BD**), PEARSON 5701 (BOL, K, MO). − **3018** (Kamiesberg): Beaem Hill, NW of Liliefontein (**-AA?**), PEARSON 6334 (BOL, K, MO; atypical); nr. Leliefontein (**-AC**), LEVYNS 4032 (BOL). − **3019** (Loeriesfontein): Kubiskow Mt. (= Kubiskouberge) (**-CD**), MARLOTH 12866 (PRE). − **3118** (Vanrhynsdorp): Vredendal D., Strandfontein (**-CC**), ACOCKS 24076 (K, LISU, PRE, US); Vanrhynsdorp D., Zandkraal (**-DB**), ACOCKS 14872 (K, PRE); −, Giftberg plateau (**-DC**), ACOCKS 14893 (K, PRE), PUFF 790713-1/3 (WU). − **3119** (Calvinia): Calvinia D., Farm Lokenburg, ca 34 km S of Nieuwoudtville (**-CA**), ACOCKS 17024 (K, PRE), PUFF 790711-2/1, 800901-1/10 (WU); summit of Butterkloof Pass (**-CD**), PUFF 800902-2/1 (WU). − **3218** (Clanwilliam): Leipoldtville- Elandsbaai rd., Kliphoutkop (**-AD**), PUFF 800915-3/1 (WU); Farm Knovlei, ca 15–20 km E of Graafwater (**-BA**), PUFF 800915-1/1 (WU); Pakhuis Pass (**-BB** to 3219-AA), STOKOE s.n. sub SAM 59523 (SAM); a few km SW of Redelingshuys, ascent to Olof

Berghs Pass (**-BC**), Puff 800915-4/1 (WU); Clanwilliam D., ca 37 km N of Citrusdal (**-BD**), Maguire 279 (BOL, NBG, STE); Farm Berg-en-dal, a few km N of Aurora (**-CB**), Puff 800915-6/1 (WU); above St. Helena Bay (**-CC**), Goldblatt 6026 (MO, WU); Aurora- Velddrif rd., nr. Farm Sandboskraal (formerly 'Weglopersheuwel'), Puff 800915-7/1 (WU); ca 8–12 km SE of Redelingshuys (**-DA**), Puff 800915-5/1 (WU); Grey's Pass nr. Citrusdal (**-DB**), Hafström & Acocks 1419 (S); Piketberg, [Farm] Boschkloof (**-DC**), Bond 546 (NBG). – **3219** (Wuppertal): from summit of Pakhuis Pass to Heuningvlei For. Stn. (**-AA**), Puff 790716-1/2 (WU); Cedarberge, Algeria For. Stn. (**-AC**), Bos 510 (K, M, PRE, STE, WAG); Farm Sanddrif, NE of Sederberg village (**-AD**), Puff 790717-3/2 (WU); nr. waterfall between Citrusdal and Elandskloof (**-CA**), Stokoe 7589 (BOL, PRE × 2); Clanwilliam D. Kromme River (**-CB?**), Stokoe s.n. sub SAM 64187 (SAM); Dasklip Pass, N of Porterville (**-CC**), Puff 800913-2/2, -4/1 (WU); Farm Grootvlei, E foot of Skurweberge (**-CD**), Puff 790718-2/1 (WU); Kat Bakkies, Zwart Ruggens (Swartruggens Mts.) (**-DC**), Levyns 1886a (BOL). – **3226** (Fort Beaufort): Hogsback, halfway up Tor Doone (**-DB**), Giffen 147 (PRE; ± atypical). – **3227** (Stutterheim): Victoria East D., Hogsback Mt. (**-CA**), Rattray 231 (BOL, PRE; ± atypical); Pirie (**-CC**), Sim 1328 (PRE, STE; ± atypical). – **3318** (Cape Town): Langebaan Peninsula, Farm Abrahamskraal (**-AA**), Puff 791226-3/1 (WU); nr. Hopefield (**-AB**), Bolus 12701 (BOL, NH, PRE); Slangkop near Darling (**-AD**), Puff 800910-1/1 (WU); West Coast, Farm Bokpunt (**-CB**), Puff 800909-1/1 (WU); Cape Peninsula, Camp's Bay (**-CD**), Penfold 161, 162 (NBG); Malmesbury D., Burgers Post Farm, nr. Pella (**-DA**), Boucher & Sheperd 4540 (PRE, STE); Bottelary rd., ca 5 km from Kuilsrivier (**-DC**), Puff 791230-1/1 (WU); Koelenhof (**-DD**), Rehm s.n. (BR, M, SRGH). – **3319** (Worcester): base of Witsenberg, N of Tulbagh (**-AA**), Puff 800924-4/2 (WU); Farm Die Erf, N of Gydoberg (**-AB**), Puff 840912-5/1 (WU); Tulbagh Kloof (**-AC**), Esterhuysen 6072 (BOL); W foot of Theronsbergpas (**-AD**), Puff 790719-4/3 (WU); Ceres-Calvinia rd., Karoo Poort (**-BA**), Esterhuysen 5479 (BOL, PRE), Marloth 9037 (PRE, STE), Puff 790719-3/1 (WU); nr. De Doorns (**-BC**), Bolus 13122 (BOL, K), Lam & Meeuse 4440 (W); around Darling Bridge (Breerivier) (**-CA**), Puff 800904-1/1 (WU); nr. Goudiniweg Stn. (**-CB**), Puff 800920-2/1 (WU); French Hoek Pass (**-CC**), Boucher 2384 (PRE, STE); Riviersonderend Mts., service rd. to Jonaskop SABC tower (**-CD**), Puff 800921-2/1 (WU); Karadouws Mts. (= Kwadouwsberg) nr. [SE of] Orchard (**-DA**), Esterhuysen 10294 (BOL); Montagu D., Eendracht (= Farm Eendrag?) (**-DB**), Compton 18389 (NBG); Aasvoelsberg, SW of Mowers Stn. (**-DC**), Puff 800920-1/2 (WU); Robertson D., Boesmanskloof Pass at McGregor (**-DD**), Esterhuysen 4478 (BOL). – **3320** (Montagu): Touwsrivier- Pienaarspoort rd., Karookop (**-AA**), Puff 790914-1/1 (WU); Witteberg, Farm Tweedside (**-AB**), Compton 22869 (NBG); Touwsrivier- Pienaar-spoort rd., Bierkraal se Rante (**-AC**), Puff 790913-6/2 (WU); Witteberg, Farm Elandsfontein (**-AD**), Compton 3745 (BOL); (nr.) Matjiesfontein (**-BA**), Pearson 1581 (NBG, PRE), Puff 790915-1/2 (WU); Witteberge, S side, Farm Fisantekraal (**-BC**), Puff 790914-3/1 (WU); Montagu D., Pampoenkloof, Baden (**-CA**), Walgate s.n. sub BOL 23358 (BOL, PRE); – , Kogmanskloof (**-CC**), Barker 5411 (BOL, NBG), 5416 (BOL, NBG, STE), Johnson 149 (NBG), Mitchell 110 (PRE), Puff 790913-1/1 (WU); Montagu- Barrydale rd., Wildehondskloof (**-CD**), Puff 790913-2/2 (WU); Touwsberg (**-DB**), Levyns 9088 (BOL); Warmwaterberg (**-DC**, **-DD**), Levyns 6207 (BOL). – **3321** (Ladismith): Ladismith D., Waterkloof [nr. Ladismith] (**-AD**), Gillett 1918 (STE), Hutchinson 1124 (BOL, K); 14 km W of Swartbergpas summit, on rd. to 'The Hell' (**-BD**), Puff 840908-3/3 (PRE, WU); Noukloof Nat. Res. (**-CA**), Laidler 148 (PRE, STE); Garcias Pass (**-CC**),

THOMPSON 3287 (K, PRE, STE); Rooiberg Pass between Vanwyksdorp and Calitzdorp (-DA), PUFF 791223-1/2 (K, WU); Calitzdorp D., Gamka Mtn. (-DB), BOSHOFF P 85, P 157, P 209 (STE). – 3322 (Oudtshoorn): Prince Albert side of Swartberg Pass (-AC), PUFF 790915-2/1 (WU); Prince Albert D., Kriedouw Hills (-AD), MARLOTH 11322 (PRE, STE); Robinsons Pass area, [Farm] Mosselbaai R 4 (-CC), BOND 1467 (NBG); Paardepoort (= Perdepoort, N of Camfer Stn.) (-CD), LEVYNS 10519 (BOL), PUFF 790909-3/1 (WU); Trakadakow Stn. nr. Ronnee Vallei P.O. (-DC), BURCHELL 5718 (K). – 3323 (Willowmore): Buyspoort, S of Willowmore on Uniondale rd. (-AD), PUFF 790909-1/1 (WU); Uniondale D., (nr.) Ongelee P.O. (Ongelegen)(-CB), ACOCKS 19998 (K, M, PRE, SRGH), LYNES 1687 (BM); Prince Alfred Pass (-CC), MARSH 627 (PRE, STE), PUFF 791221-3/4, 800927-4/2 (WU); Uniondale D., Misgund Hills (-CB to -CD), ESTERHUYSEN 6960 a (BOL); Baviaanskloofberge, Nuwekloof, S of Willowmore- Uniondale rd. junction (-DA), PUFF 791221-2/1 (WU); Joubertina, Kouga (-DD), GELDENHUYS 347 (STE). – 3324 (Steytlerville): Steytlerville (-AD), PEARSON sub TVL Museum no. 25795 (PRE); Baviaanskloofberge, Kouga River valley, track to Enkeldoorn (-CB), BOND 914 (NBG), PUFF 791221-1/2 (WU); pass W of Cambria (-DA), OLIVER 4519 (STE); Kleinfontein nr. Hankey (-DD), BAYLISS 8676 (M). – 3325 (Port Elizabeth): Uitenhage D., Armanuskraal, Zunga catchment (-CA), SCHARF 1391 (PRE); –, (Groot) Winterhoek Mts. (-CA to -CB), FRIES, NORLINDH & WEIMARCK 1072 (LISC, M, MO, PRE; – approach. subsp. uitenhagense), 1073 (BM, K, PRE; – approach. subsp. uitenhagense); Van Stadens Gorge (-CC), LONG 634 (K, PRE); Port Elizabeth D., Bethelsdorp (-kloof) (-CD), BOLUS 25119 (BOL), FRIES, NORLINDH & WEIMARCK 256 (PRE), 256a (BM, BOL, K, M, MO); –, (nr.) Humewood (-DC), LONG 434 (K, PRE), RODIN 1082 (K, MO, PRE, S, US). – 3326 (Grahamstown): Albany D., Brakkloof (-BA), ACOCKS 12045 (PRE); Port Alfred (-DB), ROGERS 17174 (J). – 3418 (Simonstown): Cape Peninsula, Schusters Bay (-AB), BARKER 763 (NBG); Cape of Good Hope Nat. Res., above Platboom Bay (-AD), PUFF 800907-2/1 (WU); between Eerste R. and Swartklip (-BA), PILLANS 9231 (BOL, K); Roman Rock, W of Klein Hangklip range (-BD), BOUCHER 1366 (STE). – 3419 (Caledon): summit of Shaw's Mountain Pass (-AD), PUFF 791224-1/2 (WU); 'Great Mountain of Baviaanskloof at Genadendal' (-BA), BURCHELL 7614 (K); N foot of Riviersonderend Mts., McGregor- Bonnievale rd., W of Farm Boesmansrivier (-BB), PUFF 840911-1/1 (PRE, WU); Stanford- Papiesvlei rd., SW base of Perdeberg, above Farm Flouhoogte (-BC), PUFF 800918-1/2 (WU); Franskraal se Berge, N of Gansbaai (-CB), PUFF 840919-1/2 (PRE, WU); SW Koueberge, Papiesvlei-Nuwepos rd., ca 4 km NNW of Nuwepos Farm (-DA), PUFF 800918-2/2 (WU). – 3420 (Bredasdorp): Swellendam D., National Bontebok Park (-AB), LIEBENBERG 6726 (PRE × 4); De Hoop Nature Reserve (-AD), MARSH 1471 (K, PRE, STE), 'Nature Conservation' 597 (PRE), VAN DER MERWE 121 (STE); –, Potberg (-BC), BURGERS 1878 (STE); ca 8–9 km N of Struisbaai on Bredasdorp rd. (-CA), PUFF 790912-4/1 (WU); Farm Martha, ca 16.5 km E of Skipskop on Bredasdorp- Arniston rd. (-CB), PUFF 790912-1/3 (WU); nr. Cape Agulhas Lighthouse (-CC), PUFF 790912-3/2 (WU). – 3421 (Riversdale): hill N of Korente R. dam (-AA), BOHNEN 7599 (STE; PRE – mixed with A. aethiopicum); Riversdale- Blombos rd., nr. Farm Victoriasdale (-AC), PUFF 840920-3/4 (WU); Still Bay hills (-AD), JOHNSTON 110 (NBG); N of Driefontein on Ystervarkfontein-Albertinia rd. (-BC), PUFF 790910-6/1 (WU). – 3422 (Mossel Bay): Goukamma Nature Res. (-BB), HEINECKEN T 4 (PRE), 'Nature Conservation' 91 (PRE). – 3423 (Knysna): Knysna D., Noetzie (-AA), LEVYNS 7825 (BOL). – Imprecise localities or no locality given: "CBS", AUGE s.n. (BM × 2), BANKS s.n. (herb. JACQUIN) (W), CARMICHAEL 150 (BM), [herb.] DELESSERT s.n. (P), HARVEY s.n. (K), NIVEN 106 (BM),

OLDENBURG 828 (BM), SCHOLL s.n. (W × 2, WU), SPARRMAN s.n. (S − mixed with *A. aethiopicum*), THUNBERG s.n. (G, S × 2), VERREAUX s.n. (G); 'everywhere in the Cape and Stellenbosch Prov.', ECKLON & ZEYHER 2307 γ (FI, G, LY × 2, M − mixed with subsp. *tulbaghense,* MO, PRE, S, SAM, W; p.p. mixed with *Nenax acerosa* subsp. *macrocarpa*); 'on Kasteelsberg, in the mountains near Simonstown and in the Hottentotshollandberge above the Palmietrivier', ECKLON & ZEYHER 2307 ζ (FI, G, LY, M, MO, P, PRE, SAM). − Additional collections: ACOCKS 629, 630, 4980 (S); BARKER 418 (BOL, NBG, PRE); BURTT DAVY 12511 (PRE); ANDREAE 271 (STE), 945 (PRE, STE), 949 (GH), 1232 (PRE); BOLUS 5059, 11294 (BOL); BOND 1057 (NBG); BOUCHER 600, 2384 (PRE, STE), 1366 (STE); BURCHELL 8681 (K); COMPTON 2521, 3004 (BOL, K), 4023 (BOL), 5882, 6436 (BOL, NBG), 6814 (NBG, STE), 4975, 6595, 6605, 6943, 6979, 7053, 7314, 8909, 9590, 12161, 16109, 17325, 21021, 23027 (NBG); DAHLSTRAND 416 (GRA − approach. subsp. *uitenhagense*); DAVIS s.n. sub PRE 41520 (PRE); DRÈGE 7661 a (E, G, W), 7663 a (E, G, K, LY, MO, P, W), 9550 (S), s.n. ('*A. aethiopicum*': E, G, S, W), 9551 (LY); ECKLON 1, 5, 7, 9, 11, 26, 1342 (& ZEYHER?; − mixed with *A. pumilum* subsp. *rigidum*), s.n. (all S); ECKLON & ZEYHER '58-8' (× 1 only; other sheets are *Nenax*), '64-9', s.n. (S), '77-9' (LY, S); ESTERHUYSEN 5564 (BOL, PRE), 5923 (BOL, K), 6167, 10819, 13024, 17162 (BOL); FOURCADE 22, 5581 (BOL); FRIES, NORLINDH & WEIMARCK 683 (PRE), 1326 (MO, PRE, US; − approach. subsp. *uitenhagense*); GARSIDE 1261 (K); GILLETT 53, 157, 221, 408 (STE); GOLDBLATT 1705 (MO, PRE), 2212 (M, MO, PRE); HAFSTRÖM & ACOCKS 1420 (PRE, S); HAFSTRÖM & LINDEBERG s.n. (S × 2); HALL 750 (NBG); HANEKOM 637, 638 (K, PRE); HOLLAND s.n. (MO sub 2306195, PRE sub 41505); HUMBERT 9588 (P, PRE); HUTCHINSON 612 (BOL, K); JOHNSON 238 (NBG); KEET 32 (GH, PRE, STE), 579 (PRE, STE), 599 (GRA, STE; − mixed with *A. aethiopicum*), 760 (PRE, STE); KERFOOT 5933 (NBG); LEVYNS 2071, 2944, 3962 a, 6069, 7390 (BOL), 8078 (BOL; − approach. subsp. *tulbaghense*); LEWIS 3107 (SAM), s.n. sub BOL 25131 (BOL; − approach. subsp. *saxatile*), s.n. sub SAM 64190 (PRE); LYNES 1797 (BM); MAGUIRE 1848 (NBG); MARLOTH 3292 (PRE); MARSH 1238 (PRE, STE); MARTIN 283 (NBG); MATINEAU 566 (SRGH); MICHEL 169 (PRE); MICHELL s.n. sub BOL 15586 (BOL, MO, PRE); MONTGOMERY 536 (STE); MOSS 5989 (BM); 5989 bis (J); OLIVIER 37 (K, M, PRE, STE), 435 (GRA), 3696 (PRE, STE); PAGE s.n. sub BOL 15634 (BOL); PAPPE s.n. (K); PARKER s.n. (GH); PEARSON 5284 (BOL, K, MO, PRE), 5730, 6263, 6554 (BOL, K); PENTHER 2697 (M, S, W); PETER 50172 (B); PHILLIPS 7460 (BOL, PRE; *A. dregei* in other herbaria); PILLANS 7336, 9101, s.n. sub BOL 14168 (BOL); PRIOR s.n. (K × 2); PUFF 790714-2/1, 790717-1/2, 790719-2/1, 790911-3/3, -7/1, 790914-2/3, 791220-4/3, -4/4, 791225-1/1, -2/1, 791226-4/1, -5/2, 791228-2/2, 791229-1/1, 791230-3/2, -4/2, 800903-1/1, 800913-5/5, 800920-3/2, 800921-2/1, 840908-1/1, 840909-2/2, -4/2, 840912-4/2, 840917-1/3, 840918-1/3, 840919-5/2, 840920-1/3, -3/4 (WU); REHM s.n. (?214, ?324: M); REHMANN 2053 (BM, BR), 2390 (BM, K); ROGERS 11306 (G), 16597 (BOL, G, GRA; *A. aethiopicum* in K), 22869 (PRE), 28564 (K), s.n. (J); RYCROFT 1944 (NBG); SALTER 6282, 7259 (BOL); SCHARF 1868 (K); SCHLECHTER 5104 (B, BOL, BR), 7976, 7977 (BM, BOL, E, G, GRA, K, MO, PRE, S, US, W), 7979 (PRE); SCHMIDT 324 (M); SMITH 2888 (PRE); STEPHENS 6870 (B, K), 7120, 7124 (K), 7122 (BOL, K); STEPHENS & GLOVER 8740 (K); STOKOE s.n. sub SAM 64190, 69991, 69992 (SAM); STORY 3660 (K, PRE; ± atypical); TAYLOR 802, 896 (NBG), 4883 (PRE, STE); THODE s.n. sub STE 6180, 7961, 8419 (STE); THOMPSON 314 (K, PRE, STE); VAN BREDA SKF-587 (K), 1695 (PRE); VAN BREDA & JOUBERT 1993 (PRE); VAN DER MERWE 2904, 2960 (K, PRE); VAN DER WESTHUIZEN 330 (K, MO, PRE, SRGH); WALL s.n. (S); WEST 360 (BOL); WILLEMS 36 (NBG); WILMS 3181 (P); WOLLEY DOD 246 (BOL, K), 1071 (BM); WORSDELL s.n. (K);

WRIGHT s.n. sub US 81701, 81702 (US); WURTS 1365 (NBG), 1401 (BOL, NBG, STE); YOUNG s.n. sub TVL Museum no. 26524 (PRE); ZEYHER 2714 (LY, S, SAM).

12 (b). Subsp. *uitenhagense* PUFF in Fl. S. Afr. **31,** 1 (2): 17 (1986).
Type: South Africa, Cape Prov., '... "Zwartkopsrivier" inque planitie inter "Krakakamma" et montes "Van Stadensriviersberge" altit. I, II (Uitenhage)', ECKLON & ZEYHER 2307 β (SAM, holo.!; BOL, GOET, LY, M, MO, S, iso.!).

− *A. aethiopicum* L. var. *uitenhagense* ECKLON & ZEYHER, Enum. Pl. Afr. Austr.: 365 (1836), non valide publ. (sine descr.).

Dioecious, several- to many-stemmed, mostly ± erect and cylindrical or somewhat reclining dwarf shrubs ca 20–60 (90) cm tall, with thick, woody roots; regularly burnt individuals often with massive, ± rosette-like woody bases. Stems to ca (10) 15 mm in diameter near the base and ca 1–2 mm in the midstem region, with several to many, ± erect to ascending branches (except in burnt individuals); younger parts obscurely 4-angled, reddish-brown to dark reddish-purplish tinged, ± densely covered with whitish papillae to 0.1 mm long; older parts terete, light brown-grey to dark grey, becoming glabrescent, epidermis often peeling; internodes mostly slightly longer than the leaves; plants quite densely leafy above. Leaves decussate, pseudo-verticillate; blades (3) 3.5–8 (10) × 0.5–1.2 mm, narrowly lanceolate or oblanceolate to ± linear, narrowed to the base, glabrous, mostly tough and thickish, upper surface often shiny; apices ± mucronate, acute or bluntish; margins revolute to ± flat; stipular sheaths mostly shortly hairy on the margins, (broadly) cup-shaped and conspicuous, ca 0.7–2.5 mm long, with a median seta ca 0.2–0.7 mm long, often with 1 (2) additional minute setae on either side. Inflorescences ± inconspicuous, made up of ± sessile, several- to few- (3-)flowered axillary clusters (♀ inflorescences often more-flowered than ♂). Flowers subsessile; ♂, ♀. Corolla 4-merous, greenish-yellow to yellowish, often purplish tinged on the outside, glabrous or occasionally with a few whitish papillae to ca 0.1 mm long near the tip. ♂: tube 0.4–0.9 mm long, (broadly) funnel-shaped, lobes (1.5) 1.9–2.5 × 0.5–1 mm, ± lanceolate, recurved; stamens 4, filaments exserted for (0.5) 0.7–1.2 mm, anthers 0.8–1.5 × 0.2–0.5 mm; rudimentary ovary very small and occasionally with minute calyx lobes. ♀: tube 0.2–0.5 mm long, cylindrical, lobes 0.3–0.7 × 0.1–0.2 mm, linear-lanceolate, ± erect; style ± 0–0.7 mm long; stigmas 2, (1.7) 2–5.1 mm long, greyish to greenish-grey; ovary 0.6–0.7 × 0.4 mm, glabrous, crowned by small and often ± indistinct calyx lobes. Fruits shiny reddish-brown to greyish-green-brown, supported by ± short U-shaped carpophores; each mericarp 2.2–3 × 1–1.3 mm, (narrowly) obovate, elliptic to oblong, often seemingly truncate above due to the position and shape of the calyx lobes, dorsal side (shallowly) convex, sometimes ± obscurely 3-ribbed, glabrous, ventral side plane to slightly concave and with a ± prominent median vertical ridge, commissure ± large; mericarps crowned by 2 sometimes ± joined (broadly) triangular calyx lobes ca 0.1–0.5 (0.7) × 0.4–0.5 mm. – Figs. 36 h, 70 a–b.

Chromosome Number (Table 3): n = 11, 2 n = 22.
Average Pollen Diameter (Fig. 48): 26.7–30.3 μm.
Habitat (Fig. 70 a): Frequently associated with rocky outcrops in grassland, occasionally in (rocky) grassland or on dry, rocky slopes. Sometimes in (marginal) Fynbos. Often over Witteberg quartzite or in red, loamy soil. – Ca 30–1 000 (1 200) m.
*Distribution (*Map, Fig. 69 d): S Africa, SE and E Cape Prov. From the Humansdorp and the Hankey Distr. NE to the Carthcart Distr. and the Great Kei River mouth.

Critical Remarks: Subsp. *uitenhagense* can be distinguished by the following combination of characters: several- to many-stemmed, ± erect, rather low habit (Fig. 70 a); relatively short and narrow, tough leaves; rather small ♂ flowers; comparatively small fruits which are often seemingly truncate above (Fig. 70 b) (also see Table 12). Its distribution (Fig. 69 d), moreover, is much more "E" than that of any of the other subspecies of *A. spathulatum*; the majority of its range is clearly outside the SW Cape Floristic Region.
　While in the E part of its distribution range, subsp. *uitenhagense* is very distinct (I originally even thought that it warranted specific rank), the delimitation to subsp. *spathulatum* can become somewhat problematic from the Port Elizabeth and Uitenhage Distr. W-ward. Coastal "Dune Sand Forms" of subsp. *spathulatum* (see p. 239 for details) may overlap in leaf size and shape and may somewhat resemble individuals of subsp. *uitenhagense* in habit; the ♂ flowers of this "Form" of subsp. *spathulatum* tend to be larger than those of subsp. *uitenhagense*, but a 100% certain distinction is not always possible. Fruiting material, which should allow an unambiguous identification is, unfortunately, often not available. Forms of subsp. *spathulatum*, ± approaching subsp. *uitenhagense* in some respects, occur as far W as the Knysna Distr.

Collections

South Africa: Cape – **3226** (Fort Beaufort): Ca 6 km N of Cathcart-Whittlesea rd., towards Fincham's Nek (**-BB**), Puff 790114-2/1 (J, WU); ca 21 km

Fig. 70. *Anthospermum spathulatum. a–b* subsp. *uitenhagense* (Puff 790114-2/1), *a* plants, to ca 40 cm tall, around a rocky outcrop (E Cape Prov., N of Cathcart-Whittlesea rd., towards Fincham's Nek), *b* fruiting branch *c* subsp. *tulbaghense* (Puff 791230-2/1), fruiting branch. – *b–c* herbarium material; the scale unit is 1 mm

from Cathcart on Whittlesea rd. (-**BD**), Puff 790114-3/1 (J, W, WU). – **3227** (Stutterheim): Keiskamahoek (-**CA**), Kotsokoane 128 (J); Kingwilliamstown D., Bolasse (-**CC**), Sim 1484 (PRE); – , Breidbach (-**CD**), Sim 1174 (PRE). – **3328** (Butterworth): near Kei Mouth (-**CB**), Flanagan 612 (GRA, PRE, SAM); on the Kei R., below 500′, Drège s.n. ['*A. spath*. b' (= V, b, 14), E, G × 3 – one mixed with '*A. spath*. ?b' = *A. galioides* subsp. *reflexifolium*, K, MO, W]. – **3324** (Steytlerville): Hankey- Willowmore rd., Baviaanskloofberge above Farm Kleinplaas (-**DA**), Puff 791220-4/2 (WU); Pisgoedvlakte, heights above Cambria, Goldblatt 2098 (BR, M, MO, NU, PRE, WAG); Elandsberg, S of (just opposite) Cockscomb Mt. (-**DB**), Puff 840921-6/1 (WU). – **3325** (Port Elizabeth): Suurberg, nr. Farm Zuurberg Manor (-**BC**), Puff 800929-1/2 (WU); – , Sidey 3126 (PRE, S); – , N descent of Suurbergpas, between Farms Aansvilla and Viewlands (-**BD**), Puff 800929-4/2 (WU); Uitenhage D., Otterford Forest Res. (-**CC**), Rodin 1122 (BOL, K, MO); Port Elizabeth D., Rocklands near Witteklip (-**CD**), Holland 3684 (BOL); – ca 8 km from Port Elizabeth on Cape road (-**DC**), Holland 3860 (BOL). – **3326** (Grahamstown): Swartwatersberg, SW of Riebeek-East (-**AA**), Puff 840922-1/5 (PRE, WU); nr. Riebeek-East, Puff 840922-2/2 (WU); near Coldstream (= Cold Spring?) Station (-**AD**), Killick 780 (K, PRE); Howiesons Poort area, Puff 790117-1/1 (J, WU); Grahamstown, 'Mountain Drive' (-**BC**), Puff 790117-3/2 (WU); – , road to Collingham, Britten 1535 (PRE; mixed with *A. aethiopicum*); – , Featherstone Kloof, Rennie & Rennie 215 (BOL); near 'Frasers Camp', ca 38 km E of Grahamstown (-**BD**), Puff 791220-1/1 (WU); Alexandria D., Kolsrand (Mt.) (-**CA**), Archibald 4088 (PRE, S); – , Alexandria (-**CB**), Burtt Davy 13404 (PRE). – **3327** (Peddie): ca 13 km from Kidd's Beach on rd. to Peddie (-**BA**), Puff 790116-2/1 (J, WU); East London (-**BB**), Rattray 115 (GRA). – **3424** (Humansdorp): Humansdorp D., ca 16 km from Humansdorp on Storms River rd. (-**BA**), Story 2836 (PRE); – , Humansdorp (-**BB**), Galpin 4104 p.p. (PRE). – Imprecise locality: 'On the Zwartkopsrivier and in the flats between Krakakamma and Van Stadensriviersberge', Ecklon & Zeyher 2307 β (BOL, GOET, LY, M, MO, S, SAM).

12 (c). Subsp. *saxatile* Puff in Fl. S. Afr. 31, 1 (2): 18 (1986).
Type: South Africa, Cape Prov., Hex River Mts., Milner Peak, 5 500– 6 000 ft., Jan. 1959, Esterhuysen 28089 (BOL, holo.!; WU, iso.!).

Dioecious, several- to many-stemmed, often decumbent cushion- or ± mat-forming dwarf shrubs. Stems to ca 25 cm long, to ca 6 mm in diameter near the base and ca 1–3 mm in the midstem region, occasionally rooting at the lowermost nodes, mostly much-branched, branches ± ascending to erect; younger parts obscurely 4-angled, reddish-brown or occasionally greyish-brown, ± densely covered with whitish papillae to 0.1 mm long to (rarely) subglabrous; older parts terete, greyish-brown to dark grey, becoming glabrescent, epidermis peeling; internodes often shorter than or as long as the leaves; plants densely leafy. Leaves decussate, pseudo-verticillate; blades (3.5) 5–8 (12) × 1–2 mm, lanceolate to oblanceolate or narrowly obovate, narrowed to the base, glabrous, tough and thickish, upper surface shiny and with conspicuous but small epidermis cells; apices acute to ± obtuse; margins ± flat to somewhat revolute; stipular sheaths mostly shortly hairy, cup-shaped, ca 0.5–1 mm long, with a median narrowly triangular seta ca 0.2–0.7 (0.9) mm long. Inflorescences inconspicuous, made up of ± sessile, few- (3–1-)flowered axillary clusters. Flowers subsessile; ♂, ♀. Corolla 4-merous, yellow to yellowish-green, occasionally purplish tinged on the outside, glabrous. ♂: tube 0.8–1.4 mm long, (broadly) funnel-shaped, lobes 2.2–2.5 × 0.8–1 mm, lanceolate, recurved; stamens 4, filaments exserted for 1.4–1.7 mm, anthers 1.5–2.1 × 0.3–0.5 mm; rudimentary ovary minute but well discernible, mostly crowned by 4 distinct calyx lobes. ♀: tube 0.2–0.5 mm long, ± cylindrical, lobes 0.5–1 (1.4) × 0.2–0.4 mm, linear-lanceolate, erect to ± spreading; style ± 0–1 mm long; stigmas 2, 2.7–9.2 mm long, greyish or greyish-green; ovary 1.4–1.7 × 0.5–0.8 mm, glabrous, shiny, crowned by 4 conspicuous calyx lobes. Fruits shiny reddish-brown, supported by long, thin U-shaped carpophores, mostly at least half as long as the fruits; each mericarp 2.4–3 (3.4) × 1.2–1.6 mm, obovate to elliptic, dorsal side shallowly convex to slightly angled (mericarps then ± triangular in section), ± indistinctly 3-ribbed, glabrous, ventral side ± plane and with a ± obscure median vertical ridge, commissure large; mericarps crowned by 2 ± triangular calyx lobes 0.3–1.5 (1.9) × 0.5–0.7 mm.

Chromosome Number (Table 3): 2 n = 44.

Average Pollen Diameter (Fig. 48): 31.5–32.6 μm.

Habitat: Only in the highest parts of the SW Cape mountains from ca 1 500–1 800 (1 900) m and growing on rocky slopes, ledges or cliffs, in steep rocky gullies and amongst, at the base of or in cracks of (TMS) rocks. Mostly in damp to moist, ± shady and sheltered places.

Distribution (Map, Fig. 71 a): S Africa, SW Cape Prov. Following the Cape folded mountain belt from the Cedarberg Mts. S to the N Caledon Distr. and E to the E-most part of the Tsitsikamma Mts. in the Humansdorp Distr.

Critical Remarks: The distinct cushion-forming or ± matted habit of subsp. *saxatile*, the rather tough, closely spaced leaves, the reduced inflorescences with few-flowered partial inflorescences and the medium-sized fruits with distinct calyx lobes in combination with its restricted occurrence to the highest parts of the SW Cape mountains and to sandstone areas justify its distinction as a subspecies. Certain individual morphological characters may, however, overlap with those of subsp. *ecklonianum* and some "Forms" of subsp. *spathulatum*. The tetraploidy in *saxatile* seems to be another argument in favour of a separation as a subspecific taxon although it should be remembered that (i) chromosome

17*

counts of only two populations are available and that (ii) the few additional pollen measurements of karyologically unknown plants do not allow definite conclusions as to their ploidy level.

Subsp. *saxatile* appears to be more closely allied to subsp. *spathulatum* than to the other subspecies of *A. spathulatum*.

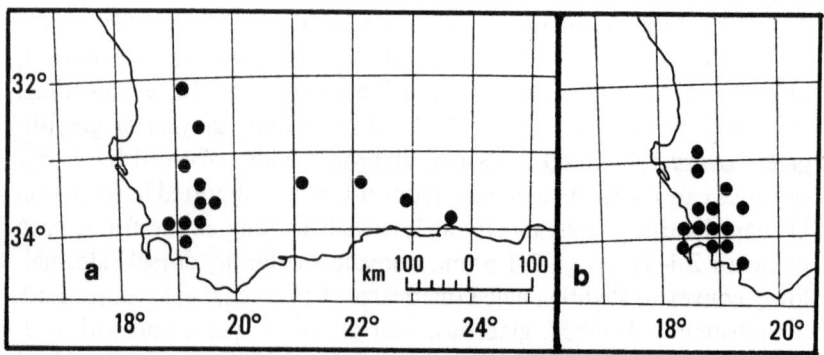

Fig. 71. Distribution of *Anthospermum spathulatum*. *a* subsp. *saxatile, b* subsp. *tulbaghense*

Collections

South Africa. Cape – **3219** (Wuppertal): Clanwilliam D., Cedarberg Mts., Krakadouwsberg (-**AA**), ESTERHUYSEN 7510 (BOL); Ceres D., S Cedarbergen, Gideons Kop (-**CB**), ESTERHUYSEN 13890 (BOL; approach. subsp. *spathulatum*). – **3318** (Cape Town): Jonkershoek State For., Third Ridge Peak (-**DD**), KRUGER 1741 (STE). – **3319** (Worcester): Witsenberge, Swartgat Peak (-**AA**), ESTERHUYSEN 27943 (BOL, WU), –, around Sneeugat Peak, PUFF 800924-3/1 (WU); Worcester D., Waaihoek Mts., Mosters Hoek Twins (-**AD**), ESTERHUYSEN 9860 (BOL), –, Hex River Mts., Milner Peak, ESTERHUYSEN 28089 (BOL, WU); –, Brandwacht Peak (-**CB**), ESTERHUYSEN 11018 (BOL, K, NBG, PRE, SAM), –, Audensberg, ESTERHUYSEN 28183 (BOL, WU); Paarl D., Wemmershoek Mts., Winterberg (-**CC**), ESTERHUYSEN 9637 (BOL, K), –, –, Wemmershoek Peak, ESTERHUYSEN 11332, 34488 (BOL); Caledon D., Stettynsberg (-**CD**), ESTERHUYSEN 11127 (BOL), –, Riviersonderend Mts., ridges W of Jonaskop, PUFF 800921-1/1, 800925-1/2 (WU); Worcester D., between Keeromsberg and Ben Heatlie (-**DA**), ESTERHUYSEN 27644 (BOL, WU). – **3321** (Ladismith): Ladismith D., Swartberg, Toorkop (-**AC**), ESTERHUYSEN 26808 (BOL). – **3322** (Oudtshoorn): Prince Albert D., nr. summit of Swartberg Pass (-**AC**), STOKOE s.n. sub SAM 67520 (PRE, SAM); Uniondale D., Mannetjiesberg (-**DB**), ESTERHUYSEN 6407 (BOL, K, NBG, PRE), 6420 (BOL × 2). – **3323** (Willowmore): Tsitsikamma Mts., Formosa Peak, S of Louterwater (-**DC**), PUFF 800928-3/1 (WU). – **3419** (Caledon): Stellenbosch D., Victoria Peak (-**AA**), ESTERHUYSEN s.n. (BOL).

12 (d). Subsp. *ecklonianum* (CRUSE) PUFF in Fl. S. Afr. 31, 1 (2): 18 (1986).
≡ *A. aethiopicum* L. var. *ecklonianum* CRUSE in Linnaea **6**: 10 (1831);
SONDER in HARVEY & SONDER, Fl. Cap. **3**: 28 (1865).
≡ *A. spathulatum* SPRENGEL var. *ecklonianum* (CRUSE) CRUSE in Linnaea **7**: 133 (1832).
Syntypes: South Africa, Cape Prov., "Baviansberg bei Genadenthal", ECKLON s.n. (S, lecto.!, selected here); −, −, "Schwarze Berg bei Caledon-Bad", ECKLON s.n. (S!, SAM!)[1].

Dioecious, single-stemmed, ± erect and typically (narrowly) cylindrical shrubs ca (0.5) 1–2 m tall. Stems to ca 10 mm in diameter near the base and ca 1–3 mm in the midstem region, ± much-branched above, branches mostly erect to ± ascending, usually regular, paired; younger parts shiny purplish-red, reddish-brown to greyish-brown, glabrous to ± densely covered with whitish papillae to 0.1 mm long; older parts brownish-grey to dark grey, becoming glabrescent, epidermis peeling; internodes mostly as long as or shorter than the leaves; plants (especially ♀) quite densely leafy. Leaves decussate, pseudo-verticillate; blades erect or ± spreading (especially in ♂), 8–18 (25) × 0.5–1.8 (2.2) mm, linear-(ob)lanceolate to ± linear, narrowed to the base, often fine and ± needle-like, glabrous, upper surface mostly ± shiny and with conspicuous but small epidermis cells; apices acute to ± acuminate; margins somewhat to strongly revolute; stipular sheaths glabrous or shortly hairy, cup-shaped, 0.9–1.2 mm long, with a median seta ca 0.6–1.2 (1.9) mm long, often with 1 (2) much smaller setae on either side. Inflorescences ± inconspicuous (♂) to conspicuous and somewhat contracted (♀; ± cylindrical inflorescence zones to a few cm long), made up of ± sessile, several- to many-flowered axillary clusters. Flowers subsessile or sometimes shortly pedicellate (♂); ♂, ♀. Corolla 4-merous, greenish-yellow to yellow, sometimes reddish-purplish tinged on the outside, glabrous. ♂: tube (0.3) 0.5–1.4 mm long, (broadly) funnel-shaped, lobes 1.9–2.9 (3.4) × 0.7–1 (1.3) mm, ± lanceolate, recurved; stamens 4, filaments exserted for (0.7) 1–1.9 mm, anthers 1.3–2 × 0.2–0.5 (0.7) mm; rudimentary ovary minute but well discernible, sometimes crowned by small calyx lobes. ♀:

---

[1] The label of the lectotype specimen in S was written by CRUSE and signed "W. CRUSE 1829". Although no reference is made on the label to ECKLON, there is no doubt that this is one of the specimens referred to by CRUSE in 1831 (p. 10–11).
It appears that at least the second-mentioned collection was later distributed as ECKLON & ZEYHER 2307 ε ['inter frutices (altit. II, III) montis "*Zwarteberg*" *haud longe a thermis* (= Caledon baths), nec non in collibus prope "Babylons Toorensberg" (Caledon)': ECKLON & ZEYHER 1836: 366]. The ECKLON & ZEYHER 2307 ε collection could, therefore, also be considered to be a syntype (BOL!, FI!, G!, GOET!, M!, MO!, P!, S!, SAM!, W!).

tube 0.1–0.3 mm long, cylindrical, lobes 0.2–0.6 (0.8) × 0.1–0.3 mm, ± linear-lanceolate, ± erect; style ± 0–0.3 mm long; stigmas 2, 2–4.4 mm long, greenish to greenish-grey; ovary ca 0.5–0.9 × 0.4–0.6 mm, glabrous, crowned by minute calyx lobes. Fruits shiny reddish-brown, supported by U-shaped carpophores to ca half as long as the fruits; each mericarp 2.5–3.1 × 1–1.5 mm, ± obovate to oblong, dorsal side ± shallowly convex, rather obscurely 3-ribbed, glabrous or, very rarely, shortly hairy, ventral side ± plane to slightly concave and with a ± prominent median vertical ridge, commissure large; mericarps crowned by 2 ± indistinct calyx lobes ca 0.5 × 0.5–0.7 mm, or calyx lobes ± lacking.

Chromosome Number (Table 3): n = 22, 2 n = 44.

Average Pollen Diameter (Fig. 48): 31.9–35.1 µm.

Habitat: Mostly growing on rocky (well drained) mountain slopes, at the base of cliffs or between rocks, along streamsides or in temporarily water-logged areas in clayey, gravelly to sandy soils. Appears to be better suited to soils derived from Cape Granite or areas where TMS is underlain by granite; in several localities (Wellington to Stellenbosch Distr.) found growing together with *Leucospermum lineare,* which is also confined to such soils (cf. ROURKE 1972: 76). Also recorded as occurring on shale plates and occasionally in sandstone soils. In the E part of its distribution range (see Critical Remarks) in areas of relatively high rainfall. – Ca 50–750 (–1 500?) m.

Distribution (Map, Fig. 69 c): S Africa, SW Cape Prov. From the N Cape Peninsula (Table Mt.) NE to the Tulbagh and Ceres Distr., and E to the Swellendam Distr. Occurring again further E from the George to the Humansdorp Distr. (See Critical Remarks, below.)

Critical Remarks: Subsp. *ecklonianum* in several respects (but especially in habit and appearance) resembles *A. aethiopicum* (11.). It seems to be somewhat intermediate between *A. spathulatum* and *A. aethiopicum.* As stated in the introduction to *A. spathulatum* (p. 233), CRUSE at first (1831) included this usually very distinct taxon in *A. aethiopicum,* but then (1832) decided that it, primarily on account of its strictly decussately arranged leaves, should be placed with *A. spathulatum.* CRUSE's arguments seem convincing, as there *is* in fact a clear line of demarcation between plants with ternately arranged leaves (*A. aethiopicum*) and plants with decussate leaf arrangement (*A. spathulatum*); individuals do not show a "mixed" leaf arrangement (i.e., opposite and 3-whorled on one plant). The principal differences between *spathulatum* subsp. *ecklonianum* and *A. aethiopicum* are summarized in Table 13.

It is tempting to speculate that *A. spathulatum* subsp. *ecklonianum* is of

Table. 13. Comparison of selected characters of *A. spathulatum* subsp. *ecklonianum*
and *A. aethiopicum*

	A. spathulatum subsp. *ecklonianum*	*A. aethiopicum*
Leaf		
arrangement	decussate	whorls of 3
size	8–18 (25) × 0.5–1.8 (2.2) mm	3–9 (13) × (0.3) 0.5–1.2 (2) mm
upper surface	"dotted"	not "dotted"
Inflorescence (♀)	condensed; infl. zones short	condensed; infl. zones long
Flowers		
♂: tube	(0.3) 0.5–1.4 mm	(0.2) 0.4–0.9 (1.2) mm
lobes	1.9–2.9 (3.4) × 0.7–1 (1.3) mm	(1.2) 1.6–2.5 (2.9) × 0.6–1.2 (1.4) mm
♀: tube	0.1–0.3 mm	0.1–0.4 (0.6) mm
lobes	0.2–0.6 (0.8) mm	0.1–0.5 (0.7) mm
Fruits		
size	2.5–3.1 × 1–1.5 mm	(1.4) 1.7–2.5 (2.7) × 0.8–1 (1.2) mm
calyx lobes	± 0; 0.5 × 0.5–0.7 mm	± 0; 0.1–0.2 × 0.2 mm
Average pollen diam.	31.9–35.1 µm	30.3–34.2 µm
Chromosome number	n = 22, 2 n = 44	n = 11, 2 n = 22

hybrid origin [*A. spathulatum* (subsp. *spathulatum*) × *A. aethiopicum*]. In favour of such a theory would be the documented tetraploidy in subsp. *ecklonianum*, a certain trend to more fertile, moister habitats (as it is typical for *A. aethiopicum*), and also the more oblong, relatively narrower and ± indistinctly ribbed fruits, which are longer but, in general, similar to those of *A. aethiopicum*. On the other hand, one may argue that subsp. *ecklonianum* has acquired a set of characters similar to that of *A. aethiopicum* through parallel evolution, and has become adapted to wetter and more favourable habitats. In my opinion, the second argument is more likely as — in at least all the populations of subsp. *ecklonianum* that I studied carefully in the field — all plants, without a single exception, had decussately arranged leaves. If *ecklonianum* were of hybrid origin, it could be expected that at least some plants would show a variation in the leaf arrangement[1]. I have, moreover, never found any concrete evidence for

---

[1] A number of herbarium sheets contain mixed collections (♂ *A. spathulatum* subsp. *ecklonianum*, ♀ *A. aethiopicum* or viceversa) but I am convinced that this must be attributed to the carelessness of collectors both in the field and when pressing the specimens. Such mixed collections can also be detected — in addition to the leaf arrangement difference — by leaf surface characters: the leaves of the subspecies of *A. spathulatum* have conspicuous (although small) upper epidermis cells which give the surface a "dotted" appearance; this feature is absent in *A. aethiopicum* (cf. Table 13).

the presence of hybrids or plants even slightly resembling subsp. *ecklonianum* in areas where I had seen *A. aethiopicum* and *A. spathulatum* (subsp. *spathulatum*) growing in nearby habitats (e.g., *A. aethiopicum* in relatively moister and fertile soils and *A. spathulatum* subsp. *spathulatum* in neighbouring drier, less fertile ground).

While collections of subsp. *ecklonianum* from the SW Cape Prov. s.str. are more "typical", rather uniform in their characters and quite distinct (except for some – especially ♂ – specimens from the N Cape Peninsula and the Worcester Distr., which are difficult to separate from subsp. *spathulatum*), the collections from the E parts of the S coast (from the George Distr. E-wards) appear to be ± atypical and often do not fully match the material from the SW. The latter occur in an area which receives a higher annual precipitation than further W. The position of these collections is not fully understood. There is some similarity to "Dune Sand Forms" of subsp. *spathulatum* (see p. 239). On the other hand, a thoroughly investigated population from the base of Formosa Peak, Tsitsikama Mts. (PUFF 800928-1/4) not only fully agreed with SW material of subsp. *ecklonianum* in its morphological characters but was also found growing in exactly the same habitats (dense fynbos along a stream). More field observations would be required.

Collections

South Africa. Cape – **3318** (Cape Town): Table Mt. (**-CD**), BOLUS 4583 (BOL, K), MEEBOLD 16740 (COI, M); – , Twelve Apostles, MCKINNON s.n. (STE); Signal Hill, N facing slopes above Sea Point, Low 965 (STE); Paarlberg (**-DB**), KRUGER M 26 (STE), PUFF 791230-2/2, 800904-2/1 (WU); Jonkershoek Forest Res. (**-DD**), PUFF 791231-1/2 (WU). – **3319** (Worcester): Witsenberge, Swartgat Peak (**-AA**), ESTERHUYSEN 27942 (BOL); Mitchell's Pass (**-AD**), PUFF 790719-5/1 (WU; ± subsp. *ecklonianum*); Klein Drakenstein Mts., Lower Kasteelkloof catchment (**-CC**), KRUGER 1409 (STE); Worcester D., Boschjesveld Mts. (**-CD?**), STOKOE s.n. sub SAM 69995 (SAM); foothills of Riviersonderend Mts., SW of McGregor (**-DC**), PUFF 840910-1/5 (WU). – **3320** (Montagu): Langeberg, below 10 o'clock Peak (**-CD**), PUFF 800919-1/1 (WU). – **3322** (Oudtshoorn): George D., Woodville (**-DC**), GALPIN 4105 (GRA, PRE; ± subsp. *ecklonianum*). – **3323** (Willowmore): Tsitsikamma Mts. S of Louterwater, Formosa Peak (**-DC**), PUFF 800928-1/4 (WU). – **3324** (Steytlerville): Kareedouw Pass (**-CD**), FOURCADE 6150 (STE). – **3418** (Simonstown): Somerset West (**-BB**), PARKER 3597 (BOL, GH, K, NBG); Kogelberg Forest Res. (**-BD**), GROBLER 6100 (PRE; *A. aethiopicum* in STE), PUFF 800101-4/1 (WU). – **3419** (Caledon): (nr.) Caledon (**-AB**), FAIRALL s.n. sub PUFF 800911-1/1 (WU), FRIES, NORLINDH & WEIMARCK 1417 (BR, PRE, WAG); Kleinmond (**-AC**), STOKOE 9513 (BOL); Vogelgat Private Nature Res. E of Hermanus (**-AD**), PUFF 800917-1/2 (WU), WILLIAMS 3070, 3071 (NBG, PRE); Baviansberg nr. Genadenthal (**-BA**), ECKLON s.n. (S). – **3422** (Mossel Bay): Goukamma Nature Res. (**-BB**), VAN DER MERWE s.n. sub STE 30967 (STE; ± subsp. *ecklonianum*). – **3423** (Knysna): Diep R. nr. Knysna (**-AA**), BOLUS 3047 (BOL, K; ± subsp. *ecklonianum*). – **3424** (Humansdorp): Witte Els Bosch

(= Witelsbos) (-AA), Fourcade 812 (BOL × 2). – Imprecise locality: "CBS", Sieber [Cape III No.] 60, 62 (P). – Additional collections: Acocks 5218 (S); Bolus 6695 (BOL); Boucher 2308 (PRE, STE; mixed with *A. aethiopicum*), 2414 (STE); Compton 15327 (NBG); Coppejans 1210 (BR; subsp. *spathulatum* in WAG); Drège s.n. ('*A. spath.* β *longifolium* E. Mey.', LY); Dümmer 183 (E); Ecklon 26 (S – mixed with subsp. *tulbaghense*), 83 (LY), s.n. (S, SAM); Ecklon & Zeyher '52-8' (S), 2307 ε (BOL, FI, G, GOET, LY, M, MO, P, S, SAM, W); Esterhuysen 13509 (BOL, K; ± approach. subsp. *saxatile*); Hafström & Lindeberg s.n. (S); Keet 857 (PRE, STE; mixed with *A. aethiopicum*; ± subsp. *ecklonianum*); Kräusel 471 (M); Moss 4471 (J); Penfold 143 (NBG); Rycroft 3286 (LMU, NBG); Stokoe 7588 (BOL), s.n. sub SAM 59527, 69994 (SAM); van der Merwe 23-13 (PRE, STE); van Rensburg 425 (K, PRE, STE × 3).

12 (e). Subsp. *tulbaghense* Puff in Fl. S. Afr. 31, 1 (2): 19 (1986).

Type: South Africa, Cape Prov., '... non procula catarractis vallis "Tulbagh" (Worcester)', Ecklon & Zeyher 2307 δ (SAM, holo.!; FI, LY, iso.!; S?).

– *A. aethiopicum* L. var. *tulbaghense* Ecklon & Zeyher, Enum. Pl. Afr. Austr.: 365 (1836), non valide publ. (sine descr.).

Dioecious, several- to many-stemmed, erect to ± reclining and spreading dwarf shrubs. Stems (10) 20–60 (75) cm long, to ca 6 mm in diameter near the base and ca 1–2.5 mm in the midstem region, mostly much-branched, branches ± irregular and rather diffuse to ± regular and paired (occasionally ♀ plants less-branched than ♂); younger parts obscurely 4-angled, reddish-purplish tinged, reddish-brown to brownish-grey, ± densely covered with whitish papillae to 0.1 mm long; older parts terete, becoming glabrescent, epidermis occasionally peeling; upper internodes as long as or shorter than the leaves. Leaves decussate, pseudo-verticillate; blades (2.2) 2.5–4.5 (5.2) × 0.5–1 (1.3) mm, (narrowly) ob-ovate, oblanceolate, lanceolate to ± linear-lanceolate, narrowed to the base, glabrous, upper surface ± shiny and with conspicuous but small epidermis cells; apices ± mucronate, acute or ± obtuse; margins flat to somewhat revolute; stipular sheaths mostly shortly hairy, cup-shaped, ca 0.5–0.9 mm long, with a median ± triangular seta ca 0.2–0.5 (0.7) mm long, seldom with an additional minute seta on either side. Inflorescences ± conspicuous, especially in ♀ and when in fruit, made up of ± sessile, several- to ± many-flowered axillary clusters (♀ inflorescences often more-flowered than ♂). Flowers subsessile; ♂, ♀. Corolla 4-merous, greenish-yellow to yellow, glabrous. ♂: tube 0.5–0.8 (1) mm long, (broad-ly) funnel-shaped, lobes 1.2–2 (2.2) × 0.5–0.9 mm, lanceolate, recurved; stamens 4, filaments exserted for 0.4–0.9 mm, anthers 0.9–1.4 × 0.3–0.4 (0.6) mm; rudimentary ovary minute. ♀: tube 0.2–0.3 mm long, cylindrical, lobes 0.2–0.5 × 0.1–0.2 mm, ± linear, erect; style ± 0; stigmas 2, 1.4–3 mm long, greenish-grey; ovary 0.5–0.8 × 0.4–0.7 mm, glabrous, papillate or, less commonly, shortly hairy, crowned by 4 ± indistinct calyx lobes. Fruits shiny, light reddish-brown to dark purplish-brown, supported by U-shaped carpophores usually less than half as long as the fruits; each mericarp (1.5) 1.7–2.3 × 0.6–0.8 (1) mm, ± oblong to (narrowly) obovate or elliptic, dorsal side convex, glabrous, papillate or, less commonly, ± densely covered with whitish papillae to 0.1 mm long, ventral side ± plane to quite strongly concave and with a prominent median vertical ridge, commissure ± small; mericarps crowned by 2 indistinct ± triangular calyx lobes ca 0.1–0.3 (0.4) × 0.3 mm or calyx lobes ± lacking. – Fig. 70 c.

Chromosome Number (Table 3): n = 11.
Average Pollen Diameter (Fig. 48): 27.8–30.9 µm.
Habitat: Often in coastal renosterveld and in degraded fynbos or in disturbed sites (roadsides, gravel quarries, etc.). Mostly growing in reddish clay soils or gravelly granite-derived soil or shales; only occasionally in sand. Predominantly in drier habitats. – Ca 60–600 (800) m.
Distribution (Map, Fig. 71 b): S Africa, SW Cape Prov. From the Piketberg Distr. S to the Cape Distr. and from there NE and E to the Worcester and the Caledon Distr.

Critical Remarks: The most striking feature of subsp. *tulbaghense* is its mericarps (Fig. 70 c), which are distinctly narrower (i.e., over two times longer than broad) and *smaller* than those of subsp. *spathulatum*, subsp. *saxatile* and subsp. *ecklonianum*. They, moreover, usually lack even ± indistinct ribs on the dorsal side. ♀ plants, especially when in fruit, differ quite markedly in habit (low dwarf shrubs; branches densely beset with fruits) from the other subspecies of *A. spathulatum*. ♂ flowers and the leaves are also rather smaller than in the other subspecies (but see comments below!).

Tulbaghense may deserve specific rank. Small-fruitedness, in combination with the distinct low habit, the small leaves and its habitat (clay!) seem to be in favour of this. The problem, however, is that it may be difficult, if not impossible, to separate herbarium specimens of ♂ plants or specimens of ♀ individuals without *mature* fruits from the other taxa of *A. spathulatum*. Leaf size and shape in particular are too variable and often do not allow an absolute separation (cf. Table 12). To a lesser extent this also holds true for ♂ corollas. Moreover, there is often dismally little detailed information on herbarium sheets as regards to habit and habitat, and specimens are often too small and fragmented to allow any conclusion as to growth form and an unambiguous identification. It is thus for "practical" reasons – more than anything else – that *tulbaghense* is at present included in *A. spathulatum* as a subspecies.

SALTER (1950) may have confused subsp. *tulbaghense* with "*A. aethiopicum* var. *montanum* SONDER" (= var. *alpinum* ECKLON & ZEYHER). He described it (1950: 733) as "smaller in all respects and usually more bushy. Leaves opposite, mucronate. Rather local (in the Cape Peninsula): Lion's Mt. and Devil's Peak." He had apparently not seen the type of var. *montanum* (cf. SALTER 1937) which differs quite markedly from subsp. *tulbaghense* in habit alone. Var. *montanum* is now included in subsp. *spathulatum*.

Collections

South Africa. Cape – **3218** (Clanwilliam): Piketberg, between Farms Kleinbegin and Goedverwag (**-DC**), PUFF 800914-5/1 (WU). – **3318** (Cape Town): Malmesbury D., [Farm] Klein Swartfontein, 6.4 km SE of Mooresburg (**-BA**), ACOCKS 24512 (MO, PRE); Cape Peninsula, Lion's Head (**-CD**), SALTER 6455 (BOL, K, NBG, SAM); Klipheuwel (**-DA**), THOMPSON 2591 (PRE); Paarlberg, below 'The Rocks' (**-DB**), PUFF 791230-2/1, 800904-2/2 (WU); N of Bottelary rd. (**-DC, -DD**), ACOCK(S) 762, 966 (S); nr. Joostenberg (**-DD**), ESTERHUYSEN 16000 (BOL). – **3319** (Worcester): Tulbagh (**-AC**), ECKLON & ZEYHER 2307 δ (FI, LY, SAM; S?), BURCHELL 1028 (GH); Worcester D., Veld reserve (**-CB**), VAN BREDA 12 (PRE), 170 (K, PRE, SAM); French Hoek summit (**-CC**), BOUCHER 2292 (STE). – **3418** (Simonstown): Wynberg (**-AB**), LEWIS GRANT 2678 (MO); Sir Lowrys Pass (**-BB**), PARKER 4261 (BOL, GH, NBG), ROGERS 24977 (PRE). – **3419** (Caledon): Lebanon [Forest Reserve] (**-AA**), KRUGER KR557 (K, PRE); around Shaw's Mountain Pass (**-AD**), PUFF 791224-1/3 (WU). – Imprecise localities: "Southwestern Districts", LEIPOLDT 9515 (BOL), "Cap and Stellenbosch", ECKLON & ZEYHER 2307 γ (M; mixed with subsp. *spathulatum*). – Additional collections: ACOCKS 4969 (S), 24493 (K, MO, PRE); BOUCHER 3386 (PRE, STE); CRUIKSHANK 7141 (PRE); DÈGE s.n. ('*A. ciliare* L. b', G); ECKLON 4 (S), 26 (GOET, S; mixed with subsp. *ecklonianum*), 26 b, 28 b (M), s.n. (GOET); ESTERHUYSEN 11953, 33058 (BOL); KRUGER 1516 (PRE, STE); LEVYNS 4227 (BOL); PARKER 4552 (MO); SCHLECHTER 7704 (BM, BOL, PRE); STREY 849 (PRE); VAN NIEKERK s.n. sub BOL 31693 (BOL); WALL s.n. (S × 2 – pathological; one mixed with *A. aethiopicum*).

13. *Anthospermum monticola* PUFF in Fl. S. Afr. **31**, 1 (2): 19 (1986).
Type: South Africa, Cape Prov., Witteberge, Lundean's Nek – Belmore rd., ca 8.5 km beyond (SSE of) New England turnoff, towards Belmore, PUFF 790113-3/1 (WU, holo.!; BR, K, NBG, NU, PRE, iso.!).

Dioecious, few- to several-stemmed, ± erect and cylindrical or roundish shrubs ca (0.2) 0.3–1.5 (2) m tall (low and ± cushion-forming if browsed excessively or occurring at unusually high altitudes), often with thick, woody tap roots. Stems to ca 20 mm in diameter near the base and ca 3–5 mm in the midstem region, mostly much-branched, branches ± spreading or curving upwards, usually paired; younger parts often obscurely 4-angled, reddish or purplish, usually densely covered with whitish papillae to 0.1 mm long; older parts terete, reddish-grey to grey, becoming glabrescent, epidermis often peeling; internode lengths variable, upper internodes often only as long as or shorter than the leaves. Leaves decussate, often pseudo-verticillate; blades (2) 3–10 (13) × (0.7) 1–2 mm, (ob)ovate, oblong, (ob)ovate-lanceolate to ± lanceolate, narrowed to the base, glabrous; apices acute, acuminate to ± mucronate; margins ± flat to (strongly) revolute; petioles subobsolete or to ca 0.3 mm long; stipular sheaths shortly hairy, cup-shaped, ca 1 × 1.5–3 mm, mostly with a median seta ca 0.2–0.5 (1) mm long. Inflorescences inconspicuous, made up of ± sessile, 6- (rarely more-) to 2-flowered axillary clusters. Flowers subsessile; ♂, ♀. Corolla 4-merous, yellowish or greenish yellow, occasionally reddish-purplish tinged on the outside, glabrous. ♂: tube 0.3–0.5 mm long, broadly funnel-shaped to (sub)campanulate, lobes 2.2–2.7 (3) × (0.5) 0.8–1.2 (1.5) mm, ± oblong-lanceolate, recurved; stamens 4, filaments ca 1.8–2 mm long, anthers 1–2 × 0.5–0.8 mm; rudimentary ovary often hardly discernible. ♀: tube 0.4–0.5 mm long, cylindrical to funnel-shaped, lobes 0.3–0.4 × 0.2–0.4 mm, ± triangular to lanceolate, mostly erect; style 0; stigmas 2, ca 3–5 mm long, greyish; ovary ca (1) 1.3–1.8 × 0.8–1 mm, glabrous, crowned by 4 conspicuous calyx lobes. Fruits reddish-brown to greyish-brown, supported by short, broadly U-shaped carpophores; each mericarp 2–3 × 1.5–2 mm, (broadly) ovate or obovate, oblong or ± elliptic, dorsal side shallowly convex, sometimes with two faint lateral ribs below the calyx lobes or faintly 3-ribbed, glabrous, ventral side ± plane to somewhat concave and with a ± prominent median vertical ridge, commissure often almost as large as the mericarp surface; each mericarp crowned by 2 ± triangular, erect to ± spreading calyx lobes 0.3–0.6 (0.8) mm long. – Fig. 35 g.

Chromosome Number (Table 3): n = 11, 2 n = 22.
Average Pollen Diameter (Fig. 48): 25.6–32.7 µm.
Habitat: On rocky slopes amongst grass or shrubs, on rocky outcrops, around rock sheets (often associated with *Cliffortia* and *Passerina*) or in cliffs; sometimes in scrub along streams; also recorded from disturbed sites (roadside cliffs, etc.). Mainly growing over sandstone or dolerite, rather infrequently over basalt. – Ca (1 200) 1 700–2 500 (3 000) m.
Distribution (Map, Fig. 72): S Africa; from the O.F.S. and Natal to the Cape Prov. Also in Lesotho. In the Cape Prov. mainly in the mountains of the interior [from the Drakensberg and Witteberg to the Sneeuwberg and

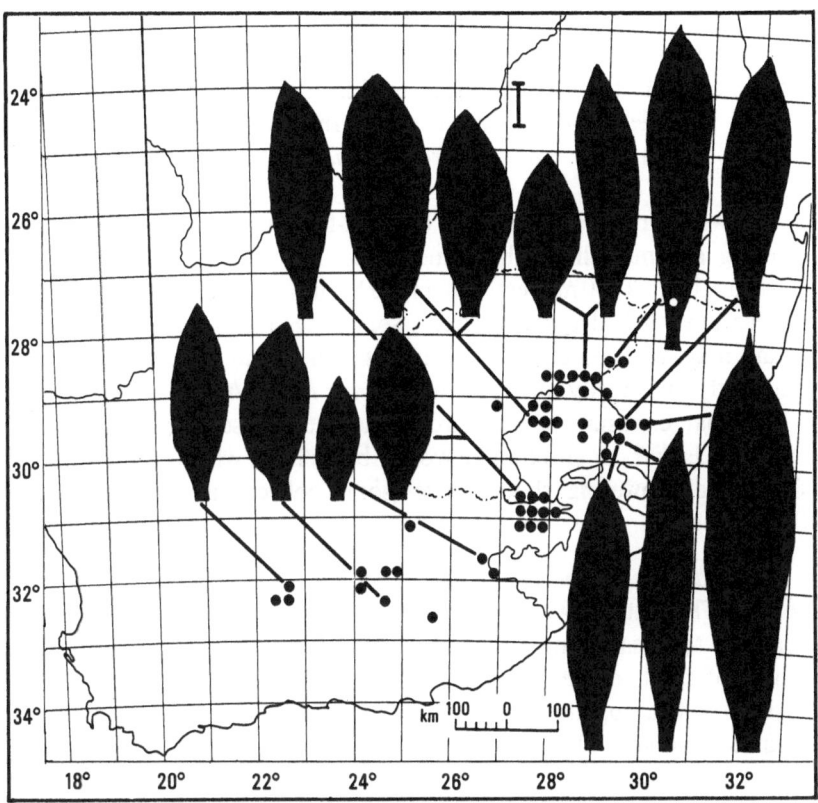

Fig. 72. Distribution and leaf silhouettes of *Anthospermum monticola*. – The scale
unit is 1 cm

as far W as the Nieuweveld Mts. (Beaufort West Distr.)]. See also Critical
Remarks, below.

Critical Remarks: The disjunct distribution range of *A. monticola* in
the interior of the Cape Prov. is remarkable (but not unique). In the area
the species "follows" the higher (and somewhat moister?) mountain
ranges as do, for example, numerous *Compositae* centered in the
Drakensberg (cf. "Group 2. Drakensberg- Sneeuwberg species",
HILLIARD 1978). Also the "temperate affinities" of plants in these
mountains are well documented (cf. ROBERTS 1966: Hangklip Mt. near
Queenstown).

The leaf shapes and sizes of *A. monticola* vary considerably and are, to
some extent, correlated with the geographical distribution: Natal collections and specimens from the Natal – O.F.S. – Lesotho border region

tend to have relatively larger and narrower leaves (cf. map, Fig. 72) and
sometimes also slightly narrower fruits which often lack even faint dorsal
ribs. These collections, furthermore, tend to differ in habit (plants ± more
densely leafy; leaf blades often ± erect). In this region, also, it often is
difficult to separate *A. monticola* from *A. galpinii* (9.; see p. 223 for further
details).

Browsing may give rise to cushion-forming plants. Also plants
growing at unusually high altitudes (Lesotho; above ca 2 400 m and
occasionally to 3 000 m. Pers. obs. in the Maluti Mts.!) may have a similar
appearance; their decumbent stems sometimes root at the nodes. Such
forms may vegetatively ± closely resemble – and were, in the past,
frequently incorrectly identified as – *Nenax microphylla*.

Regular exposure to veld fires sometimes causes the formation of
massive, ± disk- or rosette-like woody bases from which rather spindly,
short erect flowering shoots arise (some collections from the Natal
Drakensberg).

A. monticola also occurs, in part, in the same area as *A. basuticum* (14.,
below). The latter is typically found at higher altitudes and is largely
confined to basalt. There is, however, some overlap in the altitudinal
ranges of the two species, and in a few localities *A. monticola* and *A.
basuticum* occur together. Here, hybrids between the two species appear to
be quite common, although they are not always easy to detect (hybrid
plants may look like somewhat atypical *A. basuticum* with less densely
hairy leaves and fruits). In a detailed study carried out around the
terminus of the Sentinel toll road (O.F.S.; 2828-DB), concrete evidence for
the presence of hybrids, ranging from plants very closely resembling
"pure" *A. monticola* (at lower altitudes) and "pure" *A. basuticum*
(common at higher altitudes) to plants intermediate in their character
states, was found (see Fig. 56).

A. monticola appears to be allied to *A. basuticum* on the one hand and
(more closely) to the complex, SW Cape centered *A. spathulatum* (12.,
above) on the other. *A. monticola* has, in fact, often been mistaken for *A.
spathulatum* subsp. *spathulatum* (= *A. tricostatum* sensu auctt.), primarily
because of the ± similar habit and appearance, and because of the (at least
occasionally) faintly ribbed fruits with conspicuous calyx lobes. The leaves
of *A. monticola*, however, are often (relatively) smaller and broader (at
least those of Cape collections), and ♂ flowers mostly have more
campanulate and smaller corollas. The fruits of *A. monticola*, moreover,
are frequently smaller and rather broader (ovate to obovate) than those of
A. spathulatum. As is the case in several other closely allied species of
Anthospermum, a 100% clear-cut and reliable morphological separation

of the two taxa is not always possible because of the lack of sufficient "usable" distinguishing characters and the convergence of some characters. Although numerous collections do show overlaps in some morphological features, the entirely different, not overlapping distribution ranges and the differing ecological requirements of the two taxa strongly support and favour the specific rank of *A. monticola*.

Collections

Lesotho. – **2828** (Bethlehem): Leribe (-CC), DIETERLEN 681 (K, NBG, NH, P, SAM; PRE × 3 – one sheet mixed with *A. pumilum*); middle of Moteng pass (-DC), KILLICK 4446 (PRE). – **2927** (Maseru): Advance Post (-BA), JACOT GUILLARMOD 2179 (B, BM, K, PRE); Mamathes (-BB), JACOT GUILLARMOD 97 (PRE); Ha Nkoti (-BC), SCHMITZ 7483 (PRE); Blue Mountain pass (-BD), PUFF 840722-1/1 (PRE, WU), SCHMITZ 6242 (PRE); ca 67 km on Mountain Road from Maseru (-DB), JACOT GUILLARMOD 5937 (PRE). – **2928** (Marakabei): Mamalapi (-AC), COMPTON 21307 (NBG), JACOT GUILLARMOD 604, 653 (PRE); ca 3 km above (= WNW of) Likalaneng, PUFF 840722-2/1 (WU); between Sehlabathebe and Thaba Tseka Orange valley (-BC), SCHMITZ 8862 (PRE); Taung, 210 km from Maseru on rd. to Matebeng (-DA), KILLICK 4269 (K, PRE); ca 13 km beyond Likalaneng, on rd. to Marakabei, PUFF 840722-2/1 (PRE, WU). – **2929** (Underberg): Sehlabathebe National Park (-CC), BAYLISS 153 (MO, NBG), HOENER 2079 (PRE). – Additional collections: JACOT GUILLARMOD 739 (PRE); RUCH 1545, 1756 (PRE), 1782 (K, PRE); SCHMITZ 362, 4173, 4477, 6261, 6342, 6560, 6941, 7291, 7538, 7916, 8031, 8154 (PRE); STAPLES 118 (PRE).

South Africa. O. F. S. – **2827** (Senekal): ca 11 km SSE of Rosendal, E foothills of Witteberge, Farm Blydskop (-DB), PUFF 840721-1/1 (PRE, WU). – **2828** (Bethlehem): Ficksburg D., Farm Franshoek (-CA), FERREIRA F001, F013 (PRE); Farm Cornelia nr. Clarens (-CB), STAM 218 (PRE); Farm Dunblane 335, SE of Three Sisters Peaks, SCHEEPERS 1832 (K, LMU, PRE, SRGH), 1833 (K, LMU, MO, PRE, SRGH, WAG); Golden Gate National Park (-DA), PUFF 780820-1/1 (WU), ROBERTS 3121 (PRE); Qwaqwa, from 'Junior Camp' to Mopeli Hut (Kogtwane Spons), PUFF 790217-1/3 (WU); around terminus of Sentinel road, Witsieshoek Gate (-DB), PUFF 781126-1/5 (WU). – **2829** (Harrismith): Harrismith D., Platberg (-AC), HILLIARD & BURTT 11967 (E, K). – **2926** (Bloemfontein): Thaba Nchu (-BB), MOSTERT 1186 (PRE). **Natal** – **2730** (Vryheid): Utrecht D., Naauwhoek (-AD), DEVENISH 1818 (PRE). – **2829** (Harrismith): Van Reenen (-AD), FRANKS s.n. sub MEDLEY WOOD 12100 (K, NH); Cathedral Peak Forest Reserve (-CC), KILLICK 971, 990 (K, PRE; ± atypical). – **2929** (Underberg): Langalibalele pass, Giants Castle Game Reserve (-AD), PUFF 800116-2/2 (WU); Kamberg Nature Reserve, Gladstone's Nose ridge (-BC), WRIGHT s.n. (obs. in private herb. WRIGHT); Kamberg Mt. (-BD), PUFF 80015-3/1 (WU); Underberg D., Garden Castle Nature Reserve, Pillar Cave valley (-CA), HILLIARD & BURTT 10421, 10422 (E, K); Sani Pass area (-CB), HILLIARD & BURTT 15541 (E, NU, WU), KILLICK & VAHRMEIJER 3805 (PRE × 2). – Imprecise localities: "Natal, 100 miles inland", SUTHERLAND s.n. (K); "Mohlamba range" (= Quathlamba; = Zulu name for Drakensberg), SUTHERLAND s.n. (K). – Additional collections: HILLIARD & BURTT 9319 (E, NU), 9327 (E, K, NU), 13406, 13407, 13502, 13552 (E, NU, PRE, WU); HILLIARD, BURTT & MANNING 15936 (NU, PRE), 15954 (NU, WU); LAMBION & REEKMANS 82/347 (PRE); LEVYNS 8307 (BOL, ± atypical). **Cape** – **3027** (Lady Grey): Lady Grey, Jouberts pass (-CB), WERGER 1055 (K, PRE); Farm Glen

Doone, N of Lady Grey-Barkly East main rd. (-CD), Puff 840923-5/2 (PRE, WU); Witbergen (= Witteberge) (-CB to -DA), Drège 7664 (G, K, MO, W); Barkly East D., Ben McDhui, Bell River gorge (-DB), Hilliard & Burtt 16523 (NU, PRE, WU); –, Three Drifts stream below Farm Pitlochrie (-DC), Hilliard & Burtt 14737 (E, NU, WU); Lundean's Nek- Belmore rd., ca 8.5 km SSE of New England turnoff (-DD), Puff 790113-3/1 (BR, K, NBG, NU, PRE, WU). – 3028 (Matatiele): lower part of Naudes Nek pass, NE of Rhodes (-CC), Puff 790113-4/2 (WU), Rhodes to Naudes Nek, Hilliard & Burtt 16623 (NU, WU) – 3124 (Hanover): Farm Rhenosterfontein, S of Richmond, Sneeuwberg (-CC?), Acocks 15844 (K, PRE); Compasberg (-DC), Levyns 9601 (BOL), Shaw 117 (K, W); Lootsberg pass, ca 63 km from Graaff-Reinet tow. Middelburg (-DD), Puff 790908-1/1 (WU). – 3125 (Steynsburg): Kivorsch Mt. (= Kikvorsberg) nr. Naauw-poort (= Noupoort) (-AA), Hutchinson 3073 (BOL, K). – 3126 (Queenstown): Andriesberg, N of Bailey (-DA), Galpin 2177 (GRA, PRE, STE); Hangklip Mt., N of Queenstown (-DD), Roberts 1988 (PRE). – 3127 (Lady Frere): a few km S of Rossouw, on rd. to Dordrecht, nr. Farm Ponte Racina (-AB), Puff 840923-1/2 (WU); Greylings pass, above Farm Kransfontein (Roussow- Clifford rd.), Puff 840923-2/1 (WU); Otto du Plessis pass, SSE of Clifford (-BA), Puff 840923-3/1 (PRE, WU); Barkly pass area (-BB), Puff 840923-4/1 (WU). – 3222 (Beaufort West): Beaufort West D., Nieuweveld Mts., Esterhuysen 2764 (BOL – mixed with *Nenax microphylla*), Gibbs Russell, Robinson & Herman 445 (PRE), Levyns 5535 (BOL); –, around Trig. Beacon 7 (-AD), Puff 840907-1/1 (PRE, WU); –, N of TV mast (-BA), Puff 840907-4/1 (WU); –, around TV mast (-BC), Puff 840907-3/1 (WU). – 3224 (Graaff-Reinet): Koudeveldberge SE of Doornbosch (-AA), Olivier 5218 (PRE, STE); around Graaff-Reinet, nr. Valley of Desolation (-BC), Bolus 636 (BOL, STE), Galpin 11514 (K, PRE). – 3225 (Somerset East): Somerset East D., Bloukop (-DA), van der Walt 351 (PRE).

14. *Anthospermum basuticum* Puff in Fl. S. Afr. **31**, 1 (2): 20 (1986).
   Type: Lesotho, Sehlabathebe National Park, across the stream from Motsea, 2 550 m, 5. xi. 1976, Hoener 1607 (PRE, holo.!).

Dioecious, many-stemmed, cushion- or mat-forming or ± erect and cylindrical dwarf shrubs. Stems (4) 10–30 (50) cm long, erect, ascending or ± prostrate, ca 4–8 (12) mm in diameter near the base and ca 1–2.5 mm in the midstem region, mostly much-branched, branches often ascending; younger parts occasionally obscurely 4-angled, often ± reddish-brown, mostly densely covered with spreading white hairs ca 0.1–0.3 (0.6) mm long; older parts terete, greyish brown to dark grey, becoming glabrescent, epidermis often peeling; internodes, except on new growth, as long as or somewhat shorter than the leaves; plants ± densely leafy. Leaves decussate, often pseudo-verticillate; blades often curved upwards or ascending, 2.5–5 (6) × 0.8–1.2 (1.8) mm, narrowly lanceolate, narrowed to the base, often seemingly terete due to strongly revolute margins, upper surface densely covered with white spreading hairs 0.1–0.3 (0.6) mm long, lower surface (usually obscured by the revolute margins) often glabrous save for the midrib; apices ± acute; petioles subobsolete; stipular sheaths

hairy on the margins, broadly cup-shaped, ± deltoid in side view, ca 0.8–
1.5 × 1.5–3 mm, with a median hairy seta (0.5) 0.8–1.6 mm long, sometimes with an additional minute gland-tipped seta on either side.
Inflorescences inconspicuous, made up of usually paired, subsessile, ♂ or ♀
flowers. Corolla 4-merous, yellowish to yellowish-green, usually hairy
outside. ♂: tube 0.7–1 mm long, broadly funnel-shaped to subcampanulate, lobes (1.4) 1.7–2.3 × 0.6–1.1 mm, ± lanceolate to oblong,
recurved; stamens 4, filaments 0.7–1.5 mm long, anthers 1–1.6 × 0.2–
0.5 (0.7) mm; rudimentary ovary hardly discernible. ♀: tube ± 0–0.3 mm
long, ± cylindrical, lobes 0.5–1 × 0.2–0.5 mm, ± linear to lanceolate,
usually erect; style 0, stigmas 2, (2) 3–6 (7.5) mm long, white, greyish or
greenish-white; ovary 1–1.2 × 0.8 mm, mostly densely hairy, crowned by 4
calyx lobes. Fruits greyish or greyish-brown, supported by shallow,
broadly U-shaped carpophores; each mericarp 2–2.5 × 1–1.5 mm, ±
obovate, dorsal side convex, mostly densely covered with white spreading
hairs 0.1–0.3 (0.5) mm long, ventral side ± plane to slightly concave,
commissure large, almost as long as the mericarps but narrower;
mericarps crowned by 2 somewhat hairy, ± triangular erect to spreading
calyx lobes 0.5–0.8 × 0.3–0.5 (0.7) mm. – Figs. 15 b, 31 e, 35 f.

Chromosome Number (Table 3): 2 n = 22.
Average Pollen Diameter (Fig. 48): 26.4–27.8 μm.
Habitat: In cracks of rocks, in cliffs, on gravelly slopes or on loose
rocky ground; often with *Passerina, Helichrysum, Crassula* and *Lotononis.*
Mostly in coarse soil derived from basaltic rock; at lower altitudes
sometimes in Cave Sandstone-basalt derived soil mixtures. – Ca (2 150)
2 300–3 050 m.
Distribution (Map, Fig. 90 c): Lesotho and S Africa. Endemic to the
Drakensberg escarpment in the Lesotho – O.F.S., Lesotho – Natal and
Lesotho – E Cape Prov. border area.

Critical Remarks: A. basuticum has often been confused with *A.
hispidulum* (17.). It, however, differs markedly from the latter in being
strictly dioecious, in having smaller ♂ flowers with short, broad, almost
campanulate tubes and large, broader fruits, The leaves are also smaller,
although they are similar in their indumentum, in having strongly revolute
margins, and in their anatomy. Micromorphologically, however, the
leaves of *A. basuticum* are easily distinguished from those of *A. hispidulum*
by the characteristic striations on the lower epidermis cells (compare
Figs. 18 a and b). *A. basuticum* is, furthermore, well distinguished by its
distribution range and habitat (confined to high altitudes; primarily over
basalt).

I am not at all convinced that *A. basuticum* and *A. hispidulum* are very closely allied; similarities may be due to convergence. There may be a distant (or perhaps rather close?) relationship with *A. monticola* (13., above) (similar ♂ flowers, fruit structure, etc.). *A. monticola* and *A. basuticum* are known to form hybrids (for details see *A. monticola,* Critical Remarks, and Fig. 56).

Collections

Lesotho. – **2929** (Underberg): plateau above Langalibalele pass (**-AD**), Puff 800116-1/1 (WU); Sehlabathebe National Park (**-CC**), Hoener 1607, 1608 (PRE), Jacot Guillarmod, Getliffe & Mzamane 318 (K, MO, PRE).

South Africa. O. F. S. – **2828** (Bethlehem): Kogtwane Spons (Qwaqwa Skiing Center) (**-DA**), Puff 790217-2/1 (J, WU); from terminus of Sentinel road to Mount-aux-Sources, Witsieshoek Gate (**-DB**), Puff 781126-1/8, 800105-2/2 (WU). – O.F.S.-Lesotho-Natal border: summit of Drakensberg, Sentinel nr. Mont-aux-Sources (**-DB, -DD**), Killick & Marais 2196 (K, NH, NU, PRE). **Natal** – **2929** (Underberg): Giants Castle Game Res., Bushmans River (= Langalibalele) pass (**-AD**), Puff 800116-3/2, -4/1 (WU), Wright 920, 938 (NU); Underberg D., Sani pass (**-CB**), Hilliard & Burtt 7121 (E, K, MO, NU, PRE, S), 7132 (E, K, NU, PRE, S). – Additional collections: Hilliard & Burtt 13551, 13726, 14915, 15528, 15694 (NU, WU); Hilliard, Burtt & Manning 16018 (NU, WU; atypical – hybrid with *A. monticola*?); Humbert 15155, 15241 (P); Hutchinson, Forbes & Verdoorn 105 (NH, PRE); Flanagan 2002 (BOL, GRA, K, PRE, STE); Schelpe 3072 (BM, NH); Schweickerdt 30470 (PRE). **Cape** – **3027** (Lady Grey): Drakensberg, Doodmans Krans Mt. (**-DB**), Galpin 6649 (BOL, GRA, K, PRE). – **3028** (Matatiele): Maclear D., Naudes Nek pass (**-CA**), Acocks 12181 (PRE), Puff 790113-5/3 (BR, J, WU).

---

15. ***Anthospermum paniculatum*** Cruse, Rub. Cap.: 15 (1825), in Linnaea **6**: 15 (1831); Sonder in Harvey & Sonder, Fl. Cap. **3**: 31 (1865); Puff in Fl. S. Afr. **31**, 1 (2): 21 (1986).
    Type: South Africa, Cape Prov., "in terra Houtniquas ad Promontorium bonae spei mense Ianuario 1820", Mund(t) & Maire s.n. (B, holo. †; G, iso.!).

    = *A. paniculatum* Cruse var. *confertum* Ecklon & Zeyher, Enum. Pl. Afr. Austr.: 367 (1836).
    Type: South Africa, Cape Prov., ... ad "Zwartehoogdens" non longe ab urbe "Grahamstown" (Albany), in monte "Winterberg" prope fluvium "Katrivier" (Ceded Territory), Ecklon & Zeyher 2314 β (FI × 2!, G!, LY!, M!, MO!, P!, SAM!, W × 2!)[1].

---

[1] In numerous herbaria, cut-out parts from Ecklon & Zeyher's "Enumeratio" are used as labels. These often comprise the entire bottom part of page 367 ("2314. *Anthospermum paniculatum* Cruse, β. *confertum*, γ. *elongatum*"), so that it is not always certain to what variety a particular specimen is meant to belong.

= *A. paniculatum* Cruse var. *elongatum* Ecklon & Zeyher, Enum. Pl. Afr. Austr.: 367 (1836).
Type: South Africa, Cape Prov., ... prope sylvas "Krakakamma" et in montibus "Van Stadensriviersberge" (Uitenhage), Ecklon & Zeyher 2314γ (MO!, SAM!)[1] [? also Ecklon or Ecklon & Zeyher 349 (Vanstadensrivier Mts.) (BOL!, BM!, K!, M!, PRE!, SAM!), 349 bis (BM!)].

– "*A. confertum* Cruse", Walp., Rep. Bot. Syst. **2**: 462 (1843). Probably an error [*A. paniculatum* Cruse var. *confertum* Ecklon & Zeyher].

Dioecious or ± dioecious (♂ plants occasionally with odd ⚥, rarely with numerous ⚥ flowers), few- to several-stemmed dwarf shrubs or, if burnt regularly, subshrubs with numerous erect, mostly unbranched flowering stems arising from a woody base. Stems (10) 15–40 (70) cm long, ca (1) 1.5–2.5 (3) mm in diameter near the base and ca 1–1.5 (2) mm in the midstem region, erect to ascending, less commonly decumbent and sometimes basal portions rooting at the nodes, branches erect to ± spreading; younger parts obscurely 4-angled, often reddish-brown to purplish, usually densely covered with short whitish papillae to ca 0.1 mm long; older parts terete, greyish-brown to dark grey, glabrescent, epidermis often peeling; internodes often somewhat longer than to as long as the leaves in plants not exposed to fires, usually much longer in fire-exposed individuals. Leaves decussate, pseudo-verticillate; blades 5–12 (17) × (0.3) 0.5–2.5 (4) mm, linear, lanceolate to narrowly obovate, seldom (in fire-exposed individuals) ± broadly oblanceolate or obovate, narrowed to the base, glabrous or sometimes ± sparsely covered with white papillae to 0.1 mm long on the upper surface and/or the midrib below; apices acute to ± acuminate; margins flat to strongly revolute; petioles subobsolete or occasionally to 0.5 mm long; stipular sheaths subglabrous to papillate, ± cup-shaped, ca 0.6–1.2 × (1) 1.5–2.5 mm, with a ± triangular to subulate gland-tipped median seta ca (0.3) 0.7–1.5 mm long. Inflorescences terminal, thyrsic or thyrso-paniculate, ± narrowly cylindrical, ca (3) 6–20 (25) × 1–5 cm (in vigorous individuals also additional smaller inflorescences terminal on lateral branches arising below); partial inflorescences ascending, several-flowered; peduncles (0.8) 1.2–3 mm long, pedicels to 0.7 mm long, slightly elongating in fruit. Flowers ♂ (occasionally also ⚥), ♀; corolla 4-merous, creamy yellow to yellow or greenish, glabrous. ♂ (⚥): tube 0.8–1.2 (1.7) mm long, funnel-shaped, lobes 2.2–2.7 × 0.5–1.1 mm, oblong to narrowly obovate, recurved; stamens 4, filaments exserted for ca 0.8–1.3 mm, anthers 1.5–2.2 × 0.3–0.5 mm; ♂:

---

[1] See footnote on previous page.

18*

rudimentary stigmas often present, rudimentary ovary ± prominent, ca 0.3–0.7 mm long, crowned with 4 ± massive calyx lobes; ♂: gynoecium as in ♀. ♀: corolla sometimes absent or tube to 0.3 mm long, cylindrical, lobes 0.5–0.6 × 0.1–0.2 mm, ± lanceolate, erect to ± spreading; style 0–0.6 mm long; stigmas 2, (3) 4–8 mm long, reddish to reddish-purplish; ovary 1.5–2.4 × 0.9–1.6 mm, glabrous, crowned by 4 calyx lobes. Fruits brown, supported by short, broadly U-shaped carpophores; each mericarp 2.5–3.5 (4) × 1.7–2 mm, oblong to ± rectangular or ± obovate, dorsal side somewhat convex and often with 2 lateral faint ribs below the calyx lobes, ventral side plane to strongly concave and with a prominent median vertical ridge; each mericarp crowned by 2 ± triangular, often massive calyx lobes 0.5–1.2 (2.2) × 0.4–1 mm. – Figs. 15 c, 29 b, 31 i–k.

Chromosome Number (Table 3): n = 11, 2 n = 22.
Average Pollen Diameter (Fig. 48): 26.1–29.2 µm.
Habitat: Mostly amongst grass in grassveld, in bush-clump veld or in rocky grassland and on rocky ridges. Occasionally in disturbed sites such as roadsides, fire-breaks or in open ground in overgrazed and trampled grassland; seems prefer to sandy soils. – Ca 15–1 220 (1 370) m.
Distribution (Map, Fig. 95 a): S Africa, Cape Prov. (including Transkei). Occurring from the Transkei to the George Distr.; centered in the E Cape Prov.

Critical Remarks: Var. *elongatum* and var. *confertum* are certainly not worth retaining. They merely refer to plants in different stages of development (inflorescences more congested in bud or early flower, becoming elongated in fruit) and/or variations in the extensiveness of the inflorescences.

(Regular) exposure or non-exposure to veld fires seems to be the main factor responsible for the sometimes striking growth form differences.

Its typical inflorescence (cf. Fig. 29 b) makes it easy to distinguish *A. paniculatum* from all other *Anthospermum* species. Subsequently, however, this unique feature puts *A. paniculatum* into a somewhat isolated position within the genus, as all other *Anthospermum* species have variously reduced and modified thyrsic inflorescences ("axillary flower clusters"!). This isolated position is, furthermore, supported by the unusual, relatively large fruits which anatomically resemble those of *Nenax* species rather more than those of *Anthospermum* (cf. Fig. 41). Also the *reddish* stigmas of *A. paniculatum* are a character shared with numerous *Nenax* species (the vast majority of *Anthospermum* species has greenish, greyish to whitish stigmas). A position of *A. paniculatum* ± intermediate between *Anthospermum* and *Nenax* should, however, not be deduced from these similarities − the inflorescences of *A. paniculatum* are in strong disagreement with the always much reduced inflorescences of *Nenax* species.

In parts of the E Cape Prov., *A. paniculatum* and *A. herbaceum* (25.) may occur side by side in nearby or the same habitats. In such areas, odd hybrids between the two species were detected occasionally (detailed documentation in Fig. 57).

Collections

South Africa. Cape − **3127** (Lady Frere): Mt. Cala (**-BC**), PEGLER 1613 (BOL, PRE; K − as KOLBE & PEGLER). − **3225** (Somerset East): Somerset [East] (**-DA**), BOWKER s.n. (K). − **3226** (Fort Beaufort): Winterberge, 49 km from Adelaide, tow. Tarkastad (**-AD**), PUFF 790117-6/1 (J, W, WU); summit of Katberg pass (**-BC**), PUFF 790114-8/1 (WU); Hogsback- Seymour rd., ca 2 km beyond Happy Valley turnoff (**-DB**), PUFF 790114-6/1 (J, W, WU). − **3227** (Stutterheim): Cathcart D., Thomas R. (**-AB**), BARKER 3495 (NBG); − , Cathcart (**-AC**), KUNTZE s.n. (K); Fort Cunyngham (**-AD**), SIM 2784 (NU, PRE); Dohne Peak (**-CB**), FLANAGAN 2284 (BOL, K, P, PRE, SAM), PUFF 790115-5/3 (J, W, WU), SIM 1893 (PRE); Pirie (**-CC**), SIM 19609 (NU; PRE − mixed with *A. herbaceum*), 19428 a (NU, PRE × 2). − **3228** (Butterworth): Kentani D., coast (**-AD**), PEGLER 830 (BOL × 2, PRE × 2); Elliotdale D., The Haven (**-BB**), GORDON GRAY 1062 A (NU); nr. Kei Mouth (**-CB**), FLANAGAN 215 (BOL, NH, PRE, SAM × 2). − **3322** (Oudtshoorn): nr. George (**-CD**), SCHLECHTER 2296 (B, BOL, BR, J); Kammanassie Mts. (**-DB?**), ZINN s.n. sub SAM 56003 (SAM); ca 7 km W of Karatara (**-DD**), PUFF 790910-2/1 (J, K, W, WU). − **3323** (Willowmore): Prince Alfred's pass (**-CC**), WALL s.n. (S); between Kirby and The Crags (**-CD**), FOURCADE 5518 (BOL); Coldstream (**-DC**), SCHÖNLAND 1513 (PRE); N foot of Outeniquas nr. Joubertina, Farm Die Hoek (**-DD**), ESTERHUYSEN 13592 (BOL). − **3325** (Port Elizabeth): Van Staden Flower Reserve (**-CC**), DAHLSTRAND 2523 A, B (GRA, STE); Uitenhage (**-CD**), PENTHER 2072 (W × 2); Port Elizabeth (**-DC**), KEMSLEY 240 (GRA). − **3326** (Grahamstown): Swartwatersberg, SW of Riebeek-East (**-AA**), PUFF 840922-1/3 (PRE, WU); Grahamstown and surroundings (**-AD, -BC**), BRITTEN 6494 (GRA), DALY & SOLE 236 (BOL, GRA), PUFF 790117-1/3 (WU); Trapps (= Trappe's) Valley (**-BD**), DALY 655 (BOL, GRA); Alexandria (**-CB**), JOHNSON 1106 (NH, PRE × 2); Southwell (**-DA**), SCHÖNLAND 3330 (GRA, PRE). − **3327** (Peddie): Port Alfred-Peddie rd., nr. Great Fish Point/Trappe's Valley turnoff (**-AC**), PUFF 790116-4/1 (BR, J, W, WU); nr. East London (**-BB**), COMPTON 17844 (BOL, NBG). − **3422** (Mossel Bay): Goukamma Nat. Res. (**-BB**), HUGO 2026, 2028 (STE). − **3423** (Knysna): Knysna (**-AA**), BREYER 23901 (PRE). − **3424** (Humansdorp): Witte Els Bosch (= Witelsbos) (**-AA**), FOURCADE 908 (BOL); Humansdorp (**-BB**), BRITTEN 1205 (PRE). − Imprecise localities or no localities given: "Albany D.", BOLUS 25120 (BOL, BR × 2), s.n. (MO × 2, US), PRIOR s.n. (K); "Zwartehoogdens nr. Grahamstown and Winterberge nr. Katrivier", ECKLON & ZEYHER 2314 β (FI × 2, G, LY, M, MO, P, SAM, W × 2); "Krakakamma and Van Stadensrivierberge", ECKLON & ZEYHER 2314 γ (MO, SAM), as "349" (BOL, BM, K, M, PRE, SAM), as "349 bis" (BM); TYSON s.n. sub PRE 41633 (PRE). − Not traced: Tembuland, Kivenkive Mt., tow. Maclear, BOLUS 10084 (BOL). − Additional collections: ACOCKS 9269, 11067 (PRE); ARCHIBALD 5357 (PRE); BARKER 1429 (NBG); BRITTEN 1037 (GRA, PRE); BURCHELL 4476 (GH, K); COMPTON 4480 (BOL); DUTHIE 984 (STE), 984 a (MO); DRÈGE 7662 (E, G × 2, GH, K, LY, MO, P × 2, W), s.n. ("*A.*

spathulatum SPRENG. b." – P: obviously a label mix-up); ECKLON s.n. (S); ECKLON & ZEYHER 2314 (GOET, M, S), "*A. aeth.* 85.10" (GH – probably a label mix-up); FOURCADE 6300 (STE); GALPIN 3092 (GRA, PRE × 2), 10440 (PRE); GLASS 232 (SAM); KEET 1004 (PRE, STE), 1073 (GRA, PRE, STE), s.n. sub STE 15193 (STE); HILNER 267 (COI, GRA, PRE); MUND(T) s.n. (G); MUND(T) & MAIRE s.n. (G); PEGLER 2083 (PRE × 2); PENTHER 2073 (M, S, W), 2074 (W); SCHÖNLAND 3607 (GRA); SIDEY 3610 (PRE, US); SIM 1484 (NU), 19607 (NU, PRE); STORY 3300 (PRE); WALL s.n. (S); ZEYHER 2715 (FI, G, K, LY × 2, PRE, S, SAM, W; WU as "4.10").

16. *Anthospermum whyteanum* BRITTEN in Trans. Linn. Soc., ser. 2, **4**: 16 (1894); BRENAN in Mem. N.Y. Bot. Gard. **8**: 455 (1954); VERDC. in Fl. Trop. E. Afr., *Rubiaceae* **1**: 331 (1976).
Type: Malawi, Mt. Mlanje, WHYTE 48 (BM, holo.!).

= *A. albohirtum* MILDBR. in Notizbl. Bot. Garten Berlin **15**: 637 (1941).
Type: Tanzania, Matengo Highlands, Lupembe Hill, ZERNY 269 (W, holo.!).

Non-dioecious (♂, ♂ + ♀ or ♀), several- to many-stemmed, ± erect or ± rounded and low dwarf shrubs. Stems ca 20–80 (125) cm long, erect or ascending, ca (2) 3–6 mm in diameter near the base and ca 1–2.5 mm in the midstem region, mostly much-branched, branches ± erect or ascending; younger parts often reddish-brown, densely covered with whitish or reddish-brown hairs ca 0.2–0.9 (1.2) mm long, older parts reddish-brown to dark grey, becoming glabrescent, epidermis often peeling; internodes mostly as long as or shorter than the leaves, only on new growth sometimes longer. Leaves in whorls of 3, pseudo-verticillate; blades spreading to ascending, (2) 3–11 (15) × 0.8–3 (5) mm, oblanceolate, narrowly ovate-lanceolate to linear-lanceolate, narrowed to the base, mostly both surfaces densely covered with whitish spreading hairs ca 0.2–0.5 (0.9) mm long, but sometimes lower surface less hairy except for the midrib; margins mostly strongly revolute; apices acute to acuminate; petioles subobsolete; stipular sheaths usually covered with whitish hairs, cup-shaped, ca (0.5) 0.7–1 mm long, with a median seta ca (0.6) 1–2.5 mm long. Inflorescences ± inconspicuous to occasionally ± conspicuous and somewhat contracted (♀), made up of ± sessile, ca 9-3-flowered axillary clusters. Flowers subsessile, ♂, ♀, ♀. Corolla 4-merous (very rarely 5-merous in ♂ or ♀), greenish-yellow to creamy yellow, at least near the tips of the lobes with whitish spreading hairs 0.1–0.5 (0.7) mm long. ♂, ♀: tube (0.7) 1.2–2.2 mm long (in ♂ often somewhat smaller than in ♀), (narrowly) funnel-shaped, lobes (1.7) 2–2.7 (3.4) × (0.5) 0.7–0.9 mm, ± lanceolate,

recurved; stamens 4, filaments ca 1.5–3.7 mm long, anthers 1.3–1.9 × 0.3–
0.6 mm (in ♂ often smaller than in ♀); ♂: rudimentary ovary small but well
discernible, sometimes with rudimentary stigmas less than 1 mm long; ♀:
stigmas 2, ca 2–3.4 mm long, ovary as in ♀. ♀: tube 0.3–0.7 mm long,
cylindrical, lobes 0.3–0.7 (1) × 0.2–0.3 mm, linear-lanceolate or ± subulate, ± erect; style ca 0.3–0.8 mm long; stigmas 2, ca 2.7–12.2 mm long,
greyish; ovary ca 0.7–1.2 × 0.5–0.8 mm, hairy, crowned by 4 sometimes
not very distinct calyx lobes. Fruits reddish-brown, supported by small U-
shaped carpophores; each mericarp ca (1.6) 1.8–2.2 × 0.7–1 mm, ±
oblong, dorsal side convex, quite densely covered with whitish, often
upwardly directed hairs ca 0.2–0.7 mm long, ventral side ± plane and with
a median vertical ridge, commissure large; mericarps crowned by 2 small,
± triangular calyx lobes ca (0.2) 0.3–0.7 × 0.2–0.3 mm (often hidden
amongst hairs) or calyx lobes sometimes ± lacking.

Chromosome Number (Table 3): n = 11, 2 n = 22.
Average Pollen Diameter (Fig. 49): 25.6–28.7 μm.
Habitat (cf. CHAPMAN 1962: pl. 9): In mountain (upland) grassland, but
almost always associated with rocky areas. Also occurring on rocky
outcrops amongst rocks, in crevices or in shallow soil over rock slabs (then
not uncommonly growing with *Xerophyta, Iboza riparia* and *Myrothamnus flabellifolius*). – Ca 1 250–2 500 m.
Distribution (Map, Fig. 73 a): From S Tanzania, NE Zambia, Malawi
and adjacent parts of Mozambique S to Zimbabwe.

Critical Remarks: A. whyteanum is very closely allied to *A. hispidulum*
(17., below) and several morphological characters of the two species
overlap. A fundamental difference between the two species, however, lies
in the leaf arrangement, i.e., strictly *ternate* in *A. whyteanum* but without
exception decussate in *A. hispidulum. A. whyteanum*, furthermore, often
has shorter and more closely spaced leaves (shorter internodes) and tends
to be more robust and larger. It also shows a clear trend towards dioecy
(some populations observed appear to be strictly dioecious!), whereas *A.
hispidulum* exhibits quite a wide range of sex distributions. Finally, the
distribution ranges of *A. whyteanum* and *A. hispidulum* are strictly
separate: the former occurs only N of the Limpopo R., *A. hispidulum* only
S of it. For all these reasons it is, at present, thought best to maintain the
two taxa as separate species.

EYLES 343 from Zimbabwe (N) closely resembles *A. hispidulum* in
having rather long internodes and relatively broad and long leaves; its
leaves, however, are arranged in whorls of 3 as is typical of *A. whyteanum*.

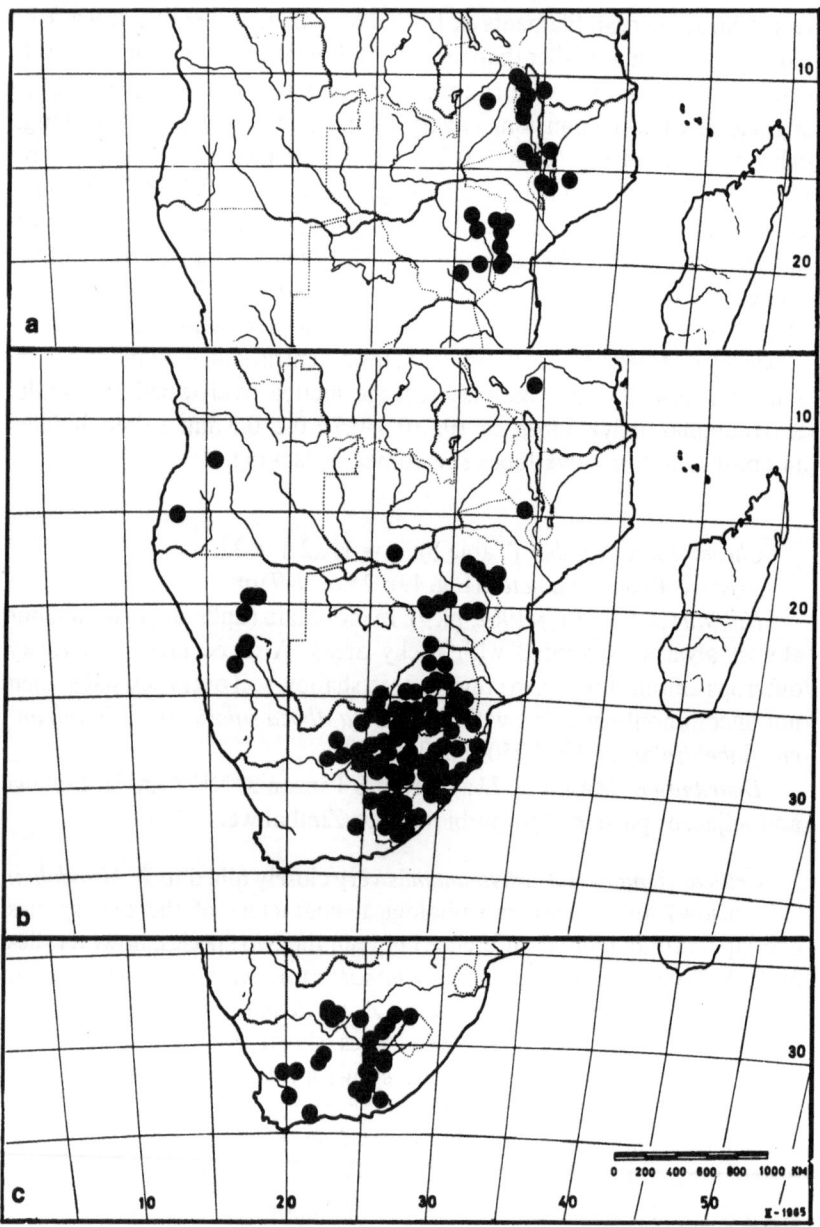

Fig. 73. Distribution of *Anthospermum. a A. whyteanum, b–c A. pumilum, b* subsp.
pumilum, c subsp. *rigidum*

The collection may represent new aerial flowering shoots produced just after a fire.

The following collections may represent an odd state of the species:

LEACH & RUTHERFORD-SMITH 11036 from Mozambique (N) in habit and its hairy fruits matches *A. whyteanum* but differs in having almost glabrous leaves.

Also TORRE & PAIVA 10309 (LISC) from Mozambique [N: Ribáuè, serra de Ribáuè (Mepalué)] and HUTCHINSON & GILLETT 4082 (K, SRGH) from Zambia (C: 32 miles NE of Serenje Corner) have rather glabrescent foliage. As mature fruits are absent, their true identity cannot be established with certainty; these two collections are, therefore, neither listed in Collections nor mapped.

Various collections differ from "typical" *A. whyteanum* in that they appear to exhibit some characters of other taxa. They are likely to be of hybrid origin[1].

PHIPPS 409 (K, SRGH) from the Chimanimani Mts. (Stonehenge Plateau) may be a hybrid *A. whyteanum* × *A. ammannioides*. Especially the specimen in K has rather large, ternately arranged oblanceolate leaves (to 20 × 4 mm) which match those of *A. ammannioides* (much larger than the leaves of "typical" *A. whyteanum* from the Chimanimanis!); also they are less densely hairy than "typical" *A. whyteanum* and the ♂ flowers have relatively short corolla tubes. "Pure" *A. ammannioides* has also been recorded from the Chimanimani Mts. (NOEL 2029 and others). Hybrids between *A. ammannioides* and *A. whyteanum*, however, seem to be very rare, although in the E highlands of Zimbabwe the two species are not uncommonly found growing in the same immediate vicinity.

In addition to the above-mentioned *A. ammannioides* (5.), *A. why-teanum* may, in E Zimbabwe, occur together with (a) *A. zimbabwense* (3.) (quite often: Inyanga National Park area, e.g., PUFF 790124- and 790125-collections; Vumba Mts., Castle Beacon, e.g., PUFF 790128- collections), (b) *A. pumilum* subsp. *pumilum* (29 a.) (Vukutu Farm, e.g., PUFF 790124-collections), (c) *A. ternatum* subsp. *randii* (19 b.) (Odzani valley, e.g., PUFF 790127- collections), (d) *A. vallicola* (6.) (Vumba Mts., Castle Beacon, e.g., PUFF 790128- collections) or (e) *A. herbaceum* (25.) (Inyanga National Park, e.g., PUFF 790124- collections). Occasionally no less than four species of *Anthospermum* (*A. whyteanum*, *A. zimbabwense*, *A. am-mannioides* and *A. vallicola*) may grow together in an area of hardly 100 × 100 m (e.g., Vumba Mts., Castle Beacon). While there is no evidence

---

[1] Specimens which quite closely resemble *A. whyteanum* but which are believed to be not "pure" *A. whyteanum* are, for practical reasons, grouped with *A. whyteanum* and are marked with an asterisk in Collections.

whatsoever for a hybridization between *A. whyteanum* and *A. herbaceum* and between *A. whyteanum* and the large shrubby *A. vallicola* and not much proof for suspected odd hybrids *A. whyteanum* × *A. ternatum* subsp. *randii* [as regards *A. ammannioides*, see above], there are strong indications that *A. whyteanum* may relatively frequently hybridize with *A. zimbabwense*. Putative hybrids quite often approach *A. zimbabwense* in habit and leaf size and shape, their leaf indumentum is less dense than that of *A. whyteanum* and the hairs may be somewhat shorter (*A. zimbabwense* has glabrous leaves!), the fruits may be partially glabrous, and ♂ and ♀ corolla tubes are rather shorter than in "typical" *A. whyteanum* and somewhat longer than the ♂ corolla tubes of *A. zimbabwense*. Collections from frequently visited localities (such as World's View or the Castle Beacon in the Vumba Mts.) often exhibit a wide range a morphological intermediates ranging from "pure" *A. whyteanum* to "pure" *A. zimbabwense*. Even duplicates from one and the same collection (e.g., BAMPS, SYMOENS & VANDEN BERGHEN 505) are sometimes either "pure" *A. whyteanum* or show certain *A. zimbabwense* characters.

At Vukutu Farm (PUFF 790124-1/3), I strongly suspect that *A. whyteanum* forms hybrids with *A. pumilum* subsp. *pumilum*. The plants are less hairy than "typical" *A. whyteanum* and have the more lanceolate and longer leaves of *A. pumilum* subsp. *pumilum* (particularly obvious on new aerial shoots).

The Mlanje Mts. in S Malawi are not only the type locality of *A. whyteanum* but also a well known collecting site of *A. welwitschii* (1.). On the mountains, a fair number of hybrids *A. whyteanum* × *A. welwitschii* appear to be present next to "typical" *A. welwitschii* [usually occurring in (forest edge) scrub] and "typical" *A. whyteanum* [typically growing on rocky outcrops (cf. CHAPMAN 1962: pl. 9)]. Putative hybrids, however, in habit often more closely resemble *A. whyteanum* (due to back crosses?) than *A. welwitschii*; they tend to have ternately arranged, but less densely hairy leaves (*A. welwitschii* has strictly decussate, glabrous leaves). I even suspect that the type of *A. whyteanum* (WHYTE 48) is not entirely "pure" *A. whyteanum*. One of my collections from Mt. Mlanje (between Chambe Plateau and Chambe Saddle, PUFF 780211-1/1; not in Collections) is truly intermediate: some plants/branches have decussately, others ternately arranged leaves. The leaves are intermediate in size and only have a few hairs; the internodes are longer than in *A. whyteanum* and shorter than in typical *A. welwitschii* collections from Mt. Mlanje.

As regards the FANSHAWE collections (nos. 7307, 9747) from the Zambian side of the Nyika Plateau, finally, I am quite convinced that they are not "pure" *A. whyteanum* but show certain characteristics of *A.*

usambarense (2.). I myself collected both *A. usambarense* and *A. whyteanum* on the Nyika Plateau and have, in one locality near the
Malawi/Zambia border, found them growing virtually side by side
(Chowe Rock; *A. usambarense* at the edge of scrub at the base of a rocky
outcrop, *A. whyteanum* in crevices on the same outcrop; PUFF 781223-3/1
and -3/2, respectively). I have not detected hybrids (although, at the time,
I did not specifically search for them), but I can at least conclude from a
comparison of my collections that FANSHAWE's specimens are similar to *A.
whyteanum* in their indumentum but resemble *A. usambarense* in size
(FANSHAWE 7303 is supposedly ca 2.4 m tall, while *A. whyteanum* hardly
exceeds 1 m in height!) and in habit and leaf shape (cf. Fig. 58).

Collections

Tanzania. – T 8: Songea D., Matengo Highlands, Lupembe Mt., MILNE-
REDHEAD & TAYLOR 8177 (B, BR, K), ZERNY 269 (W).

Zambia. – N: Mpika D., Danger Hill, 29 km NNE of Mpika, DRUMMOND &
WILLIAMSON 9708 (K, SRGH); –, Muchinga escarpment, 45 km NNE of Mpika,
Poćs &KORNÁS 6630/5 (K). – E: Mafinga [Mts.], CHISUMPA 38 (K); –, S end, PUFF
781222-2/6 (WU); Nyika plateau, Zambia side, FANSHAWE 7307*, 9747* (K),
RICHARDS 14397 (K).

Malawi. – N: Chitipa D., Mafinga Mt. above Chisenga, DRUMMOND &
WILLIAMSON 9891 (K, SRGH); Nyika Plateau, PHILLIPS 232 A (K, MO, SRGH),
RICHARDS 22491 (K), WHYTE 111, s.n. (K); –, Chelinda Bridge (Rock), PAWEK 7935
(K, MAL, MO, SRGH), PUFF 781224-1/1 (WU); –, Chosi Hill, PAWEK 2134
(MAL; K); –, Chowe (= Chowo) Rock, PAWEK 9219 (K, MO), PUFF 781223-3/1
(WU), RICHARDS 22702 (K); N Viphya, Chimaliro Hill, PUFF 781228-2/1 (WU);
Mzimba D., Viphya Plateau, ca 65 km SW of Mzuzu, Chimpayi valley, PAWEK
9155 (BR, MO, K, PRE, SRGH, WAG); S Viphya, Lumono Hills, ca 30 km N of
Chikangawa, PUFF 781229-1/1 (WU); Mzimba D., ca 3 km WSW of Chikangawa,
PHILLIPS 4553 (WAG); S Viphya, Mt. Champhila, PUFF 781231-1/1 (WU). – C:
Lilongwe D., Mkhoma Mt., CHAPMAN 1798 (MAL, SRGH); Dedza D., Chongoni
Forest Reserve, Chencherere Hill, BRUMMITT 10056 (K, SRGH × 2); –, –, Chiwau
(= Ciwao) Hill, CHAPMAN 707 (BM, K, MAL, PRE, SRGH), PUFF 780215-2/1
(WU), ROBSON & JACKSON 1270 (BR, K, SRGH), SALUBENI 4 (MAL); Ncheu D.,
Kirk Range, YOUNG 223 (MAL). – S: Blantyre, Shire Highlands, BUCHANAN 188
(?166) (E), 963 (K); Blantyre D., Ndirande Mt., BRUMMITT 10325 (K, SRGH);
Mulanje (= Mlanje) Mt., BLACKMORE, BRUMMITT & BANDA 367 (K), BRASS 16430*
(BM, BR, GH, K, MO, PRE, SRGH, US), BRUMMITT 9674 (K), HILLIARD & BURTT
4597 (E, K, NU, SRGH), 6084* (E, LMU, NU), HUMBERT 17170 (P), PUFF 780211-
2/2 (J, WU), SALUBENI 1591 (K, MAL, SRGH), SEAGRIEF s.n. sub SRGH 88248*
(SRGH), WHYTE 48 (BM).

Mozambique. – N: Niassa D., Massungula Mts., ca 65 km N of Mandimba,
LEACH & RUTHERFORD-SMITH 11036 (SRGH; atypical). – Z: Gúruè (= Vila
Junqueiro), Nàmuli, ANDRADA 1858 (COI, LISC). – T: Angónia, N part of serra
Dómué, MACUÁCA & MATEUS 1099 (LMA). – MS: Báruè, serra de Choa, 26 km
from Vila Gouveia tow. the border, TORRE & CORREIA 15401 (LISC); Chimanimani
Mts., BUTTON 101 (LMA), –, Moribane (?spelling), JOHNSON 234 (K).

Zimbabwe. – N: Mazoe, Iron Mask Hill, EYLES 343 (BM, BOL, SRGH;
atypical). – C: Wedza D., Mt. Wedza, WILD 6352 (K, LISC, SRGH). – E: W of

Juliasdale, Vukutu Farm, Puff 790124-1/3* (J, WU); − , 2 km W of Punch Rock, Biegel 4162 (B, BR, K, LISC, MO, PRE, SRGH); Inyanga D., around 'World's View', Bamps, Symoens & Vanden Berghen 322 (BR, MO, PRE, SRGH), 505* (BR, MO, SRGH), 532 (BR), Puff 790125-6/4 (J, WU), West 7081 (SRGH); Rhodes Inyanga National Park, 2 km NNE of R. I. Experimental Stn., Burrows 614 (SRGH); Stapleford Forest Reserve, North Patrol, Puff 790127-1/1 (J, WU); Vumba Mts., Chase 6038 (BM, BR, K, LISC, PRE, SRGH), Ferrar 3958 (PRE); − , Castle Beacon, Corner s.n.* (E), Puff 790128-1/2 (WU); Mt. Binga [Banti Forest Res.], Pereira, Sarmento & Marques 44* (LMU × 2, WAG); Melsetter D., Chimanimani Mts., Goodier 142 (K), Hall 245* (BM, BOL, SRGH), 326* (BM, BOL, LMA, SRGH), Noel 2024 (SRGH), Phipps 409* (K, SRGH); − , ca 33 km S of Melsetter, nr. Silverstream, Fisher 1250 (E*, NU, PRE, SRGH). − Additional collections: Chase 7894 (K, SRGH); Eyles 8519 (K, SRGH), 3113* (BOL); Gilliland 1693* (BM, K, PRE), 1930 (BM, K); Munch 358* (K, SRGH); Puff 790124-2/3 (WU), -3/2* (J, WU); Rattray 1037 (PRE, SRGH); West 4394 (SRGH); Wild 1542* (BR, K, SRGH). − S: Belingwe D., Mt. Buhwa, Biegel, Pope & Simon 4236 (K, LISC, SRGH; ± atypical); Bikita D., Mt. Horzi, Biegel 3078 (K, LISC, LMU, PRE, SRGH), Mpofu 28 (B, K, LMA, MO, PRE, SRGH).

17. Anthospermum hispidulum E. Meyer [in Drège in Flora **26**, Bes. Beigabe: 164 (1843), nom. nud.] ex Sonder in Harvey & Sonder, Fl. Cap. 3: 29 (1865); Puff in Fl. S. Afr. **31**, 1 (2): 21 (1986).

Syntypes: South Africa, Cape Prov. (Transkei), Omsamwabo to Omsamcabo[1], Drège s.n. (S, lecto.!, selected here; G!, K!, LY!, MO!, W!); − , Natal (Zululand), Gerrard [& "M'K" (= McKen)] 1361 (BM!, K!, W!).

= *A. burkei* Sonder in Harvey & Sonder, Fl. Cap. 3: 29 (1865); Compton, Fl. Swaziland: 587 (1976).

Type: South Africa, Transvaal, Magaliesberg, ["Burke & Zeyher" =] Burke 86 (BM!, K!, PRE!, SAM!), Zeyher 770 (S, holo.!; BM, G, LY, K, P, iso.!)[2].

---

[1] Omsamwabo [= Omsamvubo on map in Drège (1843)] is without doubt the Mzimvubu River; its mouth is at Port St. Johns on the Wild Coast (3129-AC). Omsamcabo is the Msikaba River; its mouth is at Mkabati (3129-BD).

[2] It appears to be a mere oversight that Sonder did not list the collection numbers of the Burke and Zeyher specimens. Collections from their joint trip to and in the O.F.S. and Transvaal were numbered separately (cf. for example, the type of *Nenax microphylla,* or that of *Otiophora calycophylla*: Puff 1981). It may be that the collections were numbered arbitrarily at a later date (perhaps even by Sonder). A strong indication for this is that the collection numbers cannot be correlated with their travelling route (cf. Drège 1847 a and b, and Gunn & Codd 1981).

Following the recommendation of Nordenstam (1980), the collection in herbarium S (Zeyher 770), bearing Sonder's handwriting, is considered to be the holotype.

= *A. arenicola* GREVES in J. Bot. (London) **63**: 203 (1925).
Type: South Africa, Transvaal, Witpoortjie Kloof nr. Johannesburg,
Moss 9806 (BM, holo.!; PRE, iso.!).

— *A. rubricaule* K. SCHUM., nom. nud. (non valide publ., since descr.)

Non-dioecious [♀̂, ♀̂ + ♀, ♀, ♀̂ + ♂ or (rarely) ♂], usually several- to
many-stemmed dwarf shrubs with often thick, woody (tap-)roots. Stems
ca (10) 15–40 (60) cm long, ascending to erect, ca 1.5–5 (8) mm in diameter
near the base and ca 1–3 mm in the midstem region, few- to much-
branched, branches ascending to ± erect; younger parts often reddish-
brown, usually densely covered with white to reddish-brown spreading
hairs ca (0.2) 0.4–0.9 mm long; older parts reddish-brown to dark grey,
becoming glabrescent, epidermis often peeling; upper internodes often as
long as or shorter than the leaves. Leaves decussate, usually pseudo-
verticillate; blades often ascending, 5–12 (20) × 1–3.5 (5) mm, oblanceolate, ovate-lanceolate to narrowly lanceolate, less commonly ±
obovate, narrowed to the base, mostly both surfaces densely covered with
whitish spreading hairs ca (0.1) 0.3–0.5 mm long, but sometimes lower
surface less hairy except for the midrib; margins somewhat or, more
commonly, strongly revolute; apices acute to acuminate or sometimes ±
mucronate; petioles subobsolete; stipular sheaths mostly densely covered
with whitish to reddish-brown spreading hairs, cup-shaped, ca 0.5–
0.9 mm long, with a median seta (0.5) 0.8–1.4 mm long. Inflorescences ±
inconspicuous to occasionally ± conspicuous and somewhat contracted
(♀), made up of ± sessile, ca (9-) 6-2-flowered axillary clusters. Flowers
subsessile, ♂, ♀̂, ♀. Corolla 4-merous (very rarely also 5-merous in ♀̂),
greenish-yellow to yellowish, at least the lobes covered with whitish
spreading hairs ca 0.1–0.4 mm long on the outside. ♂, ♀̂: tube 1.4–1.6 mm
(♂) or –2 (2.4) mm (♀̂) long, (narrowly) funnel-shaped, lobes (1.7) 2–
2.7 × 0.6–0.9 mm, ± lanceolate, recurved; stamens 4, filaments ca 1.7–
3 mm, anthers 1.4–1.9 × 0.2–0.4 mm (occasionally smaller, 1 mm or less,
and pollenless: transitional ♀̂ → ♀); ♀̂: stigmas 2, ca 2.9–4.1 mm long, ovary
ca 0.7–1.2 × 0.5–0.9 mm, hairy, crowned by 4 calyx lobes (occasionally
stigmas shorter and ovaries smaller: transitional ♀̂ → ♂); ♂: rudimentary
ovary small but well discernible, sometimes rudimentary stigmas present.
♀: tube 0.3–0.7 mm long (0.9–1.2 mm long if rudimentary anthers are
present), ± cylindrical, lobes 0.2–0.7 × 0.1–0.4 mm (1.2–1.8 × 0.4–0.5 mm
if rudimentary anthers are present), lanceolate to linear-lanceolate, ±
erect; style ± 0–0.8 mm long; stigmas 2, 3–6.5 mm long, greyish to
greenish-grey; ovary as in ♀̂. Fruits reddish-brown, supported by small, U-
shaped carpophores to ca half as long as the fruits; each mericarp ca 1.7–

2×0.7–1 mm, $\pm$ oblong, dorsal side convex, $\pm$ densely (less commonly $\pm$ sparsely) covered with whitish spreading hairs ca 0.2–0.5 mm long, ventral side $\pm$ plane and with a median vertical ridge, commissure large; mericarps crowned by 2 often hairy, $\pm$ triangular calyx lobes ca $(0.2)\,0.4$–$0.7\,(0.9) \times 0.1$–$0.3\,(0.4)$ mm. – Figs. 8 c–d.

Chromosome Number (Table 3): $n = 11$, $2n = 22$.
Average Pollen Diameter (Fig. 48): 23.9–$26.3\,\mu$m.
Habitat: Always associated with rocky areas. On rocky ridges, in cracks of rock sheets, on rocky outcrops in grassland, on koppies, on rock ledges or krantzes, or amongst boulders; less commonly in rocky grassland. In Natal and in the Transkei mostly over sandstone, in the Transvaal frequently over quartzite; occasionally also over old granite. – Ca 300–$2\,000\,(2\,300)$ m.
Distribution (Map, Fig. 74): In S Africa from the Transvaal through Natal to the E Cape Prov. (Transkei); also on the O.F.S. side of the Vaal R. and in W Swaziland.

Critical Remarks: A. hispidulum varies to some extent in leaf size and shape. Specimens from Natal and the Cape often tend to have wider leaves than those from the Transvaal. Plants may occur in fire-exposed habitats (but this is not very common!) and then behave as "sprouters" (see C.12.8.); leaves on the newly produced aerial shoots of recently burnt plants may be considerably wider than those on older, woody stems (also cf. Figs. 8 c–d). New growth, in general, has "a much more slender and graceful habit" (comments following the description of *A. arenicola,* whose type specimen represents such a state). The density of the indumentum (especially on leaves and fruits) and its colour also vary (*A. rubricaule* refers to plants with reddish-brown rather than whitish stem hairs).

It is suspected that, particularly in the Transvaal, *A. hispidulum* may form hybrids with *A. pumilum* subsp. *pumilum* (29.), and that there may be extensive hybrids populations. The two not uncommonly grow in the same vicinity (although *A. pumilum* subsp. *pumilum* is more commonly found in more open grassland habitats). Hybrids, however, are not easily detected and there is little *concrete* evidence for the hybrid nature of numerous collections (see also C.12.5.). It may be that a large number of the more *narrow*-leaved Transvaal collections of *A. hispidulum* are not "pure" *A. hispidulum* and may, at one stage, have hybridized with *A. pumilum*. It is also suspected that plants characterized by less hairy fruits with less distinct (smaller) calyx lobes and a generally less dense indumentum may not be "pure" *A. hispidulum*. The occurrence of hybrid plants is only $\pm$ well documented from a few places: On the Blaauwberg, for example, $\pm$

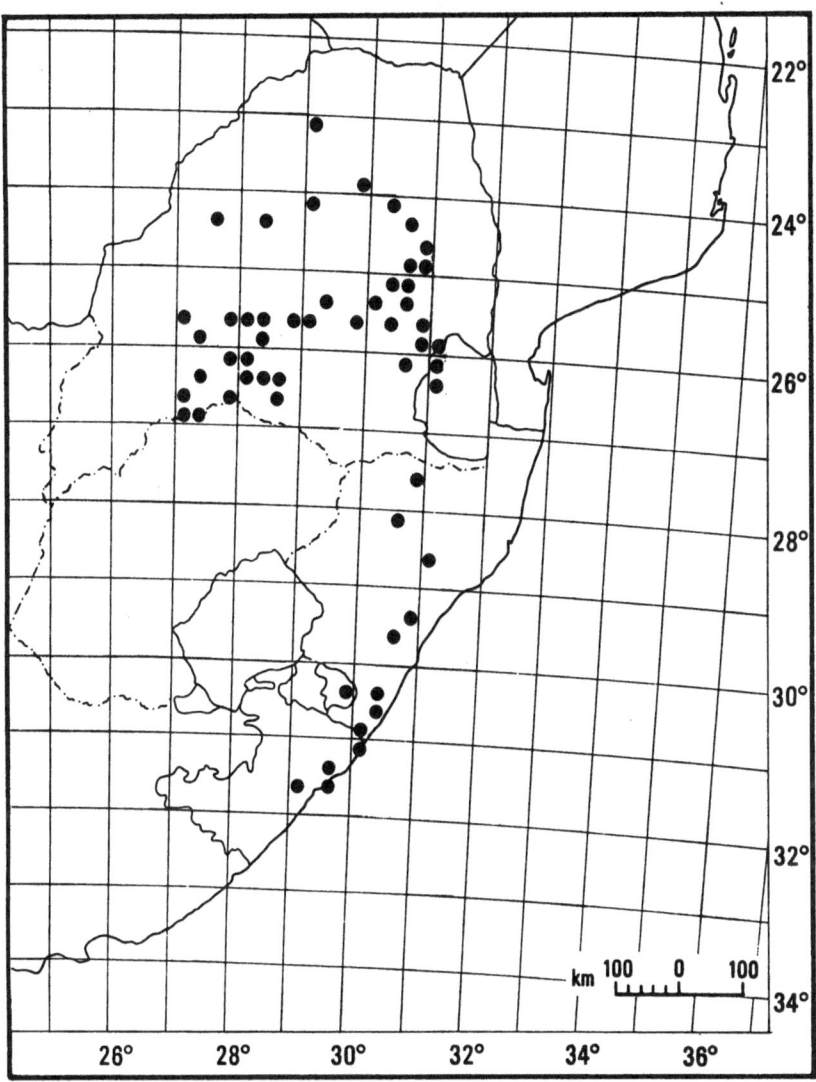

Fig. 74. Distribution of *Anthospermum hispidulum*

"pure" *A. hispidulum* (e.g., STREY & SCHLIEBEN 8498), ± "pure" *A. pumilum* subsp. *pumilum* (e.g., STREY & SCHLIEBEN 8618) and several collections intermediate in their characters, i.e., without much doubt hybrid plants[1], were collected. For practical reasons, plants with the

---

[1] E. g. CODD & DYER 9052 (NH, PRE), ESTERHUYSEN 21475 (BOL, PRE); ESTERHUYSEN 21494 (BOL, MO, PRE) is more similar to "typical" *A. hispidulum* but is, in my opinion, not "pure". These collections are not included in the specimen citations for *A. hispidulum*.

typical long hairs of *A. hispidulum* but with a generally less dense
indumentum are cited in Collections although they may not be entirely
"pure" *A. hispidulum*.

It appears that *A. hispidulum* may, occasionally, also form odd hybrids
with *A. herbaceum* (25.). Putative hybrid plants have considerably wider
leaves than "pure" *A. hispidulum*, are less woody and have fruits with ±
distinct calyx lobes (whereas the fruits of *A. herbaceum* are mostly not
crowned by calyx lobes). It is suspected that there is a (± extensive?)
hybrid population of *A. hispidulum* × *A. herbaceum* in the Hluhluwe Game
Reserve [Natal, Zululand; WARD 2638 (K, NU, NU, PRE), PUFF 790304-
2/1 (WU); neither mapped nor in Collections) but, again, there is no
definite proof. In general, however, extensive hybrid populations *A.
hispidulum* × *A. herbaceum* appear to be very exceptional.

The close alliance between *A. hispidulum* and *A. whyteanum* is
discussed above (p. 269).

Collections

Swaziland. – **2631** (Mbabane): Ngwenya Mt. (**-AA**), COMPTON 26677 (K, NBG,
PRE), PUFF 790225-1/2 (WU); Hhohho D., 5 km NW of Mbabane (**-AC**), KEMP
698 (BR, MO, US, WAG). – Additional collections: BRENAN & VAHRMEIJER 14285
(K, PRE); COMPTON 25722, 25902, 26714, 26810 (NBG, PRE), 27555, 31904
(NBG), 32021 (K, NBG, PRE).
South Africa. Transvaal – **2329** (Pietersburg): Blaauwberg (Blouberg)
plateau (**-AA**, or 2328-BB?), STREY & SCHLIEBEN 8498 (K, M, PRE); Houtbosch
(**-DD**), REHMANN 6026, 6035, 6036 (BM, K). – **2427** (Thabazimbi): Kransberg
(**-BC**), PUFF 781202-2/1 (WU). – **2428** (Nylstroom): Geelhoutkop (**-AD**), BREYER
18019 (PRE). – **2429** (Zebediela): Potgietersrus, Farm Plank Nek 154 (J,
PRE). – **2430** (Pilgrim's Rest): Shilouvane (**-AB**), JUNOD 1074 (G); Burke's Luck
Potholes (**-BC**), VAN WYK, DAHLGREN & VOK 5490 (PRE); Mariepskop (**-DB**), VAN
DER SCHIJFF 4880 (PRE, W), 4829, 5995, 6345 (PRE), 5996 (GH, K, PRE, W); Mt.
Sheba (**-DC**), PUFF 781029-1/1 (WU); 'Fairyland' nr. Graskop (**-DD**), PUFF
790704-1/2 (WU). – **2527** (Rustenburg): Rustenburg, Tierkloof (**-CA**), VENTER
646, 649 (PRE); Magaliesberg, Farm Kloofwaters (**-CD**), PUFF 790307-2/2 (WU);
Brits D., Hartebeestpoort dam (**-DB**), FAIRALL 1610 (NBG), VAHRMEIJER 407
(PRE). – **2528** (Pretoria): Wonderboom Reserve (**-CA**), REPTON 2105 (PRE);
Baviaan's poort (**-CB**), MOSS 14144 (J); Zwawelpoort (= Swawelpoort), E of
Pretoria (**-CD**), KINGES 1770 (PRE); ca 24 km N of Bronkhorstspruit on
Groblersdal rd. (**-DB**), PUFF 790209-1/1 (WU). – **2529** (Witbank): Loskopdam
(**-AD**), THERON 1292 a (PRE); Middelburg D., Olifants R. gorge, Farm Slaghoek 126
(**-CA**), MOGG 22475 (PRE); Farm Pineglades, 15 km from Belfast on rd. to
Dullstroom via Kwaggaskop (**-DB**), DU TOIT 319 (PRE). – **2530** (Lydenburg):
Spitzkop nr. Lydenburg (**-AB**), WILMS 459 (BM, E, G, K, WU); Farm Paarde-
plaats nr. Dullstroom (**-AC**), GALPIN 13079 (K, PRE, US, W), s.n. (BOL); Mt.
Anderson, W of Sabie (**-BA**), HUMBERT 10993 (P), SMUTS & GILLETT 2437 (PRE);
Wonderkloof Nature Reserve (**-BC**), ELAN-PUTTICK 294, 303 (PRE); Elands-
hoogte, nr. Machadadorp (**-CB**), HILLIARD & BURTT 14184 (E, NU, WU); Kaapse
Hoop (**-DB**), ROGERS 21565 (GRA), THORNCROFT 2218 (K); Twello Estate,
'Schoongezicht' section, W of Barberton (**-DD**), PUFF 780903-1/4 (WU). – **2531**

(Komatipoort): Saddleback Mt. (-CC), GALPIN 1310 (BOL, K, PRE, SAM), PUFF 780902-1/2 (WU), THORNCROFT 971, 972 (NH), 1198 (PRE), 2185 (K). — **2627** (Potchefstroom): Potchefstroom D., Elandsrand, near Deelkraal (-AD), BOTHA 2555 (PRE, PUC); ca 6 km NE of Krugersdorp (-BB), ACOCKS 18720 (K, M, PRE, SRGH); Potchefstroom, Mimosa Park Farm (-CA), POTTS 2697 (PRE); (nr.) Venterskroon (-CC), VAN WYK 274 (PRE); ca 8–10 km NNW of Vereeniging, Langerand Hill, Farm Houtkop 3 (-DB), MOGG 21030 (BOL, J, PRE). — **2628** (Johannesburg): Johannesburg, Melville Koppies (Nature Reserve) (-AA), FORBES 730 (J), MACNAE 1459 (J); ca 3 km W of Alberton, Klipriviersberg Farm (-AC), MOGG 18263 (J); Suikerbosrand Nature Reserve, Boschhoek 358 section (-AD), BREDENKAMP 923 B (PRE), PUFF 790516-2/1 (J, WU); ca 16 km SE of Nigel, Uitkyk Farm (-BC), MOGG 18923 (B, J); ca 10 km SE of Heidelberg, Farm Houtpoort 309 (-CB), MOGG 18474 (J); ca 66 km SE of Johannesburg, Farm Kuilfontein 289 (-DA), MOGG 20732 (J). — **2630** (Carolina): nr. The Brook (-BA), STREY 8017 (K, NH, PRE). — Imprecise localities: Magaliesberge, BURKE 86 (BM, K, PRE, SAM), LEENDERTZ 10291 (PRE), POTT 3708 (NU), ZEYHER 770 (BM, G, K, LY, P, S); "N Tvl", LE DOUX 22 (GRA). — Not traced: Buffelshoek, GOOSSENS 1503 (PRE); Zuurfontein, ROGERS 1370 (GRA). — Additional collections: BALKWILL s.n. sub PUFF 790513-1/1 (WU); BOUSMA 12718 (PRE); BREDENKAMP 862, 878, 888 (PRE); CODD 787 (K, PRE); COLLINS & SCHOLES s.n. sub PUFF 790520-1/2 (WU); DE FEIJTER 126 (PRE); DE JONGH sub GALPIN 1467 (PRE); GALPIN 1467 (PRE, US × 2), 14448 (K, PRE), s.n. (BOL); GILFILLAN 1467 (STE); GILMORE 463 (PRE); GOLDBERG s.n. (J); GONNY s.n. (J); JACOBSEN 793 (PRE); JENKINS 9235 (PRE); JUNOD 79 (PRE); KEET ?1339 (sub STE 15198); KOLBE s.n. sub BOL 25128 (BOL); LAMB 3823, 3880 (SAM); LAMBRECHTS 324 (PRE); LAVRANOS 4961 (P); LEENDERTZ 6051 (PRE); LIEBENBERG 2416 (PRE); LOUW 700 (GRA, PRE), 1412 (PRE); MAGUIRE s.n. sub J 30665 (J); McCLEAN s.n. sub PRE 41542 (PRE); MEEBOLD 14517 (M); MOGG 17623, 20141, 22049, 22535, 25521, 25561 (J), 16446 (PRE), 18144, 24980 (SRGH), 18594 (BR), 18822, 21714 (M), 20248, 20290, 23094 (J, PRE), 20481 (J, MO), 21714, 25062 (LMU); MOSS 3921, 9807, 14203, 19591 (J), 6310 (BM, J), 9806 (BM, PRE), 6746 (BOL, J); OBERMEYER 243, 27907, 27924 (PRE); PATTERSON s.n. sub J 37836 (J); PUFF 780312-1/1, -1/2, 780716-1/2 (WU); RAND 1266, 1267 (BM); REHM s.n. (M); REPTON 815 (PRE), 3777 (PRE, SRGH); SMITH 6142 (BM, PRE), 6199 (K, PRE), 6143, 6200 (PRE); STORY 992 (PRE); THODE s.n. sub STE 5045 (STE); THORNCROFT 534 (NH); VAN VUUREN 543 (K, PRE); WELLS 2340 (K, PRE); WILLIAMSON s.n. sub PRE 41535 (PRE).

O.F.S. — **2627** (Potchefstroom): Brakfontein nr. Parys (-CD), PHILLIPS s.n. sub J 32262 (J).

Natal — **2730** (Vryheid): Vryheid D., Zungeni Pk. (= Zungwini Mt.) (-DB), ACOCKS 11554 (NH, PRE × 2). — **2830** (Dundee): Nqutu D., ca 3 km E of Nqutu (-BA), CODD 7673 (K, NH, NU, PRE). — **2831** (Nkandla): ca 12 km S of Nkandla, Nomang(c)i Hill (-CA), HILLIARD 2651 (NU), PUFF 790303-3/1 (WU). — **2930** (Pietermaritzburg): New Hanover D., Little Noodsberg, Laager Farm (-BD), HILLIARD & BURTT 15476, 15477 (E, NU, WU); Table Mt. (-DA), KILLICK 77 (PRE), McLEAN 159 (PRE). — **3030** (Port Shepstone): Campbellton, [W of] Dumisa Stn. (-AD), RUDATIS 1884 (G, S, W); Oribi Gorge Nature Reserve, Inkonka Point (-CB), PUFF 790429-1/2 (WU); Port Shepstone D., Otterburn halt (-CC), HILLIARD 2754 (NU). — **3130** (Port Edward): ca 3 km inland from Port Edward (-AA), STREY 7441 (K, NH, PRE). — Imprecise locality: Natal, Zululand, GERRARD (& McKEN) 1361 (BM, K, W). — Not traced: Lebundini, Zululand, HAYGARTH sub MEDLEY WOOD 7539 (BOL, K, NH, SAM). — Additional collections: CODD 6985 (K, NH, NU, PRE); EDWARDS 1333 (NU, PRE); MEDLEY WOOD

1983 (BOL, K, NH); Meebold 14518 (M); Nicholson 988, 1727 (PRE); Puff
790423-2/2 (WU); Rump s.n. (E, NU); Strey 4928A (NH), 7505 (B, BR, K, M,
NH, NU, PRE, S), 9731 (K, NH, NU, PRE, SRGH); Thode s.n. sub STE 4404
(STE); (Wylie sub) Medley Wood 8993 (K, NH, PRE), 9058 (NH). **Cape** – **3029**
(Kokstad): Clydesdale (**-BD**), Pegler s.n. (G). – **3129** (Port St. Johns): Ntsubane
Forest (**-BC**), Strey 9015 (K, NH, PRE); Lusikisiki D., Ngogwana R. falls, Galpin
11025 (PRE); track from Old Bunting to Tombo and Pt. St. Johns (**-CA**), Puff
790415-1/3 (WU); Port St. Johns, Devil's Peak (**-DA**), Galpin 3439 (PRE). – **3130**
(Port Edward): Port Edward- Bizana rd., Sea View Farm (**-AA**), Ward 218
(NU). – Imprecise locality: between Mzimvubu and Msikaba River, Drège s.n.
(G, K, LY, MO, S, W).

18. *Anthospermum perrieri* Homolle ex Puff, spec. nova, *A. whyteano* et
A. hispidulo affine sed mericarpiis lobis calycis conspicuis destitutis
differt; foliis decussatis ab *A. whyteano* distinguitur.
Type: Madagascar, Ankaratra massif, 1 500 m, July 1912, Perrier de
la Bâthie 3924 (P, holo.).

Non-dioecious [♂ + ♀, ♀ or (seldom) ♂ + ♂ + ♀], ± rounded to erect,
several- to many-stemmed dwarf shrubs with sometimes massive, woody
tap roots. Stems ca 25–40 cm long, ascending to erect, ca 3–5 mm in
diameter near the base and ca 1.5–2 mm in the midstem region, with few to
many, ± ascending branches; younger parts terete, reddish-brown,
densely covered with whitish to reddish-brown spreading hairs ca 0.5–
0.9 mm long, older parts brownish-grey, glabrescent, epidermis peeling;
internode lengths variable. Leaves decussate, pseudo-verticillate; blades
(6.5) 8–12 (15) × (1) 1.5–2.5 (3.5) mm, (ob)lanceolate, densely covered with
whitish to reddish-brown spreading hairs ca 0.3–0.7 mm long; margins
mostly strongly revolute; apices acute to apiculate; petioles subobsolete;
stipular sheaths hairy, (broadly) cup-shaped, ca 1–1.4 mm long, with a
median hairy seta ca 1.3–1.7 mm long, sometimes with an additional
smaller seta on either side. Inflorescences inconspicuous, made up of ±
sessile, ca 6-2- (rarely more-) flowered axillary clusters. Flowers subsessile,
♂ (seldom also ♂), ♀. Corolla 4-merous, greenish(-yellow), at least the
lobes densely hairy on the outside. ♂ (♂): tube 0.9–1.7 mm long, ±
cylindrical to funnel-shaped, lobes 1.5–2.1 × 0.6–0.7 mm, lanceolate,
recurved; stamens 4, filaments ca 1–1.8 mm long, anthers 1.3–1.4 × 0.3–
0.4 mm; ♂: gynoecium as in ♀ but stigmas often somewhat shorter. ♀: tube
0.2–0.5 mm long, cylindrical, lobes 0.5–0.7 × 0.2 mm, linear-lanceolate, ±
erect; style 0–0.7 mm long; stigmas 2, ca 2.4–4.8 mm long, greyish; ovary
ca 0.8–1.2 × 0.6–0.9 mm, hairy, mostly not crowned by distinct calyx
lobes. Fruits reddish-brown, supported by small but broad ± U-shaped
carpophores; each mericarp 1.7–2.4 × 1–1.2 mm, ± ovate to elliptic,

dorsal side convex, covered with whitish, ± spreading hairs to ca 0.2 mm long, ventral side plane to ± concave and with a median vertical ridge; commissure ± large; mericarps not crowned by distinct calyx lobes.

Chromosome Number (Table 3): n = 11, 2 n = 22.
Average Pollen Diameter (Fig. 48): 23.1–24.8 μm.
Habitat: Occurring in rocky areas (granite, gneiss), mostly amongst rocks or in rock crevices, in "bois de Tapia" (*Uapaca bojeri* forest) or remnants thereof. Sometimes also in disturbed sites (edge of plantations, road banks, etc.). – Ca 1 200–2 000 m.
Distribution (Map, Fig. 67 f): Endemic to central Madagascar.

Critical Remarks: A. perrieri, without doubt, is closely allied to the species pair *A. whyteanum* (16.) and *A. hispidulum* (17.) from S Central and SE Africa. It agrees with them in habit and habitat, in the characteristic, often reddish- to rusty-brown indumentum on the young stems and the leaf surfaces, in the stipular sheaths with one hairy seta, and in the hairy corollas and fruits. *A. whyteanum* has ternately arranged leaves, while both *A. perrieri* and *A. hispidulum* have strictly decussate leaves. *A. perrieri* has somewhat smaller ♀ (♂) corollas and marginally broader (and larger) fruits than the two species from the African mainland; the fruits, furthermore, differ from *A. hispidulum* and *A. whyteanum* in that they always lack distinct calyx lobes.

Collections

Madagascar. – **1847:** rd. from Antananarivo to Arivonimamo, Pk. 26, 27, 28, BOSSER 8034, 13391, 14919 (TAN); rd. to Miarinarivo, Pk. 60, DESCOINGS 2894 (TAN). – **1947:** nr. Arivonimamo, PERRIER DE LA BÂTHIE 17602 (P × 2), SCOTT ELLIOT 1936 (K); above Antananarivo- Antsirabe rd., around Pk. 36 (between Behenjy and Ambalavao), PUFF, IGERSHEIM & RAJEMISOA 850827-2/3 (TAN, WU); Ankaratra massif, PERRIER DE LA BÂTHIE 3524, 3924 (P); 19 km N of Ambohimiadana, on rd. to Miadanandriana, PUFF, IGERSHEIM & RAJEMISOA 850824-3/1 (TAN, WU); around Antsirabe, PERRIER DE LA BÂTHIE 4012 (P). – **2047:** 54 km E of Finandrohana (= Ambatofinandrahana), along Rt. 35, CROAT 29654 (BR).

19. *Anthospermum ternatum* HIERN, Cat. Afr. Pl. Welw. 2: 499 (1898); VERDC. in Fl. Trop. E. Afr., *Rubiaceae* 1: 328 (1976).
Syntypes: Angola, Huila, around Eme, WELWITSCH 5339 (BM, lecto.!, selected here; G!, K!, P!), 5340 (BM!, G!, K!)[1].

---

[1] According to VERDCOURT (1976), both WELWITSCH collections should also be in herbarium LISU. They, however, were not amongst the material sent to me on loan.

19*

Non-dioecious (♂̣, ♀, occasionally also ♂̣ + ♀, ♂̣ + ♂ or ♂), erect to ±
straggling, few- to ± much-branched, short-lived (sometimes biennial?)
herbs [1], "woody herbs" or subshrubs with a somewhat woody base. Stems ca
20–100 (150) cm long, to 8 mm in diameter near the base and ca 1–2 mm in
the midstem region; branches often ascending to ± erect; younger parts ±
terete, reddish or sometimes reddish-grey, densely covered with white
spreading hairs ca 0.1–0.3 mm or, occasionally, to 0.5 (0.8) mm long, older
parts reddish-brown to dark grey, becoming glabrescent, epidermis
sometimes peeling; internodes often as long as or shorter than the leaves.
Leaves in whorls of 3, decussate or, occasionally, in whorls of 4,
often pseudo-verticillate; blades spreading to ± erect, (4.5) 6–
45 (55) × (0.5) 0.8–5 (8) mm, ovate- or oblong-lanceolate to ± linear,
narrowed to the base, glabrous or variously hairy, but most commonly
with short hairs (less than 0.1 to 0.3 mm long) above and with longer hairs,
to ca 0.5 mm, at least on the midrib below; upper surface sometimes with
conspicuous, large epidermis cells; margins ± flat to revolute; apices
mostly acute; petioles subobsolete or to 3 mm on the largest leaves;
stipular sheaths shortly hairy, (broadly) cup-shaped, ca (0.7) 1–1.5 mm
long, with a median, hairy, subulate seta ca (0.7) 1–3 (3.7) mm long, often
with an additional, much shorter seta on either side. Inflorescences ±
inconspicuous, made up of ± sessile to slightly elongated, several- to few-
flowered axillary clusters. Flowers subsessile, ♂, ♂̣, ♀. Corolla 4-merous,
yellowish to yellowish-green, at least near the tips of the lobes covered with
whitish hairs ca 0.1–0.3 mm long, rarely glabrous altogether. ♂̣, ♂: tube
(1.7) 1.9–2.5 mm long, cylindrical, lobes 2–2.7 × 0.5–0.7 mm, lanceolate,
recurved; stamens 4, filaments exserted for ca 1–1.7 mm, anthers 1–
2.2 × 0.3–0.5 mm (occasionally smaller, 1 mm or less, and partially
pollenless: transitional ♂̣ → ♀); ♂ with well discernible ovary and sometimes also with rudimentary stigmas to 1 mm long; ♂̣: stigmas 2, (much)
shorter than in ♀, ca 1.9–4.7 mm long. ♀: tube 0.3–0.7 mm long (ca 1.2–
1.7 mm long if rudimentary anthers are present), cylindrical, lobes 0.4–
1 × 0.1–0.3 mm (1–1.7 × 0.2–0.5 mm if rudimentary anthers are present),
± lanceolate, ± erect to recurved; style ± 0–1 mm long; stigmas 2, 4–
10.2 mm long, greyish(-green); ovary ca 0.8–1.4 × 0.5–0.7 mm, glabrous or
hairy, not crowned by distinct calyx lobes or, seldom, with indistinct, ±
triangular calyx lobes ca 0.2 × 0.2 mm. Fruits reddish-brown, supported
by short, broad U-shaped carpophores; each mericarp 2–3 × 0.8–1.2 mm,
oblong to narrowly obovate, dorsal side convex, covered with whitish

---

[1] Young plants usually come into flower in their first growing season and may
thus look like annuals; unless burnt they will, however, continue growth in the
following season.

spreading hairs ca (0.1) 0.3–0.7 mm long at least near the apex or entirely glabrous, ventral side concave and with a median vertical ridge; commissure narrow, ± linear; mericarps not crowned by distinct calyx lobes. – Figs. 31 f, 35 c–d.

Habitat: Primarily occurring in miombo (*Brachystegia- Julbernardia-Isoberlinia*; often with *Uapaca*) and related woodlands (cf. WERGER & COETZEE 1978: vegetation map, Fig. 1, and compare with the distribution of *A. ternatum*!), where it grows in sandy to gravelly soil between rocks or in more open, grassy areas. Sometimes also found in disturbed sites (gravel pits, road sides, eroded slopes, etc.). – Ca (650) 800–2 100 (2 400) m.

Distribution: From E Zaire and Tanzania through Zambia, Malawi and Mozambique S to Zimbabwe; also in Angola. It, thus, is confined to the Zambezian Domain of the Sudano-Zambezian Region as defined by WERGER (1978 a) but does not extend to S of the Limpopo R.

19 (a). Subsp. *ternatum*

Plants mostly erect, sparsely branched; stems to ca 70 (100) cm long. Leaves mostly in whorls of 3 but occasionally decussate; blades ca (4.5) 6–25 (30) × (0.5) 0.8–1.2 (1.5) mm, ± linear and sometimes ± needle-like due to strongly revolute margins, glabrous altogether or sometimes scabrous and/or with longer hairs on the lower surface (leaves then usually smaller and narrower than in subsp. *randii*); petioles subobsolete. Fruits mostly hairy, at least near the apex, rarely glabrous altogether.

Chromosome Number (Table 3): 2 n = ca 22.
Average Pollen Diameter (Fig. 49): 24.9–26.3 µm.
Habitat: As for species as a whole.
Distribution (Map, Fig. 75 a): From W of Lake Tanganyika (Zaire) to Tanzania and S to N and NE Zambia, Malawi and neighbouring parts of Mozambique; also in N Zimbabwe and Angola.
Critical Remarks: See after subsp. *randii*.

Collections

Angola. – Benguela: 48 km from Nova Lisboa, tow. Alto Hama, around Chitatamera, BARBOSA 11672 (LISC; ± intermed.). – **Huila:** around Eme, WELWITSCH 5339 (BM, G, K, P), 5340 (BM, G, K); between Humpata and Caholo, TEIXEIRA 3491 (LISC; ± intermed.); Tchivinguiro, CORREIA 1228 (LISC; atypical); Sá da Bandeira, BECQUET 1178 (BR); 20 km SE of Sá da Bandeira, on road to Joao de Almeida, KERS 3185 (LISC, PRE); Monnyino (= Munhino Mission), DEKINDT 296 (LISC). – **Bié:** N'Jaio, nr. Munongue (= Menongue = Vila Serpa Pinto?), GOSSWEILER 3136 (BM; ± intermed.).

Zaire. – Haut-Katanga: 'Tanganyika See' [probably nr. Baudouinville = Moba], KASSNER 3044 (BM, K).

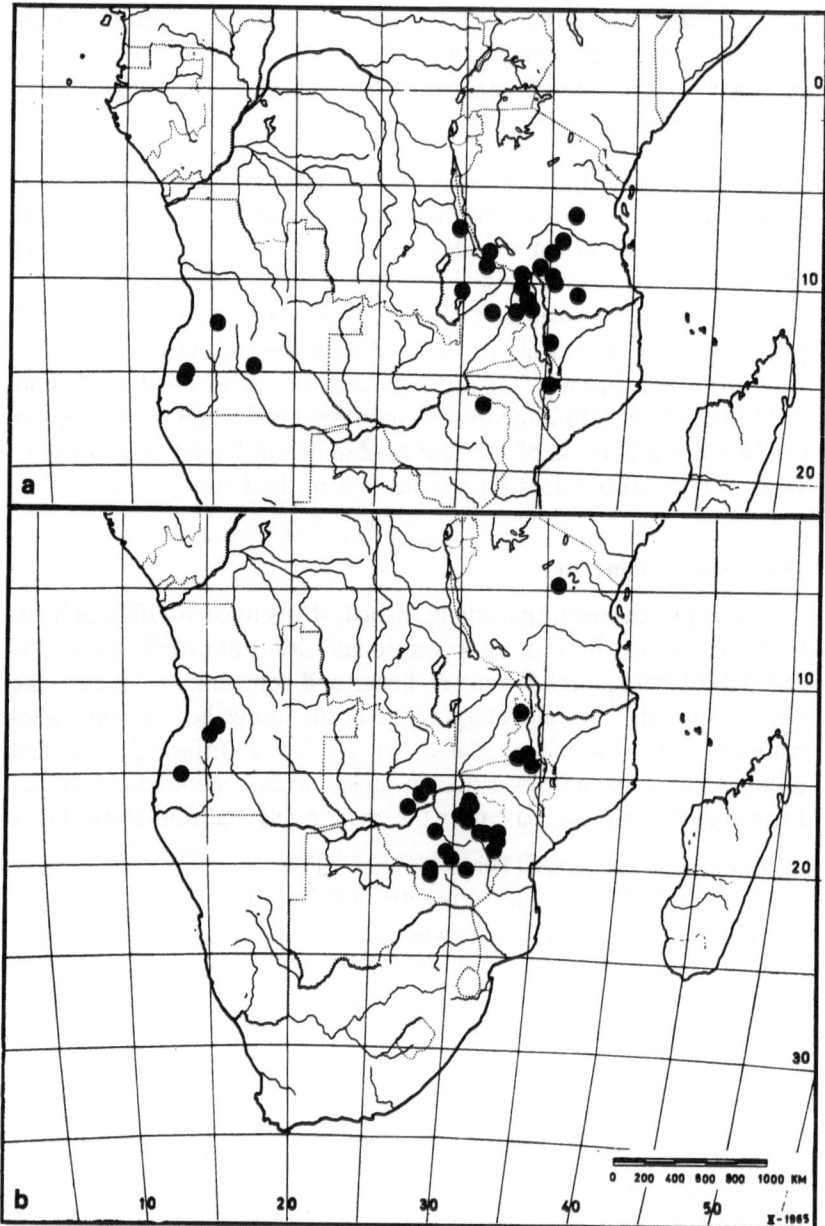

Fig. 75. Distribution of *Anthospermum ternatum. a* subsp. *ternatum, b* subsp.
randii

Tanzania. – T 4: Ufipa D., new Sumbawanga- Abercorn (= Mbala) rd., RICHARDS 11058 (K, SRGH). – T 5: Mpwapwa D., Mpwapwa, Kiboriani Mt., BURTT 3890 (K), HORNBY 936 (K), LINDEMANN 247 (BM). – T 7: Iringa D., Iringa, hill NE of town, POLHILL & PAULO 1696 (B, BR, K, LISC, P, PRE, SRGH); – , Mbeya- Sao Hill rd., BRENAN & GREENWAY 8235 (K; atypical); – , nr. Sao [Hill], GREENWAY 3441 (K); Njombe D., Ukinga, VON PRITTWITZ 5₁ (E); – , ca 4 km N of District boundary [on Songea rd.], MILNE-REDHEAD & TAYLOR 10760 (K). – T 8: Songea D., ca 44 km N of R. Hanga bridge on Njombe rd., MILNE-REDHEAD & TAYLOR 10934 (K); Tunduru D., just E of Songea D. boundary, MILNE-REDHEAD & TAYLOR 10602 (B, BR, K, LISC).

Zambia. – N: Mbala (= Abercorn) D., Mbala (= Abercorn), SIAME 186 (BR, K), 186 A (BM); – , Chilongolwelo escarpment, RICHARDS 16255 (K, LISC, SRGH); – , Sunzu Mt., RICHARDS 21387 (K); – , ca 13 km NE of Mbala, HUTCHINSON & GILLETT 4002 (BM, K, SRGH); Luwingu, FANSHAWE 8710 (K); Mpika, FANSHAWE 9250 (K, SRGH). – E: Mafingas (= Mafinga Mts.), FANSHAWE 11937 (K); – S end, on Zambian side, PUFF 781222-2/5 (W, WU); Makutus (= Makutu Mts.), FANSHAWE 11575 (K).

Malawi. – N: Chitipa D., Misuku Hills, path from Kaseye Mission to Matipa Forest, PUFF 781221-1/1 (BR, J, W, WU); – , S end of Mafinga Mts., ca 54 km from Chitipa, PAWEK 2032 (K, MAL); – , – , ca 3 km from Mulembe School and border post, PUFF 781222-1/1 (J, W, WU); Rumphi D., Nyika Plateau, ca 17.5 km N of M 1, PAWEK 11771 (K); Mzimba D., Marymount, tow. Lunyangwa, PAWEK 5453 (K, MO, SRGH), 6870 (MO, PRE, WAG). – S: Zomba Rock, WHYTE s.n. (K). – Imprecise localities: "Mpata and commencement of Tanganyika Plateau", WHYTE s.n. (K); "Nyasa Plateau", WHYTE 233 (G, K).

Mozambique. – N: Vila Cabral, TORRE 68 (LISC).

Zimbabwe. – N: Darwin D., Mt. Bandilonbidzi (= Banirembizi), WATMOUGH 136 (SRGH).

19 (b). Subsp. *randii* (S. MOORE) PUFF, stat. nov.

≡ *A. randii* S. MOORE in J. Bot. (London) **40**: 253 (1902).
Type: Zimbabwe (Rhodesia), Salisbury (= Harare), RAND 475 (BM, holo.!).

= *A. erectum* SUESSENG. in Trans. Rhod. Sci. Assoc. **43**: 54 (1951).
Type: Zimbabwe (Rhodesia), Marandellas, March 18, 1941, DEHN 55 (M, holo.!)[1].

Plants often more-branched than in subsp. *ternatum* and not uncommonly straggling; stems to 150 cm long. Leaves in whorls of 3, decussate or, occasionally, in whorls of 4; blades ca (10) 12–45 (55) × 1.5–

---

[1] In herbarium SRGH, there is another DEHN 55 collection from Marandellas; it, however, was collected on May 7, *1942*. This is not mentioned by SUESSENGUTH (in SUESSENGUTH & MERXMÜLLER 1951) and must, therefore, not be considered type material. The DEHN 55 collections from Rusape, made in *1952* ["55/52" (K!, LISC!, MO!, S!, SRGH!) and "55'/52" (BR!, K!, M!)], also must not be confused with the type.

5 (8) mm, ovate- or oblong-lanceolate to ± linear-lanceolate, variously hairy but most commonly with short hairs on the upper surface and longer hairs on the midrib below, rarely subglabrous (leaves then larger and broader than in subsp. *ternatum*); the largest leaves with petioles to 3 mm long. Fruits glabrous or hairy. – Figs. 31 f, 35 c–d.

Chromosome Number (Table 3): n = 11, 2 n = 22.
Average Pollen Diameter (Fig. 49): 24.3–27.2 µm.
Habitat: As for species as a whole.
Distribution (Map, Fig. 75 b): From Malawi to Zimbabwe and nearby areas of SE Zambia; also in Angola. A record from Tanzania is very doubtful (see Critical Remarks).

Critical Remarks (for the species as a whole): While I, on the basis of field observations and studies in Malawi and Zimbabwe, had originally thought that *A. ternatum* and *A. randii* are closely allied but easily separable, well defined species (*A. ternatum* in N Malawi always has very narrow, small, glabrous leaves, mericarps with hairs near the apex and a characteristic erect habit. *A. randii*, in central and S Malawi and in Zimbabwe, has much broader, larger, hairy leaves, glabrous fruits and often longer, straggling stems), a survey of all available material revealed that the plants are so variable that a clear separation into two species is not possible.

Subsp. *ternatum* is fairly uniform in habit and in leaf size and shape in Tanzania, N Malawi and NE Zambia[1]. It, however, shows variation in the pubescence: "typical" subsp. *ternatum* has glabrous leaves but a number of specimens has hairs on the midrib below (e.g., SIAME 186, RICHARDS 11058) or leaves hairy all over (Malawi: PAWEK 5453, 6870, and E of Lake Malawi: TORRE 68). "Typical" subsp. *ternatum* has mericarps which have hairs at least near the apex but some collections have entirely glabrous fruits (e.g., WHYTE 233).

The type of *A. ternatum* (from Angola) differs somewhat from E material in habit (longer internodes, less crowded leaves), in its trend towards slightly longer leaves, in its shorter, less coarse stem pubescence and in its slightly shorter but broader fruits. Other Angolan specimens differ (such as some S Central African material) in having hairy leaves (e.g., DEKINDT 296, KERS 3185). The KERS 3185 specimens are interesting in that they show variation in the leaf indumentum; some plants on the

---

[1] Subsp. *randii* does not seem to occur in that area. A sterile collection from Kondoa Distr. (Tanzania, T 5; BURTT 1303, K) resembles *randii* but the fragments, in my opinion, do not allow a definite identification; plotted with a question mark on map, Fig. 75 b.

sheet in PRE have glabrous leaves, while others on the same sheet have hairy leaves; the plants on the sheet in LISC are also hairy (in addition, the leaves are somewhat longer than usual, i.e., to 30 mm long). Provided the specimens originated from a single population, this seems to indicate that leaf pubescence can be discounted as a useable differentiation characters.

The type of *A. randii* has linear-lanceolate leaves (to 35 mm long and 3 mm wide) which are scabrid above (hairs ca 0.1–0.2 mm long) and hairy on the midrib below (hairs to ca 0.5 mm long). In subsp. *randii,* however, leaf indumentum is by no means uniform: Both surfaces can be covered with equally long (to 0.5 mm) or short (to ca 0.2 mm) hairs and, on the lower surface, the hairs are not necessarily confined to the midrib region; or the leaves are almost glabrous (e.g., BINGHAM 1211). Leaf sizes, shapes and arrangement are equally variable: sizes range from hardly more than 10 mm to occasionally over 50 mm, shapes from ovate-lanceolate to ± linear. The arrangement can be decussate, ternate or, occasionally, in whorls of 4. The mericarp indumentum ranges from hairy (the type, for example) to entirely glabrous on the dorsal side. There are no obvious correlations between a particular kind of indumentum and a specific leaf shape or size or between leaf indumentum and fruit hairiness. In short, subsp. *randii* is extremely variable and collections may, at least in some characters, overlap with subsp. *ternatum.* Not without reason has MOORE (1902) stated when describing *A. randii* that "... this comes nearest *A. ternatum* ...".

Because of the frequent occurrence of flowers ± transitional between ☿ and ♀ (i.e., with increasingly reduced but partially still pollen-producing anthers and, correspondingly, increasingly smaller corollas) next to "pure" ☿, ♂ and ♀ flowers, flower size differences cannot be used as differential characters.

The present subdivision into two subspecies seems more reasonable than the recognition of two separate species on the one hand, or the lumping of *A. randii* and *A. ternatum* into a single species on the other, but this solution is still not entirely satisfactory. Especially Angolan material is problematic. Some specimens appear ± intermediate and cannot be clearly assigned to one or the other subspecies. Field studies would be essential for a better understanding – especially in Huila, where all four taxa of *Anthospermum* recorded from Angola [*A. ternatum* subsp. *ternatum* and subsp. *randii; A. welwitschii* (1.); *A. pumilum* subsp. *pumilum* (29.)] occur in the same general area. These should also include field observations of *A. pumilum,* in particular in view of possible hybridogenous contact with subsp. *ternatum* (? CORREIA 1228).

A further problem concerns the delimitation of subsp. *randii* from *A.*

whyteanum (16.) (especially in E Zimbabwe): Long-hairy and rather short-leaved forms of subsp. *randii* with ternately arranged and rather closely spaced leaves may look rather similar to *A. whyteanum*. Particularly if a herbarium sheet only contains small branches or fragments of plants, a distinction between the two may become very difficult. Sheets containing larger portions of plants can be easily separated: *A. whyteanum* is a distinctly woody dwarf shrub, while subsp. *randii* generally is much less woody. Subsp. *randii*, furthermore, has flowers with distinctly cylindrical and longer (♀, ♂) corolla tubes, whereas *A. whyteanum* has smaller, (narrowly) funnel-shaped corolla tubes.

BRENAN & GREENWAY 8235 from the Iringa Distr. (T 7) appears to be an atypical, rather robust, small-leaved collection of subsp. *ternatum* but is most likely not, as suspected by VERDCOURT (note on the herbarium sheet), a hybrid *A. usambarense* × *A. whyteanum*. The collection VON PRITTWITZ 5₁ from Ukinga, also T 7, is quite similar.

The rather disjunct distribution range of subsp. *ternatum* and also of subsp. *randii* can most likely be attributed to the poor botanical knowledge of the vast and fairly uniform (miombo and related) woodlands which cover large parts of E Angola, W Zambia and adjacent parts of Zaire (see Map of the Extent of Floristic Exploration in Africa South of the Sahara; BRENAN 1965). All of these woodland areas are likely to provide suitable habitats for *A. ternatum*.

As regards the close alliance between *A. ternatum* (subsp. *randii*) and *A. rosmarinus* see p. 290.

Collections

Angola. – **Benguela**: Huambo D., Nova Lisboa, Belém do Huambo, MARTINS 80 (LISC; ± intermed.); – , – , Chianga, DA SILVA 1994 (LISC; ± intermed.); – , – , nr. Pedra do Alemao, EXELL & MENDONÇA 1685 (BM); Vila Flor- Cacoma, Chavaca, BAMPS & MARTINS 4400 (BR, PRE, WAG; ± intermed.). – **Huila**: Lubango D., Munhino Mission, MENEZES 3327 (BM, K, LISC, P, PRE, SRGH); Humpata, nr. Nene R., GOSSWEILER 10775 (K); – , Buraco do Bimbe, MENDES 3763 (LISC); – , encosta da Leba, BORGES 350 (LISC, PRE, SRGH); Serra da Chela, around Humpata, HUMBERT 16548 (P); – , around Sá da Bandeira, HUMBERT 16206 (P); 15 km from Humpata towards Tchivinguiro, BAMPS, RAIMUNDO & MATOS 4026 (BR, K, LISC, WAG; atypical); 20 km from Sá da Bandeira, tow. Vila Arriaga, BAMPS, MARTINS & MATOS 4594 (BR, WAG; ± intermed.); Quilemba, EXELL & MENDONÇA 2513 (BM, LISC), TEIXEIRA 2807 (LISC).

Tanzania. – **T 5**: Kondoa D., Kolo, BURTT 1303 (K; very uncertain – see Critical Remarks).

Zambia. – **C**: Lusaka D., Livingstone- Lusaka dr., RICHARDS 14907 (K). – **S**: Mazabuka D., Mazabuka, FANSHAWE 6583 (K, LISC), 6594 (SRGH); ca 8 km E of Choma, ROBINSON 1215 (K, SRGH).

Malawi. – N: Mzimba D., Hora Mt., PUFF 781230-1/1 (BR, J, W, WU). – C: Dowa- Ntchisi rd., nr. Golong'ozi, Ntambala Mt., PUFF 781213-2/1 (WU); – , Kongwe Forest Res., N of Dowa, PUFF 781212-2/2 (BR, J, W, WU); Lilongwe D., Dzalanyama Forest Res., BRUMMITT, SEYANI & PATEL 14947 (K); Dedza D., Dedza-Mphunzi rd., SALUBENI 1018 (K, MAL); – , Chongoni Forest Res., PUFF 780215-3/1 (BR, J, W, WU), SALUBENI 1339 (MAL, SRGH).

Zimbabwe. – N: Centenary D., Great Dyke above Mwari Palm Reserve, DRUMMOND 6819 (BR, K, SRGH); Mazoe D., Umvukwes, Horseshoe Mine, LEACH & BRUNTON 9859 (K, SRGH); – , ca 3 km W of Concession- Umvukwes rd., on rd. to Mtoroshanga pass, PUFF 790123-3/1 (BR, J, W, WU); Lomagundi D., Dyke nr. Road Camp Mine, RUTHERFORD-SMITH 571 (BR, K, LISC, M, PRE, SRGH; ± approach. subsp. *ternatum*); – , Trelawney Stn., JACK 251 (LMA, MO, PRE); – , 'Great Dyke Pass', Banket- Salisbury (= Harare) rd., PUFF 790123-5/1 (WU); Gokwe D., Gokwe, BINGHAM 1211 (BR, K, SRGH). – C: Salisbury (= Harare) D., Christonbank Nature Res., LOVERIDGE 1047 (SRGH), PUFF 790123-1/3 (BR, J, W, WU); – , Salisbury and surroundings, BRAIN 9200 (MO, PRE, SRGH), EYLES 5219 (K), 6183 (BR, K, S), 6999 (K, SRG), FLANAGAN 3196 (BOL, PRE), GILLILAND 123 (BM, K, SRGH), KOLBE 4248 (BOL, K), PUFF 790122-1/1 (BR, J, W, WU), RAND 475 (BM); Goromonzi D., Mermaid's Pool (as 'Salisbury D.'), WILD 1104 (K, SRGH); Marandellas, CORBY 89 (K, SRGH), DEHN 55 [1941] (M), 55 [1942] (SRGH); – , Ruzawi R., EYLES 4370 (K, SRGH); Makoni D., ca 20 km E of Headlands on Rusape rd., PUFF 790129-2/2 (W, WU); – , Rusape, DEHN 55/52 (K, LISC, MO, S, SRGH), 55'/52 (BR, K, M); Gwelo D., Mlezu School Farm, ca 29 km SSE of Queque, BIEGEL 992 (SRGH); – , ca 12 km S of Gwelo, BIEGEL 1948 (K, MO, SRGH); Selukwe D., S of Selukwe, on Great Dyke, WILD 6376 (K, LISC, SRGH; ± intermed.). – E: Inyanga D., nr. Inyanga village, FRIES, NORLINDH & WEIMARCK 2524 (PRE, SRGH), NORLINDH & WEIMARCK 4599 (B, MO, SRGH, WAG; COI – atypical), s.n. (BR, K, M, SRGH); – , ca 6 km from Inyanga village on Troutbeck rd., RUSHWORTH 737 (MO, SRGH); Umtali D., Odzani River valley, TEAGUE 31 (BOL, K, STE), 58 (BOL, K); – , – , ca 1 km from Penhalonga-Watsomba rd., on rd. to Stapleford, PUFF 790127-4/2 (BR, J, W, WU); – , Odzi Ardwell Mine, WILD 7558 (K, LISC, SRGH), 7560 (K, LISC, PRE); Melsetter D., Farm Silverstream, FISHER 1483 (NU, PRE; ± atypical). – S: 'Victoria' (Distr.?), MONRO 326 (BM); Victoria D., Kyle National Park, GROSVENOR 646 (SRGH). – W: Bulawayo D., Bulawayo, ROGERS 13681 (BOL, J, PRE); Matobo D., Matopos, KOLBE 4351 (BOL, K). – No locality given: HISLOP 57 (K).

20. ***Anthospermum rosmarinus*** K. SCHUM. in Bot. Jahrb. **30**: 417 (1901); VERDC. in Fl. Trop. E. Afr., *Rubiaceae* 1: 329 (1976).

Type: Tanzania, Njombe Distr., "Kingagebirge: Landschaft Ussangu, ... am Muigi-Berg", GOETZE 1010 (B, holo. †; BM, BR, E, K, iso.!).

Non-dioecious or ± dioecious (?; ♂, sometimes ♂ + odd ⚥, ♀; also ⚥ or ⚥ + ♀?), stout, erect shrubby herbs or shrubs ca 70–200 cm tall. Stems woody at least near the base and ca 10–? mm in diameter, unbranched or ± sparsely branched; younger parts ± terete, greyish, densely covered with whitish-grey spreading hairs ca 0.3–0.7 mm long, older parts reddish-grey to dark grey, becoming glabrescent, epidermis peeling; internodes (usually) shorter than the leaves, plants densely leafy above. Leaves in whorls of 3 or (less commonly) decussate, pseudo-verticillate; blades ± erect to spreading, (20) 30–50 × 3–7 (10) mm, linear-lanceolate to linear-oblong, narrowed to the base, both surfaces, but particularly the midrib below, densely covered with whitish(-grey) spreading hairs ca 0.2–0.6 mm long; margins ± flat to strongly revolute; apices ± acute; petioles subobsolete; stipular sheaths hairy, broadly cup-shaped, ca 1–2 mm long, with (5) 3 (1) hairy, ± filiform setae, the median ca (1.4) 2.3–6 mm long. Inflorescences ± inconspicuous, made up of ± sessile, several- to ± few-flowered axillary clusters. Flowers subsessile, ♂, ⚥, ♀. Corolla 4-merous, yellowish(-green), (densely) pubescent on the outside. ♂, ⚥: tube 1–2 mm long, narrowly funnel-shaped, lobes (1.7) 2–2.7 × 0.6–0.8 mm, lanceolate, recurved; stamens 4, filaments 2–3 mm long, anthers 1.7–2.2 × 0.4–0.7 mm; whitish; ♂ with minute rudimentary ovary and stigmas. ♀: tube 0.5–0.9 mm long, ± cylindrical, lobes 0.3–0.8 × 0.2–0.4 mm, ± lanceolate, erect to ± recurved; style 0.5–1.5 mm long; stigmas 2, 3–5.4 mm long (often shorter in ⚥), greyish; ovary ca 0.5–1 × 0.3 mm, hairy, not crowned by distinct calyx lobes. Fruits reddish(-brown), supported by narrow, U-shaped carpophores ca 1.2–1.7 mm long; each mericarp 1.9–2.4 × 0.8–1.2 mm, narrowly oblong to obovate, dorsal side shallowly convex, ± densely covered with whitish, spreading hairs ca 0.3–0.5 mm long, ventral side plane to ± concave and with a median vertical ridge; commissure large and broad, narrowed to the base and ± pointed; mericarps crowned by 2 very inconspicuous, ± triangular calyx lobes ca 0.3 × 0.1 mm or calyx lobes lacking.

Chromosome Number: Unknown.

Average Pollen Diameter (Fig. 48): 28–28.6 μm.

Habitat: In rocky grassland; also on bracken (*Pteridium*) hillsides and in roadside vegetation. – Ca 2 000–2 400 m.

Distribution (Map, Fig. 76 d): SW Tanzania and SE Zaire. From the Njombe Distr. N to Mpanda Distr. and – across (W of) Lake Tanganyika – in the Baudouinville (Moba) Terr.

Critical Remarks: A. rosmarinus is doubtlessly closely allied to *A. ternatum* subsp. *randii* (19 b., above). It differs in having smaller (♂, ⚥) corolla tubes, in its trend to dioecy, in its more shrubby habit and, above all, in its smaller mericarps which have large, rather broad commissures. Micromorphologically, *A. rosmarinus* differs from *A. ternatum* subsp. *randii* in having conspicuous cuticular striations on the lower leaf surface (compare Fig. 18 c and d). Some other morphological characters (such as leaf sizes and shapes), however, overlap with subsp. *randii*. For this

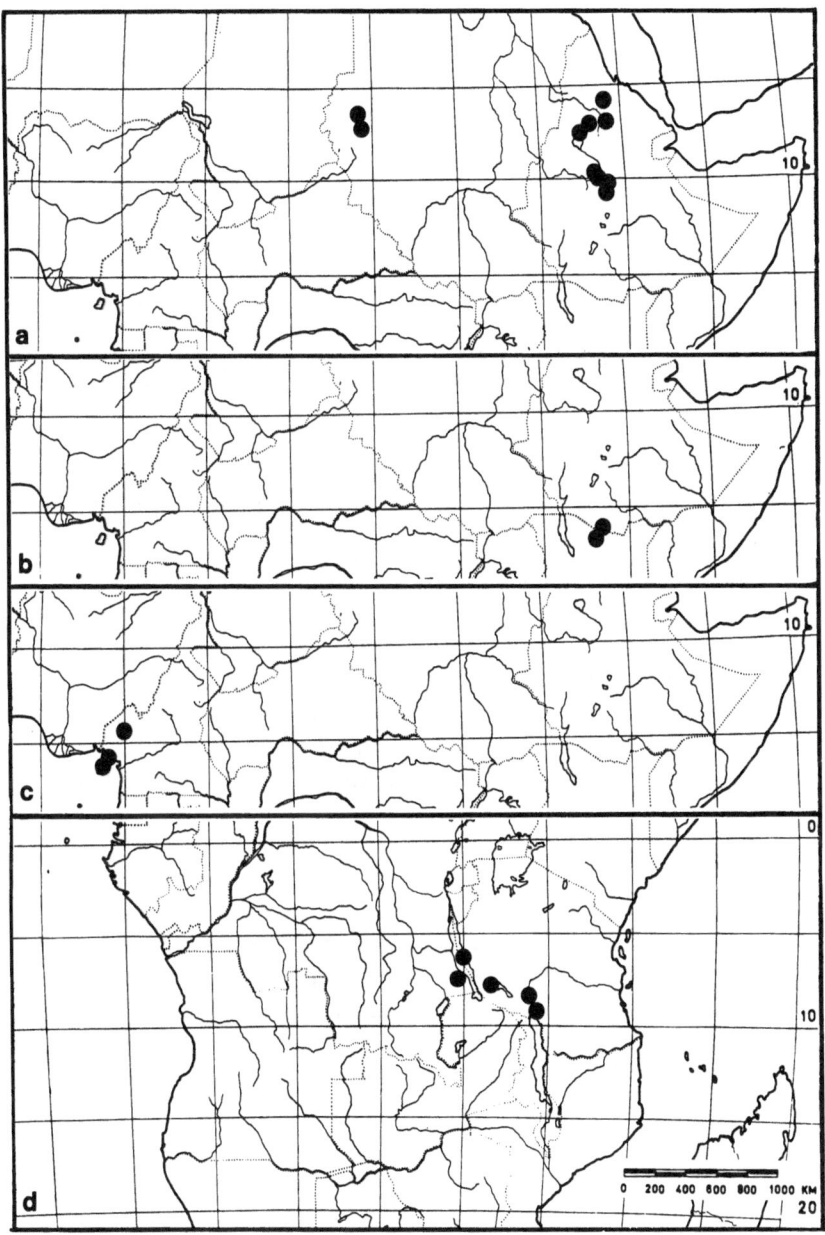

Fig. 76. Distribution of *Anthospermum. a A. pachyrrhizum, b A. villosicarpum, c A. asperuloides, d A. rosmarinus*

reason, the latter ("*A. randii*") had, in the past, been confused with *A. rosmarinus*. VERDCOURT (1976) placed the collections from the Mpanda and Ufipa Distr., (T4; HARLEY 9529, RICHARDS 8666) with "*A. randii*" although, as regards the RICHARDS collection, he must have had some doubts (p. 329: "... RICHARDS 8666 I had referred to *A. rosmarinus* but I now believe it is merely a large-leaved *A. randii*. ..."). I am, however, quite convinced that this was not correct.

    A. rosmarinus is rather uniform in its characters and has a well defined distribution range which does not overlap with that of *A. ternatum* subsp. *randii*[1]. It is considered a distinct, separate species which is believed to have evolved in isolation in SW Tanzania and adjacent parts of Zaire.

Collections

    Tanzania. – **T4**: Mpanda D., below Kungwe Mt., HARLEY 9529 (BR, K); Ufipa D., Mbisi forest, RICHARDS 8666 (K). – **T7**: Makete D., 8 km from Matamba on Inyala rd., IWARSSON, ABDALLAH, MACHA & MAGOGO 1146 (UPS); Njombe D., Kinga Mts., Muigi (= Mwigi) Mt., GOTZE 1010 (BM, BR, E, K).
    Zaire. – **Haut-Katanga**: Mt. Morumbe, KASSNER 2954 (BM, K, P).

21. *Anthospermum villosicarpum* (VERDC.) PUFF, stat. nov.

    ≡ *A. herbaceum* L. f. var. *villosicarpum* VERDC. in Kew Bull. **30**: 299 (1975), in Fl. Trop. E. Afr., *Rubiaceae* 1: 328 (1976).
    Type: Kenya, Northern Frontier Province, Furroli, GILLETT 13946 (K, holo.!).

    Non-dioecious (⚥ or ⚥ + ♀), few-stemmed, ± narrowly cylindrical dwarf shrubs or subshrubs with a woody base. Stems to 60 cm long, ca 3–4 mm in diameter near the base and ca 2 mm in the midstem region, erect or ± ascending, with few, usually short lateral branches; younger parts terete, densely covered with white spreading hairs ca 0.3 mm long; internodes frequently shorter than to as long as the leaves. Leaves in whorls of 3, pseudo-verticillate; blades 10–13 × 1–3 mm, narrowly elliptic-lanceolate, with bristly hairs to ca 0.3 mm long above and particularly on the midrib below; margins strongly revolute; apices acuminate to mucronate; petioles subobsolete; stipular sheaths sometimes slightly hairy, cup-shaped, ca 0.5–0.7 mm long and 1–1.6 mm wide, with a median seta ca 1.1–1.5 mm long, occasionally with an additional smaller seta on either side. Inflorescences inconspicuous, made up of ± sessile, several- to

---

    [1] The distribution of *A. rosmarinus*, however, overlaps with the E part of the distribution range of *A. ternatum* subsp. *ternatum*, but in the area in question the two are so markedly different that they cannot be confused.

± few-flowered axillary clusters. Flowers subsessile, ♂, ♀. Corolla 4-merous, greenish or greenish-yellow, somewhat hairy or ± glabrous. ♂: tube 0.8–1 mm long, ± narrowly funnel-shaped, lobes 1.8–2 × 0.4–0.5 mm, lanceolate, recurved; stamens 4, filaments 0.8–1.4 mm long, anthers 0.8–1.1 × 0.2 mm; gynoecium as in ♀ but stigmas only 1.3–1.4 mm long. ♀: tube 0.2–0.3 mm long, ± cylindrical, lobes ca 0.4 × 0.2–0.3 mm, ± linear-lanceolate, ± erect; style 0–0.3 mm long; stigmas 2, –2.7 mm long, greyish; ovary 1.3–1.7 × 1–1.2 mm, densely hairy, not crowned by distinct calyx lobes. Fruits brown to reddish-brown, supported by small, ± U-shaped carpophores; each mericarp 2.4–2.7 × 1.2–1.3 mm, ± obovate, dorsal side convex, densely covered with white spreading hairs ca 0.2–0.3 mm long, ventral side concave and with a median vertical ridge; commissure large and broad; mericarps not crowned by distinct calyx lobes.

Chromosome Number: Unknown (but presumably diploid).
Average Pollen Diameter (Fig. 49): 24.6–25.4 µm.
Habitat: Growing on granite or sandstone ridges and occurring in *Olea- Juniperus* scrub or in "ericoid" associations with *Struthiola, Lobelia* etc. – Ca 1 800–2 000 m.
*Distribution (*Map, Fig. 76 b): N Kenya and S Ethiopia. Only known from two collections from the Kenya – Ethiopia border area.

Critical Remarks: The type collection is a rather atypical, stunted specimen which must have been collected in a heavily grazed, trampled area. The second known collection comprises well developed (sub)shrubby plants which clearly show that *A. villosicarpum* can neither be associated with nor be considered closely related to *A. herbaceum*. VERDCOURT, when describing "*A. herbaceum* var. *villosicarpum*" (1975), was misled by the stunted habit of the type specimens (the only collection known to him). The species either is very rare or has become rare – possibly due to overgrazing by domestic stock (pers. observations on the Mega plateau; in spite of an intensive search, no plants could be detected).

It is believed that *A. villosicarpum* is, on account of its habit, its hairy leaves and its hairy mericarps with broad commissures and without distinct calyx lobes, allied to *A. ternatum* (19.) and *A. rosmarinus* (20., above), and perhaps also to *A. asperuloides* (see comments, p. 297).

Collections

Ethiopia. – **Sidamo:** Mega, GILLETT 14448 (BR, K).
Kenya. – **K 1:** Northern Frontier Prov., Furroli, GILLETT 13946 (K).

22. Anthospermum asperuloides HOOK. f. in J. Linn. Soc. (Bot.) **6**: 11 (1862)
and **7**: 197 (1864); HIERN in OLIVER, Fl. Trop. Afr. **3**: 230 (1877);
HEPPER in HEPPER, Fl. West Trop. Afr., ed. 2, **2**: 223 (1963).
Type: Fernando Po, in cacumine Clearance Peak, alt. 10 000 ped.,
MANN ("593", K, holo.!)[1].

= *A. cameroonense* HUTCH. & DALZ. in HUTCHINSON & DALZIEL, Fl.
West Trop. Afr., ed. 1, **2**: 136 (1931); HEPPER in HEPPER, Fl. West Trop.
Afr., ed. 2, **2**: 223 (1963).
Type: Cameroon Mt., 12 000 ft., MANN 1290 (K, holo.!; GH × 2,
GOET, S, W, iso.!).

Non-dioecious (⚥ or ⚥ + ♀) subshrubs with a slightly woody base or
(short-lived?) shrubs ca (30) 60–120 cm tall. Stems to ca 5 mm in diameter
near the base and ca 1.5–3 mm in the midstem region, ± erect to
ascending, sparsely to much-branched above, branches ± ascending;
younger parts ± terete, reddish-brown, densely covered with papillae or
spreading whitish hairs to 0.2 mm long, older parts reddish-brown to dark
grey, glabrescent, epidermis peeling; internodes mostly shorter than to as
long as the leaves. Leaves in whorls of 3 or sometimes also decussate (on
one and the same plant), pseudo-verticillate; blades 5–10 (15) × 0.8–
1.5 (2) mm, (narrowly) oblanceolate to ± linear-lanceolate, often the
revolute to ± flat margins and sometimes also the midrib below with
whitish hairs ca 0.1–0.2 (–0.6) mm long, occasionally also some scattered
hairs on the upper surface; apices acuminate to apiculate, often curved;
petioles subobsolete; stipular sheaths broadly cup-shaped, ca (0.7) 1–
1.5 mm long, with (1) 3–5 (6) gland-tipped setae[2] ca 1.4–2.4 (3) mm long,
the lateral ones often much shorter than the others. Inflorescences
inconspicuous, made up of ± sessile, few- to several-flowered axillary
clusters. Flowers subsessile, ⚥, ♀. Corolla 4-merous, pale yellow-green,
creamy yellow or purplish-green, somewhat hairy outside or glabrous. ⚥:
tube 0.5–0.7 mm long, ± narrowly funnel-shaped, lobes 1.4–2 × 0.6–
0.7 mm, ± lanceolate, recurved; stamens 4, filaments ca (0.7) 1–1.4 mm
long, anthers 0.9–1.7 × 0.3–0.6 mm; gynoecium as in ♀. ♀: tube ca 0.2 mm

---

[1] HOOKER (1862) did not list MANN's collection number (it appears that the
numbers were added later by someone else), but "593" with the label "growing on
the very top of Clearance Peak, Fernando Po, December 1860" is without doubt
the type. In 1864, HOOKER cited two MANN collections (the before-mentioned
Fernando Po collection and one from "Cameroons Mts., at 7–12 000 ft."), again
giving no collection numbers. This additional collection (MANN "1290") became
the type of *A. cameroonense*.

[2] The setae on the stipular sheaths of the short shoot leaves clustered in the
axils of the long shoot leaves may give the impression that more setae are present!

long, cylindrical, lobes ca 0.6–0.7 × 0.2–0.3 mm, linear-lanceolate, ±
erect; style 0; stigmas 2, 1.4–2 mm long, greyish; ovary ca 1.2–1.4 × 0.6–
1 mm, not crowned by distinct calyx lobes. Fruits reddish-brown,
supported by ± broad, U-shaped carpophores less than half as long as the
fruits; each mericarp 1.9–2.7 × 1–1.4 mm, oblong, dorsal side convex, ±
densely covered with spreading whitish hairs 0.1–0.3 mm long, papillate or
± glabrous, ventral side ± plane and with a median vertical ridge;
commissure large; mericarps crowned by 2 indistinct, ± triangular calyx
lobes to ca 0.3 × 0.2 mm or calyx lobes lacking. – Figs. 35 i, 37 f, 77.

Chromosome Number: Unknown.
Average Pollen Diameter (Fig. 48): 27.6–28.8 µm.
Habitat: In ericaceous scrub at the edge of gullies or in gully woodland,
often associated with *Phillipia, Nepeta robusta, Hypericum revolutum*
(= *H. lanceolatum*) and *Blaeria*; sometimes also on grassy slopes ("upland
grassland"). For a more detailed description of the vegetation in which *A.
asperuloides* occurs see BOUGHEY (1955: 146). – Ca 2 300–3 050 (3 650) m.
Distribution (Map, Fig. 76 c): Fernando Po (Clearance Peak) and in
the Cameroon (Cameroon Mt. and Dschang Distr. in E Cameroon).

Critical Remarks: It is, in my opinion, not justified to maintain *A.
asperuloides* and *A. cameroonense* as two distinct species. The characters
used by HEPPER (1963) to distinguish the two taxa are not reliable: leaf size
and shape overlap, and the leaves are not – as HEPPER states – always
glabrous in *A. cameroonense* (e.g., BRENAN 9531 or BRETELER & al. MC-
108, with unusually long hairs). There is no difference in stipule structure,
A. asperuloides does not typically have stipular sheaths only having two
"segments" (setae). Considering the variability of the fruit indumentum in
other *Anthospermum* species (e.g., *A. welwitschii, A. ternatum*, and others),
the characters "fruits glabrous" vs. "fruits densely and softly tomentose"
alone are hardly adequate to separate the two species. The fruit
indumentum, furthermore, varies: While the type of *A. cameroonense*
(MANN 1290) has fruits which are densely covered with hairs ca 0.3 mm
long, the fruits of PREUSS 782 have much shorter hairs (hardly 0.1 mm
long), and KEAY FHI 28592 (like the previous two collections originating
from Cameroon Mt.) has glabrous-papillate fruits. The papillae of the
latter appear to be epidermis cells which have not elongated quite enough
to become hair-like (compare Figs. 77 a–c); in addition, some fruits of that
collection do have mericarps with odd hair-like epidermal cells which in
length approach those of PREUSS 782. Note that there are also papillae
amongst the mericarp hairs of MANN 1290 (Fig. 77 a)!

The affinities of *A. asperuloides* require some comments: It is well

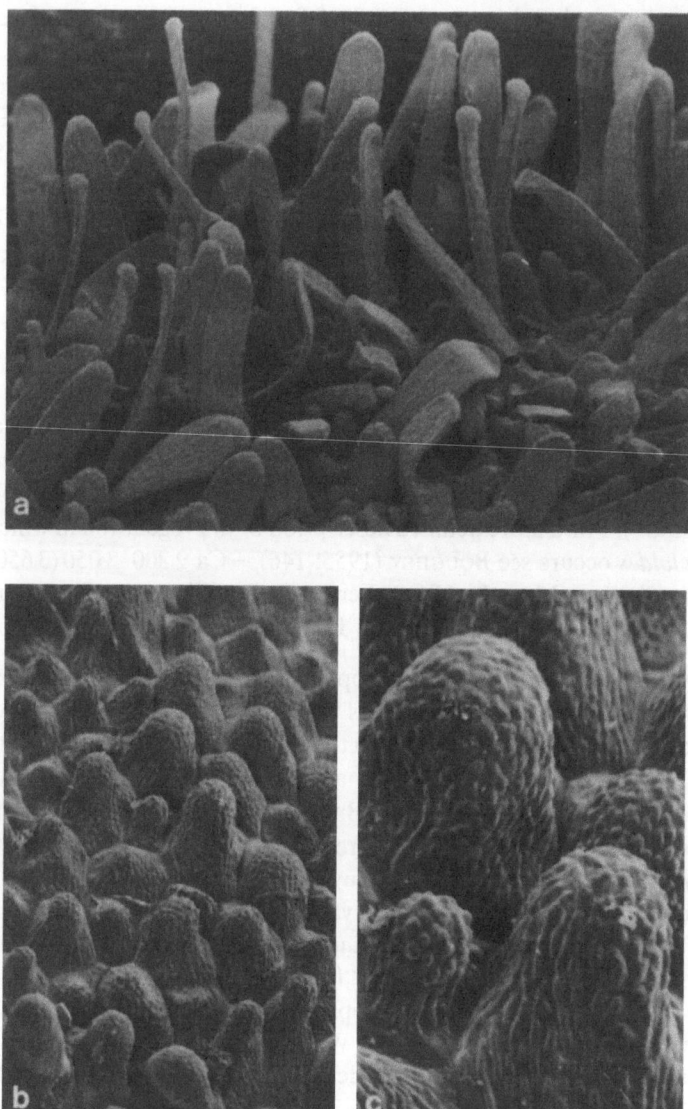

Fig. 77. *Anthospermum asperuloides*, SEMgraphs of mericarp surfaces. *a* hairs
(flattening artifical?) and papillae (MANN 1290). *b–c* papillae with surface
sculpturing similar to that of hairs (KEAY FHI 28592). – Further explanations in
the text. Herbarium material; *a* ×200, *b* ×300, *c* ×900

known that numerous afromontane taxa which are centered in tropical E
Africa and often also extend into S Africa, frequently have "W extensions" as well and occur on Cameroon Mt., Fernando Po and other higher
mountains of W Africa (see, for example, WHITE 1978). In some of these
plant groups, races which occur in the more isolated "West African
Regional Mountain System" sensu WHITE (1978) are in the process of
evolving into distinct species (cf. for example, *Galium simense*: PUFF
1979 a, 1979 b: map 581; PUFF & MANTELL 1982 b) or have become W
African endemic species. Taking this into consideration it is very tempting
to associate *A. asperuloides* with *A. usambarense* (2.) and its allies which
are widely distributed in afromontane areas of the African mainland and
comparable regions in Madagascar. There is, however, strong evidence
against such an affinity: The *A. usambarense* group is characterized by a
large shrubby habit and *dioecy,* while plants of *A. asperuloides* are rather
short-lived (?) and hardly true shrubs; the plants, furthermore, are either ⚥
or ⚥ + ♀. While odd ⚥ flowers may occur in the *A. usambarense* group (on
♂ plants, some flowers may "revert" to ⚥), it appears unlikely that purely ⚥
or − even more unlikely − gynomonoecious plants may result (see in this
context C.12.2.). Thus a link between *A. asperuloides* and the *A.
usambarense* group is hardly imaginable. Instead, there may well be a
relationship to the group of species around *A. ternatum* (*A. villosicarpum,
A. rosmarinus*; 19.–21.). That group is characterized by having, next
to pure ♀, ⚥ flowers which have − and this may also be
significant − ± cylindrical to narrowly funnel-shaped corolla tubes (as
opposed to broadly funnel-shaped tubes in the *A. usambarense* group). It,
furthermore, mostly comprises rather short-lived and not distinctly
shrubby species with frequently ternately arranged and hairy leaves. None
of these species, however, has a distinctly afromontane distribution range
and, in that respect, they differ from *A. asperuloides*. *A. villosicarpum* in
particular appears to agree with *A. asperuloides* in numerous characters:
its hairy fruits and its carpophores are almost identical to those of *A.
asperuloides* in size and shape. Also the somewhat hairy ⚥ flowers with
their relatively short but narrow corolla tubes are very similar.

Collections

Fernando Po. − Clearance Peak, MANN 593 (K).
   Cameroon. − Cameroon Mt., MANN 1290 (GH × 2, GOET, K, S, W), PREUSS
782 (BM, COI, M); −, Mann's Spring, BRENAN 9531 (BM, BR, K, P); −, around
Hut II (= Johann-Albrechts Hütte), BOUGHEY G.C 10679 (K), BRETELER, DE
WILDE, LEEUWENBERG & LETOUZEY MC-108 (G, K, LISC, P × 2, WAG), KEAY FHI
28592 (K), MORTON K 796 (K, WAG); Dschang D., Mt. Bamboutos nr. Dschang,
SANFORD 5590 (K, MO).

   20*

23. Anthospermum ibityense PUFF, spec. nova, mericarpiis calycis lobis conspicuis plus minusve taeniatisque coronatis a ceteris speciebus Anthospermi (praeter *A. longisepalo*) differt. Foliis fructibusque glabris ab *A. longisepalo* preclare distinguitur.
Type: Madagascar, Mt. Ibity, PUFF 800730-1/2 (WU, holo.; BR, K, P, iso.).

Non-dioecious [♀ or ♀ (perhaps also ♀ + ♀ and ♀ + ♂?)], ± erect and cylindrical, several-stemmed dwarf shrubs ca 25–40 (50) cm tall. Stems ca 2–5 (8) mm in diameter near the base and ca 1–1.5 mm in the midstem region, much-branched, branches ± erect; younger parts ± 4-angled, often reddish tinged, scabrid, older parts terete, dark grey, glabrescent; internodes shorter than the leaves. Leaves decussate, usually pseudo-verticillate; blades 8–12 × 0.7–1 mm, linear(-lanceolate), glabrous, margins flat or ± revolute in dried material, midrib below inconspicuous; apices acute; petioles subobsolete; stipular sheaths ± scabrid, cup-shaped, ca 0.7–1.4 mm long, usually with a median seta ca 0.2–1 mm long. Inflorescences inconspicuous, made up of ± sessile, ca 6-2-flowered axillary clusters. Flowers subsessile, ♀ (♂?), ♀. Corolla 4-merous, yellowish(-green) or creamy yellow, glabrous, often reddish(-purplish) tinged on the outside. ♀ (♂?): tube 1.2–1.8 mm long, ± cylindrical, lobes 1.3–1.5 × 0.7–0.8 mm, (linear-)lanceolate, recurved; stamens 4, filaments ca 1.4–2.5 mm long, anthers 1.1–1.4 × 0.2–0.3 mm; ♀: gynoecium as in ♀ but stigmas shorter. ♀: tube 0.5–0.7 mm long, cylindrical, lobes 0.7–0.9 × 0.4–0.5 mm, linear(-lanceolate), ± erect; style ± 0; stigmas 2, ca 3–3.5 mm long, greyish; ovary ca 1.2–1.8 × 0.6–0.8 mm, glabrous, crowned by 4 conspicuous calyx lobes. Fruits reddish-brown to greyish, supported by small, ± U-shaped carpophores; each mericarp 2–2.4 × 0.8–1 mm, ± oblong to narrowly elliptic, dorsal side convex, glabrous, ventral side plane to ± concave and with a median vertical ridge; commissure ± large; mericarps crowned by 2 narrow, ± band-like spreading to recurved calyx lobes ca 1.2–1.7 (2) × 0.2–0.3 mm.

Chromosome Number (Table 3): 2 n = 22.
Average Pollen Diameter: Unknown.
Habitat: Apparently confined to quartzite and always found amongst rocks or in crevices. In the type locality growing, inter alia, with *Pentachlaena latifolia, Pachypodium brevicaule, Cynanchum perrieri, Kalanchoe* spp., *Nematostylis anthophylla* (= *N. loranthoides*), *Aloe compressa, A. capitata* var. *quartziticola, Angraecum* sp., *Bulbophyllum* sp., *Cussonia bojeri, Agauria polyphylla* and *Helichrysum ibityense*. – Ca 1 900 m.

Distribution (Map, Fig. 67 d): Central Madagascar. So far only known from Mt. Ibity.

Critical Remarks: See *A. longisepalum* (24., below).

Collections

Madagascar. – **2047**: Mt. Ibity, Puff 800730-1/2 (BR, K, P, WU).

24. *Anthospermum longisepalum* Homolle ex Puff, spec. nova, mericarpiis calycis lobis conspicuis plus minusve taeniatisque a ceteris speciebus *Anthospermi* (praeter *A. ibityensi*) differt. Foliis fructibusque papillatis vel pilis brevissimis tectis ab *A. ibityensi* praeclare distinguitur.
Type: Madagascar, Isalo massif, W of Ranohira, Puff 800814-2/1 (WU, holo.; BR, K, P, iso.).

Dioecious, many-stemmed dwarf shrubs, ca 7–15 (25) cm tall, with thick, woody tap roots. Stems ascending to erect, ca 2–4 mm in diameter near the base and ca 1–2 mm in the midstem region, much-branched; younger parts ± 4-angled, often reddish, scabrid, older parts terete, usually dark grey, glabrescent, epidermis peeling; internodes shorter than to as long as the leaves. Leaves decussate, sometimes pseudo-verticillate; blades 7–10 (15) × 1 mm, linear-lanceolate, densely covered with whitish-grey papillae or very short hairs; margins revolute and, in dried material, closely adpressed to the prominent and broad midrib (blades thus ± 2-furrowed below); apices acute; petioles subobsolete; stipular sheaths scabrid, cup-shaped, ca 0.7–1 mm long, with a median seta ca 0.2–0.7 mm long. Inflorescences inconspicuous, made up of ± sessile, (very) few-flowered axillary clusters (in ♂ plants flowers sometimes solitary at the nodes). Flowers subsessile, ♂, ♀. Corolla 4-merous, greenish-yellow, glabrous, sometimes reddish-tinged on the outside. ♂: tube 0.3–0.5 mm long, ± cylindrical, lobes 1.3–1.6 × 0.4–0.5 mm, (linear-)lanceolate, recurved; stamens 4, filaments 0.9–1.2 mm long, anthers 0.8–1.4 × 0.2–0.3 mm; rudimentary ovary small but well discernible and crowned by 4 ± conspicuous calyx lobes. ♀: tube 0.2–0.5 mm long, cylindrical, lobes 0.2–0.3 × 0.1 mm, linear(-lanceolate), erect; style ± 0; stigmas 2, 1.7–2.9 mm long, greyish; ovary ca 0.8–1.5 × 0.4–0.7 mm, densely papillose, crowned by 4 (very rarely 5) conspicuous, long calyx lobes. Fruits reddish-brown, supported by small, ± U-shaped carpophores; each mericarp 1.7–2 × 0.6–0.8 mm, ± oblong to narrowly elliptic, dorsal side convex, densely covered with whitish papillae or very short hairs, ventral side ± plane and with a median vertical ridge; commissure ± large; mericarps crowned by 2 (or, very rarely, 3) scabrid, narrow, band-like, variously curved calyx lobes ca 2.2–3.7 × 0.2–0.3 mm. – Figs. 36 e–f.

Chromosome Number (Table 3): 2 n = 22.
Average Pollen Diameter: Unknown.
Habitat: Apparently confined to sandstone and always found amongst rocks or in crevices. In the type locality growing, inter alia, with *Euphorbia millii* (-group; *"E. splendens"*), *Mundulea phylloxylon, Xerophyta dasy-lirioides, Kalanchoe synsepala, Pachypodium rosulatum* var. *gracilis, Cynanchum* sp. and *Ceropegia dimorpha.* – Ca 800–1 000 m.
Distribution (Map, Fig. 67 e): Madagascar. So far only known from the Isalo massif.

Critical Remarks: A. ibityense (23., above) and *A. longisepalum* form a closely related (vicarious) species pair. The former is confined to quartzite, whereas *A. longisepalum* occurs on sandstone. The two species seem to occupy a rather isolated position within the genus and do not appear to have any apparent close or more distant allies in either Madagascar or on the African mainland. The species pair is characterized by having fruits which are crowned by long and narrow band-like persistent calyx lobes (cf. Fig. 36 e; unique in the genus!) and by corollas with cylindrical tubes [as in, for example, *A. herbaceum* (25.) or *A. ternatum* (19.), although not as long as in the latter].
A. ibityense is distinguished from *A. longisepalum* by its more erect habit, its glabrous leaves and glabrous fruits (with calyx lobes which are shorter than in *A. longisepalum*). In addition, the anatomy of the leaves of the two species is strikingly different (compare sections Figs. 20 a and 21 d). *A. ibityense* has larger flowers and also ⚥ plants, whereas *A. longisepalum* appears to be dioecious[1].

Collections

Madagascar. – **2245**: Isalo massif, PERRIER DE LA BÂTHIE 3884 (P); –, W of Ranohira, HUMBERT & SWINGLE 5016 (P), PUFF 800814-2/1 (BR, K, P, WU).

25. *Anthospermum herbaceum* L. f., Suppl. Pl.: 440 (1781); MURRAY, Syst. Veg., ed. 14: 919 (1784); BRENAN in Mem. N.Y. Bot. Gard. **8**: 455 (1954); AGNEW, Upl. Kenya Wild Fl.: 407, Fig., p. 406 (1974); VERDC.

---

[1] All the available material is either fruiting or in the late flowering stage so that it is not entirely certain whether the species is strictly dioecious. There are some indications that ♂ plants may also bear odd (?) ⚥ flowers (dried up remnants of a few flowers with stigmas were found on a plant with otherwise ♂ flowers).

in Fl. Trop. E. Afr., *Rubiaceae* 1: 325, Fig. 46 (1976); PUFF in Fl. S Afr.
31, 1 (2): 22 (1986).
Type: [South Africa, Cape Prov.:] "Cape of Good Hope", THUNBERG
s.n. (LINN 1233.5, holo.!).

= *A. lanceolatum* THUNB., Prod. Pl. Cap.: 32 (1794), Fl. Cap., ed.
SCHULTES: 157 (1823); CRUSE, Rub. Cap.: 12 (1825), in Linnaea **6**: 12
(1831); SONDER in HARVEY & SONDER, Fl. Cap. **3**: 30 (1865); DE WILD.,
Pl. Bequaert. **2**: 301 (1923); ROBYNS, Fl. Spermatoph. Parc Nat. Albert
2: 371 (1947); COMPTON, Fl. Swaziland: 587 (1976).
Type: South Africa, Cape Prov., THUNBERG (sheet 23314[1], UPS,
holo.!, K, PRE, WU – photo!; G, S × 3, iso.!).

= *A. muriculatum* HOCHST. ex A. RICH., Tent. Fl. Abyss. **1**: 345 (1847);
HIERN in OLIVER, Fl. Trop. Afr. **3**: 229 (1877).
Syntypes: Ethiopia, "Prov. Ouedgerate", [QUARTIN-DILLON &] PETIT
s.n. (P, lecto.!, selected here); – , "Koubi", SCHIMPER 732 (BR!, G!, K!,
S!; P! – mixed with *A. pachyrrhizum*)[2].

= *A. ferrugineum* ECKLON & ZEYHER, Enum. Pl. Afr. Austr.: 366
(1836).
Type: South Africa, Cape Prov., '... "Van Stadensriviersberge"
(Uitenhage), prope "Grahamstown" (Albany) et ad flumen "Maka-
sanirivier" prope montem "Chumiberg" (Adelaide)', ECKLON &
ZEYHER 2309 (SAM, holo.!; BOL, FI, G × 2, GOET, M, MO, S, W × 2,
iso.!).

= *A. hedyotideum* SONDER in HARVEY & SONDER, Fl. Cap. **3**: 30 (1865).
Syntypes: South Africa, Cape Prov., "Caffraria, Kreili's Country",
BOWKER [595] (S, lecto.!, selected here; K!, PRE-photo!); – , Keis-
kamma, DRÈGE s.n. ("Herb. DRÈGE, SONDER"; not located).
≡ *A. lanceolatum* THUNB. var. *hedyotideum* (SONDER) O. KUNTZE, Rev.
Gen. Pl. **3** (2): 117 (1898).

= *A. lanceolatum* THUNB. var. *latifolium* SONDER in HARVEY & SONDER,
Fl. Cap. **3**: 30 (1865).
[– *A. latifolium* E. MEYER in DRÈGE in Flora **26**, Bes. Beigabe: 164
(1843), nom. nud.]
Syntypes: South Africa, Cape Prov. and Natal: "In dist. Uit [enhage],

---

[1] = *A. lanceolatum* 'α'; *A. lanceolatum* 'β' (sheet 23315) is *Carpacoce
spermacocea*!

[2] "Prov. Quedgerate" (also spelled Wodjerat, Wogera, Wodgera of Oughera)
is probably today's Wegera Awraja in Gonder Admin. Region; "Koubi" is Mt.
Kubbi, E of Axum, in Tigre Admin. Region (see GILLETT 1972).

Albany and in Caffraria, E. & Z., Zᴇʏʜ. 2716 ex parte; near Natal, Dʀèɢᴇ, Gᴜᴇɪɴᴢɪᴜs, 468. Oct. (Herb. Tʜ., Sᴅ., D.)"[1].

— *A. nodosum* E. Mᴇʏᴇʀ in Dʀèɢᴇ in Flora **26**, Bes. Beigabe: 164 (1843), nom. nud. [South Africa, Natal, "Port Natal" (Durban), Dʀèɢᴇ s.n. (E!, G × 3!, GH!, K!, LY!, S!)].

= *A. mildbraedii* K. Kʀᴀᴜsᴇ in Wiss. Ergebn. Dt. Zentral-Afr. Exped. **2**: 341 (1914).
Type: Rwanda, W of Lake Mohasi, Mɪʟᴅʙʀᴀᴇᴅ 497 (B, holo. †; BR, iso.!).

Non-dioecious (⚥, ⚥ + ♀, ♀, ⚥ + ♂, ♂ + ⚥ + ♀, less commonly ♂ or ♂ + ♀), several- to many-stemmed perennial herbs, sometimes rather short-lived (?) herbs, or ± subshrubs, somewhat woody near the base and with often ± thick, woody roots. Habit variable: plants scrambling, straggling or trailing and with stems to 2.5 (3) m long; sometimes ± erect, cylindrical to rounded, to ca 30 (50) cm tall and with ± thick, almost rosette-like bases; occasionally low and ± mat- or cushion-forming with shoots ca (2) 7.5–15 (20) cm long. Stems to ca 5 mm in diameter near the base in the more robust forms and to ca 0.7–2 mm in the midstem region, distinctly 4-angled to ± terete; stems unbranched to much-branched, branches often ± regular and paired, spreading to ascending, frequently with short branches of a higher order; younger parts greyish-brown, reddish-brown to dark purplish-brown, ± densely covered with short hairs or papillae less than 0.1 to 0.3 mm long or with longer, whitish spreading hairs to 0.6 mm long, or sometimes entirely glabrous and shiny; older parts glabrescent, epidermis peeling; internodes mostly (much) longer than the leaves; long shoots often with much-contracted to slightly elongated short shoots bearing rather small leaves. Leaves decussate;

---

[1] Sᴏɴᴅᴇʀ considered *A. ferrugineum* and "*A. latifolium* E. Mᴇʏ." to be synonyms of his var. *latifolium*. "E. & Z." must, therefore, refer to Eᴄᴋʟᴏɴ & Zᴇʏʜᴇʀ 2309 (see *A. ferrugineum*). The collection Dʀèɢᴇ s.n. "*A. latifolium* E. Mᴇʏ." (E!, G × 3!, GH!, MO!, P × 2!, S × 2!, SAM!, W × 3!) is from the Cape Prov. ("Zuurebergen, nr. Bonjesrivier"; locality "V, a, 27" according to Dʀèɢᴇ 1843) and *not* from Natal as stated by Sᴏɴᴅᴇʀ. Sᴏɴᴅᴇʀ may have confused the locality with that of Dʀèɢᴇ s.n. "*A. nodosum* E. Mᴇʏ." (from Port Natal) which he, however, grouped with the "typical" *A. lanceolatum*. Zᴇʏʜᴇʀ 2716 (LY!, S!, SAM × 2!) is from "Vanstadesmountain". The Gᴜᴇɪɴᴢɪᴜs collection in S (here selected as lectotype!), cited as being from "near Natal", is from Port Natal (= Durban) – Sᴏɴᴅᴇʀ himself wrote the label! The collection no. [Gᴜᴇɪɴᴢɪᴜs] "468" has been added in someone else's handwriting. The Gᴜᴇɪɴᴢɪᴜs *s.n.* specimens in G and SAM (sub no. 16062), clearly identical to the plants in S, must also be considered as types.

blades 5–55 × (1) 2–25 mm, ± ovate, ovate-lanceolate, lanceolate to ± linear-lanceolate, cuneate to rounded at the base, ± densely covered with whitish papillae less than 0.1 mm long on both surfaces, with short hairs above and somewhat longer, spreading whitish hairs to 0.2 (0.4) mm long below (especially along the main veins), with whitish (sometimes ± forwardly directed) hairs on the margins and main veins only, with a few hairs near the base only or blades sometimes entirely glabrous; blades often distinctly discolourous; margins flat or sometimes ± revolute; apices acute to ± acuminate; petioles 0.7–6.5 mm long; stipular sheaths at least a little (shortly) hairy, seldom glabrous altogether, cup-shaped, ca 0.9–2.2 mm long, typically with (3–) 5 (–7) filiform setae[1], seldom (especially in stunted forms) with only 1 ± triangular seta, the longest setae (0.3) 0.7–6.1 mm long. Inflorescences inconspicuous to sometimes ± conspicuous (♀ plants); partial inflorescences sometimes ± congested on short lateral branches, made up of ± sessile to somewhat elongated, many- to 6- (less commonly only 2-)flowered axillary clusters. Flowers subsessile or occasionally with pedicels to ca 0.5 mm long (♀); ♂, ⚥, ♀. Corolla 4-merous, greenish, greenish-yellow or yellowish, sometimes reddish-purplish tinged, mostly with whitish papillae to 0.1 (0.3) mm long at least near the tip. ♂, ⚥: tube (1.5) 2–3.7 mm long (in ⚥ often longer than in ♂), cylindrical, lobes (1.5) 2–2.7 (3.4) × 0.3–0.7 mm, ± lanceolate, recurved; stamens 4, filaments exserted for 0.8–1.6 (2.2) mm, anthers (0.9) 1.2–2 (2.5) × 0.2–0.5 mm; ♂: rudimentary ovary small but well discernible, rudimentary stigmas usually also present but hidden in the corolla tube; ⚥: gynoecium as in ♀ but stigmas only ca (2) 2.4–5 (6.4) mm long; flowers occasionally transitional ⚥ → ♀ with rudimentary and ± pollenless anthers to ca 0.5 mm long and corolla tubes ca 0.8–1.7 (2) mm long and lobes 0.7–1.4 (1.7) × 0.2–0.4 mm. ♀: tube 0.3–0.7 (1.2) mm long, cylindrical, lobes 0.3–0.7 (1) × 0.1–0.3 mm, lanceolate to linear-lanceolate, erect to ± recurved; style ± 0–0.8 mm long; stigmas 2, (2) 4–10.2 mm long, greyish-white to greenish; ovary 0.5–1.2 × 0.4–0.7 mm, ± tuberculate, shortly hairy or ± glabrous, crowned by 4 often indistinct calyx lobes, or calyx lobes ± lacking. Fruits yellowish-brown or reddish-brown, mostly with ± broad and conspicuous longitudinal grooves between the mericarps, supported by small, ± broad U-shaped carpophores, less than a quarter as long as the fruits; each mericarp (1.5) 1.7–2.5 (2.8) × 0.9–1.6 mm, elliptic, oblong to ± obovate, dorsal side convex, ± densely covered with ± tuberculate structures or with whitish papillae less than 0.1 mm long or subglabrous, ventral side plane, but with age often

---

[1] The setae should not be confused with small, sometimes ± linear short shoot leaves crowded in the axils of long shoot leaves!

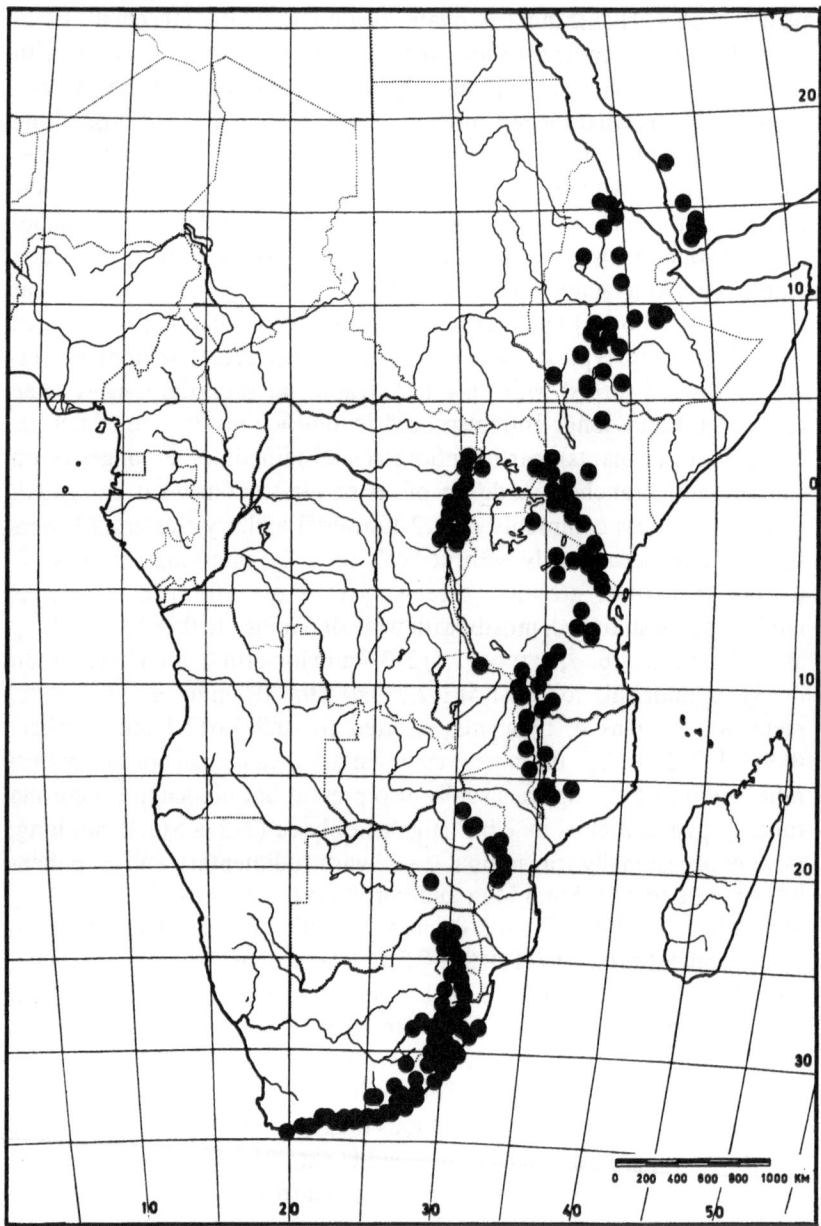

Fig. 78. Distribution of *Anthospermum herbaceum*

becoming hollowed out on both sides of the prominent median vertical ridge, commissure ± large; mericarps crowned by 2 obscure, ± triangular to rounded calyx lobes ca 0.2–0.3 × 0.2–0.3 mm, but more commonly calyx lobes ± lacking. – Figs. 30 a, 32, 35 j–k, 80–82.

Chromosome Number (Table 3): n = 11, 2 n = 22.
Average Pollen Diameter (Fig. 48): 24.6–29 µm.
Habitat: Primarily in wet, moist or damp, ± shady places (gullies, kloofs, valley bottoms, etc.). In afromontane forest, in forest edge vegetation, in secondary scrub replacing forest, at the edge of marshes or bogs, in riverine thicket, or in damp grassland. In E Africa, moreover, extending into moister, ± shady places of heathlands (with *Phillipia, Erica arborea, Hypericum, Adenocarpus, Anthospermum usambarense,* etc.). Sometimes also found on rocky slopes between boulders or in rock crevices and in (fire-prone) grassland. In SE Africa extending to the Indian Ocean coast and occasionally occurring in littoral scrub over dune sand and also near the sea in the salt spray zone. Not uncommonly growing in disturbed sites (pathsides, road-cuttings, bare spots in grassland, abandoned cultivated lands, coffee or pine plantations, etc.); sometimes in thin soil over ± recent lava flows (tropical E Africa). – Ca 0–2 900 (3 250) m.
Distribution (Maps, Figs. 78 and 79): From Ethiopia S to S Africa (Transvaal to the SW Cape Prov.), Swaziland and Lesotho; also in SW Arabia (Yemen Arab Republic and adjacent parts of Saudi Arabia).

Critical Remarks: A. herbaceum, the most widely distributed species within the genus and the subtribe, is highly variable in virtually all of its vegetative characters. Its (♂ and ♀) flowers, however, with their often very long cylindrical corolla tubes, and its fruits, having conspicuous and ± broad longitudinal grooves between the mericarps (Fig. 35 j), are rather uniform; they distinguish *A. herbaceum* from all the other species of the genus.

A. herbaceum contains a number of "Forms" (± in the sense of ecotypes) which in places are quite distinct and well discernible. There are, however, numerous morphological intermediates occurring over the entire distribution range which are often impossible to assign to one or another of these.

Seven of the more conspicuous Forms of *A. herbaceum* are circumscribed below. Natal, offering the necessary wide array of habitats, ranging from sea level to ca 3 000 m in altitude, appears to be the meeting point of virtually all of these. Populations from Natal are, for this reason, used for a photographic documentation (Figs. 80–82):

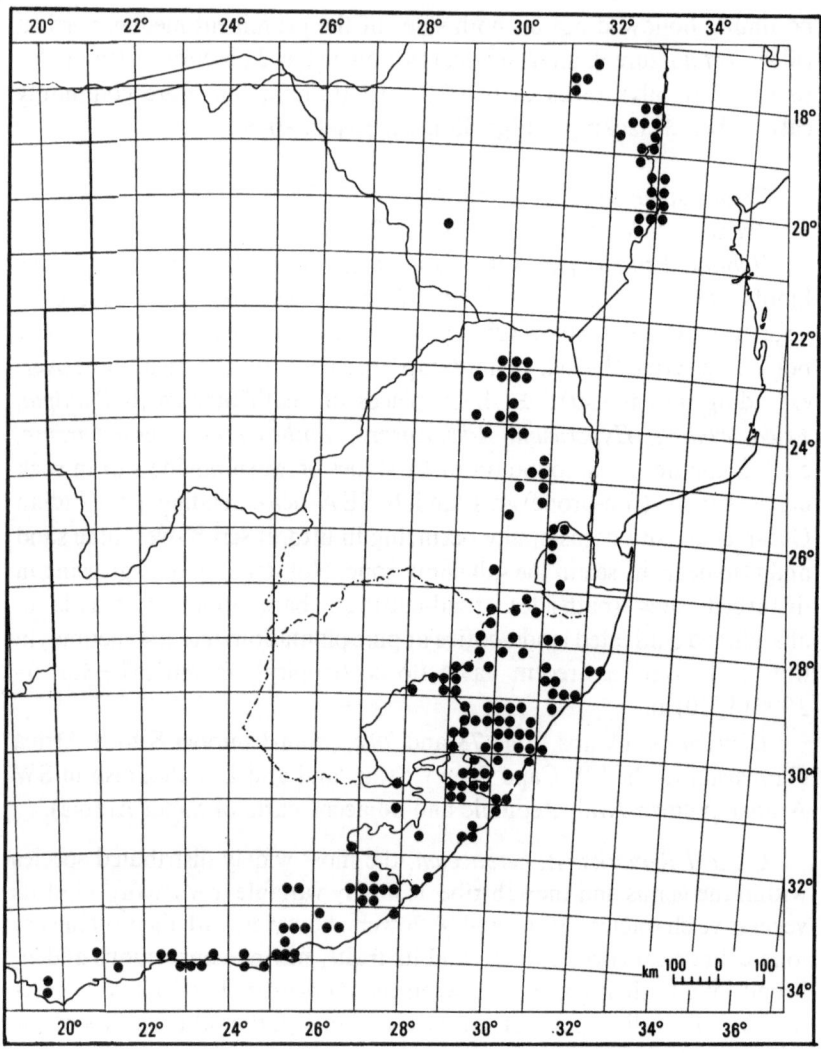

Fig. 79. Distribution of *Anthospermum herbaceum* in S Africa

(a) *"Afromontane Forest (Edge) Form"* ("typical" *A. herbaceum*) (Fig. 80 a):

The most "robust" plants of *A. herbaceum* with the longest (scrambling, straggling) stems and the largest leaves belong here. The hairiness of the plants varies; the inflorescences are relatively extensive, the partial inflorescences are comparatively many-flowered.

Growing mostly in shady and moist to wet places.

Fig. 80. *Anthospermum herbaceum.* *a* "Afromontane Forest Form" (Royal Natal National Park, Sigubudu valley, ca 1 400–1 500 m; PUFF 800106-1/2); *b* "High Altitude Form" (Natal, Giant's Castle Game Res., Langalibalele Pass, ca 2 800 m; PUFF 800116-2/1); *c–g* "Trampled Grassland Form", plants from population PUFF 800115-1/1, collected in much disturbed grassland (overgrazed and trampled by cows; Natal, S side of Kamberg Mountain, ca 1 800 m). – Further explanations in the text

Fig. 81. *Anthospermum herbaceum. a* "Burnt Grassland ('*A. hedyotideum*')
Form"; note numerous unbranched stems arising from a common woody base and
stumps of burnt off stems (Natal, Farm Dalcrue, ca 10 km from Nottingham Road
on road to Himeville, ca 1 500 m; PUFF 800115-4/1); *b* plant collected amongst
rocks at the seashore in the salt spray zone, ca 5 m above the high tide level on the
seaward side (Natal, "Black Rocks", S of Marina Beach; PUFF 790427-
2/2). – Further explanations in the text

It is perhaps the most common Form of *A. herbaceum*, found ± throughout the species' distribution range.

The following four Forms occur in more open and occasionally ± sunny localities:

(b) *"Wet to Damp Scrub/Grassland Form"*:
Found at the edge of swamps, marshes, along streams, etc.
The plants are smaller than in (a), the ± ascending to erect stems are weaker and are supported by the surrounding vegetation; the leaves are also smaller, variable in shape and hairiness, and are occasionally quite tough and almost ± leathery.
Occurring over most of the species' distribution range.

(c) *"(Burnt) Grassland ("A. hedyotideum") Form"* (Fig. 81 a):
The plants are ± erect, several- to many-stemmed and hardly more than 30 cm tall; the stems are often unbranched; the leaves are occasionally quite crowded and variable in shape (broadly ovate-lanceolate to lanceolate), relatively large short shoot leaves are frequently present; the partial inflorescences are normally very few- to one-flowered; the plants develop ± massive, almost rosette-like, ± woody bases if burnt regularly.
Mainly found in S Africa (Transvaal to the Transkei); also some collections from Zimbabwe.

(d) *"High Altitude (Grassland) Form"* (Fig. 80 b):
Often found at high altitudes (above ca 2 500 m) at the base of boulders or small rocks in grassland or, in rocky areas, on ledges or in crevices.
The plants are low, prostrate and often form mats or cushions; the stems are occasionally only a few cm long; the often relatively broad leaves are smaller than in any of the other Forms; the inflorescences are quite reduced, the flowers are often only paired at the nodes.
Mainly occurring in SE Africa, but occasionally also in tropical E Africa.

(e) *"Trampled Grassland Form"* (Figs. 80 c–g):
Found in grassland areas trampled by cattle, sheep or goats (i.e., in 'Trittrasengesellschaften'), at lower altitudes than (d).
Similar in appearance to (d), but the plants do not usually form dense cushions.
Mainly observed in S Africa (Transvaal, Natal).

The following two Forms are confined to the Indian Ocean Coastal Belt:
(f) *"Coastal Dune (Scrub) Form"* (Fig. 82 b):

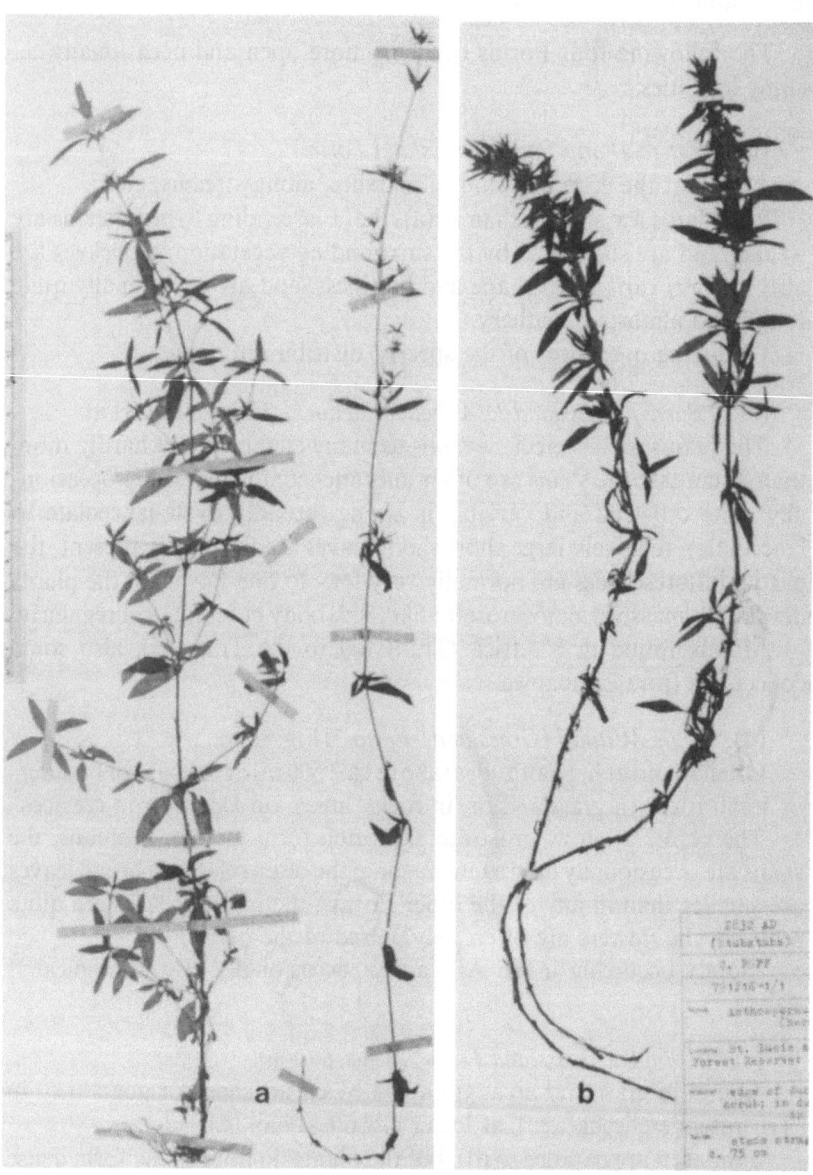

Fig. 82. *Anthospermum herbaceum. a* short-lived, thin-stemmed plants with few-
(often 1-)flowered partial inflorescences from the Natal S Coast (Umtamvuna
Nature Reserve, marshy area; Puff 790423-3/1); *b* parts of plants collected in sand
at the edge of dune forest (Natal, St. Lucia area, Eastern Shores Forest Res., nr.
Mission Rocks; Puff 791216-/1)

Found in ± dry areas in sandy soil (dune sand); associated with dune forest or scrub; often in open and ± sunny places.

The plants are ± erect, the stems hardly reach 30 cm in length; the leaves are lanceolate, relatively large and tough (almost leathery); the inflorescences are relatively many-flowered.

From Zululand to the E Cape Prov.

(g) *"Salt Spray Zone Form"* (Fig. 81 b):
Mostly found in cracks of rocks near the sea.

The ± prostrate and glabrous plants have quite broad leaves which tend to be rather succulent; the stipular sheaths bear a reduced number of setae (sometimes only a single seta is present); the partial inflorescences are often only 1-flowered.

Apparently not very common. From the Natal S Coast to the Transkei (Wild Coast).

[Another unusual Form of *A. herbaceum* from the Natal S Coast (cf. Fig. 82 a) is discussed below].

In tropical E and S Central Africa, *A. herbaceum* is relatively more uniform and does not show as much variation in habit and morphology as it does in SE Africa. "Burnt Grassland Forms" [(c), above] and low, matted ecotypes occurring in trampled areas [(e), above] are fairly uncommon; a few collections from Kenya and Tanzania, recorded from ca 2 700 m and upwards, represent dwarfed and stunted "High Altitude Forms" [(d), above].

In Ethiopia, there appears to be a trend to unisexual plants and dioecy, and to a slightly more woody habit; in some characters, plants may approach *A. pachyrrhizum* (26., below), which is probably allied to *A. herbaceum*.

In the SW-most part of the distribution range of *A. herbaceum*, subglabrous (and rather short-lived?) forms with relatively narrow, lanceolate leaves and rather reduced inflorescences with few- to one-flowered partial inflorescences (*"A. ferrugineum"*, ± *"A. lanceolatum"*) are predominant in places. They are, however, by no means confined to that area and occasionally also (although not often) occur in tropical E Africa.

Similar Forms are recorded from the Natal S Coast (e.g., GORDON-GRAY 6214, STREY 4907, 7115; Fig. 82 a). They differ slightly from the above mentioned SW Cape forms in their more slender stems, very long internodes and lateral branches with rather small leaves, but agree with "*A. ferrugineum* Forms" in being ± glabrous and having only few- to one-flowered partial inflorescences. Some collections of this Form are so distinct that I was originally tempted to recognize them taxonomically, but there are so many collections from the Natal S Coast area which are intermediate between "typical" *A. herbaceum* and this Form, that a clear morphological separation becomes impossible.

In SW Cape localities, *A. herbaceum* often occurs in habitats similar to those of *Carpacoce spermacocea*. The two taxa, moreover, show certain morphological resemblances (leaf size and shape, stipular sheaths with several setae; sometimes they ± resemble each other in habit). Subsequently, they have often been confused – especially by earlier collectors (the "*A. lanceolatum*" specimens in the THUNBERG herbarium, for example, are comprised of both *A. herbaceum* and *Carpacoce spermacocea*!). In flower or fruit, the two taxa are, however, easily separable: *Carpacoce spermacocea* has flowers with only one fertile carpel and one stigma, "hooded" corolla lobes, fruits which are crowned by conspicuous calyx lobes and which do not split into two mericarps, and different inflorescences. Even dried vegetative material can be kept apart: Herbarium specimens of *Carpacoce spermacocea* are typically uniformly darkly coloured, and the leaves are not discolourous as in *A. herbaceum*. In the field, a simple test can be used to separate the two taxa: *Carpacoce spermacocea* (syn. *Anthospermum foetidum*!) emits a foetid smell when the leaves are crushed, whereas *A. herbaceum* does not.

Some ecotypes of the SW Cape species *A. galioides* (31.) with reddish stems and relatively broad, discolourous leaves may also superficially resemble *A. herbaceum* (e.g., ECKLON & ZEYHER "2310. *A. lanceolatum*" is, in fact *A. galioides*!). They can be distinguished from *A. herbaceum*, however, by their smaller (♀) flowers, a differing fruit morphology and a different habit (they occur, moreover, in different habitats and in localities clearly outside the distribution range of *A. herbaceum*).

A. herbaceum may, occasionally, hybridize with *A. hispidulum* (17.) (in the Transvaal and Natal), *A. pumilum* subsp. *pumilum* (29.) (in the Transkei and Zimbabwe) and *A. paniculatum* (15.) (in the E Cape; Fig. 57) (also cf. Table 8). For details see the Critical Remarks sections of the respective taxa.

Collections

Saudi Arabia. – N 1842: 35 km S of Khamis Mushayt, HASSAN, KÖNIG & KÜRSCHNER 82-2343 (herb. VO 11422) (BSB).

Yemen Arab Republic. – N 1344: Jebel Sabir, SE of Taizz, RADCLIFFE-SMITH & HENCHIE 4398 (K); Ibb D., 35 km N of Taizz, around Dhisufal, HEPPER & WOOD 5866 (K); Wadi Banna above Saddah, IRONSIDE WOOD 72/122 (BM). – N 1444: ascent to Sumara pass, LAVRANOS & NEWTON 16002 (E; atypical, approach. *A. pachyrrhizum*); (around) Yarim, ACRES 13 QQ (K), IRONSIDE WOOD 72/47 a (BM); 30 km S of Dhamar, HENDY 17 (K). – N 1543: just W of Manakhah, on Hajarah rd., RADCLIFFE-SMITH & HENCHIE 4659 (E, K); Gebel Schibam (= Jebel Shibam), Menacha (= Manakhah), SCHWEINFURTH 1427 (BR, G, K). – Undecipherable: S end of Quettaql ..., ACRES 24 A/R (?247/R) (K). – Not traced: Hadie Mts., Mokhaja, FORSSKAL s.n. (C – mixed with type collection of *Galium aparinoides*).

Ethiopia. – **Eritrea**: Mt. Bizen, PAPPI 4831 (FI), SCHWEINFURTH & RIVA 1853 (BR, FI, G, K, S, US), 1882 (G); Mensa, Mt. Ira, TERRACIANO & PAPPI 1429 (FI); Amasen (= Hamasen) reg., colle Letta, FIORI 1668 (FI × 3); –, Mt. Lesa, PAPPI 4636 (FI, G, MO, S, W), 4888 (COI, FI); Oculé Cusai (= Akale Guzai Awr.), Soyrà (= Soira) Mts., PAPPI 1243 (FI); –, Halai (Mt.), PAPPI 1655 (FI). – Not traced: Jaghená (?spelling), BALDRATI 3709 (FI); Goló, DAINELLI & MARINELLI 172 (FI); Mensa, R. Ualicane (?spelling), TERRACIANO & PAPPI 1767 (FI). – **Tigre**: Mt. Kubbi (= Koubi) [E of Axum], SCHIMPER 732 (BR, G, K, S; P – mixed with *A. pachyrrhizum*); Mt. Scholada [W of Adua = Adwa], SCHIMPER 194 (B, BM, BR × 2, FI × 2, G × 3, GH, GOET, K, LISU, M × 2, MO, P × 3, W × 3, WU; S × 2, one as '224'); 19 km S of Maichew, on Dessie rd., DE WILDE 6956 (WAG). – **Gonder**: 'Prov. Ouedgerate' (= Wodjerat, Wodgera, Wogera or Uoghera; prob. today's Wegera Awr.), QUARTIN-DILLON & PETIT s.n. (P, W); Semien: Bälägäs valley nr. Schoada, SEBALD 1401 (STU); Dembià [around Gonder], Bambelo, CHIOVENDA 1035, 1054 (FI). – **Welo** (= **Wollo**): 12 km N of Dessie, tow. Lake Haik, PUFF, MANTELL & ENSERMU 810926-2/3 (ETH, K, UPS, WU); Azewagedel Mts., 2 km E of Dessie, SUTHERLAND 186, 382 (MO). – **Shewa** (= **Shoa**): Addis Ababa, Embassy Hill, MOONEY 7042, 7087 (K); Mt. Yerer nr. Bishoftu (= Debre Zeyt), MOONEY 5758 (ETH, FI, K); Menagesha (= 'Menagacio') State Forest, Mt. Wuchacha (= 'Uociacia'), DE WILDE & DE WILDE-DUYFJES 6407 (WAG), 10280 (WAG), NEGRI 468 (FI), PUFF, MANTELL & ENSERMU 810920-1/5 (ETH, K, UPS, WU); Lake Wonchi, GILBERT & TEWOLDE 3248 (ETH, K); Mt. Zuquala, ASH 2349 (K), SCOTT s.n. (K); ca 46 km from Hosaina, tow. Butajira, PUFF, MANTELL & ENSERMU 810917-1/1 (ETH, WU); ca 48 km from Wolkite, tow. Hosaina, PUFF, MANTELL & ENSERMU 810916-3/1 (ETH, K, UPS, WU). – Additional collections: ASH 2537 (K); DE WILDE & DE WILDE-DUYFJES 6105 (BR, WAG), 8209 (BR, ETH, K, MO, WAG), 9984 (B, BR, K, MO, PRE, WAG); MESFIN, HEDBERG, EDWARDS, TEWOLDE & ENSERMU 2870 (ETH); MOONEY 6496 (K), 7938 (ETH); NEGRI 237, 305 (FI); PIOVANO 157 (FI). – **Kefa** (= **Kaffa**): Mt. Maigudo, MOONEY 6163 (K); Magi (= Maji), GILBERT 350 (ETH, K). – **Gamo Gofa** (= **Gamu-Gofa**): Gughe highlands above Arba Minch, GILBERT, THULIN & AWEKE 514 (ETH, K, MO); around Chencha (= 'Cencia'), PUFF & ENSERMU 821219-1/4, -2/8 (ETH, WU), VÀTOVA 2070 (FI); –, Ociollo, VÀTOVA 2092 (FI); mts. above and S of Gidole, DE WILDE 5665 (BR, LMU, WAG). – **Arsi** (= **Arussi**): Mt. Cilalo nr. Asella, DE WILDE & DE WILDE-DUYFJES 8037 (BR, WAG); on the Coriftu R., SMEDS 1363 (FI; K – mixed with *Galium scioanum*). – **Harerge** (= **Harar**): Jifara, ca 25 km S of Assabot (= Asebot), MOONEY 6324 (ETH, FI, K); between Kuni and Bedessa, PUFF, ENSERMU, DAWE & EDWARDS 820921-2/3 (ETH, WU); W flank of Gara Mullata Mt., DE WILDE 6503 (BR, WAG); Bedeno, AMARE GETAHUN H-86 (ETH; K – as IECAMA H-86); around Kulubi, DE WILDE 4412, 4431 (BR, WAG), IECAMA RS 189 (K); nr. Carsa (= Kersa), ROBERTSON 1325 (ETH, K); Mt. Kondudu, DE WILDE 5885 (BR, WAG); Sarerta Mt., GILLETT 5188 (FI, K). – **Sidamo**: 8 km SE of Hagere (= Agere) Selam, SE of Wendo, DE WILDE & DE WILDE-DUYFJES 8415 (BR, WAG); 20 km E of Adola (= Kibre Mengist), MOONEY 9991 (WAG); Mega, CORRADI 2712 (FI, W), 2726 (FI), GILLETT 14232 (B, BR, FI, K), 14233 (BR, FI, K), PUFF & ENSERMU 821224-4/2 (ETH, WU). – Imprecise locality: between Harrar and Addis Ababa, WELBY s.n. (K). – Not traced: Passo Sià (?spelling), VÀTOVA 2463, 2488 (FI); May Deghera (?spelling), RIVA 165 (FI).

314   D. Systematic Part

Uganda. – U 2: Bunyoro d., Hoina (= Hoima), Bagshawe 393 (BM); Toro D., (nr.) Fort Portal, Bagshawe 989 (BM), Hazel 94 (K); –, nr. Kyanyewara For. Stn., Puff, Katende & Grabherr 830706-1/8 (WU); Ruwenzori, Scott Elliot 7572 (BM, K, MO); –, from Ibanda to Nyabitaba, Puff, Katende & Grabherr 830716-1/1 (WU); Ankole D., Igara, Kyamahungu, Purseglove 687 (BR, K); Kigezi D., Maziba, Purseglove 1607 (K); –, Kachwekano Farm, Purseglove 2739 (K); –, Mabungo, Rogers (& Gardner) 324 (BM, BR, K), Thomas 1149 (K); –, Virunga Mts., between Mgahinga and Sabinio, Taylor 2093 (BM, S); –, –, Muhavura, Snowden 1527 (BM), Stauffer 906 (BR, K, P, PRE, WAG). – U 3: Mbale D., Bulago, Thomas 317 (K); –, Kapchorwa, Lind 479 (K).

Kenya. – K 1: Samburu D., Uaraguess (= Warges Mt.), Newbould 3102 (K). – K 2: West Suk D., Sekerr Mt., Agnew & al. 10345 (EA). – K 2/3: S Cherangani, Symes 211 (K). – K 3: Trans-Nzoia D., (E)NE Mt. Elgon, Lewis 5959 (K, MO), Tweedie 1997 (K); –, Saiwa Swamp National Park ('Sitatunga Swamp'), N of Kitale, Hooper & Townsend 1451 (K), Puff & Weber 780116-1/2 (WU); Elgeyo D., Kapsowar, Kipkunurr Forest Res., Hepper & Field 5001 (K); Leikipia plateau, Gregory 71 (BM); Kaptagat, Agnew & Agnew 9021 (MO); Eldama ravine, Whyte s.n. (K); Nakuru D., W Mau, Ndoinet River valley, Gillett 19030 (BR, K, M, MO, WAG); –, NE Menengai Crater, Mwangangi 193 (BR, FI, K, MO); –, E Mau Forest Res., camp 11, Maas-Geesteranus 6196 (BR, COI, G, K, PRE, S, WAG); Naivasha D., Kinangop, Albrechtson 14 (K). – K 4: Kiambu D., Muguga, Verdcourt 728 (FI, K, MO, PRE, S); Ukamba (= Ukambani), Scott Elliot 6527 (BM, K). – K 5: Kericho D., SW Mau Forest Res., camp 7, Maas-Geesteranus 5683 (BR, COI, G, K, PRE, S, WAG); Nakuru D., Tinderet Forest Res., nr. Malaget For. Stn., Perdue & Kibuwa 9194 (BR, K). – K 6: Narok D., Ol'Pusimoru sawmill, ca 14 km from Olokurto, Glover, Gwynne & Samuel 1395 (K, PRE); –, Ndunyangerro area, nr. Njura settlement, Glover, Gwynne & Samuel 1613 (K, PRE); Chyulu N (K 6 or K 4?) Bally 7815 (K). – K 7: Teita Hills, Yale Peak, Drummond & Hemsley 4302 (B, BR, FI, K, PRE), Gillett 17260 (BR, FI, K, PRE); –, Mt. Vuria, Lewis 5932, 5937 (K, MO, US). – Additional collections: Bally 4740 (G, K); Brodhurst-Hill 259, 270 (K); Chandler 2340 (B, BR, K); Dümmer 5178 (GH, US); Fries & Fries 330 (BR, K, MO, US); Irwin 57 (FI, K); Kerfoot 2778 (K), 2897 (PRE); Maas-Geesteranus 5040 (B, BR, G, K, PRE, US, WAG), 6019 (B, BR, COI, G, K, MO, PRE, S, US, WAG); Synge 1144 (BM); Taylor 1445, 1538 (BM); Verdcourt 1046 (K, PRE); Whitall 153 (K).

Tanzania: – T 1: North Mara D., Kadesha [Ukiriguru Research stn.] 4597 (EA). – T 2: Mangati, from Mdumgara to Dareda, Peter 44006 (B, K); 'Winterhochland: Lager Lemunge' (= Munge), Peter 43064 (B, K); Mt. Meru, Arusha (Ngurdoto) National Park, Burtt 4093 (K), Cooper 30, 119 (BM), Greenway & Kanuri 12291 (BR, K, PRE), Puff & Kibuwa 820826-1/1 (WU), Richards 25604 (BR, K); Mt. Kilimanjaro, Peter 41777 (B, K), Thulin 313 (K), Volkens 2001 (BM, G). – T 3: N Pare Mts., from Kilomeni to Kissangara, Peter 55747 (K); S Pare Mts., Chome Forest, Puff & Kibuwa 820901-1/4 (WU); Lushoto D., W Usambara Mts., SW Shagai Forest, Kwegoka, Drummond & Hemsley 2717 (B, BR, K, LISC, LMA, S); –, –, Mazumbai, Puff & Kibuwa 820828-5/1 (WU). – T 6: Kilosa D., Mandege For. Project, 50 km N of Kilosa, Harris, Pócs & Mwanjabe BJH 6412 (K); Morogoro D., Nguru Mts., Mabega Mt. nr. Maskati Mission, Thulin & Mhoro 3067 (K); –, Morogoro, Schlieben 4191 (MO); –, Uluguru Mts., Schlieben 3834 (B, BM, BR × 3, G, GH, K, LISC, M, MO, P, PRE, SRGH). – T 7: Iringa D., Dabaga Highlands, Kilolo, Polhill & Paulo 1414 (B, BR, K, LISC, PRE, SRGH); Mufindi, Paget-Wilkes 810 (K, MO); Njombe D., Msima Stock Farm, Emson 277 (K); –, Milo, Archbold 2436, 2563 (K); Rungwe

D., Kiejo Volcano, HEPPER, FIELD & MHORO 5393 (K); upper Ruhudje, Lupembe are, SCHLIEBEN 477 (B, BM, BR, G, K, M, P, S). – T 8: Songea D., ca 7 km E of Songea, MILNE-REDHEAD & TAYLOR 8803 (B, BR, K, LISC); – , Matengo Hills, ca 3 km NE of Mpapa by Luhekea R., MILNE-REDHEAD & TAYLOR 10452 (B, BR, K, LISC). – Additional collections: ARCHBOLD 1182 (K); FAULKNER 4843 A (K, MO, SRGH); GEILINGER 1837, 2167 (K); HAARER 1561, s.n. (K); MILNE-REDHEAD & TAYLOR 9399, 9400 (B, BR, K, LISC, SRGH); MSHANA 7, 54 (BR, K); PETER 46741, 55456 (B), 55637, 55639, 55656 (B, K), K 725 (K), K 730 (B, BR, K); PUFF & KIBUWA 820829-3/1 (WU); RENVOIZE & ABDALLAH 1770 (K); ROUNCE 422 (K); SCHLIEBEN 4196 (B, BR, GH, K, LISC, MO, SRGH); TANNER 886 (K); VOLKENS 437 (BM, BR, E, G, K, WU); ZERNY 615 (W); ZIMMER 73 (BM).

Zaire. – Lac Albert: Nioka, GERMAIN 4103 (BR, J, US), TATON 864 (BR, J); Djugu (Kibali, Ituri), LEBRUN 3975 (BR × 2, K); Bogoro, BEQUAERT 4933 (BR). – Lacs Edouard et Kivu: [W of foot of] Ruwenzori, Kalonge, HAUMANN 225 (BR); Lubero Terr., Luofu, LÉONARD 5295 (BR, J, US); mts. SW of L. Edouard, Kisale (= Kasali) Mts., HUMBERT 8241 (B, BR, P): Rutshuru, BEQUAERT 6129 (BR); Parc National Albert, E Niiagongo (= Nyiragongo) Volcano, LEBRUN 8755 (BR); – , between Rumoka Volcano and Sake, LOUIS 4844 (BR, K); Kivu Terr., Mt. Kahuzi, nr. Mukaba, LISOWSKI 60291 (BR); – , Tshitshiangutu, SCAETTA 1745 (BR, K), 1746 (BR); Fendula, Nyakolonge, SCAETTA 672 (BR). – Not traced: Kiloti, DE WITTE 1299 (BR); Luherera, HENDRICKX 194 (BR). – Undecipherable: Bushuk..., MEURILLON 899 (BR). – Additional collections: BEQUAERT 3570 (BR); DERU 188 (BR); FICHE 215 (BR); GERMAIN 3303 (BR, K); GILON 181 (BR); HEINE 378 (BR); HENDRICKX 3757, 3950, 5569 (BR), 7848 (BR, P); LAURENT 557 (BR), 614 (K); LEBRUN 6951 (BR), 8461 (BR, K), 8541 (BR, K, PRE); LISOWSKI 48215 (BR × 2); OSMASTON 2096 (BM); SPITAELS 336 (BR).

Rwanda. – Préf. Ruhengeri, Ruhengeri, SCAETTA 350, 370 (BR); – ,L. Bulera, AUQUIER 4489 (BR); Rutare, TATON 965 (BR, J); W of L. Mohasi, MILDBREAD 497 (BR); Préf. Kibuye, Rutsiro, Gakoma- Kibuye rd., TROUPIN 15289 (BR); Kigali Terr., Remera, BECQUET 315 (BR); Rubona, MICHEL 5936 (BR, J); Préf. Gikongoro, Mata, TROUPIN 15140 (BR). – Not traced: Nyakabande, LOVERIDGE 401 (B, GH, K, MO).

Burundi. – Mukarumbi [nr. Kibira], ROBYNS 2385 (BR, GH).

Zambia. – N: Mbala (= Abercorn) D., Mbala, FANSHAWE 10620 (BR, K, SRGH); – , Ndundu, RICHARDS 961, 4869, 11377 (K); – , Zombe plain, SANANE 908 (K, SRGH); – , Tunduma rd., Kamuswazi R., SANANE 1280 (K, SRGH). – E: Nyika plateau, below Rest House, tow. N Rukuru waterfall, ROBSON 394 (BM, BR, K, LISC, PRE, SRGH).

Malawi. – N: Misuku Hills ('Masuku Plateau'), WHYTE 263 (K); – , Mugesse Forest, PUFF 781219-1/3, -1/6 (WU); – , Wilindi Forest Res., PUFF 781220-1/10 (WU); – , border of Matipa Forest Res., PUFF 781221-2/2 (WU); Chitipa D., around Chisenga, foot of Mafinga Mts., BRUMMITT 12036 (K, SRGH), TYRER 529 (BM, BR); Nyika Plateau, PUFF 781225-1/3 (WU), WHYTE s.n. (K); – , nr. Nganda (Hills), BRUMMITT & SYNGE WC 91 (K, MAL, SRGH), TYRER 859 (BM, SRGH); – , 7 km SW of Chelinda camp, BRUMMITT 11806 (K, SRGH); between Kondowe (= Livingstonia) and Karonga, WHYTE 322 (K); N Viphya Plateau, ca 52 km N of Mzuzu, PAWEK 5575 (K, MAL, MO, SRGH); Mzimba D., Mzuzu, Marymount [Mission], PAWEK 65, 1345 (MAL), 1765 (SRGH), 6870 (MO, PRE); Nkhata Bay D., ca 8 km E of Mzuzu, PAWEK 7315 (K); – , ca 35–37 km SE of Mzuzu, Lusangazi Forest, PAWEK 8739 (BR, K, MO, WAG); – , – , Mpamphala Mt., GROSVENOR & RENZ 1091 (BR, K, SRGH); S Viphya Plateau, Champayo, SALUBENI 1684 (K, MAL, SRGH); – , Luwawa Dam, PAWEK 8538 (K, MO). – C: Ntchisi (= Nchisi)

Mt., Brass 17107 (BM, BR, GH, K, MO, PRE, SRGH, US), Puff 781213-1/4 (WU); Dedza D., Dedza Mt., Salubeni 1086 (K, MAL, SRGH). – S: Zomba D., Zomba Plateau, Brass 16189 (BM, BR, GH, K, MO, PRE, SRGH, US), Brummitt 12236 (BR, K, SRGH), Puff 780208-2/3 (WU); Shire Highlands, Buchanan 412 (E, K); Blantyre D., Ndirande Mt., Brummitt 8864 (K, SRGH); Mt. Mulanje (= Mlanje, Milanji), Whyte s.n. (BM); –, Chambe plateau, Puff 780213-1/1 (WU). – Imprecise localities ("Nyasaland"): Buchanan 229 (BM, MO, SAM), 247 (GH, US), 257 (K), 861 (BM, E, G, US), s.n. (MO); Whyte 375 (G).

Mozambique. – N: Vila Cabral, Torre 196 (COI × 2, LISC). – Z: Namuli Peaks, Leach & Schelpe 11477 (K, SRGH). – T: NE of Vila Coutinho, Mozambique/Malawi border, W of Ncheu- Dedza rd., Puff 780216-1/1 (WU). – MS: Vila de Manica, Pedro & Pedrógão 6999 (LMA); Manica, Mavita, Barbosa 913 (LISC), Pereira, Sarmento & Marques 1156 (LMU, WU); Mt. Maruma, Swynnerton 2148 (BM, K, SRGH).

Zimbabwe. – N: Sipolio (= Sipolilo) D., Semhenheka, ca 11 km S of Sipolilo, Mavi 1208 (SRGH); Mazoe D., Suri Suri Dam, Gibbs-Russell 1144 (SRGH); Shamva rd., Mansala R., Rutherford-Smith 332 (BR, LMU, MO, SRGH). – C: Salisbury- Mazoe rd., Christonbank Nature Res., Puff 790123-2/1 (WU); Salisbury, Arthurs Seat, Greatrex 395 (SRGH); –, Rua R., Eyles 8785 (BR, K, SRGH); Umvukwe Hills, Mermaid's Pool, Gilliland 635 (BM, K); Goromozi D., Chin(d)amora Nature Res. [? Mazoe D.], Rutherford-Smith 45 (K, SRGH); Makoni D., Chiduku Sacred Forest, West 6182 (SRGH); –, Rusape, Hislop 222 (K); –, ca 37 km E of Rusape, Miller 4751 (SRGH). – E: Inyanga D., World's View, Edwards 948 (SRGH), Puff 790125-6/2 (WU); –, Rhodes Inyanga National Park, Fort Nyangwe, Puff 790124-2/4 (WU); –, Pungwe R. (Falls), Fries, Norlindh & Weimarck 3839 (BM, BR, K, M, MO, SRGH), Loveridge 1104 (BR, K, LISC, SRGH), Puff 790126-3/3 (WU); Stapleford Forest Res., Gilliland 309 (BM, K, MO, PRE, SRGH), Puff 790127-3/2 (WU); Umtali D., Odzani R. valley, Teague 181 (BOL, K); – Vumba Mts., Obermeyer 2148 (PRE); –, Banti Forest Res., Mavi 556 (B, LISC, SRGH); –, Engwa, Exell, Mendonça & Wild 170 (BM, LISC, SRGH); Melsetter D., Melsetter, Walters 2759 (K); –, Martin Forest Res., 'Rocklands', Chase 2923 (BM, MO, SRGH); –, Chimanimani Mts., 'Stonehenge', Hall 285 (BM, SRGH); –, Tarka Forest Res., Goldsmith 15/70 (K, LISC, PRE, SRGH); –, 2 km below Haroni gorge, Biegel 2792 (K, PRE, SRGH); Chipinga, Noel 2018 (SRGH); Chirinda (Forest), Chase 631 (BM, K, SRGH), Hack 203/50 (SRGH). – Additional collections: Allen 3584 (BM); Biegel 1680 (K, SRGH); Brain 7061 (SRGH); Burrows 269 (SRGH); Crook M 166 (K, LMU, MO, SRGH); Davies 2721 (MO, SRGH); Fries, Norlindh & Weimarck 2140 (NBG), 2410 (PRE, SRGH), 2595 (E), 3237 (B, COI, PRE, SRGH, WAG); Johnson 171 (K); Ngoni 26 (B, SRGH); Obermeyer 3587 (PRE); Phipps 1174 (BR, PRE, SRGH); Puff 790125-5/2, 790129-1/1 (WU); West 6403 (LMU, SRGH); Whellan 714 (MO, SRGH); Wild 478 (K, SRGH), 1526 (BR, K, SRGH). – W: Matobo D., Matopo Hills, Gibbs 93 (BM); –, Farm Chesterfield, Miller 5887 (BR, LMA, MO, NU, SRGH); –, Farm Besna Kobila, Miller 3665 (K, LMA, SRGH); –, Metabezi (? spelling) valley, Meara 48 (MO, SRGH).

Swaziland. – 2531 (Komatipoort): Havelock (-CC), Compton 30647 (NBG, PRE); Piggs Peak, King's Forest (-CD), Compton 31732 (NBG). – 2631 (Mbabane): Forbes Reef (-AA), Compton 31808, 32459 (NBG), Puff 780409-1/1 (WU); Mbabane (-AC), Compton 23249, 23789 (NBG), 23765 (BOL, NBG); Hlatikulu (-CD), Compton 26227 (NBG, PRE). – Additional collections: Compton 24588, 27736, 31094 (NBG), 24752, 25280, 26693 (NBG, PRE), 25052 (K, PRE,

SRGH), 25394 (BOL, K, M, PRE, SRGH); DLAMINI s.n. sub NBG 67546, 71751, 77710 (NBG), s.n. sub PRE 41606, 41607 (PRE); PUFF 790224-2/1, -4/1, 790225-1/5 (WU); ROGERS 11615 (SAM).

Lesotho. – **2828** (Bethlehem): Leribe D., Malaoaneng (**-CC**), DIETERLEN 908 (P, PRE, SAM). – **2929** (Underberg): Sehlabathebe National Park (**-CC**), BAYLISS LES 40, 40 a (K), LES 160 (MO), BEVERLEY 224, 297 (PRE), JACOT GUILLARMOD, GETLIFFE & MZAMANE 201 (K, MO), SCHMITZ 7079 (PRE).

South Africa. Transvaal – **2229** (Waterpoort): Zoutpansberg, Wyllie's Poort (**-DD**), HAFSTRÖM & ACOCKS 1422 (PRE, S). – **2230** (Messina): Entabeni Forest Res. at Muchindudi Falls (**-CC** to **-CD?**), CODD 4474 (PRE); Venda, Tate Vondo Forest Res. (**-CD**), PUFF 791201-1/2, -6/1 (WU). – **2329** (Pietersburg): Louis Trichardt D., nr. Mara Bergplaas, nr. Lejuma (**-AB**), SCHLIEBEN & STREY 8293 (BR, G, K, LMA, M, PRE, SRGH, W); – , Louis Trichardt (**-BB**), YOUNG 26884, 26928 (PRE); Pietersburg (**-CD**), ROGERS 16005 (PRE); Haenertsburg (**-DD**), POTT 4715 (TVL Museum no. 13707) (PRE). – **2330** (Tzaneen): Elim (**-AA**), OBERMEYER 29232 (PRE); Tshakoma [= Tshakhuma (Mission)] (**-AB**), OBERMEYER 30375 (PRE), s.n. sub NH 26979 (NH); Westfalia Estate nr. Duiwelskloof (**-CA**), BOS 1157 (M, PRE, STE, WAG), PUFF 790211-2/1 (WU); New Agatha [Forest Res.] (**-CC**), MCCALLUM s.n. sub PRE 41613 (PRE). – **2429** (Zebediela): Chuniespoort (**-BA**), WALL s.n. (S). – **2430** (Pilgrim's Rest): The Downs, Mamotswiri (= 'Mamotsuiri') Peak area (**-AA**), JUNOD 2046 a (G), PUFF 791208-1/3 (WU); Shilovane (= Shiluvane Mission) (**-AB**), JUNOD 1704 (TVL Museum no. 5532) (PRE); Mariepskop (**-DB**), VAN DER SCHIJFF 5596 (K, PRE), 6341 (PRE); Mt. Sheba Nature Res. (**-DC**), FORRESTER & GOOYER 61 (MO, PRE), PUFF 781029-2/1, -4/1 (WU); Graskop (**-DD**), GALPIN s.n. (BOL), ROGERS 23593 (G). – **2530** (Lydenburg): Lydenburg (**-AB**), WILMS 590 (G); around Sabie, tow. Nelspruit (**-BB**), HUMBERT 10734 (P); Witklip (**-BD**), KLUGE 665, 698 (PRE); Belfast (**-CA**), LEENDERTZ 9190 (PRE). – **2531** (Komatipoort): Barberton D., ca 40 km SE of Barberton (**-CC**), LEWIS 6334 (K, MO). – **2629** (Bethal): Farm Spitskop, [N of] Ermelo (**-DB**), POTT 4881 (BOL, sub TVL Museum no. 15164 in PRE). – **2730** (Vryheid): Wakkerstroom D., Farm Oshoek nr. Wakkerstroom (**-AD**), DEVENISH 757 (K, M, PRE), PUFF 791213-1/1, -2/1 (WU); Kwa-Mandlangampisi, Farm Groothoek (**-BA**), DU TOIT 107 (PRE). – Imprecise localities: between Louis Trichardt and Wyllie's Poort (2329-BB to 2229-DD), GILLETT 2934 (K), WALL s.n. (S); between Pilgrim's Rest and Sabie (2430-DD to 2530-BB), ROGERS 14880 (GRA, J, K, SAM); "TVL", JUNOD 4318 (G), ROGERS 18045 (BOL, PRE). – Additional collections: GALPIN 9715 (K, PRE), 10947, 14429 (PRE); JUNOD 50 (TVL Museum no. 25578) (PRE), 2422 (G); MEEUSE 9382 (K, M, PRE); MOGG 9538 (K); MOSS & ROGERS 258 (J), 281 (BM, J); MUDD s.n. sub PRE 41613 (PRE); OBERMEYER 362 (PRE); PUFF 781028-3/1, -3/2, -4/1, 790210-3/2, 790211-1/2, -4/2, 790701-1/2, 791201-4/1, 791202-1/1, -3/1, -4/2, 791212-1/1, -2/1 (WU); REHMANN 6028 (K); ROGERS 18261 (J, sub TVL Museum no. 15998 in PRE), 21276 (PRE), 22008 (G), 23202 (G, J); SCHEEPERS 366 (K, M, PRE × 2, SRGH); SCHEEPERS & HAASBROEK s.n. sub PRE 41603 (PRE); VAHRMEIJER (?, as 'O.S. V.') 362 (PRE); VAN DER SCHIJFF 7354 (PRE); WAGER 23126 (PRE); WALL s.n. (S). **O.F.S.** – **2828** (Bethlehem): Golden Gate National Park (**-DA**), LIEBENBERG 6873 (K, PRE); Qwaqwa, from 'Junior Camp' to Mopeli Hut (Kogtwane Spons), PUFF 790217-1/4 (WU); Witsieshoek (**-DB**), JUNOD s.n. (G, sub PRE 17451); from terminus of Sentinel rd. to Mont-aux-Sources (**-DB, -DD**), PUFF 800105-2/3 (WU). – **2829** (Harrismith): Harrismith (**-AC**), JANKEY (or SANKEY; ? spelling) 80 (K), SMIT 129 (PRE); – , Platberg, JACOBSZ 3038 (PRE); Kerkenberg, PUFF 781125-4/1 (WU); Rensburgskop, [SW of] Swinburne (**-AD**), JACOBSZ 205 (K, PRE); (Farm) Paulina, [NE

of] Van Reenen, THODE s.n. sub STE 4606 (STE). – No locality given ("OFS"):
COOPER 1071 (BM, BOL, E, W × 2). **Natal** – **2729** (Volksrust): Majuba (**-BD**),
ROGERS s.n. (GRA); Mullers pass summit (**-DC**), PUFF 780425-1/1 (WU);
Drakensberg, Mountainprospect (= Mt. Prospect?) (**-DB**), REHMANN 7011 (BM,
K). – **2730** (Vryheid): [Farms] Tweekloof, Altemooi [ESE of Wakkerstroom]
(**-AD**), THODE A 1162 (K, NH, PRE; STE sub 4590); Paulpietersburg D.,
Paulpietersburg (**-BD**), GALPIN 10877 (K, PRE); Utrecht D., Farm Klipspruit
(**-CD**), BREYER 16990 (PRE). – **2731** (Louwsburg): Farm Leeuwnek [23 km SE of
Vryheid] (**-CC**), THODE s.n. sub STE 5046 (STE); Ngome Forest, ca 15 km from
Forest Stn. on rd. to Vryheid (**-CD**), GERMISHUIZEN 2125 (PRE). – **2828** (Beth-
lehem): Royal Natal National Park, path from Tendele Camp to Bushman's
paintings (**-DB**), PUFF 800106-3/1 (WU); O.F.S./Lesotho/Natal border: summit
of Drakensberg, Beacon Buttress nr. Mont-aux-Sources (**-DB, -DD**), GALPIN 9422
(K, PRE). – **2829** (Harrismith): Natal side of Van Reenen pass, Farm Nolens
Volens (**-AD**), PUFF 781125-2/1 (WU); Collin's pass (**-BA**), PUFF 840819-1/3 (PRE,
WU); Kliprivier D., Elandslaagte (**-BD**), SHIRLEY 310 (NU); Oliviershoek pass
(**-CA**), THODE s.n. sub STE 4593 (STE); Drakensberg, Cathedral Peak area (**-CC**),
ESTERHUYSEN 12884 (BOL). – **2830** (Dundee): Dundee D., Indumeni (= Ndumeni)
Mt. (**-AA**), ACOCKS 10289 (NH, PRE); ca 1.5 km S of Nqutu (**-BA**), PUFF 790303-
1/3 (WU). – **2831** (Nkandla): Babanango (**-AC**), KING 253 (PRE); Nkandla Forest
Res., ca 19 km S of Nkandla (**-CA**), PUFF 790303-5/1 (WU); Mtonjaneni D., ca
7.5 km E of Melmoth (**-CB**), VENTER 3373 (PRE); Eshowe (**-CD**), GERSTNER 2787
(K, MO, NH, PRE), LAWN 138, 557, 1905 (NH); Mtunzini D., Ngoye (= Umgoye)
Forest Res. (**-DC**), HILLIARD & BURTT 5626 (E, K, NH, NU), PUFF 791217-2/1
(WU), VENTER 3052 (PRE); –, Umlalazi Nature Res. (**-DD**), WARD 4316 (K, NH,
NU, PRE). – **2832** (Mtubatuba): Palm Ridge Farm, 7 km N of Mtubatuba (**-AC**),
HARRISON 484 (K, NH, PRE, WAG); Hlabisa D., N coast of St. Lucia estuary
(**-AD**), FEELY, TINLEY & WARD 4 (K, NH, NU, PRE). – **2929** (Underberg):
Drakensberg, Cathkin Peak (**-AB**), GALPIN 11891 (K, PRE); Giant's Castle Game
Res., Langalibalele pass (**-AD**), PUFF 800116-2/1, -3/1 (WU); Estcourt (**-BB**),
SCHLECHTER 3350 (GRA, PRE); Estcourt D., Kamberg Nature Res., Stillerust Vlei
(**-BC**), HILLIARD & BURTT 8731 (E, K, MO, NU, PRE); Fort Nottingham
commonage (**-BD**), PUFF 800115-5/1 A, -5/1 B (WU); Underberg D., Garden
Castle Forest Res., Mlambonja valley (**-CA**), HILLIARD & BURTT 14967 (NU,
WU); –, Bamboo Mt. (**-CB**), GRICE s.n. (NU); Bulwer (**-DD**), BAYER 329 (NH,
NU). – **2930** (Pietermaritzburg): Mooi River Stn. (**-AA**), KUNTZE s.n. (K); Umvoti
D., Seven Oaks-Rietvlei (**-BA** to **-AB**), GORDON-GRAY 4601 (NU); Lidgetton
(**-AC**), MOGG 6689 (PRE); New Hanover D., Impolweni (= Mpolweni) (**-AD**),
RUMP s.n. (NU; NH sub 20335); Greytown (**-BA**), WYLIE s.n. (K, S), sub NH 20466
(NH), sub NH 22393 (BOL, NH, PRE); Noodsberg, Farm Kronsberg (**-BD**),
THODE s.n. sub STE 5234 (STE); Dargle (**-CA**), BEWS 3335 (NU); Pietermaritzburg,
Hawthorn's Hill (**-CB**), ALLSOPP 805 (NH, NU, PRE), 943 (NU); Richmond D.,
Keerom, (nr.) Byrne (**-CC**), STREY 10851 (E, NH, NU, PRE, WAG); Table Mt.
(**-DA**), KILLICK 323 (NU, PRE, SRGH), McCLEAN 191 (PRE); Inanda (**-DB**),
MEDLEY WOOD 433 (BM, K, SAM), 695 (? 645) (BOL, K, NH), STREY 4848 (K, NH,
PRE); Umbumbulu (**-DC**), COLEMAN 139 (NH); nr. Pinetown (**-DD**), MEDLEY
WOOD s.n. (G, GRA, MO, US). – **2931** (Stanger): Tugela Beach (**-AB**), JOHNSON
377 (NBG); Durban D., Wentworth (The Bluff) (**-CC**), SANDERSON 928 (K, NH),
WARD 6098 (NH, PRE), 6464 (E, NU, PRE). – **3029** (Kokstad): Natal/Transkei
border: Riverside, The Valley (**-BA**), PHILLIPS s.n. sub J 35684 (J); –, Zuurberg
(**-BC**), PUFF 790416-1/2 (WU); Weza, Belfast outlook (**-DA**), STREY 10925 (E, NU,
PRE), 10945 (NH); Harding, Bedford Farm (**-DB**), STREY 9158 (BR, E, K, LISU,

NH, NU, PRE). – **3030** (Port Shepstone): Maxwell, Ixopo (**-AA**), Evans 280 (NH); Dumisa Stn., Enadale (? spelling) (**-AD**, or -BC?), Rudatis 1297 (BM, E, K, PRE); Umbogintwini (**-BB**), Medley Wood 12437 (US); Umdoni Park (**-BC**), Ward 1308 (NU); Beacon Hill, Port Edward- Izingolweni rd. (**-CC**), Puff 790423-2/1 (WU); Shelly Bay (Beach) (**-CD**), Mogg 11916, 12184 (PRE). – **3130** (Port Edward): Port Edward (**-AA**), Strey 7450 (B, BR, K, M, NH, NU, PRE, S). – No locality given: "Zululand", Gerrard & Mc'Ken 318 (BM, K, P), "Natal", Gueinzius 138 (P). – Additional collections: Acocks 10677, 11401 (NH, PRE); Barker 5157 (NBG); Bayer & McClean 12, 139 (PRE), 161 (GRA, K, NU, PRE); Coleman 230 (NH); Drège s.n. (E, G × 3, GH, K, LY, S, W × 3); Edwards 3134 (K, NH, NU, PRE); Evans 91 (NH); Fisher 1652 (PRE); Franks s.n. sub Medley Wood 12160 (K, NH), sub NH 12975 (NH); Galpin 10380 (PRE); Gordon-Gray 118 (NU), 6214 (E, NU); Gueinzius 468 (S), s.n. (G; SAM sub 16062); Hafström & Acocks 1423, 1424, 1425 (PRE, S); Hilliard & Burtt 7897 A (E, NU), 11721 (E, K, NU, WU), 13756, 13757 (E, NU, WU); Humbert 15130 (P); Huntley 299 (PRE); Johnston 175, 552 (E); Junod 247 (? 241) (G); Killick 1083 (K, LMA, PRE), 1668 (K, NH, PRE); Killick & Vahrmeijer 3639 (PRE × 2); Kuntze s.n. (K, US); McClean 859 (K, NH, PRE), 884 (MO, NH, PRE); McKnown 83, 105 (NU); Medley Wood 230 (BM, E, MO, NU), 4531 (K, NH), 6691 (SAM, US), 7406 (GOET, LY, PRE); Mogg 4073, 4615, 5578, 7106 (PRE), 13520 (K, PRE, SRGH); Moll 1164 (PRE), 1454 (K); Nicholson 1193 (PRE); Pole-Evans 3902 (K, PRE); Puff 790423-1/3, -3/1, 790427-1/1, -2/2, -4/2, 791216-1/1, 791217-3/1, 800106-1/2, 800115-1/1, -2/1, -3/2, -4/1, 800116-6/2 (WU); Ruch 2058 (PRE); Rump s.n. (NU; NH sub 20243); Schlechter 6764 (GRA); Shirley 239 (NU); Smith 5629, 5743 (PRE); Strey 4907 (K, NH, PRE), 6358 (NH, PRE), 7104 (K, NH, NU, PRE), 7115 (BR, K, NH, NU, PRE, S), 10616 (NH); Trauseld 754 (NU, PRE); Venter 402 (NH); Wall s.n. (S × 3); West 29 (PRE); Wright 1686 (NU).

Cape – **3027** (Lady Grey): Barkly East D., summit of Ben McDhui (**-DB**), Galpin 6648 (BOL, K, PRE); Hilliard & Burtt 16378 (NU, WU). – **3029** (Kokstad): Mt. Currie (**-AD**), Goosens 351 (PRE); nr. Clydesdale (**-BD**), MacOwan 733 (PRE); Cedarville- Mt. Frere rd., Farm Milkfontein (= Melkfontein) (**-CA**), Story 544 (PRE); Kokstad (**-CB**), Mogg 5217 (PRE); Mt. Ayliff D., Insiswa Mt. (**-CC, -CD**), Hilliard & Burtt 6567 (E, K, MO, NU, PRE), Schlechter 6435 (B, BOL, BR; in PRE sub TVL Museum no. 3298), Strey 10775 (E, K, MO, NH, NU, PRE, SRGH, WAG). – **3126** (Queenstown): Fincham's Nek (**-DD**), Galpin 2547 (COI, GRA, PRE × 2, S). – **3127** (Lady Frere): Elliot- Maclear D. boundary, Baster-voetpad (**-BB**), Hilliard & Burtt 16670 (NU, WU). – **3128** (Umtata): Bazija (**-CB**), Baur 166 (E, K), 266 (K, MO, SAM). – **3129** (Port St. Johns): Ntsobane (**-BC**), Strey 9010 (K, NH, PRE, SRGH); Port St. Johns D., Umgazi (= Mngazi) R. mouth (**-CB**), Wells 3462 (GRA, K, PRE, S); –, Port St. Johns (**-DA**), Wager s.n. sub PRE 41634, 41635 (PRE). – **3225** (Somerset East): upper part of Bruintjes-hoogte Berg above the Station (**-CB**), Burchell 3033 (K); Boschberg nr. Somerset East (**-DA**), Bolus 1759 (BOL), Levyns 5569 (BOL), Scott Elliot 136 (PRE), 623 (E). – **3226** (Fort Beaufort): Alice- Seymour rd., nr. Brambledene (**-DB**), Barker 2907 (NBG). – **3227** (Stutterheim): Stutterheim D., Fort Cunyngham (**-AD**), Schönland 8 (GRA), Sim s.n. (NU); Hogsback (**-CA**), Rattray 77 (GRA); Dohne Peak (**-CB**), Puff 790115-5/1, -5/5, -5/8 (WU); Pirie (**-CC**), Sim 19609 (PRE– mixed with *A. paniculatum*); Breidbach nr. King William's Town (**-CD**), Sim 1173 (PRE); Amabele (**-DA**), Moss 5017 (BM, J); nr. Komgha (**-DB**), Flanagan 24 (GH, K, PRE). – **3228** (Butterworth): Willowvale D., Qora River mouth (**-BC**), Hilner 472 (GRA, NBG, PRE); Butterworth- Qolora Mouth rd., just S of Kei Mouth turnoff (**-CB**), Puff 791219-2/1 (WU). – **3320** (Montagu): Langeberg,

Farm Strawberry Hill, between Lemoenshoek and Naauwkrantz (**-DD**), Stokoe s.n. sub NBG 99658 (NBG). – **3322** (Oudtshoorn): George (**-CD**), Michell s.n. (GH), sub PRE 41636 (PRE), Moss 5989 (J); George D., Woodville (**-DC**), Galpin 4106 (PRE). – **3323** (Willowmore): Coldstream (**-DC**), Daly & Sole 261 (GRA); throughout Tsitsikamma- Storms River commonage (**-DD**), Laughton s.n. sub BOL 25129 (BOL). – **3324** (Steytlerville): Zuuranys pass above Kareedouw (**-CD**), Thompson 1814 (PRE, STE). – **3325** (Port Elizabeth): Zuurebergen (= Suurberg), nr. Bonjesrivier (**-AD**?), Drège s.n. (E, G × 3, GH, MO, P × 2, S × 2, SAM, W × 3); –, top of Zuurberg range (**-BC**), Archibald 5105 (PRE); –, East ridge (**-BD**?), Archibald 5206 (K, PRE × 2, W); Uitenhage D., Winterhoek Mts., Woodfield's Krantz (**-CB**), Fries, Norlindh & Weimarck 694 (E, S, SRGH, STE); Loerie Plantation (**-CC**), Dix 138 (GRA); Theescomb nr. [W of] Port Elizabeth (**-CD**), Paterson 2069 (BOL); by the Backens River nr. Algoa Bay (**-DC**), Burchell 4349 (GH, GOET, K, M, W). – **3326** (Grahamstown): Swartwatersberg, SW of Riebeek-East (**-AA**), Puff 840922-1/4 (PRE, WU); Grahamstown and surroundings (**-AD, -BC**): Barker 733 (NBG), Comins 973 (GRA, K, PRE, W), Dyer 440 (NBG, PRE), Glass 233 (SAM), MacOwan 275 (BM, K), Puff 790117-2/2 (WU). – **3327** (Peddie): East London (**-BB**), Hilner 243 (PRE), Rattray 161 (GRA), Wood 2692 (K, PRE). – **3419** (Caledon): Salmonsdam Nat. Res., E of Stanford (**-BC**), Puff 840918-1/7 (PRE, WU); Groenkloof, inland (NE) of Pearly Beach (**-DA**), Puff 840919-3/1 (PRE, WU); Klein Hagelkraal, Thompson 3900 (STE). – **3421** (Riversdale): nr. Riversdale (**-AB**), Schlechter 1999 (BM, BOL, G × 2, GRA, J, K, LY, M, W, WU × 2). – **3422** (Mossel Bay): Knysna- George rd., Groenvlei (**-BB**), Martin 4545 (PRE). – **3423** (Knysna): Concordia nr. Knysna (**-AA**), Keet 706 (GH, NBG, PRE, STE); Plettenberg Bay (**-AB**), Compton 4485 (BOL), Pappe s.n. (K, S). – **3424** (Humansdorp): Clarkson (**-AB**), Thode A 855 (GH, K, NH, PRE); Humansdorp (**-BB**), Compton 23349 A (NBG). – Not traced: Hofman's Bosch, Britten 1370 (GRA); Mossel Bay D., above Longfontein, Muir 2192 (BOL, PRE). – Imprecise localities: "CBS", collection Burman s.n. [herb. Delessert] (G), Mund(t) & Maire [herb. Bergius] s.n. (G); Oldenburg (? Masson) s.n. (BM); [herb.] Thunberg s.n. (G, GOET, S × 3); "Kaffraria", Cooper 304 (BM, BOL, W × 2); –, "Kreilis country", Bowker 595 (K, PRE, S); "Faku's Territory", Sutherland s.n. (K); Van Stadensriviersberge, nr. Grahamstown and on the Makasaniriver nr. the Chumiberg, Ecklon & Zeyher 2309 (BOL, FI, G × 2, GOET, M, MO, S, SAM, W × 2; GH, PRE as '4.10'). – Additional collections: Acocks 9178 (PRE); Archibald 4972 (BM, K, PRE); Barker 974 (NBG); Bolus s.n. (BOL); Compton 4485, 10463, 14372, 17031, 19165, 19179 (NBG); Drège '4.10.' (E); Dyer 306 (GRA, PRE); Ecklon & Zeyher (Ecklon?) 596 (BM, E, K, PRE, SAM, W), '47-5' (S), s.n. (S); Flanagan 1218 (BOL, GRA, PRE, SAM), 1818 (BOL, NBG, NH, PRE, SAM); Gillett 1214 (BOL, K); Keet 423 (GRA, PRE, STE); Kotsokoane 45 (J); Krauss s.n. (1205?) (M, W); Leighton 2702 (BOL, PRE); Pegler 790 (BOL, PRE), 1254 (BOL, PRE, SAM); Puff 790114-6/3, 790115-2/1, 791219-4/1 (WU); Rennie 184 (BOL); Rogers 12722 (GRA); Salter 6943 (K); Schönland 3220 (GRA, PRE); Story 2262 (GRA, PRE), 3336 (PRE); Tyson 576 (BOL, G, GH, K, PRE, SAM, W), 733 (BM, BOL, G, GH, K, SAM, W), 1318 (BOL, G × 2, PRE, SAM, W × 2), 2131 (PRE × 2; SAM – mixed with *Galopina crocyllioides*), s.n. (BR); Zeyher 2716 (LY, S, SAM).

26. *Anthospermum pachyrrhizum* HIERN in OLIVER, Fl. Trop. Afr. **3**: 26 (1877).

Syntypes: "Abyssinia, QUARTIN-DILLON & PETIT, SCHIMPER". Ethiopia, Tigre Admin. Region, from Maigoigoi (Mai-Gouagoua) to Dobra (Debra) Sina, QUARTIN-DILLON & PETIT s.n. (P, lecto.!, selected here; W!); –, (Mt.) Scholoda, QUARTIN-DILLON & PETIT s.n. (BR!, K!, P!); –, Hedscha (Mt.), SCHIMPER 327 (BM × 2!, WU!; B†); –, Bellaka, SCHIMPER 538 (FI × 2!, G!, W!); Gonder Admin. Region, without locality [Semien], SCHIMPER 128 (BR!, G!, K!, P × 3!); –, Debre Eski, SCHIMPER 167 (FI!, G!, P!, S!), 2063 (G!, S!, W!)[1].

– *A. hirsutum* sensu A. RICH., Tent. Fl. Abyss. **1**: 346 (1847), non DC. [Prod. **4**: 580 (1830); see 36. *A. hirtum*].

– *A. unisetum* HOCHST., nom. nud.

Dioecious or ± dioecious (♂ plants occasionally with odd ⚥ or ♀ flowers), ± rounded or sometimes ± cylindrical many-stemmed dwarf shrubs or subshrubs to 40 cm tall, with a thickened base and a long, woody tap root; woody base massive, ± disk-like and to 5 cm in diameter in individuals exposed to fire. Stems ca (15) 20–40 cm long, ca 2–4 mm in diameter near the base and 0.5–2 mm in the midstem region, ascending to erect, sparsely to much-branched; younger parts terete, often reddish-purplish tinged, ± densely covered with whitish papillae 0.1 mm or less long, older parts reddish-brown to greyish, glabrescent, epidermis peeling; internodes shorter than to as long as the leaves (seldom longer). Leaves decussate or, rarely, in whorls of 3, often pseudo-verticillate; blades (5) 9–18 (25) × (0.5) 1–2 (3) mm, narrowly elliptic-lanceolate to linear-lanceolate, the flat or ± revolute margins often with short, forwardly directed whitish prickle-like hairs, surfaces mostly glabrous, upper surface often with large, conspicuous epidermis cells; apices acute; petioles subobsolete; stipular sheaths slightly hairy, cup-shaped, ca 0.7 mm long and 1.3–1.4 mm wide, with a gland-tipped seta ca 0.5–1 (2.5) mm long, rarely with an additional, smaller seta on either side. Inflorescences ± inconspicuous, made up of ± sessile, several- to ± few-flowered axillary clusters (♀ inflorescences often more-flowered and ± more conspicuous than ♂). Flowers subsessile or sometimes with pedicels to ca 0.7 mm long (♀, especially when in fruit); ♂, ♀ (occasionally also odd ⚥). Corolla 4-merous, greenish-yellow to pale yellow, outside sometimes purplish tinged and with a few, short whitish hairs. ♂ (⚥): tube 0.7–1 mm long, funnel-shaped,

---

[1] For the exact and up-to-date spellings and grid references of the above collecting sites see GILLETT (1972).

Fig. 83. *Anthospermum pachyrrhizum. a* plant, ca 30 cm in diam., on basalt cliff, N of Gonder, Gonder Admin. Reg., N Ethiopia (PUFF 811005-4/1). *b* mericarp, dorsal side. *c* mericarp, ventral side, upper part with persistent calyx lobes. *d* carpophore; note pedicel below. – *b–d* SEMgraphs; herbarium material (ROBERTSON 127); *b, d* × 16; *c* × 29

lobes 1.8–2.5 × 0.3–0.7 mm, lanceolate, recurved; stamens 4, filaments 1–1.4 mm, anthers 2–2.4 × 0.5–0.8 mm; ♂: rudimentary ovary well discernible. ♀: tube 0.3–0.5 mm long, lobes 0.3–0.7 × 0.2–0.3 mm, ± linear-lanceolate, ± erect; style 0; stigmas 2, 6–9 mm long (in odd ♀ 1.5–2.2 mm long), greyish(-green); ovary 0.8–1 × 0.6–0.8 mm, (densely) hairy, crowned by 4 often unequal calyx lobes. Fruits reddish-brown, supported by small, ± U-shaped carpophores; each mericarp 1.8–2.1 × 0.7–0.9 mm, ± oblong, dorsal side convex, densely covered with white spreading hairs ca 0.3–0.7 mm long, ventral side concave and with a median vertical ridge; commissure large; mericarps crowned by 2 calyx lobes ca (0.4) 0.6–1.3 (1.7) mm long, one often shorter than the other. – Fig. 83.

Chromosome Number (Table 3): n = 11.
Average Pollen Diameter (Fig. 48): 24.4–26.8 μm.

Habitat (Fig. 83 a): Mostly growing in cracks of rocks in cliffs or on steep, rocky road banks; in Ethiopia usually over basalt. In the Sudan also recorded from upland (rocky?; occasionally or ± regularly burnt?) grassland. – Ca 1 700–3 100 m.

Distribution (Map, Fig. 76 a): Ethiopia (from the Tigre Admin. Region S to the N Shewa Admin. Region) and Sudan (Jebel Marra and N outlier Jebel Gourgeil; cf. WICKENS 1976 and QUÉZEL 1969).

Critical Remarks: A. *pachyrrhizum* appears to be (± remotely?) allied to the widely distributed, essentially afromontane species *A. herbaceum* (25., above). In Ethiopia (from where both species are recorded), *A. pachyrrhizum* and *A. herbaceum* have, in the past, often been confused. *A. pachyrrhizum* differs primarily in its fruits (covered with long hairs, crowned by often unequally long, conspicuous persistent calyx lobes; absence of a broad, longitudinal groove between the mericarps. Compare Figs. 83 b and 35 j–k) and in its shorter, funnel-shaped (δ) corolla tubes.

A few collections from the Yemen Arab Republic (Y.A.R.), at present grouped with *A. herbaceum* (e.g., LAVRANOS & NEWTON 16002; E), seem to closely approach *A. pachyrrhizum* in having somewhat hairy fruits with rather distinct calyx lobes. As an E extension of the (E–W) distribution range of *A. pachyrrhizum* (cf. map, Fig. 76 a) to the mountains of the Y.A.R. would, at least theoretically, be plausible, it might be worthwhile to pay special attention in the field to the *A. herbaceum* – *A. pachyrrhizum* complex in that area.

Collections

Ethiopia. – **Tigre:** from Maigoigoi (= Mai-Gouagoua) to Debra Sina, QUARTIN-DILLON & PETIT s.n. (P, W); (Mt.) Scholada, QUARTIN-DILLON & PETIT s.n. (BR, K, P); Hedscha (Mt.), SCHIMPER 327 (BM × 2, WU); Mt. Kubbi [E of Axum], SCHIMPER 732 (P – mixed with *A. herbaceum*); Bellaka, SCHIMPER 538 (FI × 2, G, W). – **Gonder:** Gondar and surroundings, CHIOVENDA 2401 (FI); – , Mt. Quatelé, PICHI-SERMOLI 760 (FI, K); – , Mt. Iabéc and Mt. Cocô (= Zoz), PICHI-SERMOLI 761 (FI); – , Bambelo, CHIOVENDA 1051 (FI); – , 1.5–16 km N of Gondar, SEEGELER 2199 (WAG), MESFIN & KAGNEW 1846 (ETH), PUFF, MANTELL & ENSERMU 811005-4/1 (ETH, K, UPS, WU); Simien, SCHIMPER 128 (BR, G, K, P × 3); – , Debre Eski, SCHIMPER 167 (FI, G, P, S), 2063 (G, S, W); – , Wolkefit pass, SCOTT 336, 337 (BM, K). – **Gojam:** Debre Markos Awraja, 15 km N of Dejen, MESFIN & KAGNEW 1613 (ETH); – , 5 km N of Dejen, MESFIN & KAGNEW 1641 (ETH). – **Shewa** (= **Shoa):** Bole gorge, upper part of Mugher River system, FRIIS, GILBERT, RASMUSSEN & VOLLESEN 1156 (BR, C, ETH, K); nr. 'Portuguese Bridge', nr. Debre Libanos, DE WILDE & DE WILDE-DUYFJES 8661 (BR, WAG), PUFF, MANTELL & ENSERMU 810930-2/1 (ETH, K, UPS, WU); ca 2 km from Goha Tsion, edge of Blue Nile Gorge, PUFF, MANTELL & ENSERMU 811011-2/1 (ETH, K, UPS, WU).

Sudan. – **Darfur Prov.:** Jebel Marra massif, DE WILDE & DE WILDE-DUYFJES 5532 (BR, K, MO, WAG), JACKSON 2655 (B, K), LYNES s.n. (BM), 147 b (BM, K, US), 150 (BM, K), ROBERTSON 127 (BM, K), SAHNI 435 (K), SANDISON 58, 59, 60 (all BM), WICKENS 1734, 2654 (K); Jebel Gourgeil (no specimens seen; cf. QUÉZEL 1669).

27. *Anthospermum palustre* HOMOLLE ex PUFF, spec. nova, *A. thymoidi* affine sed praecipue mericarpiis glabris differt.
Type: Madagascar, around Antsirabe, 1 900 m, PERRIER DE LA BÂTHIE 3809 (P, holo.).

♀(?) perennial herbs. Stems to ca 60 cm long, prostrate to ± ascending, sometimes rooting at the lower nodes, diffusely branched, ca 1.5 mm in diameter in the midstem region; younger parts ± 4-angled, reddish-brown, scabrid particularly below the nodes, older parts reddish- to greyish-brown, glabrescent; internodes (much) longer than the leaves. Leaves decussate, ± pseudo-verticillate; blades 8–13 × 1.5–2.5 mm, (linear-)lanceolate to narrowly oblanceolate, glabrous, margins flat to (slightly) revolute; apices acute to ± acuminate; petioles subobsolete; stipular sheaths scabrid, cup-shaped, ca 0.7–2 mm long, with a narrowly triangular, frequently recurved median seta ca 0.9–1.2 mm long, some-times with an additional, smaller seta on either side. Inflorescences inconspicuous, made up of ± sessile, several- to few-flowered axillary clusters. Flowers subsessile, ♂. Corolla 4-merous, tube ca 0.9–1.5 mm long, ± cylindrical to narrowly funnel-shaped, lobes 1–1.7 × 0.5–0.7 mm, lanceolate, recurved; stamens 4, filaments 1.4–2 mm long, anthers 1–1.2 × 0.2–0.3 mm; style 0–0.3 mm long; stigmas 2, 1.7–2.5 mm long, greyish; ovary 0.7–0.9 × 0.7 mm, glabrous, not crowned by distinct calyx lobes. Fruits reddish-brown, supported by small but broad, ± U-shaped carpophores, mericarps divergent, held at angles of ca 45–60° to each other, free (at least) above the middle; each mericarp 1.5–1.9 × 0.8–

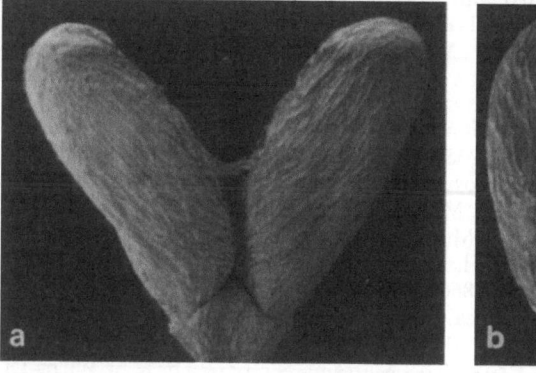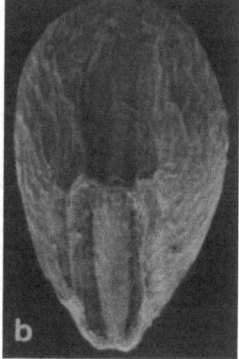

Fig. 84. *Anthospermum palustre* (PERRIER DE LA BÂTHIE 3809). *a* fruit in side view. *b* mericarp, ventral side. – SEMgraphs; herbarium material; *a–b* × 35

0.9 mm, oblong to ± elliptic, dorsal side convex, glabrous, ventral side with a prominent, short and broad median vertical ridge; commissure very short, ± narrow; mericarps not crowned by calyx lobes. – Fig. 84.

Chromosome Number: Unknown (but presumably diploid).
Average Pollen Diameter (Fig. 48): 25.0 µm.
Habitat: Growing in (possibly confined to?) peat bogs ("tourbières")[1]. [No further detailed habitat information available]. – Ca 1 900 m.
Distribution (Map, Fig. 85 c): Central Madagascar. So far only known from around Antsirabe and Ambositra and from Reserve Nat. No. V.

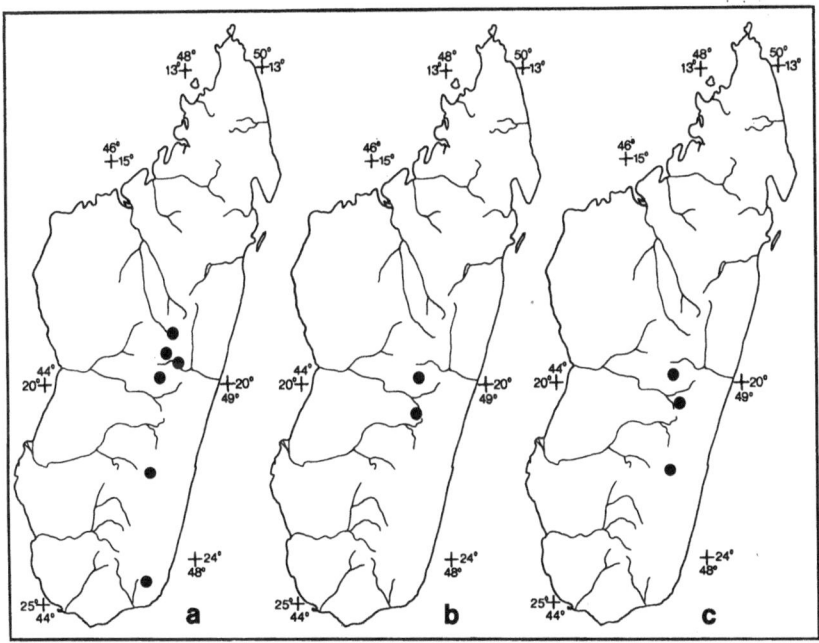

Fig. 85. Distribution of *Anthospermum* in Madagascar. *a–b A. thymoides, a* subsp. *thymoides, b* subsp. *antsirabense, c A. palustre*

Critical Remarks: A. palustre is doubtlessly very closely allied to *A. thymoides* (28., below); leaf size and shape, inflorescences and flowers are similar. It, however, differs from the latter in its often more prostrate habit, in its habitat (peat bogs!) and, most of all, in its differently shaped,

---

[1] See STRAKA (1960) for details on peat bogs in Madagascar.

glabrous fruits (compare Figs. 84 and 37 c). Also the anatomy of the fruits of *A. palustre* is unusual in that the mericarps have two (air-filled?)"pockets" (cf. section, Fig. 40 c), a feature not observed in any other species of the genus.

I have serious doubts as to whether the plants of *A. palustre* are always ⚥. Too little material of the species is available at present.

The collection DEANS COWAN s.n. (BM) is provisionally placed here. It agrees with *A. palustre* in its vegetative features, but the absence of flowers and fruits make a definite determination impossible.

For a discussion of the possible relationships and affinities of the species pair *A. palustre* and *A. thymoides* to other species see p. 328.

Collections

Madagascar. – **1947**: around Antsirabe, PERRIER DE LA BÂTHIE 3809 (P). – **2047**: around Ambositra, PERRIER DE LA BÂTHIE 12180 (P). – **2246**: Reserve Nat. No. V, Sendrisoa Canton, Ambalavao Distr., RAZAFINDRAKOTO RN 3550 (TAN). – Not traced: Ankafana, DEANS COWAN s.n. (BM).

28. *Anthospermum thymoides* BAKER in J. Linn. Soc. (Bot.) **20**: 171 (1883). Type: Central Madagascar, BARON 2005 (K, holo.!; P, iso.!).

Non-dioecious [⚥ + ♀ (and perhaps also ♂ and/or ♀?)] subshrubs or perennial herbs with ± woody rootstocks. Stems ca 10–30 (40) cm long, ascending to prostrate, sometimes rooting at the lowermost nodes, usually ± much-branched, ca 1–2 mm in diameter in the midstem region; younger parts ± 4-angled to terete, reddish-brown, scabrid or with whitish spreading hairs ca 0.1–0.2 mm long, older parts reddish to greyish-brown, glabrescent; internodes (much) longer than the leaves. Leaves decussate, pseudo-verticillate; blades 4–8 (12) × 1–2.5 (4) mm, ± linear-lanceolate or (narrowly) oblanceolate to obovate, glabrous or ± densely covered with whitish spreading hairs ca 0.2–0.3 mm long; margins usually revolute; apices acute to acuminate; petioles subobsolete; stipular sheaths scabrid or shortly hairy, (broadly) cup-shaped, ca (0.3) 0.5–1.4 mm long, with a narrowly triangular, erect to spreading-recurved median seta ca 0.6–1.4 mm long, sometimes with an additional, minute seta on either side. Inflorescences ± inconspicuous, made up of ± sessile to slightly elongated, several- to ± few-flowered axillary clusters. Flowers subsessile, ⚥, ♀. Corolla 4-merous, yellowish(-green), glabrous or hairy on the outside. ⚥: tube 0.7–1.4 mm long, ± cylindrical to narrowly funnel-shaped, lobes (1) 1.2–2.4 × 0.2–0.7 mm, ± lanceolate, recurved; stamens

4, filaments 1.4–2 mm long, anthers (0.7) 1–1.7 × 0.2–0.3 mm; gynoecium as in ♀ but stigmas usually shorter. ♀: tube 0.3–0.5 mm long, cylindrical, lobes 0.3–0.7 × 0.1–0.2 mm, ± linear-lanceolate, ± erect; style 0–0.5 mm long; stigmas 2, 2–4.7 mm long, greyish; ovary 0.7–0.9 × 0.5–0.9 mm, very shortly hairy, not crowned by distinct calyx lobes. Fruits reddish-brown, supported by very small, ± U-shaped carpophores, mericarps ± kidney-shaped in side view, touching each other only in the lower quarter; each mericarp ca 1.4–1.7 × 1–1.5 mm, broadly obovate to ± round in dorsal view, ± densely covered with whitish, often curved and flattened papillae or very short hairs usually less than 0.1 mm long, ventral side with a very short prominent median vertical ridge; commissure very short, narrow; mericarps not crowned by calyx lobes. – Fig. 37 c.

Two subspecies are recognized:

28 (a). Subsp. *thymoides*

Stems and stipular sheaths scabrid; leaves to 8 (12) mm long, glabrous; corolla glabrous.

Chromosome Number (Table 3): 2 n = 22.
Average Pollen Diameter (Fig. 49): 27.3–27.4 µm.
Habitat: Growing on (dry) rocky hills, in grassland or in riverbank vegetation; sometimes also in secondary vegetation ("savoka") and in disturbed sites (waste ground, plantations, edge of rice fields, etc.). – Ca 1 400–2 500 m.
Distribution (Map, Fig. 85 a): Madagascar. From around Antananarivo S to the Andringitra massif and to Mt. Itrafanaomby (= Trafonaomby), NW of Ft. Dauphin.
Critical Remarks: See below.

Collections

Madagascar. – **1847**: around Tananarive (= Antananarivo), GOUDOT s.n. (G), SCOTT ELLIOT 1814 (BM, K). – **1947**: around Miantsoarivo, S of Imerintsiatosika, DESCOINGS 2942 (TAN); Ankaratra massif, slopes below Tsiafajavona, PUFF, IGERSHEIM & RAJEMISOA 850826-1/1 (TAN, WU); –, pine plantation (Manjaka-tompo For. Stn.), PUFF, IGERSHEIM & RAJEMISOA 850826-2/3 (WU); –, route Ambatolampy-Faratsiho, BOSSER 10914 (TAN); Onive basin, Tsinjoarivo, PERRIER DE LA BÂTHIE 3925 (P); around Antsirabe, PERRIER DE LA BÂTHIE 4018 (P). – **2246**: Andringitra massif, PERRIER DE LA BÂTHIE 1426 (P × 2). – **2446**: Mt. Itrafanaomby (= Trafonaomby), HUMBERT 13431 (P). – Imprecise localities ("Central Madagascar", "Madagascar"): BARON 1335, 1395 (galled), 3318 (K), 2005 (K, P); BENOIST s.n. (P).

28 (b). Subsp. **antsirabense** Puff, subspec. nova, caulibus, stipulis laminisque pilosis a subsp. *thymoidi* differt.
Type: Madagascar, around Antsirabe, 1 500 m, Perrier de la Bâthie 4019 (P, holo.).

– *A. antsirabense* Homolle, nom. nud.

Stems, stipular sheaths and leaf blades (densely) covered with whitish spreading hairs ca 0.1–0.3 mm long; leaves ca 4–6 × 1–2 mm; corolla hairy on the outside.

Chromosome Number: Unknown (but presumably diploid).
Average Pollen Diameter: Unknown.
Habitat: "On slopes protected from fire" (further notes quite illegible).
Distribution (Map, Fig. 85 b): Central Madagascar. So far only known from around Antsirabe and a locality further S.

Critical Remarks (for the species as a whole): The two subspecies are easily distinguished by the presence or absence of the characteristic indumentum of whitish spreading hairs. There are, however, no other reliable characters that can be used to separate the two taxa. They are, therefore, provisionally considered as subspecies rather than as two distinct species as was originally suggested by Mme Homolle on herbarium sheets.
Subsp. *thymoides* is very variable in leaf size and shape, and possibly also in habit. In the localities studied on the Ankaratra, plants of subsp. *thymoides* tended to form dense mats between tufts of grasses, while it appears that specimens from other localities are straggling plants with (longer) stems which are supported by the surrounding vegetation.

The fruits of *A. thymoides* and *A. palustre,* characterized by their very small commissures (similar those of *Galopina tomentosa* – compare Figs. 37 c and 84 with Fig. 39 c), are very unusual in *Anthospermum.* This is certainly one of the main reasons for the two species' somewhat isolated position within the genus. The habit (not distinctly shrubby!), certain floral characters (the ± cylindrical corolla tubes in particular), the sex distribution (non-dioecious!) and their main occurrence in wetter habitats (at least *A. palustre*) may, however, point to and suggest a ± distant (?) affinity to *A. herbaceum* (25.) from the African mainland.

Collections

Madagascar. – **1947:** around Antsirabe, Perrier de la Bâthie 4019, 16728 (P). – **2047:** ca 300 km [from Antananarivo] on rd. to Fianarantsoa [= N of Ambatofitorahana], Descoings 2126 (TAN).

29. *Anthospermum pumilum* Sonder in Harvey & Sonder, Fl. Cap. **3**: 31 (1865); Puff in Fl. S. Afr. **31**, 1 (2): 25 (1986).
Type: South Africa, O.F.S., Caledon River[1], Zeyher s.n. (S, holo.!; LY, SAM sub no. 16064, iso.!).

[− *A. rigidum* sensu auctt. Afr. austr., non Ecklon & Zeyher (1836)]

= *A. humile* N.E.Br. in Kew Bull. **1895**: 145 (1895).
Type: South Africa, Natal, Ulundi [2929-BC][2], Evans 370 (K, holo.!; NH, J, NU, PRE, WU − photo!).

= *A. ericoideum* K. Krause in Bot. Jahrb. **39**: 570 (1907); Engler, Pflanzenwelt Afr. **1**: 574, Fig. 508 (1910); Launert & Roessler in Merxm., Prod. Fl. SWA **115**: 8 (1966).
Type: South West Africa (Namibia), Auasberge, Damara-Namaland, Dinter 291 (B, holo. †?).

= *A. pumilum* Sonder var. *pilosum* Phillips in Ann. S. Afr. Mus. **16**: 112 (1917).
Syntypes: Lesotho, Leribe Plateau, A. Dieterlen 629 (SAM, lecto.!, selected here, J, NU, WU − photo!; GRA!, K!, NH!, PRE × 2!, STE × 2!); −, Qoqolosi Peak, Phillips 619 (SAM!), 947 (SAM!)[3].

= *A. spicatum* Suesseng. in Trans. Rhod. Sci. Assoc. **43**: 55 (1951).
Type: Zimbabwe (Rhodesia), Marandellas, Dehn 547 (M, holo.!)[4].

− *A. frutescens* Dinter, nom. nud. [South West Africa (Namibia), Auros, Otavi, Dinter 5659 (B!, BOL!, G!, PRE!, SAM!)].

Two subspecies are recognized. For a discussion of occasional problems concerning the distinction and separation of the subspecies and for details about the alliance of *A. pumilum* to other taxa see Critical Remarks, p. 332 and 344.

---

[1] The type specimen most likely originated from "Trans-Garipina, Umgebung des Caledonrivier, bei Komissiedrift" (collecting site no. 149: cf. Drège 1847 a: 595). Farm Komissiedrift, still existing today, is located near the Rouxville-Smithfield road (3026-BC).

[2] It should be noted that this is not the well-known Ulundi in Zululand. See Hilliard & Burtt (1985) for details on Evans' collecting localities in the Drakensberg.

[3] Phillips also lists "Dieterlen 172 partly" as a syntype, but with the exception of one duplicate in SAM (containing a mixture of plants with hairy and glabrous fruits) all other duplicates (K!, NH!, P!, PRE × 2!, STE × 2! and another sheet in SAM!) only contain plants with glabrous fruits (= "var. *pumilum*" sensu Phillips).

[4] Not to be confused with the *1952* Dehn collections 547/52 and 547'/52.

29 (a). Subsp. *pumilum*
  Synonyms as above.

Non-dioecious (⚥, ⚥ + ♀, ♀, occasionally ♂ or ⚥ + ♂), rounded to cylindrical subshrubs with numerous unbranched stems, ca (5) 8–20 (30) cm long, arising from a massive, often ± rosette-like woody base (if exposed to fire) or much-branched dwarf shrubs to 40 cm tall with thick, woody roots. Stems to 3 (5) mm in diameter near the base in shrubby plants, otherwise stems hardly more than 1 mm in diameter; younger parts reddish-brown to greyish-brown, densely covered with whitish papillae to 0.1 mm long; in shrubby plants older parts glabrescent, epidermis peeling; internodes as long as or shorter than the leaves, less commonly somewhat longer; plants mostly ± densely leafy. Leaves decussate, pseudo-verticillate; blades (4) 6–12 (22) × (0.5) 0.8–1.5 (2) mm, linear, linear-(ob-)lanceolate or narrowly lanceolate, narrowed to the base, ± membranaceous, mostly with a few whitish papillae to 0.1 mm long on the flat to somewhat revolute margins, midrib below often reddish-brown and prominent; apices acute; petioles subobsolete; stipular sheaths covered with whitish hairs to 0.1 mm long, cup-shaped, ca 0.7–1.2 mm long, with 1 (rarely 3) gland-tipped setae, ca 0.2–0.5 (0.7) mm long. Inflorescences inconspicuous, made up of ± sessile, several- to 2-flowered axillary clusters. Flowers subsessile, ♂, ⚥, ♀. Corolla 4-merous, greenish to yellowish, often with whitish hairs or papillae to 0.1 mm long near the tip. ⚥, ♂: tube (0.5) 0.7–1.4 (1.7) mm long, narrowly funnel-shaped, lobes 1.2–1.9 (2.5) × 0.3–0.7 mm, ± lanceolate, recurved; stamens 4, filaments exserted for (0.5) 0.7–1.2 mm, anthers 1–1.8 (2) × 0.2–0.5 mm; ⚥: gynoecium as in ♀, but stigmas often only ca (1.2) 1.7–5 mm long; flowers occasionally transitional ⚥ → ♀ with rudimentary, ± pollenless anthers less than 1 mm long and with smaller corollas. ♀: tube 0.2–0.5 mm long, cylindrical, lobes 0.2–0.5 × 0.1–0.2 mm, linear-lanceolate, ± erect; style 0–0.5 mm long; stigmas 2, 2.4–9.8 mm long, whitish-grey; ovary ca 0.5–0.9 × 0.3–0.8 mm, ± glabrous or a little hairy, crowned by 4 ± indistinct calyx lobes. Fruits reddish-brown, shiny, supported by short, U-shaped carpophores less than ⅓ as long as the fruits; each mericarp (1.5) 1.8–2.4 × 1–1.5 mm, elliptic to obovate, dorsal side convex, ± glabrous or with a few whitish, sometimes ± hook-like hairs less than 0.1 to 0.3 mm long, ventral side plane or shallowly concave and with a ± indistinct median vertical ridge, commissure mostly ± small; mericarps crowned by 2 small, broadly triangular to rounded calyx lobes ca 0.1–0.3 × 0.3–0.4 mm, or calyx lobes ± lacking. – Figs. 7 c, 8 a–b, 87, 89 a.

Chromosome Number (Table 3): n = 11, 2 n = 22.

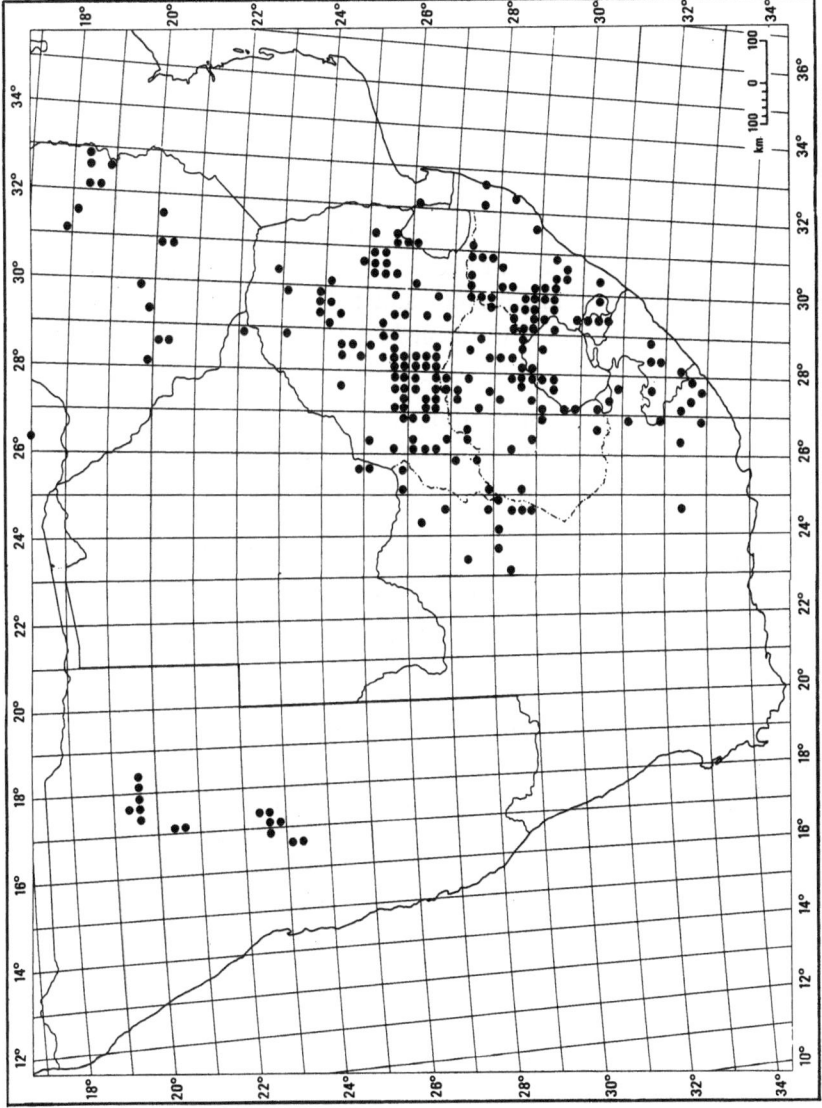

Fig. 86. Distribution of *Anthospermum pumilum* subsp. *pumilum* in S Africa

Average Pollen Diameter (Fig. 48): 23.3–25.8 μm.

Habitat: Primarily in grassveld or in grassy areas in thorn veld, bush-clump veld or open woodland; also in grassy areas on rocky outcrops or in rocky grassland. Mostly in fire-prone habitats. Growing in sandy soil or sandy loam or red clay, somewhat less commonly in coarse gravelly to stony soil. Not uncommonly also found in open, bare patches and disturbed sites (abandoned fields, fallow lands or roadsides). Usually in ± dry habitats, but occasionally in moist ground, in ditches, near dams or ponds (Namibia), at the edge of dambos (S Zambia), or in seepage areas (Natal, Transvaal). – Ca (50) 250–2 300 (2 800) m.

Distribution (Maps, Figs. 73 b and 86): Widely distributed in the E part of S Africa and also occurring in the neighbouring parts of Botswana, in Lesotho, Swaziland, Zimbabwe and S Zambia; in addition, known from a few localities in Mozambique (T) and S Tanzania (T 7). In the W, extending from Namibia N to the highlands of Angola.

Critical Remarks: A. pumilum doubtlessly belongs to the SW and W Cape centered group of species around *A. galioides* (species 31.–34.). Certain ± erect ecotypes of *A. galioides* appear very similar to subsp. *pumilum* and various morphological characters overlap, but a *direct,* close alliance between the two taxa is rather unlikely (entirely different distribution ranges!). The Natal S Coast sandstone endemic, *A. streyi* (30., below), also forms part of this alliance (see there for further details).

Considering the wide range of distribution and the frequency of subsp. *pumilum,* it is not surprising that the subspecies shows considerable variation in some of its characters: Fruit hairiness, for example, varies throughout the range of (both subspecies of) *A. pumilum,* and mericarps with ± hooked hairs ("*A. pumilum* var. *pilosum*") are not uncommon; similarly, the size of the persistent calyx lobes (from small but distinct to ± lacking) is variable. Also leaf width, ranging from linear (leaves almost needle-like) to narrowly lanceolate, is not constant.

A few Angolan collections, in habit and in their slightly broader and more widely spaced leaves, may somewhat approach *A. ternatum* (subsp. *randii*) (19.) (e.g., TEIXEIRA & ALMEIDE 7764, LISC). Also see comments, p. 287.

In Namibia, plants of subsp. *pumilum* fairly often become quite shrubby. "*A. frutescens*" refers to such collections from that territory; they resemble subsp. *rigidum* in some respects but have the smaller flowers typical of subsp. *pumilum.*

In the E part of S Africa (Transvaal, Natal), some populations of subsp. *pumilum* are characterized by an unusual ± cushion- or mat-forming habit (stems ± prostrate and branched, giving rise to numerous

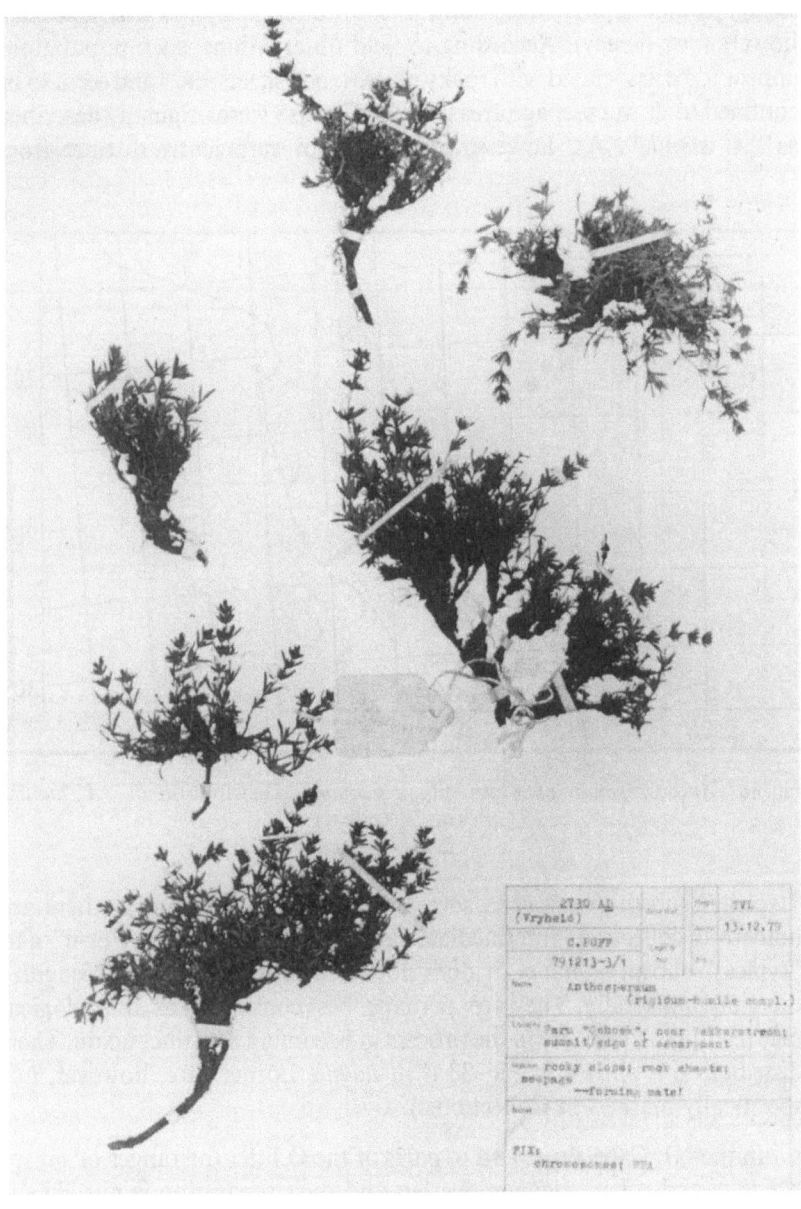

Fig. 87. *Anthospermum pumilum* subsp. *pumilum*. "*A. humile* Form" (Transvaal, Farm Oshoek near Wakkerstroom; Puff 791213-3/1)

short ascending to erect branches) (Fig. 87). Plants from these populations
sometimes have flowers which are slightly larger than those of "typical"
subsp. *pumilum*; plants also seem to show a tendency towards unisexual
flowers (i.e., dioecy). According to field observations, such populations
appear to be associated with rocky places (rock sheets, etc.) and seem to be
confined to ± wet seepage areas. Such "Forms" were originally described
as "*A. humile*". As, however, they are not sufficiently distinct from

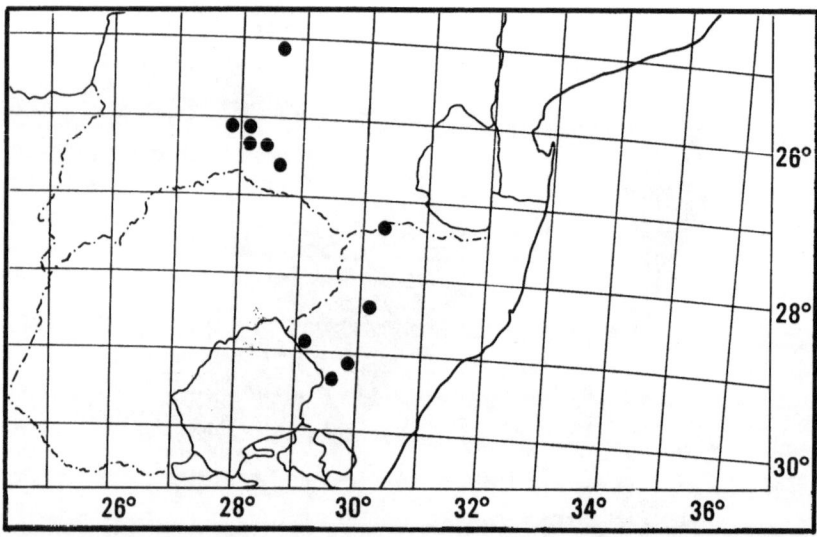

Fig. 88. *Anthospermum pumilum* subsp. *pumilum*. Distribution of "*A. humile*
Forms" (see text)

"typical" subsp. *pumilum* in several other characters, and as there are
numerous collections intermediate between this "*A. humile* Form" and
"typical" subsp. *pumilum,* it does not seem to be justified to recognize
them taxonomically. They are, perhaps, best considered as an ecological
race (ecotypes), possibly in the process of becoming a distinct taxon. Their
distribution is shown in Fig. 88 ("*A. humile* Forms" are, however, not
specifically marked in Collections!).

In the NE Cape Prov. and in parts of the O.F.S., the ranges of subsp.
pumilum and subsp. *rigidum* overlap and their separation is not always
clear-cut. It may also be that in Lesotho (especially in the Berea Distr. and
on the Leribe Plateau) both subspecies, rather than subsp. *pumilum* only,
are present; further field observations, however, would be needed to reach
a final decision on the matter.

Environmental factors, in particular fire-exposure and browsing, often appear to complicate the separation of subsp. *pumilum* and subsp. *rigidum*:

Subsp. *pumilum* typically occurs in fire-prone habitats and plants – due to ± regular burning – remain low (Figs. 7c and 8a). If

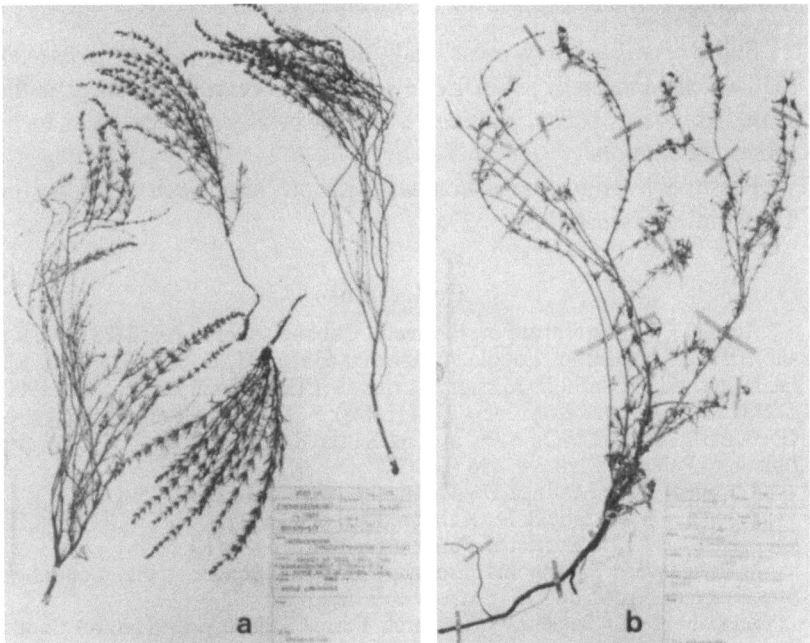

Fig. 89. *Anthospermum pumilum. a* subsp. *pumilum* (O.F.S., ca 10–12 km E of Parys; Puff 800109-1/1), *b* subsp. *rigidum* (O.F.S., N of Bloemfontein, near Glen; Puff 790111-2/1b)

individuals of subsp. *pumilum,* however, are allowed to grow undisturbed for several seasons, they can become quite woody (distinct dwarf shrubs; cf. Fig. 89 a) and then may somewhat resemble plants of the typically quite robust, woody subsp. *rigidum* (usually occurring in less fire-prone habitats). Shrubby plants of subsp. *pumilum,* however, are less robust, are more densely leafy and have thinner, less "tough" and often longer and relatively narrower leaves than subsp. *rigidum* (compare Figs. 89 a and b).

Subsp. *rigidum,* on the other hand, is quite often browsed (while browsed plants were not observed in subsp. *pumilum*). Browsed individuals also remain low and produce thin, unbranched shoots with relatively larger and broader leaves; they may resemble plants of subsp. *pumilum*.

In such cases, a separation of the two subspecies may become difficult and requires some experience. "Typical" subsp. *rigidum,* however, has larger (♀, ♂) flowers. In field studies, moreover, a clear trend to dioecy was observed in populations of subsp. *rigidum*. I suspect that some populations of the latter may be strictly dioecious, while in subsp. *pumilum,* on the other hand, such a trend to the separation of the sexes is absent, and ♂ flowers (individuals) appear to be uncommon.

Subsp. *pumilum* may occasionally form hybrids with *A. herbaceum* (25.) which – especially in S Africa – sometimes occurs in nearby or ± the same habitats as subsp. *pumilum*[1]. Such hybrids, however, seem to be rather uncommon.

Possible hybridogenous contact with *A. hispidulum* (17.) in the Transvaal is discussed on p. 276.

Collections

Angola. – Benguela: at Fort P. Amelia, Cubango, GOSSWEILER 2488 (BM, K), 4027 (BM). – **Huila:** nr. Lopolo, WELWITSCH 5337 (BM, G, K), 5338 (BM, K); Huila, around the Catholic Mission, SANTOS 41 (LISC); Humpata, TEIXEIRA 1907, 2121 (LISC), TEIXEIRA & ALMEIDA 7764 (LISC); – , Serra da Chela, Tchivinguiro, GOSSWEILER 12689 (LISC); – , – , around Sá da Bandeira, HUMBERT 16269 (P); Lubango, Palanca, MENDES 3636 (LISC).

Tanzania. – T7: Mafinga D., Mafinga, KIBUWA 5514 (NHT, WU).

Zambia. – S: Mazabuka D., Kalomo, ROBINSON 2559 (K, M, SRGH), ROGERS 8419 (BOL); – , Tara Protected Forest area, WHITE 6414 (K).

Mozambique. – T: Angónia, around Ulongwe (Ulongué = Vila Coutinho), STEFANESCO & NYONGANI 270 (LMA).

Zimbabwe. – C: Salisbury D., Garch Farm, Salisbury (= Harare) South, STRANG 2324 (SRGH); – , Rua (= Ruwa) R., E of Salisbury airport, PUFF 790122-2/1 (WU); Marandellas D., Marandellas, DEHN 547 (M); – , Diggleford [School], CORBY 560 (K, SRGH); – , Pasture Station, STENT s.n. sub SRGH 5454 (SRGH); Makoni D., Rusape, DEHN 547/52 or 547'/1952 (BR, K, M, MO, SRGH); – , Rusape- Headlands rd., ca 20.5 km E of Headlands, PUFF 790129-2/3 (WU); – , between Rusape and Maidstone, FRIES, NORLINDH & WEIMARCK 4038 (BR, K, M, MO, SRGH, WAG); Gwelo D., ca 10 km S(E) of Gwelo, BIEGEL 413, 1354 (SRGH). – **E:** W of Juliasdale, Vukutu Farm, PUFF 790124-1/2 (WU); Inyanga D.,

---

[1] Putative hybrid collections are:

SCHLECHTER 6400 [Mt. Ayliff (Cape 3029-CD); B, BOL, BR, GRA, PRE × 2]; SCHLECHTER 6435 from the same locality is typical *A. herbaceum.*

COLEMAN 769 [Farm Thornman (Cape 3029-BD); NH]; the plants have stipular sheaths bearing several small setae (rather than one seta as in *A. pumilum*) and leaves covered with short hairs on the surfaces.

WILD 4324 [Belingwe Distr., Mt. Buhwe (Zimbabwe); K, LISC, MO, SRGH], described as "tufted bush to 1 foot", has ♂ (♀?) corollas and leaves which are intermediate in size and shape between subsp. *pumilum* and *A. herbaceum.*

Inyanga, BOUGHEY 453 (SRGH), FRIES, NORLINDH & WEIMARCK 3011 (BR, SRGH); −, Rhodes Estate, FISHER 1156 (NU, PRE); Pungwe Hills, HOPKINS 7156 (SRGH); Umtali D., Odzani River valley, TEAGUE 421 (BOL, K, STE); −, Penhalonga, ROBINSON 1840 (K, MO, SRGH). − S: Victoria D., Zimbabwe National Park, CHIPARAWASHA 619 (K, SRGH); Bikita D., Bikita, FRIES, NORLINDH & WEIMARCK 2122 (BM, K, MO, PRE, SRGH). − W: Shangani, FEIERTAG s.n. sub SRGH 45550 (K, SRGH); Nyamandhlovu, PARDY 4916 (K); Bulawayo D., Bulawayo, RAND 338 (BM), ROGERS 13507 (BOL); −, Hope Fountain Mission, NORRGRANN 421 (K, SRGH); Matobo D., Matobo, WEST 2451 (SRGH); −, Matopos, RATTRAY 265 (K, SRGH); −, Matopo Hills, GIBBS 186 (BM); −, Farm Besna Kobila, MILLER 2535 (K, SRGH), 3118 (LISC, PRE × 2, SRGH), 5767 (K, SRGH), 7699 (SRGH), 8170 (K). − Not traced: Shanawe River, ca 67 km from Salisbury (= Harare), EYLES 8901, 8929 (K). − Imprecise localities: "S Rhodesia", HISLOP 21 (K); Victoria (Distr.?), MONRO 884 (BM).

Botswana. − 2425 (Gabarone): ca 14 km S of Ramotswa (Ramoutsa) (-DC), LEACH & NOEL 215 (K, LISC, SRGH). − 2525 (Mafeking): ca 3 km S of Lobatsi, E of railroad, Farm Springfield (-BA), LEACH & NOEL 148 (SRGH). − 2228 (Maasstroom): Mashatu Nature Res., Tuli Block (-BB), PUFF 781009-1/5 (WU).

Namibia (South West Africa). − 1917 (Tsumeb): Grootfontein D., Farm Elandhoek, GR 771 (-BC), GIESS 15149 (M, WIND); Ovati (-CB), DINTER 785 (NH, SAM), 5436 (NU), 5438 (B, BOL, G, PRE, SAM); Grootfontein D., Farm Askevold, GR 525 (-DA), GIESS & SMOOK 10618 (M, PRE, S, WIND); Guchab (-DB), DINTER 7661 (B, K). − 1918 (Grootfontein): S of Grootfontein (-CA), SCHÖNFELDER 942 (PRE); Grootfontein D., Farm Philadelphia (-CB), WALTER & WALTER 720 (WIND; in M as Voigtland, Auas, 2217-CB!). − 2017 (Waterberg): Otjiwarongo D., Waterberg, Omuverume Plateau (-AC), RUTHERFORD 308 (WIND); −, Okosongomingo-Nord (-CA), VOLK 548 (M). − 2216 (Otjimbingwe): Khomas Hochland, Farm Friedenau (-DB), SASSNER 57 (M). − 2217 (Windhoek): Windhoek D., Farm Voigtskirch, WIN 135, 45 km NE of Windhoek on Steinhausen rd. (-AD), WANNTORP & WANNTORP 196 (K, S); −, Farm Regenstein, WIN 32, Großherzog Friedrichberg (-CA), GIESS 11679 (M, PRE, WIND); −, Farm Finkenstein, WIN 71 (-CB), SEYDEL 3887 (B, BR, COI, GH, K, M, MO, US, WAG, WIND); −, Lichtenstein (-CC), DINTER 4318 (B). − 2316 (Nauchas): Rehoboth D., Farm Verdwaal, REH 41 (-BA), MERXMÜLLER & GIESS 28089 (M); −, Farm Göllschau (-BC), WALTER & WALTER 1770 (WIND). − Not traced: Gobabis- Kihors (? spelling), DINTER 2757 (SAM); Farm Hoffnung, DINTER s.n. sub SAM 7458 (SAM). − Additional collections: Boss s.n. sub TVL Museum no. 36263 (PRE); DE WINTER 2361 (K, M, PRE, WIND); DINTER 1881 (SAM), 5659 (B, BOL, G, PRE, SAM), 6115 (GH); GIESS 8699 (M, WIND); KERS 782 (S, WIND), 2877 (S); LEIPPERT 4381 (WIND); MERXMÜLLER & GIESS 30189 (M, PRE, WIND); MEYER 71 (M, WIND); REHM s.n. (S × 2); RUTHERFORD A 32 (PRE); SCHÖNFELDER 943 (PRE), 946 (K, PRE); SEYDEL 4104 (B, GH, K, MO, SRGH, US); VOLK 611 a, 11044 (M), 796 (PRE), 1034 (LISU); WALTER & WALTER 1609 (M); WANNTORP & WANNTORP 579 (PRE, S).

Swaziland. − 2631 (Mbabane): Mbabane D., Ngwenya Mt. (-AA), PUFF 790225-1/1 (J, WU); −, Bomvu Ridge, COMPTON 31195 (K, NBG, PRE); −, Komati pass [between Forbes Reef and Piggs Peak], COMPTON 31041 (K, NBG, PRE). − 2632 (Bela Vista): Lebombo Mts., Blue Jay Ranch, ca 4.5 km SW of entrance to Umbuluzi Gorge (-AA), CULVERWELL 701 (PRE), PUFF 780408-1/3 (WU).

Lesotho. − 2828 (Bethlehem): Leribe (-CC), DIETERLEN 172 (K, NH, P, PRE × 2, SAM × 2, STE × 2), 629 (GRA, K, NH, PRE × 2, SAM × 2, STE × 2),

681 (PRE – mixed with *A. monticola*), 922 (PRE, SAM), 970 (P, PRE, SAM), PHILLIPS 619, 947, 2877 (SAM); Butha Buthe D., Moteng, JACOT GUILLARMOD 3687 (PRE). – **2927** (Maseru): Berea D., Mamathes (**-BB**), JACOT GUILLARMOD 2432, 2567, 4818 (PRE), LAWSON 846 (NH); Mapotong valley, SCHMITZ 585 (PRE); Masoeling, JACOT GUILLARMOD 1585 (PRE); Maseru D., Ha Khotso, SONGCA 17 (PRE); Roma D., Roma (**-BC**), RUCH 1 (PRE), SCHMITZ 555 (PRE); Berea plateau, Botsabelu area, WILLIAMSON 817 (K; approach. subsp. *rigidum*); 'Mountain Road', between Rual and Mpao (**-BD**), SCHMITZ 7328 (PRE); around Molimo Nthuse lodge, PUFF 840721-5/1 (WU); Mafeteng (**-CC**), WATT & BRANDWYK 1244, 1927 (PRE). – **2928** (Marakabei): Thaba Basiu (**-BA**), JUNOD 1810 (G). – **2929** (Underberg): Mohkotlong (**-AC**), COMPTON 21519 (NBG), 21525 (NBG, PRE), JACOT GUILLARMOD 1125 (PRE), LIEBENBERG 5853 (PRE). – No locality given ("Basutoland"), COOPER 2487 (E, K).

South Africa. Transvaal – **2230** (Messina): Tate Vondo (**-CD**), VAN WYK 5568 (PRE). – **2328** (Baltimore): Pietersburg D., Blouberg, Lenare (**-BB**), STREY & SCHLIEBEN 8618 (B, BM, K, M, PRE, SRGH). – **2329** (Pietersburg): Louis Trichardt, Zoutpansberg (**-BB**), ROGERS 18136 (J, K, S); Pietersburg (**-CD**), MOSS & ROGERS 967 (J); ca 28 km E of Pietersburg on rd. to Tzaneen (**-DC**), VAN VUUREN 1401 (K, PRE); Haenertsburg (**-DD**), MOSS & ROGERS 892 (J), POTT 13669 (PRE). – **2427** (Thabazimbi): Kransberg, off Bakkerspas (**-BC**), PUFF 781202-1/1, -2/2 (J, WU). – **2428** (Nylstroom): Swaershoekberge, Farm Geelhoutkloof 195 (**-AD**), PUFF 781203-2/1 (J, WU); Waterberg above Warmbaths (= Warmbad) (**-CD**), GILLETT 2760 (BOL, K); Waterberg D., Naboomspruit, Farm Mosdene (**-DA**), GALPIN M 141 (PRE). – **2429** (Zebediela): Percy Fyfe Nature Res. (**-AA**), HUNTLEY 1552, 1576 (PRE); Mogotokloof (**-AD**), ROBERTSE 575 (PRE); Donkerkloof nr. Chuniespoort (**-BA**), VAHRMEIJER 2449 (K, MO, PRE, SRGH). – **2430** (Pilgrim's Rest): The Downs, Mamotswiri Peak area, Farm Haffenden Heights (**-AA**), PUFF 791208-1/4 (WU); Ohrigstad Dam Nature Res. (**-DC**), JACOBSEN 1344 (PRE). – **2526** (Zeerust): Marico D., Rooderand Farm (**-AB**), CARTER 918 (PRE); Zeerust (**-CA**), BURTT-DAVY 110 (PRE), LEENDERTZ 11290 (PRE), ROGERS 22685 (GRA), THODE A 1421 (NH, PRE); Koster Stn. (**-DD**), BURTT-DAVY 4145 (7175?) (PRE). – **2527** (Rustenburg): Rustenburg Nature Res. (**-CA**), JACOBSEN 877 (PRE); Waterkloof, Tierkloof (**-CC**), GERMISHUIZEN 676 (PRE); Magaliesberg, Farm Kloofwaters (**-CD**), PUFF 780312-1/3, 780716-1/3 (J, WU); Brits D., Jacksontuin (**-DA**), VAN VUUREN 386 (PRE); Bokfontein (**-DB**), JENKINS 6918 (PRE); Farm Uitkomst 499 J.Q. (**-DC**), COETZEE 584 (PRU); Farm Uitkomst (**-DD**), COETZEE 384, 386 (PRE). – **2528** (Pretoria): Roodeplaatdam Nature Res., NE of Pretoria (**-AD**), VAN ROOYEN 2083, 2818 (PRE); Leeuwkraal nr. Roikop (**-BA**), GILLETT 249 (K); Kwa-Ndebele, Farm Gemsbokfontein (**-BD**), DU TOIT 220 (PRE); Pretoria D., Hornsnek (**-CA**), SCHLIEBEN 7842 (B, BR, G, K, M, US); – , Silverton (**-CB**), OBERMEYER 110, 2769 (PRE); – , Irene (**-CC**), OBERMEYER s.n. sub TVL Museum no. 27737 (PRE); – , Fairy Glen Alsation St. (**-CD**), SAGP/SAAB 1/69, 4/17 (PRE); – , Premier Mine (**-DA**), MOSS 3921 bis (J), ROGERS 19035 (19835?) (G); ca 24 km N of Bronkhorstspruit on Groblersdal rd. (**-DB**), PUFF 790209-1/2 (WU). – **2529** (Witbank): Bronkhorstspruit-Groblersdal rd., ca 16 km S of Dennilton (**-AC**), PUFF 790209-2/1 (WU); Middelburg, Farm Doornkop 243 (**-CB**), DU PLESSIS 1032 (PUC); nr. Middelburg (**-CD**), KASSNER 124 (BR), RUDATIS 132, s.n. sub STE 2458, 2547 (STE); Farm Pineglades, 15 km from Belfast on rd. to Dullstroom via Kwaggaskop (**-DB**), DU TOIT 313 (PRE). – **2530** (Lydenburg): Farm Zwagershoek nr. [SW of] Lydenburg (**-AB**), OBERMEYER 242 (PRE); Belfast D., Schoemans Kloof (**-AD**), YOUNG A 352 (PRE); Pilgrim's Rest D., Mt. Anderson (**-BA**), SMUTS & GILLETT 2374 (MO, PRE); Sabie, Tweefontein Exp. area

(-BB), Wager B 164 (PRE); Wonderkloof Nature Res. (-BC), Elan-Puttick 245 (PRE); Nelspruit (-BD), Rogers 4781 (PRE); Waterval Boven (-CB), Pole-Evans 2614 (PRE). – 2531 (Komatipoort): Kruger National Park, (nr.) Pretorius Kop (-AB), Codd & de Winter 4926 (PRE, SRGH), van der Schijff 1082, 2086, 2402 (PRE); Eertegeluk 16, Uitkyk, 18 km S of Nelspruit (-CA), Buitendag 1272 (PRE); Maid of the Mist Mt. [nr. Louws Creek] (-CB), Hutchinson 2463 (BOL, K); Saddleback Mt. (-CC), Galpin 1309 (BOL, K, NH, PRE, SAM). – 2626 (Klerksdorp): Lichtenburg (-AA), Burtt-Davy 60 (PRE), Jenkins 11234 (PRE); – D., Witstinkhoutboom (-AB), Liebenberg 37 (PRE); Farm Morgenzon 42, ca 21 radial km N of Ventersdorp (-BB), Mogg 22204 (PRE, W); Ventersdorp (-BD), Taylor 5087 (NBG); Doornpoort (-CA), Pole-Evans H 13253 (PRE); Klerksdorp, Wolwerand (-CD), Hanekom 1816 (K, LISC, LMA, SRGH, WAG). – 2627 (Potchefstroom): Ventersdorp D., Rysmierbult (-AC), Louw 1247 (PRE); Potchefstroom D., Welverdient (-AD), Rogers 22691 (G); Krugersdorp D., Sterkfontein Caves and Farm Swartkrans 67 (Isaac Stegmann Nature Res.) (-BA), Mogg 35226 (LISC, SRGH), 36008 (MO); (nr.) Witpoortjie (-BB), Moss 7976, 8464, 11589, 13731 (J), 9538 (BM, J); Potchefstroom (-CA), Leendertz 9484 (PRE), Liebenberg 1015 (PRE); Farm Klipdrif, E of Potchefstroom (-CB), Theron 1105 (PRE), 1274 (NH, PRE); Potchefstroom D., Losberg, Farm Elandsfontein (-DA), Theron 481 (781?) (PRE), 1077 (NH); nr. Vereeniging (-DB), Burtt-Davy 17138 (MO, PRE). – 2628 (Johannesburg): Johannesburg, Bryanston (-AA), Puff 781105-1/1 (WU); Benoni (-AB), Bradfield 345 (PRE); Klipriviersberg Farm, ca 3 km W of Alberton (-AC), Mogg 18303 (J); Suikerbosrand (Nature Res.), Boschhoek (-AD), Bredenkamp 919, 920, 925 (PRE); –, Keyterskloof (-CA), Lambrechts 77 (PRE); –, entrance by Nolte (-CB), Bredenkamp 388 (PRE); Nigel D., Farm Vrisgewaag 337 (-DA), Mogg 22576 (J, SRGH), 22606A (SRGH). – 2629 (Bethal): Bethal (-AD), Leendertz 9399 (PRE), Moss 16317 (J); Standerton (-CD), Burtt-Davy 898 (PRE), Leendertz 11053 (PRE), Rehmann 9798 (K), Weiss s.n. sub STE 16987 (STE); Ermelo, Nooitgedacht Res. Stn. (-DB), Balinhas 2922 (K, MO, PRE, SRGH, WAG). – 2630 (Carolina): Carolina (-AA), Moos & Rogers 1107 (J), Rademacher 7279 (PRE); Rogers 19555 (BOL). – 2725 (Bloemhof): Wolmaranstad (-BB), Rogers 18471 (K); Christiana (-CC), Burtt-Davy 14410 (PRE); Bloemhof D., Verlatenkloof (-DB), Lamplough 21 (NBG). – 2729 (Volksrust): Volksrust (-BD), Jenkins 10684 (PRE). – 2730 (Vryheid): Wakkerstroom (-AC), Beeton 119 (PRE, SAM); Farm Oshoek, E of Wakkerstroom (-AD), Puff 791213-2/2, -3/1 (WU). – Imprecise localities: Waterberg, Galpin 9168 (NBG, PRE); Palala River, Breyer 25244 (PRE). – Not traced: Lydenburg D., Sekukuniland, (Farm?) Groenlands, Barnard & Mogg 1063 (PRE); Buffelshoek, Goosens 1485 (PRE); 'Hogge Veld', Trigardsfontein, Rehmann 6706 (BM, K). – Additional collections: Aaron 4982 (PRE); Acocks 11328 (PRE); Acocks & Naude 48 (PRE); Botha & Ubbink 1898 (PRE); Bredenkamp 868 (PRE); Bremekamp & Schweickerdt 95 (PRE); Bryant D 89 (PRE); Buitendag 208 (NBG, PRE); Burtt-Davy 1987, 9545 (K), 5392, 7099 (PRE); Carter 887 (PRE); Chippindale 136 (PRE); Clarke 388 (PRE); Codd 2152 (K, PRE); Collins & Scholes s.n. sub Puff 790520-1/1 (WU); Conrath 357, 1200 (K); Fairall 1646 (NBG); Flugge-de-Smidt s.n. sub PRE 41656 (PRE); Galpin 1292 (GRA, PRE, STE), 1415 (PRE), 9800 (K, NBG, PRE); Gilfillan 1414, 1415 (PRE); Gilmore 460 (PRE); Haagner s.n. sub Conrath 190 (K); Hanekom 1660 (LISU, PRE, SRGH, WAG); Hutchinson 2379 (K), 2578 (BOL, K); Jacobsen 2087 (PRE); Jenkins 10277 (PRE); Junod 4273 (PRE × 2; one as TVL Museum no. 20410); Kings 1446 (PRE); Kuntze s.n. (K); Leeman 5 (PRE); Leipoldt s.n. sub PRE 41600 (PRE); Leendertz 1018 (BOL), 3739, 6068, 6551, 7820, 8657 (PRE);

LIEBENBERG 85, 173, 3454 (PRE), 3231 (K, PRE); Louw 1247, 2477, 2698, 3331, 3801 (PUC), 1861, 2092 (PRE); MACNAE 1403 (BOL, J, K); MERXMÜLLER 391 (LISC, M); MOGG 18438, 19151, 19356, 22995 (J), 18647 (B, SRGH), 18827 (BR), 18951 (LMA), 19751 (COI), 19884 (B), 20265, 20558 (J, PRE), 20493 (J, MO), 21715 (MO), 25260 (M, PRE); MOGG & PEDRO 657 (LMA); MOSS 3922, 6164, 6309, 8406, 8625, 9537, 10562, 11274, 11427, 13385, 13555, 13783, 14358, 14359, 16216, 16676, 18207 (J), 6311, 13819, 18277 (BM, J); MOSS & ROGERS 1855 (BM, GRA); NATION 250, 342 (BOL, K); PHILLIPS s.n. sub PRE 41638 (PRE); PIENAAR 815 (PRE); POLE-EVANS 348 (PRE, W); POTTER 1698 (PRE); PROSSER 1649 (PRE); PUFF 790210-4/1, 790516-1/4 (WU); 'Pupils of [Ermelo] Convent' 125 (PRE); RAND 901 (BM); REHMANN 4150, 4240, 4242, 4518, 4613, 6037 (BM, K); REPTON 101, 165, 2039, 4335 (PRE); ROGERS 1408 (GRA), 1855 (J), 14115 (BOL), 20629, 25291 (PRE); SCHLECHTER 3583 (B, BR); SCHOLES s.n. sub PUFF 781212-1/1 (WU); SCOTT ELLIOT 1312 (E); SMITH 1195 (NBG), 1177, 1195, 1225, 1235, 1265, 1300, 1774, 1823, 2168 (PRE); SMUTS & GILLETT 3073 (PRE, STE); STEYN 961 (NBG); STREY 3069 (B, K, M, PRE); SUTTON 367, 368, 369 (PRE); TAYLOR 5062 (NBG); THODE A 428 (NH, PRE); VAN NIEKERK & WASSERFALL 28 (K, PRE); VENTER 645 (PRE); WAGER s.n. (NU); WALFORD 6252 (PRE); WALL s.n. (S); WATT & BRANDWYK 2221 (J); WEDERMANN & OBERDIECK 1270 (B × 2, K, PRE, US); WILMS 475 (BM, E, G, GOET, K, M, NU, P, W, WU), 475a (BM, E, G, K), 476, 477 (BM), sub TVL Museum no. 6476 (PRE); ZEYHER s.n. (LY). **O. F. S.** — **2627** (Potchefstroom): ca 10–12 km E of Parys, Farm Palmietfontein ('Glendal') along Vaal R. (-DC), PUFF 800109-1/1 (WU); Coalbrook (-DD), GILMORE 2126 (G, PRE); Sasolburg, THERON 608 (PRE). — **2726** (Odendaalsrus): Farm Sandfontein, ca 24 km W of Bothavilla (-AD), SCHWEICKERDT 1089 (PRE); Bothaville (-BC), GOOSENS 1184 (PRE). — **2727** (Kroonstad): Leeuw Spruit and Vredefort (-AB?), BARRETT-HAMILTON s.n. (BM); Farm Schuttesrust, 40 km from Heilbron on rd. to Lindley (-BA), STADLER s.n. (PRE); Heilbron (-BD), GOOSENS 477 (PRE), 568 (PRE; ± approach. subsp. *rigidum*); Kroonstad D., Kroonstad (-CA), PONT 424 (PRE); ca 6 km N of Kroonstad on Vredefort (Johannesburg) rd., PUFF 790111-1/1 (J, WU), SCHEEPERS 1315 (BR, K, M, MO, PRE); ca 3–5 km N of Steynsrus, on rd. to Edenville (-DC), PUFF 840723-2/1 (WU). — **2728** (Frankfort): Farm Rietfontein, 23 km S of Frankfort on gravel rd. to Tweeling (-BC), RETIEF 1063 (PRE); Sweet Home, 20 km W of Reitz on Bulthoek rd. (-CD), RETIEF 967 (PRE); Susannaskop, 26 km from Reitz on Vrede rd. (-DB), RETIEF 996 (PRE). — **2825** (Boshof): Boshof, Olifant's Rug (-CA?), FERRAR s.n. (K; approach. subsp. *rigidum*); — , Smits Kraal, BURTT-DAVY 10334 (PRE). — **2826** (Brandfort): Bultfontein, Sandveld Nature Res. (-AC), VILJOEN 58 (PRE; ± approach. subsp. *rigidum*); Landboukollege Glen (-CD), V. D. BERG 3911 (PRE), School of Agriculture 3432 (PRE). — **2827** (Senekal): ca 22 km from Ventersburg on new rd. to Senekal (-AB), WARD 6007 (PRE); just N of Rosendal (-BD), PUFF 840721-3/1 (WU); Clocolan D., Mequatling (-CD), STAM 100 A (PRE); Senekal D., Doornkop (-DA?), GOOSENS 739 (K, PRE, PUC); E foothills of Witteberge, SSE of Rosendal, Farm Bothasberg (-DB), PUFF 840721-2/1 (PRE; WU; ± approach. subsp. *rigidum*); just W of Ficksburg, base of Mpharane Mt. (-DD), PUFF 840721-4/1 (WU). — **2828** (Bethlehem): Bethlehem (-AB), PHILLIPS 3173 (PRE), POTGIETER 21849 (PRE); ca 6 km SE of Bethlehem, on Kestell rd. (-AD), WERGER 66 (K, PRE); ca 9 km from Bethlehem on Clarens rd., SCHEEPERS 1388 (M, PRE); ca 17 km from Fouriesburg on Ficksburg rd. (-CA), PUFF 800104-5/1 (WU); Golden Gate National Park (-DA), COMPTON 22512 (NBG), LIEBENGERG 6945 (PRE), ROBERTS 3024 (PRE); Qwaqwa, above 'Junior Camp' on path to Mopeli Hut, PUFF 790216-1/1 (WU); Witsieshoek (-DB), JUNOD s.n. (G); rd. from Witsieshoek from Sentinel, PUFF 800105-1/1 (WU). — **2829**

(Harrismith): hills W of Harrismith (-AC), MEDLEY WOOD 5406 (BM); Rensburgskop, [S of] Swinburne (-AD), JACOBSZ 226 (PRE); –, Manyenyeza Mt., JACOBSZ 356 (PRE). – 2926 (Bloemfontein): Thaba Nchu Mt., Wilgerboomnek (-BB), ROBERTS 2322, 2960 (PRE). – 2927 (Maseru): ca 10–15 km from Tweespruit on Excelsior rd. (-AA), PUFF 790118-3/1 (J, WU); O.F.S./Lesotho border, Tsupane Gate above Wepener (-CA), PUFF 790118-2/1 (WU). – 3026 (Aliwal North): Caledon River [nr. Kommissiedrift] (-BC), ZEYHER s.n. (LY, S; SAM sub 16064). – 3027 (Lady Grey): Aasvoelberg above Zastron (-AC), REID 196 (PRE). – Not traced: W. P. W. KLOOF, KOK s.n. (PRE). – Imprecise locality: between Winburg and Vrerefort (2827-CA to 2727-AB), ACOCKS 8326 (S). – No locality given ('O.F.S.'): COOPER 2494 (K), RICHARDSON s.n. (K), STOCKDALE s.n. (BM), ZEYHER 773 (BM). **Natal** – 2729 (Voksrust): nr. Charlestown (-BD), MEDLEY WOOD 5674 (G); Newcastle D., Ingogo (-DB), COMPTON 23844 (NBG); Mullerspas (-DC), PUFF 780425-1/2 (J, WU); nr. Newcastle (-DD), WILMS 1982 (BM). – 2730 (Vryheid): Utrecht D., Farm Naauwhoek (-AD), DEVENISH 857 (PRE × 3), HILLIARD & BURTT 9174 a (E, NU); Paulpietersburg (-BD), SIDEY 3332 (US); Vryheid D., Hlobane (-DB), JOHNSTONE 503 (NU); –, Vryheid, Lancaster Hill (-DD), GALPIN 9750 (K, PRE). – 2731 (Louwsburg): Louwsburg D., Itala Nature Reserve (-AC), BROWN & SAPIRO 146 (PRE). – 2732 (Ubombo): Ubombo D., Mkuzi Game Reserve (-CA), WARD 3538 (PRE); –, Bazwana [= Mbazwana (W of Sodwana Bay)] (-DA), STREY 5119 (K, NH, PRE). – 2828 (Bethlehem): Royal Natal National Park, Sigubudu valley (-DB), PUFF 800106-1/1 (WU). – 2829 (Harrismith): O.F.S./Natal border, Van Reenen (-AD), BEWS 55 (NU), KROOK 2069 (W), MEDLEY WOOD s.n. (G); Klipriver, between Van Reenen and Ladysmith (-BC), SCHWEICKERDT 916 (PRE); Oliviers Hoek pass (-CA), THODE s.n. sub STE 4592 (STE); Bergville D., Cathedral Peak Forest Res. Stn. (-CC), KILLICK 1624 (NH, PRE); Farm Bergveld [5595; 11 km W of Winterton] (-CD), THODE s.n. sub STE 10422 (STE); Bergville D., Spioenkop area (-DA), VAN RENSBURG 16 (E, NU); nr. Colenso (-DB), MEDLEY WOOD 909 (E, GRA, MO × 2, SAM, US); Grantleigh, [E of] Winterton (-DC), KING 2, 3 (PRE); Ennersdale (-DD), GILL s.n. sub NH 23448 (NH). – 2830 (Dundee): Dundee D., Dundee (-AA), SHIRLEY s.n. sub NU 31822 (NU); –, Waschbank (-AC), SHIRLEY 217 (NU); ca 1.5 km S of Nqutu (-BA), PUFF 790303-1/2 (J, WU); Umlilumba (= Umlumba) Nek, Muden- Weenen rd. (-CC), EDWARDS 3210 (PRE). – 2831 (Nkandla): upper Umhlatuzi (= Mhlatuze) River (-DC?), GERSTNER s.n. sub NH 23118 (NH). – 2832 (Mtubatuba): St. Lucia Estuary Game Park (-AD), POOLEY 1876 (E, K, NU). – 2929 (Underberg): Estcourt D., ca 19 km from Giants Castle Game Rest Camp (-AB), TINLEY 626 (NH, NU); Estcourt, Highmoor Forest Stn. (-BB), KILLICK & VAHRMEIJER 3600 (PRE × 4); Giants Castle Game Res., Boesmans River, nr. Cave Museum (-BC), PUFF 800116-6/1 (WU); Fort Nottingham commonage (-BD), WRIGHT 1777 (NU); Underberg region (-CD?), MCCLEAN 659 (NH, PRE). – 2930 (Pietermaritzburg): (nr.) Mooi River (-AA?), MEDLEY WOOD 3459 (K, NH); Tweedie (-AC), FORBES 286 (NH); Albert Falls (-AD), COMINS 276 (NU); New Hanover D., ca 3 km W of Noodsberg (-BD), MOLL 1498 (K, NH, NU, PRE); Pietermaritzburg (-CB), ALLSOPP 566 (NU), 953 (NH, NU); Drummond (-DA), RUMP s.n. sub NH 20263 (NH). – 3030 (Port Shepstone): Dusmisa Stn., [Farm] Ellesmere (-AD), RUDATIS 1459 (BM, E, G, K, S, US, W). – No locality given ('Natal, Zululand'): GERRARD 363 (BM, K, W). – Additional collections: BARKER 4428, 5163 (NBG); EVANS 370 (K, NH, PRE – photo); HILLIARD & BURTT 15434 (E, NU, WU); HUMBERT 15033, 15137 (P); HUNTLEY 121 (NU); JOHNSTON 308, 475 (E); MOGG 7105 (PRE); RUMP s.n. sub NH 20334 (NH); WEST 733 (BOL, PRE, SRGH). **Cape** – 2525 (Mafeking): Molopo, Plaas Louvain (-CC), KARS 48 (PRE);

Mafeking (-DC), Duparquet 7 (P). – 2624 (Vryburg): Vryburg D., Farm Palmyra, ca 90 km NW of Vryburg (-AD), Rodin 3514 (BOL, PRE, S; ± approach. subsp. *rigidum*); – , Vryburg (-DC), Brueckner 465 (K, PRE), Sharpe 7185 (PRE); – , Farm Asmoedsvlakte, Bisschop s.n. sub PRE 41719 (PRE), Rogder 2266 (KMG). – 2723 (Kuruman): Kuruman D., Esperanza (-AD), Esterhuysen 2255 (BOL). – 2724 (Taung): Barkly West D., Groot Boetsap Kuil (-DC), Marloth 986 (STE). – 2823 (Griekwastad): Postmasburg D., ca 60 km SW of Olifantshoek, ca 3 km NW of Chrisanto P.O. (-AC?), Leistner 2133 (K, KMG, M, PRE, SRGH); Barkly West D., Asbestos Hills, Daniels Kuil (-BA), Esterhuysen 822 (BOL), Lewis s.n. sub SAM 53490 (SAM); – , Farm Ouplaas, Ferrar s.n. sub KMG 6642 (KMG). – 2824 (Kimberley): Barkly West D., Farm Cristaalfontein (-AA), Acocks 1488 (K, PRE); Kimberley D., Warrenton (-BB), Adams 178 (GRA, NBG; ± intermed.), 2733 (PRE; ± intermed.); – , Kamfers Dam (-DA), Acocks 2265 (PRE); – , ca 6 km S of Kimberley (-DB), Leistner 2021 (K, KMG, PRE, SRGH); – , Boshof rd., Wilman s.n. sub BOL 25130 (BOL); – , Farm Bultfontein (-DC), Acocks 260 (PRE). – 3027 (Lady Grey): Herschel D., Sterkspruit (-CB), Hepburn 3 (GRA); ca 31 km from New England on rd. to Lundean's Nek/Rhodes (-DC), Puff 790112-11/1 (WU). – 3029 (Kokstad): Mt. Currie D., Swartberg, Farm Vaalfontein (-AB), Comins 1905 (GRA, K, PRE); Mt. Currie (-AD), Goosens 344 (PRE); Franklin, Coles 8409 (PRE); nr. Clydesdale (-BD), Tyson 818 (BM, BOL, G, GH, K, SAM, STE, W), 2587 (BOL, G, K, PRE, SAM); Kokstad (-CB), Mogg 1572 (PRE), Tyson s.n. (E). – 3126 (Queenstown): Aliwal North D., Jamestown (-BB), Barker 2226 (NBG); Fincham's Nek, S of Queenstown (-DD), Galpin 1790 (K, PRE), Puff 790114-1/1 (J, WU); Queenstown, nr. Komani River, Galpin 2158 (GRA, PRE). – 3127 (Lady Frere): Xalanga D., Cala (-DA), Bolus 25123 (BOL), Flanagan 2675 (PRE × 2), Pegler 1665 (K, PRE × 2). – 3128 (Umtata): Baziya (-CB), Baur 474 (K); Umtata-Idutywa rd., ca 1.5 km above Bashee valley (-CD), Puff 791219-1/1 (WU); tow. Umtata waterfall (-DB), Schönland 3767 (GRA). – 3224 (Graaff-Reinet): Tàndjesberg nr. Graaff-Reinet (-BC), Bolus 809 (BOL, K, SAM). – 3226 (Fort Beaufort): Winterberge, 37 km from Adelaide on rd. to Tarkastad (-AD), Puff 790117-5/1 (WU); upper Gaga valley (-DD), Giffen s.n. (PRE × 2). – 3227 (Stutterheim): ca 4–6 km outside Cathcart on Hogsback rd. (via Happy Valley) (-AC), Puff 790114-4/1 (WU); Stutterheim D., just outside Stutterheim (-CB), Puff 790115-6/1 (WU); – , Dohne, Sim 19428 (E, NU, PRE); Komgha (-DB), Compton 17749 (NBG), Flanagan 1217 (BOL, GRA, PRE, SAM); King William's Town D., Berlin (-DC), Barker 2752 (NBG). – 3228 (Butterworth): Butterworth (-AC), Pegler 1804 (BM, BOL, K, PRE). – Not traced: E Griqualand, Amalfi (? spelling), 'Vielsalm', Forbes 1115 (NH). – Imprecise localities: Albert Distr., Cooper 634 (BM, E, K, W × 2; in part ± intermed.); Ceded Territory, nr. Phillipstown, on Katrivier, Ecklon & Zeyher 33-6 (S).

29 (b). Subsp. *rigidum* (Ecklon & Zeyher) Puff in Fl. S. Afr. **31**, 1 (2): 26 (1986).

≡ *A. rigidum* Ecklon & Zeyher, Enum. Pl. Afr. Austr.: 367 (1836). Type: South Africa, Cape Prov., 'in collibus Karro similibus (alt. II) prope fluvium "Gauritzrivier" (Swellendam)'[1], Ecklon & Zeyher 2315 (S, holo.!; LY, SAM, iso.!).

---

[1] For comments on this locality see Critical Remarks!

– *A. ciliare* sensu auct., non L.: HOBSON & JESSOP, Veld. Pl. S. Afr.: 216 & pl. 21 (1975).

Non-dioecious [♂, ♀̂ (occasionally ♀̂ + odd ♂ or ♀̂ + odd ♀) or ♀], several- to many-stemmed, ± erect and cylindrical to ± rounded dwarf shrubs with often thick, woody roots; if browsed, low and ± cushion-forming or with long, thin shoots from a thick woody base. Stems (10) 20–40 (55) cm long, ca (2) 3.5–6 mm in diameter near the base and ca 1–2 mm in the midstem region, ± terete, ± sparsely to much-branched; branching becoming irregular if browsed; younger parts brownish-grey or occasionally reddish tinged, densely covered with whitish papillae to 0.1 mm long; older parts greyish to blackish, glabrescent, epidermis peeling, bark sometimes with long fissures; internodes mostly longer than the leaves; plants not densely leafy. Leaves decussate, sometimes ± pseudo-verticillate (short shoot leaves often distinctly smaller than long shoot leaves); blades 3–6 (8) × 1–1.5 mm (occasionally to 22 × 3 mm on newly developed shoots), narrowly ovate- or obovate-lanceolate, lanceolate to linear-lanceolate, narrowed to the base, ± tough and thickish, glabrous or with whitish papillae less than 0.1 mm long on the flat to ± revolute margins, upper epidermis cells often ± large and conspicuous; apices acute to ± acuminate; petioles subobsolete; stipular sheaths covered with whitish papillae to 0.1 mm long, cup-shaped, 0.5–1.2 mm long, with a median, gland-tipped seta ca 0.1–0.3 mm long or seta ± lacking. Inflorescences inconspicuous, made up of ± sessile, 2- to 6- (rarely more-) flowered axillary clusters. Flowers subsessile or occasionally with pedicels to 0.5 mm long (♀, especially when in fruit); ♂, ♀̂, ♀. Corolla 4-merous (very rarely 5-merous in ♀̂), creamy yellow to greenish-yellow, sometimes with a few papillae to 0.1 mm long near the tip. ♂, ♀̂: tube 0.8–1.5 (1.9) mm long (in ♀̂ often longer than in ♂), narrowly funnel-shaped to ± cylindrical, lobes (1.7) 2.2–3 (3.2) × 0.5–0.8 mm, ovate-lanceolate to ± lanceolate, recurved; stamens 4, filaments exserted for (1) 1.2–1.9 mm, anthers 1.5–2.2 (2.5) × 0.3–0.6 mm; ♂: rudimentary ovary small but well discernible, often with rudimentary stigmas (flowers occasionally transitional ♀̂ → ♂); ♀̂: gynoecium as in ♀ but stigmas often only to ca 3 mm long; flowers occasionally transitional ♀̂ → ♀ with rudimentary anthers less than 0.5 mm long, tubes to 0.8 mm and lobes to 1.5 mm long. ♀: tube 0.2–0.5 mm long, cylindrical and narrow, lobes 0.3–1 × 0.1–0.3 mm, ± linear-lanceolate, ± erect to spreading; style ± 0–1 mm long; stigmas 2, ca 3.4–7.5 mm long, greyish-white; ovary ca 0.6–0.9 × 0.5–0.7 mm, covered with curved, ± hook-like hairs or ± glabrous, crowned by 4 often ± indistict calyx lobes. Fruits reddish-brown, shiny, supported by short, broadly U-shaped carpophores less than ⅓ as long as the fruits; each mericarp 2–

2.6 (2.8) × 1–1.5 mm, elliptic to obovate, dorsal side convex, mostly ±
sparsely covered with whitish, curved, ± hook-like hairs ca 0.1–0.3 mm
long or (less commonly) glabrous, ventral side plane to very shallowly
concave, without a prominent median vertical ridge, commissure ± large;
mericarps crowned by 2 ± indistinct, ± trapezioidal, rounded or
triangular calyx lobes ca 0.1–0.4 × 0.3–0.4 mm, or calyx lobes ±
lacking. – Fig. 89 b.

Chromosome Number (Table 3): n = 11.
Average Pollen Diameter (Fig. 48): 23.5–24.8 μm.
Habitat: Mostly associated with rocky areas (koppies, rocky hillsides)
and growing in rock crevices, in gravelly soil between rocks and boulders
and occasionally also in red, sandy loam. Usually in relatively dry, karroid
or "false karroid" (sensu ACOCKS 1975) vegetation. – Ca (?–) 800–1 500 m.
Distribution (Maps, Figs. 73 c, 90 a): S Africa. From the O.F.S. to the
Cape Prov., but not extending into the SW Cape Floristic Region.

Critical Remarks: At first glance, the type locality of *A. rigidum* [SW
Cape Prov.: "in collibus Karro similibus, prope fluvium Gauritzriver
(Swellendam)"] seems to be suspect as it appears to be clearly outside the
distribution range of the taxon. The reference to Swellendam, however, is
misleading; the Gauritzriver (= Gouritz R.) is approximately 100 km E of
Swellendam and forms the border between the Riversdale and the Mossel
Bay Distr. The Gouritz R. valley both N and S of the Langeberge is indeed
characterized by karroid vegetation rather than SW Cape Fynbos (obs.
PUFF). Already MUIR (1929), in his "Vegetation of the Riversdale Area",
notes (p. 48) that "on the banks and bed of the Gauritz River south of the
mountains [= the Langeberge] there is found a curious concentration of
species which might not be excpected in this region" and, further on, lists a
number of species centered in the Karoo and Namaqualand (amongst
those "*A. rigidum*"!) as also occurring in the Gouritz River System.
ACOCKS (1975) lists the Gouritz River Scrub as a subtype of Veld Type 23
(Valley Bushveld), which is grouped with "Karoo and karroid" Veld
Types. The specimen may be from 3321-DC (mapped as such with
question mark in Fig. 90) or from further S (3421-BA).

In a number of respects, subsp. *rigidum* is ± intermediate between *A.
dregei* (subsp. *dregei*) (33.) and *A. pumilum* subsp. *pumilum*. It was
originally tempted to group "*A. rigidum*" with *A. dregei* subsp. *dregei*, as it
resembles the latter in its more woody and robust habit, in its tougher,
relatively shorter and broader leaves and in its trend to dioecy. Some W
collections of subsp. *rigidum* (e.g., from Houthoek, Southerland, or from
Calvinia) are indeed very similar to *A. dregei*. In its E and NE range of

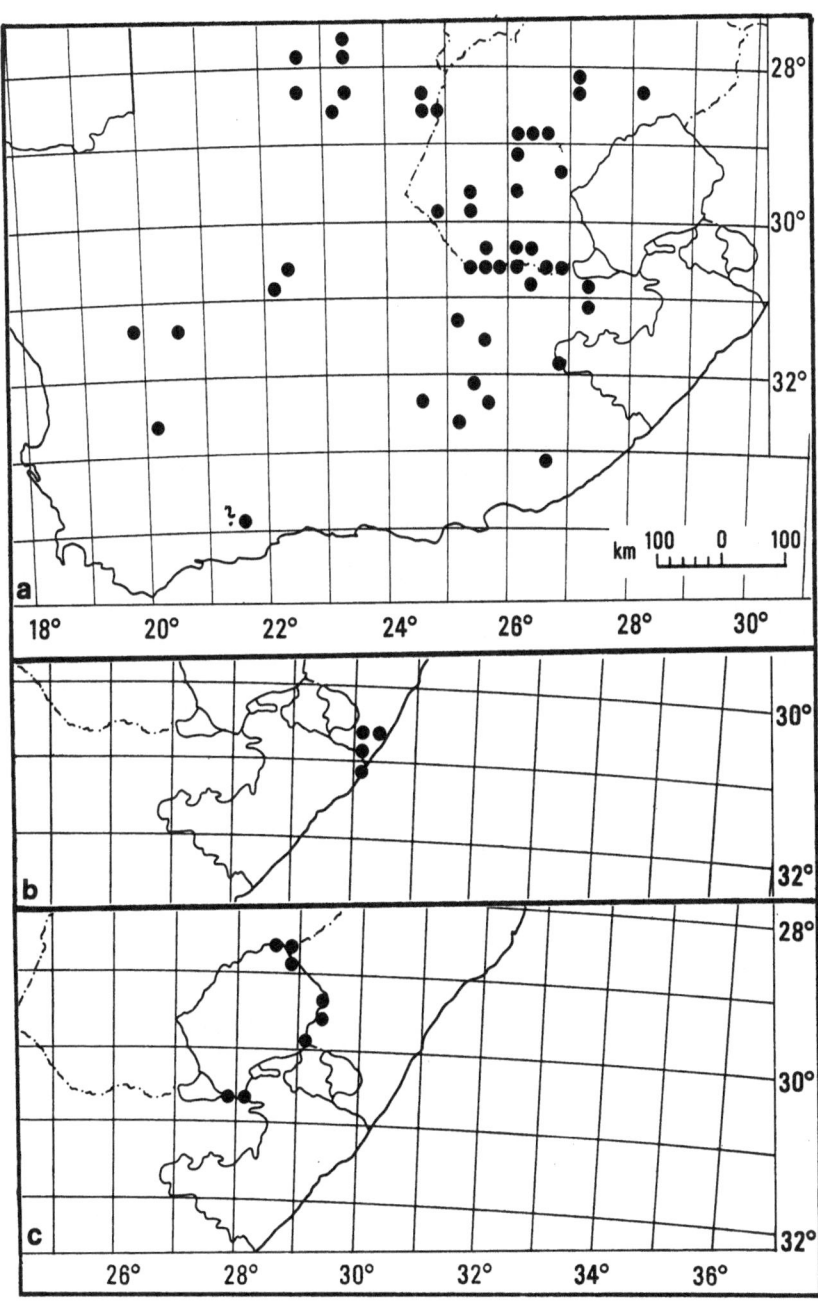

Fig. 90. Distribution of *Anthospermum. a A. pumilum* subsp. *rigidum, b A. streyi, c
A. basuticum

23*

distribution, however, *rigidum* is more similar to *A. pumilum,* shares more characters with it and is sometimes morphologically difficult to separate, so that I prefer to consider it a subspecies of *A. pumilum.*

In some parts of the O.F.S. and the Cape Prov. (e.g., around Bloemfontein or around Graaff-Reinet) both subsp. *rigidum* and subsp. *pumilum* are recorded from the same general area, but then they always seem to occur in different habitats: subsp. *rigidum* is confined to drier, rocky areas, whereas subsp. *pumilum* occurs mainly in grassy, open places.

Collections

South Africa. O. F. S. − **2826** (Brandfort): Bloemfontein D., Krugersdriftdam Nature Reserve (-CC), MÜLLER 1732 (PRE); −, Glen, N of Bloemfontein (-CD), PUFF 790111-2/1, -2/1 b (J, WU); 80 km S of Winburg on N 1, tow. Bloemfontein (-DC), PUFF 800930-3/1 (WU). − **2827** (Senekal): ca 4 km from N 1, on rd. to Willem Pretorius Game Res., Farm Paradys, along Sand R. (-AA), PUFF 800930-5/1 (WU); Willem Pretorius Game Res. (-AC), VAN ZINDEREN BAKKER 1169 (PRE). − **2828** (Bethlehem): Fouriesburg D., Farm Help My (-AC), STAM 165 (PRE). − **2924** (Hopetown): 15 km from Luckhoff on rd. to Phillipolis, Farm Knoffelfontein (-DD), HERMAN 348 (PRE). − **2925** (Jagersfontein): Sannalis Poort, E end of Fauresmith (-CB), SMITH 434 (PRE); S of Fauresmith (-CD), PUFF 790112-4/2 (WU). − **2926** (Bloemfontein): Bloemfontein, Naval Hill (-AA), COMPTON 15673 (NBG); 3 km N of Dewetsdorp (-CA), SMOOK & GIBBS RUSSELL 2210 (PRE). − **3025** (Colesberg): ca 7 km W of Springfontein on Phillipolis rd. (-BC), PUFF 790112-7/1 (J, WU). − **3026** (Aliwal North): ca 14 km E of Bethulie on Aliwal North rd. (-AC), WERGER 310 (PRE); ca 4 km SE of Caledon River bridge on Bethulie- Aliwal North rd. (-AD), PUFF 790112-8/1 (J, WU); foot of Witbergen, Nieuwejaarspruit, between Garip [Orange] and Caledon River [= E of Beestekraal Nek Stn., between Aliwal North and Rouxville] (-DB), ZEYHER '114-10' (LY, S). − Additional collections: BOLUS 11085 (BOL); BOUWER 2190 (PRE); BRUECKNER 948 (KMG); HANEKOM 605 (LISC, PRE); HENRICI 1725, 1841, 2779 (PRE); MOSS 3924 (BM, J); MULLER 63 (NBG), 371, s.n. sub PRE 41688 (PRE); PHILLIPS 26341 (J); POLE-EVANS 1598 (PRE, W), 1599, 1791 (PRE); PUFF 790112-2/1 (J, WU); ROBERTS 5391 (PRE); SMITH 426 A, 5201 (PRE); VAN BREDA 31 (PRE); VERDOORN 862 (PRE), 1175 (NBG, PRE); WASSERFALL 843 (NBG). **Cape** − **2722** (Olifantshoek): Postmasburg D., ca 3 km S of Olifantshoek (-DC), LEISTNER & JOYNT 2734 (K, KMG, PRE, SRGH); ca 15 km N of Olifantshoek, Langeberg, PUFF 780415-1/2 (WU). − **2723** (Kuruman): Kurumanheuwels, a few km S of Farm Mansfield (Kuruman- Grypoort gravel rd.) (-CB), PUFF 840905-1/1 (PRE, WU); −, ca 3 km S of Farm Newcastles (-CD), PUFF 840905-2/1 (WU). − **2822** (Glen Lyon): Hay D., Farm Dunnmurr(a)y (-BC), VIGNE 7 (NH), s.n. sub NBG 23235 (NBG), WILMAN 2406 (GRA, K, KMG). − **2823** (Griekwastad): Hay D., Smuts (-AD), ACOCKS 2429 (B, BOL, K); −, Farm Boven Ongeluk (-CA), ACOCKS [or H(afström)] 1058 (KMG, GH, PRE, S). − **2824** (Kimberley): Kimberley D., Farm Rietputs (-BC), ACOCKS 945 (KMG, PRE); Barkly West D., Farm Klipvlei (-DA), ESTERHUYSEN 2099 (BOL); Kimberley D., Farm Wit Pan (-DB), ESTERHUYSEN 25117 (BOL, K); −, Farm Riet Pan, WIOMAN 15918 (BOL). − **3022** (Carnarvon): Karrebergen (-CB), BURCHELL 1561 (K); Carnarvon, Schietfontein (-CC?), BURCHELL 1551 (GH, K). − **3025** (Colesberg): 19 km from Colesberg on Norvalspont rd. (-CB), PUFF 800104-1/1 (K, WU); Oviston Reserve (-DA), FOURIE ORFS 139 (PRE); KP 5 Hills, ANDERSON ORFS 253 (PRE). − **3026** (Aliwal North):

'Burghersdorp D.' (= Albert D.), (Farm?) Nooitgedacht (-**CA**?), THORNE s.n. sub SAM 51932 (SAM); Burghersdorp (= Burgersdorp) (-**CD**), GUTHRIE 2313 (NBG); Aliwal North (-**DA**), GERSTNER 256 (PRE); Elands Hoek, nr. Aliwal North, BOLUS 230 (STE). – **3027** (Lady Grey): Farm Glen Doone, N of Lady Grey- Barkly East main rd. (-**CD**), PUFF 840923-5/1 (PRE, WU). – **3119** (Calvinia): Calvinia, Railway Camp (-**BD**), HENRICI 3375 (PRE); – , Ekerdam (= Akkerendam), ACOCKS 17746 (PRE × 2), TAYLOR 2769 (BOL, NBG). – **3120** (Williston): ca 30 km from Williston on Calvinia rd. (-**BC**), STORY 4264 (GRA, PRE). – **3125** (Steynsburg): Middelburg D., Bangor Farm (-**AC**?), BOLUS 14054 (BOL); – , Farm Grootfontein, ARCHIBALD 3350 (GRA); – , Middelburg road, KUNTZE s.n. (K; approach. subsp. *pumilum*); ca 27.5 km S of Hofmeyer, tow. Cradock (-**DA**), PUFF 800929-10/1 (WU). – **3126** (Queenstown): Queenstown D., Lesseyton Nek (-**DD**), GALPIN 2013 (PRE). – **3127** (Lady Frere): a few km S of Rossouw, on rd. to Dordrecht, nr. Farm Ponte Racina (-**AB**), PUFF 840923-1/1 (PRE, WU). – **3220** (Sutherland): Houthoek (-**CA**), HANEKOM 1558, 1558 A (K, LISU, PRE, WAG). – **3224** (Graaff-Reinet): nr. Graaff-Reinet (-**BC**), BOLUS 362 (BOL × 2, K, S); – , Valley of Desolation, PUFF 790908-3/1 (WU). – **3225** (Somerset East): National Mountain Zebra Park (-**AB**), BRYNARD 132 (K, PRE), LIEBENBERG 7383 (PRE); Cradock D., nr. Mortimer (-**BC**), KENSIT s.n. sub BOL 25121 (BOL); Wildebeestkuil nr. Pearston (-**CA**), HOBSON 81 (PRE). – **3321** (Ladismith): nr. Gauritzrivier (-**DC**, or 3421-B?), ECKLON & ZEYHER 2315 (LY, S, SAM). – **3326** (Grahamstown): Bothasberg (Botha's Hill), near the Great Fish R. (-**BA**), (? ECKLON &) ZEYHER 1342 (S – mixed with *A. spathulatum* subsp. *spathulatum*). – Imprecise locality: Albert Distr., COOPER 589 (G, K, NH, W; E – mixed with *Nenax microphylla*).

30. *Anthospermum streyi* PUFF in Fl. S. Afr. **31**, 1 (2): 28 (1986).

Type: South Africa, Natal, Port Shepstone Distr., Beacon Hill East, January 1, 1967, STREY 7248 (NH, holo.!; K, NU, PRE, iso.!).

Non-dioecious [♂ or ♀ (always?)], procumbent and ± straggling or cushion-forming dwarf shrubs. Stems ca (10) 15–30 (40) cm long, ca 2.5–4 mm in diameter near the base and ca 1–2.5 mm in the midstem region, ± sparsely to much-branched; younger parts ± 4-angled, often reddish to reddish-brown, ± densely covered with whitish papillae to 0.1 mm long; older parts terete, yellowish-grey to grey, becoming glabrescent, epidermis peeling; internodes as long as or shorter than the leaves; plants mostly densely leafy. Leaves decussate, pseudo-verticillate; blades often curved, (5) 8–12 × (0.3) 0.5–0.8 (1) mm, linear to linear-lanceolate, in dried material often seemingly terete and needle-like due to strongly revolute margins, glabrous, midrib below (and sometimes also upper surface) reddish-brown; apices acute to ± acuminate; petioles subobsolete; stipular sheaths often with a median line of short hairs, broadly cup-shaped, ca 2–2.5 × 1 mm, with a median seta ca 0.4–0.6 (1) mm long, sometimes with an additional, minute gland-tipped seta on either side. Inflorescences inconspicuous, made up of ± sessile, 2- to 6-flowered axillary clusters. Flowers subsessile, ♀̂, ♀. Corolla 4-merous, yellowish, often with a few short hairs near the tip. ♂: tube 1–1.3 mm long, narrowly funnel-shaped, lobes (1.3) 1.5–2.5 (3) × 1 mm, ± lanceolate, recurved; stamens 4, filaments ca 1.8–2 mm long, anthers 1.3–2 × 0.3–0.5 mm; style 0; stigmas 2, 3–5 mm long, greyish; ovary ca 0.8–1.2 × 1 mm, covered with short, curled whitish hairs, crowned by 4 indistinct calyx lobes. ♀: tube ca 0.3–0.5 mm long, cylindrical, lobes 0.2–0.5 × 0.2 mm, ± linear, erect; gynoecium as in ♂. Fruits reddish-brown, ± heart-shaped in side view, supported by shallow, broad U-shaped carpophores less than ⅓ the length of the fruits; each mericarp 2–2.5 × (1) 1.2–1.5 mm, ovate to obovate, dorsal side convex, ± sparsely covered with curled whitish hairs ca 0.1–0.2 mm long, ventral side ± plane and with a ± indistinct median vertical ridge, commissure considerably smaller than the ventral surface; mericarps crowned by 2 indistinct ± triangular calyx lobes ca 0.1–0.2 mm long or calyx lobes lacking. – Figs. 4 a, 37 a.

Chromosome Number (Table 3): n = 11, 2 n = 22.

Average Pollen Diameter (Fig. 49): 25.8–26.9 μm.

Habitat (Fig. 4 a): On rocky outcrops in grassland, amongst rocks, at the edge of krantzes; confined to sandstone areas. – Ca. 100–600 m.

Distribution (Map, Fig. 90 b): S Africa. Narrowly endemic to S Natal (Port Shepstone Distr.).

Critical Remarks: A. streyi "replaces" the closely allied and widely distributed *A. pumilum* subsp. *pumilum* (29., above) in the sandstone areas of Natal's S border (an interesting region, notable not only for its endemics but also for its connections with the SW Cape Flora; cf. HILLIARD 1978: 422).

Morphological similarities between *A. streyi* and *A. pumilum* (subsp. *pumilum*) are numerous and undeniable. Several characters overlap to some extent but, apart from being a sandstone endemic, *A. streyi* differs in habit, in its more needle-like and often curved leaf blades and in having larger ♂ flowers and somewhat larger and broader (rounder) fruits.

The sex distribution in *A. streyi* (♂ and ♀ plants) is rather unusual for an *Anthospermum* species. It would not be surprising if, in addition, ♂ + ♀ plants (as in numerous other taxa of *Anthospermum*) also occur; further field observations would be required.

Collections

South Africa. Natal – **3030** (Port Shepstone): Highlands Farm, NNW of Paddock, heights above Umzimkulwana R. (-CA), NICHOLSON 1145 (PRE), PUFF 790426-3/1 (WU); Oribi Gorge Nature Reserve, 'The Rocks', MANTELL & VASSILATOS 221 (J); –, Inkonka Point (-CB), PUFF 790426-1/3 (J, WU); Port

Shepstone D., Beacon Hill East (-CC), Strey 7248 (K, NH, NU, PRE). — 3130 (Port Edward): Umtamvuna Nature Res., southernmost section (-AA), Puff 840821-1/3 (PRE, WU); Port Edward- Izingolweni rd. (-AA to 3030-CC), Huntley 752 (NH, NU).

31. *Anthospermum galioides* Rchb. in Sprengel, Syst. Veg. **4**, 2: 338 (1827)[1]; Puff in Fl. S. Afr. **31**, 1 (2): 28 (1986).
Type: "C.B.S.", no collector given[2].

 — *A. ciliare* sensu auctt., non L.: Cruse, Rub. Cap. 13 (1825), in Linnaea **6**: 11 (1831); Ecklon & Zeyher, Enum. Pl. Afr. Austr.: 366 (1836); Sonder in Harvey & Sonder, Fl. Cap. **3**: 28 (1865); Salter in J. S. Afr. Bot. **3**: 110 (1937), in Adamson & Salter, Fl. Cape Penins.: 733 (1950)[3].

Non-dioecious (♀̣, ♀̣ + ♀, ♀, ♂, or less commonly ♀̣ + ♂), several- to many-stemmed, erect and ± cylindrical, rounded or prostrate and ± mat- or cushion-forming subshrubs, or dwarf shrubs with often thick, woody roots. Stems ca (5) 8–30 (50) cm long, ca 2–4 (5) mm in diameter near the

---

[1] It is not clear to me from Sprengel's publication whether the description was actually made by Ludwig Reichenbach and handed over to Sprengel to be included in "Curae Posteriores" (Syst. Veg. **4**, 2) or whether the name was merely suggested by Reichenbach on (a) herbarium sheet(s) that were passed on to Sprengel who eventually draw up the descriptions (in which case the authority should be "Rchb. ex Sprengel"). Also the "Index auctorum in hoc opere citatorum" Sprengel (l.c.: 373) added to his "Curae posteriores" does not provide conclusive evidence as to L. Reichenbach's extent of contribution. It, furthermore, is not possible to draw further conclusions from Cruse's publication (1831); see also the footnote below.
   The same applies to the authority of *A. rubiaceum* (see 36. *A. hirtum*), *Cliffortia spicata* (see 37. *A. bergianum*) and *A. spermacoceum* (see *Carpacoce spermacocea*).
   As Reichenbach is accepted as authority by Ecklon & Zeyher (1836), Sonder (1865) and Salter (1950), I at present retain 'Rchb. in Sprengel'.

[2] I have not been able to trace a specimen which, without doubt, could be identified as the type. For the following reasons the specimen "*A. ciliare* L., Pr. b. Sp., Bergius s.n." (B †; G, iso.!), seen and mentioned by Cruse (1825, 1831), is chosen as a *neotype:* Cruse (1831) considered *A. galioides* to be a synonym of "*A. ciliare* L." (see also the following footnote). He reached this conclusion after studying "Reichenbach's specimen" (which, however, could not have been collected by Reichenbach himself, as he had never visited the Cape). Hence any specimen cited by Cruse under "*A. ciliare* L." and studied by him (1825, p. 14: "in promontorio bonae spei ... legit Bergius", and 1831, p. 11: "In monte Tabulari ... legerunt Bergius, C. F. Ecklon, Zeyher") can be considered to be identical to *A. galioides.*

[3] The type of *A. ciliare* L. (1763) corresponds to *A. bergianum* Cruse (1825). Since the name *A. ciliare* has consistently been misapplied, it has been proposed to reject the name (Puff 1982 b). See also 37. *A. bergianum.*

base and ca 0.8–2 mm in the midstem region, unbranched to much-branched, not rooting at the nodes in prostrate forms; younger parts obscurely 4-angled to terete, often reddish-purplish tinged or shiny reddish-brown, mostly densely covered with whitish papillae to 0.1 mm long; older parts terete, often greyish, glabrescent, epidermis peeling; internodes often ca as long as the leaves. Leaves decussate, pseudo-verticillate; blades ascending to spreading, but often recurved at least near the tip, (2.2) 2.7–10 (12) × (0.5) 0.8–2.5 (3) mm, linear to lanceolate or ovate-lanceolate (on young plants or shoots produced after a fire blades to 20 × 4 mm and ovate-lanceolate to lanceolate), narrowed to the base, ± membranaceous to ± tough and thickish, glabrous above and margins (sometimes also midrib below) with whitish spreading hairs ca 0.1–0.4 mm long, or blades entirely glabrous or only minutely scabrid on the margins; blades sometimes distinctly discolourous, upper surface often shiny and with large, conspicuous epidermis cells; apices acuminate to acute; margins flat to strongly revolute; petioles subobsolete or to ca 1 mm long; stipular sheaths ± densely covered with whitish papillae to 0.1 mm long to subglabrous, cup-shaped, ca (0.5) 0.7–1 mm long, with a (± broadly) triangular to subulate median seta ca (0.5) 0.7–1.5 (2) mm long, sometimes with an additional minute seta on either side. Inflorescences inconspicuous, made up of ± sessile, 2- to 6- (rarely more-) flowered axillary clusters. Flowers subsessile, ♂, ☿, ♀. Corolla 4-merous (very rarely also odd 5-merous ♂ and 3-merous ♀), greenish-yellow to yellowish, sometimes reddish-purplish tinged on the outside, mostly with whitish papillae to 0.1 (0.3) mm long at least near the tip. ♂, ☿: tube 0.7–1.7 (2) mm long, ± cylindrical to (narrowly) funnel-shaped, lobes 1.4–2.4 (2.9) × 0.3–0.8 (1) mm, ± lanceolate, recurved; stamens 4, filaments exserted for (0.4) 0.7–1.5 (1.7) mm, anthers (0.7) 1–1.8 (2.1) × 0.2–0.5 mm; ☿: gynoecium as in ♀, but stigmas often only to ca 5 mm long; flowers occasionally transitional ☿ → ♀ with rudimentary, ± pollenless anthers to 0.5 mm long and corolla tubes ca 0.5–0.8 mm long and lobes ca 1–1.7 × 0.4–0.6 mm. ♀: tube 0.2–0.5 mm long, cylindrical, lobes 0.2–0.8 × 0.1–0.3 mm, lanceolate to linear-lanceolate, erect to somewhat spreading; style ± 0–0.6 mm long, stigmas 2, 2–8.1 mm long, whitish grey, greyish or (seldom) purplish; ovary ca 0.3–0.7 × 0.3–0.5 mm, glabrous, shortly hairy or papillate, crowned by 4 ± indistinct calyx lobes or calyx lobes ± lacking. Fruits reddish-brown, shiny, supported by short but relatively broad U-shaped carpophores; each mericarp (1.5) 1.7–2.4 × 0.9–1.3 (1.5) mm, elliptic to obovate, dorsal side convex, glabrous, ± sparsely covered with whitish papillae to 0.1 mm long or densely papillate, ventral side plane to shallowly concave and with a ± indistinct median vertical ridge, commissure ± small; mericarps crowned by 2 indistinct calyx lobes to 0.2 × 0.2 mm, but more commonly calyx lobes ± lacking. – Figs. 31 g–h, 37 b, 92–94.

Chromosome Number (Table 3): n = 11, 2 n = 22.

Average Pollen Diameter (Fig. 48): 23.5–28.4 μm.

Habitat: Found in all principal vegetation types of the SW Cape Floristic Region – i.e., Coastal Fynbos, Mountain Fynbos (incl. Arid Fynbos) and Coastal Renoster(bos)veld (cf. TAYLOR 1978, KRUGER 1979) – except for the Strandveld (Veld Type 34 of ACOCKS 1975). Occurring on rocky slopes, in sandy flats or gravelly areas and occasionally also in disturbed sites such as roadsides or plantations. Found in moderately dry habitats but also in moister areas along streams, in kloofs, on well drained mountain slopes, etc. Growing over a variety of substrates: in TMS derived sandy to gravelly soil, in clayey soil of shale bands, in reddish sandy loams derived from Cape Granite, in gravelly or coarse-sandy clay loams formed from the Malmesbury Beds and Bokke-veld Series, over Witteberg quartzite and in limestone or limestone sand. – Ca 10–1 600 (1 800) m.

Distribution (Maps, Figs. 91 a and b): S Africa, widely distributed in the SW Cape Floristic Region and, in the E, slightly extending beyond its limits to the Alexandria and Albany Distr.

Critical Remarks (for the species as a whole): *A. galioides* is highly variable in habit, leaf size, shape, indumentum and consistency, and in fruit indumentum and fruit size. It, nevertheless, is possible to subdivide the species into two subspecies, which – although overlapping in their distribution range to a considerable extent – are (at least in part) ecologically distinct and are normally easily separated morphologically. Within each of these two subspecies, there are a number of distinct ecotypes which, however, are connected by a series of intermediates so that it seems better to refrain from further subdivision and from recognizing these "Forms" taxonomically.

Species 29.–34. are very closely allied. Some Forms of *A. galioides* may, in fact, rather closely resemble *A. pumilum* subsp. *pumilum* (29.) in general morphology (leaf size and shape, general fruit and carpophore structure, flowers) and in the rather variable sex distributions of the plants. Other Forms may be similar to *A. prostratum* (32.) or approach *A. dregei* subsp. *ecklonis* (33.) in certain characters. In terms of absolute measurements, several characters often overlap considerably between the taxa and – on top of it – some characters are variable within a given taxon (e.g., fruit surface characters!) and, thus, cannot be used to distinguish species.

D. Systematic Part

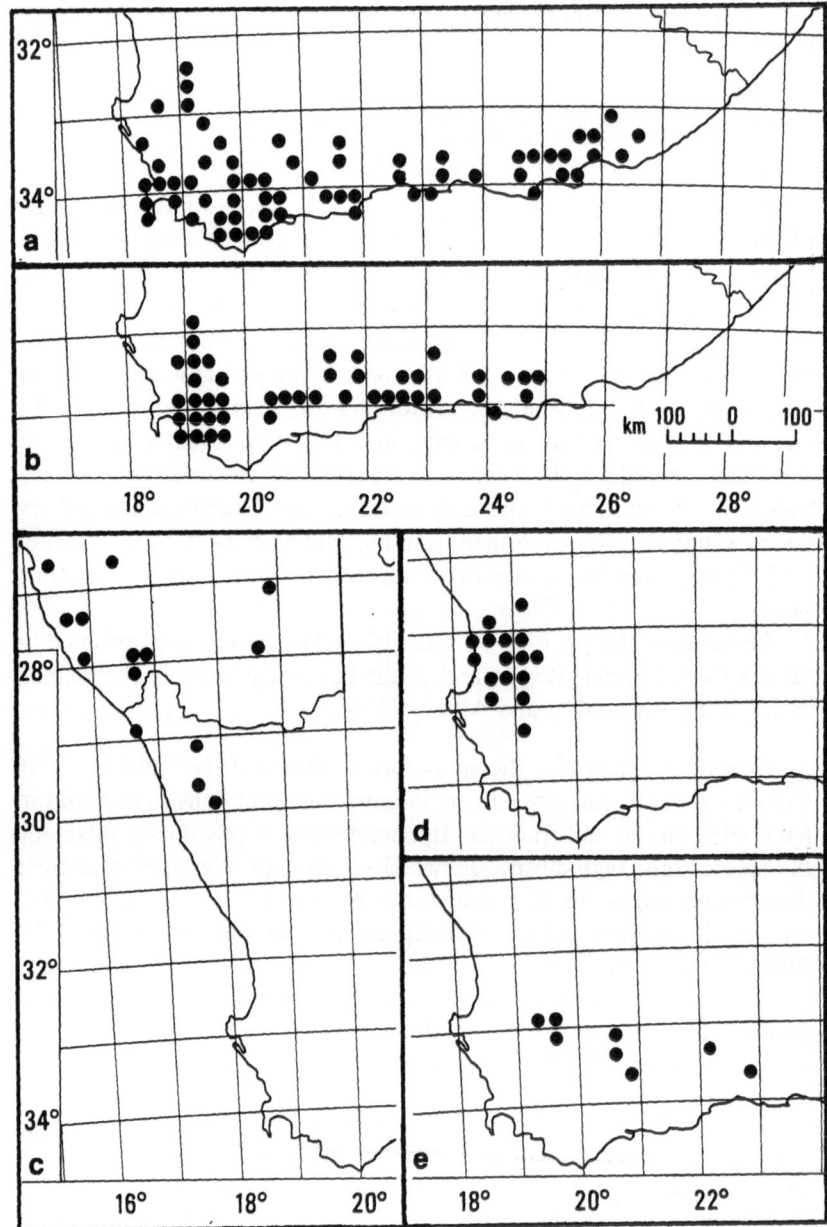

Fig. 91. Distribution of *Anthospermum. a–b A. galioides, a* subsp. *galioides, b*
subsp. *reflexifolium. c–d A. dregei, c* subsp. *dregei, d* subsp. *ecklonis. e A. comptonii*

Differences between the taxa sometimes only become evident in the field and if their growth habit and habitat is known.

Individuals of *A. galioides* with rather broad leaves and with flowers having comparatively long and cylindrical corolla tubes may somewhat resemble *A. herbaceum* (25.), which does extend into the S part of the SW Cape Floristic Region (for this reason, collections of these two species from the SW Cape were often confused in the past; see comments, p. 312). In my opinion, these similarities between the two taxa may be an indication that *A. galioides* and related species have some links to — i.e., are (± distantly) related to — the *A. herbaceum* alliance.

There is no concrete evidence for the occurrence of hybrids between *A. galioides* and other species.

31 (a). Subsp. *galioides*

= *A. ciliare* L. var. *angustifolium* ECKLON & ZEYHER, Enum. Pl. Afr. Austr.: 366 (1836).
Type: South Africa, Cape Prov., ... ad radicem montis "Duyvelsberg", in monte prope "Tokay" (Cap), in montibus "Hottentottshollands- and Hauhoeksberge" (Stellenbosch), prope "Caledon", ECKLON & ZEYHER 2308 γ (FI!, LY!, M!, MO!, S!, SAM!, W!; also "51-8: Caledon": S!).

= *A. ciliare* L. var. *latifolium* ECKLON & ZEYHER, Enum. Pl. Afr. Austr.: 366 (1836).
Type: South Africa, Cape Prov., ... prope "Somerset" in "Hottentottsholland" (Stellenbosch), ECKLON & ZEYHER 2308 β (FI!, GOET!, LY!, M!, MO!, S × 2!, SAM!, W × 2).

= *A. ciliare* L. var. *scabrum* ECKLON & ZEYHER, Enum. Pl. Afr. Austr.: 366 (1836), *pro parte*.
Type: South Africa, Cape Prov., ... inter "Coega- et Zondagsrivier", in planitie prope "Krakakamma" inque alveo lapidoso fluminis "Zwartkopsrivier" (Uitenhage), ECKLON & ZEYHER 2308 δ (B!, FI!, G × 2!, GOET × 2!, P − one sheet!, S!, SAM × 2!, W × 2!) [specimens mixed; subsp. *reflexifolium* in other herbaria].

= *A. ciliare* L. var. *glabrifolium* SONDER in HARVEY & SONDER, Fl. Cap. 3: 29 (1865).
Type: none given.

= *A. ciliare* L. var. *papillatum* SONDER in HARVEY & SONDER, Fl. Cap.

3: 29 (1865); SALTER in J. S. Afr. Bot. **3**: 110 (1937), in ADAMSON & SALTER, Fl. Cape Penins.: 733 (1950).
Type: South Africa, Cape Prov., Simon's Bay and Rietvalley; no collector given.
≡ *A. aethiopicum* L. ("β *ciliare*") var. *papillatum* (SONDER) O. KUNTZE, Rev. Gen Pl. **3**(2): 127 (1898), quoad typum tantum.

Sex distributions and growth form as for species as a whole. Leaf blades linear to ovate-lanceolate, at least margins ciliate (if glabrous, blades ascending or spreading and straight, not distinctly recurved). Fruits glabrous, distinctly papillate or very shortly hairy (if glabrous, mericarps mostly not longer than 2 mm). – Figs. 31 g–h, 37 b, 92–93.

Average Pollen Diameter (Fig. 48): 23.5–26.6 µm.
Habitat: As for species as a whole (but normally not in moister areas). See also Critical Remarks below.
Distribution (Map, Fig. 91 a): As for species as a whole.

Critical Remarks: Three "Forms" (in the sense of morpho- and/or ecotypes) can be distinguished which may occasionally be quite distinct:
(a) *"Typical"* subsp. *galioides*:
Characterized by having ovate-lanceolate to linear-lanceolate (due to revolute margins!) leaves with ciliate margins. Plants are often many-stemmed and ± densely tufted. The fruits are either glabrous and shiny or papillate/shortly hairy. On younger shoots, on plants resprouting after fire, etc. the leaves are often rather broad, i.e., ovate-lanceolate (to 10 × 4 mm in the extreme!; "*A. ciliare* var. *latifolium*"), and widely spaced.
This "typical" Form seems to be centered in the Cape Peninsula, although it appears again in the E parts of the distribution range of subsp. *galioides* and as far N as the Piketberg.

(b) *"Papillatum Form"* (Figs. 92 b–d):
Characterized by having few to several, ± erect to ascending, and often quite woody stems; the plants are not densely tufted. The straight, (narrowly) lanceolate to linear (due to strongly revolute margins) leaf blades [ca 5–10 × (0.5) 0.8–1.5 mm] are spreading or held at an angle of ca 45° to the stems; they are glabrous (except for, sometimes, minute prickles on the margins) and have large, conspicuous upper epidermis cells. The fruits are distinctly papillate (to very shortly hairy). In habit, the "*Papillatum* Forms", may somewhat resemble *A. pumilum* subsp. *pumilum* (which, however, does not extend into the SW Cape), especially if the plants are quite densely leafy due to rather short internodes.
The "*Papillatum* Form" is apparently ± confined to Mountain

Fig. 92. *Anthospermum galioides* subsp. *galioides*. *a* "Prostrate Form" (Cape Prov., ca 8–9 km N of Struisbaai, on road to Bredasdorp; PUFF 790912-4/2); *b–d* "*Papillatum* Form", three plants from population PUFF 800913-4/2 (Cape Prov., top of Dasklip Pass). – Further explanations in the text

Fynbos and TMS areas. Its distribution is fairly restricted: it is recorded from the Cape Peninsula E to the Caledon Distr. and NE to the Worcester, Ceres (Cold Bokkeveld Mts.), Piketberg (Porterville Mts.) and the Clanwilliam Distr. (Cedarberge).

SALTER (1937, 1950), concentrating on the Cape Peninsula, distinguished this Form as "*A. ciliare* var. *papillatum*" and maintained that it "is very distinct ... It is possibly worthy of specific rank". As far as the Cape Peninsula is concerned, there is some truth to his observations. Considering the whole range of variation over the entire distribution area of subsp. *galioides*, however, it is no longer valid. It often only differs very slightly from more robust and ± erect individuals of "typical" subsp. *galioides* with relatively narrow, ciliate-margined leaves and with papillate/shortly hairy fruits.

(c) *"Prostrate Form"* (Fig. 92 a):

Characterized by an often entirely prostrate habit: several stems radiate out from a common base and may form ± dense mats; occasionally, however, the upper parts of the stems are ± ascending. In more luxurious forms, stems may be to ca 30 (40) cm long and much-branched; in plants growing under less favourable conditions or in disturbed sites, such as in trampled areas, etc., they may not be longer than 5 cm and are hardly branched. The prostrate stems may be partly covered in soil but have never been found to root at the nodes. The leaves are often relatively broad (ovate-lanceolate to lanceolate) but small [ca 4–8 (10) × 1.5–2.5 (3) mm]; their margins (and, rarely, also the upper surfaces) are ciliate or – more commonly – glabrous. The fruits are relatively smaller than in the other forms (hardly reaching 2 mm in length), glabrous or papillate/shortly hairy.

The "Prostrate Form" occurs from the Caledon Distr. E to as far as the Suurbergpas N of Addo (e.g., PUFF 800929-1/3), and N to the Ceres and Laingsburg Distr. Whether it also occurs in the Cape Peninsula is somewhat doubtful (possibly ESTERHUYSEN 32955, clay loam slopes above the UCT buildings in Rondebosch, belongs here). It is noteworthy that this form does not seem to occur in TMS areas. It is, perhaps, most frequently found in the relatively fertile soils formed from the Bokkeveld Series of the Cape System (often growing in association with the renosterbos, *Elytropappus rhinocerotis*!), but it also is not uncommon – along the S Coast – in limestone areas. In inland habitats, it may also be found – although not so frequently – over Witteberg quartzites.

These "Prostrate Forms" may superficially resemble (and have often been confused with) *A. prostratum* (32., below). The latter, however, has rather longer trailing stems which root at the nodes and bear short, erect

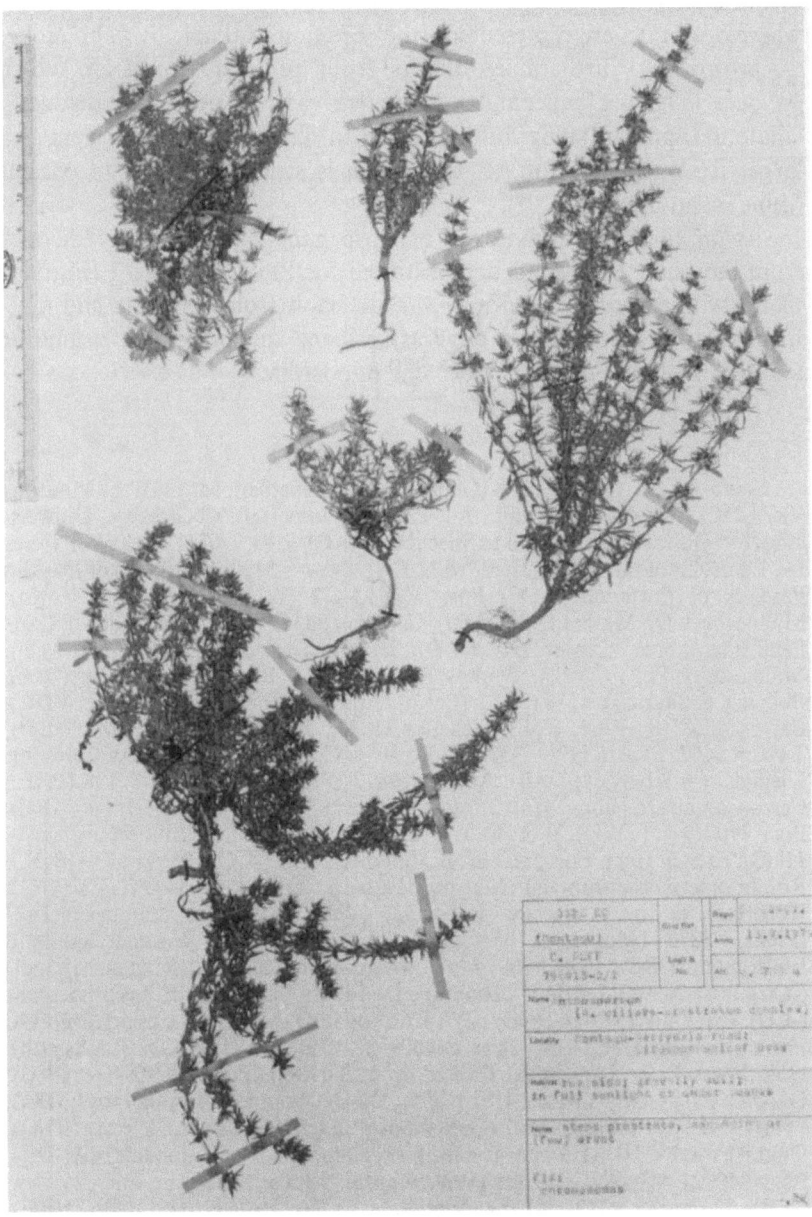

Fig. 93. *Anthospermum galioides* subsp. *galioides*. Plants from population Puff 790913-2/1; note growth form differences (stems ± prostrate → ascending → ± erect). – Further explanations in the text

lateral branches (Figs. 3 b–c); plants of *A. prostratum,* moreover, do not form mats. When observed in the field – e.g., in the Bredasdorp Distr., where both taxa are represented – they are unmistakable by habit alone. *A. prostratum,* furthermore, differs from prostrate forms of subsp. *galioides* in having longer and narrower leaves which never have distinctly ciliate margins, in being dioecious and in having larger ♂ flowers and larger fruits. In addition, *A. prostratum* is strictly confined to coastal (dune) sand areas.

Again, this "Prostrate Form" of subsp. *galioides* is not always distinct from the other Forms. As demonstrated for PUFF 790913-2/1 (Fig. 93), plants within a population may vary in habit from prostrate and mat-forming to erect so that a clear assignment and placement to one or another of these Forms may become impossible.

Collections

South Africa. Cape – **3218** (Clanwilliam): Piquetberg Mt., NW of Moutons Vlei (-DC), PILLANS 7326 (BOL, K). – **3219** (Wuppertal): Cedarberge, Duiwels-kloof (= Duiwelsgat), on path to Sneeuberg (-AC), PUFF 790717-1/1 (WU); Ceres D., Elandskloof [Citrusdal- Ceres rd.] (-CA), LEVYNS 5120 (BOL); top of Dasklip Pass, N of Porterville (-CC) PUFF 800913-2/1 (WU). – **3318** (Cape Town): Malmesbury D., Mamre Hills (-AD?), COMPTON 9434 (NBG; very atypical); Cape Peninsula, above Camps Bay (-CD), ESTERHUYSEN 33992 (BOL, M, WU); Malmesbury D., nr. Pella, Fynbos Research Site, Burgers Post Farm (-DA), BOUCHER & SHEPHERD 4538 (STE); [Farm] Langverwacht above Kuilsrivier (-DC), OLIVER 4368 (K, PRE, STE); Stellenbosch D., Jonkershoek For. Res. (-DD), TAYLOR 5467 (PRE, STE). – **3319** (Worcester): Ceres D., ca 40 km N of Ceres, nr. Kleinvlei (= Klynvley) (-AB), ACOCKS 19872 (PRE), SCHLECHTER 10198 (PRE); Ceres-Calvinia rd., above Hottentotskloof Outspan (tow. Theronsbergpas) (-BC), PUFF 790719-4/1 (WU); Worcester, at the waterfall (-CB), ECKLON & ZEYHER '1– 11' (S); French Hoek For. area, nr. Bushman's Castle (-CC), SALTER 5739 (BOL); Rooihoogte Pass (Montagu- Matroosberg rd.) (-DB), PUFF 840911-3/1 (WU); Robertson D., nr. Robertson aerodrome (-DD), VAN BREDA & JOUBERT 1927 (PRE). – **3320** (Montagu): Witteberge, between Farms Fisantekraal and Eselsfontein (-BC), PUFF 790914-4/1 (WU); Montagu Pass (Kogmans Kloof) (-CC), PUFF 790913-1/2 (WU); Montagu D., foot of Langeberg nr. Leeurivierberg (-CD), ESTERHUYSEN 24560 (BOL, WU); nr. top of Ouberg Pass, around turnoff to Farm Heintjesvlakte (-DB), PUFF 840909-4/1 (PRE, WU). – **3321** (Ladismith): Huis River Pass (Ladismith- Calitzdorp rd.) (-BC), PUFF 840909-1/1 (WU); Garcia's Pass (-CC), GALPIN 4101 (GRA, PRE); summit of Rooibergpas (-DA), PUFF 791223-1/1 (WU). – **3322** (Oudtshoorn): Farm Die Krans, De Rust (-DA?), DAHLSTRAND 2429 (PRE – very atypical); Eseljagpoort (N of Uniondale rd.) (-DC), PUFF 840921-4/2 (WU). – **3323** (Willowmore): between Misgund and Haarlem (-CB), FOURCADE 4339 (K); nr. Walletjies (-CD), BAYLISS 6461 (BR, NBG); Uniondale D., hill N of Joubertina (-DD), FOURCADE 2681 (BOL). – **3324** (Steytlerville): E end of Baviaanskloofberge, between Grootrivierpoort and Farm Hadley (-DA), PUFF 840921-5/2 (WU); Elandsberg, just opposite (S of) Cockscomb Mt. (-DB), PUFF 840921-6/2 (PRE, WU); Humansdorp D., [Farm] Rietvlei (-DC), ESTERHUYSEN 6611 (BOL). – **3325** (Port Elizabeth): Suurberg Pass, nr. Farm Zuurberg Manor (**-BC**), Puff 800929-1/3 (BOL); Addo National Park, Zuurkop (= Suurkop) (**-BD**), Archibald 3866 (K, PRE, S); Uitenhage D., Winterhoek Mts. (**-CA, -CB?**), Fries, Norlindh & Weimarck 1189 (BOL); Uitenhage (**-CD**), Prior s.n. (K); Alexandria D., Nanaga (**-DB**), Archibald 5944 (K, PRE×2, S); Zwartkops River mouth (**-DC**), Long 808 (GRA, K, PRE). – **3326** (Grahamstown): base of Swartwatersberg, nr. Riebeek-East (**-AA**), Puff 840922-2/1 (PRE, WU); Grahamstown commonage, nr. Sugar Loaf (**-BC**), Rennie & Rennie 111 (BOL); Alexandria (**-CB**), Burtt-Davy 14232 (PRE). – **3418** (Simonstown): Muizenberg Mt. (**-AB**), Salter 6274 (BOL); Cape of Good Hope Nat. Res., above Platboom Bay (**-AD**), Puff 800907-2/1 (WU); Sir Lowrys Pass (**-BB**), Penther 2071 (W). – **3419** (Caledon): (around) Caledon (Baths) (**-AB**), Bolus 9149 (BOL, K), Henderson 1779 (NBG), Purcell s.n. sub SAM 46162 (SAM); between Bot River and Kleinmond (**-AC**), Esterhuysen 3122 (BOL); N foot of Riviersonderend Mts., W of Boesmansrivier Farm (McGregor- Bonnievale rd. (**-BB**), Puff 840911-1/2 (PRE, WU); Stanford- Papiesvlei rd., Perdeberg above Farm Flouhoogte (**-BC**), Puff 800918-1/3 (WU); just S of Napier (**-BD**), Esterhuysen 32962 (BOL, WU); Groenkloof, NE (inland) of Pearly Beach (**-DA**), Puff 840919-3/2 (PRE, WU); ca 1–1.5 km E of Elim (**-DB**), Puff 800918-3/1 (WU). – **3420** (Bredasdorp): Swellendam (**-AB**), Burchell 7292 (K); ca 4.5 km E of Wydgelee P.O. (Ouplaas) (**-AD**), Puff 790911-6/1 (WU); Buffeljagsrivier-Malgas Pont rd., ca 5 km N of Farm Swartklip (**-BA**), Puff 790911-5/1 (WU); Bredasdorp D., Potberg (**-BC**), Pillans 9284 (BOL); –, (ca 1 km inland from) Arniston (= Waenhuiskrans) (**-CA**), Acocks 22612 (K, PRE), Puff 790912-2/2 (WU), Taylor 3805 (K, PRE, STE); Farm Martha, ca 16.5 km W of Skipskop on Bredasdorp rd. (**-CB**), Puff 790912-1/2 (WU). – **3421** (Riversdale): N of Riversdale, on rd. to Farm Klein-Kruisrivier (**-AB**), Puff 790911-1/1 (WU); Albertinia- Gouritsmond rd., ca 5.5 km from main road (R 2), SE of Aasvoelberg (**-BA**), Puff 800927-1/1 (WU); W of Kleinberg (**-BB**), Burchell 6360 (K); ca 3.5 km NW of The Fisheries (= Gouritsmond) (**-BD**), Puff 790910-4/1 (WU). – **3422** (Mossel Bay): Knysna D., Belvedere (**-BB**), Duthie 1096 (BOL, STE). – **3423** (Knysna): Knysna (**-AA**), Duthie 646 (BOL, LMA, STE). – **3424** (Humansdorp): Jeffreys Bay (**-BB**), Duthie 1063 (BOL, STE). – Imprecise localities ("CBS") or no locality given: Alexander s.n. (BM, K × 2); Bergius s.n. (G); Brehm s.n. (M); coll. Burman [herb. Delessert] s.n. (G × 3 – one mixed with *Galium capense*); Carmichael 250 (BM); herb. Jacquin (W × 5); Lehmann s.n. (P); Oldenburg s.n. (BM); Scholl s.n. (W × 4); Sieber [Flora Cap.] 88 (G, K, M, MO, S, W), [Flora mixta] 24 (E, S); Sparrman s.n. (S); Thunberg '22', '29', s.n. (S); herb. Ventenat s.n. (G – mixed with *A. aethiopicum*); Verreaux s.n. (G – mixed with *A. spathulatum*). – Additional collections: Acock(s) 537, 627, 2775, 4645 (S), 12794 (PRE); Archibald 5246 (K, LISC, PRE, W); Barnard 458 (SAM); Bond 394 (NBG); Boucher 2491 (PRE, STE); Bunbury 113 (BM); Burchell 1029, 6381 (GH, K); Burgers 2426 (STE); Burtt Davy 12157 (PRE; ± atypical); Compton 15749 (NBG), 22710 (NBG; appr. subsp. *reflexifolium*); de Vos 1038 (STE); Drège 3065 (PRE), 7661 b (E, G, K, W), 9548 (LY), 9549 (LY); Dümmer 20 ε, 63 b, 1395 (E); Duthie 515 (BOL); Ecklon 14, 29 (S), 27 (GOET, S), 27 b (M, MO), 28 (GOET), 28 b (MO), 191 (S, W); Ecklon & Zeyher '44-5', '56.7' (M, LY – one sheet), '89-9' (S), '51-8' (?G sub herb. Verreaux, S), 2308 (B, G × 2, LY, M, MO, SAM, W), 2308 β (FI, GOET, LY, M, MO, S × 2, SAM, W × 2), 2308 γ (FI, LY, M, MO, S, SAM, W), 2308 δ (B, FI, G × 2, GOET × 2, P – one sheet, S, SAM × 2, W × 2; mixed collections p.p.; subsp. *reflexifolium* in some herbaria); Esterhuysen 447 (PRE), 32960 (BOL, K), 32078 (BOL, WU; approach. subsp. *reflexifolium*), 1082, 21274, 32955 (BOL); Fries, Norlindh & Weimarck 111 (MO, PRE, US,

WAG), 1642 (K, MO, PRE); GALPIN 4101 (GRA, PRE); HANEKOM 1273 (K, PRE); HUTCHINSON 105 (BM, BOL, K, PRE); JOHNSON 731 (PRE × 2); KRUGER 1406 (STE); LEIPOLDT 4450 (BOL); LEVYNS 5350, 8041 (BOL); LIEBENBERG 6385 (PRE, STE); MARLOTH 455 (PRE); MARSH 1356 (PRE, STE); MOFFETT 2797 (STE); MOSS 8465 (J); MUIR s.n. (GRA); PARKER 4463 (BOL, K, MO, NBG); PATERSON 779 (BOL); PENTHER 2700 (W); PUFF 790912-4/2, 790913-2/1, 791225-3/1, 791226-5/3, 800903-2/1, 800908-2/2, 800919-4/1, 800928-4/1 (WU); PURCELL s.n. sub SAM 90335, 90336, 90339 (SAM); REHMANN 1211 (BM, BR); ROGERS 24893 (PRE); SALTER 6188, 6266 (BOL, K), 6222 (BM, BOL, K), 6239, 6706, s.n. sub BOL 25125 (BOL); SCHÖNLAND 3502 (PRE); TAYLOR 309 (NBG), 3764 (K, PRE, STE), 5406 (BR, K, PRE, STE), 6525 (STE); THODE s.n. sub STE 7962 (STE), 8407 (STE × 3); VAN BREDA 660 (K × 2); VAN DER MERWE 1124 (PRE); WOLLEY DOD 1073 (BM), 1106 (BM, BOL), 1224 (K), 2660 (BM, BOL, K); WRIGHT s.n. (GH, US × 2 – all mixed with *A. spathulatum* subsp. *spathulatum*); ZEYHER 2718 (LY, S), 2719 (LY, SAM), s.n. (K, P), s.n. sub SAM 16057, 16058, 16063 (SAM).

31 (b). Subsp. ***reflexifolium*** (O. KUNTZE) PUFF in Fl. S. Afr. **31**, 1 (2): 30 (1986).

≡ *A. aethiopicum* L. var. *reflexifolium* O. KUNTZE, Rev. Gen. Pl. **3** (2): 117 (1898).

Type: South Africa, Cape Prov., Swellendam, January 30, 1894, O. KUNTZE s.n. (US sub 554666, holo.!; GH, K, iso.!).

= *A. ciliare* L. var. *scabrum* ECKLON & ZEYHER, *pro parte*.
Type: ECKLON & ZEYHER 2308 ♂ (LY!, M!, MO!, P – one sheet!) [see also subsp. *galioides*].

Plants mostly ♂, ⚥ + ♀ or ♀, densely leafy, ± erect, cylindrical and quite woody to less densely leafy, ± rounded, weaker and ± subshrubby. Young stems often only with two rows of very short hairs or papillae. Leaf blades linear-lanceolate to lanceolate, recurved (at least near the tip) to strongly recurved, mostly firm, thickish and almost leathery, glabrous (except for, sometimes, minute prickles on the margins), upper surface shiny, epidermis cells large, conspicuous; margins often revolute. Flowers mostly only in pairs at the nodes. Fruits glabrous, epidermis cells often large and conspicuous, but mericarps never distinctly papillate or very shortly hairy; mericarps relatively large, ca 2–2.4 mm long. – Fig. 94.

Average Pollen Diameter (Fig. 48): 24.9–28.4 μm.

Habitat: Apparently confined to Mountain Fynbos (incl. Arid Fynbos) and exclusively (?) in TMS areas. In moderately dry habitats but also in moister places (well drained mountains slopes, along streams, in wooded kloofs, at the edge of riparian communities, etc.).

Distribution (Map, Fig. 91 b): From the Porterville Mts. S to the

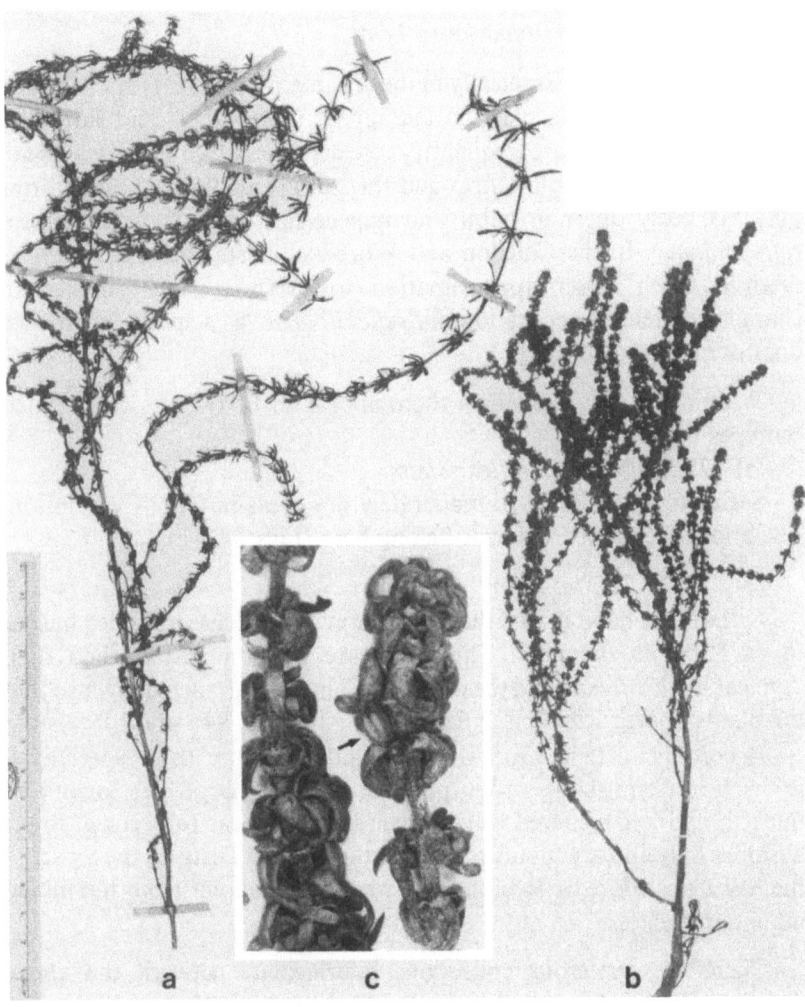

Fig. 94. *Anthospermum galioides* subsp. *reflexifolium. a* branch of rounded bush, ca 30 cm in diam., from a moist, shady, steep slope at the edge of forest (Cape Prov., ca 7 km W of Karatara, ca 150–200 m; PUFF 790910-3/2); *b* part of few-stemmed, erect plant collected in ± dry fynbos (Cape Prov., Prince Alfred Pass, nr. Donkerhoek-se-Nek, ca 800–900 m; PUFF 791221-3/1); *c* enlargement from *b*, showing distinctly recurved, toughish, shiny leaf blades; the arrow points to a mature fruit. – Further explanations in the text

Caledon Distr. (but not in the Cape Peninsula). E-wards following the
mountain ranges to the Groot Winterhoekberge and, further S, to the
Kouga Mts. and the Humansdorp Distr.

Critical Remarks: Especially in the mountains in the E part of the SW
Cape Floristic Region, plants often appear more robust and shrubbier
than elsewhere. They are frequently quite densely leafy, and the leaves are
small (to 5 mm long; quite firm and thickish) (Figs. 94 b–c). Such forms
may markedly differ in habit and appearance from "typical" subsp.
reflexifolium[1]. In the Caledon and Worcester Distr., however, there is
often a much closer approximation to subsp. *galioides* in certain
characters. Hence a separation of "*reflexifolium*" as a species of its own
does not seem justified.

Within subsp. *reflexifolium* there appear to be two ± well defined
ecotypes ("Forms"):
(a) "*Typical*" subsp. *reflexifolium:*
Seems to be confined to moderately dry areas in fynbos vegetation.
The above mentioned E forms belong here (Figs. 94 b–c).

(b) "*Moist Habitat Form*" (Fig. 94 a):
Differing in habit in that the plants form rather dense, rounded bushes
to ca 50 cm in diameter. The stems are often more-branched than
"typical" *reflexifolium* and the internodes tend to be (much) longer so that
the plants appear less densely leafy (compare Figs. 94 a and b).

According to field observations, plants of Form (b) – especially if
growing in (semi)shade – often produce relatively large, less tough and
thickish and less recurved leaves. Such plants appear to occur at lower
altitudes only and are found as far E as the Knysna Distr. In the S part of
the SW Cape Floristic Region, they seem to be absent from the inland
mountain ranges.

There are numerous collections intermediate between the above
described Forms (a) and (b), so that a clear separation of herbarium
specimens, which often neither give a clue to habitats nor growth forms, is
frequently impossible. It, furthermore, must be noted that the "curliness"
of the leaves was often found to vary with age. Leaves on new flowering
shoots (coming up after a fire, etc.) and on plants flowering for the first
time may have relatively thinner and larger blades (to 20 × 4 mm) which
are less distinctly recurved and have less prominently revolute margins

---

[1] Acocks 23082 may either be such an "extreme" E Form or represent a
closely allied new species. It, at present, is tentatively grouped with subsp.
reflexifolium.

than older woody plants of the same population (the 13 duplicates of SCHLECHTER 9149, for example, document this situation!). Finally, attention is drawn to the fact that the degree of the leaf-"curliness" may (markedly) increase when the plants are dried.

In contrast to subsp. *galioides*, subsp. *reflexifolium* exhibits a clear trend to dioecy. According to field observations, some populations are strictly dioecious, while in others, unisexual (predominant) and ♂ + ♀ individuals (occasional) occur together.

Some collections, primarily from the Bredasdorp Distr., may represent a new taxon, allied to subsp. *reflexifolium*[1]. The plants have tough, leathery leaves with conspicuous upper epidermis cells; they appear to be unisexual and seem to have fruits which differ quite markedly from those of *A. galioides* (larger, broader). As neither good flowering nor fruiting material is available, I, at present, refrain from a taxonomic recognition.

Collections

South Africa. Cape — 3219 (Wuppertal): plateau of Porterville Mts. (Skurweberge) (-CC), PUFF 800913-5/6, 800914-1/1 (WU). — 3318 (Cape Town): Malmesbury D., Riebeeks Kasteel (-BD), PILLANS 6093 a (BOL, K); Stellenbosch D., Jonkershoek For. Res. (-DD), ESTERHUYSEN 33137, 33750 a (BOL), PUFF 791231-1/3 (WU). — 3319 (Worcester): Witsenberg, around Sneeugat Peak (-AA), PUFF 800924-3/4 (WU); Tulbagh waterfall (-AC), COMPTON 12428 (NBG, STE); Waaihoek Mts., Mosters Hoek Twins (-AD), ESTERHUYSEN 9841 (BOL); Paarl D., Bains Kloof (-CA), ESTERHUYSEN 26310 (BOL, WU), SCHLECHTER 9149 (BM, BOL, BR, COI, E, G, K, MO, P, PRE, S, US, W); Klein Drakenstein Mts., upper Kasteelkloof catchment (-CC), KRUGER 1423 (PRE, STE); Riviersonderend Mts., ridges W of Jonaskop (-CD), PUFF 800921-1/2 (WU); Worcester D., Keeromsberg (-DA), ESTERHUYSEN 9216 (BOL); Riviersonderend Mts., service rd. to Jonaskop SABC tower (-DC), PUFF 800921-2/2 (WU). — 3320 (Montagu): base of 10 o'Clock Peak, N of Swellendam (-CD), PUFF 800919-2/1 (WU); Tradouwspas area (-DC), PUFF 790911-3/2, 790913-3/1, -4/1, -4/2 (WU); Swellendam D., Lemoenshoek Peak (-DD), ESTERHUYSEN 10430 (BOL). — 3321 (Ladismith): S slopes of Swartbergen nr. Ladismith (-AD), ESTERHUYSEN 13957 (BOL); 14 km W of Swartberg Pass summit, on rd. to The Hell (-BD), PUFF 840908-3/1 (WU); Rooiberg, 4.1 km above North Gate on rd. to Bailey Peak (-CB), TAYLOR 9664 (PRE, STE); Gysmanshoek Pass between Brandrivier and Heidelberg (-CC), PUFF 791223-3/1 (WU); Calitzdorp D., Gamka Mountain [Zebra] Reserve (-DB), BOSHOFF P 324 (STE); Mosselbay D., Langeberg (-DC), MUIR 1373 (BOL × 2, PRE, SAM). — 3322 (Oudtshoorn): Robinson Pass (-CC), PUFF 791222-3/1 (WU); ca 6.5 km WNW of Camfer Stn. (-CD), ACOCKS 23082 (K, PRE; very atypical); Kammanassie Mts.,

---

[1] SW Koueberge, on Papiesvlei- Nuwepos rd. (3419-DA), PUFF 800918-2/4 (WU); hills above (N of) Viljoenshof (3419-DA), PUFF 800918-6/1 (WU); De Hoop- Potberg Nature Res., Potberg, top of mt. above Boskloof (3420-BC), BURGERS 1444 (STE); de Hoop (3420-CA), VAN DER MERWE 2014 (STE). Also BOUCHER 1134 (STE) from the Kogelberg Nature Res., peak above Cascades (3418-BD), may belong here.

Kleinberg Bosreservat, Kleinplass R. valley (-DA), Matthews 1207 (NBG); Uniondale D., Mannetjiesberg (-DB), Esterhuysen 6491 (BOL, K); Wilderness-George (-DC), Levyns 5079 (BOL); 6 km W of Karatara (-DD), Puff 790910-3/2 (WU). – 3323 (Willowmore): Uniondale D., Slypsteenberg (-AC), Esterhuysen 6297 (BOL, K); Prince Alfred Pass, nr. Donkerhoek se Nek (-CC), Puff 791221-3/1, -3/2 (WU); Uniondale D., Kouga Mts. nr. Smutsberg (-DB), Esterhuysen 10716 (BOL); – , N slopes of Outeniquas nr. Joubertina (-DD), Esterhuysen 10659 (K, BOL). – 3324 (Steytlerville): Baviaanskloof, Eilandvlakte (-CB), Bond 1315 (NBG); Baviaanskloof Mt., above Couties Kraal, NW of Cambria (-DA), Thompson 1907 (PRE, STE); Uitenhage D., Winterhoek Mts., Cockscomb (-DB), Esterhuysen 27496 (BOL, PRE); Humansdorp D., [Farm] Rietvlei (-DC), Esterhuysen 6617, 6644 (BOL). – 3418 (Simonstown): Sir Lowrys Pass (-DB), Schlechter 7802 (BM, G, K, PRE); Kogelberg Forest Res. (-BD), Haynes 565 (STE), Puff 800101-1/1, -2/2 (WU). – 3419 (Caledon): Lebanon [For. Res.], Grabouw (-AA), Kruger KR 270 (K, PRE); Ezelsjagt (= Eselsjagsberg) or Donkershoek (= Donkerhoekberge) nr. [S of] Villiersdorp (-AB), Bolus 5060 (BOL); Caledon D., Kleinmond Kloof (-AC), de Vos 1146 (PRE); Vogelgat (Private) Nature Res., E of Hermanus (-AD), Puff 800917-2/2 (WU); Caledon D., Riviersonderend Mts. (-BA?), Stokoe s.n. sub SAM 59514, 64184 (SAM); Salmonsdam Nature Res., E of Stanford (-BC), Puff 840918-1/3 (PRE, WU). – 3420 (Bredasdorp): Swellendam (-AB), Kuntze s.n. (GH, K, US sub 554666). – 3424 (Humansdorp): Witte Els Bosch (= Witelsbosch) (-AA), Esterhuysen 6741 (BOL). – Additional collections: Boucher 1197 (STE); Drège s.n. ['*A. spath.* ?a': E, G × 3, K, LY, MO, S × 2, SAM, W × 3; '*A. spath.* ?b': E, G – mixed with *A. spathulatum* subsp. *uitenhagense* ('*A. spath.* b'), K, MO, W]; Ecklon s.n. (S); Ecklon & Zeyher '56-6' (S), '56.7' (E × 2- p.p., GH × 2, GOET, GRA, LY-p.p., MO, PRE × 2, W, WU), 2308 δ (mixed – LY, M, MO, P; subsp. *galioides* in many other herbaria), 2310 (LY, S, SAM); Esterhuysen 9557, 11877 (BOL, K), 2684, 6532, 10836, 10916, 11110, 11423, 11479, 11510, 12341, 12477, 12562 (BOL); Keet 770 (GRA, STE); Levyns 6000 (BOL); Lewis s.n. sub BOL 25132 (BOL); Masson s.n. (BM); Parker 4690, 4895, 4896 (BOL, GH, K, NBG); Phillips 1158 (SAM); Puff 791223-2/1, 800921-3/1 (WU); Rehm s.n. (M); Rycroft 2899 (NBG, STE); Stokoe 7587 (BOL), s.n. sub SAM 59515, 59516, 59517, 59521 (SAM), sub SAM 64185 (PRE, SAM); Taylor 5422 (PRE, STE), 8568 (STE); Thompson 1465, 1847 (PRE, STE); Williams 2697 (NBG); Zeyher s.n. sub SAM 16063 (SAM).

32. **Anthospermum prostratum** Sonder in Harvey & Sonder, Fl. Cap. 3: 28 (1865): Salter in J. S. Afr. Bot. 3: 110 (1937), in Adamson & Salter, Fl. Cape Penins.: 733 (1950); Puff in Fl. in Fl. S. Afr. 31, 1 (2): 30 (1986).
Type: only given for the varieties cited below.

= *A. prostratum* Sonder var. *glabrum* Sonder in Harvey & Sonder, Fl. Cap. 3: 28 (1865).
Type: South Africa, Cape Prov., Cape Flats, 1841, Ecklon s.n. (S, holo.!, LY, iso.!)[1] [selected as lectotype of *A. prostratum*].

---

[1] The label on the specimen in LY only says "*A. prost.* Sond." in Sonder's handwriting.

= *A. prostratum* Sonder var. *velutinum* Sonder in Harvey & Sonder, Fl. Cap. **3**: 28 (1865).
Type: South Africa, Cape Prov., Cape Flats, C. Wright (S, holo.!)[1].

Dioecious (very seldom also ♂ plants with odd ⚥ flowers), prostrate dwarf shrubs or subshrubs with basally woody stems, often with ± thick, woody tap roots. Stems trailing, rooting at the nodes, to ca 1 m long, ca 2–2.5 mm in diameter near the base and ca 1 mm in the midstem region, with ascending to ± erect short lateral branches, ca 1–5 (10) cm long; younger parts ± 4-angled, often reddish-purplish tinged, densely covered with whitish papillae less than 0.1 mm long to subglabrous; older parts terete, (reddish-)grey, glabrescent, epidermis often peeling off; internodes to 30 mm long on prostrate long shoots and usually shorter than the leaves on the lateral branches. Leaves decussate, pseudo-verticillate; blades (5) 7.5–12 × 0.7–2 (2.3) mm, lanceolate, oblanceolate to ± linear-lanceolate, glabrous or with a few very short whitish hairs or papillae on (the lower half of) the margins; apices acute to ± mucronate; margins revolute to ± flat; petioles subobsolete; stipular sheaths shortly hairy to subglabrous, cup-shaped, small, 1 mm or less long, with a ± broadly triangular median seta ca 0.3–0.7 mm long, sometimes with an additional, minute gland-tipped seta on either side. Inflorescences inconspicuous, made up of subsessile, solitary or paired flowers (seldom paired 3-flowered cymes) on short lateral branches. Flowers ♂ (very rarely also ⚥), ♀; corolla 4-merous (very seldom also 5-merous in ♂), frequently reddish-purplish tinged outside and yellowish or greenish-yellow inside, glabrous. ♂ (odd ⚥): tube 1.2–2.3 mm long, narrowly funnel-shaped, lobes (2) 2.4–3 (3.7) × 0.7–1 mm, lanceolate, recurved; stamens 4, filaments 1.9–2.7 mm long, anthers 1.5–2.2 × 0.2–0.4 mm; rudimentary ovary reddish-brown, small but well discernible. ♀: tube (0.5) 0.7–0.9 (1.1) mm long, cylindrical, lobes 0.5–1 × 0.1–0.2 mm, linear(-lanceolate), ± erect; style 0–1.3 mm long; stigmas 2, 4.8–10 mm long (shorter in odd ⚥), greyish; ovary 0.7–1.4 × 0.5–1 mm, glabrous or shortly hairy, not crowned by distinct calyx lobes. Fruits reddish(-brown), shiny, supported by small, ± broad U-shaped carpophores; each mericarp (2) 2.3–2.9 × 1.2–1.7 mm, (ob)ovate, dorsal side convex, glabrous or with short whitish hairs or papillae to 0.1 mm long, ventral side ± plane and with an indistinct median vertical ridge or a vertical ridge absent; commissure relatively large; mericarps not crowned by calyx lobes. – Figs. 3 b–c.

---

[1] No collection number is given on the label handwritten by Sonder although, according to Sonder in Flora Capensis, the specimen is Wright 491.

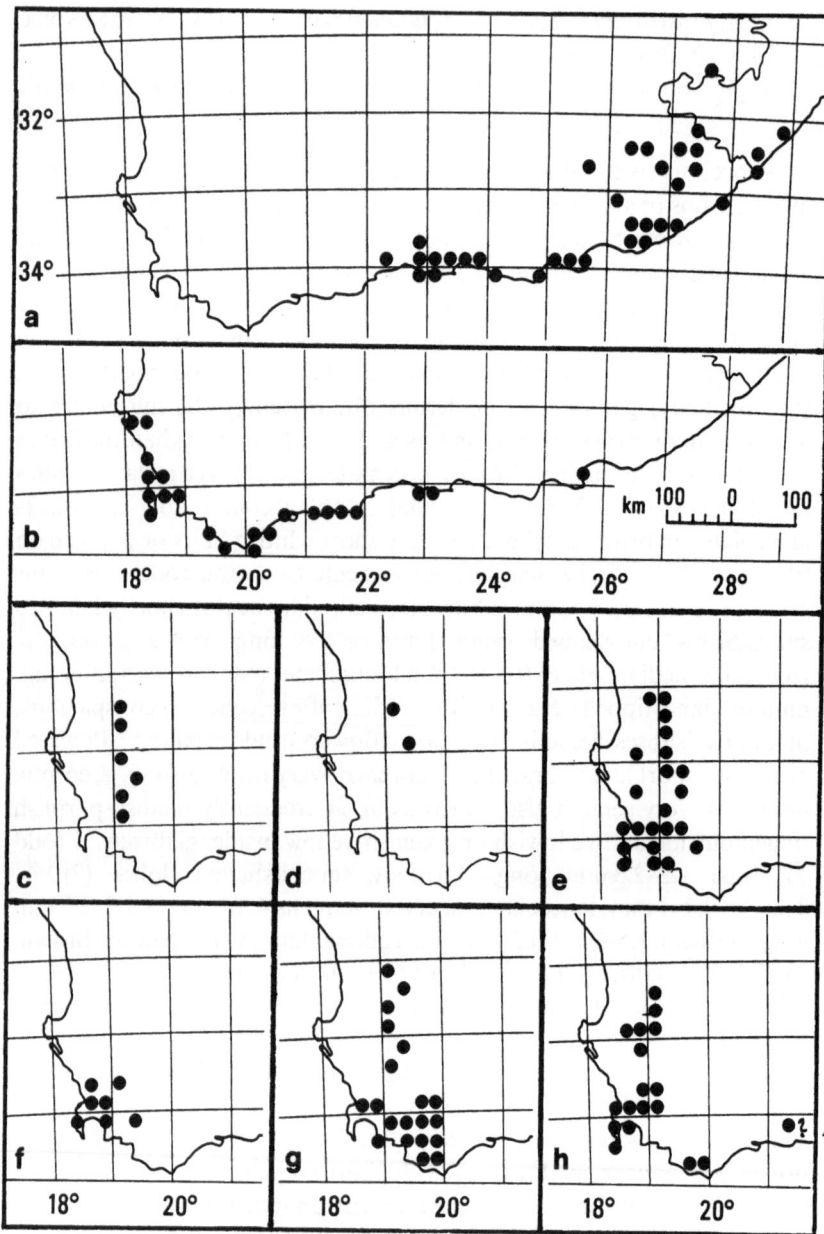

Fig. 95. Distribution of *Anthospermum* . *a A. paniculatum, b A. prostratum. c–d A. esterhuysenianum, c* var. *esterhuysenianum, d* var. *hirsutum. e A. bergianum, f A. ericifolium, g A. bicorne, h A. hirtum*

Chromosome Number (Table 3): n = 11, 2 n = 22.

Average Pollen Diameter (Fig. 48): 24.5–27.8 µm.

Habitat (Fig. 3 b): Confined to coastal areas and occurring in sandy flats and dunes; typically in Coastal Fynbos. – Ca 0–50 (80) m.

Distribution (Map, Fig. 95 b): S Africa, SW Cape Prov. From the Cape Peninsula N to the Malmesbury Distr., and E to the Knysna Distr.; also around Port Elizabeth.

Critical Remarks: A. prostratum is well distinguished by its growth form and habitat. It is relatively uniform in its characters except for the fruits. Considering the great variability in the fruit indumentum in numerous other *Anthospermum* taxa (including in those allied to *A. prostratum*) it, however, does not seem to be justified to uphold var. *velutinum* for hairy-fruited forms of *A. prostratum*.

A. prostratum is without doubt closely allied to *A. galioides* (31., above), and plants may superficially resemble prostrate ecotypes of subsp. *galioides* (see comments, p. 356). An alliance to *A. littoreum,* another species confined to coastal sand dunes (cf. Fig. 3 a; Tongaland to the E Cape Prov.), is – on account of major morphological differences (flowers, fruits, growth form) – highly improbable.

Collections

South Africa. Cape – **3318** (Cape Town): Langebaan peninsula, Oude Post Private Nature Reserve (**-AA**), Boucher 2989 (PRE, STE); ca 2.5 km ESE of Hopefield (**-AB**), Acocks 20675 (K, M, PRE, SRGH); around Melkbosstrand (**-CB**), Leistner 5 (STE), Puff 791226-1/1 (WU); ca half way between Melkbosstrand and Bloubergstrand (**-CD**), Puff 791226-1 A/1 (WU); nr. Modder Dam, ca 4.5 km SW of Bellville (**-DC**), Salter 6245 A–C (BM, BOL, K). – **3325** (Port Elizabeth): around Port Elizabeth (**-DC**), Fries, Norlindh & Weimarck 166 (SRGH). – **3418** (Simonstown): Simon's Bay (**-AB**), Wright s.n. (K, GH, US sub 81704, 81705); [Cape of Good Hope Nature Reserve], Buffels Bay (**-AD**), Compton 6037 (NBG); Cape D., nr. Swartklip (**-BA**), Compton 18135 (NBG), Pillans 1775 (BOL); Lower Eerste Rivier area at Macassar below Kramat (**-BB**), Boucher 3483 (PRE, STE). – **3419** (Caledon): Bredasdorp D., beyond Strandkloof (**-CB**), Compton 14754 (NBG); around Buffelsjag (**-DA/-DC**), Puff 840919-5/1 (PRE, WU). – **3420** (Bredasdorp): De Hoop- Portberg Nature Res., Potberg Farm, 0.3 km from coast (**-BC**), Burgers 1008 (PRE); ca 1 km inland from Arniston (**-CA**), Puff 790912-2/1 (J, W, WU); ca 4–5 km N of Skipskop (**-CB**), Puff 790911-8/1 (J, K, W, WU); nr. Cape Agulhas lighthouse (**-CC**), Puff 790912-3/1 (J, W, WU). – **3421** (Riversdale): Riversdale- Blombos rd., nr. Farm Victoriasdale (**-AC**), Puff 840920-3/1 (PRE, WU); Stillbay Hoogte (**-AD**), Bohnen 7605 (PRE, STE); Stillbay East-Ystervarkfontein rd., just W of Ystervarkpunt (**-BC**), Puff 840921-1/1 (PRE, WU); Vleesbay, Fransmanshoek (**-BD**), Puff 840921-3/1 (PRE, WU). – **3422** (Mossel Bay): Groenvlei (= 'Groene vallei') (**-BB**), Burchell 5663 (GH, GOET, GRA, K, M, P, W), Levyns 10280 (BOL). – **3423** (Knysna): Buffalobay, Knysna (**-AA**), Keet 854 (STE). – Imprecise locality ("CBS"): Masson s.n. (BM). – Additional collections: Compton 13437 (NBG); Dümmer 63 b

(GH), 380 (BM, E, NH); Ecklon s.n. (S, LY); Gillett 1024 (BOL, K); Hutchinson 9 (BOL, K); Lamb 1278 (SAM); Marloth 5679 a, b (PRE); Martin 4421, 4552 (PRE); Nature Conserv. herb. no. 192 (PRE); O'Callaghan 525 (STE); Purcell 343, 344 (SAM); Salter 274/15 (BM, 6238 (BM, BOL, K); Taylor 7029 a (STE); Tölken 35 (PRE, STE); Wolley Dod 1343 (BM, K); Wright s.n. (S); Zeyher s.n. sub SAM 16056 (SAM).

33. *Anthospermum dregei* Sonder in Harvey & Sonder, Fl. Cap. **3**: 29 (1865); Launert & Roessler in Merxm., Prod. Fl. SWA **115**: 7–8 (1966); Puff in Fl. S. Afr. **31**, 1 (2): 31 (1986).
Type: South Africa, Cape Prov. [Namaqualand]: between Koussie and Zilverfontein, 2 000 ft., Aug., Drège 3016 (S, holo.![1]; E, G × 3, K, LY × 2, MO, P × 3, S[1], SAM, W × 5, iso.!).

Non-dioecious (♂, ♀̄, ♀̄ + ♀, or ♀) or sometimes ± dioecious, diffusely to ± regularly branched, many-stemmed, often ± rounded dwarf shrubs ca 15–40 cm tall. Stems to ca 7 mm in diameter near the base and ca 1–3 mm in the midstem region, ± terete, branches ± irregular (especially if browsed) and often divaricate; younger parts often reddish-purplish-brown, densely covered with whitish papillae or hairs to 0.2 mm long; older parts reddish-brown to greyish, glabrescent, epidermis peeling; internodes usually longer than the leaves. Leaves decussate, sometimes ± pseudo-verticillate; blades 4–14 (20) × 1–4 (4.5) mm, ± ovate, oblong-lanceolate to ± lanceolate, narrowed to the base, ± membranaceous to thickish, glabrous or with a few papillae to 0.1 mm long on upper and/or lower surface, but especially on the margins; upper surface sometimes distinctly purplish-brown, lower surface (light) green; apices acute to acuminate; margins ± revolute, convolutely rolled or ± flat; petioles subobsolete or to ca 1 mm on the largest leaves; stipular sheaths covered with papillae or hairs to 0.2 mm long, cup-shaped, ca 0.5–1.2 mm long, with a median seta ca 0.1–0.4 mm long or seta ± lacking. Inflorescences inconspicuous, made up of ± sessile, 2- to 6-flowered axillary clusters. Flowers subsessile or occasionally with distinct pedicels to 1 mm long (♀); ♂, ♀̄, ♀. Corolla 4-merous (in ♀ occasionally also 5-merous), greenish-yellow to creamy yellow or reddish-purple at least on the outside, glabrous or with a few papillae or hairs to 0.2 mm long. ♂, ♀̄: buds 2.5–4 mm long;

---

[1] There are two sheets in S. On one sheet there are two specimens ('3016 ♂' and '3016 ♀'), on the other one specimen ('3016 ♀'). On the first mentioned sheet, the label "*Anthospermum* 3016. ♂" in E. Meyer's handwriting (?) contains the addition "*Dregei* Sond." and "8" (= species number in Flora Capensis) in Sonder's handwriting. This sheet must, therefore, be considered the holotype. For details of O. W. Sonder's herbarium, his types in herbarium S, and for samples of his handwriting the reader is referred to Nordenstam (1980).

tube 0.7–1.4 (1.8) mm long, (narrowly) funnel-shaped to ± cylindrical, lobes 1.5–3.2 × 0.5–1.2 mm, ± lanceolate, recurved; stamens 4, filaments ca (1) 1.2–1.7 mm long, anthers 1.3–2.8 × 0.3–0.6 mm; ⚥: gynoecium as in ♀, but stigmas often only to 3 mm long; flowers occasionally transitional ⚥ → ♀ with smaller, rudimentary anthers and smaller corollas or transitional ⚥ → ♂ with shorter stigmas and smaller ovaries; ♂: rudimentary ovary small but well discernible. ♀: tube ± 0 or 0.1–0.7 mm long, cylindrical, lobes 0.2–1.3 (2) × 0.1–0.5 mm, linear-lanceolate, ± erect; style 0–0.7 mm long, stigmas 2, (2.5) 3–8.5 mm long, greyish-white or (seldom) reddish-purplish; ovary ca 0.7–1.5 × 0.5–1 mm, glabrous or a little hairy, crowned by 4 indistinct calyx lobes or calyx lobes ± lacking. Fruits reddish-brown, shiny, supported by short, broad U-shaped carpophores ca 0.3–0.7 mm long; each mericarp (1.5) 1.8–2.7 × 0.8–1.5 mm, ± oblong, dorsal side convex, glabrous or with a few whitish papillae to 0.1 mm long, ventral side ± plane and with a median vertical ridge; commissure ± small and often only ca half as large as the mericarp surface; mericarps crowned by 2 indistinct ± triangular calyx lobes to ca 0.2–0.3 mm long or calyx lobes ± lacking.

Chromosome Number (Table 3): n = 11, 2 n = 22.
Average Pollen Diameter (Fig. 48): 23.3–25.1 µm.

Two subspecies are recognized:

33 (a). Subsp. *dregei*

– *A. thymifolium* DINTER & K. KRAUSE in Feddes Repert. **15**: 91 (1917), nom. non valide publ. (sine descr.) [South West Africa (Namibia), Aus, DINTER 1088 (SAM!)].

Robust, thick-stemmed dwarf shrubs to 40 cm tall, often irregularly (diffusely) and divaricately branched (especially if browsed); browsed plants sometimes ± cushion-forming and low. Internodes normally much longer than the leaves. Leaves thickish and tough, often broad and large, to 14 (20) mm long and to 4 (4.5) mm wide, ovate, ovate-lanceolate or oblong-lanceolate, mostly papillate on lower and/or upper surface and margins, distinctly discolourous, upper surface often very dark; leaves frequently shortly petiolate. Plants often ♂, ♀; ⚥ flowers fairly uncommon. ♂ (⚥): buds ca 3.5–4 mm long; corolla lobes 2.3–3.2 × 0.7–1.2 mm, glabrous. Fruits glabrous, carpophores ca 0.6–0.7 mm long; each mericarp (1.8) 2.2–2.7 × 1–1.5 mm.

Chromosome Number (Table 3): 2 n = 22.

Average Pollen Diameter (Fig. 48): 23.7–25.1 µm.

Habitat: In semi-desert and desert areas, usually occurring on granite hills and outcrops and growing in rock crevices, amongst rocks or sheltered under boulders. Occasionally also over mica-slate (Glimmerschiefer) and sandstone (?; Great Karasberge: PEARSON 8553). – Ca 500–1 830 m.

Distribution (Map, Fig. 91 c): S Africa, W Cape Prov. (Namaqualand Distr.), and S Namibia (South West Africa), Warmbad and Keetmanshoop Distr. and S Namib desert (Lüderitz-Süd Distr.).

Critical Remarks: See after subsp. *ecklonis.*

Collections

South Africa. Cape – 2816 (Oranjemund): Alexander Bay, Boegoeberg [= Buchuberg] South (**-DC**), LE ROUX & RAMSEY 176 (PRE). – **2917** (Springbok): Steinkopf, flats of Anenous (= Aninaus) (**-BA**), MARLOTH 12225 (PRE, STE); nr. Spektakel (**-DA**), BOLUS 9421 (BOL, K); Buffels R., COMPTON 17255 (NBG); ca 16 km N of Komaggas, COMPTON 22802 (BOL, NBG, STE); 32 km S of Springbok (**-DD**), GOLDBLATT 2397 (BR, MO, PRE, WAG); Misklip (= Farm Mesklip?), ESTERHUYSEN 5843 (BOL, PRE). – Imprecise locality: between Koussie (= Buffels R.) and Zilverfontein (2917-CB to 2918-CC), DRÈGE 3016 (E, G × 3, K, LY × 2, MO, P × 3, S, SAM, W × 5). – Specimens mixed: DRÈGE 7665 a (*A. dregei* in W; *Nenax microphylla* in other herbaria).

Namibia (South West Africa). – 2615 (Lüderitz): Kovisberge (**-CB**), MERXMÜLLER & GIESS 28458 (M). – **2616** (Aus): Aus and surroundings, Farm Kubub (**-CB**): DINTER 1088 (SAM), 4159 (B, BM, G, SAM), 6115 (B, BM × 2, BOL, E, G, K, M, PRE, S, STE), GIESS & van VUUREN 868 (K, M × 2, PRE, WIND), PUFF 780810-2/1 (WU), SCHÄFER 365 (B). – **2715** (Bogenfels): Klinghardtberge (**-BC, -BD**), DINTER 3965 (B, BOL, PRE), MERXMÜLLER & GIESS 32124 (M), MÜLLER 678 (M, PRE, WIND); Buchuberge (**-DD**), DINTER 6453 (B, M), 6458 (B). – **2716** (Witpütz): Lüderitz-Süd D., around Farm Spitskop (**-DC**), MERXMÜLLER & GIESS 3397 (M, WIND), 3426 (BR, M, PRE, WIND); Farm Namuskluft (LU 88) (**-DD**), GIESS 12934 (WIND). – **2718** (Grünau): Great Karasberg, Lord Hill (**-BB**), PEARSON 8553 (BM, BOL, GRA, K, SAM); Kanus (**-DC**), RANGE 560 (SAM). – **2816** (Oranjemund): SW of Farm Spitskop, tow. Obib (**-BA**), MERXMÜLLER & GIESS 3426 (PRE). – Imprecise locality: Diamond Area 1, DINTER 3937 (PRE).

33 (b). Subsp. *ecklonis* (SONDER) PUFF in Fl. S. Afr. 31, 1 (2): 32 (1986).
≡ *A. ecklonis* SONDER in HARVEY & SONDER, Fl. Cap. 3: 32 (1865). Type: [South Africa, Cape Prov., Clanwilliam Distr.] "On the Olifantsriver and near Villa Brakfontein, ECKLON (Herb. SD.)" (S, holo.!; LY, iso.!)[1].

---

[1] The original label on the sheet in S reads "*Anthosp. ciliare*, 77-9". SONDER crossed out the epithet "*ciliare*", replaced it with the name "*Ecklonis* SOND." and

Rather weak, thin-stemmed, often rounded dwarf shrubs to ca 20 cm tall (less commonly more erect and to 40 cm tall), ± regularly branched, branches often numerous and ascending to erect. Internodes often not much longer than the leaves (except on new growth). Leaves thinnish, ± membranaceous, often narrow and short, 4–9 (12) × 1–2.5 (3) mm, (narrowly) oblong-lanceolate to lanceolate, glabrous or papillate (mainly on the margins), ± discolourous; margins often revolute; petioles usually subobsolete. Plants ♂̦, ♂̦ + ♀ or ♀, rarely ♂. ♂̦(♂): buds ca 2.5–3.2 (3.8) mm long; corolla lobes 1.5–2.5 (3) × 0.5–0.8 mm, mostly with a few short hairs or papillae on the outside. Fruits glabrous or covered with some whitish papillae, carpophores ca 0.3–0.5 mm long; each mericarp (1.5) 1.8–2.2 × 0.8–1.1 mm.

Chromosome Number (Table 3): n = 11, 2 n = 22.

Average Pollen Diameter (Fig. 48): 23.3–25 μm.

Habitat: Confined to TMS areas and typically growing in crevices and amongst rocks; sometimes in gravelly soil and sandy places between rock sheets and in depressions of large open rock surfaces. Often growing in very dry fynbos (not Arid Fynbos sensu TAYLOR 1978) and not uncommonly associated with succulents (various *Mesembryanthemaceae, Anacampseros, Senecio (Kleinia)* spp., or *Pelargonium fulgidum*) and kliphout (*Heeria argentea*). – Ca 65 (200)–1 230 m.

Distribution (Map, Fig. 91 d): S Africa, W part of the SW Cape Prov. From the SW Calvinia Distr. through the Clanwilliam and Piketberg Distr. S to the Tulbagh Distr. (and parts of the Ceres Distr.). In this area its distribution range closely follows the distribution of Table Mountain Sandstone (cf. LAMBRECHTS 1979: map, Fig. 1).

Critical Remarks (for the species as a whole): Several morphological characters of the two taxa overlap to some extent. For this reason it is, in my opinion, not justified to maintain them as separate species. The two subspecies, however, differ in their ecology [sandstone vs. granite; annual precipitation in habitats of subsp. *ecklonis* certainly (much) higher than in those of subsp. *dregei*] and their distribution range. Subsp. *dregei* is the only taxon in the genus *Anthospermum* which extends into desert areas.

---

corrected the number to "76-9". The numbers are not collection numbers but locality numbers coupled with the month in which the collection was made (9 after the dash stands for September). These locality numbers can be looked up in DRÈGE's "Standörter-Verzeichniss" (1847 a). No collector is given on the original label, but there can be no doubt that the specimen is ECKLON's, as SONDER had his specimens at his disposal (see also GUNN & CODD 1981). The duplicate in LY bears a label to which "16" (= species number in Flora Capensis) is added in SONDER's handwriting.

Typically, the two taxa differ in their habit (subsp. *dregei* is a much more robust plant); this often, however, only becomes evident in the field. Subsp. *dregei*, moreover, has much tougher leaves with darkly coloured upper surfaces (the exceptionally large epidermal cells are filled with a dark substance – cf. leaf section, Fig. 20 c; adaptation to hotter, drier climate?). Frequently, plants are browsed (both subspecies, although probably more often in subsp. *dregei*; sheep were observed, but possibly also browsed by goats). Browsing causes a more diffuse and divaricate branching and often a ± cushion-like habit. Herbarium sheets only consisting of new shoots of browsed plants (characterized by rather thin leaves and long internodes) are difficult to identify and are very similar in both subspecies.

The trend to the complete separation of the sexes (i.e., to dioecy) in subsp. *dregei* appears to be quite significant and may point to the more "derived" or "advanced" state of that taxon. It may be in the process of becoming – in isolation and under (semi-)desert conditions – a species of its own. Similar situations, although at a more "progressed" state, are also encountered in other taxa of the *A. galioides* alliance (to which *A. dregei* is thought to belong; see below): *A. prostratum* (32.) and *A. comptonii* (34.), for example, have become dioecious and may well have, in the (relatively) not too distant past, "split off" the ancestral stock (with variable sex distributions!) that gave rise to the modern *A. galioides*.

A. dregei is without doubt closely allied to the SW Cape endemic and very variable *A. galioides* (31.) and may resemble some "Forms" of the latter in leaf size and shape and in fruit and floral morphology. It also appears to be quite closely related to *A. pumilum* (29.) which, however, does not extend into the SW Cape Region (or into the winter rain fall area respectively). Subsp. *dregei* and *A. pumilum* subsp. *rigidum* are sometimes difficult to separate morphologically, but the distribution ranges of the two taxa do not overlap (see p. 344 for further details). *A. dregei* (subsp. *ecklonis*) appears to be rather closely allied to *A. comptonii* (34., below; see there for comments).

Collections

South Africa. Cape – **3118** (Vanrhynsdorp): Gfitberg (-**DC**), Drège 7660 (E, G, K, W), Phillips 7460 (K, SAM; *A. spathulatum* in other herbaria), Puff 790713-1/2 (WU). – **3119** (Calvinia), Farm Lokenburg, ca 33 km S of Nieuwoudtville (-**CA**), Acocks 17321 (PRE), 17479 (K, PRE), Puff 790712-1/1, 800901-1/2 (WU). – **3218** (Clanwilliam): Clanwilliam D., Lamberts Bay, Nortier Reserve (-**AB**), Acocks 15194 (K, PRE); Leipoldtville- Elandsbaai rd., Kliphoutkop (-**AD**), Puff 800915-3/2 (WU); Rietvleiberg, SE extension (-**BA**), Oliver 3836 (K, PRE, STE); Clanwilliam, Olifants River dam (-**BB**), Steyn 512 (NBG; STE); ca 19 km S of Clanwilliam, Khamiesberg (-**BD**), Hardy 447 (PRE × 2); Piketberg D., ca 35 km SE of Redelingshuys (-**DA**), Pillans 7684 (BOL); pass overlooking Citrusdal

(**-DB**), Marsh 772 (PRE, STE); Piketberg, lower half of Versveld pass (**-DC**), Puff 800914-3/1 (WU). – **3219** (Wuppertal): Clanwilliam D., Welbedacht, Bidouw (**-AA**), Maguire 1850 (NBG), Stokoe s.n. sub SAM 59520 (SAM); Cedarberge, ca 6 km NW of Sederberg village tow. Algeria Forestry Stn. (**-AC**), Puff 790717-2/1 (WU); – , E of Sederberg village, Farm Sanddrif (**-AD**), Puff 790717-3/1 (WU); Elandskloof [E of Citrusdal] (**-CA**), Levyns 5742, 5776 (BOL); base of Dasklip pass, N of Porterville (**-CC**), Puff 800913-2/3, 2/3 A (WU). – **3319** (Worcester): Mts. nr. Tulbagh kloof (**-AC**), Davis s.n. sub SAM 67519 (SAM). – Not traced: Ceres D., De Straat, Compton 6437 (NBG). – Additional collections: Bodkin s.n. sub Guthrie 2599 (NBG – mixed with *A. hirtum*); Bolus 9053 (BOL, K); Compton 4966, 4983, 6872, 6920 (NBG), 9633 (BOL, NBG); Ecklon "76-9" (S), s.n. (LY); Esterhuysen 3256 (BOL), 12066 (BOL, K, PRE); Levyns 3971, 1237 (BOL); Maguire 115 (NBG); Puff 790716-1/1, 800902-4/1, -6/5 (WU); Rehm s.n. (M); Salter 7506 (BM, BOL, K); Stephens 7119 (BOL); Stephens & Glover 8738 (K); Stokoe 8218 (BOL, PRE), s.n. sub SAM 59522 (PRE × 2, SAM).

34. *Anthospermum comptonii* Puff in Fl. S. Afr. **31**, 1 (2): 32 (1986).
Type: South Africa, Cape Prov., S side of Witteberg, Farm "Fisante-kraal", ca 1 200–1 300 m, Puff 790914-3/2 (WU, holo.!; BOL, NBG, PRE, STE, iso.!).

Dioecious, ± regularly to diffusely branched, many-stemmed, rounded to ± cylindrical or, if browsed, low cushion-like dwarf shrubs to ca (20) 30 cm tall with thick, woody, sometimes twisted (tap-)roots to ca 10 (20) mm in diameter. Stems to ca 4 mm in diameter near the base and ca 1–2.5 mm in the midstem region, ± terete, branches few to numerous, ascending to ± erect but often irregular and becoming ± spine-tipped if browsed; younger parts often reddish-purplish-brown, covered with whitish papillae to 0.1 mm long; older parts greyish to dark grey-blackish, glabrescent; internodes usually shorter than or as long as the leaves [frequently (much) longer on newly developed shoots of browsed plants]. Leaves decussate; blades 5–8 (10) × 0.8–2 mm, narrowly ovate-lanceolate to linear-lanceolate (broader, to 3 mm, on new growth of browsed plants), thickish and ± tough, glabrous except for odd papillae less than 0.1 mm long on the margins; upper surface shiny, with large conspicuous epidermis cells, midrib below often ± prominent and broad; apices acute; margins strongly to ± revolute; petioles subobsolete; stipular sheaths subglabrous or covered with whitish papillae to 0.1 mm long, cup-shaped, ca 0.8–1.7 mm long, with an often indistinct median seta ca 0.1–0.3 mm long. Inflorescences inconspicuous; flowers solitary or paired at the nodes, subsessile or occasionally with pedicels ca 0.2–1 mm long (♀, especially when in fruit), ♂, ♀. Corolla 4-merous, (greenish-)yellow or (dark) purplish at least on the outside, glabrous or occasionally with a few whitish papillae less than 0.1 mm long near the tip. ♂: buds 3.5–5.5 mm long; tube 1–2 mm long, (narrowly) funnel-shaped, lobes 2.7–3.5 × 0.7–1 mm, ± lanceolate, recurved; stamens 4, filaments 1.7–2.2 mm long, anthers 1.9–2.4 × 0.4–0.5 mm; rudimentary ovary well discernible and often with broad, ± rounded calyx lobes to 0.3 mm long, ♀: tube 0.1–0.2 mm long or ± 0, lobes 0.7–0.8 (1) × 0.2–0.3 mm, ± linear-lanceolate, erect; style ± 0–0.7 mm long; stigmas 2, 2.7–5.6 mm long, greyish; ovary 1–1.4 × 0.5–0.9 mm, usually glabrous, crowned by ± indistinct calyx lobes. Fruits reddish-brown, shiny, supported by small U-shaped carpophores; each mericarp 2.7–3 × 1.5–1.7 mm, dorsal side convex, glabrous or (rarely) with odd whitish papillae to 0.1 mm long, ventral side ± plane and with a prominent median vertical ridge, commissure often ± small and narrow, ca half as long as the mericarp; mericarps crowned by 2 ± indistinct, rounded to ± triangular calyx lobes ca 0.1–0.3 × 0.3 mm. – Figs. 96, 97 a–c.

Chromosome Number (Table 3): n = 11, 2n = 22.

Average Pollen Diameter (Fig. 48): 26.8–27.3 µm.

Habitat: Typically found in Arid Fynbos (sensu TAYLOR 1978 and KRUGER 1979), but also intruding into (open, succulent) Karoo vegetation. Growing in rock fissures and crevices, among large boulders, at the base of rocks or in the shelter of large bushes in coarse, sandy soil. Usually in Witteberg quartzite areas, but sometimes also over TMS. – Ca 650–1 700 m.

Distribution (Map, Fig. 91 e): S Africa, SW Cape Prov. Mostly inland of the major SW Cape folded ranges (i.e., on the inland margin of the SW Cape Floristic Region) from the S Schurweberge and the Swartruggens to the Witteberg and the Touwsberge (Ceres, Laingsburg and Ladismith Distr.), and also further E in the Cango Caves region and in the W Uniondale Distr.

Critical Remarks: Like several other taxa S African *Anthosperminae* occurring in drier areas, *A. comptonii* is frequently browsed by domestic animals (sheep and goats were observed). Browsing appears to account for most of the variability in the species. Compare in this context Figs. 96 a–d, showing plants taken from a single population (the type locality!). The plants in Figs. 96 a and b have obviously been untouched by animals, whereas the plants in Figs. 96 c and d have been browsed to varying degrees. During the growing period, browsed plants continuously produce new growth, comprised of thin branches with longer internodes and thinner leaves than in non-browsed plants; these are often eaten as soon as they are developed, resulting eventually in low cushion-like growth forms (Fig. 96 d).

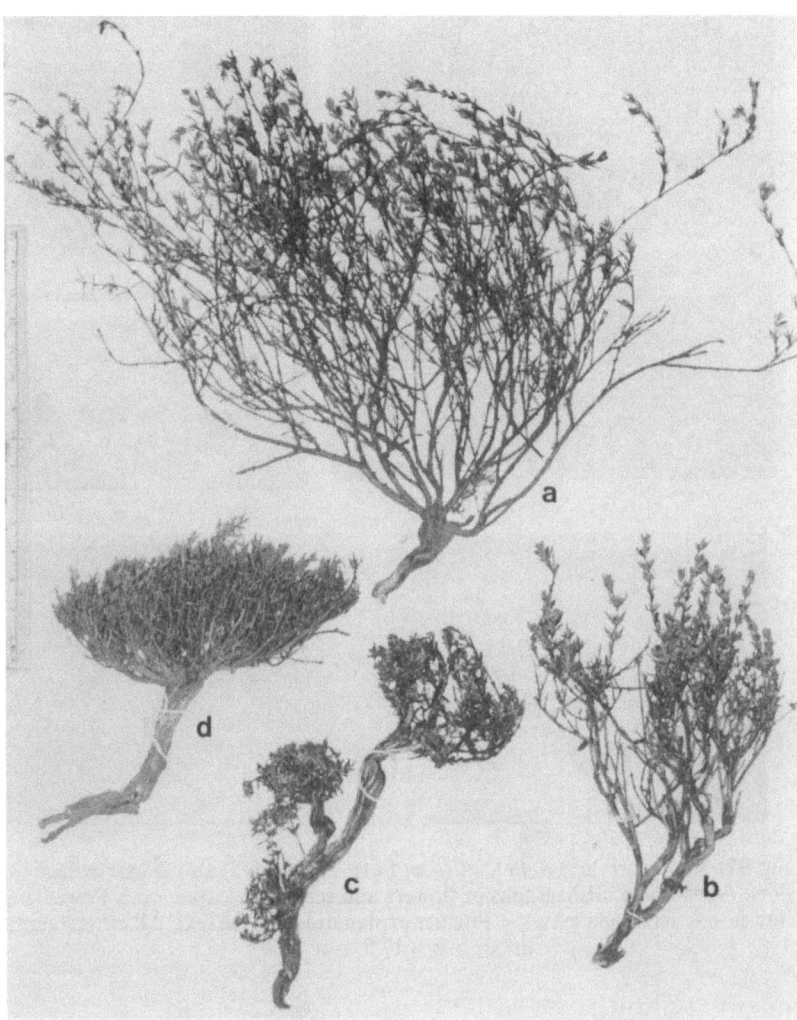

Fig. 96. *Anthospermum comptonii*, plants from population PUFF 790914-3/2 (Cape Prov., Witteberg, S side, Farm Fisantekraal). *a–b* plants from a sheltered place; *c–d* plants browsed to varying extents (goats, sheep). – Further explanations in the text

The habit of *A. comptonii* is similar to that of certain *Nenax* species. *A. comptonii* should not be confused with *N. elsieae* which – in part – occurs in the same general area. ♀ flowering and fruiting specimens are easily kept apart (cf. flower and fruit of *N. elsieae*, Figs. 33 c and 38 e–f and compare with Fig. 97 b), but ♂ and vegetative plants of the two species are more

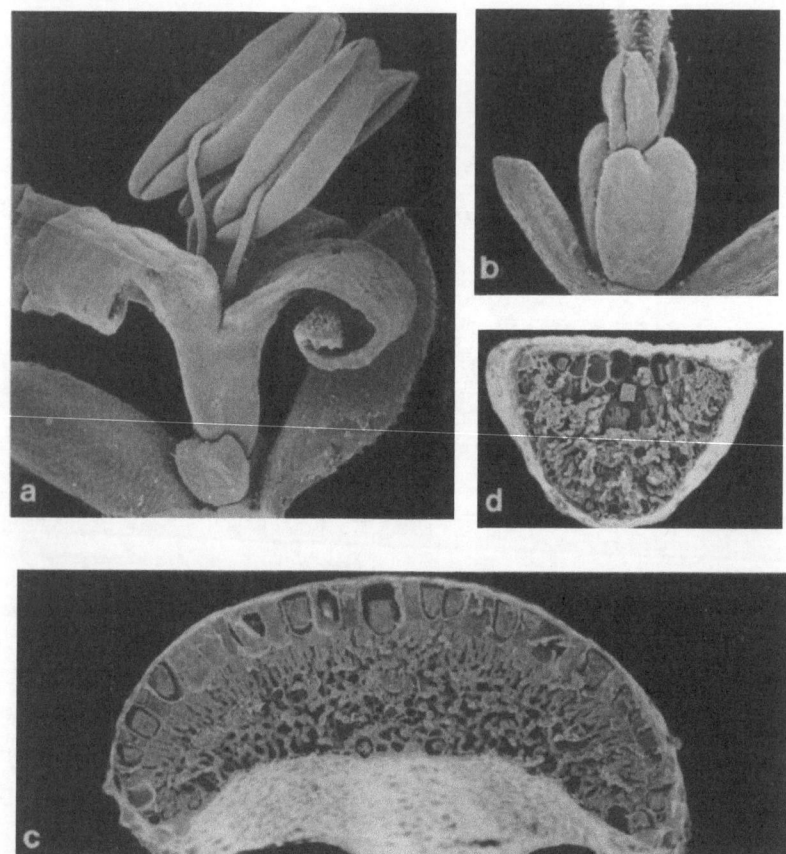

Fig. 97. *Anthospermum comptonii* (*a–c*; Puff 790914-3/2) and *Nenax elsieae* (*d*; Puff 790913-6/1). SEMgraphs of flowers and sectioned leaves. – *a* ♂ flower. *b* ♀ flower. *c–d* sectioned leaves. – Further explanations in the text. All critical point dried; *a–b* × 17.5, *c–d* × 50

difficult to separate (e.g., TAYLOR 5871: *A. comptonii,* and TAYLOR 5874: *N. elsieae,* both from the Swartruggens). The ♂ flowers of the two species are rather similar in size and shape, although the rudimentary ovaries with their calyx lobes provide a distinguishing character: In ♂ *N. esieae,* the calyx lobes are relatively large and ± trapezoid, whereas in *A. comptonii* they are smaller, rounded and less conspicuous (cf. Fig. 97 a). This difference, however, is not easily seen as the flowers are (sub)sessile and subtended by a pair of leaf-like bracts which often ± enclose and obscure the base of the flower. The leaves provide a more reliable character for the separation of the two taxa: in section, the leaves of *N. elsieae* are ±

triangular to semiterete (Fig. 97 d), while those of *A. comptonii* are broader and have revolute margins (Fig. 97 c). Finally, the branching is also different: in *N. elsieae* the branches usually arise singly at a node, while in *A. comptonii* the branching is more regular, i.e., the branches are arranged decussately (provided that the plant is not browsed).

The similarities pointed out above should not be taken as evidence for a close affinity between the two taxa. Rather it is thought that *A. comptonii* may – in adaptation to similar environmental conditions – merely have acquired certain characters similar to those of *N. elsieae*. I, personally, am of the opinion that *A. comptonii* is rather closely allied to *A. dregei* (subsp. *ecklonis*), with which it has often been confused (e.g., in the Witteberg area: COMPTON 1931). *A. comptonii* differs from the latter in (a) having tougher, thickish leaves which often appear narrower due to their revolute margins, (b) usually having larger ♂ flowers, and (c) having much larger mericarps (fruits). *A. comptonii,* furthermore, is strictly dioecious, whereas *A. dregei* subsp. *ecklonis* is not. Finally, *A. comptonii* is usually found in more arid areas than *A. dregei* subsp. *ecklonis* and occurs along the inner margins of the SW Cape Floristic Region (i.e., shows a distribution trend away from the SW Cape Region into the surrounding karroid areas); also it is mostly found growing on Witteberg quartzites, whereas *A. dregei* subsp. *ecklonis* is confined to TMS areas.

Collections

South Africa. Cape – **3219** (Wuppertal): Ceres D., Zoo Ridge ('Suurvlakte'), on Groot River rd. from Koue Bokkeveld to Cedarberg, ca 22 km N of Farm Excelsior (**-CD**), TAYLOR 5917 (PRE, STE); – , Swartruggens (**-DC**), TAYLOR 5871 (PRE × 3); –, –, ca 15 km W of Ceres- Calvinia rd., PUFF 790719-1/1 (WU). – **3319** (Worcester): rd. to Farm Hoop en Uitkoms and Bavianshoek, NNW of Hottentotskloof (**-BA**), PUFF 840912-4/1 (PRE, WU). – **3320** (Montagu): Laingsburg D., [abandoned] 'Karro Garden', Whitehill (**-BA**), COMPTON 3662 (BOL), 11255 (NBG); – , hill S of Matjiesfontein, PUFF 790915-1/1 (WU); – , Witteberg (**-BC**), COMPTON 2566, 2648, 2792, 3007 (BOL), 15220 (NBG); –, –, S side, Farm Fisantekraal, PUFF 790914-3/2 (BOL, NBG, PRE, STE, WU), between Farms Fisantekraal and Ezelsfontein, PUFF 790914-4/3 (WU); Ladismith D., Touwsberg (**-DB**), LEVYNS 7482 (BOL). – **3322** (Oudtshoorn): Swartberg Mts., Cango Caves region (**-AC**), VLOK 1248 (SAAS, WU); WSW of Barandas, Farm Kookooboo (**-DB**), PUFF 790909-2/2 (WU). – No locality given, TAYLOR s.n. sub PRE 57580 (PRE).

35. *Anthospermum esterhuysenianum* PUFF in Fl. S. Afr. **31,** 1 (2): 33 (1986). Type: South Africa, Cape Prov., Ceres- Tulbagh Distr., Witzenbergen, Swartgat Peak, 1 520–1 830 m, ESTERHUYSEN 27926 (BOL, holo.!).

Non-dioecious (♂̦̦♀, ♂̦̦♀ + ♀, ♀, or occasionally ♂), mat-forming or trailing

25*

subshrubs ± woody near the base and with woody tap roots or ± dwarf shrubs. Stems ca (6) 12–30 cm long, prostrate, to ca 3 mm in diameter near the base and ca 0.5–1 mm in the midstem region, mostly much-branched, branches often single at a node or, if paired, frequently of unequal length; younger parts terete, ± reddish-brown, glabrous or sometimes with whitish papillae less than 0.1 mm long; older parts glabrous, greyish-black, epidermis often peeling; internodes usually longer than the leaves; leafy short shoots, if present, mostly single at a node. Leaves decussate, sometimes ± pseudo-verticillate, occasionally slightly anisophyllous; blades 4–6 (7.5) × (1.4) 2–3 (4) mm, oblong, elliptic, ovate, ovate-lanceolate to lanceolate, narrowed to the base, glabrous or margins and often also the upper surface covered with white spreading hairs ca 0.2–0.4 mm long, margins flat; apices ± obtuse, acute or mucronate; petioles subobsolete or to 0.5 mm long; stipular sheaths glabrous, cup-shaped, ca 0.6–1 × 0.8–1.2 mm, with a gland-tipped median seta ca 0.4–0.8 mm long, sometimes with an additional, minute seta on either side. Inflorescences very inconspicuous; flowers solitary, terminal on the shoots, often overtopped by lateral branches arising below, with distinct pedicels ca 0.2–0.3 (1) mm long, often slightly elongated and curved in fruit; ♂, ⚥, ♀. Corolla 5-merous (very rarely 4-merous), yellowish to greenish-yellow, often dark purplish tinged on the outside, glabrous or hairy. ♂, ⚥: tube 1–1.8 (2) mm long, ± cylindrical or narrowly funnel-shaped, lobes (2) 2.2–3 (3.5) × 0.5–1 (1.3) mm, lanceolate to ± linear, recurved; stamens 5 (4), filaments (1.3) 1.7–2.5 mm long, anthers 1.4–1.9 × 0.3–0.5 mm; ♂: rudimentary ovary minute; ⚥: style 0; stigmas 2, (4.5) 5.5–7.5 (9) mm long, whitish-grey; ovary ca 0.8–1.2 × 0.7–1 mm, mostly densely papillate, crowned by 5 minute calyx lobes. ♀: tube ± 0–0.3 mm long, cylindrical, lobes 0.6–0.9 × 0.4 mm, ± lanceolate, erect; gynoecium as in ⚥. Fruits brownish, supported by short, broad U-shaped carpophores; each mericarp 1.9–2.5 × 1.3–1.5 mm, ovate, ± obovate or oblong, dorsal side convex, ± densely covered with papillae to ca 0.1 mm long, ventral side concave and with a prominent median vertical ridge, commissure long and narrow, ca 1.7–2 × 0.3–0.4 mm; mericarps crowned by 2 or 3 indistinct calyx lobes (0.2) 0.3–0.5 × 0.2 mm, or calyx lobes ± lacking. – Fig. 4 b.

Two varieties are recognized:

35 (a). Var. *esterhuysenianum*

Leaves and corollas glabrous; leaf blades oblong, elliptic or ovate, apices not distinctly mucronate.

Chromosome Number (Table 3): 2 n = 22.

Average Pollen Diameter (Fig. 48): 25.3–26.4 µm.

Habitat (Fig. 4 b): Higher parts of the SW Cape mountains. Forming mats at the base of rocks or in bare patches between higher vegetation or between rocks on steep rocky slopes, sometimes trailing over rock ledges. Apparently confined to shale bands and shale-derived soils. Seems to be more common on S-facing aspects. – Ca 1 350–2 000 m.

Distribution (Map, Fig. 95 c): S Africa, SW Cape Prov. Confined to the high mountains in the triangle Clanwilliam – Stellenbosch – Worcester.

Critical Remarks: See after var. *hirsutum.*

Collections

South Africa. Cape – **3219** (Wuppertal): Clanwilliam D., S Cedarberg, base of Sneeuberg, nr. Bakleikraal (-AC), ESTERHUYSEN 34148 (BOL); –, –, Apex Peak (-CA), ESTERHUYSEN 28943 (BOL); – **3319** (Worcester): Tulbagh D., Witsenberg, Swartgat Peak (-AA), ESTERHUYSEN 27926 (BOL); –, –, around Sneeugat Peak, ESTERHUYSEN s.n. (BOL), PUFF 800924-3/2 (WU); Worcester D., Waaihoek Mts., N end (-AD), ESTERHUYSEN 18204 (BOL); –, –, Waaihoek Peak, ESTERHUYSEN 22599 (BOL, K, PRE), 22599 a (BOL, K × 2, NH, PRE × 2); –, –, Mosters Hoek Twins, ESTERHUYSEN 9878 (BOL); Paarl D., Hoelhoek (Haelhoek) Sneeukop (-CA?), ESTERHUYSEN 34159 a (BOL, PRE); Worcester D., between Goudini Sneeukop and Du Toits Peak (= DuToitskloofpiek) (-CC), ESTERHUYSEN 24734 (BOL).

35 (b). Var. ***hirsutum*** PUFF in Fl. S. Afr. **31**, 1 (2): 33 (1986).

Type: South Africa, Cape Prov., Ceres Distr., N Cold Bokkeveld, Schurweberg Peak, 1 370–1520 m, ESTERHUYSEN 29458 (BOL, holo.!).

The outside of the corollas, the leaf margins and often also the upper surfaces covered with white spreading hairs ca 0.2–0.4 mm long; leaf blades lanceolate to ovate-lanceolate, apices mostly ± distinctly mucronate.

Habitat: As for var. *esterhuysenianum.*

Distribution (Map, Fig. 95 d): S Africa, SW Cape Prov. Only known from the Central Cedarberg and the Schurweberg (N Cold Bokkeveld).

Critical Remarks (for the species as a whole): *A. esterhuysenianum* is a very distinct and unmistakable SW Cape mountain endemic. Its growth form and inflorescences are rather unique for *Anthospermum*: the terminal, solitary flowers are overtopped by one – or two un-equal – branches arising from below. If two unequal lateral branches arise below a flower, the branching may become ± pseudo-dichotomous (as in some *Nenax* and *Carpacoce* species; see also C.2., p. 19 and 24, Fig. 7 d). A slight trend to anisophylly (marginally larger leaf blades associated with the promoted lateral axes; cf. Fig. 29 a) is also remarkable. These peculiarities make it difficult to place the species; there appear to be no close allies. A closer affinity to the few other *Anthospermum* species with 5-merous corollas is unlikely as there is major disagreement in most other characters. It is also uncertain whether there is some distant relationship to the widely distributed and variable *A. herbaceum* (25.) which, primarily in patches of kloof forests, etc., extends into the SW Cape. Forms of *A. herbaceum* occasionally show a trend to overtopping in the inflorescence region, and growth forms of some ecotypes approach *A. esterhuysenianum* (i.e., perennial herbs which become rather woody near the base and ± subshrubby). The corollas of *A. herbaceum* with their long, cylindrical tubes also agree to some extent, but numerous other characters (stipule structure, fruits, etc.) differ markedly from *A. esterhuysenianum*.

Some ecotypes of *A. galioides* (31.) may superficially resemble *A. esterhuysenianum* (e.g., leaf size and shape, low, mat-forming habit), but inflorescence differences (axillary flower clusters in *A. galioides*!) and a dissimilar branching pattern provide strong evidence against a close affinity between the two.

At present, too little herbarium material is available and not enough field observations have been made to be certain about the taxonomic status of the hairy and glabrous plants of *A. esterhuysenianum*. For this reason they are provisionally separated as varieties. So far no individuals ± intermediate in their character states have been found.

Collections

South Africa. Cape – 3219 (Wuppertal): Clanwilliam D., Central Cedarberg, plateau between Scorpioonspoort and Klein Koupoort (-AC), ESTERHUYSEN 34826 (BOL, K, WU); Ceres D., N Cold Bokkeveld, Heidveldberg (Schurweberg) (-CD?), ESTERHUYSEN 29431 (BOL × 2); –, –, Schurweberg Peak, ESTERHUYSEN 29458 (BOL).

36. *Anthospermum hirtum* CRUSE, Rub. Cap.: 11 (1825), in Linnaea 6: 13 (1831); SONDER in HARVEY & SONDER, Fl. Cap. 3: 30 (1865); SALTER in J. S. Afr. Bot. 3: 110 (1937), in ADAMSON & SALTER, Fl. Cape Penins.: 733 (1950); PUFF in Fl. S. Afr. 31, 1 (2): 34 (1986).
≡ *A. hirsutum* DC., Prod. 4: 580 (1830), nom. illeg.
Type: [South Africa, Cape Prov.:] ad promontorium bonae spei in monte Diaboli Orientem versus, mense Augusto 1816, C. W. BERGIUS [non P. J. BERGIUS; see VON SCHLECHTENDAL (1826)] s.n. (B, holo. †; G, iso.!).

= *A. rubiaceum* RCHB. in SPRENGEL, Syst. Veg. **4**, 2: 338 (1827).
Type: "C.B.S.", no collector given[1].

Non-dioecious (♂, ⚥, ⚥ + ♀, ♀, or ⚥ + ♂), few- to several-stemmed, ±
cylindrical, rounded or low, ± matted, weak subshrubs somewhat woody
near the base or distinct dwarf shrubs with often thick, woody (tap-)roots.
Stems ca 20–100 (150) cm long, ca 2–5 (7) mm in diameter near the base
and ca 1–2.5 mm in the midstem region, often ± much-branched,
branches ± erect to spreading; younger parts reddish-brown to dark
purplish-red, ± densely covered with whitish spreading hairs ca (0.2) 0.5–
1.4 mm long or hairs sometimes in two distinct rows on the ± obscure
angles, rarely hairs around the stipular sheaths only; older parts reddish-
brown-grey to dark grey, becoming glabrescent, epidermis often peeling;
internode lengths variable. Leaves decussate, occasionally ± pseudo-
verticillate; blades (6) 8–25 (30) × (0.5) 0.8–3 (4.5) mm, narrowly oblanceolate to ± linear-lanceolate, narrowed to the base; margins, at least near
the base or on the lower half, and often also upper surfaces ± densely
covered with white spreading hairs ca 0.4–1 (1.5) mm long, seldom also
with a few hairs on the midrib below; blades often distinctly discolourous,
upper surface shiny, epidermis cells large, conspicuous, lower epidermis
distinctly papillate; apices acute to acuminate or terminating in long,
white hairs; margins ± revolute; petioles subobsolete or to ca 1 mm on the
largest leaves; stipular sheaths reddish, covered with whitish spreading hairs (0.2) 0.5–0.8 (1) mm long, (broadly) cup-shaped, ca 0.7–
1.7 (2) × (2) 2.7–3.5 (4) mm, with a median, subulate, hairy seta ca (0.5) 1–
2.5 (3) mm long. Inflorescences inconspicuous, made up of ± sessile, 6- to
2-flowered axillary clusters. Flowers subsessile or occasionally with
pedicels to ca 0.7 mm long (♀, especially when in fruit); ♂, ⚥, ♀. Corolla 5-
merous (occasionally also 4-merous in ♀), greenish-yellow or (dark)
purplish-red, covered with whitish spreading hairs ca 0.2–0.5 (1) mm long
at least near the tip. ♂, ⚥: tube 0.5–1.5 (2) mm long, (narrowly) funnel-
shaped, lobes (1.7) 2–3.4 (3.7) × 0.5–0.8 (1) mm, ± lanceolate, recurved;
stamens 5, filaments (1) 1.4–3 mm, anthers (1) 1.4–2 (2.2) × 0.3–0.5 mm; ⚥:

---

[1] CRUSE (1831) claims having seen "the author's" specimen when putting *A.
rubiaceum* into synonymy with his *A. hirtum*, but he did not cite the collection. I am
quite convinced that SIEBER "Fl. Cap. No. 413" (G!, MO!, W!) is the type specimen.
In an "advertisement" published in Flora 7 (2): 717–720 and dated "Dresden d. 4.
Nov. 1824" SIEBER offered for sale (amongst other items) his "Herbarium Florae
Capensis" and also noted (p. 719) that "Die Bestimmungen der Phanerogamen
sind von Hrn. Prof. Reichenbach und dem Herausgeber; ..." The specimen in
W bears a label "*Anthospermum rubiaceum*, Cap, REICHENB." (likely to be in
REICHENBACH's handwriting) next to SIEBER's distinct printed "Fl. Cap." label
which is directly attached to the plant.

gynoecium as in ♀, but stigmas often only to 3 (5.5) mm long; flowers occasionally transitional ⚥ → ♀ with smaller, less than 1 mm long, ± pollenless anthers and smaller corollas or transitional ⚥ → ♂ with shorter stigmas and smaller ovaries; ♂: rudimentary ovary small but well discernible, sometimes also rudimentary stigmas present. ♀: tube 0.3–0.8 (1.2) mm long, ± cylindrical, lobes (0.1) 0.3–1 × 0.1–0.3 mm, ± linear-lanceolate, erect to ± spreading; style 0–1 mm long; stigmas 2, 3.7–7.5 mm, greenish-white, greyish or (seldom) reddish-purple; ovary ca (0.5) 0.9–1.2 × 0.5–1 mm, usually glabrous, not crowned by distinct calyx lobes. Fruits reddish-brown, supported by small, U-shaped carpophores to 1 mm long and 0.6–0.9 mm wide; each mericarp 1.9–2.7 × (1) 1.2–1.7 mm, ± oblong, dorsal side convex, with large conspicuous epidermis cells or (rarely) with odd papillae less than 0.1 mm long, ventral side ± concave and with a median vertical ridge, commissure small and narrow; mericarps not crowned by distinct calyx lobes.

Chromosome Number (Table 3): 2 n = 22.

Average Pollen Diameter (Fig. 48): 31.1–35.4 μm.

Habitat: Typically growing in ± moist to damp sites (along streams or near trickles of water at the base of overhangs or cliffs, or on well drained steep slopes) and in partial shade. Occasionally also found in drier habitats – i.e., in dry mountain fynbos (not Arid Fynbos sensu TAYLOR 1978) or Coastal Fynbos – and growing amongst rocks or boulders in sandy to gravelly soil. Sometimes recorded from disturbed areas (clearings, neglected fields, etc.). It seems to occur mainly on TMS derived substrates but is occasionally also found over Granite (clay) soils. – Ca 50–700 (850) m.

Distribution (Map, Fig. 95 h): S Africa, SW Cape Prov. Occurring from the Cape Peninsula N to the Clanwilliam Distr. and, to the SE, to the Bredasdorp Distr. Also in the Riversdale Distr.? (see Critical Remarks.)

Critical Remarks: Although the characters which distinguish *A. hirtum* from all other SW Cape taxa of *Anthospermum* (i.e., the characteristic indumentum of the leaf blades, the stipules and the 5-merous hairy corollas) are ± constant, there is nevertheless a fair amount of variation. The variation primarily concerns the habit (woodiness, branching, internode lengths and leaf size and shape) and the extensiveness of the inflorescence. "Dwarf" or stunted quite woody forms with relatively short internodes and rather small leaves appear to occur in drier habitats and, in general, under less favourable environmental conditions than "normal". This is particularly obvious around Elim, where *A. hirtum* (e.g., SCHLECHTER 9708, PUFF 800914-4/1) grows on gravelly soil amongst peculiar low fynbos vegetation[1]. Similar forms were also observed (next to "typical" *A. hirtum*) on the plateau of the Porterville Mts. (TMS; PUFF 800913-5/4) and collected in the Paarlberg area (granite soils; e.g., TAYLOR 5440, GROBLER 334) and even in the SW Cape Peninsula (e.g., SALTER 6621). As many collections are, in habit, intermediate between such "extreme" stunted forms and "typical" longer and weaker-stemmed forms, it does not appear justified to recognize them taxonomically.

Yet another "Form" occurs occasionally in the Piketberg Distr. (Piketberg area): The plants are ± weak and thin-stemmed and the characteristic indumentum of long, whitish hairs is confined to the stipules and the base of the leaf blades while even the youngest stems are glabrous or nearly so; in addition, the flowers are always only in pairs at the nodes. It may be worthwhile to separate these plants as a variety[2].

In his "Vegetation of the Riversdale Area, Cape Prov.", MUIR (1929) does not list *A. hirtum* as occurring in that area. It is, in my opinion, not unlikely that the MUIR specimen in MARLOTH's herbarium (no. 5655), supposedly from the Riversdale Distr., was in fact collected somewhere else. The Riversdale Distr. is clearly outside the species' distribution range (i.e., much further E).

The collection SALTER 5686 (BOL) may be a hybrid *A. hirtum × A. spathulatum* (more robust habit, smaller leaves, glabrous ♂ corollas with shorter tubes); otherwise no putative hybrids between *A. hirtum* and other taxa have been detected.

A. hirtum appears to occupy a rather isolated position within the Cape species of *Anthospermum* and in the genus as a whole. There may be a distant alliance to the SW Cape endemic *A. bergianum* (37., below; see there for details). *A. hirtum* has also sometimes been compared and associated with *A. herbaceum* (25.). The latter has similarly variable floral sex forms but differs from *A. hirtum* in so many other characters (e.g., indumentum, stipules with several setae, markedly different fruits, small

---

[1] It is, in this context, noteworthy that ACOCKS (1975: 87) speculates that the dwarf Fynbos of the Elim flats should probably be regarded as a distinct veld type.

[2] "Var. *subglabrum*". The following collections from 3218-DC (not included in the general specimen citations) belong here: BOLUS 8505 (BOL), DRÈGE 7667 (E, FI − as '64.9', G, K, LY, MO, P × 2, S, W × 2), PILLANS 7317 (BOL, K), 8004 (BOL); also two collections of unknown origin in herb. THUNBERG (S) and in herb. PALLAS (BM).

I, at present, hesitate to formally recognize this variety. Further field observations would be required. Some collections from the Piketberg (e.g., Versveld pass, PUFF 800914-3/2, or from "Sandleegte" on path to Sebrakop, ESTERHUYSEN s.n. sub PUFF 800914-4 A/3) are "typical" *A. hirtum* and others are ± intermediate between "typical" *A. hirtum* and "var. *subglabrum*" in their character state (e.g., BOLUS 13562, BM, BOL, GRA, K, PRE).

average pollen diameters, etc.) that, in my opinion, a close relationship between the two species is unlikely.

Collections

South Africa. Cape — **3218** (Clanwilliam): Piketberg, The Rest (-DC), GILLETT 3742 (BOL, STE); — plateau, "Sandleegte" (-DC, -DD), ESTERHUYSEN s.n. sub PUFF 800914-4 A/3 (WU). — **3219** (Wuppertal): Clanwilliam D., Cedarberg, Duiwel-skloof (-AC), STOKOE s.n. sub SAM 64188 (PRE, SAM); ca 15–17 km E of Citrusdal, base of Buffelshoek pass, Farm Kleinplaas (-CA), PUFF 800903-1/2 (WU); Dasklip pass, N of Porterville (-CC), PUFF 800913-3/1 (WU), THOMPSON 1481 (PRE, STE). — **3318** (Cape Town): nr. Porterville (-BB), SCHLECHTER 10722 (BM, BOL, BR, COI, E, G, GRA, K, MO, PRE × 2, S, US, W); Kirstenbosch, around Window stream (gorge) (-CD), ESTERHUYSEN 12022 (BOL), 26751 (BOL, MO), SALTER 6236, 6237 (BM, BOL, K); Paarl D., Paarlberg (-DB), GROBLER 334 (PRE, STE), KRUGER M 38 (PRE, STE), M 131 a (STE); Bellville D., Tygerberg (Nature Reserve) (-DC), COMPTON 20749 (NBG), LOUBSER 3839 (MO); Stel-lenbosch, Onderpapegaaiberg (-DD), TAYLOR 6883 (PRE, STE). — **3319** (Wor-cester): Bains Kloof nr. Wellington (-CA), LEWIS GRANT 2254 (BOL, MO); French Hoek Forest Reserve, nr. Bushman's Castle (-CC), SALTER 25126 (BOL). — **3418** (Simonstown): above Hout Bay (-AB), WHITE 5215 (PRE); [Cape of Good Hope Nature Reserve], Smitswinkel Bay (-AD), SALTER 6221 (BOL, K), 274/18 (BM); Phesantekraal (-BA?), VAN NIEKERK 168 (BOL). — **3419** (Caledon): around Elim (-DA, -DB), SCHLECHTER 9708 (BM, BOL, G, K, PRE). — **3421** (Riversdale): Riversdale D., MUIR s.n. sub MARLOTH 5655 (PRE). — Imprecise localities ("CBS") or no locality given: ALEXANDER s.n. (herb. PRIOR) (BM, K); AUGE s.n. (BM); F. BAUER "94" (W); BUNBURY s.n. (BM); DRÈGE s.n. ("*A. hispidulum*", obviously a label mix-up) (E); GARSIDE 1441 (K); KRAUSS s.n. (W); PAPPE s.n. (K); SCHOLL s.n. (W × 2); SIEBER [Flora Cap.] 90 (BR, M, MO, P, S, W × 2), 413 (G, MO, W), [Flora mixta] 23 (E); VERREAUX s.n. (G × 2). — Additional collections: BERG(IUS) s.n. (G); BODKIN s.n. sub GUTHRIE 2599 (NBG; approach. 'var. *subglabrum*'; mixed with *A. dregei*); BOLUS 3944 (BOL, K), 13562 (BM, BOL, GRA, K, PRE; appr. 'var. *subglabrum*'), 25118 (BOL), s.n. sub STE 25757 (STE); BOS 461 (K, M, WAG); BURCHELL 450 (K); DÜMMER 903 (E); ECKLON 19 (S), 29 (GOET, S); ECKLON (& ZEYHER?) "85" (S); ECKLON & ZEYHER 2311 (B, BOL, G × 2, GH, GOET, LY, M, MO, P, S, SAM, W × 2, WU); GAMBLE 22173 (K); HUMBERT 9681 (P, PRE); PENFOLD 89, 90 (NBG); PRIOR s.n. (K); PUFF 791228-1/1, 800906-1/1, 800913-4/4, -5/4, 800914-3/2, 800918-4/1 (WU); SALTER 6248 A–C (BM, BOL, K); SCHMIDT 343 (M); TAYLOR 5440 (PRE, STE), 7512 (STE); THODE s.n. sub STE 6181, 9269 (GRA, STE); WOLLEY DOD 245 (BM, BOL, K); ZEYHER s.n. sub SAM 16061 (PRE, SAM).

37. ***Anthospermum bergianum*** CRUSE, Rub. Cap.: 9 (1825), in Linnaea 6: 7 (1831); SONDER in HARVEY & SONDER, Fl. Cap. 3: 29 (1865); SALTER in J. S. Afr. Bot. 3: 110 (1937), in ADAMSON & SALTER, Fl. Cape Penins.: 733 (1950); in Fl. S. Afr. 31, 1 (2): 35 (1986).
Type: South Africa, Cape Prov., "in planitie capensi versus Tyger-berg", August 1816, C. W. BERGIUS s.n. (B †). Neotype: Cape Flats, August 1818, MUND 91 (S)[1].

---

[1] The specimen was seen by CRUSE and the label bears the inscription "*A.*

= *A. ciliare* L., Sp. Pl., ed. 2: 1512 (1763), nom. rej.[2]
Type: [South Africa, Cape Prov.] "CBS", no collector given (LINN 1233.4, holo.!).
≡ *A. aethiopicum* L. β *ciliare* (L.) O. KUNTZE, Rev. Gen. Pl. **3**(2): 117 (1898), quoad typum.

−? *Cliffortia spicata* RCHB. in SPRENGEL, Syst. Veg. **4**, 2: 209 (1827).
Type: "C.B.S.", no collector given[3].

Mostly dioecious (seldom also ♂ plants + odd ♀ flowers or ♀ plants), single- to few-stemmed, usually erect subshrubs or ± short-lived dwarf shrubs with woody tap roots to 22 cm long. Stems (5)8–15 cm long in younger, occasionally to 45 (75) cm long in older flowering plants, ca 1.5–5 (6) mm in diameter near the base and ca 1–1.5 (2.2) mm in the midstem region, sparsely to much-branched, branches ascending to erect; younger parts terete, often reddish(-brown), densely covered with white spreading hairs ca 0.3–0.5 mm long; older parts greyish-brown to dark grey, becoming glabrescent, epidermis peeling; internodes, except on new growth, shorter than to as long as the leaves; shoots densely leafy above. Leaves in whorls of 3 or occasionally in whorls of 4 or decussate (sometimes on one and the same plant), often pseudo-verticillate; blades mostly erect or ascending, frequently ± imbricate, (4)5–12 (15) × (1.2) 1.5–3 (3.5) mm, linear-lanceolate to ovate-lanceolate, narrowed to the base; margins, sometimes also the midrib below or the entire lower surface, covered with white spreading hairs (0.3) 0.5–1 mm long, upper surface glabrous, often reddish-brown; apices acute to acuminate;

---

bergianum mihi. CRUSE". CRUSE may, in error, have attributed the specimen to ECKLON, because in 1831 he wrote "in planitie capensis legerunt d. BERGIUS nec non C. F. ECKLON" (p. 7), but he did not cite MUND. MUND (his name is often − as on the label of the type collection − misspelled 'MUNDT'), after his arrival in S Africa, was introduced to some interesting collecting localities in the Cape by CARL BERGIUS (cf. GUNN & CODD 1981).
   In 1979, I revised the collections ECKLON & ZEYHER 2306 as neotypes, but the above mentioned MUND collection (of which I was then not aware) seems to be the more preferable neotype.

   [2] In a letter, dated January 28th, 1986, Dr. R. K. BRUMMITT, Secretary of the Nomenclature Committee for *Spermatophyta*, informed me that the "Proposal to reject the name *A. ciliare* L. (1763) (*Rubiaceae*)" (proposal 685; PUFF 1982 b) was accepted; the official report of the Nomenclature Committee will appear in Taxon in 1987.

   [3] In his *Cliffortia* monograph, WEIMARCK (1934: 160) states that "REICHENBACH's type specimen [of *C. spicata*] corresponds ... to *Anthospermum bergianum* CRUSE ..." but I could not check the correctness of this statement as I was unable to trace the type specimen.

petioles 0; stipular sheaths ciliate on the margins, shallowly cup-shaped, ca 0.4–0.8 × 1–2 mm, with a median seta ca 0.2–0.4 mm or occasionally to 1.2 mm long or seta absent. Inflorescences inconspicuous to ± conspicuous [♀; ± dense cylindrical inflorescence zones ca (2) 3–12 (17) × 1–2 cm], made up of ± sessile, 3- to 9-flowered axillary clusters (sometimes more-flowered in ♀). Flowers subsessile, ♂ (occasionally also ⚥), ♀. Corolla 5-merous (sometimes also 4-merous in ♀), yellowish, sometimes reddish tinged outside, covered with white hairs ca 0.3–0.5 mm long at least near the tip or glabrous (especially ♀). ♂ (odd ⚥): tube 0.7–1 mm long, funnel-shaped, lobes 1.9–3 × 0.6–1.1 mm, ± ovate-lanceolate, recurved; stamens 5, filaments 1.5–2.5 mm long, anthers 1–1.6 × 0.3–0.5 mm; ♂: rudimentary ovary to ca 0.2–0.5 mm long. ♀: tube (0.1) 0.3–0.6 mm long, cylindrical, lobes 0.5–1.2 × 0.2–0.4 mm, ± oblong, erect; style 1–2.8 mm long; stigmas 2, (6) 8–13.5 mm long, greyish; ovary 0.8–1.5 × 0.6–1 mm, papillate or occasionally shortly hairy, not crowned by distinct calyx lobes. Fruits brown or reddish-brown, often ± shiny, supported by short, ± cup-shaped carpophores, ca 0.3–1.2 mm long; each mericarp 1.8–2.5 × 0.9–1.1 mm, ± obovate to oblong, dorsal side convex, ± densely papillate or sometimes with white spreading hairs to ca 0.2 mm long, ventral side (strongly) concave and with a prominent median vertical ridge, commissure long and narrow, ca 1.7–2.2 × 0.2–0.4 mm; mericarps not crowned by distinct calyx lobes. – Figs. 12, 98.

Chromosome Number (Table 3): n = 11, 2 n = 22.

Average Pollen Diameter (Fig. 48): 31.1–33.8 µm.

Habitat: On (dry) rocky slopes in coarse to fine sandy soil, on rock ledges, or sometimes in sandy patches; often (mostly?) associated with TMS, occasionally in shale bands. Frequently found in fire-prone areas. – Ca 0–1 800 m.

Distribution (Map, Fig. 95 e): S Africa, SW Cape Prov. From the Clanwilliam Distr. (Pakhuis Pass) S to the Caledon Distr. and the Cape Peninsula.

Critical Remarks: A. bergianum often occurs in areas that are burnt quite regularly. Plants sometimes start resprouting from the base after a fire, but individuals are more commonly killed by fire. *A. bergianum,* thus, frequently behaves as a "seed regenerator" (cf. C.12.8.). For this reason, young flowering plants are common in areas that were burnt some months ago. They are often only a few cm tall and could be mistaken for annuals (cf. Figs. 12 a and 98). A search in isolated patches not reached by or protected from fires will, however, often prove that older, somewhat woody and ± shrubby plants are also present in the population.

Fig. 98. *Anthospermum bergianum.* Young ♂ plants, flowering for the first time, on a TMS slope burnt in the previous dry season (SW Cape Prov., summit of Buffelshoek Pass on Citrusdal- Ceres rd.; PUFF 800903-2/3). Note that branching is more extensive than in equally aged ♀ plants (compare Fig. 12a). – The plant in the foreground is ca 10 cm tall

A. bergianum, although variable in the leaf arrangement, the degree of hairiness of the leaves and in the fruit indumentum (in addition to the above mentioned environment-dependent growth form differences), is one of the most easily recognizable *Anthospermum* species and can hardly be confused with any other SW Cape species of the genus.

The relationships of *A. bergianum* remain somewhat obscure, and the species appears to be ± isolated; there do not seem to be any apparent or obvious *close* allies in the SW Cape Floristic Region or in the surrounding areas. Noteworthy are the 5-merous corollas (only in ♀ occasionally also 4-merous) and also the average pollen diameters, which are unusually large for diploid, slightly woody (i.e., non large-shrubby) *Anthospermum* species (cf. Fig. 48). *A. bergianum* shares these two characters with *A. hirtum* (36., above). Also the fruits of the two species are similar (no distinct calyx lobes; mericarps ± oblong, often papillate; commissure narrow) and, furthermore, there is some agreement in the stem and leaf indumentum. Apart from these similarities, however, there are considerable differences between *A. bergianum* and *A. hirtum* (sex distributions, habit, leaf arrangement, leaf size and shape and micromorphology, and stipule structure) so that a *close* alliance seems somewhat improbable.

Collections

South Africa. Cape – **3218** (Clanwilliam): Clanwilliam D., Pakhuis pass (-**BB** to 3219-AA), Compton 6955, 9600 (NBG); Piquetberg (= Piketberg), nr. Goedverwacht (-**DC**), Bolus 25122 (BOL). – **3219** (Wuppertal): Cedarberg Mts., between Pakhuis and Heuningvlei (-**AA**), Esterhuysen 7441 (BOL); –, around Sneeuberg (-**AC**), Esterhuysen 13059 (BOL, PRE), Puff 790718-1/1 (WU); summit of Buffelshoek pass on Citrusdal- Ceres rd. (-**CA**), Puff 800903-2/3 (WU); plateau of Porterville Mts. (Skurweberge) (-**CC**), Puff 800913-5/3, 800914-1/2 (WU). – **3318** (Cape Town): Mamre Rd. (Mamreweg) (-**BC**), Esterhuysen 12983 (BOL); Cape Town (-**CD**), Prior s.n. (K); Kraaifontein (-**DC**), Dümmer 1825 (E, NBG); ca 25 km from Cape Town, S of Bottelary rd. (-**DD**?), Acocks 307, 521 (S). – **3319** (Worcester): Witzenberg, Inkruip (-**AA**?), Esterhuysen 23432 (BOL, WU); Cold Bokkeveld, NW of village Op die Berg (-**AB**), Thompson 1556 (PRE, STE); Ceres Peak (-**AD**), Acocks 1850 (S); Paarl D., Bailey's Peak (-**CA**), Esterhuysen 8527 (BOL); Klein Drakenstein Mts., Upper Kasteelkloof, Franschhoek (-**CC**), Kruger 723 (STE); Caledon D., Stettynsberg (-**CD**), Esterhuysen 11078 (BOL). – **3418** (Simonstown): Scarborough (-**AB**), Acocks 1418 (S), Goldblatt 2652 (MO, PRE, WAG); above Buffels Bay [Cape of Good Hope Nature Reserve] (-**AD**), Bolus 17194 (BOL); Sir Lowrys pass (-**BB**), Parker 4599 (BOL, K, MO); Kogelberg Forest Reserve (-**BD**), Boucher 406 (K, PRE, STE). – **3419** (Caledon): Caledon D., Palmiet River Mts. (-**AC**), Stokoe 975 (PRE); "Great Mountain" of Baviaanskloof at Genadendal (-**BA**), Burchell 7821 (K). – No locality given ("CBS"): Alexander (herb. Prior) (K); Forbes s.n. (BM, G, K); Grey s.n. (K); Masson s.n. (BM); Scholl s.n. (herb. Jacq.) (W × 4). – Imprecise locality: Cape Flats, Zeyher 644 (LY). – Additional collections: Bolus 4745 (BM, BOL, GH, NBG, K); Bond 396 (NBG); Boucher 2190 (PRE, STE); Compton 5369, 7047, 9357, 13508 (NBG); Drège 7668 (E, FI, G × 2, K × 2, LY, MO, P, W); Dümmer 65 a (E, GH); Ecklon & Zeyher 2306 (BOL, M, MO, SAM, W; S? – 'Cruse'); Esterhuysen 3257, 3258, 4208, 12955, 24635, 33169 (BOL), 3452 (PRE), 7379 (BOL, K, NBG); Fries, Norlindh & Weimarck 1673 (K, SRGH); Goldblatt 2580 (MO, NBG, PRE, WAG); Meebold 11168 (M); Michell s.n. sub NBG 23181 (NBG); Moss 9372 (BM × 2, J); Mund(t) 91 (S); Purcell 337–341 (SAM); Rogers 11233 (GRA, PRE); Salter 6265 A–E (BOL, K); Schlechter 8758 (BM, BOL, G, K), 1099 (B, G, GRA, P, W × 2, WU); Stephens 7315 (BOL, K); Stokoe 8216, 9514

(BOL), s.n. (PRE sub 41552, SAM sub 27421, 61866); TAYLOR 1547 (SAM), 2536 (STE); WALL s.n. (S); WALLICH 360, s.n. (BM); WOLLEY DOD 1394 (BM, K); WRIGHT s.n. (GH, MO; US sub 81699, 81700); (?ECKLON &)ZEYHER s.n. (GOET, SAM × 3); ZEYHER 10 (S, SAM), s.n. (S).

38. *Anthospermum ericifolium* ('ericaefolium') (LICHTENSTEIN ex ROEMER & SCHULTES) O. KUNTZE, Rev. Gen. **3**: 117 (1898); PUFF in Fl. S. Afr. **31**, 1 (2): 35 (1986).

≡ *Spermacoce ericaefolia* LICHTENSTEIN [Spicileg. Fl. Cap., MS.] ex ROEMER & SCHULTES, Syst. Veg. **3**: 281 (1818).

Type: South Africa, Cape Prov., nr. Rivier Zonderend (= Riviersonderend), LICHTENSTEIN s.n. (? B †)[1].

= *Anthospermum lichtensteinii* CRUSE, Rub. Cap.: 15 (1825), in Linnaea **6**: 16 (1831); SONDER in HARVEY & SONDER, Fl. Cap. **3**: 32 (1865).

Type: as above[1].

Non-dioecious (⚥, ♀, seldom ⚥ + ♂ or ♂), few- to many-stemmed, ± erect dwarf shrubs. Stems (15) 25–40 (60) cm long, ca 2.5–4 mm in diameter near the base and ca 1–2 mm in the midstem region, sparsely to much-branched, branches ± erect or ascending; younger parts terete, reddish-brown to purplish, ± densely covered with spreading to ± curled whitish hairs to 0.2 mm long; older parts dark grey, glabrous, epidermis often peeling; internodes somewhat shorter to slightly longer than the leaves; shoots densely leafy above. Leaves decussate, pseudo-verticillate; blades ascending to erect, 4–7 × (0.6) 0.8–1.2 (1.4) mm, linear-lanceolate, often broadly triangular and shallowly concave above in section; margins

---

[1] CRUSE must have studied LICHTENSTEIN's "*Spermacoce ericaefolium*" specimen and have come to the conclusion that it needed to be transferred to *Anthospermum*; subsequently, he introduced the name *A. lichtensteinii* for it. According to CRUSE's publications (1825, 1831), the LICHTENSTEIN specimen in question – i.e., the type – was in WILLDENOW's herbarium ('Willd. herb. No. 2631'). Neither this nor any other LICHTENSTEIN collection, however, could be traced. The LICHTENSTEIN collection supposedly in "herb. SD." (SONDER 1865: 32) was not amongst the material of the herbaria containing specimens from SONDER's herbarium (especially S, LY).

For the following reason the collection ECKLON & ZEYHER 2316 (LY!, M!, S!, SAM!, W!) seems to be the most appropriate choice of a *neotype*: CRUSE, when presenting a more detailed redescription of *A. lichtensteinii* (1831), also cites an ECKLON specimen from 'prope Lautenbach ad Riedvalley'. It is quite likely that this collection was later distributed as ECKLON & ZEYHER 2316 and quoted by ECKLON & ZEYHER (1836: 368) as '2316 ... *prope "Rietvalley"* (Cap), in montibus "Hottentottshollandberge" prope "Somerset".'

(at least lower half) with white spreading hairs ca 0.3–0.4 mm long, blades otherwise glabrous, often reddish-brown above and on the midrib below; apices ± mucronate; petioles 0; stipular sheaths with white hairs ca 0.3–0.4 mm long on the margins, shallowly cup-shaped, 0.3 – 1 × 1 – 2 mm, with or without a minute, gland-tipped median seta to 0.1 mm long. Inflorescences ± inconspicuous, made up of ± sessile, (2-) 6- to 10- (14-)flowered axillary clusters. Flowers subsessile; ♂, ♀̂, ♀. Corolla 5-merous, yellowish (?), ± densely to sparsely covered with white spreading hairs ca 0.2–0.3 mm long at least near the tip. ♂, ♀̂: tube 0.7–1 mm long, (narrowly) funnel-shaped to ± cylindrical, lobes 1.2–1.7 × 0.5–0.8 (1) mm, ± ovate-lanceolate, recurved; stamens 5, filaments exserted for 0.8–1 mm, anthers 1–1.2 × 0.3–0.4 mm; ♂: rudimentary ovary minute, hardly discernible. ♀̂: style 0; stigma *one*, 3–7 mm long, greyish (?); ovary with only *one* fertile carpel, ca 1–1.3 × 0.5–0.8 mm, densely hairy; the reduced carpel crowned by 2, the fertile carpel by 3 calyx lobes. ♀: tube ca 0.2 mm long, cylindrical, lobes 0.4–0.6 × 0.1–0.2 mm, ± oblong, erect; gynoecium as in ♀̂. Fruits greyish-brown to greyish, supported by asymmetrical U- or fork-shaped carpophores ca 2.2–3 mm long; the fertile mericarps (2.5) 2.8–3.5 (4) × (1) 1.3–1.8 mm, ± ovate to oblong, often curved lengthwise, dorsal side convex, densely covered with straight white hairs 0.4–0.6 mm long, ventral side strongly concave and with a prominent median vertical ridge, commissure long and narrow, ca 2–3.2 × 0.5–0.7 mm; the reduced carpels ± strap-like, 2.5–3.5 × 0.5–0.7 mm, occasionally somewhat less hairy than the fertile ones; the fertile carpels crowned by 3 triangular-lanceolate calyx lobes 0.3–0.7 (1) × 0.2–0.4 mm, the reduced carpels crowned by 2 sometimes marginally larger calyx lobes. – Figs. 15 d, 36 d.

Chromosome Number: Unknown (presumably diploid).
Average Pollen Diameter (Fig. 48): 26.1–27.6 µm.
Habitat: Mostly in sandveld fynbos; in sandy or gravelly soil [no additional detailed habitat information available]. – Ca 50 (?)–300 m.
Distribution (Map, Fig. 95 f): S Africa, SW Cape Prov. Occurring from the Cape Flats to the W Worcester Distr. and the Caledon Distr.

Critical Remarks: A. ericifolium, a rather rare but unmistakable species (ovaries with only one fertile carpel, and flowers with only one stigma!) is allied to *A. bicorne* (39., below), the only other *Anthospermum* species in which one of the two carpels is reduced and modified into a strap-like structure bearing the calyx lobes[1]. It differs from *A. bicorne* in

---

[1] The reduction of one of the two carpels in these two *Anthospermum* species should *not* be interpreted as evidence for a close alliance between *Anthospermum* and *Carpacoce* (most species with only one fertile carpel). Detailed morphological

having ovaries and fruits which are crowned by 5 ± equally sized calyx lobes (vs. 2 large and 3 minute calyx lobes in *A. bicorne;* compare Figs. 36 a, c and d), and in having only one stigma per flower (according to field observations, *A. bicorne* has always two stigmas, although one of the two ovaries is invariably reduced). Finally, *A. ericifolium* differs in having more-flowered partial inflorescences and in having flatter (i.e., not markedly needle-like) and more closely spaced leaves with distinctly ciliate margins.

The affinities of *A. ericifolium* and *A. bicorne* to other (SW Cape) species of *Anthospermum* are not entirely understood and are somewhat uncertain. There may be some relationship to *A. bergianum* (37., above) and *A. hirtum* (36.), two other species which are also characterized by (unusual for *Anthospermum*) having 5-merous corollas (the high SW Cape mountain endemic, *A. esterhuysenianum,* also has 5-merous corollas, but major morphological differences make an alliance highly unlikely). Also the size and shape of the ♂ and ♀ corollas of the four species are similar (relatively long, narrowly funnel-shaped to ± cylindrical tubes). The fruits of *A. bergianum* and *A. hirtum,* however, differ – apart from having two fertile mericarps – in being crowned by considerably smaller, or by having no obvious calyx lobes. The two (diploid!) species *A. bergianum* and *A. hirtum,* furthermore, differ in having unusually large average pollen diameters (cf. Fig. 48). *A. bergianum* is similar to *A. ericifolium* in leaf size, shape and basic indumentum type (ciliate margins!) and stipular sheaths with much reduced or absent setae, but the leaves of *A. bergianum* are typically arranged in whorls of 3.

Collections

South Africa. Cape – 3318 (Cape Town): Riverlands (formerly Michiel Heyns Kraal) (-**DA**), ESTERHUYSEN 35553 (BOL, WU); nr. Rietvalley (-**DC**) [and Hottentottsholland (3418-BB)], ECKLON & ZEYHER 2316 (LY, M, S, SAM, W); Paarl D., nr. Paarl, on Cape Town rd. (-**DD**), ESTERHUYSEN 28054 (BOL, WU). – 3319 (Worcester): nr. Bainskloof (-**CA**), SCHLECHTER 10242 (BM, BOL, BR, COI, E, G, GRA, K, MO, P, PRE × 3, S, US, W). – 3418 (Simonstown): Table Mt., nr. Constantia (-**AB**), ECKLON & ZEYHER "85" (S); Sir Lowrys pass (-**BB**), HAFSTRÖM s.n. (S); Hottentottsholland, ZEYHER 275 (S). – 3419 (Caledon): Donkerhoek Berg (-**AB**), BURCHELL 7966 (K × 2). – Imprecise localities: "between Stellenbosch and Somerset [West]", "Klein-Drakenstein and Dal Josaphat", DRÈGE s.n. (FI, G, K, LY, MO, W).

---

and anatomical investigations of ovaries and fruits (see C.7.) reveal considerable structural differences between the two genera.

39. *Anthospermum bicorne* PUFF in Fl. S. Afr. **31**, 1 (2): 36 (1986).
Type: South Africa, Cape Prov., Caledon Distr., Houw Hoek, GILLETT
829 (BOL, holo.!; STE, iso.!).

Non-dioecious (⚥, ⚥ + ♀ or ♀), many-stemmed, usually ± erect dwarf
shrubs or subshrubs. Stems (10) 20–55 cm long, ca 2–3 (4) mm in diameter
near the base and 0.5–1 mm in the midstem region, sparsely to ± much-
branched, branches ascending to erect; younger parts terete, purplish or
reddish-brown, sparsely to densely covered with often upwardly directed
whitish hairs to 0.2 mm long or glabrous; older parts grey-brown to dark
grey, glabrous; internodes mostly longer than the leaves. Leaves decussate, pseudo-verticillate; blades erect or ascending, (6.5) 8–16 (19) × 0.4–
0.5 mm, linear, needle-like, ± semiterete or ± triangular and shallowly
concave above in section; margins (often lower third only) with white hairs
ca 0.1–0.2 mm long, or blades glabrous altogether; apices acute to ±
mucronate; petioles 0; stipular sheaths glabrous or with white hairs ca 0.1–
0.2 mm long, cup-shaped and often ± inflated, 1–1.2 × 2 mm, with a
minute median seta ca 0.1–0.2 mm long, sometimes with an additional,
obscure seta on either side. Inflorescences inconspicuous, made up of ±
sessile, 2- to 6-flowered axillary clusters. Flowers subsessile, ⚥, ♀. Corolla
5-merous, yellowish or purplish on the outside, covered with white
papillae or short hairs to 0.2 mm long mostly near the tip. ⚥: tube 0.8–
1.3 mm long, funnel-shaped, lobes 1.7–2.5 × 0.4–0.7 mm, ± oblong to
lanceolate, recurved; stamens 5, filaments 1.2–1.7 mm long, anthers 1.6–
2.4 × 0.2–0.4 mm; style ca 0.7–0.8 mm long; stigmas 2, ca 2.5–4 mm long,
whitish to greyish; ovary with only *one* fertile carpel, 0.8–1.5 × 0.5–
0.7 mm, densely hairy; the reduced carpel strap-like and with 2 large calyx
lobes, the fertile carpel crowned by 3 minute, indistinct calyx lobes. ♀: tube
(0.1) 0.3–0.4 mm long, cylindrical, lobes 0.3–0.6 × 0.1–0.2 mm, ± linear,
erect; gynoecium as in ⚥, but stigmas to 6.5 mm long. Fruits reddish-
brown to greyish, supported by asymmetrical U- or fork-shaped carpophores ca 1.8–2.5 mm long; the fertile mericarps 2.5–3.5 × 1–1.7 mm, ±
ovate, sometimes distinctly curved lengthwise, dorsal side convex, densely
covered with white spreading hairs 0.3–0.5 mm long, ventral side strongly
concave and with a prominent median vertical ridge, commissure long and
narrow, ca 2.2–3 × 0.5 mm; the fertile carpels sometimes crowned by 3
minute calyx lobes ca 0.1–0.4 mm long; the reduced carpels ± strap-like,
ca 2.5–3.5 × 0.6–0.8 mm, less hairy than the fertile ones or glabrous, with 2
large, ± divaricate calyx lobes 1.4–2 × 0.2–0.4 mm. – Figs. 36 a–c.

Chromosome Number (Table 3): 2 n = 22.
Average Pollen Diameter (Fig. 48): 27.4–29.8 μm.

Habitat: In dry to moist (coarse) sandy soil; mostly over TMS, sometimes over shale. – Ca. 90–1 070 m.

Distribution (Map, Fig. 95 g): S Africa, SW Cape Prov. Occurring from the Clanwilliam Distr. (Cedarberg Mts.) S to the Cape Flats, the Caledon Distr. and the Bredasdorp Distr.

Critical Remarks: A. bicorne is easily recognized by its ovaries with only one fertile carpel and the two greatly enlarged calyx lobes borne on the reduced and modified sterile carpel. It is allied to and was often confused with the preceding species (see p. 390 for further details).

Collections

South Africa. Cape – **3219** (Wuppertal): Clanwilliam D., Cedarberg Mts., between Pakhuis and Heuningvlei (**-AA**), ESTERHUYSEN 7428 (BOL, NBG, PRE, SAM); –, Grootberg (**-AD**), ESTERHUYSEN 4168 (BOL, PRE); Elands Kloof, Cold Bokkeveld (**-CA**), LEVYNS 4881 (BOL); plateau of Porterville Mts. (Skurweberge) (**-CC**), PUFF 800913-5/2 (WU). – **3318** (Cape Town): nr. Kraaifontein (**-DC**), DÜMMER 427, 1805 (E); N of Kanonberg, ACOCK(s) 4298 (S); between Bottelary rd. and Main Line (**-DD**), ACOCK(s) 75 (S). – **3319** (Worcester): Rocklandplaas, ca 30 km N of Ceres (**-AB**), HUGO 2516 (PRE, STE); ca 1 km N of Romansrivier (**-AC**), PUFF 840915-2/1 (PRE, WU); Riviersonderend Mts., service rd. to Jonaskop SABC tower (**-DC**), PUFF 800925-2/1 (WU); Riviersonderendberge (**-DD**), ROBBERTSE and students 1028 (PRU). – **3418** (Simonstown): Kogelberg Forest Reserve (**-BD**), BOUCHER 1238 (STE), DURAND 109 (STE), PUFF 800101-5/1 (WU). – **3419** (Caledon): Houwhoek (**-AA**), BOLUS 341 (BM, G, GH, K, PRE, SAM, US, W; as 5377 in BOL), GILLETT 829 (BOL, STE), GUTHRIE 2252 (NBG), SCHLECHTER 7356 (BM, COI, E, G, K, MO, P, S, SAM, US, W; sub TVL Museum no. 1992 b in PRE); Caledon D., S of [Caledon] baths (**-AB**), PURCELL s.n. sub SAM 46161 (SAM × 3); –, Shaw's pass (**-AD**), BARKER 6118 (NBG), LEWIS 4244 (SAM), PUFF 791224-1/1 (WU); Riviersonderend Mts., rd. to P.O. tower, 'Die Galg' area (**-BA**), PUFF 840910-2/6 (WU); River Sondereinde (= Riviersonderend) (**-BB**), GUTHRIE 3226 (NBG); E of Farm Lusern, W of Salmonsdam Nature Res. (**-BC**), PUFF 840917-4/2 (PRE, WU); Die Skeiding, between Elandsberge and Normanskop (**-BD**), HUGO 907 (K, PRE, STE); ca 7 km SW of Elim, tow. Viljoenshof, between Farms Bruinklip and The Springs (**-DA**), PUFF 800918-5/1 (WU); ca 8–10 km from Viljoenshof, tow. Bredasdorp, Farm Vlooikraal (**-DB**), PUFF 800918-7/1 (WU). – No information given: ZEYHER s.n. (K).

Nenax GAERTN., Fruct. **1**: 165, t. 32, f. 7 (1788); HOOK. f. in BENTH. & HOOK. f., Gen. Pl. **2** (1): 140 (1873); K. SCHUM. in ENGLER & PRANTL, Nat. Pflanzenfam. **IV, 4**: 129 (1891); SALTER in J. S. Afr. Bot. **3**: 111 (1937), in ADAMSON & SALTER, Fl. Cape Penins.: 734 (1950); DYER, Gen. S. Afr. Flow. Pl. **1**: 622 (1975); PUFF in Fl. S. Afr. **31**, 1 (2): 37 (1986).

Type species: *N. acerosa* GAERTN., Fruct. **1**: 165, t. 32, f. 7 (1788).

26*

= *Ambraria* CRUSE [non HEISTER ex FABRICIUS], Rub. Cap.: 16 (1825), in Linnaea **6**: 18 (1831), SONDER in HARVEY & SONDER, Fl. Cap. **3**: 33 (1865).
Type species: *A. glabra* CRUSE, Rub. Cap.: 17 (1825) (Lectotypus) [= *Nenax acerosa* GAERTN.].

Dioecious, many-stemmed, often much- and intricately branched dwarf shrubs with thick, woody roots. Leaves decussate or (rarely) in whorls of 3, often in seemingly larger numbers at the nodes due to much-contracted, leafy short shoots ("pseudo-verticillate"), blades small and often ± ericoid, (sub)sessile, with small, ± cup-shaped stipular sheaths with or without 1 (–3) minute setae on either side. Inflorescences frequently leafy and inconspicuous, made up of ± sessile, mostly 3–1-flowered cymes, arranged in pairs or single at the nodes or flowers solitary and terminal on shoots. Flowers subsessile, subtended by a pair of leafy bracts, ♂ or ♀, 4–5 (6)-merous; ♂: corolla tubes ± cylindrical to broadly funnel-shaped, short, lobes ± lanceolate, recurved, anthers yellow to whitish, exserted, dangling on long slender filiform filaments; ♀: corollas much smaller, tubes cylindrical, lobes erect to ± spreading, linear to ± lanceolate; style 0; stigmas 2, long exserted, hairy, often puplish-red; ovary bicarpellate and biovulate, crowned by 4–5 small calyx lobes. Fruits crowned by the persistent calyx lobes, sometimes supported by small U-shaped carpophores, dehiscent, occasionally ± inflated, or hard, indehiscent and without carpophores.

Chromosome Number: $x = 11$; $n = 11, 22$, $2n = 22, 44$.
Average Pollen Diameter (Fig. 49): 23.1–35.7 μm.
Distribution (Maps, Figs. 1 c and 99 e): S Africa; centered in the SW and W Cape Prov. with one species extending into Namibia (South West Africa) and one species widely distributed from the central Cape Prov. to the O.F.S. and W Lesotho.

Critical Remarks: Nenax is doubtlessly very closely allied to *Anthospermum*. If compared one by one, most characters overlap to some extent, but there is a conspicuous trend to more "derived" character states in *Nenax* (e.g., fruits dehiscent → indehiscent; di- → tetraploidy; dioecy in the entire genus; trend to very reduced, few-flowered inflorescences). An essential difference between the genera lies in habit and appearance: *Nenax* consists exclusively of distinctly woody (and presumably *secondarily* woody, cf. KOEK-NOORMAN & PUFF 1983) long-lived dwarf shrubs, while *Anthospermum* has either tall shrubby (and presumably *"primarily"* woody) species or (at least in the SW Cape) shorter-lived

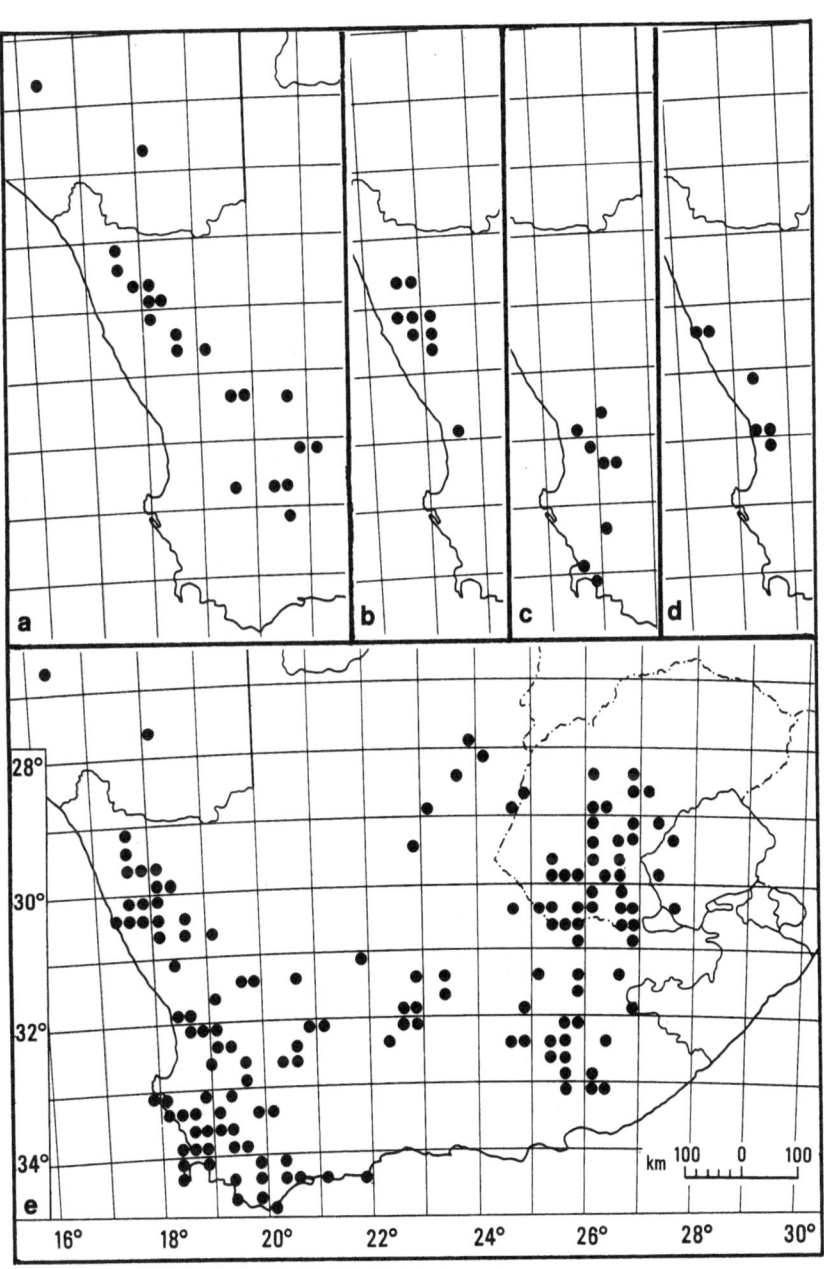

Fig. 99. Distribution of *Nenax*. *a N. cinerea*, *b N. namaquensis*, *c N. divaricata*, *d N. arenicola*, *e* all taxa

dwarf shrubs, subshrubs or perennial herbs. It is, however, the combination of characters rather than the individual characteristics that allow a more reliable distinction between the genera. The combination of the characters (a) distinctly woody, dwarf shrubby habit, (b) needle-like leaves, (c) reduced, few-flowered inflorescences and (d) dioecy, is unique to *Nenax*. As long as these distinctions – as feeble as they may seem – remain, it appears most feasible to keep the genera apart (last but not least for convenience), rather than to include *Nenax* in the (already huge, "unwieldy" and difficult) genus *Anthospermum*.

Except for the two distinct species, *Nenax microphylla* and *N. cinerea*, taxa may be extremely difficult to distinguish, and some experience, including field knowledge, is often required to ensure a definite identification. ♂ specimens of different taxa can be quite similar to each other and do, in general, not provide easy-to-use key characters (♂ individuals should not be confused with *Anthospermum*!). The following key is, therefore, based on (♀) fruiting material; only occasionally ♂ flowers are referred to. If no mature fruits are present on a ♀ individual, older stems should be searched for carpophores (cf. Fig. 38 g): the presence of carpophores indicates the presence of dehiscent (not inflated) fruits (species 3.–6.); the absence of carpophores indicates the presence of (a) inflated, dehiscent fruits (species 1. and 2.) or of (b) indehiscent fruits (species 7.–11.). It is recommended that *always* both ♂ and ♀ specimens of a given taxon are collected.

Nenax is presumably a (relatively) "young" genus (see E.2. for a detailed discussion) and probably in the process of actively and rapidly producing new species. This may explain the rather frequent occurrence of "Forms" (mostly in the sense of geographically isolated and morphologically anomalous populations or distinct geographical or ecological races) which in some respects do not fully match typical material of a given taxon. They may be extremely difficult to place. A number of such "Forms" are mentioned in the Critical Remarks section of the respective taxa. Unlike in *Anthospermum*, I virtually have no evidence for hybridogenous contact between SW Cape taxa of *Nenax*. Hybridization, in my opinion, is to be discounted as a possible explanation for these "Forms".

Key to Species and Subspecies

1. Fruits dehiscent .... 2.
1*. Fruits indehiscent .... 7.

2. Fruits inflated, ± round in outline, not supported by obvious carpophores .... 3.

2*. Fruits not inflated, mericarps much longer than wide, supported by U-shaped carpophores.. 4.

3. Fruits ca 1.7–3.2 mm in diameter, subglobose to ± ellipsoidal; leaves glabrous or sometimes with a few papillae on the margins. .... **1. *N. microphylla***

3*. Fruits large, ca (5) 6–8 mm in diameter, laterally compressed, often ± round or broadly obovate in outline; leaves silvery-grey to greyish(-green) due to a dense cover of very short, often curled whitish hairs or papillae.. **2. *N. cinerea***

4. Mericarps covered with ± curled, whitish hairs, ca 0.2–0.5 mm long; leaves needle-like, small, 3–5 × 0.5 – 0.8 (1) mm; in drier parts of the Worcester and Ceres Distr. .... **6. *N. elsieae***

4*. Mericarps glabrous or papillate; leaves not distinctly needle-like or, if needle-like, to 9 (11) mm long... 5.

5. Leaves not distinctly needle-like, 2.5–4.5 (6) × 0.9–1.4 mm; mericarps small, ca 2.1–2.6 mm long; corollas 5-merous, tube (♂) relatively long and ± cylindrical; mainly in Namaqualand .... **3. *N. namaquensis***

5*. Leaves needle-like, (3) 3.5–9 (11) × 0.4–1 mm; mericarps (2.4) 2.7–5 mm long.. 6.

6. Mericarps large, 3–5 × 1.9–2.5 (3) mm, crowned by relatively large calyx lobes, ca 0.6–1.2 mm long; ♂ corolla 5-merous; dwarf shrubs with often pseudo-dichotomous and ± irregular branching; narrowly confined to the S Vanrhynsdorp Distr. and adjacent Clanwilliam Distr. .... **4. *coronata***

6*. Mericarps smaller, ca (2.4) 2.7–3.4 × 1.2–1.7 mm, crowned by small calyx lobes ca 0.3–0.4 mm long; ♂ corolla 4-merous; dwarf shrubs with mostly ascending, regular, opposite branches; ± widely distributed from the Calvinia Distr. S to the Sir Lowry's Pass...... .... **5. *N. divaricata***

7. Fruits ± soft, easily squashed between two fingers .... **10. *Nenax* sp. A**

7*. Fruits too hard to be squashed between two fingers................. 8.

8. Fruits small, ca 2–2.8 mm long... 9.

8*. Fruits larger, 3–8 mm long.. 10.

9. Fruits reddish, shiny, glabrous, obscurely ribbed; leaves decussate, needle-like.. **11. *Nenax* sp. B**

9*.   Fruits greyish, densely covered with whitish, spreading hairs ca 0.1–0.2 mm long; leaves in whorls of 3 .... **7 (a). _N. hirta_ subsp. _hirta_**

10.   Fruits 3–5 mm long, shortly hairy or ± glabrous .... 11.

10*.   Fruits 5–8 mm long, entirely glabrous; flowers paired or single at the nodes, widely spaced; W Coast Strandveld........... **9. _N. arenicola_**

11.   Leaves to 3.5 mm long, decussate or occasionally in whorls of 3, widely spaced; fruits ca 3–3.7 mm long, shortly hairy; intricately branched, robust dwarf shrubs; only on the W Coast (Langebaan Peninsula and immediately S) over limestone............................. .... **7 (b). _N. hirta_ subsp. _calciphila_**

11*.   Leaves to 12(15) mm long, strictly decussate, often pseudo-verticillate; fruits 3–5.5 mm long and often ± crowded on the shoots .... 12.

12.   Stems reddish (at least when young), densely leafy, leaves usually pseudo-verticillate, needle-like; plants often only distinctly woody below; fruiting inflorescences ± conspicuous and spike-like; flowers in clusters of 6(–2) at the nodes; fruits reddish, glabrous or ± papillate.. **8 (a). _N. acerosa_ subsp. _acerosa_**

12*.   Stems grey(ish), mostly with dimorphic leaves (long, needle-like and smaller, broader); distinctly woody dwarf shrubs; flowers mostly paired at the nodes; fruits greyish(-brown), mostly ± densely covered with short whitish spreading hairs................................. .... **8 (b). _N. acerosa_ subsp. _macrocarpa_**

1. **_Nenax microphylla_** (SONDER) SALTER in J. S. Afr. Bot. **3**: 113 (1937); HOBSON & JESSOP, Veld Pl. S. Afr.: 220, pl. 21 (1975); PUFF in Fl. S. Afr. **31**, 1 (2): 38 (1986).
≡ _Ambraria microphylla_ SONDER in HARVEY & SONDER, Fl. Cap. **3**: 34 (1865).
Types: South Africa, O.F.S., Sandrivier, BURKE 506 (BM, K, PRE, SAM, syn.!), ZEYHER 769 or s.n. (BM, G, K, PRE, syn.!; SAM, lecto.!) [1].

---

[1] In Flora Capensis, SONDER did not state the province from which the type collections originated. According to DRÈGE's (1847 a) "Standörter-Verzeichniss der von C. L. ZEYHER in Südafrika gesammelten Pflanzen", the collections probably originated from the locality "156. – felsige Höhen am Zandrivier und Grasfläche bis zu den Abhängen vom Falsrivier, ungefähr 5000′ Höhe". The Sandrivier in question (in "Southern African Place Names", LEISTNER & MORRIS

Dioecious, rounded to ± cylindrical, many-stemmed, intricately branched dwarf shrubs ca 10–30 (40) cm tall and ca 10–70 cm in diameter; lower, cushion- to ± mat-forming if browsed; shoots sometimes arching downward, rooting at the nodes and capable of producing new plantlets. Stems ca 3–11 mm in diameter at the base and ca 1.5–4 mm in the midstem region, ascending or erect, sometimes arching downward or ± prostrate, much-branched, often ± spine-tipped if browsed; younger parts ± 4-angled, greenish to reddish-purple, papillate; older shoots terete, glabrescent, light to dark grey or brownish-grey, epidermis often peeling; longest internodes ca 1–2 mm long on old shoots, to ca 10 mm long on new shoots. Leaves decussate, often pseudo-verticillate; blades (1.5) 2–4 (5.4) × 0.7–1.2 mm, (broadly) ovate to ± elliptic (often longer, to 7 mm, and ± linear-lanceolate or lanceolate on new growth), frequently recurved, glabrous or sometimes with a few papillae on the margins; apices ± obtuse to acute; petioles subobsolete; stipular sheaths cup-shaped, ca 0.3–0.7 mm long, with or without a minute, median gland-tipped seta. Inflorescences inconspicuous to ± conspicuous ($\female$, especially when in fruit). Flowers ± sessile, in clusters of 2 (seldom 4–6) at the nodes; $\male$, $\female$. Corolla 4-merous, pale yellow to greenish-yellow, sometimes reddish(-purplish) tinged on the outside. $\male$: tube 0.5–0.7 mm long, broadly funnel-shaped, lobes 1.5–2.2 × 0.6–0.8 mm, ± lanceolate, recurved; stamens 4, filaments 1–1.7 (2) mm long, anthers 1.2–1.9 × 0.3–0.5 mm; rudimentary ovary mostly minute. $\female$: tube 0.2–0.4 mm long, lobes 0.2–0.5 × 0.1–0.2 mm, ± linear, corolla sometimes absent altogether; style 0; stigmas 2.8–4.4 mm long, whitish-grey (and purplish-red?); ovary 1.3–1.7 × 0.6–0.9 mm, green at first, later reddish, ellipsoidal, ± glabrous or papillate, crowned by 4 ± triangular calyx lobes, ca 0.3–0.4 × 0.2–0.3 mm. Fruits (bright) reddish-brown or reddish, inflated, (sub)globose or ± ellipsoidal, ca 1.7–3.2 mm in diameter, dehiscent, not supported by distinct carpophores, crowned by the sometimes obscure persistent calyx lobes; each mericarp ca 2–3.2 × 1.7–2.8 (3.2) mm, dorsal side convex, glabrescent or covered with whitish papillae less than 0.1 mm long, ventral side concave, with lateral ledges and a prominent vertical median ridge. – Figs. 6 b, 9, 38 a–b, 59 c.

Chromosome Number (Table 3): n = 11 (Fig. 47 d), 2 n = 22.

Average Pollen Diameter (Fig. 49): 23.1–28.1 μm.

Habitat (Fig. 6): Usually in Karoo or karroid vegetation (including "false" Karoo and pure grassveld veld types sensu AcOcks 1975) with few

---

1976, there are no less than 32 entries for 'Sandrivier' covering all provinces!) must be in the Senekal Distr., O.F.S. (2827-B). The 'Falsrivier' (= Vals R.) mentioned in Drège must have been reached by Zeyher and his companion Burke somewhere between Lindley and Kroonstad (2727-C, -D) (see also Gunn & Codd 1981).

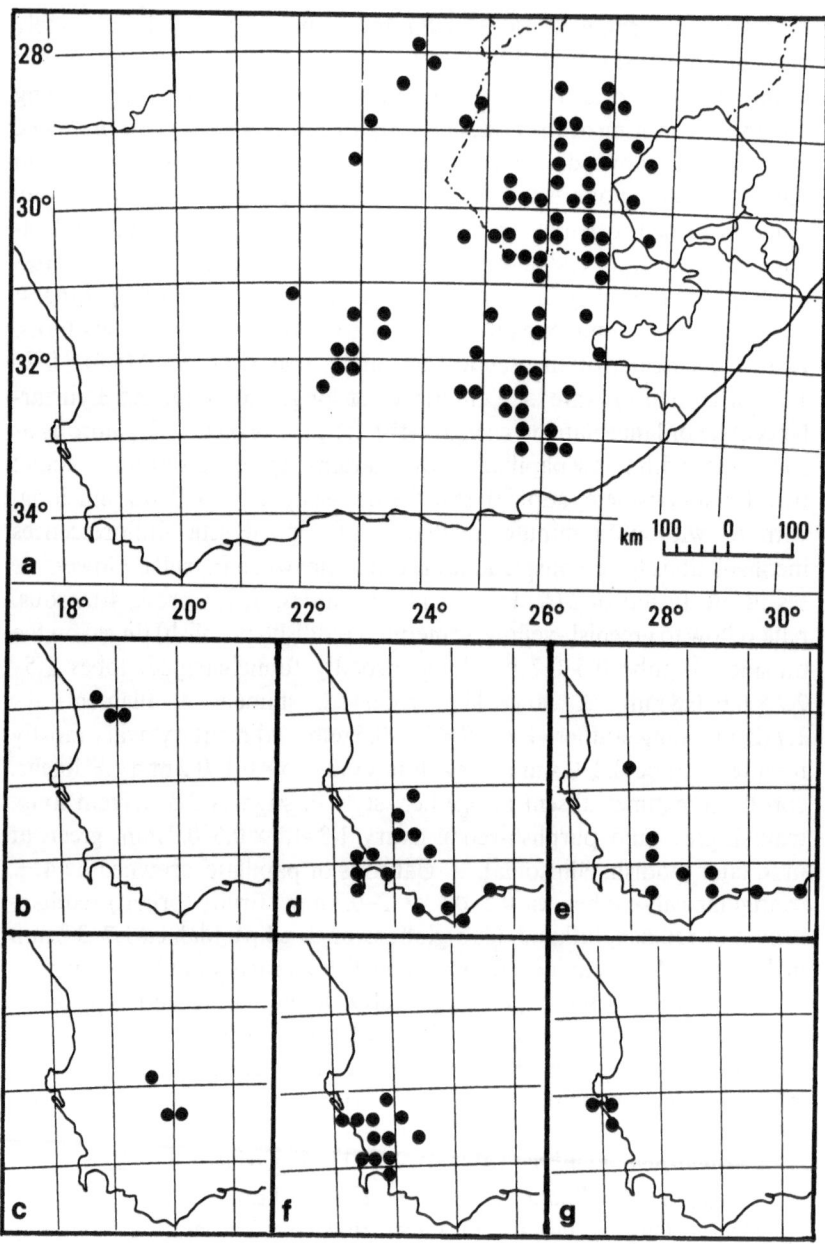

Fig. 100. Distribution of *Nenax. a N. microphylla, b N. coronata, c N. elsieae. d–e
N. acerosa, d* subsp. *acerosa, e* subsp. *macrocarpa. f–g N. hirta, f* subsp. *hirta, g*
subsp. *calciphila*

or no trees or tall shrubs. Between rocks on stony hills and rocky slopes, in crevices of rock sheets, in dry grassland between rocks in sandy to gravelly soil; according to HOBSON & JESSOP (1975), in level ground where the soil is deep. – Ca 500–1 500 (2 250) m.

Distribution (Map, Fig. 100 a): S Africa. From the E, central and N (E) Cape Prov. E to the central O.F.S.; also extending into the lower-lying parts of Lesotho.

Critical Remarks: Leaves on newly produced shoots are much longer and (relatively) narrower than on old shoots. Such shoots, particularly if they bear ♂ flowers, may closely resemble *Anthospermum pumilum*, whose range of distribution, in part, overlaps with that of *N. microphylla*. Both taxa sometimes even grow side by side. Hybrids between the two, however, appear to be very uncommon. The collections HANEKOM 605 and 606 [O.F.S., 2926-AA: Heuwelsig, W of Dan Pienaar (605: PRE, LISC; 606: PRE, SRGH; not in Collections)] may include hybrids *N. microphylla* × *A. pumilum* subsp. *rigidum*. The 606 sheets bear ♀ shoots with *A. pumilum*-like leaves, but ± inflated fruits, intermediate in size and shape between *N. microphylla* and *A. pumilum*. The plants on sheets 605 (with fruits and ♂ flowers) look more like *A. pumilum* subsp. *rigidum*.

In Lesotho, *N. microphylla* should not be confused with habitually similar, cushion-forming plants of *Anthospermum monticola*; see comments, p. 260.

In general, however, *N. microphylla* is easily distinguished from the remaining *Nenax* species and *Anthospermum* by its inflated, ± globose fruits and small leaves.

The species is probably much more widely distributed than the available herbarium material suggests. I have little doubt that the plants were often ignored by collectors because the species is frequently one of the most common plants in karroid areas (s.l.), and because it is often deformed and difficult to collect (due to browsing). To prove the point I have, on my collecting trips in the Cape Prov. and the O.F.S., specifically searched for plants in quarter-degree squares from which they had not yet been recorded and have, for the most part, been successful in finding them in suitable habitats.

N. microphylla is an excellent fodder bush (primarily for sheep) and, according to HOBSON & JESSOP (1975), regrowth is excellent when the rains are reasonable. The authors, furthermore, state that the species is characterized by an outstanding drought resistance. The common name for *N. microphylla* is "Daggaput karoo" because it was supposedly used by the Griquas and Hottentots as a "Dagga"-substitute[1].

---

[1] In S Africa, the term "dagga" is nowadays used for *Cannabis sativa*. The

It is noteworthy that ♂ plants of *N. microphylla* were detected which bore odd ♀ flowers (e.g., in population PUFF 800104-3/1). This, however, is an extremely uncommon and rare situation in the species (and in the genus *Nenax* as a whole).

For details on the kind of vegetative reproduction in *N. microphylla* (presumably unique in the genus) see C.2.2. and Fig. 6 b.

Collections

Lesotho. − **2927** (Maseru): nr. Roma (**-BC**), SCHMITZ 192 (PRE); Mafeteng D., Likhoele (**-CD**), DIETERLEN 1326 (PRE × 2) − **3027** (Lady Grey): Quthing (Moyeni) D., Leloaleng (**-BC**), DIETERLEN 1224 (PRE). − Imprecise locality: Maluti Mts. (2927-BD to 2928-AA, -AB), STAPLES 102 (PRE).

South Africa. O. F. S. − **2826** (Brandfort): 5 km from Bultfontein on Brandfort rd. (**-AC**), HERMAN 257 (PRE); Erfenisdam (**-BD**), WIPPLINGER 34 (PRE); Krugersdriftdam Nature Reserve (**-CC**), MULLER 1490 (K, PRE); Glen, N of Bloemfontein (**-CD**), POLE-EVANS 19619 (PRE), PUFF 790111-2/3 (WU), VAN DEN BERG 3879 (PRE); 29 km S of Winburg on N 1 (**-DB**), PUFF 800930-4/1 (WU). − **2827** (Senekal): Sandrivier (**-B**), BURKE 506 (BM, K, PRE, SAM), ZEYHER 769 (BM, G, K, PRE, SAM); a few km SE of Winburg on rd. to Excelsior (**-CA**), PUFF 840924-8/1 (WU). − **2925** (Jagersfontein): Fauresmith (**-CB**), VERDOORN 2099 (PRE); ca 33 km from Phillipolis on Fauresmith rd. (**-CD**), PUFF 790112-5/1 (WU); ca 10.5 km from Jagersfontein on Bloemfontein rd., nr. Trompsburg turnoff (**-DC**), PUFF 790112-1/1 (WU); between Trompsburg and Edenburg, along N 1 (**-DD**), PUFF 790720-1/1 (WU). − **2926** (Bloemfontein): Bloemfontein (**-AA**), Moss 3923 (J); Kafferrivier (**-AC**), SAUNDERS 6 (PRE); Thaba Nchu D., Farm Glamorgan-Wilgeboomnek boundary (**-BB**), ROBERTS 2964 (PRE); ca 7.5 km NE of Meadows Stn., on rd. to Thaba Nchu (**-BC**), PUFF 840924-6/1 (WU); Victoria Nek, 15 km S of Thaba Nchu (**-BD**), PUFF 840924-7/2 (WU); ca 67 km S of Bloemfontein on N 1, at Wurasoord- Reddersburg turnoff (**-CA**), PUFF 800930-1/1 (obs.); Reddersburg-Smithfield rd., nr. Sonop turnoff, base of Klein-Boesmansberg (**-CD**), PUFF 790413-1/1 (BR, J, WU); Dewetsdorp (**-DA**), PUFF 840924-5/1 (obs.); nr. Helvetia, Smithfield- Dewetsdorp rd. (**-DC**), PUFF 840924-4/1 (obs.) − **2927** (Maseru): just outside Ladybrand, on Clocolan rd. (**-AB**), PUFF 800104-4/1 (WU). − **3025** (Colesberg); 9 km from Phillipolis on rd. to Draaikloof (**-AC**), REID 262 (PRE); ca 7 km from Phillipolis on Springfield rd. (**-AD**), PUFF 790112-6/1 (WU); ca 19 km beyond Trompsburg on rd. to Phillipolis, nr. Blaauheuwel (**-BD**), WERGER 222 (K, PRE); ca 12 km W of Bethulie, on Donkerpoort rd. (**-DB**), PUFF 800104-2/1 (WU). − **3026** (Aliwal North): base of Boesmansberg (**-AA**), REID 214 (PRE); ca 19 km E of Bethulie, tow. Aliwal North (**-AC**), WERGER 194 (K, PRE, SRGH); Wepener- Smithfield rd., ca 15 km SW of Vanstadensrus turnoff (**-BA**), PUFF 800104-3/1 (WU); Smithfield- Rouxville rd., nr. Caledon R. (**-BC**), PUFF 840924-2/1 (obs.); ca 27 km N of Aliwal North on Rouxville rd. (**-BD**), PUFF 790413-2/1 (obs.); ca 6 km from Goedegemoed on rd. to Aliwal North (**-DA**), PUFF 790112-9/1 (WU); nr. Beestekraalnek Stn., Aliwal North- Rouxville rd. (**-DB**), PUFF 840923-1/1 (PRE, WU). − Additional collections: BURTT-DAVY 11794 (PRE); Grey College Herb. 88 (BOL); HENRICI 2543

---

word is most likely of Hottentot origin and also refers to any of several species of *Leonotis* which were chewed and used medicinally by the Hottentots (see BRANFORD 1978: 52).

(PRE); Phillips 26344 (J); Potts 88 B (BM, K, PRE), 4182 (PRE, STE); Puff 790112-2/2, -4/1 (WU); Rabie s.n. sub PRE 41793 (PRE); Roberts 5364 (PRE); Smith 3888, 4037, 4129, 5196, 5289 (PRE); Verdoorn 917, 943, 1339 (PRE). **Cape** − **2723** (Kuruman): Barkly West D., ca 4.5 km NW of Blikfontein (**-DD**), Leistner 962 (K, KMG, PRE). − **2823** (Griekwastad): Hay D., Papkuil (**-BC**), Wilman s.n. (sub KMG 1300, GH, sub PRE 41805); Griquatown (**-CC**), Burchell (K). − **2824** (Kimberley): Barkly West D., Cristaalfontein Farm (**-AA**), Acocks 1483 (KMG, M, PRE); Kimberley D., nr. Wesselton Mine (**-DB**), Wilman s.n. (sub KMG 1177, sub PRE 41798); −, Mauritzfontein Farm (**-DC**), Acocks 61 (KMG, PRE). − **2922** (Prieska): Hay D., Niekerks Hoop (**-BD**), Wilman s.n. sub PRE 41797 (PRE). − **3024** (De Aar): Philipstown, Grasfontein Farm (**-BC**), Vahrmeijer 1385 (PRE). − **3025** (Colesberg): Oviston Nature Reserve (**-CB, -DA**), Fourie 158, 409 (PRE), van Schoor 91 (PRE); ca 22 km S of Venterstad, on gravel rd. to Steynsburg, nr. Farm Dwarshoek (**-DD**), Puff 800929-12/1 (WU). − **3026** (Aliwal North): Aliwal North (**-DA**), Gerstner 256 (K, PRE); − D., Braamspruit (**-DD**), Burtt-Davy 10451 (PRE × 2). − **3121** (Fraserburg): ca 4 km W of Carnarvon, Taandjies Nek area (**-BB**), Puff 790710-1/1, 800831-3/1 (WU). − **3122** (Loxton): Victoria West D., Melton Wold (**-BD**), Thorne s.n. sub SAM 51931 (SAM); nr. Farm Hoveldene, Wagenaarskraal- De Jagers Pass rd. (**-DC**), Puff 840906-3/1 (obs.); Aasvoelberg, SW of Wagenaarskraal (**-DD**), Puff 840906-2/1 (PRE, WU). − **3123** (Victoria West): Richmond- Hutchinson rd., ca 15 km W of R 2, Farm Roggenfontein (**-AD**), Puff 800103-3/1 (WU); Uitvlugt (**-CB**), Drège 7665 a (E, G, K × 2, MO, P; W − mixed with *Anthospermum dregei*). − **3124** (Hanover): nr. summit of Lootsberg pass, Graaff-Reinet- Middelburg rd. (**-DD**), Puff 790908-1/1 (WU). − **3125** (Steynsburg): Middelburg D., Grootfontein (**-AC**), Verdoorn 1497 (K, PRE); ca 5–7 km S of Steynsburg on Middelburg rd. (**-BD**), Puff 800929-11/1 (WU); Maraisburg D. (**-DB**), Archibald 2687 (GRA). − **3126** (Queenstown): Penhoek pass area (**-BC**), Puff 790118-1/3 (obs.); Bradford Nek, Queenstown (**-DD**), Galpin 5643 (PRE), 8290 (GRA, PRE). − **3222** (Beaufort West): Karoo National Park, Nieuweveld Mts., between Trig. Beacon and TV mast (**-AD**), Puff 840907-2/1 (PRE, WU); Beaufort West D., Nieuweveld Mts. (**-BA**), Esterhuysen 2764 (BOL − mixed with *Anthospermum*); nr. summit of De Jagers pass (**-BB**), Puff 840906-4/1 (PRE, WU). − **3224** (Graaff-Reinet): Graaff-Reinet D., Melrose (**-BC**), Burtt-Davy 12311 (PRE); −, Klipfontein (**-BD**), Burtt-Davy 13455 (PRE). − **3225** (Somerset East): Somerset East- Craddock via Swaershoek pass, ca 14 km N of Swaershoek village (**-AD**), Puff 800929-8/1 (obs.); Wapadskloof, N of Bracefield (**-BA**), Puff 800929-5/1 (WU); Craddock D., nr. Rietfontein (**-BB**), Sim 5643 (K, PRE); 32 km NNE of Swaershoek village, tow. Craddock (**-BC**), Puff 800929-9/1 (obs.); nr. Bruintjies Hoogte (**-CB**), Bolus 1778 (BOL, K, MO); Somerset [East] (**-DA**), Bowker s.n. (K × 2); ca 36 km SSW of Somerset East (**-DC**), Acocks 15672 (PRE). − **3226** (Fort Beaufort): Winterberge, Springs Valley area (**-AD**), Puff 790117-7/1 (WU); "Fish River Rand", 33 km from Bedford on Grahamstown rd. (**-CC**), Puff 790117-4/1 (WU). − **3325** (Port Elizabeth): Albany D., ca 10 km along Craddock rd. (**-BA**), Martin s.n. sub GRA 9091 (GRA). − **3326** (Grahamstown): Carisle Bridge, Grahamstown (**-AA**), Forward 1590 (PRE); ca 16 km NW of Grahamstown, Farm Hilton (**-AB**), Story 4535 (GRA, K). − Imprecise localities: between Papkuil and Postmasburg (2823-BC to 2823-AC), Hutchinson 3014 (BOL, K); Albert D., Cooper 589 (E, K, NH, PRE, W; mixed with *Anthospermum*). − Additional collections: Acock(s) 5523 (S); Anderson 226 (PRE); Burtt-Davy 9712 (PRE); Gill 84 (PRE); Lynes 139 (BM); Puff 800929-6/1, -7/1 (WU); Story 74 (PRE); Tölken 1015 (K, PRE); van der Walt 68 (PRE).

2. Nenax cinerea (THUNB.) PUFF in Fl. S. Afr. **31**, 1 (2): 40 (1986).
   ≡ *Cliffortia cinerea* THUNB., Prod. Pl. Cap.: 93 (1800)[1].
Type: [South Africa, Cape Prov.]: "CBS", THUNBERG (sheet 23686, UPS, holo.!; S, iso.!).

   = *N. dregei* L. BOLUS in Ann. S. Afr. Mus. **9**: 215 & pl. VI, A (1917); LAUNERT & ROESSLER in MERXMÜLLER, Prod. Fl. S.W.A. **115**: 19 (1966).
Types: South Africa, Cape Prov., Bot Riverbed, between Calvinia and Holle River, PEARSON 3966 (BOL, lecto.!; K, NBG, iso.!); −, −, ca 24 km N of Alewyn's Fontein (= Aalwynsfontein), PEARSON 3930 (BOL, syn.!); −, −, between Klipplaat and Bitterfontein, PEARSON 3295 (BOL, syn.!, STE, isosyn.!); −, −, between Anenous and Chubiessis Outspan, PEARSON 5979 (BOL, syn.!, K, isosyn.!).

   − *N. hantamensis* SCHLECHTER, nom. nud. [South Africa, Cape Prov., Calvinia Distr., Hantam Mts., MARLOTH 10444 (PRE!)]

Dioecious, many-stemmed, intricately branched, rigid (dwarf) shrubs ca 50–100 cm tall; much lower, ca 10–30 (40) cm tall, and occasionally cushion-forming if browsed. Stems ca 10–20 mm in diameter at the base and ca (2) 4–8 (10) mm in the midstem region, erect, ascending or sometimes ± prostrate, much-branched; branches often single at the nodes or, if in pairs, of unequal length; shoots frequently spine-tipped if browsed; younger parts often purplish, densely papillate, older parts glabrescent, light to dark grey, less commonly reddish-brown, bark often with long fissures; longest internodes ca 5–11 mm. Leaves decussate; blades (2) 3–7 (10) × (1.2) 1.5–2 mm, ovate- or linear-lanceolate, shallowly concave above, convex below, appearing silvery-grey, greyish or greenish-grey due to a dense cover of whitish, often curled short hairs or papillae; apices ± acute; petioles 0; stipular sheaths broadly cup-shaped, less than 1 mm long, hairy-papillate, sometimes with a minute, median gland-tipped seta. Flowers subsessile, single or, less commonly, paired (seldom in clusters of 4–6) at the nodes; ♂, ♀. Corolla 5-merous, yellowish to greenish-yellow, sometimes reddish tinged on the outside and usually densely papillate. ♂: tube 0.8–1.3 (1.7) mm long, broadly funnel-shaped, lobes 2.5–3.2 (3.5) × (0.6) 0.8–1 mm, ± lanceolate, recurved; stamens 5, filaments (0.8) 1.2–1.7 (2.2) mm long, anthers (1.5) 1.8–2.5 × 0.4–0.7 mm; rudimentary ovary minute. ♀: tube (0.1) 0.3–0.6 mm long, lobes (0.1) 0.3–0.7 (0.9) × 0.1–0.2 (0.4) mm, ± linear, erect, corolla sometimes absent

---

[1] In his *Cliffortia* revision, WEIMARCK (1934: 15) removed the species from *Cliffortia* and considered it identical with *Ambraria hirta* (= *Nenax hirta*).

altogether; style 0; stigmas (1.9) 2.5–7.5 mm, dark purple, seldom greyish; ovary 1.4–2 × 0.8–1.2 (1.7) mm, obovoidal, densely papillate, crowned by 5 (4) rounded calyx lobes. Fruits light reddish-brown, inflated, laterally compressed, ca (5) 6–8 mm in diameter, dehiscent, not supported by distinct carpophores, crowned by the persistent rounded to ± triangular calyx lobes, ca 0.2–0.4 × 0.3 mm; each mericarp ca (5) 6–7.5 × (5) 6–8 mm, ± round, broadly obovate or ± heart-shaped in outline, dorsal side shallowly convex, papillate to glabrescent, ventral side concave, with a prominent vertical median ridge. – Figs. 10 a, 15 e–f, 37 g.

Chromosome Number (Table 3): n = 11, 2 n = 22.

Average Pollen Diameter (Fig. 49): 26.6–28.8 μm.

Habitat: In sandy soil of dried up river beds and along ephemeral watercourses; in cracks of rocks, between rocks or at the edge of rock sheets. Appears to be confined to localities with a reasonably good water supply in generally hot, arid areas. – Ca (600) 800–1 200 (1 400) m.

Distribution (Map, Fig. 99 a): S Africa; in the interior of the W Cape Prov. from the Laingsburg Distr. N to Namaqualand. Also extending into S Namibia (South West Africa) as far N as Aus.

Critical Remarks: N. cinerea is easily recognized by its large, inflated, laterally compressed fruits (Fig. 10 a). Flowering or vegetative material may, however, in habit, branching, leaf size and shape closely resemble forms of *N. coronata* (4.) but the leaf indumentum of *N. cinerea* is unique (cf. Figs. 15 e–f). *N. coronata* may have papillate leaf margins, but papillae are never found on both leaf surfaces. Attention is, furthermore, drawn to the strikingly different leaf anatomy of the two species (compare Figs. 24 a and b). See also *N. namaquensis,* Critical Remarks (3., below).

Collections

South Africa. Cape – **2917** (Springbok): between Anenous and Chubiessis Outspan (**-BA**), PEARSON 5979 (BOL, K); nr. Steinkopf (**-BC**), SCHLECHTER 11428 (BOL, GRA); Modderfontein, ca 15 km W of Springbok (**-DB**), VAN DER WESTHUIZEN 328, 329 (K, PRE). – **2918** (Gamoep): Hester Malan Nature Reserve nr. Springbok (2917-DB to 2918-CA), ROESCH & LE ROUX 197, 242 (WIND); Silwerfontein (**-CC**), THORNS s.n. sub NBG 23287 (NBG); Koppieskraal, ca 4.5 km S of Gamoep (**-CD**), VAN DER WESTHUIZEN 157 (PRE). – **3018** (Kamiesberg): Kamieskroon D., Pedroskloof, ca 18 km NO of Kamieskroon (**-AA**), VAN DER WESTHUIZEN 156 (PRE); ca 24 km N of Alewyn's Fontein (= Aalwynsfontein) (**-BD**), PEARSON 3930 (BOL); nr. Nieuwefontein (**-DA**), PEARSON 3355 (BOL, PRE). – **3019** (Loeriesfontein): between Klipplaat and Bitterfontein (**-CA**), PEARSON 3295 (BOL, STE). – **3119** (Calvinia): Bot River, between Calvinia and Holle River (= Holrivier) (**-BC**), PEARSON 3966 (BOL, K, NBG); Akker(en)dam Nature Reserve, N of Calvinia (**-BD**), PUFF 790711-1/1 (WU), 800831-5/1 (BR, J, WU). – **3120** (Williston): Williston D., ca 37 km W of Williston (**-BC**), ACOCKS 14703 (K, PRE). – **3219** (Wuppertal): nr. Gansfontein (**-DA**), LEVYNS 1767

(BOL). – **3220** (Sutherland): ca 45 km NE of Sutherland, along Prinshof rd. (**-BB**), VAN BREDA 2019 (PRE); Sutherland D., Houthoek (**-CB**), HANEKOM 1560, 1560 A (STE); ca 30 km S of Sutherland, Verlatekloof, Farm Klipbanksrivier (**-DA**), PUFF 800103-1/1 (WU). – **3221** (Merweville): between Sutherland and Fraserburg, Farm Blomfontein (**-AA**), PUFF 800103-2/1 (WU). – **3320** (Montagu): Laingsburg D., Matjiesfontein (**-BA**), MARLOTH 10767 (PRE). – No locality given ("CBS"): Boos 1106 (WU); herb. THUNBERG s.n. (BM, S × 3; UPS 23686). – Additional collections: ACOCKS 16947 (BOL, K, PRE); ARCHER 119 (BOL); BARKER 9507 (NBG); COMPTON 2939, 3661, 9248 (BOL), 5615, 5616, 7823 (NBG), 14919 (NBG, PRE); LEISTNER 374 (K, PRE); MARLOTH 10444 (PRE); PILLANS 5438 (BOL, K); PUFF 800102-4/1 (WU); SCHMIDT 132 (PRE); TAYLOR 2672 (NBG); VAN BREDA 4200 (K, MO).

Namibia (South West Africa). – **2616** (Aus): Aus (**-CB**), DINTER 4132 (B, BM, BOL, G, J, K, PRE, S, STE; GH as '432'), GIESS & VAN VUUREN 869 (K, M, PRE, WIND). – **2718** (Grünau): Klein-Karas (**-CA**), DINTER 3178 (SAM × 2), 4909 (B, BOL, GH, PRE × 2, STE).

3. *Nenax namaquensis* PUFF in Fl. S. Afr. **31**, 1 (2): 41 (1986).

Type: South Africa, Cape Prov., Namaqualand, a little N of Middelkraal, PEARSON 5615 (BOL, holo.!; K, iso.!).

Dioecious, many-stemmed, intricately branched, ± rounded to spreading dwarf shrubs, ca 15–60 cm tall; often ± cushion-forming if browsed. Stems ca 5–10 mm in diameter at the base and ca 1–3 mm in the midstem region, often much-branched; branching irregular, branches mostly single at the nodes, usually spine-tipped if browsed; younger parts ± densely covered with whitish papillae to 0.1 mm long, greyish-brown to dark grey, sometimes reddish tinged; older parts glabrescent, grey to blackish, epidermis often peeling and bark with long fissures; internodes mostly (much) longer than the leaves except near shoot tips or on short lateral branches. Leaves decussate; blades ascending to ± spreading, often slightly curved, 2.5–4.5 (6) × 0.9–1.4 mm, narrowly ovate-lanceolate to linear-lanceolate, shallowly concave above and convex below to ± triangular in section, fresh green in nature, glabrous except for some whitish papillae to 0.1 mm long on the margins; apices ± acute; petioles 0; stipular sheaths broadly cup-shaped, ca 0.7–1.4 mm long, glabrous to papillate, sometimes with a minute, median gland-tipped seta to ca 0.1 mm long. Flowers subsessile, single or (rarely) paired at the nodes, sometimes also terminal and overtopped by lateral shoots arising below; ♂, ♀. Corolla 5-merous (in ♂ very rarely, in ♀ occasionally also 4-merous), yellowish (?), reddish tinged on the outside. ♂: tube 1–2.1 mm long, funnel-shaped or occasionally ± cylindrical, lobes 2.4–3.4 × 0.7–1 mm, ± lanceolate, recurved; stamens 5 (4), filaments 1.4–1.7 mm long, anthers 1.5–2.2 × 0.4–0.6 mm; rudimentary ovary minute, often with conspicuous calyx lobes. ♀:

tube 0.2–0.4 (0.7) mm long, cylindrical, lobes 0.4–0.7 (1.2) × 0.1–0.2 (0.3) mm, lanceolate to linear-lanceolate, ± erect to spreading; style 0; stigmas 1.4–2.4 mm; ovary 1–1.2 × 0.8–1 mm, glabrous or sparsely to densely covered with short whitish hairs, crowned by 5 (4) rounded to ± triangular calyx lobes. Fruits dark reddish-brown, dehiscent, supported by distinct, sometimes large U-shaped carpophores or (seldom) carpophores ± absent; each mericarp 2.1–2.6 × 1–1.4 mm, ± obovate, dorsal side ± convex, glabrous or sparsely to densely covered with whitish papillae to 0.1 mm long, ventral side ± plane and with a ± indistinct vertical median ridge; each mericarp crowned by 3 (2) ± triangular calyx lobes, ca 0.4–0.7 × 0.4–0.7 mm.

Chromosome Number (Table 3): 2 n = 22.

Average Pollen Diameter (Fig. 49): 28.1–28.8 µm.

Habitat: Growing in rocky areas (mainly granite) in sandy to gravelly soil amongst rocks or in rock crevices. Primarily occurring in Namaqualand Broken Veld and, in the Kamiesberg area, in Mountain Renosterbosveld [Acocks (1975) Veld Type Nos. 33 and 43]; no habitat notes are available for the S-most locality (Nardouw). – Ca 450–1 200 m.

Distribution (Map, Fig. 99 b): S Africa. Confined to the W Cape Prov. and occurring from the Namaqualand Distr. S to the N Clanwilliam Distr.

Critical Remarks: Although, in habit and appearance, *N. namaquensis* may superficially resemble *N. cinerea* (2., above) (and has, for this reason, often been identified as such), Bolus (1917) quite correctly kept the type collection (Pearson 5615, listing it as "? *Nenax sp.*") separate from her *N. dregei* (= *N. cinerea*), when describing the *Rubiaceae* collected on the Percy Sladen Memorial Expeditions. Her comments as regards to the "immature fruits" on the ♀ type collection are, however, incorrect. The fruits can readily be split into two mericarps, which could not be done with ease if they were immature. The collection Hutchinson 836 from "between Garies and Kamieskroon" can be considered a ♂ topotype [the holotype locality – (Farm) Middelkraal – is roughly 15–20 km N of Garies on the road to Kamieskroon].

Vegetatively, *N. namaquensis* differs from *N. cinerea* primarily in lacking the distinct leaf indumentum and anatomy of the latter. In fruit, the two species cannot be confused (*N. cinerea* has large, inflated, laterally compressed fruits). The relatively small, "*Anthospermum*-like" dehiscent fruits of *N. namaquensis* seem to point to a close affinity to both *N. coronata* (4., below) and *N. divaricata* (5.). The latter, however, differs in growth form (± regular branches, branches mostly opposite) and in having 4-merous flowers. *N. coronata* is perhaps the closest ally of *N.*

namaquensis. The two species have in common their 5-merous flowers and a similar growth form (unequal promotion of lateral shoots, or branches produced singly at the nodes); *N. namaquensis,* however, differs in having smaller, relatively wider and less needle-like leaves, smaller fruits and narrower, occasionally ± cylindrical, longer ♂ corolla tubes.

The ♂ collection HAFSTRÖM & ACOCKS 1441 (PRE, S) from near Sutherland (3220-BC), appears to be allied to *N. namaquensis.* It is an obviously browsed, intricately branched dwarf shrub with spine-tipped, mostly opposite, grey branches and 5-merous ♂ flowers. It differs in having short, rather broadly funnel-shaped corolla tubes and ± glabrous, flat leaves, which are ± oblong in section. The absence of fruiting (♀) material makes it impossible to place it with certainty; it may be a new species.

Collections

South Africa. Cape – **2917** (Springbok): top of Spektakelberg, ca 1.5 km S of Naries homestead (**-DA**), VAN DER MERWE 190 (PRE, STE); Hester Malan Nature Reserve nr. Springbok (**-DB**), ROESCH & LE ROUX 728 (PRE). – **3017** (Hondeklip-baai): ca 22 km W of Kamieskroon (**-BA**), ACOCKS 16443 (PRE × 5); Bowesdorp (**-BB**), STOKOE s.n. sub SAM 59518 (PRE, SAM); a little N of Middelkraal (**-BD**), PEARSON 5615 (BOL, K); between Garies and Kamieskroon, HUTCHINSON 836 (BOL, GRA, K, PRE). – **3018** (Kamiesberg): Farm Bovlei, ca 7–8 km E of Kamieskroon (**-AA**), PUFF 790714-1/1 (BR, J, WU); Farm Welkom, Kamiesberg (**-AC**), ESTERHUYSEN 23731 (BOL); Skilpad (= Farm Skilpadrug, SE of Garies and E of Klein-Kamiesberg?) (**-CA**?), WINKLER 133 (NBG, STE). – **3118** (Van-rhynsdorp): Nardouw (**-DC**), STOKOE 8217 (BOL). – Imprecise locality: 'Little Namaqualand', STOKOE 9512 (BOL).

4. *Nenax coronata* PUFF in Fl. S. Afr. **31**, 1 (2): 41 (1986).

Type: South Africa, Cape Prov., W side of Pakhuis Pass, a little W of Leipoldt grave, PUFF 800902-6/4 (WU, holo.!; BOL, NBG, PRE, STE, iso.!).

Dioecious, many-stemmed, ± erect dwarf shrubs, often with thick, woody roots. Stems ca 25–40 cm long, ca 3–5 mm in diameter at the base and ca 1–2 mm in the midstem region, much-branched; branches usually single at the nodes, branching often pseudo-dichotomous[1] and ± irregular; younger parts covered with whitish papillae to 0.1 mm long, ± terete, grey to dark grey or occasionally reddish tinged; older parts glabrescent, grey to blackish, epidermis often peeling; internodes much longer than the leaves except near shoot tips. Leaves decussate; blades

---

[1] See Critical Remarks!

(3) 5–9 (11) × 0.7–1 mm, linear, needle-like and ± rigid, semiterete to ± triangular in section, glabrous or with a few papillae on the margins, upper surface and midrib region sometimes (dark) brownish-red in dried material; apices ± acute to acuminate; petioles 0; stipular sheaths broadly cup-shaped, ca 0.5–0.8 mm long, sometimes covered with a few short, whitish hairs, with 1–3 minute, median setae to 0.2 mm long or setae ± absent. Flowers subsessile, mostly single at the nodes, sometimes also terminal and overtopped by short lateral branches arising below; ♂, ♀. Corolla 5- (rarely 4-)merous, yellowish (?), glabrous. ♂: tube ca 0.8 mm long, funnel-shaped, lobes 2.5–3.4 (4) × 0.6–1 mm, lanceolate to ± linear-lanceolate, recurved; stamens 5 (4), filaments 1.2–1.5 mm long, anthers 1.4–2.4 × 0.3–0.7 mm; rudimentary ovary minute, sometimes with well discernible calyx lobes. ♀: tube 0.1–0.2 mm long, lobes 0.3–0.8 × 0.1–0.2 mm, ± linear, erect; style 0; stigmas (1.5) 2–4 mm; ovary 1.2–2 × 0.7–1 mm, glabrous or shortly hairy, crowned by 5 conspicuous calyx lobes. Fruits reddish-brown, dehiscent, either supported by small carpophores or carpophores sometimes ± absent; each mericarp 3–5 × 1.9–2.5 (3) mm, oblong to obovate, dorsal side convex, glabrous or covered with short whitish hairs or papillae to 0.1 mm long, ventral side ± concave and with a ± prominent vertical median ridge; each mericarp crowned by 3 (2) triangular calyx lobes ca 0.6–1.2 × 0.3–0.5 mm. – Fig. 37 h.

Chromosome Number (Table 3): 2 n = 22.

Average Pollen Diameter (Fig. 49): 27.0 μm.

Habitat: In dry Mountain Fynbos (Arid Fynbos sensu TAYLOR 1978, KRUGER 1979); growing in sandy patches between rocks or in cracks of rocks. – Ca 600–1 000 m.

Distribution (Map, Fig. 100 b): S Africa, W Cape Prov. Confined to the S Vanrhynsdorp Distr. (Giftberg) and the adjacent Clanwilliam Distr.

Critical Remarks: N. coronata is allied to the remaining *Nenax* species with non-inflated, dehiscent fruits (species 3., 5. and 6.).

N. *elsieae* (6.) differs primarily in its smaller leaves, its habit, its somewhat smaller fruits which are densely covered with often curled whitish hairs and crowned by somewhat smaller but broader calyx lobes (compare Figs. 37 h and 38 e–f) and its 4-merous corollas; it has a much more SE distribution. A very close alliance between the two species, therefore, seems improbable.

The distribution range of N. *divaricata* (5., below), on the other hand, covers the entire range of the much more narrowly distributed N. *coronata* (Giftberg, Pakhuispas area). It appears to be much rarer than N. *divaricata,* which I have found to be quite common in both localities,

27*

whereas it took several visits and an extensive search to discover *N. coronata* in the Pakhuispas area. In the field, the two species are easily separable by growth form alone: *N. divaricata* is much more regularly branched, the branches are mostly opposite and divaricate on the ± straight and erect stems, whereas in *N. coronata,* the branches mostly arise singly at the nodes. As they continue growth, they frequently push the main axes to the side, so that a Y-like appearance of the branchings may result ("pseudo-dichotomous" branching in the description, above). Consequently, the main axes often also show a distinct zig-zag pattern. *N. divaricata,* furthermore, differs in having 4-merous ♂ flowers and smaller fruits with much less conspicuous, minute calyx lobes and often smaller, thinner leaves.

The affinities between *N. coronata* and *N. namaquensis* (3., above) are discussed on p. 407.

Collections

South Africa. Cape – **3118** (Vanrhynsdorp): Vanrhynsdorp D., Giftberg plateau (-**DC**), ESTERHUYSEN 22044 (BOL). – **3218** (Clanwilliam): W side of Pakhuispas, a little W of Leipoldt grave (-**BB**), PUFF,800902-6/4 (BOL, NBG, PRE, STE, WU); Pakhuis, Pakhuisberg or -pas (-**BB** to 3319-AA), ESTERHUYSEN 3254 (BOL), SALTER 8130 (BM, BOL × 2, K, NBG), SCHLECHTER 8652, 10805 (BOL, GRA, PRE). – **3219** (Wuppertal): Clanwilliam D., Brandewyns River (-**AA**), COMPTON 9649 (NBG).

5. *Nenax divaricata* SALTER in J. S. Afr. Bot. **3**: 113 (1937); PUFF in Fl. S. Afr. **31**, 1 (2): 43 (1986).

= *Ambraria acerosa* SONDER in HARVEY & SONDER, Fl. Cap. **3**: 34 (1865), *pro parte* [♂; see *N. acerosa,* Critical Remarks (p. 417)].
Type: South Africa, Cape Prov., Worcester Distr., near Tulbaghs-kloof, ECKLON & ZEYHER 2319 (or "1.9") (SAM, holo.!; E × 2, FI, GOET, LY, M, MO, NBG, PRE, S, US, W × 2, WU, iso.!).

– *N. acerosa* sensu ECKLON & ZEYHER, Enum. Pl. Afr. Pl. Afr. Austr.: 368 (1836), non GAERTN.

Dioecious, several-stemmed, rounded to ± cylindrical, usually diva-ricately branched dwarf shrubs ca (15) 30–60 (100) cm tall. Stems ca 3–7.5 mm in diameter at the base and ca 1.5–3 mm in the midstem region, erect to ascending, much-branched, principal branches usually paired at the nodes, held at an angle of ca 45°; younger parts greyish, papillate; older parts glabrescent, branches grey to dark grey, bark often with long fissures; internodes mostly longer than the leaves except near shoot tips. Leaves decussate; blades (3) 3.5–8 (9.5) × 0.4–0.5 mm, linear, needle-like, semiterete to ± triangular in section, glabrous except for a few papillae or short hairs on the margins; apices ± acute to mucronate; petioles 0; stipular sheaths small, ± cup-shaped, ca 0.5–0.7 mm long, with or without a minute median seta. Flowers subsessile, mostly single at the nodes, sometimes also terminal and overtopped by short lateral branches arising below; ♂, ♀. Corolla 4-merous, yellowish-green, often purplish or purplish-brown tinged or streaked on the outside. ♂: tube (0.8) 1–1.5 mm long, funnel-shaped, lobes 2.7–3.2 × 0.7–1 mm, ± lanceolate, recurved; stamens 4, filaments (0.8) 1–1.5 mm long, anthers 1.4–1.7 × 0.3–0.5 mm; rudimentary ovary very small. ♀: tube 0.2–0.3 mm long, lobes 0.3–0.5 × 0.1–0.2 mm, ± linear, erect; style 0; stigmas 3.4–4.6 mm long, purplish-red to dark purple; ovary 0.7–1.2 × 0.4–0.8 mm, often papillate, crowned by 4 small calyx lobes. Fruits reddish-brown, greyish to dark grey, tardily dehiscent, supported by small but distinct, narrowly U-shaped carpophores; each mericarp (2.4) 2.7–3.4 × 1.2–1.7 mm, ± obovate, dorsal side convex, glabrescent to densely covered with papillae to 0.1 mm long, ventral side ± plane to somewhat concave and with a vertical median ridge; each mericarp crowned by 2 rounded to ± triangular calyx lobes ca 0.3–0.4 mm long. – Figs. 33 g, 38 g.

Chromosome Number (Table 3): n = 11, 2 n = 22.

Average Pollen Diameter (Fig. 49): 30.1–32.6 μm.

Habitat: Primarily in dry Mountain Fynbos (Arid Fynbos sensu TAYLOR 1978, KRUGER 1979) associated with TMS; growing in sandy to gravelly soil. – Ca. (150) 500–900 (1 250) m.

Distribution (Map, Fig. 99 c): S Africa. Confined to the (S)W Cape Prov. and occurring from the Sir Lowry's Pass N to the S Vanrhynsdorp and the Calvinia Distr.

Critical Remarks: Primarily because of its similar, small, dehiscent fruits supported by carpophores and its similar branching pattern, *N. divaricata* has often been confused with dioecious *Anthospermum* species (*A. spathulatum* in particular). ♀ flowering specimens of *N. divaricata* are, however, easily distinguished by their purplish-red stigmas (in SW Cape *Anthospermum* species stigmas are, with very few exceptions, greenish-white to greyish!) and their few-flowered inflorescences. ♂ specimens, especially if only small branches are available and the habit of the plant is not known (*N. divaricata:* several-stemmed dwarf shrubs; dioecious SW Cape *Anthospermum* species: mostly single-stemmed larger shrubs), may be more problematic. Leaf sections, nevertheless, allow a certain distinction: *N. divaricata* differs in having distinctly needle-like, ericoid leaves which are semiterete to ± triangular in section (cf. Figs. 25 g–h).

Plants from the S-most locality, Sir Lowry's Pass (STOKOE s.n. sub SAM 69993, ACOCKS s.n., SCHLECHTER 1138), differ somewhat in habit from the remaining collections. ACOCKS s.n. in particular is a much more densely leafy and more branched plant than "typical" *N. divaricata,* although the fruits fully match other collections. Whether these forms deserve taxonomic recognition cannot be determined without further field observations and additional herbarium material.

Collections

South Africa. Cape – **3118** (Vanrhynsdorp): Giftberg (-DC), PUFF 790713-1/1 (WU), THOMPSON 2090 (K, PRE, STE). – **3119** (Calvinia): Boskop area, ca 45–50 km S of Nieuwoudtville (-CA), PUFF 790712-2/1, 800901-1/1 (WU). – **3218** (Clanwilliam): Clanwilliam side of Pakhuis pass, nr. Leipoldt grave (-**BB**), PUFF 790716-1/5, -2/1 (WU). – **3219** (Wuppertal): Clanwilliam D., Nieuwoudt (= Nieuwhoudt) pass (-AC), COMPTON 4780 (NBG), ESTERHUYSEN 8145 (BOL); E of Sederberg village, Farm Sanddrift (-AD), PUFF 790717-3/3 (WU). – **3318** (Cape Town): Bottelary hills, nr. Stellenbosch rd. (-DC), ACOCK(s) 2423 (S). – **3319** (Worcester): Tulbaghskloof (-AC), ECKLON & ZEYHER 2319 (or E. & Z. or ZEYHER "1.9") (E × 2, FI, GOET, LY, M, MO, NBG, PRE, S, SAM, US, W × 2, WU). – **3418** (Simonstown): Sir Lowry's pass (-**BB**), STOKOE s.n. sub SAM 69993 (SAM; ± atypical). – Not traced: Brakfontein (Clanwilliam), ECKLON s.n. (S). – No locality given: SIEBER 42 (W). – Additional collections: ACOCKS 17445, 17458 (K, PRE), 18544 (PRE), s.n. (S, very atypical); SCHLECHTER 8767 (BOL, GRA, PRE × 2 – one sheet as TVL Museum no. 88), 1138 (BM, BOL, G × 2, GRA, K, W; ± atypical).

6. *Nenax elsieae* PUFF in Fl. Afr. **31**, 1 (2): 43 (1986).

Type: South Africa, Cape Prov., Worcester Distr., Bonteberg, Eikenbosch Hoek, 1 070–1 220 m, ESTERHUYSEN 3656 (BOL, holo.!).

Dioecious, many-stemmed, rounded to ± cylindrical low dwarf shrubs with thick woody roots. Stems ca 8–20 (30) cm long, ca 2–4 mm in diameter at the base and ca 1–1.5 mm in the midstem region, ascending to erect, much-branched; branches usually single at the nodes, ± erect; younger parts ± densely papillate, often reddish-purple tinged; older parts glabrescent, greyish-brown to dark grey, epidermis often peeling; longest internodes (3) 5–8 (10) mm long. Leaves decussate, sometimes ± pseudo-verticillate; blades 3–5 × 0.5–0.8 (1) mm, linear(-lanceolate), needle-like, semiterete to ± triangular in section, margins often papillate, midrib region sometimes reddish-brown on the lower side; apices ± acute; petioles 0; stipular sheaths much reduced, broadly cup-shaped, less than 1 mm long, with or without a minute median seta. Flowers subsessile, usually single at the nodes; ♂, ♀. Corolla 4-merous, yellowish-green. ♂: tube ca 0.8–1 mm long, (broadly) funnel-shaped, lobes (1.7) 2–3 × 0.8–

1 mm, ± lanceolate, recurved; stamens 4, filaments 1.2–1.8 mm long, anthers 1.7 × 0.5–0.8 mm; rudimentary ovary very small, usually crowned by 4 conspicuous, ± rounded calyx lobes. ♀; tube 0.2–0.5 mm long, lobes 0.3–0.8 × 0.2–0.3 mm, linear-lanceolate, erect to ± spreading; style 0; stigmas 3.4–4.4 mm long, (dark) purplish-red; ovary 1–1.4 × 0.7–0.8 mm, hairy, crowned by 4 conspicuous calyx lobes. Fruits reddish-brown, dehiscent, supported by minute, shallowly U-shaped carpophores; each mericarp 2.7–3.5 × 1.8–2.2 mm, obovate, dorsal side convex, densely covered with (±) curled whitish hairs, ca 0.2–0.5 mm long, ventral side plane, commissure relatively small; each mericarp crowned by 2 glabrous, broad, rounded to ± trapeziform calyx lobes ca 0.5–0.7 × 0.5–0.9 mm. – Figs. 33 c, 38 e–f.

Chromosome Number: Unknown.

Average Pollen Diameter (Fig. 49): 32.3 µm.

Habitat: Growing in very dry fynbos over Witteberg Quartzite together with *Protea* and *Leucodendron* spp., *Restionaceae, Passerina* sp., *Polygala* sp., *Elytropappus rhinocerotis* and various *Mesembryanthemaceae*; occurring on rocky slopes between rocks (data based on field notes from PUFF 790913-6/1). – Ca 1 000–1 220 m.

Distribution (Map, Fig. 100 c): S Africa, interior of the SW Cape Prov. Only known from N and NE of Touws River and from the Swartruggens Mts. (Worcester and Ceres Distr.).

Critical Remarks: This apparently rather rare species is easily and well distinguished by its dehiscent fruits with longish, ± curled hairs, and by the conspicuous, relatively large calyx lobes (Figs. 33 c and 38 e–f), which are also found on rudimentary ovaries of ♂ flowers (although they are – relatively – much smaller in ♂ than in ♀ flowers).

N. elsieae (♂ specimens in particular) may be confused with *Anthospermum comptonii* if not studied with care. See *A. comptonii,* Critical Remarks (p. 375), for details.

Collections

South Africa. Cape – **3219** (Wuppertal), Ceres D., Swartruggens (-DC), TAYLOR 5874 (PRE, STE). – **3319** (Worcester): Worcester D., Bonteberg, Eikenbosch Hoek (-BD), ESTERHUYSEN 3656 (BOL). – **3320** (Montagu), Touws River- Pienaarskloof rd., Bierkraal se Rante (-AC), PUFF 790913-6/1 (WU).

7. *Nenax hirta* (CRUSE) SALTER in J. S. Afr. Bot. **3**: 113 (1937), in ADAMSON & SALTER, Fl. Cape Penins.: 735 (1950); PUFF in Fl. S. Afr. **31**, 1 (2): 44 (1986).

≡ *Ambraria hirta* CRUSE, Rub. Cap.: 17 (1825), in Linnaea **6**: 19 (1831); SONDER in HARVEY & SONDER, Fl. Cap. **3**: 34 (1865).

Types: [South Africa, Cape Prov.] "C.B.S.", MUND(T) & MAIRE s.n. (B, syn. †), at base of Lion's Mt. towards Drieanckerbay, BERGIUS s.n. (B, syn. †). Neo- and Topotype: slopes of Signal Hill above Three Anchor Bay, SALTER 6407 (BOL!).

Two subspecies are recognized:

7(a). Subsp. *hirta*

Dioecious, many-stemmed, rounded to ± cushion-forming, diffusely branched dwarf shrubs to ca 30 cm tall and 30–50 cm in diameter. Stems ca 3.5–5 mm in diameter at the base and ca 1–2 mm in the midstem region, ascending, prostrate or ± erect, much-branched; branches usually single at the nodes; younger parts often purplish, densely covered with whitish hairs ca 0.1–0.2 mm long; older parts glabrescent, greyish(-brown); longest internodes 3–6(9) mm long. Leaves in whorls of 3, pseudo-verticillate; blades (1.2) 1.5–3.5 × 0.4–0.9 mm, linear(-lanceolate) (often longer, to 6.4 mm, on new growth), margins and sometimes also midrib below with whitish, ± straight hairs ca 0.1–0.2 (0.4) mm long; apices ± acute to mucronate; petioles 0; stipular sheaths minute, cup-shaped, ca 0.5–0.6 mm long, with or without a minute median seta. Flowers subsessile, usually in clusters of 3 at the nodes, ♂, ♀; inflorescences sometimes ± conspicuous in fruiting ♀. Corolla 4-merous, yellowish-(-green), sometimes with a few hairs on the outside. ♂: tube 0.7–1.2 mm long, broadly funnel-shaped, lobes (1.5) 1.7–2.1 × 0.6–0.9 mm, ± lanceolate, revolute; stamens 4, filaments (1.1) 1.4–2 mm long, anthers 1–1.7 × 0.2–0.5 (0.6) mm; rudimentary ovary minute. ♀: tube 0.3–0.4 mm long, lobes (0.3) 0.5–1 × 0.1–0.3 mm; style 0; stigmas (2.2) 2.5–4.8 mm long, purplish-red; ovary 1–1.2 × 0.6–1 mm, ± obovoidal, hairy, crowned by 4 conspicuous, broad, rounded calyx lobes. Fruits 2–2.8 × (1.5) 1.7–2.2 mm, obovoidal to spheroidal, indehiscent, hard, not supported by carpophores, greyish, covered with whitish, ± spreading hairs ca 0.1–0.2 mm long, crowned by 4 glabrous, rounded calyx lobes ca 0.3–0.4 × 0.5 mm. – Fig. 38 d.

Chromosome Number (Table 3): 2n = 22.
Average Pollen Diameter (Fig. 49): 28.5–30.7 µm.
Habitat: Mostly in Coastal Renosterveld and in Coastal Fynbos [ACOCKS (1975) Veld Type Nos. 46 and 47] in sand dunes; sometimes in disturbed sites (roadsides, cultivated lands). Growing in white sand,

gravelly or red, clayey (granite-derived) soil or over shale. – Ca 0–300 (600) m.

Distribution (Map, Fig. 100 f): S Africa, SW Cape Prov. Occurring from the N end of the Cape Peninsula E and NE to the Worcester and the Piketberg Distr.

Critical Remarks: Easily recognized by its small indehiscent hairy fruits, by its small ♂ flowers and by its small leaves arranged in whorls of 3.

The collection Acocks 14523, originating from "a patch of *Themeda* veld on white sand", has the same habit and the characteristic, small fruits of subsp. *hirta* but somewhat approaches subsp. *calciphila* in having some leaves arranged decussately rather than in whorls of 3.

See subsp. *calciphila* for further comments.

Collections

South Africa. Cape – **3318** (Cape Town): Malmesbury D., ca 12 km NE of Yzerfontein (= Ysterfontein) (**-AC**), Acocks 14523 (G, K, PRE; ± approach. subsp. *calciphila*); Slangkop nr. Darling (**-AD**), Compton 6778 (NBG – mixed *Anthospermum spathulatum*), Puff 800910-1/2 (WU); nr. Porterville (**-BB**), Schlechter 10724 (BM, BOL, BR, E, G, K, MO, PRE × 2, S, US, W); Malmesbury D., nr. Oude Post (= Oupos) Hotel (**-BC**), Acocks 15226 (K, PRE); Cape Town, Bantry Bay (**-CD**), Salter 6812 (BM, BOL, K); Klipheuwel [on Durbanville- Malmesbury rd.] (**-DA**), Thompson 2560 (PRE, STE); nr. Wellington via Hermon rd. (**-DB**), Hafström & Acock(s) 1433 (PRE, S); Cape [Bellville] D., De Grendel, Tigerberg (**-DC**), Esterhuysen 23077 (BOL); Paarl D., Joostenberg's Kloof (**-DD**), Esterhuysen 33882 (BOL, WU). – **3319** (Worcester): between Tulbagh and "The Drostdy" (**-AC**), Burchell 1028 (K); Worcester D., Chavonnes (**-CB**), van Breda 256 (PRE). – **3418** (Simonstown): Faure (**-BB**), Strey 846 (K). – Imprecise localities or no locality given: "Hopefield- Saldhana Bay- Darling trip", Bolus 12764 (BOL); "CBS", Alexander s.n. (K), Harvey s.n. (BM); Jordaan s.n. sub STE 30970 (STE). – Additional collections: Acocks 2641 (S); Barker 2540 (NBG); Bolus 4998 (BOL); Compton 6778, 13425, 18119, 20065 (NBG); Drège s.n. (FI; S, as "64.9"); Duthie 18430 (BOL); Ecklon & Zeyher 2318 (FI, G, GOET, M, MO, NBG, S, SAM, W), "1-11" (S), "62.9" (E × 2, GH, GRA, GOET, PRE, US, WU); Harvey 3105 (GH); Leighton 1759 (BOL); Liebenberg 4204 (STE); Niven 127 (BM); Pappe s.n. (S); Puff 800910-2/1 (WU); Rogers 17049, 17090 (J); Salter 6407, 6452 (BOL); Schlechter 1050 (G, GRA, W × 2); Smith 2900 (PRE); Zeyher s.n. (K; SAM sub 16066).

7 (b). Subsp. *calciphila* Puff in Fl. S. Afr. **31**, 1 (2): 44 (1986).

Type: South Africa, Cape Prov., Langebaan Peninsula, Oude Post Private Nature Reserve, Boucher 2964 (STE, holo.!; PRE, iso.!).

Much more robust dwarf shrubs than subsp. *hirta*, ca 30–100 cm tall, intricately branched; branches often spine-tipped. Leaves much less crowded than in subsp. *hirta*, ± widely spaced, less distinctly pseudo-verticillate, mostly decussate (but some leaves occasionally in whorls of 3); blades to 3.5 mm long. Flowers unknown. Fruits with somewhat shorter hairs, larger, ca 3–3.7 × 2.7 mm, not as crowded and numerous as in subsp. *hirta.*

Chromosome Number and *Average Pollen Diameter:* Unknown.

Habitat: Apparently confined to exposed limestone ridges and sandy limestone soils. In West Coast Strandveld [ACOCKS (1975) Veld Type No. 34] and characteristic for *Nenax-Maytenus-Zygophyllum* Limestone Evergreen Shrubland (see BOUCHER & JARMAN 1977 for a detailed description of this plant community in the Langebaan area). Often occurring together with *Aloe distans* which has a distribution range similar to that of subsp. *calciphila* (cf. VENTER & BEUKES 1982: map 1). – Ca 25–50 m.

*Distribution (*Map, Fig. 100 g): S Africa, W part of the SW Cape Prov. From the Langebaan Peninsula S to Yzerfontein (Hopefield and NW Malmesbury Distr.).

Critical Remarks: Although all the available specimens were collected in the dry season and bear no flowers, their fruits, habit and vegetative morphological features are sufficiently distinct to allow their separation as (at least) a subspecies. Subsp. *calciphila* appears to be restricted to and may have evolved in the relatively "young" and rather narrowly confined limestone areas of the W Cape.

The Langebaan specimens originated from areas with heavily trampled and grazed vegetation, but browsing alone is not likely to be responsible for the characteristic growth form of subsp. *calciphila,* as obviously grazed specimens of subsp. *hirta* differ markedly in habit.

Flowering specimens collected after the rains would be needed for comparison with subsp. *hirta.*

Collections

South Africa. Cape – **3317** (Saldanha): Langebaan Peninsula, above Juttenbaai [just NW of Postberg (= Oupos) Private Nature Res.] (**-BB**), PUFF 791226-2/1 (WU). – **3318** (Cape Town): Oude Post Private Nature Reserve (**-AA**), BOUCHER 2694 (PRE, STE); Malmesbury D., Yzerfontein (= Ysterfontein) (**-AC**), ACOCKS 14514 (PRE).

8. *Nenax acerosa* GAERTN., Fruct. **1**: 165, t. 32, f. 7 (1788); SALTER in J. S. Afr. Bot. **3**: 112 (1937), in ADAMSON & SALTER, Fl. Cape Penins.: 734 (1950); PUFF in Fl. S. Afr. **31**, 1 (2): 46 (1986).
   Type: [South Africa, SW Cape Prov.] "CBS", MASSON s.n. in herb. BANKS (BM, holo.!).

– *Cliffortia acerosa* MS. [in herb. BANKS (BM!)].

= *Ambraria glabra* CRUSE, Rub. Cap.: 17 (1825), in Linnaea **6**: 18 (1831); SONDER in HARVEY & SONDER, Fl. Cap. **3**: 33 (1865).
≡ *Nenax glabra* (CRUSE) O. KUNTZE, Rev. Gen. **3**: 121 (1898).
Type: [South Africa, SW Cape Prov.] "CBS", BERGIUS s.n. (B, holo. †).

= *Ambraria glabra* CRUSE var. *papillata* SONDER in HARVEY & SONDER, Fl. Cap. **3**: 34 (1865).
Type: none cited, but almost certainly "Capfläche", 1841, ECKLON s.n. (S, holo.!) [label bearing SONDER's handwriting].

= *Ambraria glabra* CRUSE var. *tulbaghica* SONDER in HARVEY & SONDER, Fl. Cap. **3**: 34 (1865).
Type: South Africa, Cape Prov., (near) waterfall, Tulbagh, no collector cited but without doubt PAPPE s.n. (S, holo.!; K, iso.!) [label bearing SONDER's handwriting].

– *Ambraria acerosa* SONDER in HARVEY & SONDER, Fl. Cap. **3**: 34 (1865), *pro parte* (♀; see below), excl. type.

Critical Remarks: SALTER (1937) clarified that *Ambraria glabra* CRUSE is synonymous with *Nenax acerosa* GAERTN. and pointed out that *Ambraria acerosa* SONDER (= "2319. *Nenax acerosa*": ECKLON & ZEYHER 1836: 368) is a very different species from *Nenax acerosa* GAERTN. He thus proposed the name *N. divaricata* (5.) in place of *Ambraria acerosa* SONDER. While this is correct if one looks at the numerous duplicates of ECKLON & ZEYHER 2319, mostly bearing as labels cut-outs from their "Enumeratio …" (ECKLON & ZEYHER 1836) or printed labels "*Nenax acerosa* 1.9" (1.9 = locality number taken from DRÈGE's "Standörterverzeichniss …", 1847 a), the situation becomes more complicated if the specimens studied by SONDER (now in herbarium S) are reinvestigated. One sheet with ♀ specimens, bearing the label "*Ambraria acerosa* SOND. *Nenax acerosa*, 77-9, E & Z, Tulbaghskloof, Worcester" in SONDER's handwriting and with a pouch containing entire, indehiscent and sectioned fruits and drawings of the "4-locular" fruits, is *without doubt Nenax acerosa* GAERTN. To a ♂ specimen, bearing the label "*Anthospermum* n.sp. … bei Brakfontein (Clanwilliam), Juni" in ECKLON's handwriting, SONDER added "*Ambraria acerosa* SOND., *Nenax acerosa* E & Z"; this specimen is *Nenax divaricata*. There can be no doubt that SONDER (1865) based the description of *Ambraria acerosa* SOND. in part on this specimen ("male flowers tetrandrous, exactly as in *Anthospermum*"). The description of *Ambraria acerosa* SOND. thus is based on *Nenax acerosa* GAERTN. (♀) and *Nenax divaricata* (♂).

Why SONDER (1865) did not consider the above mentioned ♀ specimen ("... 77-9, E & Z, ..."; S) to be identical with his *Ambraria glabra* var. *tulbaghica,* the type of which came from the same area (near waterfall, Tulbagh), will remain a mystery. His descriptions of the fruits of both *Ambraria glabra* var. *tulbaghica* and *Ambraria acerosa* are virtually identical ("fr. 3- or 4-locular, 2.5 lines long, obsoletely angular, crowned by the short acute or obtuse calycine teeth" versus "Fr. 2 lines long, obsoletely angular, 4-locular, terminated by the nearly obsolete calyx teeth").

Two subspecies are recognized:

8. (a) Subsp. *acerosa*

Synonyms as above.

Dioecious, many-stemmed, erect to ± diffuse dwarf shrubs or, if burnt regularly, subshrubs with often massive, woody bases and numerous, thin aerial stems. Stems ca (10) 15–30 (40) cm long, to ca 3 mm in diameter at the base and ca 1–2 mm in the midstem region, unbranched to ± much-branched, branches ascending to erect; younger parts often ± 4-angled, glabrous to ± densely covered with whitish papillae or short hairs, reddish(-brown), shiny; older parts glabrescent, reddish-brown to greyish, epidermis sometimes peeling; internodes mostly shorter than to as long as the leaves. Leaves decussate, pseudo-verticillate; short shoot leaves sometimes slightly longer and narrower than the long shoot leaves; blades (5) 8–12 (15) × 0.5–1 mm, linear (sometimes linear-lanceolate on long shoots), ± needle-like and ± triangular in section or, if broader, shallowly concave above, glabrous or margins with whitish papillae or hairs less than 0.1 to 0.2 (0.3) mm long, upper surfaces and midrib below often dark brown; apices acute to ± mucronate; petioles 0; stipular sheaths cup-shaped, 1.2–1.6 mm long, with or without 1–2 minute, gland-tipped setae. Flowers subsessile, in clusters of 6-2 at the nodes, ♂, ♀; inflorescences spike-like, sometimes ± conspicuous (♀, especially when in fruit). Corolla 5-merous (occasionally 4-merous in ♀, very rarely in ♂), greenish-yellow to creamy yellow, often reddish-purple tinged, occasionally with a few whitish hairs near the tip. ♂: tube 0.7–1.5 mm long, broadly funnel-shaped, lobes 2.4–3.7 × 0.8–1 mm, ± lanceolate, recurved; stamens 5, filaments 1.7–2.2 mm long, anthers 1.4–2 × 0.5–0.7 mm; rudimentary ovary minute but well discernible, sometimes crowned by calyx lobes.

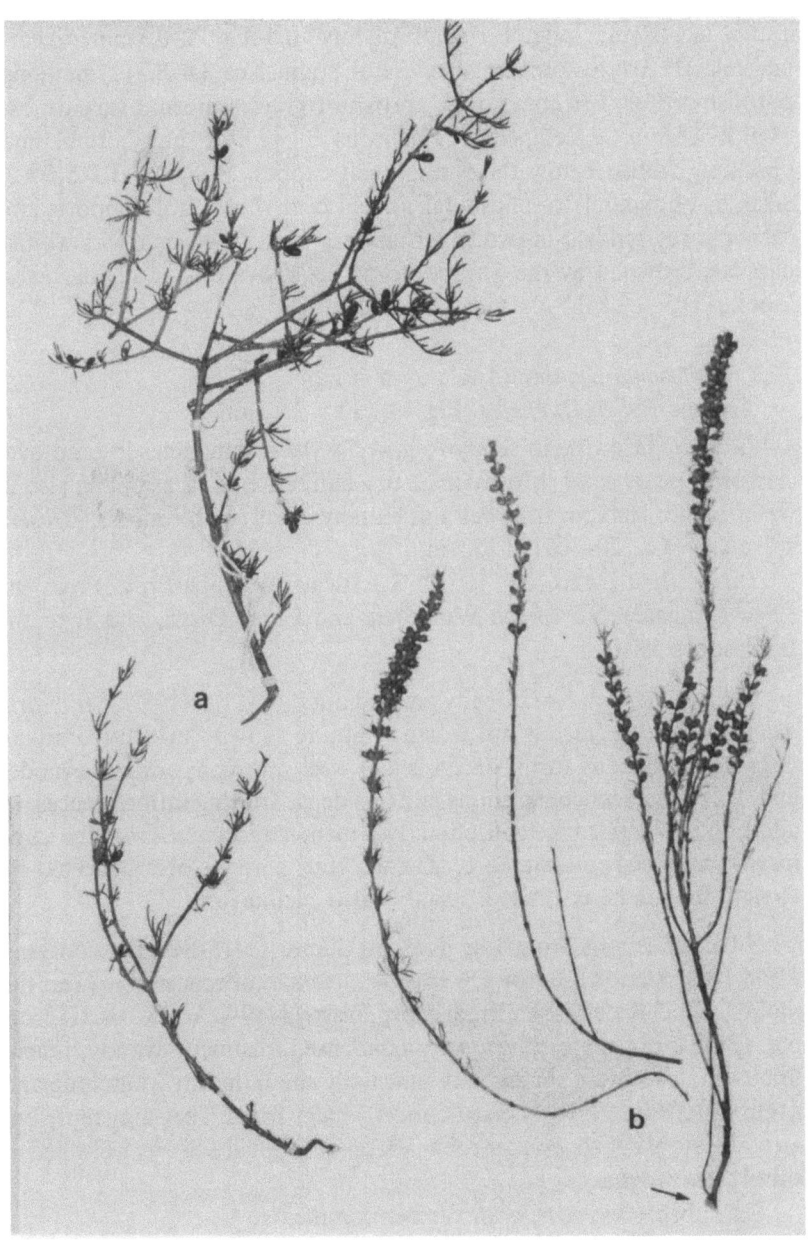

Fig. 101. *Nenax. a N. arenicola* (Cape Prov., ca 3 km SE of Graafwater- Lambert's Bay road, on road to Leipoldtville; PUFF 800915-2/1), note widely spaced fruits, ± irregular branching and some spine-tipped branches (browsed!); *b N. acerosa* subsp. *acerosa* (Cape Prov., ca 1.5–2 km NE of Darling Bridge; PUFF 800920-3/1), note ± many-fruited, spike-like inflorescences; the arrow points to the apex of the massive, ± disk-like woody base. – Further explanations in the text

♀: tube 0.3–0.7 mm long, ± cylindrical, lobes 0.5–1 × 0.2–0.3 mm, (linear-) lanceolate, erect to spreading; style ± 0; stigmas ca 3.4–6.3 (7) mm long, purplish-red or, less commonly, greyish or greyish-green; ovary ca 1.4–2 × 0.9–1.5 mm, ± ellipsoidal, glabrous or a little hairy, sometimes obscurely ribbed below the 5 small calyx lobes. Fruits ca 3.2–5.5 × 2–3.4 mm, obovoidal to ellipsoidal, indehiscent, hard, not supported by carpophores, reddish-brown, glabrous or ± sparsely covered with whitish papillae, crowned by the 5 trapeziform, ± obovate to triangular calyx lobes ca 0.5–0.9 × 0.7–0.8 mm. – Fig. 101 b.

Chromosome Number (Table 3): n = 22.

Average Pollen Diameter (Fig. 49): 30.1–35.7 μm.

Habitat: Growing in sandy to gravelly flats; sometimes in sand over clay or in peaty sand. Mostly in ± dry habitats but occasionally also in water-logged, temporarily wet and marshy areas (with restiads, *Drosera* sp., etc.). – Ca. 20–300 (? 1 150) m.

Distribution (Map, Fig. 100 d): S Africa, SW Cape Prov. From the Cape Peninsula NE to the Worcester and Ceres Distr., and E to the Bredasdorp Distr.

Critical Remarks: As already pointed out by SALTER (1937), [*Ambraria glabra*] var. *tulbaghica* is not worth retaining as it is "evidently only an ungrazed luxurious form" (SALTER l.c) with unusually long internodes and, therefore, less conspicuous and less dense fruiting inflorescences. In addition to PAPPE's type collection, two further specimens from the same locality in SAM, attributed to ZEYHER (but perhaps also collected by PAPPE?) belong here; LEVYNS 10800 is also such a form.

Two collections from [the flats at] Faure [3418-BB; ESTERHUYSEN 11932 (BOL), and COMPTON 15998 (NBG)] and a collection from [near the mouth of the] Ratel River, Bredasdorp Distr. [3419-DA?; BOLUS 21875 ex pte. (BOL); not mapped] are very condensed, distinctly woody, much-branched, low dwarf shrubs with unusually small (hardly 3 mm long and 2 mm wide) but seemingly mature shortly hairy fruits. They may represent an odd state of subsp. *acerosa* (or a new taxon?) but are clearly not allied to subsp. *macrocarpa*.

See subsp. *macrocarpa* for further comments.

Collections

South Africa. Cape – **3318** (Cape Town): Rondebosch Common (-**CD**), ESTERHUYSEN 26583 (BOL, WU); ca 22 km from Cape Town on Koeberg rd. (-**DC**), BOLUS 21813 (BOL). – **3319** (Worcester): Ceres D., Gydo (-**AB**), COMPTON 18716 (NBG); Tulbagh (Tulbaghskloof) (-**AC**), ECKLON & ZEYHER '77-9' (S), PAPPE s.n.

(K, S), Zeyher s.n. sub SAM 16065 (SAM × 2); (nr.) Darling Bridge (over Breerivier) (-CA), Compton 9909 (NBG), Esterhuysen 3786 (BOL), Puff 800920-3/1 (WU); Worcester (-CB), van Breda 169 (BOL, PRE); Robertson D., Poesjenels River (-DC), Levyns 10800 (BOL). – **3418** (Simonstown): Constantia, Bergvliet Farm (-AB), Purcell s.n. sub SAM 90340- 90342 (SAM); Cape of Good Hope Nature Reserve, Platboom rd., nr. windmill (-AD), Puff 800907-3/1 (WU), Taylor 6656 (K, PRE, STE); Stellenbosch D., Van der Stel (-BB), Smith 5146 (PRE). – **3419** (Caledon): Caledon D., nr. Elandskloof (-BD), Schlechter 9747 (BM, BOL – mixed with *Anthospermum hirtum*, BR, E, G, K, MO, PRE × 2, S, US, W); –, Danger Point, (-CB), Compton 10261 (NBG); Bredasdorp D., Elim, sandveld tow. Agulhas (-DB), Marloth 10008 (PRE). – **3420** (Bredasdorp): Potberg range, Hamerkop (-BC), Levyns 8389 (BOL); nr. Cape Agulhas (-CC), Esterhuysen 4420 (BOL). – Imprecise localities ('CBS' or 'Cape Flats'): Bunbury 114 (BM); Carmichael 238, 239 (BM); Ecklon 187, 188 (S); Harvey 425 (BM, E, K), 455 (E, K); Wallich 361 (BM), s.n. (BM, G, K, S). – Additional collections: Acocks 2051, 3397, 3399 b (S); Bolus 4997 (BOL, NH, PRE); Burchell 185 (K); Compton 6025 (NBG); Ecklon & Zeyher 2317 (M, S × 2, SAM, W); Esterhuysen 32348 (BOL, M, WU), 33327 (BOL, WU), s.n. (BOL); Levyns s.n. sub Salter 5664 (K); Hafström & Acocks 1432 (PRE, S); Puff 800926-1/2 (WU); Salter 5655 (BM, BOL, K, SAM), 5669 (K), 5689 (BOL, SAM), 6376 (BM, BOL, K).

8 (b). Subsp. ***macrocarpa*** (Ecklon & Zeyher) Puff in Fl. S. Afr. **31**, 1 (2): 46 (1986).

= *Ambraria hirta* Cruse var. *macrocarpa* Ecklon & Zeyher, Enum. Pl. Afr. Austr.: 368 (1836).

Type: South Africa, Cape Prov., on the Breederivier, Swellendam, Mund(t) s.n. sub Ecklon & Zeyher 2318 β (SAM, holo.!; S, iso.!).

Dioecious, many-stemmed, often ± rounded or low, ± cushion-forming dwarf shrubs ca 10–50 cm tall and to ca 50 cm in diameter; often with thick, woody roots. Stems ca 20–50 cm long, to ca 5 (8) mm in diameter at the base and ca 2–3 mm in the midstem region, ascending to ± prostrate or erect, mostly much-branched; younger parts obscurely 4-angled to ± terete, ± densely covered with whitish papillae to 0.1 mm long, mostly greyish or greyish-brown, not distinctly reddish; older parts glabrescent, grey to blackish, epidermis often peeling and bark with long fissures; internodes often as long as or shorter than the leaves. Leaves decussate, pseudoverticillate, often distinctly dimorphic; blades 2–6 × 0.4–1 mm, narrowly ovate-lanceolate and oblong to ± elliptic in section on old long shoots, linear-lanceolate to linear, needle-like, relatively longer and ± round in section on short shoots and/or new growth, glabrous except for a few papillae to ca 0.1 mm long on the margins; apices ± acute; petioles 0; stipular sheaths cup-shaped, ca 0.3–1 mm long, occasionally with a minute, gland-tipped median seta. Flowers subsessile, usually paired (seldom in clusters of 4–6) at the nodes, ♂, ♀; inflorescences sometimes ± conspicuous in ♀. Corolla 4–5-merous (in ♀ more commonly 4-merous), yellowish to greenish-yellow, often reddish-purplish tinged, glabrous. ♂: tube 0.7–1.2 mm long, broadly funnel-shaped, lobes 2.1–3 × 0.8–1 mm, ± lanceolate, recurved; stamens 4–5, filaments 1.4–1.8 mm long, anthers 1.7–2.1 × 0.5–0.7 mm; rudimentary ovary minute but well discernible, often crowned by conspicuous calyx lobes. ♀: tube ± 0 or 0.2–0.3 mm long, cylindrical, lobes 0.5–1 × 0.2–0.3 mm, linear-lanceolate, ± erect; style 0; stigmas 3–5 mm long, purplish-red or, less commonly, greenish-white; ovary 1.2–1.6 × 1–1.4 mm, shortly hairy, crowned by 5 triangular or rounded calyx lobes. Fruits ca 3–5 × 2.8–3.5 mm, ellipsoidal, obovoidal to ± subglobose, indehiscent, hard, not supported by carpophores, greyish or greyish-brown, densely covered with whitish spreading hairs to ca 0.2 mm long (if ± glabrescent, fruits ± subglobose), crowned by 5 ± erect to spreading, triangular to ± rounded calyx lobes, ca 0.5–1 × 0.4–0.7 mm. – Figs. 24 c–d, 102.

Chromosome Number (Table 3): n = (ca) 22.
Average Pollen Diameter (Fig. 49): 33–35.7 μm.
Habitat: Hillsides, sandy flats; in Coastal Fynbos over sand. Also recorded from limestone areas (dune limestone, limestone banks). – Ca 20–300 m.
Distribution (Map, Fig. 100 e): S Africa, SW Cape Prov. From the Caledon Distr. N to the S Clanwilliam Distr. and E to the Riversdale Distr.

Critical Remarks: Although subsp. *macrocarpa* has fruits which, at least in their indumentum, are similar to *N. hirta* (subsp. *hirta*) (7., above), it is grouped with *N. acerosa* rather than *N. hirta* (with which it was associated previously – "*Ambraria hirta* var. *macrocarpa*"). The latter is diploid, while all other *Nenax* taxa with indehiscent fruits are tetraploid like "*macrocarpa*".

"Typical" subsp. *acerosa* and "typical" subsp. *macrocarpa* are quite strikingly different in both habit and appearance. Subsp. *macrocarpa* is much more woody and robust, it does not have the characteristic reddish stems of subsp. *acerosa,* is less densely leafy and shows an often rather conspicuous leaf dimorphism. It, furthermore, has less conspicuous inflorescences with mainly paired flowers (versus the more often more-flowered and more conspicuous inflorescences, especially in fruit, of subsp. *acerosa*) and shortly hairy fruits (typical subsp. *acerosa* has glabrous or nearly glabrous fruits). A number of specimens are, however, ± intermediate in their character set, so that it, at present, seems most feasible to assign subspecific rank to the two taxa.

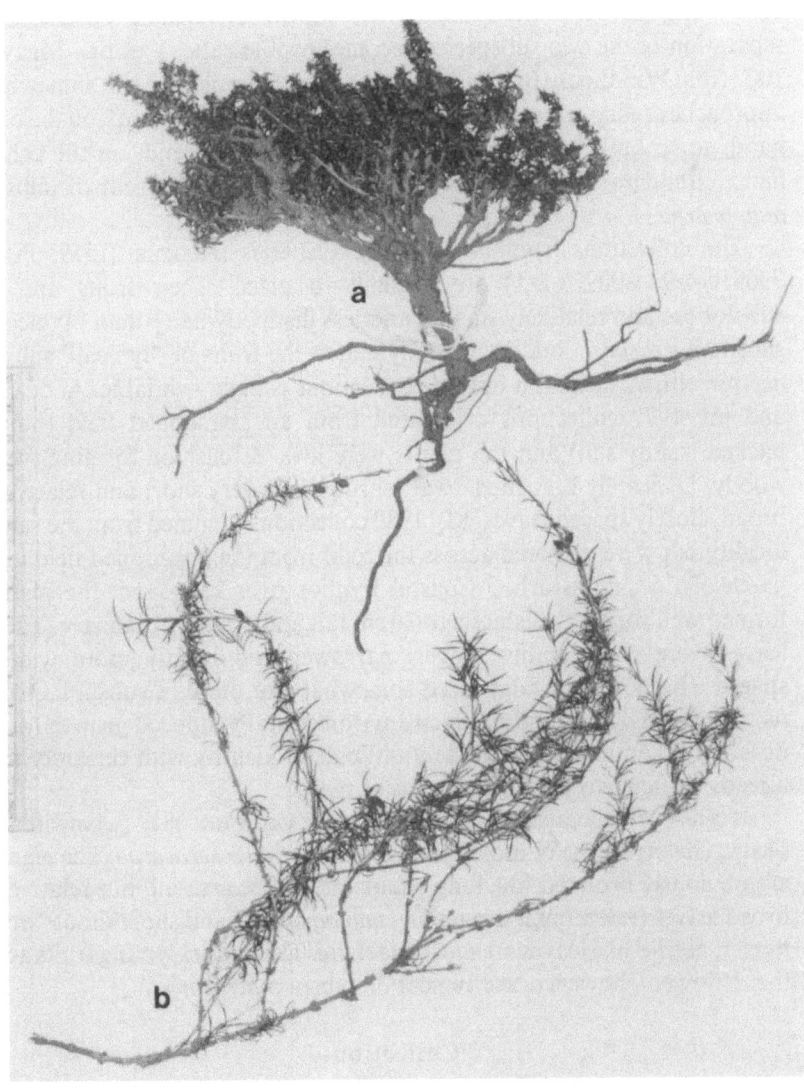

Fig. 102. *Nenax acerosa* subsp. *macrocarpa* from NW of The Fisheries (Cape Prov.). *a* plant from an abandoned field growing in hard-packed sand (trampled by domestic stock?; Puff 790910-4/2); *b* plant from a patch of undisturbed Coastal Fynbos over kalksand across the road from the abandoned field (Puff 800927-2/1). – Further explanations in the text

It appears that, especially in the SE-most part of the distribution range of *N. acerosa* (E Bredasdorp Distr. to the Riversdale Distr.), the separation of the two subspecies becomes problematic. VAN DER MERWE 1007 from Windhoek (Bredasdorp Distr.), for example, in habit somewhat approaches subsp. *acerosa,* and the herbarium sheet consists of, on the one hand, some branches bearing almost glabrous fruits and, on the other hand, branches with very shortly hairy fruits more typical of subsp. *macrocarpa.*

The collections from NW of The Fisheries (ACOCKS 21559; PUFF 790910-4/2, 800927-2/1) are difficult to place. Their fruits are ± subglobose and relatively smaller and less distinctly hairy than "typical" subsp. *macrocarpa,* but do not fully match the fruits of "typical" subsp. *acerosa* either. The habit of these collections is highly variable. ACOCKS's and my 1979 collections originated from an abandoned field (hard-packed, sandy soil) and the plants were low, ± cushion-forming, very woody, intricately branched dwarf shrubs with very short and relatively broad, closely spaced leaves. My 1980 collections stemmed from the same locality but were gathered across the road from the abandoned field in a patch of fine, undisturbed Coastal Fynbos over kalk-sand; the plants formed well rounded cushions to 30 cm tall and 50 cm in diameter. Their leaves were considerably longer, narrower and mostly more widely spaced – branches thus appeared somewhat like those of subsp. *acerosa* (compare Figs. 102 a and b). Such environmentally induced growth form differences can make it extremely difficult to identify with certainty the already difficult to distinguish taxa of *Nenax.*

It is also with some hesitation that I place COMPTON 6819 (Clanwilliam Distr., Grootvlei; the N-most locality) with subsp. *macrocarpa.* The plants are obviously browsed; the long shoots mostly bear small, but relatively broad leaves (resembling those of *N. namaquensis*), and short shoots with longer, needle-like leaves are mostly lacking. The ovaries, young fruits and the ♂ flowers, however, are typical of subsp. *macrocarpa.*

Collections

South Africa. Cape – **3218** (Clanwilliam): Clanwilliam D., Grootvlei [N of Citrusdal, on old Citrusdal- Clanwilliam rd.?] (**-DB** or 3219-AC?), COMPTON 6819 (NBG × 2). – **3319** (Worcester): Worcester- Ceres rd., nr. (NW of) Goudiniweg Stn. (**-CB**), PUFF 800920-2/2 (WU); Boschjesveld nr. Mordkuil, on the Doornrivier (**-CD**), DRÈGE s.n. (E, G, K, W). – **3419** (Caledon): Caledon D., Kleinriviersberg (**-AD**, **-BC**), ECKLON & ZEYHER '58-8' (S × 2, third sheet: *Anthospermum spathulatum;* as '85.10' in E, PRE; as 2307γ in LY, MO, PRE – p.p. mixed with *A. spathulatum*); Robertson D., Bushmans River (**-BB**), COMPTON 11898 (NBG). – **3420** (Bredasdorp): on the Breederivier (Swellendam) (**-AB**?), MUND(T) s.n. sub ECKLON & ZEYHER 2318 β (S, SAM); Windhoek plateau (**-AD**), VAN DER MERWE 1007 (PRE; p.p. ± atypical). – **3421** (Riversdale): Riversdale D., ca 25 km

S of Riversdale (-AC), Acocks 23998 (PRE); −, ca 3.5 km NW of The Fisheries
(= Gouritsmond) (-BD), Acocks 21559 (K, PRE), Puff 790910-4/2, 800927-2/1
(WU).

9. Nenax arenicola Puff in Fl. S. Afr. **31**, 1 (2): 47 (1986).
   Type: South Africa, Cape Prov., ca 3 km SE of Graafwater- Lambert's
   Bay rd., on Leipoldtville rd., Puff 800915-2/1 (WU, holo.!; BOL,
   NBG, PRE, iso.!).

Dioecious, many-stemmed, intricately branched, often rounded dwarf
shrubs, ca 30–50 cm in diameter and ca 20–75 (100) cm tall, frequently
with thick, woody roots. Stems ca 0.9–2 mm in diameter in the midstem
region, ascending to erect, ± much-branched; branches typically paired at
the nodes, spreading to ascending, branching very irregular if browsed;
younger parts papillate to glabrescent, greyish or sometimes reddish-brown tinged; older parts glabrous, greyish-brown, grey or dark
grey; internodes mostly longer than the leaves, branches not densely
leafy. Leaves decussate, sometimes pseudo-verticillate; blades (5) 7–
13 (16) × 0.5–1 mm, linear, often shallowly concave above and convex
below or ± semiterete in section, glabrous or with a few papillae on the
margins; apices ± acute; petioles 0; stipular sheaths much reduced,
shallowly cup-shaped, to ca 1 mm long, mostly with 1 (–3) gland-tipped
setae, ca 0.1–0.2 mm long. Flowers subsessile, paired or, more often, single
at the nodes, widely spaced; ♂, ♀. Corolla 4(5)-merous, greenish-yellow,
sometimes purplish tinged. ♂: tube 0.4–0.8 mm long, broadly funnel-
shaped, lobes 2–2.9 × 0.6–0.8 mm, ± lanceolate, recurved; stamens 4 (5),
filaments 1.4–1.8 mm long, anthers 1.6–1.9 × 0.5–0.7 mm; rudimentary
ovary to 0.3 mm long, often crowned by ± conspicuous calyx lobes. ♀:
tube ± 0 or 0.1–0.3 mm long, lobes 0.5–1 (1.3) × 0.1–0.3 mm, linear-
lanceolate, erect to ± spreading; style 0; stigmas 3–5.4 mm long, purplish-
red or creamy-grey tinged purplish; ovary ca 2.2–2.9 × 1–1.4 mm, ellipsoidal, bright green, glabrous, crowned by 4–5 rounded calyx lobes. Fruits
5–8 × 2–3.5 mm, ellipsoidal, indehiscent, hard, not supported by carpophores, reddish-brown to dark grey, glabrous, crowned by 4–5 small,
often ± indistinct, rounded calyx lobes, to ca 0.4−0.5 mm long.
− Figs. 33 d−f, 38 c, 101 a.

Chromosome Number (Table 3): n = 22 (Fig. 47 e), 2 n = ca 44.
Average Pollen Diameter (Fig. 49): 33.7–34.3 µm.
Habitat: In sandy coastal plains and occasionally in the adjacent
foothills in heavier, stony soil, i.e., in West Coast Strandveld and
Succulent Karoo of the Namaqualand Coast Belt [Acocks (1975) Veld

28*

Type Nos. 34 and 31 a]; rainfall ca 50–300 mm per annum, in winter or throughout the year. – Ca 30–500 (650) m.

Distribution (Map, Fig. 99 d): S Africa, W Cape Prov. Occurring from the SW Namaqualand Distr. S to the W Clanwilliam Distr.

Critical Remarks: Although clearly allied to *N. acerosa* (8., above) (hard, dehiscent fruits; tetraploidy), *N. arenicola* is a well and easily distinguished species which is characterized by a less leafy appearance, a different habit (much more woody, intricately branched dwarf shrubs; cf. Fig. 101 a), much more reduced inflorecences with often only solitary flowers at the nodes, a more NW distribution range and drier, sandy W Coast Strandveld habitats.

N. arenicola is often browsed by domestic animals and – as in other *Nenax* species – browsing results in a much more irregular branching pattern (Fig. 101 a) and more compact, denser plants which, in habit, may differ considerably from unbrowsed specimens.

Collections

South Africa. Cape – 3017 (Hondeklipbaai): ca 8 km E of Hondeklipbaai, on Garies rd. (**-AD**), Puff 790714-3/1 (J, WU); Farm Naries, ca 38 km SW of Kamieskroon (**-BC**), van der Westhuizen 114 (PRE). – **3118** (Vanrhynsdorp): Karee Bergen (**-AB**), Schlechter 8295 (BM, BOL, G, GRA, K, PRE sub TVL Museum no. 11279); Vanrhynsdorp D., ca 35 km S of Vredendal (**-CD**), Gentry, Barclay & van Breda 18762 (PRE, US); – , ca 16 km SSW of Vanrhynsdorp, base of Giftberg (**-DC**), Acocks 19635 (K, M, PRE, SRGH). – **3218** (Clanwilliam): ca 3 km SE of Graafwater- Lambert's Bay rd., on Leipoldtville rd. (**-BA**), Puff 800915-2/1 (BOL, NBG, PRE, WU). – Imprecise locality: "CBS", Masson s.n. (BM). – Additional collections: Acocks 14936 (K, PRE); Leipoldt 3669 (BOL, NBG, PRE, SAM); Pillans 18035 (BOL); Puff 790714-4/1 (WU); Rourke 1585 (NBG, PRE); van Breda 4075 (PRE).

10. *Nenax* sp. A

South Africa, SW Cape Prov., Laingsburg Distr., [Farm] Cabidu [N of Constabel, 3320-AB], Compton 22209 (NBG).

Dwarf shrub with greyish stems, ± much-branched; branches mostly paired at the nodes, ± ascending. Leaves decussate, widely spaced, dimorphic; blades of long shoot leaves ca 2–3 × 1 mm, ovate-lanceolate to linear-lanceolate, those of short shoot leaves to 6 (8) mm long and narrower. Flowers unknown. Fruits indehiscent, ± soft, easily squashed between two fingers, not supported by carpophores, ca 3.5–4.5 mm in diameter, subglobose, greyish to greyish-brown, shortly hairy, crowned by minute to subobsolete calyx lobes.

The fruits are not as hard as those of other *Nenax* species with indehiscent fruits (species 7.–10.). They lack the thick, continuous ring of sclerenchyma which is characteristic of the indehiscent fruits of *N. acerosa* and allies (cf. Figs. 41 c and g).

It is perhaps closest to *N. acerosa* subsp. *macrocarpa* (8.). As the (♀) fruiting collection cited above is the only one known it is probably best to wait until more material comes to light before describing it as a new species.

11. *Nenax* sp. B

South Africa, SW Cape Prov., Zonder Einde Mts. (= Riviersonderend Mts.), Boesmanskloof [3419-BA], LEVYNS 9200 (BOL).

Dwarf shrubs with reddish stems, ± much-branched; branches mostly paired at the nodes, ± spreading. Leaves decussate, pseudo-verticillate; blades ca 4–6 × 0.5 mm, ± needle-like, glabrous, shiny. Flowers unknown. Fruits indehiscent, hard, not supported by carpophores, ca 2–2.5 mm in diameter, subglobose, shiny red or reddish-brown, obscurely ribbed, crowned by minute calyx lobes.

Probably allied to *N. acerosa* (8.). Most likely a good new species but so far only known from the (♀) fruiting collection cited above.

Galopina THUNB., Nov. Gen. Pl. **1**: 3 (1781); CRUSE, Rub. Cap.: 18 (1825); SONDER in HARVEY & SONDER, Fl. Cap. **3**: 26 (1865); HOOK. f. in BENTH. & HOOK. f., Gen. Pl. **2** (1): 139 (1873); K. SCHUM. in ENGLER & PRANTL, Nat. Pflanzenfam. **IV, 4**: 128 (1891); DYER, Gen. S. Afr. Flow. Pl. **1**: 622 (1975); COMPTON, Fl. Swaziland: 586 (1976); PUFF in Fl. S. Afr. **31**, 1 (2): 48 (1986).
Type species: *G. circaeoides* THUNB., Nov. Gen. Pl. **1**: 3 (1781).

= *Oxyspermum* ECKLON & ZEYHER, Enum. Pl. Afr. Austr.: 365 (1836). Type species: *O. asperum* ECKLON & ZEYHER, Enum. Pl. Afr. Austr.: 365 (1836) [≡ *Galopina aspera* (ECKLON & ZEYHER) WALP.].

– *Phyllis* sensu CRUSE in Linnaea **6**: 19 (1831), non L.

☿, ☿ + ♀, ♀, or occasionally ♂, ☿ + ♂, ♂ + ☿ + ♀ or ♂ + ♀ perennial herbs with branched, often ± woody rhizomes or rootstocks. Leaves

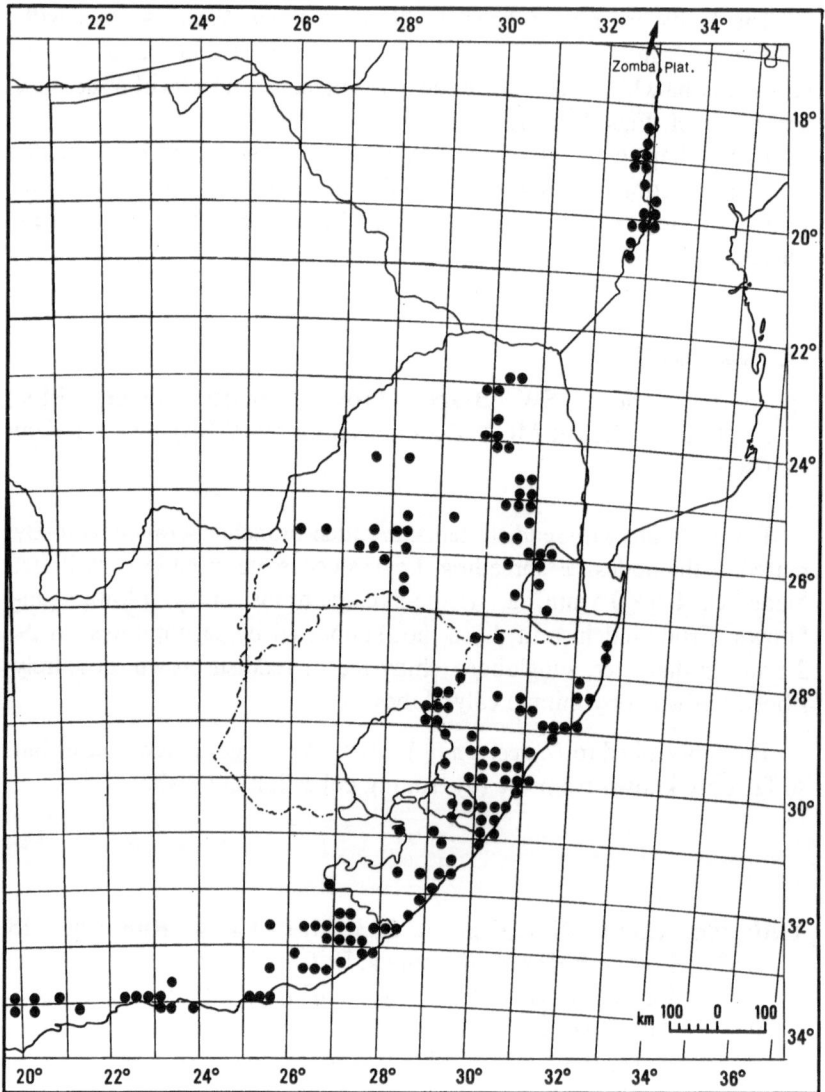

Fig. 103. Distribution of *Galopina* (all species)

decussate, blades broadly ovate to lanceolate, relatively large, membranaceous, with ± prominent lateral veins, distinctly petiolate and with
stipular sheaths bearing 3–5(–7) setae. Inflorescences terminal, paniculate
to thyrso-paniculate, bracteate, lax and ± spheroidal to narrowly
cylindrical. Flowers with distinct, (±) filiform peduncles and pedicels; ♂,
⚥ or ♀; calyx 0; corolla 4(5)-merous. ♂, ⚥: corolla tubes (very) short,

broadly funnel-shaped to campanulate, lobes ± lanceolate, recurved; anthers yellowish to whitish, exserted, dangling on long slender filiform filaments. ♀: corollas much smaller, tubes sometimes ± 0, lobes ± linear, erect to spreading; ⚥, ♀: style 0; stigmas 2, long exserted, in ⚥ often shorter and thinner than in ♀, hairy, greyish-white, yellowish-grey or greenish; ovary bicarpellate and biovulate. Fruits dehiscing into two mericarps, not supported by carpophores.

Chromosome Number (Table 3) x = n = 11, 2 n = 22.
Average Pollen Diameter (Fig. 49): 23.3–26.2 μm.
Distribution (Maps, Figs. 1 b and 103): Centered in SE Africa and occurring from the E Transvaal through Swaziland, Natal and Transkei to the E Cape Prov. One widely distributed species extends N into Zimbabwe and S Malawi (Zomba Plateau) and, in S Africa, into the SW Cape Prov.

Key to Species

1. Ovaries and fruits villous; stems usually without short shoots .... **4. G. crocyllioides**
1*. Ovaries and fruits smooth, wrinkled or tuberculate, without hairs; stems often with leafy short shoots .... 2.

2. Inflorescences ± narrowly cylindrical; peduncles and pedicels not conspicuously divergent in fruit; fruiting pedicels (0.8) 1–2 (3) mm long; short shoot leaves conspicuous, often almost as long as the long shoot leaves; petioles to 2 (3) mm long .... **3. G. aspera**
2*. Inflorescences pyramidal, ellipsoidal or ± spheroidal; peduncles and pedicels divergent in fruit; fruiting pedicels (2) 3–26 (37) mm long; short shoot leaves much smaller than the long shoot leaves; petioles to 14 mm long.. 3.

3. Peduncles and pedicels (2) 3–6 (7) mm long in fruit, ± stiff; inflorescences pyramidal to ellipsoidal; leaves typically ± densely tomentose; blades (27) 35–50 (62) × (15) 20–35 mm, broadly ovate-lanceolate or ovate.. **2. G. tomentosa**
3* Peduncles and pedicels (7) 12–26 (37) mm long in fruit, filiform and slender; inflorescences lax, broadly pyramidal to ± spheroidal; leaves glabrous or sparsely hairy; blades (28) 40–80 (105) × (8) 12–30 (37) mm, ± lanceolate to ovate-lanceolate....... **1. G. circaeoides**

1. _Galopina circaeoides_ THUNB., Nov. Gen. Pl. **1**: 3 (1781); CRUSE, Rub. Cap.: 18 (1825); SONDER in HARVEY & SONDER, Fl. Cap. **3**: 26 (1865); BRENAN in Mem. N.Y. Bot. Gard. **8**: 454 (1954); COMPTON, Fl. Swaziland: 586 (1976); PUFF in Fl. S. Afr. **31**, 1 (2): 48 (1986).
Type: [South Africa, Cape Prov.] in sylvis Hautniquas, Groot Vaders-Bosch, aliisque. THUNBERG in herb. THUNBERG (sheet 23313, UPS, holo.!; BOL, PRE, WU – photo!), in herb. MONTIN (S, iso.!) and in herb. GASSTRÖM (S, iso.!).

= _Anthospermum galopina_ THUNB., Prod. Fl. Cap.: 32 (1794), nom. illeg.
≡ _Phyllis galopina_ CRUSE in Linnaea **6**: 20 (1831), nom. illeg.

= _G. circaeoides_ THUNB. var. _glabra_ O. KUNTZE, Rev. Gen. Pl. **3**: 120 (1898).
Types: [South Africa, Cape Prov.] Swellendam, 300 m, 1984, O. KUNTZE s.n. (K, lecto.!); Caffraria, Perie-Wald, 600 m, O. KUNTZE s.n. (K, syn.!).

= _G. circaeoides_ THUNB. var. _pubescens_ O. KUNTZE, Rev. Gen. Pl. **3**: 120 (1898).
Type: [South Africa, Cape Prov.] Caffraria, Cathcart, 1 400 m, March 2, 1894, O. KUNTZE s.n. (K, holo.!; US sub 554842, iso.!).

♀, ♀ + ♀, ♀, or occasionally ♀ + ♂ or ♂ + ♀ + ♀, several- to many-stemmed perennial herbs with often branched, ± woody rootstocks or rhizomes. Stems (30) 40–120 (150) cm long, ca (1) 1.5–3 (4) mm in diameter at the base and ca (0.8) 1–2.5 (3) mm in the midstem region, ascending to erect or occasionally decumbent and rooting at the nodes; terete to obscurely 4-angled, glabrous or occasionally sparsely covered with whitish spreading hairs to ca 0.2 mm long; internodes mostly longer than or as long as the leaves; nodes sometimes ± swollen; stems often with much-contracted short shoots bearing small leaves. Leaves decussate; blades (28) 40–80 (105) × (8) 12–30 (37) mm, ± lanceolate to ovate-lanceolate, narrowed to the base, glabrous or with some whitish spreading hairs ca 0.1–0.2 (0.3) mm long along the midrib and principal veins and/or scattered on both surfaces; apices acute; petioles (3) 4–14 mm long; stipular sheaths ± cup-shaped, ca 3–4 × (0.8) 1.2–2 mm, glabrous or slightly hairy, with 3 or 5 (–7) gland-tipped setae, fused basally for ca 0.6–1.5 mm, the free portion of the longest (3.5) 4–8 (10.8) mm. Inflorescences terminal, paniculate, broadly pyramidal to ± spheroidal, lax, ca (9) 15–30 × (8) 13–28 cm, in vigorous individuals often additional, smaller inflorescences arising in the upper leaf axils; bracts leaf-like below, minute

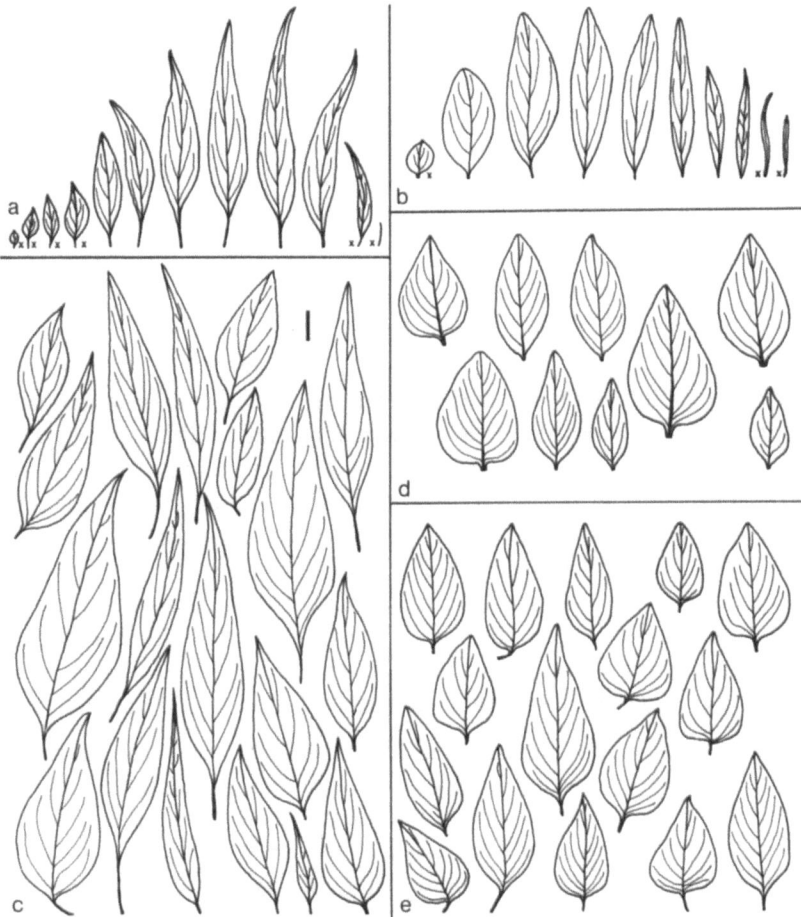

Fig. 104. Leaves of *Galopina*. *a–b* leaf sequence of an individual flowering shoot (lowermost stem leaves on the left, bracts of lowermost inflorescence branches on the right; both marked with ×), *a G. circaeoides* (Puff 790516-1/3), *b G. crocyllioides* (Puff 790416-1/1). *c–e* midstem leaves, *c G. circaeoides, d G. aspera, e G. tomentosa. – a–e* same scale. The scale unit is 1 cm

and linear-lanceolate to subulate above; peduncles and pedicels filiform, slender, glabrous, strongly divergent in fruit, elongating to (7) 12–26 (37) mm. Flowers ♂, ♀̄, ♀; corolla whitish, creamy white, yellowish, greenish and occasionally reddish-purplish tinged outside, glabrous or papillate. ♂, ♀̄: tube 0.3–0.5 (0.6) mm long, broadly funnel-shaped to campanulate, lobes (1.1) 1.3–1.7 (1.9) × (0.3) 0.4–0.6 mm, ± lanceolate, recurved; filaments (0.6) 0.8–1.2 (1.4) mm long, anthers 1.1–1.6 × 0.2–

0.3 mm; ♂: gynoecium as in ♀, but stigmas often shorter and thinner. ♀:
tube ± 0–0.2(0.3) mm long, cylindrical, lobes 0.6–0.8(1) × 0.1–
0.2(0.3) mm, ± linear, erect to spreading; stigmas (2.3) 3–5(6.5) mm long,
whitish to greyish-white; ovary ca 0.5–0.8(1) × (0.5) 0.8–1.3 mm, ±
densely warty. Fruits blackish to black; each mericarp ca 1.5–2 × (0.7) 1–
1.4 mm, oblong to ± obovate, dorsal side convex, ± densely warty,
ventral side ± plane or slightly concave; commissure relatively large, ca
0.9–1.2 × 0.5–0.7 mm. – Figs. 13 a, 15 h, 28 a, 33 h–k, 39 a, 104 a, c.

Chromosome Number (Table 3): n = 11, 2 n = 22 (Fig. 47 b).
Average Pollen Diameter (Fig. 49): 23.3–25.9 µm.
Habitat: Growing on the forest floor in afromontane, kloof or gallery
forests, in scrub along streams, bushclumps around and between rocky
outcrops and in gullies. Sometimes found in plantations (pine, *Eucalyptus*)
and in disturbed sites in indigenous forest (roadsides, banks, etc.). Mostly
in (wet to) moist or damp places, in sandy or clayey soil, in leaf-mould or
humus. Usually found growing in deep shade or in places which are in
shade for most of the day. – Ca 25–1 700 (1 800) m.
Distribution (Map, Fig. 105): In S Africa, from the Transvaal through
Natal and the Transkei to the SW Cape Prov. (see also Critical Remarks,
below); also in Swaziland, in the E Highlands of Zimbabwe and
neighbouring parts of Mozambique and in S Malawi (Zomba Plateau).

Critical Remarks: Stem and leaf indumentum and leaf size and shape
(cf. Fig. 104 c) may vary considerably. Within populations, leaf indumentum was found to vary from absent to hairy along the principal
veins, to scattered hairs all over. It is for this reason that KUNTZE's (1898)
subdivision into a glabrous and hairy variety is not followed. *G.
circaeoides,* nevertheless, is a well defined, easily recognizable species (lax,
broad and extensive inflorescences!). Even hairier individuals are not
difficult to separate from the densely tomentose *G. tomentosa* (2., below).

It is suspected that *G. circaeoides* may hybridize with *G. tomentosa* in
areas where the two species occur in the same or nearby habitats.
According to my field observations, ± sympatric occurrence is, however,
rare, thus accounting for the rarity of hybrids. WELLS 3485 from the
Transkei Wild Coast appears to be a hybrid collection.

From a phytogeographic point of view, the distribution range of *G.
circaeoides* is rather remarkable: It is essentially a species occurring as
undergrowth in afromontane forests. At first glance, numerous
localities – especially in the Transvaal and the Cape Prov. – appear to be
markedly out of the range of afromontane vegetation. This, however, is

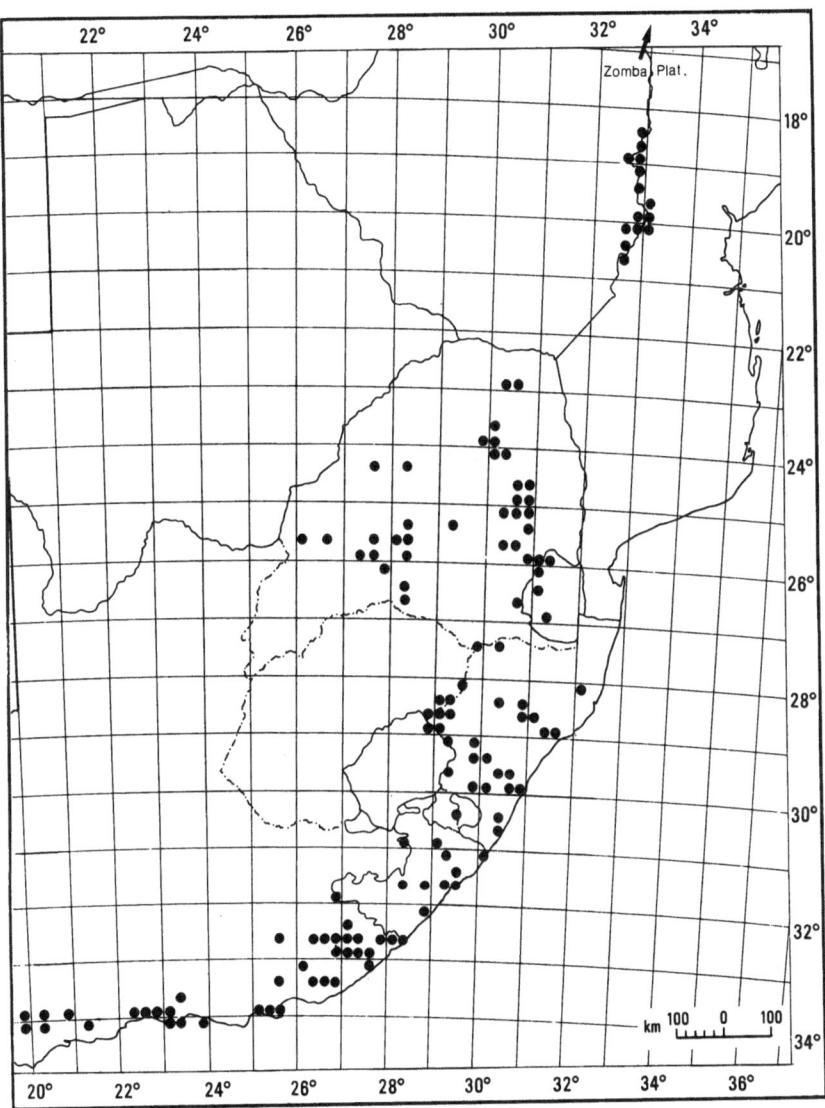

Fig. 105. Distribution of *Galopina circaeoides*

not so: The species merely "follows" afromontane forest patches (with *Ilex mitis, Halleria lucida, Podocarpus, Pittosporum,* etc.) occurring in sheltered, moist kloofs in areas which otherwise do not provide suitable habitats for afromontane vegetation ("distant satellite populations of Afromontane near-endemic species" sensu WHITE 1978: 477). Examples are various localities in the Magaliesberg and further W ("Magaliesberg extension" of WHITE, l.c.) or the Swaershoekberge (2428-AD, Farm Geelhout Kloof; geelhout = yellow wood = *Podocarpus*!). Much the same is likely to apply to the W-most localities in the Cape Prov. (e.g., Langeberge nr. Robertson or Riviersonderend; cf. WHITE 1978: 506 and Fig. 4).

Collections

Malawi. [S] − **1535** (Zomba): Zomba Plateau (-AD), BRASS 16295 (K, MO, SRGH, US), BRUMMITT 9967 (K, SRGH), PUFF 780208-2/9 (WU).

Mozambique. [MS] − **1832** (Umtali): Manica, Macequece (-DD), BARBOSA 1193 (LISC). − **1933** (Vila Pery): Mavita- Rotanda R. valley (-CA), PEDRO & PEDRÓGÃO 6603 (LMA); Chimanimani Mts., between Skeleton Pass and the Plateau, Martin Falls path (-CC), DRUMMOND 9125 (SRGH). − **2032** (Chipinga): Mafusi (-BB), JOHNSON 155 (K); serra do Espungabera, Mossurize (R.) (-DA), PEREIRA, SARMENTO & MARQUES 1343 (LMU).

Zimbabwe. [E] − **1832** (Umtali): Rhodes Inyanga National Park, along Pungwe R. (-BD), PUFF 790126-3/2 (MO, WU); Stapleford Forest Reserve, Dandy's Nek, Mt. Nuza (-DB), GILLILAND 394 (BM, K, SRGH), PUFF 790127-3/1 (MO, WU); Penhalonga (-DC), WILD 838 (PRE). − **1932** (Melsetter): Vumba Mts., Bunga Forest (-BB), PUFF 790128-2/1 (BR, J, MO, WU); Umtali D., Engwa, Himalayas (-BD), EXELL, MENDONÇA & WILD 148 (BM, LISC, SRGH), WILD 4530 (K, LISC, MO, PRE, S, SRGH); Muchira D., Mt. Peni (-DD), BAMPS, SYMOENS & VANDEN BERGHEN 857 (BR, SRGH). − **1933** (Vila Pery): Chimanimani Mts., Longgulley (-CC), NOEL 2079 (SRGH). − **2032** (Chipinga): Melsetter D., Kasipiti (-BA), LOVERIDGE 1692 (B, BR, LISC, SRGH), 1272 (SRGH); Chirinda (Forest) (-BC), GOLDSMITH 74/62 (LISC, M, MO, PRE, SRGH), SWYNNERTON 317 (BM, K), WHELLAN 420 (PRE, SRGH). − **2033** (Chibavava): Melsetter D., Haroni-Makurupini Forest (-AA), MÜLLER 1193 (PRE, SRGH). − Additional collections: CHASE 342 (BM, SRGH), 2160 (BM, BR, LISC), HALL 296 (BM, BOL, SRGH), PHIPPS 1208 (BR, PRE, SRGH), WHELLAN 1265 (MAL), WOLHUTER 24 (SRGH).

Swaziland. − **2531** (Komatipoort): Piggs Peak area (-CC, -CD), COMPTON 27836 (NBG), MILLER 7247 (M, PRE, SRGH), PUFF 790225-2/1 (WU). − **2631** (Mbabane): Ngwenya Mt. (-AA), PUFF 790225-1/4 (WU); Mbabane and surroundings (-AC), COMPTON 24960, 24999, 25827 (all NBG, PRE), KARSTEN s.n. (NBG, sub PRE 41490), PUFF 790224-3/1 (BR, J, WU); nr. Hlatikulu (-CD), COMPTON 29901 (NBG, PRE), 31341, 31343 (NBG).

South Africa. Transvaal − **2230** (Messina): Venda, Tate Vondo Forest Reserve (-CD), PUFF 791201-2/1 (J, WU); Venda, Tshaulu, Mutanzhela (-DC), VAN WYK 3854 (PRU). − **2329** (Pietersburg): Houtbosch (Woodbush) (-DD), REHMANN 6024 (BM, K), WAGER s.n. (NU, sub PRE 22970). − **2330** (Tzaneen): Westfalia Estate nr. Duiwelskloof (-CA), BOS 1332 (B, GH, LISC, M, PRE, STE), PIENAAR 867 (PRE), PUFF 790211-1/1, -2/2 (WU), SCHEEPERS 548 (M, MO, PRE, SRGH), SCHEEPERS & HAASBROEK s.n. sub PRE 41475 (PRE); New Agatha Forest Reserve, Wolkberg (-CC), MCCALLUM 137 (PRE × 2), s.n. (GH), MEEUSE 9886 (M, PRE, S), MULLER & SCHEEPERS 269 (PRE). − **2427** (Thabazimbi): between Rankins Pass and Bakkers Pass, Farm Groothoek (-BC), HARDY, RETIEF & HERMAN 5357 (PRE). − **2428** (Nylstroom): Swaershoekberge, Farm Geelhoutkloof 195 (-AD), PUFF 781203-1/1 (J, W, WU). − **2430** (Pilgrim's Rest): The Downs area (-AA), JUNOD 4234 (G, PRE, US), PUFF 790210-3/1 (W, WU); Farm Paris (-AB), VAHRMEIJER 2382 (MO, PRE); Lydenburg D., Starvation Creek Nature Res. (-DA), KLUGE 1167 (PRE); Mariepskop (-DB), MEEUSE 9937 (PRE), VAN DER

Schijff 4324 A (PRE, W), 4994 (PRE), van Son sub TVL Museum no. 30975 (BR, PRE); Ohrigstaddam Nature Reserve (-DC), Jacobsen 2659 (PRE), Puff 781028-2/2 (W, WU), Theron 3678 (PRE); Graskop, "Fairyland" (-DD), Puff 790704-2/2 (J, WU). – 2526 (Zeerust): Zeerust (-CA), Jenkins 11682 (PRE); Swartruggens, Elandsrivier (-DA), Sutton 987 (PRE). – 2527 (Rustenburg): Magaliesberg, Farm Kloofwaters (-CD), Puff 780326-2/1 (W, WU); Brits D., Farm Jacksontuin (-DA), Mogg 14977 (PRE), van Vuuren 165 (PRE); Hekpoort (-DC), Phillips 455 (PRE). – 2528 (Pretoria): Roodeplaatdam Nature Reserve (-AD), van Rooyen 2749 (PRE); Pretoria D., Fontain Valley (-CA), Repton 228, 369 (PRE), Verdoorn 720, 783 (PRE); Roodeplaatdam Nature Reserve, Gelofte … (indecipherable) (-CB), van Rooyen 2749 (PRU); Pretoria Distr., Irene (-CC), Murray s.n. sub PRE 41469 (PRE); –, Garstfontein, Wolwekloof (-CD), Mogg 15953 (PRE). – 2529 (Witbank): Loskopdam (-AD), Theron 1732 (PRE). – 2530 (Lydenburg): Lydenburg (-AB), Jenkins 10323 (PRE); Maritzbos (-BA), Mohle 263 (PRE); Vertroosting Nature Reserve, 12 km S of Sabie (-BB), Muller 2363 (PRE), 2414 (PRE); Witklip Forestry Station (-BD), Kluge 466 (MO, PRE); Waterval Boven (-CB), Rogers 14769 ("14457" added in brackets) (PRE); ca 7.5 km from Kaapsche Hoop to Ngodwana (-DA), Buitendag 819 (NBG, PRE); Cythna Letty Nature Reserve, Barberton (-DD), Muller 2198 (PRE). – 2531 (Komatipoort): ca 3–4 km from Josefsdal border post on rd. to Barberton (-CC), Puff 791212-2/2 (WU). – 2627 (Potchefstroom): Witwatersrand, Witpoortje Kloof (-BB), Moss 9529, 10801 (J), 5037 (BM, J). – 2628 (Johannesburg): Suikerbosrand Nature Reserve, Valsfontein 311 section (-AD), Puff 790516-1/3 (WU); Heidelberg D., Suikerbosrand, Evergreen Ranch, Koedoeskloof (-CB), Bredenkamp 595 (PRE). – 2630 (Carolina): 5 km from Amsterdam, on rd. to Lothair (-DA), van Wyk 2296 (PRU). – Imprecise or uncertain localities: Zoutpansberg D., McCallum 137 (PRE); Witwatersrand, Wetter, Suikerbosrand, Porter s.n. (J). – Not traced: Farm Mooihoek, Piet Rietief D., Devenish 1505 (MO, PRE, SRGH); Spelonken, Junod sub TVL Museum no. 19859 (PRE). – Additional collections: Benardi 8948 (G, K); Bredenkamp 817 (PRE); Burtt-Davy 5286 (K); Clinning s.n. (J); Forrester & Gooyer 150 (J); Jacobsen 911 (PRE); Leendertz 1615, 4743, 8663 (PRE); Mogg 16091 (PRE); Nation 222 (BOL, K); Pont 913 (PRE); Puff 781029-3/1 (J, W, WU), 790701-1/1 (WU); Thorncroft 1909 (MO), 2782 (PRE); van der Merwe 87 (PRE); van Rensburg s.n. (J). O.F.S. – 2829 (Harrismith): Plaas Rensburgskop, Manyenyeza Mt. (-AC), Jacobsz 336 (PRE); Van Reenen Station (-AD), Phillips s.n. sub PRE 112186 (PRE); O.F.S./Natal border: Van Reenen pass, Thode s.n. sub STE 4605 (STE). – No locality given ("OFS"): Cooper 3558 (K). Natal – 2729 (Volksrust): Boscobello (-BD), Jenkins 12467 (PRE). – 2730 (Vryheid): Utrecht D., Donkerhoek (-AD), Devenish 1126 (M, PRE), 1420 (NH, PRE). – 2828 (Bethlehem): Royal Natal National Park, Tugela R. below Tendele Camp (-DB), Puff 800106-2/1 (WU); –, Gudu Forest (-DD), Schelpe 1515 (NH, NU). – 2829 (Harrismith): Farm Nolens Volens, Natal side of Van Reenen pass (-AD), Puff 781125-1/1 (J, W, WU), Jabocsz 1598 (NBG, PRE); Collins Pass (-BA), Puff 840819-1/1 (NU, PRE, WU); Oliviershoek pass (-CA), Medley Wood 3560 (K, NH), Thode s.n. sub STE 4602 (STE); Bergville, "The Cavern" (-CB), Gemmell 5355 (PRE); Cathedral Peak Forest Research Station (-CC), Killick 1392 (PRE). – 2830 (Dundee): Farm Blesboklaagte, ca 13.5 km S of Dundee (-AD), Codd 2417 (PRE); Babanango (-BD), Venter 1850 (PRE); Qdeni Forest (-DB), Puff 790303-6/1 (WU). – 2831 (Nkandla): Nkandla Forest Reserve, ca 16–19 km S of Nkandla (-CA), Codd 1386 (PRE), Puff 790303-5/2 (J, W, WU); Eshowe (-CD), Gerstner 3712 (NH), Lawn 292 (NH); Mtunzini D., Ngoye Forest Reserve (-DC), Huntley 311 (MO, NH,

PRE), Puff 791217-1/1 (WU).－**2832** (Mtubatuba): Hluhluwe Game Reserve (**-AA**), Ward 2070 (NH, NU, PRE).－**2929** (Underberg): Cathkin Park, path to Grotto (**-AB**), Howlett 18 (NH, NU, PRE); Estcourt Pasture Research Station (**-BB**), Acocks 10061 (NH, PRE), West 1762 (NH, PRE), 1765 (NH); Fort Nottingham Commonage (**-BD**), Puff 800115-6/1 (WU); Underberg D., Bamboo Mt. (**-CB**), McClean 688 (NH, PRE); Bulwer (**-DD**), Badenhuizen & party s.n. (J).－**2930** (Pietermaritzburg): Karkloof (**-AC**), Evans s.n. sub NH 19969 (NH), Moll 3529 (NH, NU, PRE, S, SRGH); Pietermaritzburg (**-CB**), Moss 5159 (J), Schlechter 1993 (PRE), 6747 (GRA, NBG); Richmond D., Byrne (**-CC**), Galpin 12032 (PRE, W); Table Mtn. nr. Pietermaritzburg (**-DA**), Killick 328 (NU, PRE, SRGH); Richmond D., Tala Farm (**-DC**), Moll 3047 (NU, PRE); Cowies Hill (**-DD**), Platt 170 (NH).－**3029** (Kokstad): Natal/Transkei (Cape) border: Zuurberg (Ingeli Forest Reserve; nr. Weza) (**-BC**), Holland 225 (PRE), Puff 790416-2/1 (J, W, WU), Tyson 1169 (BOL, GRA, K, PRE, SAM).－**3030** (Port Shepstone): Hlokozi (**-AD**), Rudatis 2336 (STE × 3); Oribi Gorge Nature Reserve, along Mzumkulwana R. (**-CB**), Kerfoot 7736 (J), Puff 790426-2/1 (BR, J, WU); Izothsa (**-CD**), Strey 8082 (NH, NU, PRE).－**3130** (Port Edward): Umtamvuna Nature Reserve, Bulolo gorge (**-AA**), Puff 790422-1/2 (W, WU).－No locality given ("Natal"): Cooper 3557 (K).－Additional collections: Bayer & McClean 62 (BOL, GRA, PRE), 125 (NU, PRE); Buchanan s.n. (K); Franks sub Medley Wood 11827 (PRE, SAM), 13008 (NH); Junod 370 (G); Hutchinson 4560 (K); Hutchinson, Forbes & Verdoorn 82 (NH, PRE); Meebold s.n. (sub M 14574, sub PRE 41501); Skead 147 (NU); Thode s.n. sub STE 5277 (STE); Tyson 2548 (S). **Cape** － **3028** (Matatiele): Maclear D., Pot Rivier Berg (**-CD**), Galpin s.n. sub PRE 6647 (PRE).－**3029** (Kokstad): Insizwa Forest (**-CC**), Strey 10783 (GRA, NH, NU, PRE).－**3126** (Queenstown): Queenstown D., Gwatgu, Junction Farm (**-DD**), Galpin 8124 (PRE).－**3128** (Umtata): Baziya (**-CB**), Baur 580 (K); Umtata (**-DB**), Convent of the Holy Cross s.n. sub TVL Museum no. 25693 (PRE), Lewis Grant s.n. (MO).－**3129** (Port St. Johns): Bulembu (**-AB**), Kotze 595 (NBG, PRE); ca 3 km inland from Umgazi (= Mngazi) River mouth (**-CB**), Wells 3485 (GRA, PRE; very atypical－probably *G. circaeoides × G. tomentosa*); Port St. Johns (**-DA**), Moss 5038 (BM), 5039 (BM, J), Puff 790415-3/1 (BR, J, WU).－**3225** (Somerset East): Boschberg nr. Somerset East (**-DA**), Bolus 315 (BOL, K), Burchell 3183, 3203 (GOET, K), Macowan 216 (BM, GH, K), Scott Elliot 326 (E).－**3226** (Fort Beaufort): Fort Armstrong (**-CB**), Zeyher 31 (?41) (E); Katberg (**-DA**), Puff 790114-7/1 (J, W, WU), Schönland 4309 (GRA); nr. Hogsback village, H. Forest Reserve (**-DB**), Dahlstrand 2818 (PRE), Giffen 409 (PRE), Puff 790115-1/1 (W, WU); Seymour, Lily Bend, Wolf Ridge (**-DD**), Giffen 1524 (PRE).－**3227** (Stutterheim): Cathcart (**-AC**), Kuntze s.n. (K, sub US 554842); Dontsa pass (**-CA**), Puff 790115-4/1 (WU); Dohne Peak (**-CB**), Acocks 9489 (PRE), Puff 790115-5/7 (J, W, WU); Pirie Forest (**-CC**), Kuntze s.n. (K); King William's Town (**-CD**), Sim 19612 (PRE); Komg(h)a (**-DB**), Flanagan 546 (GRA, PRE × 2, SAM); Berlin (**-DC**), Barker 2736 (NBG).－**3228** (Butterworth): Elliotdale D., The Haven (**-BB**), Gordon-Gray 319 (NU); Kei R. (**-CA**), Krook 2047 (M, S, W); Kentani D. (**-CB**), Pegler 383 (BM, BOL, GH, GRA, PRE × 3, SAM).－**3319** (Worcester): nr. Robertson (**-DD**), Duthie 6491 (BOL).－**3320** (Montagu): Langeberge, (N of) Swellendam (**-CD**), Schlechter 2048 (BOL);－, Grootvaders Bosch (**-DD**), Burchell 7231 (K, M, W).－**3322** (Oudtshoorn): Montagu pass, at Keurrivier bridge (**-CD**), Puff 790909-6/1 (BR, J, W, WU); Wilderness (**-DC**), Compton 10703, 14302 (NBG); 6 km W of Karatara (**-DD**), Puff 790910-3/1 (W, WU).－**3323** (Willowmore): Uniondale D., Haarlem, beyond Stone's Hill (**-CB**), Schönland 3145 (GRA, PRE); Prince Alfred pass

(-**CC**), Puff 79122-1/1 (WU).-**3325** (Port Elizabeth): Suurbergpas, 3 km from Zuurberg Inn (-**BC**), Puff 800929-2/1 (WU); Vanstadens R. (-**CC**); Drège s.n. ("IV, C, c, 12") (E, G, MO, W × 3); Uitenhage D., Krakakamma (-**CD**), Zeyher 532 (K, SAM), "3-2" (S); Port Elizabeth (-**DC**), Laidley & al. 433 (G), Paterson 12964 (PRE).-**3326** (Grahamstown): Swartwatersberg, SW of Riebeek-East (-**AA**), Puff 840922-1/1 (PRE, WU); Grahamstown and surroundings (-**AD**,-**BC**), Daly 912 (GRA, PRE), Gane 17153 (PRE), Galpin 30 (PRE), Glass 234 (G), Read sub Macowan s.n. (K), Marais 133 (GRA, M, MO, PRE), Puff 790117-2/1 (WU), Stingeon s.n. sub SRGH 25259 (SRGH), Story 2198 (PRE); Grahamstown- Peddie rd., nr. Frasers Camp (-**BD**), Puff 791220-1/3 (WU).-**3327** (Peddie): East London (-**BB**), Barker 2927 (NBG, PRE), Rattray 280 (GRA).-**3419** (Caledon): Rivierzonderend Mts. nr. Rivierzonderend village (-**BB**), Esterhuysen 25338 (BOL, M, WU).-**3420** (Bredasdorp): Swellendam (-**AB**), Kuntze s.n. (K), Marloth 4965 (PRE × 2, STE).-**3421** (Riversdale): Riversdale D., Klein Kuils R. (-**AB**), Muir 3529 (PRE). -**3423** (Knysna): Knysna (-**AA**), Duthie 539 (BOL, SAM, STE × 3), Pappe s.n. (K, S); between Knysna and Plettenberg Bay (-**AA** to -**AB**), Zeyher s.n. (BR, SAM); Tzitzikama (= Tsitsikamma) National Park (-**BB**), Liebenberg 7771 A (BR, G, M, MO, PRE, S).-No localities given or localities imprecise: "Albany", Bowker s.n. (K); "CBS", Herb. Forsyth (K), Masson s.n. (BM), Prior s.n. (K), Thunberg, several sheets (S); "Riverzondereine (Swellendam), Krakakamma (Uitenhage), Grahamstown and on Katrivier (Albany)", Ecklon & Zeyher 2304 (or "89.2") (BOL × 3, E × 2, FI, G, GH, GOET × 2, LY × 3, M, MO, PRE × 2-one as "Zeyher 89.2", S, SAM, W × 3, WU).-Not traced: E Griqualand, Nalogha, Krook 2061 (M, S, W).-Additional collections: Barker 2888 (NBG × 2, PRE); Britten 1171 (PRE); Burchell 5204 (GH, K); Compton 4470 (BOL, NBG), 14302 (PRE), 19235 (NBG); Drège s.n. (FI, G, K, S); Duthie s.n. sub STE 15217 (STE); Esterhuysen 2662 (BOL); Fourcade 109 (GRA); Galpin 6301 (PRE × 2); Garside s.n. (K); Giffen s.n. (PRE); Gillett 1343 (STE); Jacot Guillarmod 7797 (PRE); Krauss 1497 (MO, W); Laidley 433 (488?) (LY); Lewis 4909 (PRE, SAM); Lucas s.n. (J); Moss 6087 (J); Oliver 18 (PRE); Prior s.n. (K); Puff 790115-3/1 (W, WU), 791220-3/1 (WU); Schönland 3440 (PRE), 3617 (GRA, PRE); Smith 3751 (PRE); Spearmann 36 (PRE); Thode A 2539 (NH, PRE); Wall 349/289, s.n. (S); Well s.n. sub GRA A1194 (GRA); West 383 (MO); Williamson A.J. 114 (GRA); Worsdell s.n. (K); Zeyher 2712 (LY, SAM).

2. **Galopina tomentosa** Hochst. in Flora **27**: 555 (1844); Puff in Fl. S. Afr. **31**, 1 (2): 50 (1986).
Type: South Africa, Natal, nr. Umlaas R., December 1839, Krauss 52 (BM!, G!, GH!, K!, M!, MO!, W!).

- *G. hirsuta* E. Meyer in Drège in Flora **26**, Bes. Beigabe: 186 (1843), nom. nud.

♀, ♂ + ♀, ♂, or occasionally ♂, ♂ + ♀ or ♂ + ♀ + ♀, few- to many-stemmed perennial herbs with branched, often ± woody rootstocks or rhizomes; sometimes with long, plagiotropous subterranean shoots. Stems (30) 40–80 (100) cm long, ca (2) 3–5 mm in diameter at the base and ca 1.5–3 mm in the midstem region, ascending to ± erect or ± prostrate and rooting at the nodes; obscurely 4-angled above and terete below, mostly densely covered with whitish or yellowish-white hairs ca (0.2) 0.3– 0.5 (0.8) mm long; internodes slightly shorter to somewhat longer than the leaves; stems often with contracted to somewhat elongated short shoots bearing small leaves. Leaves decussate; blades (27) 35–50 (62) × (15) 20– 35 mm, broadly ovate-lanceolate or ovate, cuneate, truncate or ± subcordate at the base, both surfaces, particularly the midrib and the prominent lateral veins, ± densely covered with yellowish to whitish spreading hairs ca (0.2) 0.3–0.5 (0.8) mm long, very rarely glabrescent; apices subacute to ± obtuse; petioles (3) 4–6 (9) mm long; stipular sheaths ± cup-shaped, 2.5–4 × 1–2 (3) mm, pubescent, with 3 (–5) gland-tipped setae slightly fused below, the median seta (1.5) 2–4 (5) mm long, the others (much) shorter. Inflorescences terminal, paniculate, pyramidal to ellipsoidal, (6) 10–20 (25) × (4) 7–15 (20) cm; in vigorous individuals often additional, smaller inflorescences arising in the upper leaf axils; bracts leaf-like below, small to minute and ± lanceolate above; peduncles and pedicels thin but ± stiff, hairy to subglabrous, divergent in fruit, elongating to (2) 3–6 (7) mm. Flowers ♂, ⚥, ♀; corolla greenish, greenish-yellow or creamy yellow and sometimes reddish-brown tinged outside, glabrous or papillate near the tip. ♂, ⚥: tube ± 0–0.4 (0.5) mm long, campanulate, lobes 1.1–1.5 (1.7) × 0.4–0.6 (0.9) mm, ± lanceolate, recurved; filaments 0.5–0.7 (1) mm long, anthers 0.9–1.2 × 0.2–0.4 mm; ⚥: gynoecium as in ♀, but stigmas often shorter and ± thinner. ♀: tube 0– 0.2 (0.3) mm long, cylindrical; lobes 0.6–0.9 × 0.1–0.2 (0.3) mm, ± linear, erect to spreading; stigmas (2.2) 3–4.5 (5) mm long, whitish-grey to yellowish; ovary ca 0.6–0.9 × 0.8–1.3 mm, ± wrinkled to ± warty. Fruits greenish-brown, brown to brownish-black; each mericarp (1) 1.2– 1.8 (2) × 0.8–1.4 (1.6) mm, obovate to elliptic, dorsal side convex, ± smooth, wrinkled or ± warty, ventral side ± plane; commissure small, ca 0.5–0.7 (1) × 0.3–0.5 (0.8) mm. – Figs. 13 b, 28 b, 39 c–d, h; 104 e.

Chromosome Number (Table 3): n = 11, 2 n = 22.

Average Pollen Diameter (Fig. 49): 23.6–26.2 μm.

Habitat: Growing at the edge of (coastal and dune) scrub and forest, but occasionally also in open grassland; sometimes in disturbed areas (plantations, waste places). Generally, often in ± sunny habitats. Frequently found growing in sandy to sandy-loamy soil. – Ca 20–500 m.

Distribution (Map, Fig. 106 a): S Africa. From Tonga- and Zululand extending to the Natal S Coast and the Wild Coast (Transkei).

Critical Remarks: G. tomentosa has persistently been confused with and identified as *G. aspera* (3., below), although the two species are easily

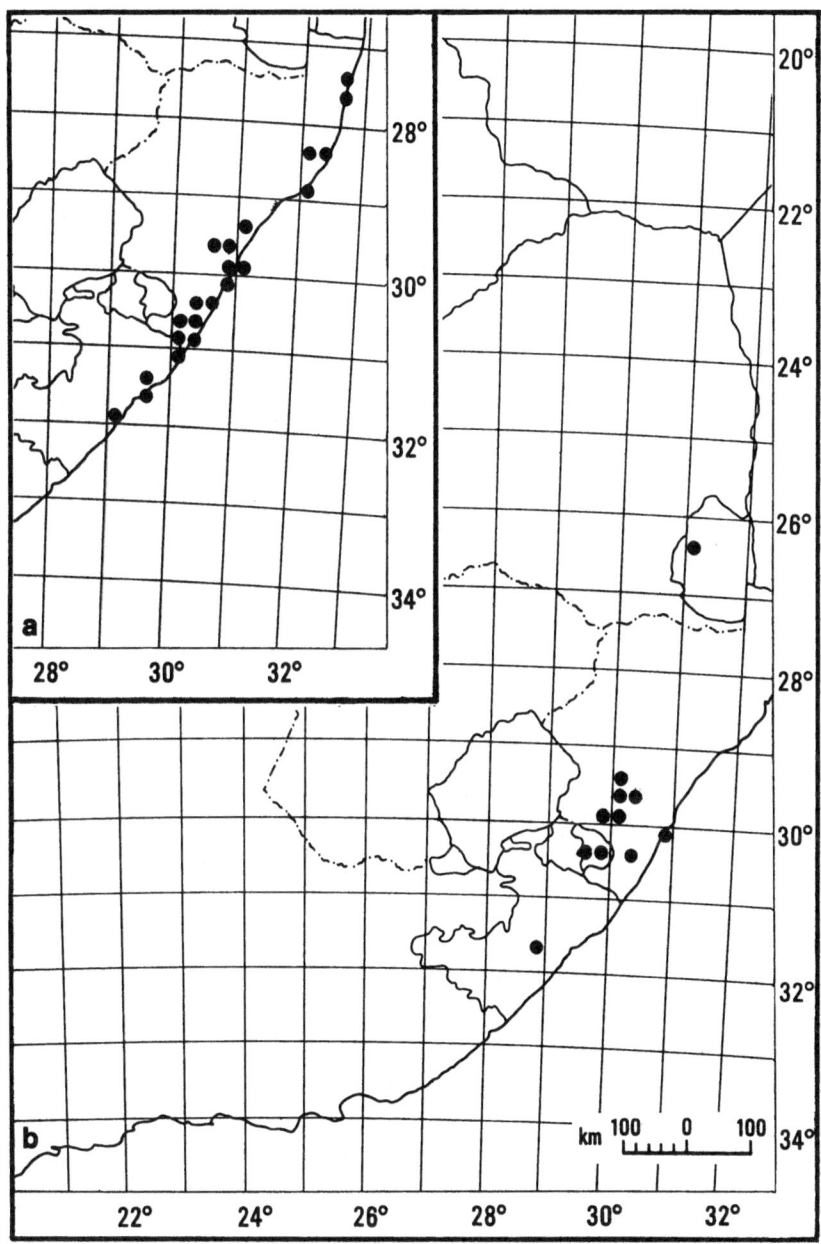

Fig. 106. Distribution of *Galopina*. *a G. tomentosa, b G. crocyllioides*

separated morphologically (inflorescences, leaf characters, mericarps!). *G. tomentosa,* in fact, is thought to be more closely allied to *G. circaeoides* than to *G. aspera.* In nature, *G. tomentosa* is easily recognized by its dull green to dark blueish-green foliage (due to the dense indumentum!), while *G. circaeoides* has shiny, bright green leaves. There are also significant ecological differences between the two species.

Some glabrescent or less densely hairy forms, such as HARRISON 412 or PUFF 790304-3/1 from N of Mtubutuba (2832-AC) and a few other collections from coastal areas of Tonga- and Zululand, appear to approach *G. circaeoides* in leaf size, shape and indumentum but have the typical, less lax inflorescences with short, stiffer peduncles and pedicels of "typical" *G. tomentosa.* As *G. circaeoides* is not recorded from any of these areas, it appears unlikely that these are hybrids.

Collections

South Africa. Natal — **2732** (Ubombo): E Ingwavuma, N shore L. Sibayi (**-BC**), AITKEN & TALE 6 (NU, PRE; atypical); Sordwana Bay (**-DA**), BALSINHAS 3209 (PRE), VAHRMEIJER & TÖLKEN 832 (PRE; atypical). — **2832** (Mtubatuba): Palm Ridge Farm, 7 km N of Mtubatuba (**-AC**), HARRISON 412 (NH, PRE), PUFF 790304-3/1 (BR, J, W, WU); St. Lucia Resort Game Park (**-AD**), POOLEY 2002 (E; atypical); Richards Bay village (**-CC**), VENTER 5391 (PRU). — **2930** (Pietermaritz-burg): Stainbank Nature Reserve (**-BD**), WARD 6378 (NU); (nr.) Umlaas Road (**-DA**), EVANS 325 (NH), KRAUSS 52 (BM, G, GH, K, M, MO, W); Inanda D., Groeneberg (**-DB**), BURTT 3022 (E, NU), STREY 5630 (PRE); Manors, Pinetown (**-DD**), COLEMAN 48 (NH, PRE). — **2931** (Stanger): Chakaskraal (**-AC**), THODE s.n. sub STE 4612 (STE); Durban (Port Natal) (**-CC**), DRÈGE s.n. (sub "*G. hirsuta* E. MEY.") (G, W), HARVEY 1840 (GH). — **3030** (Port Shepstone): Stn. Dumisa, Farm Friedenau, Umgaye and Ifafa R. (**-AD**), RUDATIS 332 (BM, E, G, K, PRE, S, STE, W), 820 (BM, E, G, S, W); Isipingo Beach (**-BB**), WARD 1378 (NU); Umzinto D.; Umgayi (**-BC**), WARD 5488 (NH, NU, PRE); Port Shepstone D., Oribi Flats, (**-CA**), McCLEAN 529 (NH, PRE); Oribi Gorge Nature Reserve (**-CB**), DAVIDSON 1189 (J), GLEN 317 (J), PUFF 790426-1/2 (J, W, WU); Umtamvuna Nature Reserve, Smedmore area (**-CC**), PUFF 840823-2/2 (PRE, WU); Uvongo River and -Beach (**-CD**), LIEBENBERG 8032 (PRE), STREY 9561 (BR, NH, PRE, S, SRGH), 11056 (E, NH, NU). — **3130** (Port Edward): Umtamvuna Nature Reserve (**-AA**), PUFF 790422-3/1 (BR, J, W, WU). — No locality given ("Natal"): GERRARD 388 (BM, K, W), SANDERSON 536 (S). — Additional collections: FRANKS sub MEDLEY WOOD 10854 (GRA, SAM); GUEINZIUS 161 (W); HAYGARTH s.n. sub STE 246 (STE); HUNTLEY 174 (MO, NU); JOHNSON & COLEMAN 1471 (NH, PRE, SRGH); KILLICK 72 (PRE, SRGH); MEDLEY WOOD 92 (BM, BOL, E, K, MO, NU, SAM), 222 (BM, BOL, MO, SAM), 1004 (BM, BOL, G, GH, K, NBG, SAM, W), 5753 (E, MO); MOBERLY 57 (NU); PHILLIPS s.n. sub PRE 41487 (PRE); PLATT 169 (NH); RUMP s.n. sub NH 21075 (NH); SANDERSON s.n. sub NH 1164 (NH); SIM 120 (NU); STREY 5206 (BR, M, NH, PRE, SRGH), 5630 (M, NH, NU, PRE); 7378 (NH), 11067 (E, NH, NU); THODE s.n. sub STE 5039 (STE). **Cape** — **3129** (Port St. Johns): Lusikisiki D., Umsikaba (= Msikaba) gorge (**-BC**), ACOCKS 13254 (PRE); around Tshani (**-CC**), PUFF 790414-1/2 (WU); Port St. Johns and surroundings (**-DA**), COMINS 1958 (GRA, PRE), GALPIN 3500 (BOL, PRE), MOSS 5038 (BM, J), PUFF 790415-2/1 (BR, J, W, WU), SCHÖNLAND 4148 (GRA).

3. Galopina aspera (ECKLON & ZEYHER) WALP., Rep. Bot. Syst. **2**: 462
(1843) pro 'Galopina ? aspera'; SONDER in HARVEY & SONDER, Fl. Cap.
3: 26 (1865); PUFF in Fl. S. Afr. **31**, 1 (2): 51 (1986).
≡ *Oxyspermum asperum* ECKLON & ZEYHER, Enum. Plant. Afr. Austr.:
365 (1836).
Type: [South Africa, Cape Prov.] Katriviersberg, above Philipstown
(Ceded Territory), and on Chumiberg (Kafferland), ECKLON &
ZEYHER 2305 (SAM, holo.!; FI, G, GOET, M, MO, S, W × 2, iso.!).

= *G. oxyspermum* STEUD., Nom., ed. 2, **1**: 662 (1841), nom. illeg.

♂, ♂̦, ♀̂ + ♀, ♀, or occasionally ♂ + ♀̂ + ♀ or (very rarely) ♂ + ♀, single-
to several-stemmed perennial herbs with often branched, ± woody
rootstocks. Stems (30) 50–120 (140) cm long, ca (2) 3–5 mm in diameter at
the base and ca 2–3 (4) mm in the midstem region, ± erect to ascending,
mostly unbranched; at least younger parts obscurely 4-angled, usually
densely covered with yellowish to whitish spreading hairs ca 0.1–
0.3 (0.5) mm long, occasionally glabrescent or papillate; internodes mostly
as long as or longer than the leaves; stems often with contracted short
shoots bearing relatively large leaves. Leaves decussate; blades (25) 30–
45 (50) × (8) 12–25 (30) mm, ovate, elliptic or ovate-lanceolate, ± cuneate
or ± truncate at the base, both surfaces, particularly the midrib and the
prominent lateral veins, ± densely covered with spreading whitish to
yellowish hairs ca 0.1–0.3 mm long or seldom ± papillate or subglabrous;
apices subacute to obtuse; petioles 0.5–2 (3) mm long; stipular sheaths ±
cup-shaped, ca 3–4.5 (5) × 1–2 mm, densely hairy to ± glabrous, with 5 or
3 (–4) gland-tipped setae, the median ca 3–4 (4.5) mm long, the others
(much) shorter. Inflorescences terminal, (thyrso-)paniculate, ± narrowly
cylindrical, ca (8) 15–30 (38) × (2) 4–8 (10) cm; in vigorous individuals
additional, smaller inflorescences arising in the upper leaf axils; bracts
relatively large and ± lanceolate below, minute and ± linear above;
peduncles and pedicels thin but stiff, hairy to subglabrous, not conspicuously divergent in fruit, peduncles elongating to ca (1.5) 2.5–
5 (6) mm, pedicels to (0.8) 1–2 (3) mm. Flowers ♂, ♀̂, ♀; corolla greenish,
greenish-yellow or purplish or brownish-red tinged, often densely papillate outside. ♂, ♀̂: tube (0.1) 0.2–0.5 mm long, broadly funnel-shaped to
campanulate, lobes 1–1.5 (1.6) × (0.4) 0.5–0.6 (0.7) mm, ± lanceolate-
oblong, recurved; filaments (0.3) 0.4–0.7 (1) mm long, anthers (0.5) 0.7–
1.1 (1.2) × (0.2) 0.3–0.4 mm; ♀̂: gynoecium as in ♀, but stigmas often
shorter and thinner. ♀: tube 0–0.2 mm long, cylindrical, lobes (0.5) 0.6–
1 × (0.1) 0.2–0.4 mm, ± linear, erect to spreading; stigmas (2) 2.5–
4 (4.5) mm long, whitish-grey or -green; ovary ca 1–1.5 × 0.8–1 mm,

29*

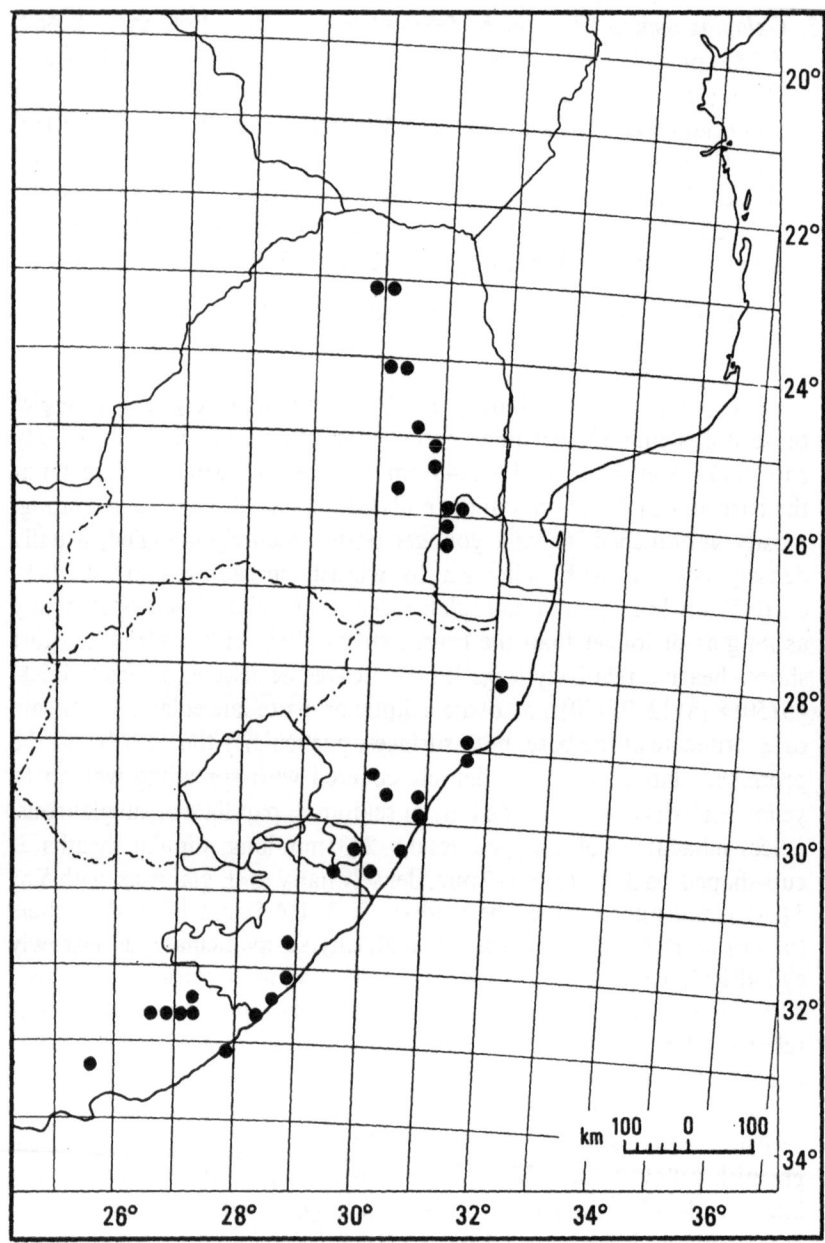

Fig. 107. Distribution of *Galopina aspera*

densely tuberculate. Fruits brownish; each mericarp 1.5–2 × (0.7) 1–1.2 (1.5) mm, dorsal side convex, conspicuously and densely warty-tuberculate, ventral side ± plane to strongly concave due to inrolled sides; commissure relatively large, ca 1–1.2 (1.4) × 0.6–0.9 (1.2) mm. – Figs. 28 c, 39 e–g, i, 104 d.

Chromosome Number (Table 3): n = 11, 2 n = 22.

Average Pollen Diameter (Fig. 49): 24.3–25.9 µm.

Habitat: Often growing amongst grass and other vegetation at the edge of forests, in forest edge scrub or in bushclumps around or between rocks; occasionally in moist places (edge of marshes or swamps). Sometimes found in disturbed sites such as roadsides, firebreaks or plantations. Often growing over clay, clayey loam or silicaceous loam. In ± sunny habitats. – Ca 25–1 400 m.

Distribution (Map, Fig. 107): S Africa and Swaziland. From the Transvaal, roughly along the Drakensberg escarpment, to Swaziland and, through Natal and the Transkei, to the E Cape Prov.

Critical Remarks: G. aspera varies in leaf size and shape (cf. Fig. 104 d) and indumentum, but its often conspicuous leafy short shoots and ± cylindrical inflorescences distinguish it from both of its presumed allies, *G. tomentosa* (2., above) (with more extensive, broader inflorescences) and *G. crocyllioides* (4., below) (less extensive inflorescences with more congested partial inflorescences). Mericarps of *G. aspera* occasionally have rather massive, ± elongated tubercles, but never produce distinct hairs as in *G. crocyllioides* (compare Figs. 39 i and j). The similar basic structure of the tubercles of *G. aspera* and the hairs of *G. crocyllioides* may be yet another feature pointing to the close alliance between the two species.

Collections

Swaziland. – **2531** (Komatipoort): Piggs Peak (**-CD**), Compton 27632 (NBG, PRE). – **2631** (Mbabane): Steynsdorp village, ca 29 km NNW of Mbabane (**-AA**), Maguire 7420 (J, WU); Black Mbuluzi valley nr. Mbabane (**-AC**), Compton 28654 (NBG), 31262 (NBG, PRE × 3), Puff 790224-4/2 (WU).

South Africa. Transvaal – **2329** (Pietersburg): Louis Trichardt (**-BB**), Breyer 22104 (PRE), Moss 16849 (J). – **2330** (Tzaneen): Elim (**-AA**), Schlechter 4549 (BM, BOL, G, K, PRE, W, WU); Duiwelskloof, Westfalia Estate (**-CA**), Puff 790211-3/1 (J, W, WU), Scheepers 1101 (BM, G, M, PRE, SRGH), s.n. sub PRE 41496 (PRE). – **2430** (Pilrim's Rest): The Downs area (**-AA**), Junod 4247 (PRE), Puff 790210-5/1 (WU); Shilouvane (**-AB**), Junod 1041 (K); Ohrigstaddam Nature Reserve (**-DC**), Jacobsen 2639 (PRE). – **2530** (Lydenburg): Vertroosting Nature Reserve, 12 km S of Sabie (**-BB**), Muller 2474 (PRE); Witklip (**-BD**), Kluge 682 (PRE); Waterval Boven (**-CB**), Rogers 14774 (PRE). – **2531** (Komatipoort): Barberton and surroundings (**-CC**), Galpin 1319 (K, PRE), Liebenberg 2399 (PRE), Thorncroft 2105 (PRE). – Uncertain locality: Thorncroft 465 (NH).

Natal – **2831** (Nkandla): Melmoth (**-CB**), Lawn 1953 (NH); Ngoye (**-DC**), Medley Wood 10433 (NH). – **2832** (Mtubatuba): Hluhluwe Game Reserve (**-AA**), Fakude 76 (PRE), Ward 2188 (GRA, NH, NU, PRE). – **2930** (Pietermaritzburg): Shafton, Howick (**-AC**), Hutton 71 (BM); Pietermaritzburg and surroundings (**-CB**), Rajah 5 (E, NH, NU), Rehmann 7568 (BM, K); Inanda (**-DB**), Medley Wood 867 (K); Field Hill (**-DD**), Evans 157 (NH); Pinetown D., Gibson s.n. (NU). – **2931** (Stanger): Tugela Mouth Hills (**-AB**), Strey 6472 (NH, PRE). – **3030** (Port Shepstone): Umzinto D., Hazelwood (**-BC**), Baijnath 400 (NU, PRE); Oribi Flats (**-CA**), McClean 529 (NH). **Cape** – **3029** (Kokstad): nr. Clydesdale (**-BD**), Tyson 734 (BM, BOL, G, GH, K, PRE, SAM, W), 2112 (G), 2158 (BOL, BR, K, PRE, WU); Gwalaweni (= Gxwaleni or Gwaleni Forests) (**-DA**), Vahrmeijer & Hardy 1704 (PRE). – **3128** (Umtata): Umtata (?[1]) (**-DB**), Drège s.n. (S). – **3226** (Fort Beaufort): Katberg (**-DA**), Hutton s.n. (K, S); Katriviersberg above Philipstown and on Chumieberg (**-DB**), Ecklon & Zeyher 2305 (or "47-5") (FI, G, GOET, M, MO, S, SAM, W × 2). – **3227** (Stutterheim): Fort Cunyngham (**-AD**), Sim 2682 (GRA, PRE); Keiskamahoek D., Dontsa pass (**-CA**), Acocks 9431 (PRE); Dohne Peak (**-CB**), Acocks 9663 (PRE); Stutterheim D., below Kabousie Forest, Hilliard & Burtt 12430 (E, NU, WU), 12431 (NU, WU). – **3228** (Butterworth): Elliotdale D., The Haven (**-BB**), Gordon-Gray 1401 (NU); Komga D., Haga-Haga (**-BC**), Hilliard & Burtt 11082 (E), 11083 (E, K); nr. Kei Mouth (**-CB**), Flanagan 444 (GRA, PRE, SAM); Kentani D., Pegler 312 (BM, K, PRE × 2). – **3325** (Port Elizabeth): Zuurberg, Afalaya valley (**-BC?**), Archibald 4978 (PRE). – **3327** (Peddie): East London (**-BB**), Wood 3403 (PRE).

4. *Galopina crocyllioides* Bär ex Schinz in Vierteljahresschr. Naturforsch. Ges. Zürich **68**: 437 (1923); Hilliard & Burtt in Notes Roy. Bot. Gard. Edinb. **32**: 387 (1973); Puff in Fl. S. Afr. **31**, 1 (2): 52 (1986). Types: [South Africa, Natal] Howick, Junod 238 (Z, lecto.!);" Alexandra Distr.", Station Dumisa, Umgaye Flats, 600 m, Rudatis 867 (BM, E, K, S, W, syn.!).

♂, ♂̄ + ♀, ♀, or occasionally ♂̄ + ♂, few- to single-stemmed perennial herbs with short, often branched, ± woody rootstocks. Stems (50) 60–100 (120) cm long, ca (2) 3–5 mm in diameter at the base and ca 2–3 (4) mm in the midstem region, erect, unbranched; obscurely 4-angled, glabrous or younger parts ± sparsely covered with whitish papillae; internodes as long as or longer than the leaves; stems usually without short shoots. Leaves decussate; blades (35) 40–60 (70) × 12–25 (30) mm, ovate, ovate-lanceolate to ± lanceolate (very variable in shape on individual plants), narrowed to the base, glabrous or ± sparsely covered with whitish or yellowish-white papillae to 0.1 mm long; petioles (0.5) 1–2 (2.5) mm long; stipular sheaths ± cup-shaped, ca 2–4 × 1.5–3 mm, glabrous to sparsely hairy, with (2) 3–4 (5) gland-tipped setae, the longest 2.5–5 mm long.

---

[1] Added in Sonder's handwriting; according to Drège (1843) from "Port Natal" (Durban; locality 'V,c,37').

Inflorescences terminal, thyrso-paniculate to thyrsic, ± narrowly cylind-rical, ca (7) 10–30 × 1.5–5 (7) cm; in vigorous individuals additional, smaller inflorescences arising in the upper axils; bracts small, lanceolate to linear below, minute and often subulate above; peduncles and pedicels thin but stiff, usually glabrous, not divergent in fruit, peduncles elongating to ca (1) 2–4 (5) mm, pedicels to 0.7–1.5 (2.5) mm. Flowers ♂, ⚥, ♀; corolla greenish-yellow or brownish tinged, usually hairy outside. ♂, ⚥: tube 0–0.5 mm long, broadly funnel-shaped to campanulate, lobes 1–1.5 × 0.4–0.7 mm, ± oblong-lanceolate, recurved; filaments 0.5–1 mm long, anthers 0.5–0.8 (1) × 0.2–0.3 mm; ⚥: gynoecium as in ♀, but stigmas often shorter. ♀: tube ± 0, lobes ca 0.5 × 0.1–0.2 mm, linear, ± erect; stigmas 2–3 (3.5) mm long, whitish-grey; ovary ca 0.8–1.2 × 0.7–1 mm, densely hairy. Fruits reddish-brown to blackish; each mericarp 1.3–2.2 (2.6) × 0.8–1.2 mm, dorsal side convex, densely covered with whitish spreading hairs ca (0.3) 0.5–0.8 mm long, ventral side plane to slightly concave; commis-sure relatively large, ca 1–1.5 (1.7) × 0.6–1 mm. – Figs. 13 e, 28 d, 39 b, j, 104 b.

Chromosome Number (Table 3): n = 11, 2 n = 22.

Average Pollen Diameter (Fig. 49): 24–24.6 µm.

Habitat: Mostly growing in full sunlight amongst tall grass in grassland or in forest edge scrub. – Ca 20–1 600 m.

Distribution (Map, Fig. 106 b): S Africa, centered in the Natal Midlands and Coastal regions and extending SE into the Transkei. Also in Swaziland.

Critical Remarks: G. crocyllioides, the least common of all *Galopina* species, is well distinguished by its hairy mericarps. Only few specimens bear some leafy short shoots; such collections, especially if they *also* have relatively shortly hairy fruits, somewhat resemble *G. aspera*. It remains uncertain whether they merely represent an odd state of the species or perhaps are hybrids *G. crocyllioides* × *G. aspera* (e.g., COMPTON 32265 from around Mbabane, an area from where *G. aspera* is also recorded).

Collections

Swaziland. – **2631** (Mbabane): Mbabane (and surroundings) (-AC), COMPTON 23278, 24995, 25036 (all NBG), 25527 (NBG, PRE), 32265 (NBG; atypical).

South Africa. Natal – **2929** (Underberg): SSE of Donnybrook Stn., Polela (-DD), ACOCKS 22130 (PRE). – **2930** (Pietermaritzburg): Howick (-AC), JUNOD 238 (Z); Swartkop (-CA), NIXON 14 (NU); Pietermaritzburg, Town Hill (-CB), HILLIARD 5275 (NU); Byrne area, Keerom (-CC), STREY 11266 (NH). – **3029** (Kokstad): Natal/Transkei (Cape) border: Zuurberg (Ingeli Forest Reserve; nr. Weza) (-BC), HILLIARD & BURTT 8072 (E × 2, K, MO, NBG, NU, PRE, S), 10184 (E, NU), PUFF 790416-1/1 (BR, J, W, WU), TYSON 1176 (K, SAM; BOL – a-typical); Alfred D., Harding, Farm Rooivaal (-DB), HILLIARD & BURTT 16735 (NU,

WU). – **3030** (Port Shepstone): Dumisa, Umgaye (**-AD**), RUDATIS 867 (BM, E, K, S, W); Isipingo (**-BB**), DE VILLIERS 34 (NU). **Cape – 3029** (Kokstad): nr. Clydesdale (**-BD**), TYSON 2131 (SAM; mixed with *Anthospermum herbaceum*). – **3127** (Lady Frere): Pondoland, Engcobo, BOLUS s.n. (BOL). – **3128** (Umtata): nr. Umtata (**-DB**), FLANAGAN 2848 (PRE × 2).

Carpacoce SONDER in HARVEY & SONDER, Fl. Cap. **3**: 32 (1865); HOOK. f. in BENTH. & HOOK. f., Gen. Pl. **2** (1): 141 (1873); K. SCHUM. in ENGLER & PRANTL, Nat. Pflanzenfam. **IV, 4**: 130 (1891); SALTER in J. S. Afr. Bot. **3**: 113 (1937), in ADAMSON & SALTER, Fl. Cape Penins.: 733 (1950); DYER, Gen. S. Afr. Flow. Pl. **1**: 623 (1975); PUFF in Fl. S. Afr. **31**, 1 (2): 52 (1986).
Type species: *Anthospermum scabrum* THUNB., Prod. Pl. Cap. **1**: 32 (1794) [= *C. scabra* (THUNB.) SONDER].

– *Lagotis* E. MEYER in DRÈGE in Flora **26**, Bes. Beigabe: 197 (1843), nom. nud.

⚥, ♀, or occasionally ♂, ⚥ + ♀ or ⚥ + ♂ dwarf shrubs, rarely perennial herbs. Leaves decussate or (rarely) in whorls of 3, often ericoid, subsessile; stipular sheaths cup- or funnel-shaped, bearing minute setae or (less commonly) longer bristles on either side. Inflorescences frequently very inconspicuous, made up of paired or solitary flowers (seldom more-flowered clusters) at the nodes or, occasionally, flowers solitary and terminal on shoots. Flowers ⚥, ♀ or occasionally ♂; calyx 3–5 (6)-merous, lobes leaf-like, often unequal in size; corollas 4–7-merous; ⚥, ♂: tubes cylindrical to ± narrowly funnel-shaped, lobes hooded, linear to ± lanceolate, spreading to spreading-recurved; stamens 4–7, anthers purplish-red to dark purplish-brown, exserted, dangling on long slender filaments; ♀: corollas much smaller, tubes cylindrical, lobes hooded, ± linear, ± erect; style 0; stigmas 1 (2), long exserted, hairy, purplish-red, greenish-grey, greyish or whitish; ovary with 1 (2) fertile ovule(s). Fruits 1- (2-)seeded, crowned by the persistent calyx lobes; exocarp dehiscing into valves, releasing endocarp with enclosed seed[1].

Chromosome Number (Table 3): x = n = 11, 2 n = 22.
Average Pollen Diameter (Fig. 49): 31.6–41.4 µm.
Distribution (Maps, Figs. 1 d and 108 j): S Africa. Endemic to the SW Cape Prov.; only one species extending as far E as the Bathurst and Albany Distr.

---

[1] Endocarp plus enclosed seed (cf. Figs. 43 f–k) are referred to as "diaspore" (dispersal unit) in the following descriptions.

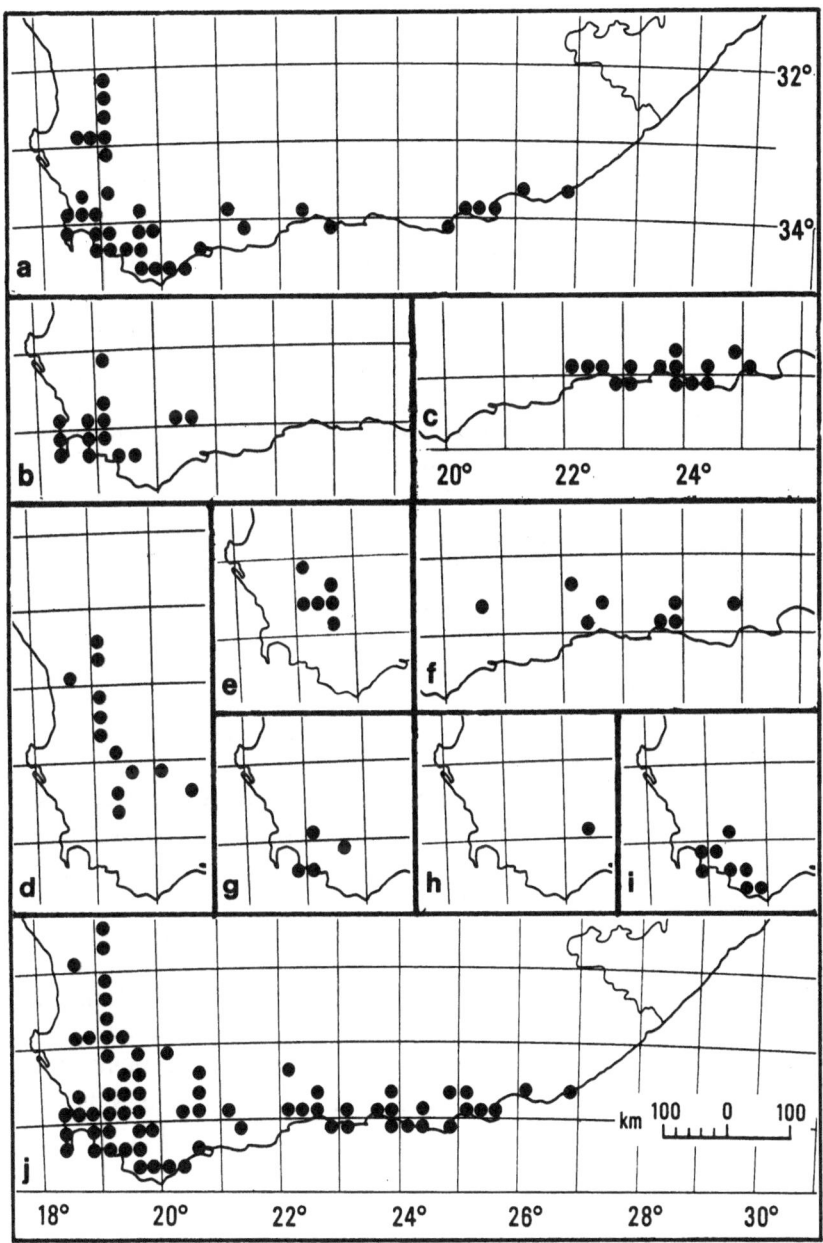

Fig. 108. Distribution of *Carpacoce. a C. vaginellata. b–c C. spermacocea, b* subsp. *spermacocea, c* subsp. *orientalis. d–e C. scabra, d* subsp. *scabra, e* subsp. *rupestris, f C. curvifolia, g C. burchellii, h C. gigantea, i C. heteromorpha, j* all taxa

Key to Species and Subspecies

1. Leaves (35) 50–70 (80) mm long; fruits crowned by 3 calyx lobes ....
    .... **2. *C. gigantea***
1*. Leaves smaller, 3–30 (35) mm long; fruits crowned by 4–6 calyx lobes
    .... 2.

2. Leaves lanceolate or ovate-lanceolate, to 5 (–8) mm wide, not dis-
    tinctly ericoid .... 3.
2*. Leaves ericoid, often ± needle-like, linear, linear-lanceolate or
    narrowly ovate-lanceolate, not wider than 2 mm .... 5.

3. Stipular sheath with a single minute seta, or seta absent; leaf blades
    rigid, conspicuously recurved, (4) 6–10 (12) mm long; corolla and
    calyx always 4-merous .... **3. *C. curvifolia***
3*. Stipular sheath with numerous bristles, 1–2.5 (3) mm long; leaf blades
    ± membranaceous, not conspicuously recurved (except for the tips),
    10–30 (35) mm long; corolla and calyx mostly 5-merous; plants foetid
    .... 4.

4. Plants slender, lax; partial inflorescences few- to 1-flowered; fruiting
    pedicels 3–12 (20) mm long; widely distributed in the SW Cape s.str.,
    E to the Heidelberg Distr. ....
    .... **1 (a). *C. spermacocea* subsp. *spermacocea***
4*. Plants more robust, distinctly woody at least near the base; partial
    inflorescences several- to ± many-flowered; fruiting pedicels 1–3 mm
    long; from the Mossel Bay Distr. E-wards...
    .... **1 (b). *C. spermacocea* subsp. *orientalis***

5. Stipular sheaths broadly cup-shaped, (4) 5–7 mm wide, 3–5 mm long;
    corolla tubes (♀) 3–5.5 mm long; fruits crowned by 4 calyx lobes ...
    .... **7. *C. heteromorpha***
5*. Stipular sheaths funnel- or cup-shaped, much smaller; corolla tubes
    (♀, ♂) 1–2.5 mm long; fruits crowned by 5–6 calyx lobes............ 6.

6. Stipular sheaths funnel-shaped, (1.5) 2–3 (4) mm long; fruits (4) 5–
    7.5 × 2–3 mm.. **6. *C. vaginellata***
6*. Stipular sheaths cup-shaped, not longer than 0.5–1.2 mm; fruits 2–
    4 × (2.5) 3–3.5 mm.. 7.

7. Flowers with 1 stigma; fruits 1-seeded; leaves small, 3–7 (9) × 0.7–
    1 (2) mm.. **5. *C. burchellii***

7*. Flowers with 2 stigmas; fruits 2-seeded; leaves (3) 5–15 (18) × (0.6) 0.8–2 mm .... 8.

8. More or less erect dwarf shrubs, internodes (except near shoot tips) frequently longer than the leaves; fruiting pedicels ca 1–7 mm long .... **4 (a). *C. scabra* subsp. *scabra***

8*. Decument, ± much-branched dwarf shrubs, internodes mostly shorter than the leaves; fruiting pedicels to ca 1 mm long .... **4 (b). *C. scabra* subsp. *rupestris***

1. ***Carpacoce spermacocea*** (Rchb.) Sonder in Harvey & Sonder, Fl. Cap. **3**: 33 (1865); Salter in J. S. Afr. Bot. **3**: 116 (1937), in Adamson & Salter, Fl. Cape Penins.: 734 (1950); Puff in Fl. S. Afr. **31**, 1 (2): 53 (1986).
    ≡ *Anthospermum spermacoceum* Rchb. in Sprengel, Syst. Veg. **4**, 2: 338 (1827); Cruse in Linnaea **6**: 17 (1831), in Linnaea **7**: 134 (1832). Type: "C.B.S.", no collector given. Neotype: "Reich[enbach] 63.9" (E!, GH!, LY!, WU!)[1].

    – *Lagotis spermacocea* (Rchb.) E. Meyer in Drège in Flora **26**, Bes. Beigabe: 197 (1843), nom. non valide publ.

    – *Anthospermum foetidum* Ecklon, nom. nud.

    Two subspecies are recognized:

1 (a). Subsp. *spermacocea*

    Synonyms as above.

    ♂, ♀ or occasionally ♂ + ♀, lax, straggling or scrambling, few- to several-stemmed (strongly) foetid perennial herbs with ± woody bases or short rhizomes. Stems ca 20–60 (90) cm long, ca 1–2.5 mm in diameter at the base and ca 0.8–1.5 (2) mm in the midstem region, sometimes rooting at the lower nodes, branching irregular, branches to 30 cm long, subglabrous or slightly scabrid below the nodes; shoots often with contracted to slightly elongated leafy short shoots. Leaves decussate; blades 10–18 (23) × (1.5) 2–3 (4) mm, narrowly ovate-lanceolate to linear-lanceolate,

---

[1] These specimens are almost certainly duplicates of Ecklon & Zeyher 2312 or "63.9" (from "below Constantia, Tafelberg and in the Caledon and Swellendam Provinces"; also see Collections) which must have been available to and studied by Reichenbach.

± spreading, membranaceous, glabrous or margins faintly scabrid; apices acuminate, often recurved; petioles 0; stipular sheaths cup-shaped, ca 1–1.5 × 1.5–2 mm, often scabrid, with 3–5 (or sometimes more) bristles, ca 1–2 (2.5) mm long. Flowers ♂ or ♀, in groups of few to 1, lateral and/or terminal, mostly on short lateral branches; fruiting pedicels elongated, 3–12 (20) mm. Calyx (4)5–6-merous, lobes (1) 2.5–5 × 1 mm, linear-lanceolate, 1 or 2 often much larger than the others, erect to spreading, or apices or upper thirds recurved. Corolla (4)5(6)-merous, greenish, greenish-yellow or dark purplish-brown to blackish; ♂: tube 1–2 × 0.3–0.5 mm, narrowly funnel-shaped, lobes 2–3.5 × 0.7–1.2 mm, ± lanceolate, hooded, ± spreading-recurved; stamens (4) 5 (6), filaments 2.5–4 mm long, anthers 1.5–2.2 × 0.3–0.5 mm; gynoecium as in ♀, but stigma only ca 5–8 mm long. ♀: tube 0.8–1.2 × 0.3–0.4 mm, ± cylindrical, lobes 0.7–1.5 × 0.2–0.4 mm, ± linear-lanceolate, hooded, ± erect (corollas somewhat larger in ♀ with rudimentary ± sessile anthers); style 0, stigma 1, 5–13 × 0.3–0.6 mm, purplish or occasionally whitish-grey; ovary green, ca 1.5–2 × 0.7–1.5 mm, with 1 fertile ovule. Fruits green, turning dark brown, ca 2–3.5 (4) × 1–1.5 mm, ± obovate to cylindrical, often ± ribbed, crowned by the persistent calyx lobes, 1-seeded; diaspore ca 2–3 (4) × 1–1.5 mm, greyish to black, ± cylindrical, hollowed out at the base (hollow sometimes filled with an easily removable "plug" of spongy tissue), ± rugose, occasionally indistinctly ribbed.

Chromosome Number (Table 3): 2 n = 22.

Average Pollen Diameter (Fig. 49): 34.5–39.3 μm.

Habitat: Always in wet or damp, often shady and sheltered localities such as in scrub near streams or at the edge of kloof forest; also in marshy or boggy areas, near trickles of water at the base of cliffs, etc. Usually in fine (TMS derived) sand mixed with black humus. – Ca 50–800 (1 000) m.

Distribution (Map, Fig. 108 b): S Africa, SW Cape Prov. From the Cape Peninsula and the Tulbagh Distr. E to the Heidelberg Distr.

Critical Remarks: See subsp. *orientalis*.

Collections

South Africa. Cape – **3318** (Cape Town): Kirstenbosch, below Window gorge (-CD), Salter 6413-6416 (BOL, K); Jonkershoek, Jakkalsvlei (-DD), Taylor 5708 a (PRE). – **3319** (Worcester): Tulbagh D., Witzenberg (-AA), Stokoe s.n. sub SAM 59553 (SAM; ± approach. subsp. *orientalis*); Dutoitskloof (-CA), Drège s.n. (E, FI, G × 3, GH, K, LY × 2, M, MO, S, SAM, W × 5); Frenchhoek (-CC), Schlechter 9225 (BM, BOL, BR, E, G, K, MO, PRE, S, US, W). – **3320** (Montagu): Duiwelsbos, N of Swellendam (-CD), Puff 800919-3/1 (WU); Heidelberg D., Tradouws Pass Kloof (-DC), Taylor 3550 (PRE, STE). – **3418** (Simonstown): Muizenberg (-AB), Compton 21216 (NBG); Buffels Bay [Cape of

Good Hope Nature Reserve] (-**AD**), Compton 6038 (BOL); Lourensford Estate (E of Somerset West) (-**BB**), Parker 4279 (BOL, GH, K, NBG); Kogelberg Forest Reserve (-**BD**), Boucher 1144 (PRE, STE), Puff 800101-2/1 (WU). – **3419** (Caledon): Lebanon Forest Reserve (-**AA**), Kruger 541 (PRE); Vogelgat (Private) Nature Reserve, E of Hermanus (-**AD**), Puff 800917-3/1, -6/4 (WU), Williams 2604 (K, NBG, PRE; approach subsp. *orientalis*). – Imprecise localities: "below Constantia, Tafelberg and in the Caledon and Swellendam provinces", Ecklon & Zeyher 2312 [or as "63.9" (= Tafelberg; cf. Drège 1847 a)] [G, GOET, M, MO – mixed with *Anthospermum* spp., PRE, S, SAM, W × 2; as "Reich. (Reichenbach) 63.9": E, GH, LY, WU]; "CBS", Harvey 482 (E × 2), 12157 (BM, K); Sparmann s.n. (S); Thunberg (herb. Gaström) s.n. (S; mixed with *Anthospermum herbaceum*); herb. Thunberg sub 23315 (UPS); Wallich & Hartman s.n. (S); Zeyher s.n. (K). – Additional collections: Boucher 1069 (STE); de Vos s.n. sub PRE 41828 (PRE); Dümmer 672 (E); Ecklon 30 (E, GOET, K, M, MO, PRE, S, W), s.n. (S); Esterhuysen 346, 448 (PRE); Goldblatt 3738 (MO); Levyns 10381 (BOL); Marloth 11459 (BOL, PRE, STE), 11459 b (BOL, STE); Martin 4605 (PRE); Moss 18991 (J); Pappe s.n. (K); Parker 4518 (BOL, K, MO, NBG, SAM); Puff 791228-2/3, -3/1, 800908-1/1 (WU); Salter 5770 (BOL); Stokoe 5770 (SAM), 7586 (NBG, PRE), 8647 (BOL, PRE), s.n. sub SAM 64193, sub SAM 67530 (SAM), sub SAM 59519 (SAM, STE), sub SAM 69998 (NBG); Taylor 7128 (PRE, STE); Wallich 425 (BM, K), s.n. (G); Wolley Dod 773 (BOL, K), 1601 (BM), 1992, 2078 (BOL), 2893 (BM, K; as s.n. in MO); Wright s.n. (US); Zeyher s.n. (GOET, SAM).

1 (b). Subsp. *orientalis* Puff in Fl. S. Afr. 31, 1 (2): 56 (1986).

Type: South Africa, Cape Prov., Montagu Pass, Puff 790909-5/2 (WU, holo.!; BOL, GRA, NBG, PRE, STE, iso.!).

♀̂, ♀, occasionally ♀̂ + ♀ or (seldom) ♀̂ + ♂̂, often quite robust, straggling, scrambling or sometimes ± erect, few- to several-stemmed (strongly) foetid perennial herbs with woody bases, or distinct dwarf shrubs. Stems ca (10) 20–150 cm long, ca 3–6 mm in diameter at the base and ca 1.5–3 mm in the midstem region, branching often quite regular, branches usually not more than 20 cm long, ascending to ± spreading; subglabrous or younger parts scabrid especially below the nodes; shoots often with contracted to slightly elongated short shoots; plants often ± densely leafy. Leaves decussate; blades (15) 20–30 (35) × (2) 3–5 (8) mm, narrowly ovate-lanceolate to linear-lanceolate, membranaceous, margins usually scabrid; apices acuminate, sometimes recurved; petioles 0; stipular sheaths cup-shaped, ca 1.5–3 × 2.5–3 mm, usually densely scabrid, with 3– 5 (–8) bristles, ca 1–2.5 (3) mm long. Flowers ♀̂, ♀ or (seldom) ♂̂, ± sessile, in groups of ± many (8–5) to few, lateral and/or terminal, mostly on short lateral branches; fruiting pedicels not longer than ca 1–3 mm. Calyx 5–6-merous, lobes 2–5 × 0.6–0.8 mm, linear-lanceolate, 1 or 2 often much larger than the others, ± erect to spreading, apices often recuved. Corolla

(4) 5 (6)-merous, greenish to dark brown; ♀ (♂): tube 0.8–1.5 (2) × 0.4–
0.8 mm, (narrowly) funnel-shaped, lobes 2–3.5 (4) × 0.8–1.2 mm, ± lanceolate, hooded, ± spreading-recurved; stamens (4) 5 (6), filaments
(2) 2.5–4 mm long, anthers 1.7–2.5 × 0.3–0.5 mm; ♂: gynoecium as in ♀, but
stigma shorter; ♂: rudimentary gynoecium well discernible. ♀: tube 0.6–
1 × 0.4 mm, ± cylindrical, lobes 0.6–1 (1.5) × 0.2–0.4 mm, ± linear-
lanceolate, hooded, erect to ± spreading (corollas somewhat larger in ♀
with rudimentary ± sessile anthers); style 0, stigma 1, 5–12 × 0.4 mm,
purplish-red; ovary green, ca 1.5–2.5 × 1–1.5 mm, with 1 fertile ovule.
Fruits green, turning dark brown, ca (3) 3.5–5 × 1.5–2.5 mm, ± obovate
to cylindrical, crowned by the persistent calyx lobes, 1-seeded; diaspore ca
3–3.5 (4) × 1.2–1.7 mm, greyish to black, ± cylindrical, hollowed out at
the base (hollow sometimes filled with an easily removable "plug" of
spongy tissue), ± rugose, occasionally indistinctly ribbed. – Figs. 34 a–e,
43 a, 44 f.

Chromosome Number (Table 3): n = 11.
Average Pollen Diameter (Fig. 49): 38.1–40.5 μm.
Habitat: As for subsp. *spermacocea.* – Ca (30) 50–1 200 (1 850) m.
Distribution (Map, Fig. 108 c): S Africa. In the E part of SW Cape
Floristic Region from the Mossel Bay Distr. to the Uitenhage and the Port
Elizabeth Distr.

Critical Remarks: C. *spermacocea* is subdivided into two geographi-
cal/ecological subspecies. Subsp. *orientalis* occurs in high rainfall E
coastal areas and some of the mountains further inland (also with rather
high precipitation?), but – more important – in areas with rainfall ±
evenly distributed throughout the year or with a slight excess of summer
rain. In contrast, subsp. *spermacocea* is confined to an area receiving only
winter rainfall.

The two subspecies are not always satisfactorily separable morphologically. Odd luxurious "forms" of subsp. *spermacocea* (e.g.,
WILLIAMS 2604 from 3419-AD) may overlap with subsp. *orientalis* in, for
example, leaf size and in having more flowers per partial inflorescence
than usual, etc. In general, however, subsp. *orientalis* tends to be more
robust, is often distinctly woody, has inflorescences with more flowers and
is generally larger (leaves, fruits, etc.).

Within subsp. *orientalis,* two ± distinct "Forms" (in the sense of
ecotypes) can be distinguished:
(a) "Typical" subsp. *orientalis:* luxurious, straggling or scrambling
plants with stems to 1.5 m long.
(b) Erect, ± stunted, distinctly shrubby and rather densely leafy plants

to ca 50 cm tall which are usually only found in the mountains (750 m and above), where they occur in rocky, wet areas (at the base of cliffs, etc.).

C. spermacocea seems to be allied to the large-leaved, shrubby and narrowly endemic *C. gigantea* (2., below), but this species pair appears to occupy a somewhat isolated position within the genus. There may be a rather remote relationship to *C. curvifolia* (3.; see there for further comments).

C. spermacocea has, in the past, often been confused with *Anthospermum herbaceum* (see comments, p. 312).

Collections

South Africa. Cape – **3322** (Oudtshoorn): Robinson pass (-**CC**), PUFF 791222-3/2 (WU); Montagu pass (-**CD**), COMPTON 7593 (NBG), ESTERHUYSEN 10862 (BOL, PRE), PUFF 790909-5/2 (BOL, GRA, NBG, PRE, STE, WU); Trakadokow Stn., nr. Ronnee Vallei P.O. (-**DC**), BURCHELL 5743 (G, GH, GOET, K, M, W). – **3323** (Willowmore): Prince Alfred pass, Ysternek Nature Reserve (-**CC**), PUFF 791222-2/1, 800927-3/1, -3/1 A (WU); Uniondale D., Kouga Mts., Kouga Peak (-**DB**), ESTERHUYSEN 10811, 16267 (BOL); –, Tsitsikamma Mts., Formosa Peak (-**DC**), PUFF 800928-1/3, -2/1 (WU); –, Outeniqua Mats. nr. Joubertina, Camel Pile (-**DD**), ESTERHUYSEN 13579 (BOL). – **3324** (Steytlerville): Humansdorp D., Kareedouw pass (-**CD**), GILLETT 1526 (STE); Great Winterhoek Mts., Cockscomb (-**DB**), ESTERHUYSEN 27492, 28003 (BOL, PRE). – **3325** (Port Elizabeth): Groendal Wilderness Res., Kroompoortkloof Catchment basin, Farm Strydomsberg (-**CA**), SCHARF 1889 (PRE); Loerie Plantation (-**CC**), DIX 78 (BOL). – **3422** (Mossel Bay): Knysna D., Belvedere (-**BB**), DUTHIE 1011 (MO, STE). – **3423** (Knysna): –, Springfield Plantation (-**AA**), Forestry Dept. herb. s.n. sub STE 15214 (STE), KEET 23 (PRE); Humansdorp D., above Storms River mouth (-**BB**), ACOCKS 21177 (K, M, PRE). – **3424** (Humansdorp): Witte Els Bosch (= Witelsbos) (-**AA**), FOURCADE 2295 (BOL, K, PRE, STE); Clarkson (-**AB**), THODE A 856 (GH, K, MO, NH, PRE). – Imprecise localities: between Knysna and Avontuur (3423-AA to 3323-CA), FRIES, NORLINDH & WEIMARCK 1612 (MO, PRE); "Cape", SIEBER "Fl. Cap. 239" (S, W). – Not traced: Hofmans Bosch, BRITTEN 1163 (GRA, PRE). – Additional collections: "ECKLON & ZEYHER" (ZEYHER?) "4.10" (PRE); FOURCADE 27 (BOL, GRA, STE); GALPIN 4103 (GRA, PRE); HOLLAND 3872 (BOL); HUMBERT 9890 (P); KAPP 68 (PRE); LONG 672 (K, PRE), s.n. (GRA); OLIVER 4470 (PRE, STE); PRIOR s.n. (K, ± atypical); ROGERS 2849 (GRA); STORY 2755 (PRE); THODE A 2540 (K, NH, PRE); TYSON 992 (BM, BOL, G, GH, K, W), 2954 (SAM).

2. ***Carpacoce gigantea*** PUFF in Fl. S. Afr. **31**, 1 (2): 56 (1986).
Type: South Africa, Cape Prov., "in the ascent of the Craggy Peak in the Great Range at Swellendam" [Langeberg N of Swellendam], BURCHELL 7320 (K, holo.!; GH, iso.!).

♂ or ♀, ± erect, lax (dwarf) shrubs to 90 cm. Stems ± much-branched, ca. 2.5–4 mm in diameter in the midstem region, glabrous; internodes shorter than the leaves. Leaves decussate; blades (35) 50–70 (80) × (2) 3–5 (7) mm, linear-lanceolate, glabrous, margins flat to slightly revolute; apices acute to acuminate; petioles 0; stipular sheaths cup-shaped, ca 2–5 mm long, with or without several short bristles to ca 1 mm long. Flowers ♂ or ♀, solitary on short lateral branches, mostly hidden amongst foliage; pedicels to 4 mm long. Calyx 3-merous, lobes (6) 8–12 × 0.5–1 mm, linear-lanceolate, ± erect, sometimes ± unequal in size. Corolla 5 (6)-merous; ♂: tube 4–5.5 × 0.4–0.6 mm, ± cylindrical, lobes 4–4.5 × 0.7–1 mm, ± lanceolate, hooded, ± spreading; stamens 5 (6), filaments 4–6 mm long, anthers 3–3.5 × 0.5 mm; gynoecium as in ♀. ♀: tube 2.5 × 0.4 mm, cylindrical, lobes 0.7 × 0.1–0.2 mm, erect, hooded; style 0, stigma 1, ca 10–15 × 1 mm; ovary 2–2.5 × 1.2–2 mm, with 1 fertile ovule. Fruits grey, 2.5–5 × 1.2–2 mm, ± ellipsoidal to cylindrical, crowned by the persistent calyx lobes, 1-seeded; diaspore (2) 2.5–3 × 1.5 mm, greyish to black, ± elliptic to ± rectangular in outline, rugose.

Chromosome Number: Unknown.
Average Pollen Diameter (Fig. 48): 36.3 μm.
Habitat: Moist, well-drained mountain slopes; confined to TMS.
Distribution (Map, Fig. 108 h): S Africa, SW Cape Prov. Only known from the mountains above Swellendam.

Critical Remarks: This distinct species (ovaries/fruits crowned by 3 calyx lobes; the largest leaves in the genus) is only known from three collections, all from the same locality (and perhaps from the same population?). I visited this locality in 1979 and 1980, but in spite of detailed habitat descriptions and sketches (kindly provided by H. C. TAYLOR, who collected the plants there in 1962) and an extensive search, no plants were found. I suspect that the species may be extinct. If the species — like the allied *C. spermacocea* — (1) can only regenerate from seed and if (2) the frequency of fires has been higher than the regeneration cycle in recent years, it is quite possible that the population may have gradually vanished since it was last collected in 1962.

Collections

South Africa. Cape – **3320** (Montagu): Swellendam D., Swellendam Mt., 9 o'Clock Peak (= Craggy Peak) (**-CD**), BURCHELL 7320 (GH, K), 7396 (K), TAYLOR 4241 (PRE, STE).

3. *Carpacoce curvifolia* PUFF in Fl. S. Afr. **31**, 1 (2): 57 (1986).
    Type: South Africa, Cape Prov., Uniondale Distr., Tsitsikamma Mts. near Joubertina, ESTERHUYSEN 27334 (BOL, holo.!; WU, iso.!).

♂ or ♀, decumbent, seldom ± erect, many-stemmed perennial herbs with woody bases or dwarf shrubs. Stems ca (10) 15–50 (60) cm long, ± sparsely to much-branched, glabrous; longest internodes 7–10 (20) mm long. Leaves decussate or (rarely) in whorls of 3; blades (4) 6–10 (12) × (1) 2–4 mm, lanceolate to ovate-lanceolate, usually strongly recurved, rigid, shiny, glabrous or margins faintly scabrid; apices acuminate; petioles 0; stipular sheaths cup-shaped, 1–2 × 1.3–3 mm, glabrous or faintly scabrid, sometimes with a median seta ca 0.1–0.2 mm long. Flowers ♂ or ♀, single or paired at the nodes, ± sessile, largely hidden in the stipular sheaths, or terminal, with pedicels up to 4 mm long. Calyx 4-merous, lobes (0.8) 1–2.4 × 0.7–1 mm, lanceolate to triangular, ± spreading, 2 often up to twice as long as the others. Corolla 4-merous; ♂: tube (1) 1.5–2 (3) × 0.3–0.4 mm, ± cylindrical, lobes ca 2–3 × 0.8–1 (1.5) mm, linear-lanceolate, hooded, ± spreading; stamens 4, filaments 3–4 mm long, anthers 0.8–1.5 (1.7) × 0.3–0.5 mm; gynoecium as in ♀. ♀: tube 0.5–1 mm, cylindrical, lobes 0.5–1 × 0.3 mm, linear, hooded, erect; style 0, stigma 1, 4–6 × 0.3–0.4 mm; ovary 1–1.2 × 0.6–0.8 mm, with 1 fertile ovule. Fruits greyish-brown, ca 2–2.5 × 1–1.5 mm, ± obovate, lower half often laterally compressed and hidden in stipular sheath, crowned by the persistent calyx lobes, 1-seeded; diaspore ca 1.2–2 × 0.8–1.2 mm, black, ellipsoidal, the base often with an appendage of (easily removable) spongy tissue, ± rugose. – Fig. 15 g.

Chromosome Number: Unknown.
Average Pollen Diameter (Fig. 49): 34.9–35.3 μm.
Habitat: On steep rocky slopes, on ledges and in crevices. Appears to favour S aspects and to prefer moist, sheltered situations. – Ca 1 000–2 000 m.
Distribution (Map, Fig. 108 f): S Africa. Endemic to the mountains in the E part of the SW Cape Floristic Region and occurring from the Uitenhage to the Ladismith Distr. (Anysberg).

Critical Remarks: Unlike *C. spermacocea* subsp. *orientalis* (1.), which is also endemic to the E part of the SW Cape Floristic Region and also largely occurs in areas with a ± even annual rainfall distribution or slight summer rainfall excess, *C. curvifolia* seems to be confined mostly to the (somewhat drier?) inland mountain ranges (i.e., belongs to the group of "Karro-Mountain" and "South-Eastern Endems" of WEIMARCK 1941).

The collection from the W-most, ± isolated locality (Anysberg; ESTERHUYSEN 25973) is a somewhat atypical, depauperate specimen, which, however, clearly belongs to this species.

Habit, similar (although often somewhat smaller) flowers and diaspores and its occurrence in moist to wet habitats seem to suggest a remote alliance to *C. spermacocea*.

Collections

South Africa. Cape – **3320** (Montagu): Ladismith D., Anysberg (**-DA**), Esterhuysen 25973 (BOL, WU). – **3322** (Oudtshoorn): Prince Albert D., nr. top of Swartberg pass (**-AC**), Bond 1539 (BOL), Esterhuysen 29576 (BOL), Stokoe s.n. sub SAM 69997 (SAM); Postberg, N of George (**-CD**), Burchell 5692 (K); Uniondale D., Mannetjiesberg (**-DB**), Esterhuysen 6400 (BOL, K). – **3323** (Willowmore) – , Kouga Mts., peak E of Smutsberg (**-DB**), Esterhuysen 7675 (BOL); – , Tsitsikamma Mts., Formosa Peak (**-DC**), Esterhuysen 27381 (BOL), Puff 800928-3/2 (WU); – , – , near Joubertina (**-DD**), Esterhuysen 27334 (BOL, WU). – **3324** (Steytlerville): Uitenhage D., Great Winterhoek Mts., Cockscomb (**-DB**), Esterhuysen 27093, 27529, 28042 (BOL, PRE).

4. *Carpacoce scabra* (Thunb.) Sonder in Harvey & Sonder, Fl. Cap. **3**: 33 (1865); Salter in J. S. Afr. Bot. **3**: 115 (1937); Puff in Fl. S. Afr. **31**, 1 (2): 57 (1986).

≡ *Anthospermum scabrum* Thunb., Prod. Pl. Cap. **1**: 32 (1794), Fl. Cap. **1**: 573 (1813), Fl. Cap. ed. Schultes: 158 (1823); Cruse, Rub. Cap.: 14 (1825), in Linnaea **6**: 14 (1831).

Types: [South Africa, Cape Prov.] "CBS", Thunberg (sheet 23317, UPS, lecto.!, BOL, PRE, WU – photo!; sheet 23318, UPS, syn.!, BOL, PRE, WU – photo!).

Two subspecies are recognized:

4 (a). Subsp. *scabra*

♀̂, ♀ or occasionally ♀̂ + ♀, ♀̂ + ♂ or ♂, many-stemmed, often rounded to ± cylindrical, sometimes slightly foetid dwarf shrubs. Stems (8) 15– 40 (45) cm long, ca (1) 2.5–5 mm in diameter at the base and ca 0.8–1.5 mm in the midstem region, sparsely to ± much-branched, branches ascending to ± spreading, glabrous to scabrid; internodes mostly longer than the leaves, plants not densely leafy. Leaves decussate; blades (3) 5– 15 (18) × (0.6) 0.8–2 (2.5) mm, linear(-lanceolate), ascending to erect, ericoid, often shallowly concave above or semiterete in section, scabrid at least on the margins; apices acuminate; petioles 0; stipular sheaths cupshaped, ca 0.5–1 mm long, glabrous to faintly scabrid, sometimes with a median seta ca 0.1 mm long. Flowers ♀̂, ♀ or occasionally ♂, solitary or paired, ± sessile to shortly pedicellate; fruiting pedicels sometimes elongated, to ca 7 mm long. Calyx 5 (6)-merous, lobes (2.5) 4–6 × 0.7– 1 mm, linear-lanceolate, subequal, spreading to ± erect. Corolla 5 (6)-

merous, yellowish-green, often with reddish-brown or brownish-purple streaks; ♀, ♂: tube 1.5–2.5 × 0.3–0.5 (1) mm, (narrowly) funnel-shaped, lobes 2.5–4 × 1–2 mm, ± lanceolate, hooded, ± spreading-recurved; stamens 5 (6), filaments 3–4 mm long, anthers 2–2.5 (3) × 0.5–1 mm; ♀: gynoecium as in ♀; ♂: small rudimentary ovary and stigmas present. ♀: tube 0.3–0.6 × 0.4–0.5 mm, ± cylindrical, lobes 0.5–1.2 × 0.4–0.5 mm, ± linear(-lanceolate), hooded, ± erect (corollas somewhat larger in ♀ with rudimentary ± sessile anthers); style 0, stigmas 2, (4) 5–9 × 0.4–0.5 mm, purplish-red; ovary green, ca 1.5–2 × 1–2 mm, with 2 fertile ovules. Fruits green, turning grey-bown, ca 2–3 × (2.5) 3–3.5 mm, ± turbinate to subglobose, crowned by the persistent calyx lobes, 2-seeded; each diaspore ca 1.5–2.5 × 1–1.5 mm, greyish-white to dark grey, ± elliptical to obovate in dorsal view, dorsal side convex, ventral side plane to slightly concave, ± rugose and ribbed. – Figs. 43 b, f, j–k.

Chromosome Number (Table 3): n = 11 (Fig. 47 c), 2 n = ca 22.
Average Pollen Diameter (Fig. 49): 34.8–39.5 µm.

Habitat: Seems to occur mainly in sandy areas on mountain slopes, but also grows on ridges and amongst rocks. Primarily found in TMS areas, occasionally also over Witteberg quartzite. Apparently confined to arid Fynbos types. – Ca (150) 650–1350 (1 600) m.

Distribution (Map, Fig. 108 d): S Africa. In drier inland areas of the SW Cape Prov. from the Calvinia Distr. SE and S to the Laingsburg and the Worcester Distr.

Critical Remarks: See subsp. *rupestris.*

Collections

South Africa. Cape – **3118** (Vanrhynsdorp): Vanrhynsdorp D., Giftberg (-DC), Esterhuysen 22105 (BOL). – **3119** (Calvinia): Farm Glennridge nr. Nieuwoudtville (-AC), van Breda 1015 (K, PRE); Farm Lokenburg, ca 33 km S of Nieuwoudtville (-CA), Acocks 17356 (K), 17377 (K, PRE), Leistner 416 (PRE), Story 4327 (GRA, PRE), 4360 (K, PRE). – **3219** (Wuppertal): from summit of Pakhuis pass to Heuningvlei For. Stn. (-AA), Puff 800914-6/1 (WU); Koudeberg nr. Wuppertal (-AC), Bolus 9054 (BOL, K, NBG, PRE); Cedarberg, Boschkloof (-CA), Pocock 689 (STE); Ceres D., Schurweberg (-CD), Esterhuysen s.n. (BOL). – **3319** (Worcester): Ceres D., Michells pass, below Castle Rocks (-AD), Esterhuysen 14148 (BOL); Ceres D., Swartruggens, ca 14 km NNE of Farm Hoop en Uitkomst (-BA), Acocks 23743 (K, PRE); Karroo Poort, Hafström & Acock(s) 1428, 1440 (PRE, S); Worcester D., Fonteintjiesberg (-CB), Esterhuysen 10961 (BOL, K); –, Gouronna (-CB?), Esterhuysen 3746 (BOL). – **3320** (Montagu): Ceres D., poort N of Pienaarskloof (-AA), Acocks 23705 (PRE); Laingsburg D., Witteberg (-BC), Compton 2791 (BOL); –, S of Bantams, Esterhuysen 30511 (BOL); –, S side, between Farms Fisantekraal and Ezelsfontein, Puff 790914-4/2 (WU). – No locality given ("CBS"): Thunberg ("herb. Swartz, herb. Gaström") (UPS 23317, 23318, S × 3).

30*

4(b). Subsp. *rupestris* Puff in Fl. S. Afr. **31**, 1 (2): 59 (1986).
Type: South Africa, Cape Prov., Ceres-Tulbagh Distr., Swartgat Peak, Witzenbergen, Esterhuysen 27930 (BOL, holo!).

♀, ♀ or occasionally ♀ + ♀ or ♀ + ♂, several-stemmed, decumbent or sometimes cushion-forming dwarf shrubs. Stems 10–20 (25) cm long, ca 3–5 mm in diameter at the base and ca 1–1.5 mm in the midstem region, ± much-branched, at least younger parts densely scabrid; internodes mostly (much) shorter than the leaves, plants densely leafy. Leaves decussate; blades (6) 8–12 (15) × 0.8–1.5 (2) mm, linear-lanceolate to narrowly oblanceolate, ± erect, ± triquete in section or ± flat above and with a prominent midrib below, scabrid at least on the margins; apices ± pungent to acuminate; petioles 0; stipular sheaths cup-shaped, 0.5–1 mm long, scabrid, usually with a median seta ca 0.1–0.3 mm long. Flowers ♀, ♀ or occasionally ♂, solitary, ± sessile; fruiting pedicels sometimes to 1 mm long. Calyx 5 (6)-merous, lobes 2.5–4 × 0.2–0.5 mm, linear(-lanceolate), subequal, erect to spreading. Corolla 5 (6)-merous; ♀, ♂: tube 1–2 (2.5) × 0.3–0.5 mm, narrowly funnel-shaped, lobes 2.5–4 (5) × 0.8–1.2 mm, ± lanceolate, hooded, ± spreading; stamens 5 (6), filaments (2) 2.5–5 mm long, anthers 1.5–2.5 × 0.5–0.7 mm; ♀: gynoecium as in ♀; ♂: small rudimentary ovary and stigmas present. ♀: tube 0.3–0.5 (1) × 0.3 mm, ± cylindrical, lobes 0.5–0.8 × 0.2–0.3 (0.5) mm, ± linear(-lanceolate), hooded, ± erect, often hidden between calyx lobes (corollas somewhat larger in ♀ with rudimentary ± sessile anthers); style ± 0, stigmas 2 (rarely 1), 5–9 × 0.3 mm; ovary green, ca 1–1.5 (2) × 1–1.7 mm, with 2 (rarely 1) fertile ovule(s). Fruits green, turning dark brown or grey-brown, 2–3.5 × (1.5) 2–3.5 mm, ± obovate, crowned by the persistent calyx lobes, 2- (rarely 1-) seeded; each diaspore ca 1.5–2.5 × 1.2–1.5 mm, black, ± obovate in dorsal view, dorsal side convex, ventral side strongly concave, ± rugose and ribbed.

Chromosome Number: Unknown.
Average Pollen Diameter (Fig. 49): 34.6–37.7 µm.
Habitat: On rocky slopes amongst rocks, on ledges, etc.; mainly over (confined to?) TMS. – Ca. 1 200–2 000 m.
Distribution (Map, Fig. 108 e): S Africa, SW Cape Prov. In the higher mountains of the Worcester Distr. and extending N to the Witsenberg range.

Critical Remarks: Subsp. *rupestris* differs from subsp. *scabra* in habit (i.e., low, densely leafy plants), even more reduced inflorescences, shorter pedicels and flowers with smaller calyces. Subsp. *rupestris* seems to be confined to rocky areas on the highest mountains and to mountain tops. As there is some overlap in several morphological characters, subspecific rank seems to be the most appropriate for the two taxa.

C. scabra appears to be allied to *C. burchellii* (5., below). The latter, however, has ovaries/fruits with always only 1 fertile ovule, whereas *C. scabra* is the only species in the genus with two fertile ovules (although odd flowers with only 1 fertile ovule also occur in plants of subsp. *rupestris!*).

Collections

South Africa. Cape – **3319** (Worcester): Witzenbergen, Swartgat Peak (**-AA**), ESTERHUYSEN 27930 (BOL, WU); – , Matroosberg (**-BC**), ESTERHUYSEN s.n. sub BOL 31694 (BOL); – , Upper Wellington Sneeukop (**-CA**), ESTERHUYSEN 26491 (BOL); – , Hex River Mts., Horseshoe Ridge Peak (**-CB**), ESTERHUYSEN 22208 (BOL); – , Kwadouws Mts., above Orchard (**-DA**), ESTERHUYSEN 10900 (BOL); – , Sawedge Peak, E of Keeromsberg, ESTERHUYSEN 31146 (BOL, WU); Caledon/Worcester D. border, Bobbejaankop nr. Jonaskop, Riviersonderend Mts. (**-DC**), TAYLOR 6567 (PRE, STE). – Not traced: Worcester D., Hex River Mts., Shale Peaks, ESTERHUYSEN 28714 (BOL).

5. *Carpacoce burchellii* PUFF in Fl. S. Afr. **31**, 1 (2): 59 (1986).
Type: South Africa, Cape Prov., Caledon Distr., S slopes of Riviersonderend Mts. near Greyton, ESTERHUYSEN 20782 (BOL, holo.!).

☿ or ♀, many-stemmed, ± erect and cylindrical dwarf shrubs or subshrubs with ± woody bases. Stems ca (5) 8–30 cm long, ca 2 mm in diameter at the base and ca 0.5 mm in the midstem region, unbranched to sparsely branched, glabrous; internodes usually somewhat longer than the leaves. Leaves decussate; blades 3–7 (9) × 0.7–1 (2) mm, linear, ericoid, glabrous or margins slightly scabrid; apices ± pungent to acuminate; base of blades ± sac-like; petioles 0; stipular sheaths cup-shaped, ca 0.5–1.2 mm long, glabrous or slightly scabrid on the rim, occasionally with a median seta ca 0.5–0.6 mm long. Flowers ☿ or ♀, in groups of 1–4 at the nodes, ± sessile, or sometimes terminal, shortly pedicellate. Calyx 5-merous, lobes (1.5) 2–3.5 × 1–1.5 mm, 1 often distinctly larger than the others. Corolla 5-merous; ☿: tube 1–2 × 0.5 mm, cylindrical to ± funnel-shaped, lobes (2) 3–4 × 0.5–1 mm, ± linear-lanceolate, hooded, spreading; stamens 5, filaments ca 3–4 mm long, anthers 1.5–2 × 0.6 mm; gynoecium as in ♀, but stigma often shorter; ♀: tube 0.3–0.6 × 0.4 mm, cylindrical, lobes 0.4–1 × 0.2–0.3 mm, ± linear, hooded, erect; style 0, stigma 1, (4) 5.5–8 × 0.4 mm; ovary (1) 1.5–2.2 × 1–1.5 mm, with 1 fertile ovule. Mature fruits not seen.

Chromosome Number: Unknown.

Average Pollen Diameter (Fig. 49): 33.5–35.1 µm.

Habitat: Occurs mainly on moist, well-drained S slopes; also known to occur in marshy situations and shady kloofs. – Ca 300–1 150 (1 350) m.

Distribution (Map, Fig. 108 g): S Africa, SW Cape Prov. Only known from a few mountains in the Paarl and Caledon Distr.

Critical Remarks: Plants are capable of resprouting after a fire, and the newly produced aerial shoots become rather long, remain unbranched and the leaves are mostly larger, wider and less distinctly ericoid (e.g., ESTERHUYSEN 13707). Plants apparently not exposed to fires (the majority of known collections) are distinctly dwarf shrubby.

STOKOE sub MARLOTH 11009 is a very dwarfed form; flowering shoots are only a few cm tall.

In habit, *C. burchellii* is very similar to *C. scabra* (subsp. *rupestris* in particular) but can be distinguished by its ovaries/fruits with only one fertile ovule.

Collections

South Africa. Cape – **3319** (Worcester): Paarl D., Wemmershoek Mts., Tierkloof (**-CC**), ESTERHUYSEN 20092 (BOL, K, PRE). – **3418** (Simonstown): Hangklip Mts. (**-BD**), STOKOE sub MARLOTH 11009 (PRE). – **3419** (Caledon): Caledon D., mts. nr. Palmiet R. mouth (**-AC**), ESTERHUYSEN 13707 (BOL, K, PRE); – , Riviersonderend Mts. nr. Greyton (**-BA**), ESTERHUYSEN 20782 (BOL); "Great Mountain of Baviaanskloof at Genadendal", BURCHELL 7736 (K).

6. *Carpacoce vaginellata* SALTER in J. S. Afr. Bot. **3**: 113, Fig. 2 (1937), in ADAMSON & SALTER, Fl. Cape Peninsula: 734 (1950); PUFF in Fl. S. Afr. **31**, 1 (2): 59 (1986).

Type: South Africa, Cape Prov., Cape Peninsula, Muizenberg Mt., SALTER 6271 (BOL, holo.!; K, NBG, iso.!).

♀̇, ♀, or occasionally ♀̇ + ♀, ♀̇ + ♂ or ♂, many-stemmed, cylindrical and suberect to ± rounded or occasionally decumbent, sometimes faintly foetid dwarf shrubs or subshrubs with ± woody bases. Stems (10) 15–40 (70) cm long, ca 2–7 mm in diameter at the base and ca 1–3 mm in the midstem region, ± much-branched (unbranched or sparsely branched in plants resprouting after a fire), branches ascending to ± erect, glabrous or younger parts occasionally scabrid; internodes frequently shorter than the leaves. Leaves decussate; blades (7) 10–25 (30) × 0.7–1.5 (2) mm, linear, ± erect, ericoid, semiterete to ± triquete in section, margins scabrid or with rigid whitish hairs to 0.2 mm long; apices often ± pungent or acuminate; petioles 0; stipular sheaths distinctly funnel-shaped, (1.5) 2–3 (4) mm long, glabrous to scabrid, sometimes with a median seta ca 0.3–0.4 mm long. Flowers ⚥, ♀ or occasionally ♂, mostly solitary, lateral and/or terminal on branches, ± sessile or shortly pedicellate. Calyx 5 (6)-merous, lobes 3–8 × 0.8–1.5 mm, linear(-lanceolate), subequal, ± erect. Corolla 5-(6-, very rarely 7-)merous, greenish-yellow, reddish-brown to dark purplish-brown; ⚥, ♂: tube (1) 1.5–2.5 × 0.6–1 mm, (narrowly) funnel-shaped to ± cylindrical, lobes (3) 3.5–5 (5.5) × (0.7) 1–2 mm, ± lanceolate, hooded, spreading to ± spreading-recurved; stamens 5 (6), filaments (2.5) 3.5–5.5 mm long, anthers (1.5) 2–3 (4) × 0.5–1 mm; ⚥: gynoecium as in ♀, but stigma always shorter and ovary sometimes smaller than in ♀; ♂: small rudimentary stigma and ovary present. ♀: tube 0.5–1.5 × 0.3–0.7 mm, ± cylindrical, lobes 0.5–1.5 × 0.1–0.5 mm, ± linear, hooded, ± erect, corolla usually hidden between calyx lobes (corollas somewhat larger in ♀ with rudimentary ± sessile anthers); style 0; stigma 1 (very rarely 2), (6) 8–12 (17) × 0.8–1.5 mm, purplish-red, seldom greyish; ovary green, ca 2–4 × 1.5–2 mm, with 1 (very rarely 2) fertile ovule(s). Fruits green, turning dark grey-brown, (4) 5–7.5 × 2–3 mm, ± obovate, crowned by the persistent calyx lobes, 1- (very rarely 2-)seeded; diaspore (3) 3.5–5 × 1.5–2 mm, grey to blackish, often laterally ± compressed, ± elliptical to obovate in dorsal view, rugose to muricate. – Figs. 10 b, 34 f, 43 d, g, 44 i.

Chromosome Number (Table 3): n = 11, 2n = 22.

Average Pollen Diameter (Fig. 49): 31.6–41.4 μm.

Habitat: In Coastal and Mountain Fynbos, mostly on sandy (or sandy-rocky) slopes, plateaux or flats. Frequently in TMS-derived sand, but occasionally also over kalksand or Cape Granite – TMS mixed sands. In the E-most part of its distribution range, it also occurs in grassveld. – Ca 20–1 000 (1 350) m.

Distribution (Map, Fig. 108 a): S Africa. Widely distributed in the SW Cape Prov. and extending as far E as the Albany and Bathurst Distr.

Critical Remarks: As opposed to *C. spermacocea* (1.), for example, plants are not usually killed by fire and new aerial shoots are produced from the base after a fire. Individuals that are frequently exposed to fire, often develop quite thick, ± disk-like woody bases.

Plants with 2-seeded or both 1- and 2-seeded fruits do occur but are uncommon. The frequency of plants with 2-seeded fruits varies from population to population. In most populations, all plants observed had the characteristic 1-seeded fruits. In a few others, however, e.g., in population Puff 800914-6/2, several plants had flowers with both 1 and 2 stigmas/fertile ovules.

"Pseudo-fruits" without well developed seeds are not uncommon.

They are (much) smaller than fertile fruits and are often reddish at first
and then brownish-grey; they could either have originated from rudimentary ovaries of ♂ flowers or be aborted fruits of ⚥ or ♀.

Superficially, *C. scabra* (4.) closely resembles *C. vaginellata* but can be
distinguished vegetatively by its shorter, cup-shaped stipular sheaths and
also by its 2-ovulate carpels (see SALTER 1937 for further details).
Although *C. vaginellata* is much more widely distributed than *C. scabra,*
both species may occur together in some areas. In my experience, the two
are easily distinguished in the field by the colour of their foliage alone: *C.
vaginellata* has fresh-green leaves, *C. scabra* a more blueish-green foliage.

Collections

South Africa. Cape − **3218** (Clanwilliam): plateau of Piketberg, "Sandleegte"
(**-DC, -DD**), ESTERHUYSEN s.n. sub PUFF 800914-4 A/1 -4 A/2 (WU). − **3219**
(Wuppertal): from summit of Pakhuis pass to Heuningvlei For. Stn. (**-AA**), PUFF
790716-1/3 B, 800914-6/2 (WU); Cedarberg, Algeria Nature Reserve, Ventersberg
(**-AC**), TAYLOR 7518 (K, PRE); summit of Buffelshoek pass (Citrusdal- Ceres rd.)
(**-CA**), PUFF 800903-2/2 (WU); top of Dasklip pass, above Porterville (**-CC**), PUFF
800913-4/3 (WU). − **3318** (Cape Town): Table Mt., above Bridle Path, S of
Wynberg Reservoir (**-CD**), ESTERHUYSEN 17290 (BOL, NBG, PRE); Malmesbury
D., Kalkboskraal (**-DA**), BOUCHER 4445 (PRE, STE); nr. Kraaifontein (**-DC**),
DÜMMER 96 b (E, GH), 1810 (E); Stellenbosch D., Jonkershoek Forest Reserve
(**-DD**), TAYLOR 5148 (K, M, PRE, STE). − **3319** (Worcester): Olifants River Mts., S
of Groen (**-AA**), ESTERHUYSEN 13494 (BOL); Worcester D., Du Toits Peak above
Du Toits Kloof (**-CA**), ESTERHUYSEN 12375 (BOL); SW of McGregor, foothills of
Riviersonderend Mts. (**-DC**), PUFF 840910-1/3 (PRE, WU). − **3321** (Ladismith):
"The Great Mountain" at "Mountain Station" (Langeberg, W of Garcia's Pass)
(**-CC**), BURCHELL 7070 (GH, K), 7124 (GH, K, P). − **3322** (Oudtshoorn): Postberg,
N of George (**-CD**), BURCHELL 5977 (K). − **3325** (Port Elizabeth): Uitenhage D.,
Vanstaadesberg (Van Stadens Berg) (**-CC**), "ECKLON & ZEYHER" (?ZEYHER) "4-7"
(S); Port Elizabeth D., Parson's Vlei (**-CD**), HOLLAND 3856 (BOL); − , Walmer
(**-DC**), PATERSON 3155 (GRA). − **3326** (Grahamstown): Dassies Klip
(= Dassieklip), between Port Elizabeth and Grahamstown (**-CA**), BOLUS 2670
(BOL, K); Bathurst D., from Riet Fontein to Kowie R. (= E of Port Alfred)
(**-DB**), BURCHELL 3991 (GH, K). − **3418** (Simonstown): Simonstown D.,
Wildevöelvlei (**-AB**), TAYLOR 6382 (PRE, STE); Hottentots Holland Mts. (**-BB**),
ZEYHER s.n. (G, sub SAM 37519); Kogelberg Forest Reserve (**-BD**), BOUCHER 420
(PRE, STE), PUFF 800101-1/2 (WU). − **3419** (Caledon): Lebanon Forest Reserve
(**-AA**), HAYNES 440 (STE), KRUGER 527 (PRE); Caledon D., Palmiet River Mt., nr.
Palmiet R. (**-AC**), STOKOE s.n. sub SAM 64194 (PRE, SAM, STE); Vogelgat
(Private) Nature Reserve, E of Hermanus (**-AD**), PUFF 800917-2/1 (WU);
Riviersonderend Mts., rd. to P.O. Tower, 'Die Galg' area (**-BA**), PUFF 840910-2/1
(WU): Caledon D., Riviersonderend Mts., nr. Riviersonderend village (**-BB**),
ESTERHUYSEN 25082 (BOL, WU); Stanford- Papiesvlei rd., Perdeberg, Farm
Flouhoogte (**-BC**), PUFF 800918-1/1 (WU); Papiesvlei- Nuwepos rd., SW
Koueberge (**-DA**), PUFF 800918-2/3 (WU); Elim (**-DB**), SCHLECHTER 7649 (BM,
BOL, E, G, GH, GRA, K, MO, NH, PRE × 2, S, US, W). − **3420** (Bredasdorp):
Bredasdorp D., Potberg (**-BC**), ACOCKS 22841 (K, PRE); − , nr. Struis Bay (**-CA**),
ESTERHUYSEN 4431 (BOL, PRE); − , Farm Martha, ca 16.5 km from Skipskop

towards Bredasdorp (-CB), Puff 790912-1/1 (WU). – 3421 (Riversdale): Riversdale D., ca 6.5 km W of Albertinia turning (-AB), Acocks 22356 (K, PRE). – 3422 (Mossel Bay): nr. Sedgefield (-BB), Levyns 10325 (BOL). – 3424 (Humansdorp): Humansdorp (-BB), Galpin 4104 (PRE). – Imprecise localities: "Hottentottshollandberge (Stellenbosch) and Van Stadensrivierberge (Uitenhage)", Ecklon & Zeyher 2313 (B, FI, G, GOET, LY, M, MO, S, SAM, W × 2); Swellendam D., Bowie s.n. (K), "CBS", Berg(ius) & Mund(t) s.n. (G × 2); Dahl s.n. (S). – Additional collections: Bolus 21875 (BOL); Bos 137 (K, STE); Burchell 5638 (K); Dümmer 99 γ, 1447 (E); Ecklon s.n. (S × 2); Esterhuysen 3255, 13706, 16137, 26684, 33504 (BOL), 20185 (BOL, PRE), 26688 (BOL, WU); Humphrey-Smith s.n. (K, sub BOL 21633); Kerfoot 6088 (PRE, STE); Levyns 6338, 9262 (BOL); Long 572 (K, PRE); Puff 791228-2/4, 800908-2/1, 800913-5/1, 840917-4/1, 840918-1/4 (WU); Salter 275/1 A, /1 B, /2 (BM, BOL), 2681 (BM, BOL, SAM; as 2861 in K), 2823 (BM, BOL, MO), 6271 (BOL, K, NBG), 6432, 6489 (BOL); Taylor 8592 (PRE, STE); Wright 530 (GH, K, P); Zeyher 622 (LY), s.n. (S).

7. Carpacoce heteromorpha (Buek) L. Bolus in J. Bot. (London) **34**: 25 (1896); Puff in Fl. S. Afr. **31**, 1 (2): 60 (1986).

≡ *Merciera heteromorpha* Buek in Ecklon & Zeyher, Enum. Plant. Afr. Austr.: 387 (1837).

Type: [South Africa, Cape Prov.] Caledon Distr., on Babylons-Tooernsberg [= "Babilonstoring", Babylons's Tower Mt.], Zwart & Marais sub Ecklon & Zeyher 2421 (SAM, holo.!).

☿ (occasionally also ☿ + ♂ and ♀?), single- to several-stemmed, ± cylindrical and erect to rounded dwarf shrubs. Stems ca 15–45 cm long, ca 5–10 mm in diameter at the base and ca 1.5–2.5 mm in the midstem region, ± sparsely to much-branched, branches often long, ascending to ± erect; at least younger parts densely covered with fine whitish hairs ca 0.2–0.4 (0.5) mm long; shoots with much contracted to slightly elongated leafy short shoots, internodes shorter than the leaves, plants very densely leafy. Leaves decussate; blades 10–25 (30) × 0.8–1.5 mm, linear, spreading-recurved to strongly recurved, ± terete, semiterete to ± triquete in section, margins (lower half) with fine white hairs ca 0.2–0.4 mm long; apices ± pungent to acuminate; petioles 0; stipular sheaths broadly cup-shaped, gradually widened into the base of the leaf blade, 3–5 mm long, (4) 5–7 mm wide, with fine, white hairs on the rim and on the outside. Flowers ☿ (always?), solitary, lateral and/or terminal on short branches, ± sessile, mostly hidden amongst foliage. Calyx 4-merous, lobes (5) 6–8 × 0.7–1.2 mm, linear, subequal, ± erect, hairy on the margins. Corolla (5) 6–7-merous, greenish-yellow to yellow, tube 3–5.5 × 0.3–0.5 mm, ± cylindrical, lobes (3) 3.5–4.5 (5) × 0.5–1 mm, ± lanceolate, spreading, hooded, sometimes with a few stiff, erect bristles near the apices; stamens

(5) 6–7, filaments 4.5–6 mm long, anthers 2–2.7 × 0.7–0.8 mm; style 0; stigma 1 (very rarely 2), 8–13 × 0.4–0.5 mm, greenish-grey, greyish or whitish; ovary green, ca 1.5 × 1 mm, with 1 (very rarely 2) fertile ovule(s). Fruits ash-grey, ca 1.5–3 mm in diameter, subglobose to ± turbinate, crowned by the persistent calyx lobes, 1-(very rarely 2-)seeded; diaspore ca 1.2–2.5 × 1.2–2.2 mm, black or dark grey, subglobose to ± pyriform, with a distinct vertical groove, muricate or ± rugose. – Figs. 10 c–d, 43 c, h–i.

Chromosome Number (Table 3): n = 11, 2 n = 22.

Average Pollen Diameter (Fig. 49): 32.8–33.6 µm.

Habitat: In Coastal and Mountain Fynbos, less commonly in Coastal Renosterveld; mostly in dry areas, in gravelly to sandy soil amongst rocks; primarily over TMS. – Ca 100–850 m.

Distribution (Map, Fig. 108 i): S Africa, SW Cape Prov. Occurring from the Worcester and Somerset West Distr. SE to the W Bredasdorp Distr.

Critical Remarks: Forms of *C. heteromorpha* with less distinctly curled/recurved leaf blades than those depicted in Figs. 10 c–d have occasionally been confused with (the doubtlessly ± closely allied) *C. vaginellata* (6., above). *C. heteromorpha,* however, is easily distinguished by having flowers with unusually long corolla tubes, by its ± round fruits with four long calyx lobes, by its ± round diaspores, and by the rather long, whitish hairs on the stipular sheath and lower half of the leaf blades (unique in the genus!).

If odd two-seeded fruits are present (very rare!), the diaspores are no longer ± round but laterally compressed.

Both flowers and fruits are usually hidden amongst the dense foliage and plants give the impression of permanently being in a vegetative state. According to field observations, *C. heteromorpha,* moreover, seems to flower very rarely. In the populations studied in the field, very few plants were seen in flower and, so far, only individuals with ⚥ flowers were detected. Herbarium studies, however, seem to indicate that there are also individuals with both ⚥ flowers and flowers transitional ⚥ → ♂, i.e., with short stigmas and comparatively very small ovaries. I am convinced that observations of populations over a longer period of time will show that there are also pure ♀ individuals as in all other species of *Carpacoce.*

Collections

South Africa. Cape – **3319** (Worcester): Worcester D., Villiersdorp area (**-CD**), STOKOE 9431 (BOL). – **3418** (Simonstown): Hottentots Holland Mts., Somerset Sneeukop (**-BB**), STOKOE s.n. (BOL, sub SAM 66477); Caledon D.,

Hangklip Estate (**-BD**), Stokoe s.n. sub SAM 64195 (SAM); Kogelberg Forest Reserve, Boucher 782 (PRE, STE), 1623 (PRE), Puff 800101-3/1 (WU). – 3418-BB/3419-AA: Hottentots Holland Mts., between Somerset Sneeukop and Dwarsberg, Stokoe s.n. sub SAM 59528 (SAM). – **3419** (Caledon): Caledon D., Nuweberg (Nieuweberg Forest Reserve) (**-AA**), Bond 483 (NBG); – , Babylons-Tooernsberg (= Babilonstoring Mt.) (**-AD**), Zwart & Marais sub Ecklon & Zeyher 2421 (SAM); Vogelgat (Private) Nature Reserve, E of Hermanus, Puff 800917-4/1 (WU); Salmonsdam Nature Reserve, E of Stanford (**-BC**), Puff 840918-1/5 (PRE, WU); from Elim to Ratels River (**-DA**), Bolus 3077 (NBG); (around) Elim (**-DB**), Bolus 6972 (BM, BOL, K, PRE), Schlechter 7648 (BM, E, K, PRE, US).

Taxa To Be Excluded

from the *Anthosperminae*

Anthospermum calycophyllum Sonder in Harvey & Sonder, Fl. Cap. **3**: 31 (1865).

≡ *Otiophora calycophylla* (Sonder) Schlechter & K. Schum. in Bot. Jahrb. **30**: 416 (1901).

Anthospermum crocyllis Sonder in Harvey & Sonder, Fl. Cap. **3**: 32 (1865).

= *Crocyllis anthospermoides* E. Meyer ex K. Schum. in Engler & Prantl, Nat. Pflanzenfam. **IV, 4**: 132 (1891).

See Puff & Mantell (1982 a).

Anthospermum holtzii K. Schum. in Bot. Jahrb. **34**: 340 (1904).

≡ *Psychotria holtzii* (K. Schum.) Petit in Bull. Jard. Bot. Brux. **36**: 90 (1966).

Anthospermum mazzocchii-alemannii Chiov. in Bull. Soc. Bot. Ital. **1924**: 40 (1924).

= *Spermacoce subvulgata* (K. Schum.) Garcia in Mem. Junta Invest. Ultram. II, **6**: 49 (1959).

See Verdcourt (1983).

Anthospermum plicatum Hilsenb. & Bojer ex Baker in J. Linn. Soc. (Bot.) **20**: 143 (1883), pro syn. *Myosurandrae moschatae*

= *Myrothamnus moschatus* (Baillon) Baillon in Adansonia **9**: 370 (1870).

Anthospermum polyacanthum Baker in J. Linn. Soc. (Bot.) **20**: 171 (1883).

≡ *Galium polyacanthum* (Baker) Puff in Plant Syst. Evol. **140**: 64 (1982).

To be Excluded from *Anthospermum*
(but *not* from the *Anthosperminae*)

Anthospermum viscosum WEBB in sched. [BURGEAU, Pl. Canar. 414 (1845),
    Pl. Canar., Iter sec. 1363 (1855)], nom. nud.
    ≡ **Phyllis viscosa** CHRIST in Bot. Jahrb. **9**: 144 (1888).

E. Phylogenetic Relationships and Evolutionary Aspects

1. Relationships Within *Anthospermum*

1.1. Introductory Remarks

If one looks at any particular species of μm, one will not experience any problem or uncertainty in recognizing it as a member of the genus. The sexually dimorphic, anemophilous flowers with their hairy, long stigmas and the large anthers dangling on long, thin filaments and the dry, dehiscent fruits with their characteristic anatomy "hold" the genus together. And yet, there is extreme variability within the genus in growth form (large shrubs . . . perennial herbs, etc.), corolla tube lengths and shapes (long and ± cylindrical . . . short and subcampanulate), in the sex distributions (non-dioecious, dioecious) and several other features.

Obviously, the range of variation of several of these characters is indicative of the presence of evolutionary "progressions" (from "primitive" to "advanced" or "derived" to very advanced). In order to clarify the status of these characters it is, at first, attempted to establish what a hypothetical ancestral form of *Anthospermum* might have looked like.

Principles for the establishment of "character phylogenies" have been widely and frequently discussed (e.g., ESTABROOK 1977, EHRENDORFER 1983: 539). On the basis of these, a hypothetical ancestor may be assumed to have had the following set of "primitive" features:

- Habit: tree or large shrub, "primarily" woody
- Foliage: decussately arranged; leaves ± large and broad
- Inflorescence: a "normal" panicle or thyrsus with discrete peduncles and pedicels
- Flowers: possibly ill-adapted to anemophily; with long, narrow and cylindrical corolla tubes; stigmas somewhat exserted; anthers relatively small, filaments relatively short
- Sex distribution: flowers hermaphrodite or perhaps plants gynodioecious ($\male\female/\female$) or $\male\female/\male\female + \female/\female$
- Fruits: (? fleshy to) dry; differentiated into exo- and endocarp, crowned by small persistent calyx lobes

Habitat and distribution: possibly forest edge inhabitants in (montane?) tropical areas.

The following "character progressions" can be recognized within *Anthospermum*:

Growth Form: 1) large shrubs (mostly single-stemmed, "primarily" woody) → 2) perennial herbs → 3a) many-stemmed (*secondarily woody!*) dwarf shrubs; 3b) subshrubs → (?) 4) short-lived shrubs or "woody herbs" (presumably secondarily woody).

Note: Wood anatomical investigations (KOEK-NOORMAN & PUFF 1983) of *Anthospermum* revealed considerable differences in the wood structure of large shrubby and dwarf shrubby taxa, strongly suggesting that the assumption of a linear progression large shrubs → dwarf shrubs → perennial herbs does – in the case of *Anthospermum*! – not apply.

Leaves:
Arrangement: 1) decussate → 2) variable (decussate and in whorls of 3 or even 4; often within individuals) → 3) strictly in whorls of 3.

Shape and Size: 1) ± large, broad and thin → 2) variously rolled, relatively small and narrow → 3) distinctly ericoid (± triangular or semiterete in section).

Indumentum: 1) ± glabrous, papillate or very shortly hairy → 2) densely (long-)hairy.

Stipules: 1) stipular sheaths with several setae or fimbriae → 2) with a single median seta → 3) seta minute or ± lacking.

Note: This is somewhat problematic. One may also argue that the presence of only one distinct median seta is the "primitive" situation, while the presence of several setae could be interpreted as a secondary increase. Thus the following may also be permissable: 1) stipular sheaths with a single median seta → 2a) with several setae; 2b) seta reduced and minute or ± lacking.

Inflorescences: 1) extensive, (paniculate or) thyrsic, with discrete peduncles and pedicels (e.g., *A. paniculatum*, Fig. 29b) → 2) partial inflorescences ± many-flowered but congested ("axillary flower clusters") → 3) partial inflorescences few- (3 – 1-)flowered and sessile → 4) solitary flowers.

Note: The distinct, ± cylindrical inflorescence zones in ♀ of several dioecious species could be considered to be a special situation of 2).

Flowers:
Calyx: 1) relatively small, but nevertheless well developed → 2a) somewhat enlarged and sometimes somewhat unequal in size; 2b) minute or lacking.

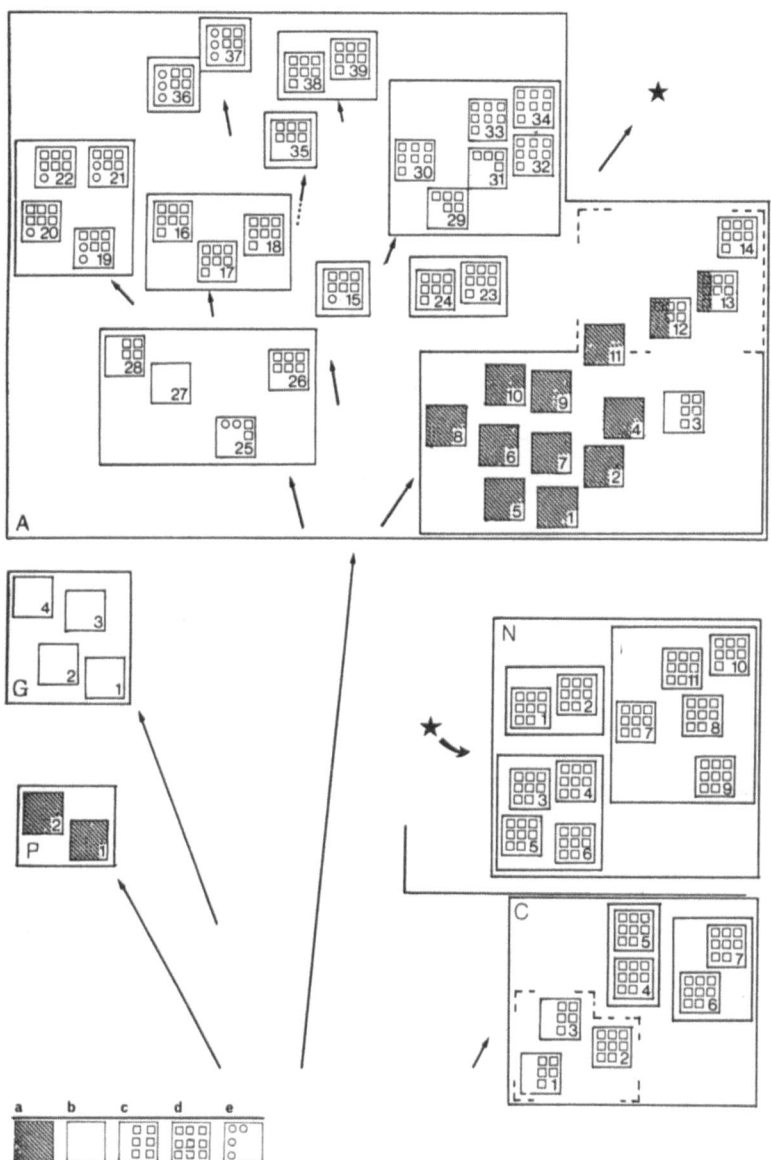

Fig. 109. Growth forms of the *Anthosperminae* (A *Anthospermum*; C *Carpacoce*; G *Galopina*; N *Nenax*; P *Phyllis*) and presumed relationships between and within genera (a square stands for a species; numbers correspond to those used in part D., except for *Phyllis*: 1 *P. nobla*, 2 *P. viscosa*). *a* large (usually single-stemmed), (presumably) primarily woody shrubs (seed regenerators); *b* perennial herbs; *c* subshrubs (woody at base); *d* many-stemmed, (presumably) secondarily woody dwarf shrubs (*c–d*: mostly resprouting from base after fire); *e* short-lived shrubs ("woody herbs")

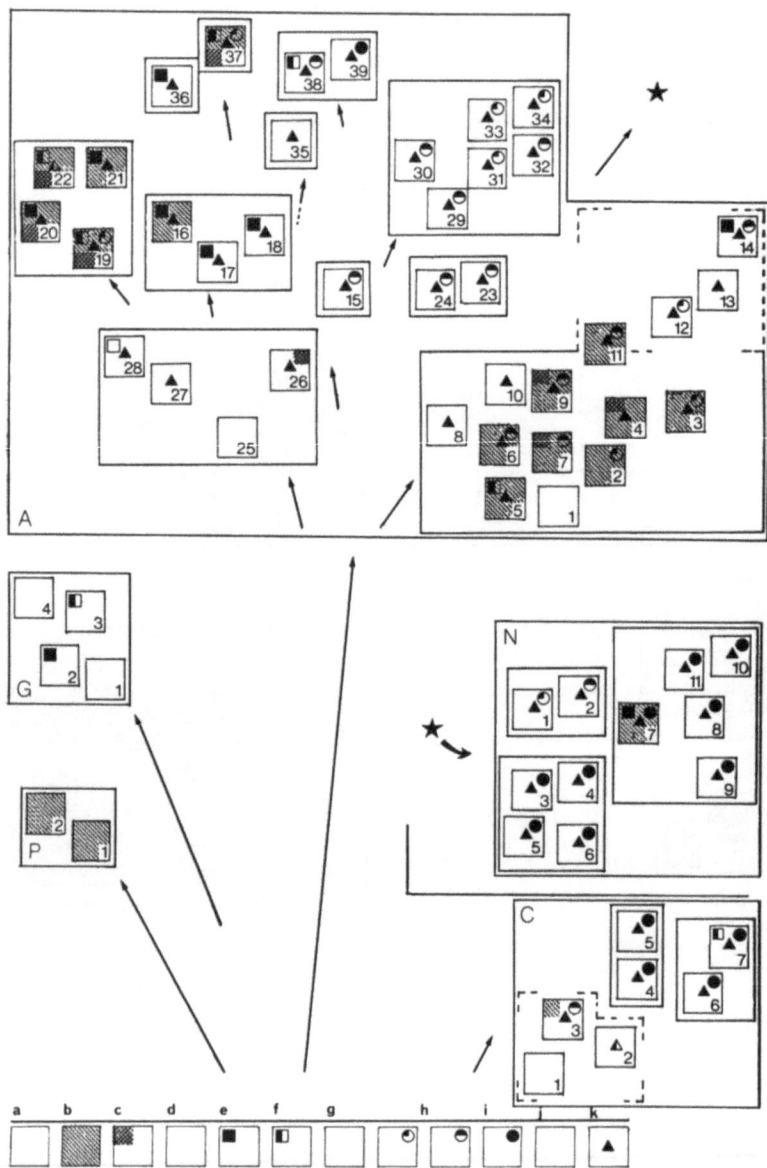

Fig. 110. Leaf and stipule characters of the *Anthosperminae*. *a–c* leaf arrangement: *a* decussate, *b* in whorls of 3, *c* variable (decussate and/or in whorls of 3 or 4); *d–f* surface: *d* glabrous or subglabrous, *e* hairy, *f* sometimes hairy or partially hairy (margin, lower half of blade, etc.); *g–i* shape: *g* broad, *h* narrow, con- or revolutely rolled to ± ericoid, *i* distinctly ericoid (semiterete or ± triangular in section); *j–k* stipules: *j* with several setae, *k* with 1 seta or setae lacking. – For further explanations see Fig. 109

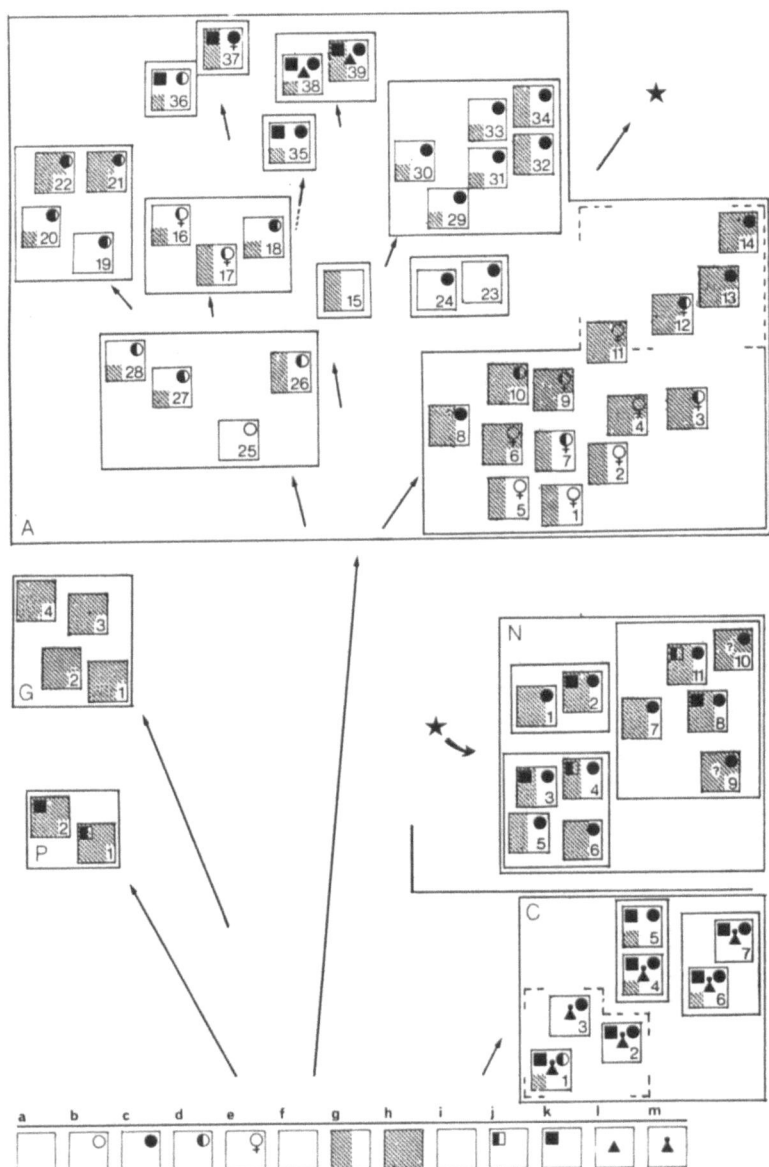

Fig. 111. Inflorescence and floral characters of the *Anthosperminae*. *a–e* inflorescence: *a* " basic type" (cf. Figs. 27a–b), *b* congested, ± many-flowered, *c* few-flowered (partial inflor. 3 – 1-flowered), *d* ± intermediate between *b* and *c, e* in dioecious taxa: distinct, ± cylindrical inflorescence zones in ♀; *f–h* corolla tube: *f* long, ± cylindrical, *g* ± short, (broadly) funnel-shaped, *h* (very) short, ± subcampanulate; *i–k* corolla lobes: *i* 4, *j* 4 or 5, *k* 5(–7); *l–m* gynoecium: *l* only 1 carpel fertile, the second strap-like (cf. Figs. 36a, c and 42a), *m* only 1 carpel fertile, the second greatly modified (cf. Fig. 44). – For further explanations see Fig. 109

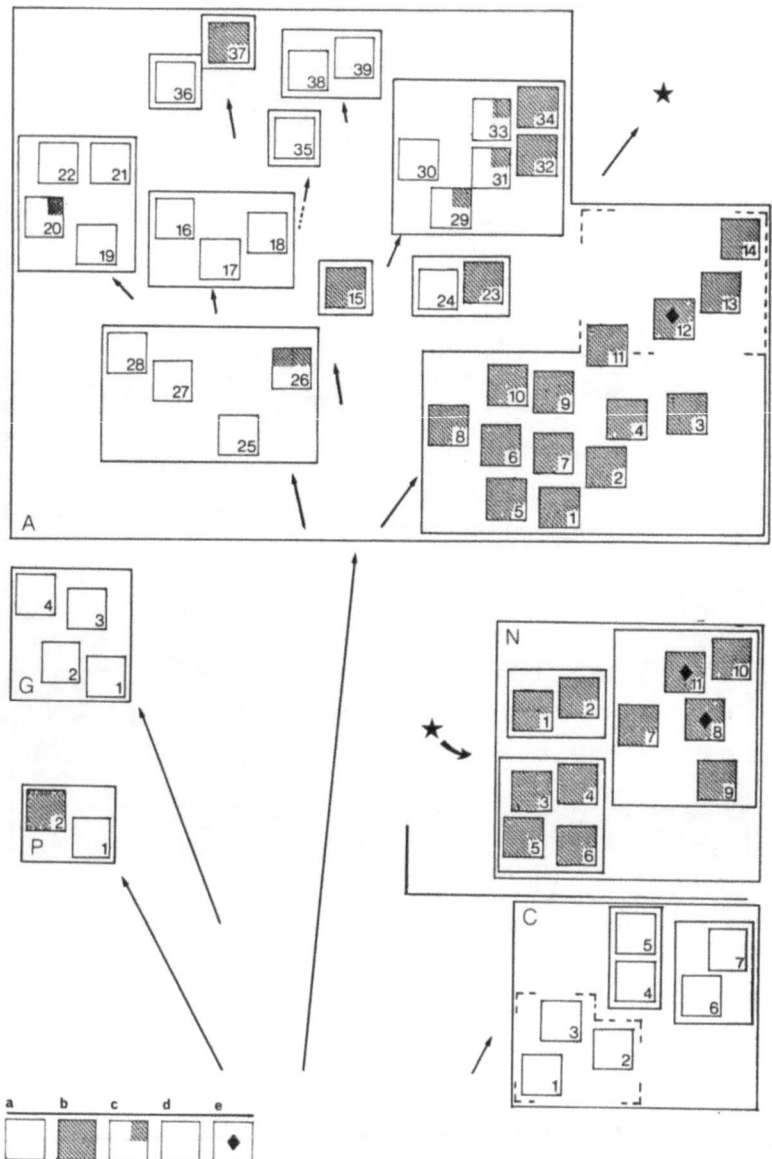

Fig. 112. Sex distributions and ploidy levels of the *Anthosperminae*. *a* non-dioecius taxa; *b* dioecious taxa; *c* trend to dioecy (some populations of a taxon dioecious, others not, etc.); *d* diploid taxa; *e* polyploidy (4x; 2x, 4x and 6x: A 12). – For further explanations see Fig. 109

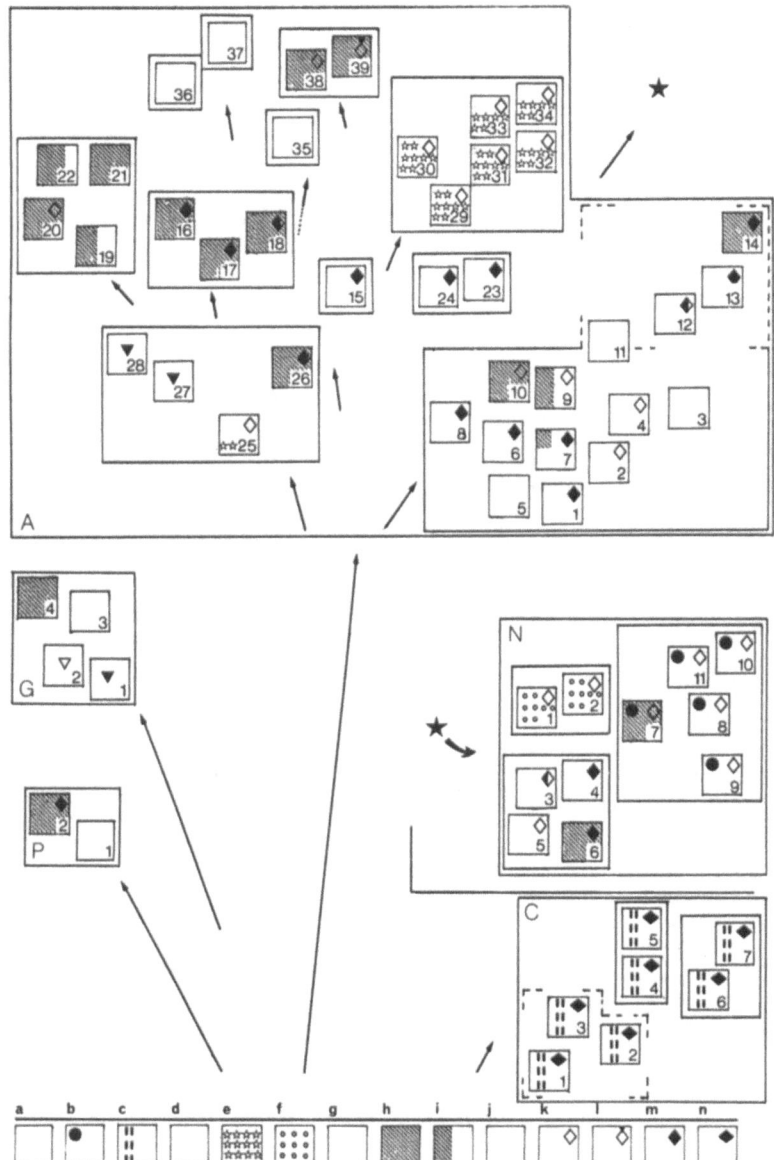

Fig. 113. Fruit characters of the *Anthosperminae*. *a–c* dehiscence: *a* dehiscing into two mericarps, *b* indehiscent, *c* exocarp valves (cf. Fig. 43f); *d–e* mericarps in side view: *d* considerably longer than wide (cf. Fig. 35c), *e* ± round (cf. Fig. 37a); *f* fruits round, ± "inflated" (cf. Figs. 38a–b); *g–i* indumentum: *g* glabrous or papillate, *h* hairy, *i* ± hairy or hairness variable (absent/present within taxa); *j–n* persistent calyx lobes on fruits/mericarps: *j* absent, *k* small but well discernible, *l* as *k* but calyx lobes of sterile carpel enlarged, *m* ± large, sometimes unequal, *n* large, leaf-like. Commissures: ca half as long to as long as ventral mericarp surface (cf. Fig. 37b) (no symbol), very short, less than half as long as ventral mericarp surface (cf. Fig. 37c) (inverted black triangles), often very short (inverted open triangle). – For further explanations see Fig. 109

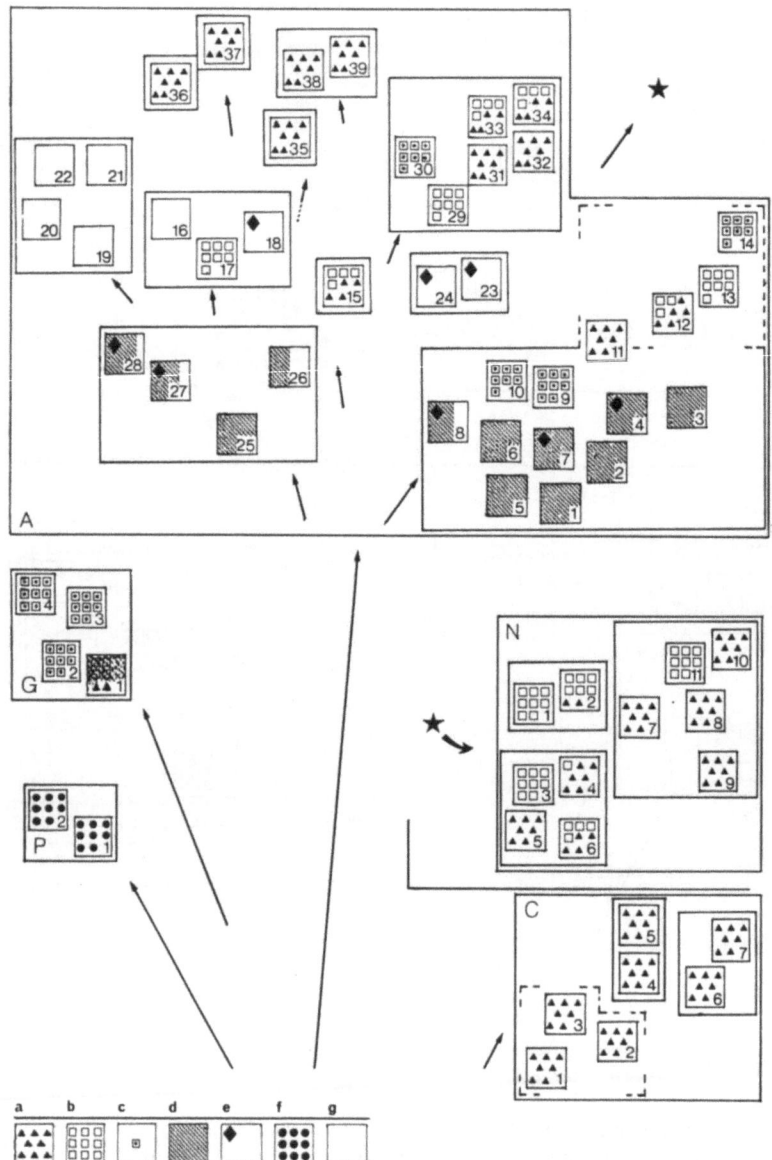

Fig. 114. Distribution of the *Anthosperminae*. *a* SW Cape Floristic Region; *b* Southern Africa (excl. SW Cape); *c* SE South Africa; *d* afromontane; *e* Madagascar; *f* Macaronesia; *g* other (S Central and tropical Africa; not afromontane). – For further explanations see Fig. 109

Number of Corolla Lobes (= number of stamens): 1) 4 → 2) 4 or occasionally 5 → 3) 5.[1]

Corolla Tube: 1) rather long, cylindrical → 2) quite short, (broadly) funnel-shaped → 3) (very) short, subcampanulate.

Number of Carpels: 1) 2 fertile → 2) 1 fertile, 1 sterile and modified.

Ovary Indumentum: 1) glabrous (or ± papillate) → 2) (long-)hairy.

Sex Distribution: 1) ⚥ (and also ⚥/♀ or ⚥/⚥ + ♀/♀?) → 2) a wide range of sex distributions (⚥, ⚥+♀, ♂+⚥+♀, ...; see C.12.2.) → 3) dioecious.

Without going into any detail at this stage, it can be said that (1) individual taxa are invariable inhomogenous in their character expression in that they always show a combination of both "primitive" and "advanced" characters, and that (2) there are some (± well definable) species groups which, in turn, are characterized by specific sets of both "primitive" and "advanced" characters. The combination of "primitive" and "advanced" characters, however, can differ markedly from group to group.

Each of these groups will be discussed separately in the subsequent subchapters. In 1.10. a synthesis of the findings presented in 1.2. to 1.9. will be attempted.

The reader is also referred to Figs. 109–114, in which selected character states are given for each species.

1.2. The *Anthospermum usambarense* Group
(Species 1.–10.; 11.?)[2]

is comprised of usually quite large, mostly single-stemmed shrubs, all of which are *dioecious*. Only *A. zimbabwense,* which is clearly adapted to fire-prone grassland habitats, deviates in growth form (its presumed affinities are discussed below). In the majority of taxa, the leaves are arranged in whorls of three rather than decussate. Members of this group, as far as they have been investigated wood anatomically, have typical rubiaceous wood, i.e., are believed to show "normal" ("primary") woodiness (KOEK-NOORMAN & PUFF 1983) (but also see *A. isaloense,* below). They do not produce innovation buds or shoots at or near the base

---

[1] Within the *Rubiaceae as a whole* 5 corolla lobes are without doubt the "primitive" situation [(3)←4←5→6, 7 (or more)]. In my opinion, this does not seem to apply within *Anthospermum*; subsequently, the presence of 5 corolla in a few (isolated and obviously "derived") *Anthospermum* species is interpreted as a "secondary" increase.

[2] Species numbers correspond to those used in the Systematic Part (D.)

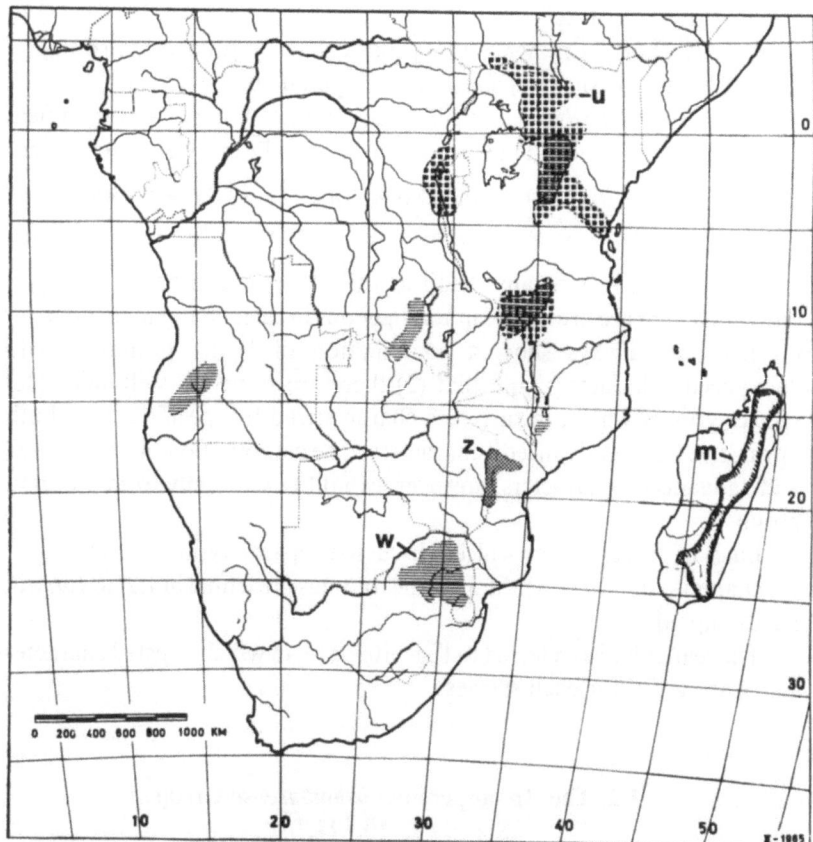

Fig. 115. Distribution ranges of *Anthospermum usambarense* (u) and presumably allied species. – w *A. welwitschii*; z species occurring in the E highlands of Zimbabwe (and, in part, in neighbouring parts of Mozambique; *A. ammannioides, A. vallicola, A. zimbabwense*); m Madagascan species (*A. emirnense, A. isaloense, A. madagascariense*)

of the stem (exception: *A. zimbabwense*) and are, therefore, killed by fire. Plants thus behave as "seed regenerators"; their seeds germinate rapidly (cf. C.12.7.) and the plants often show ± "weedy" tendencies (quickly invading recently burnt areas – cf. *A. usambarense*, Fig. 62, etc.). Their ♂ flowers have rather short and broadly funnel-shaped to ± subcampanulate corolla tubes. The fruits often have distinct, and not uncommonly unequal calyx lobes; their indumentum ranges from glabrous to long-hairy. This group is widely distributed in ± (afro-)montane areas of the African mainland and Madagascar (although not entirely confined to such regions) (Fig. 115).

I am convinced that this entire group is a natural one, derived from a common (tropical-afromontane?) ancestral stock which may have been present along forest edges or, at higher altitudes, in the "ericaceous zone". The Madagascan representatives of this group may have arrived there by Long Distance Dispersal (LDD). Whether this happened in fairly recent times or when Madagascar and the African mainland were still much closer to each other than today, is basically irrelevant as there could never have been a direct connection between the two areas which would have offered continuous and suitable habitats (cf. AXELROD & RAVEN 1978).

Also, the Madagascan species of this group may have established themselves there at different times: While *A. emirnense* and *A. madagascariense* show very close morphological affinities to the African mainland taxa, *A. isaloense* differs more markedly in several morphological features, in leaf and wood anatomical details and ± in ecological aspects (see PUFF 1986 for details). These deviations can possibly be interpreted as an indication of a greater age of that taxon or, in other words, as an indication of an earlier arrival of the stock that gave rise to *A. isaloense* (theoretically, another alternative interpretation which in my opinion, however, is less probable, could be applied: The stock that gave rise to *A. isaloense* did not arrive on the island earlier than that of the other montane species but has, in a different and new environment, evolved more quickly than the other species and acquired a "new", i.e., more deviating, set of characters).

Noteworthy, furthermore, is the fragmented distribution range of some of the species of this group – notably of *A. welwitschii*. Again, to assume a once-continuous range of suitable habitats – and thus a continuous distribution range is, in my opinion, quite unrealistic. LDD offers, in my mind, the only plausible explanation for the disjunct distribution range. Interestingly, the isolated, S ("mega"-) populations of *A. welwitschii* (plants from Angola, the Transvaal and Mt. Mlanje in S Malawi, respectively) are rather similar to each other. There are no obvious indications that these geographically isolated populations are in the process of evolving into discrete entities, which is possibly an indication of the youthfulness of the species. This situation contrasts with that of the disjunctly distributed Madagascan *A. madagascariense,* which is no longer uniform; a number of, at present, rather ill-known "Forms" are known to occur which may indicate that the species is in the process of active evolution (further details about this in PUFF 1986). Another, in my opinion, interesting point is that *A. welwitschii* "misses out" the Chimanimani Regional Mountain System sensu WHITE (1978). Instead, the highlands of E Zimbabwe (and nearby Gorongosa Mt. in Mozambique) have some endemic *Anthospermum* species which, however, are clearly allied to *A. welwitschii*. Possibly the particular geological and climatic (and perhaps other, as yet not fully understood?) factors of this area may have favoured the evolution of discrete new taxa[1] from the ancestral stock that also gave rise to *A. welwitschii*. Two of these E Zimbabwean endemics, *A. ammannioides* and *A. vallicola*, have a similar habit but differ in their ecology (*A. vallicola* at higher altitudes!) and various morphological characters (most conspicuous are the leaf shape differences). The third species, *A. zimbabwense*, it is not so close to the previous two. With regard to growth form and regeneration behaviour, it is an exception within the group: *A. zimbabwense* is typically a many-stemmed, small subshrub — obviously a growth form adaptation to the fire-prone grassland habitats in which it occurs. In all its other characters, however, it shows close affinities to the tropical E African montane *A. usambarense: A. zimbabwense* is believed to be a "S offshoot" of that species.

While the relationship between *A. ammannioides* and *A. welwitschii* is doubtlessly a rather close one, the alliance between the former and the Madagascan *A. emirnense* seems to be even closer. The two species agree with each other in more characters than with any other species of that group. Similarly, numerous agreements in character states appear to exist between *A. vallicola* and *A. madagascariense*. It is, nevertheless, believed that the similarities between the species of each of these two pairs do not reflect a direct, close relationship. It is in my opinion rather more likely that, as was stressed above, the entire *A. usambarense* group forms a natural entity, in which similar or even identical character states recur independently in several of its members.

The SE African coastal species *A. littoreum* (not in map, Fig. 115) appears, at first glance, to be misplaced in this group, but I am certain this is not so. Because of its adaptation to dune sand habitats (where most of the shoots are buried in sand), its habit may give a wrong and misleading impression. Plants of *A. littoreum*, cultivated in the greenhouse under the same conditions as other species of this group, do not differ in their growth form. The fruits, densely covered with quite long hairs (such fruits are unknown in other members of the group) may also be an adaptation to its ecology — or, more precisely, to the better dispersal of the mericarps in their habitats (the mericarps are most probably "rollers", blown about by the wind in the coastal sands). I am convinced that *A. littoreum*, in turn, is closely allied to *A. galpinii* (not in map, Fig. 115), and that both share a common ancestor. The latter occurs in roughly the same area as *A.*

---

[1] It must be stressed, however, that none of these species is a distinct Chimanimani quartzite endemic (for a list of these see WILD 1964), although they may occur over quartzite also.

littoreum but in localities further inland [subsequently, *A. galpinii* is often associated with (remnants of) afromontane vegetation]. The structure of the mericarps of *A. galpinii* is basically identical to that of *A. littoreum,* the main difference lies in the much shorter, rather closely spaced hairs covering the fruits.

Whether the widely distributed SW Cape species *A. aethiopicum* links up with this group remains somewhat uncertain. On the one hand, *A. galpinii* and *A. aethiopicum* appear to be so close to each other that their morphological separation may cause extreme difficulties where the distribution ranges of the two species overlap (in the E Cape − SW Cape Floristic Region border or "transition" zone!). On the other hand, *A. aethiopicum* deviates in certain anatomical characters (amongst these it is particularly noteworthy that *A. aethiopicum* no longer has "typical" rubiaceous wood; cf. KOEK-NOORMAN & PUFF 1983). *A. aethiopicum,* furthermore, is never in any way associated with (remnants of) afromontane vegetation which in gullies, kloofs, coastal areas, etc. intrudes into the SW Cape vegetation proper.

1.3. The *Anthospermum spathulatum* Group
(Species 12.–14.)

is comprised of strictly dioecious shrubs to dwarf shrubs with always decussately arranged leaves. The plants are glabrous (*A. spathulatum, A. monticola*) or hairy (*A. basuticum*). The ♂ flowers have short to very short, broadly funnel-shaped to ± subcampanulate tubes. The fruits mostly have rather conspicuous, relatively large calyx lobes.

The three species of this group are confined to S Africa: *A. spathulatum* is centered in the SW Cape Floristic Region; the other two occur outside the SW Cape (from the Cape interior E and NE to the O.F.S., Lesotho and Natal; cf. Fig. 116).

A. spathulatum is an extremely variable and widely distributed species which contains di-, tetra- and hexaploids. There is some possibility that it is (very remotely) linked to afromontane species (i.e., the *A. usambarense* Group, above) with which it shares the dioecy, the rather short corolla tubes and, to some degree, the shrubby habit. I am, however, not at all certain about such a relationship since, for example, in growth form *A. spathulatum* is very variable and shows a different (and heterogenous) regeneration behaviour. Plants of *A. spathulatum* can be erect, rather tall, single-stemmed shrubs to rounded, several-stemmed dwarf shrubs; the former are killed by fire and regenerate from seed, the latter have the

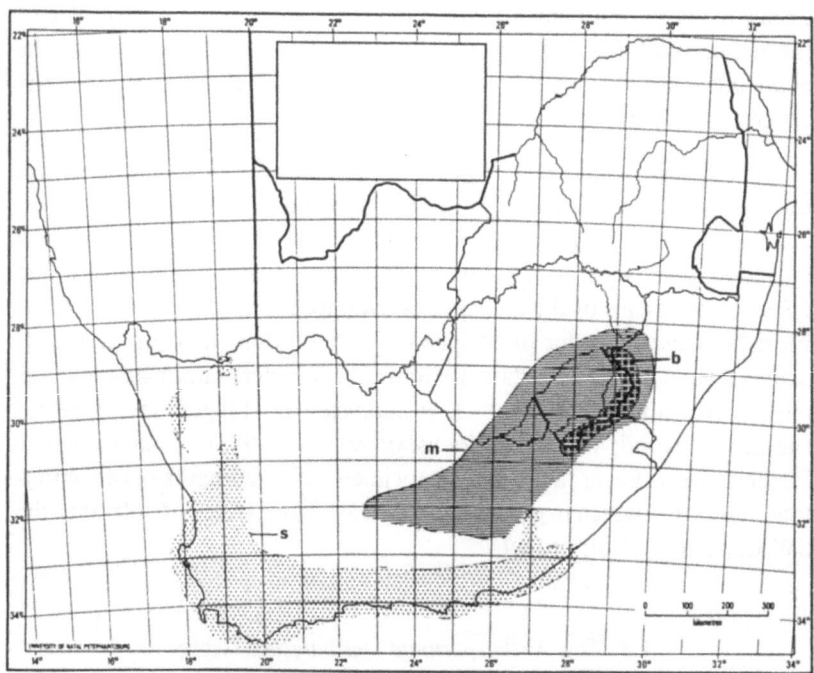

Fig. 116. Distribution ranges of *A. spathulatum* (s) and presumably allied species. – b *A. basuticum*; m *A. monticola*

ability to innovate from the base after a fire. In contrast, species of the *A. usambarense* Group are (with the exception of *A. zimbabwense*) invariably single-(or few-)stemmed, erect, tall shrubs always regenerating from seed. Furthermore, wood anatomical characters of *A. spathulatum* (cf. KOEK-NOORMAN & PUFF 1983), revealing features deviating from the "typical" rubiaceous wood structure (the latter is typical for the investigated African members of the *A. usambarense* Group), do not support a direct and close link between the two groups. As there are no obvious morphological or other relationships to any other species group, the affinities of *A. spathulatum* and its allies *A. monticola* and *A. basuticum* remain rather obscure.

It seems certain that *A. spathulatum* has, in the SW Cape, experienced extensive radiation. Five entities have evolved which have become sufficiently different morphologically, ecologically and/or in their geographical distribution to be recognized taxonomically as subspecies (see the Systematic Part for further details). At least one of these has become so distinct (e.g., subsp. *tulbaghense*) that it could almost be considered a separate species. Especially within the large and widely distributed subsp. *spathulatum*, further differentiation and diversification is obviously still in progress: a number of "Forms" (± in the sense of ecotypes or races), some of which may be rather distinct, are recognizable (details in the Systematic Part). Finally, also the establishment of tetraploidy in some of the subspecies (subsp. *ecklonianum*, subsp. *saxatile*) and the proven odd occurrence of both tetra- and hexaploids (next to diploids) in subsp. *spathulatum* are an indication for the active evolution in *A. spathulatum* (in that context it is particularly noteworthy that *A. spathulatum* is the only species in the genus in which polyploidy occurs).

The morphological similarities between *A. spathulatum* and *A. monticola*, in my opinion, leave no doubt that the two species are closely allied. The latter is disjunctly distributed in the mountains of the interior of the Cape Prov. (where it is ± confined to the rather wet mountain tops); from there, its distribution stretches E to Lesotho and the Drakensberg area. It, in turn, links up with *A. basuticum*, which is confined to the high Drakensberg in the Lesotho − E Cape, Lesotho − Natal and Lesotho − O.F.S. border area; the two species, by the way, hybridize freely where they occur in vicinity to each other (cf. C.12.5.!).

I am convinced that *A. monticola* is but an "offshoot" of *A. spathulatum* which established itself in the mountains of the Cape interior from *A. spathulatum*-like ancestors at a time when this stock (during the climatic changes and fluctuations at the end of the Pliocene and in the Pleistocene?) had expanded into the interior of the Cape. *A. monticola* could have attained its present-day distribution by LDD. Subsequently, *A. basuticum* could have evolved from part of this *A. spathulatum*- *A. monticola*-like stock in the high Drakensberg area. Hence the above speculations would imply that *A. basuticum* is relatively younger than *A. monticola* (but not necessarily younger than some of the subspecies of *A. spathulatum*). It can, furthermore, be deduced from the above that the austro-afroalpine *A. basuticum* can neither be associated with (E Africa-centered) afromontane species nor with (S)E African species of lower altitudes (such as *A. hispidulum* or *A. whyteanum* which superficially − last not least because of their similar indumentum − resemble *A. basuticum*).

1.4. The *Anthospermum whyteanum* Group
(Species **16.–18.**)

is comprised of several- to many-stemmed, presumably secondarily woody, hairy dwarf shrubs. None of the species is dioecious; individuals may be ⚥, ♀, ♂, ⚥ + ♀ or (seldom) ♂ + ⚥ + ♀ or ? ⚥ + ♂.

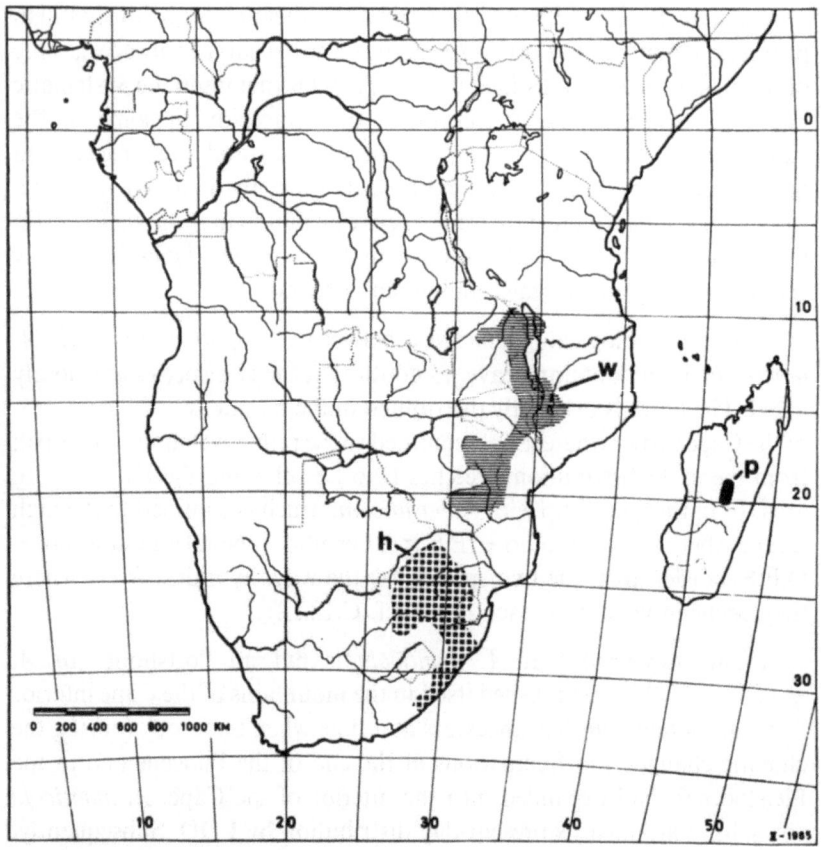

Fig. 117. Distribution ranges of *A. whyteanum* (w) and presumably allied
species. – h *A. hispidulum*; p *A. perrieri*

The three species form a close-knit and rather "homogenous" S
Central – SE African and Madagascan group (Map, Fig. 117). Species
differ from each other vegetatively mainly in their leaf arrangement
(decussate in the SE African *A. hispidulum* and the Madagascan *A.
perrieri*; strictly in whorls of three in the S Central African *A. whyteanum*)
and, in the fertile region, primarily in slight corolla tube shape differences
and in the size of the persistent calyx lobes crowning the fruits. All three
occur in comparable rocky (mountain) habitats; *A. whyteanum* and *A.
hispidulum* may extend to afromontane rocky habitats, but none of them
can really be classified as a typical afromontane species. The substrates
over which they occur are geologically divergent, none of the three species
is really substrate-specific; the best known species of the group, *A.*

hispidulum, for example, is known to occur over quartzite, sandstone and also over granite. In contrast to taxa of the *A. usambarense* Group (1.2., above), these three species show no "weedy" tendencies, and the seeds apparently do not germinate rapidly and readily at any place. Germination trials (cf. C.12.7.) indicate that either specific conditions may be essential for germination and/or that the germination rate it decreased (possibly due to the short viability of the seeds, etc.?). This may be one of the reasons for the less "aggressive" behaviour of this group of species. The dwarf shrubs of this group have the potential to regenerate from the base ("resprout") after a fire.

There can be no doubt that these three species are derived from a common stock. Part of this stock must, at one state, have reached Madagascar by LDD to give rise to *A. perrieri.*

1.5. The *Anthospermum ternatum* Group
(Species 19.–22.)

is comprised of subshrubs, rather short-lived shrubs ("woody herbs") or ± dwarf shrubs. The plants have mostly non-dioecious sex distributions (⚥, ⚥ + ♀ or ♀ individuals are most common), only *A. rosmarinus* shows a clear trend to dioecy. The leaves are generally arranged in whorls of three, but this not an entirely constant feature (considerable variation – sometimes within individual plants – may occur, especially in *A. ternatum:* decussate and in whorls of three, or even decussate, in whorls of three or four). The blades are mostly variously hairy (no taxon has glabrous leaves throughout). The ⚥ (♂) flowers have rather long, cylindrical to relatively short, funnel-shaped corolla tubes; the corollas are mostly hairy on the outside. Also the fruits are mostly hairy (but, again, there is variation in this character, even within species – e.g., *A. asperuloides,* cf. Figs. 77 a – c). The fruits either entirely lack persistent calyx lobes, or the calyx lobes are minute. The mericarps, throughout, are ± elongated, have a large commissure and appear ± flattened in side view (e.g., *A. ternatum* subsp. *randii,* Figs. 35 c – d).

The five taxa (four species) of this group have a discontinuous, ± tropical African range of distribution which, however, does not extend to S Africa (Map, Fig. 118). *A. ternatum* subsp. *ternatum* and subsp. *randii* and *A.*rosmarinus are centered in the Zambezian Domain of the Sudano-Zambezian Regional Center of Endemism; *A. villosicarpum* occurs in N Kenya and S Ethiopia, and *A. asperuloides* in W Africa. While the Zambesian taxa are associated with miombo [*Brachystegia- Julbernardia-Isoberlinia (-Uapaca)*] or related woodlands and occur at altitudes of ca

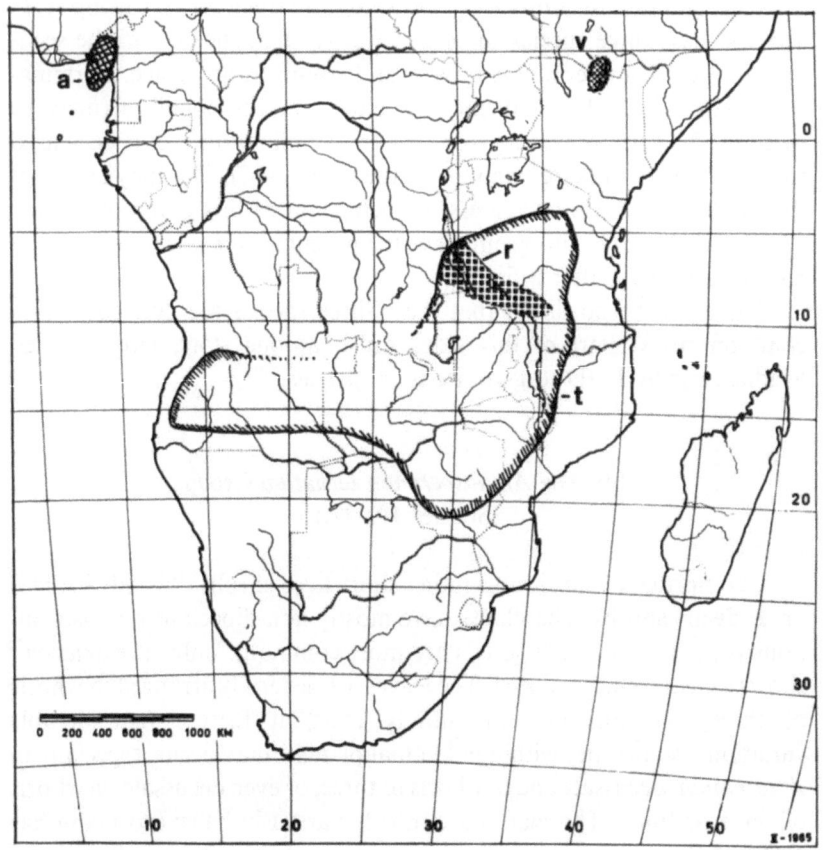

Fig. 118. Distribution ranges of *Anthospermum ternatum* (t) and presumably allied
species. – a *A. asperuloides*; r *A. rosmarinus*; v *A. villosicarpum*

(650) 800–2 400 m, *A. villosicarpum* and *A. asperuloides* are confined to
montane vegetation. The former occurs in dry forest or scrub with *Olea*
and *Juniperus*, *A. asperuloides* in the high mountains of Cameroun and
Fernando Po [2 300–3 050 (3 650) m] in ericaceous scrub at the edge of
gully woodland or in upland grassland.

It appears that the species of this group may, on the one hand, be
(loosely) linked with the *A. whyteanum* Group (1.4., above) and, on the
other hand, with the *A. herbaceum* Group (1.6., below) (also see 1.10.!).
Subsequently, I believe that all taxa of these three groups may be derived
from a common ancestral stock, part of which established itself in the vast
savanna-woodlands of S Central Africa to give rise to the modern *A.*

ternatum and *A. rosmarinus*. On account of its trend to dioecy and its often more shrubby (presumably secondarily woody!) habit, the rather narrowly distributed *A. rosmarinus* may be the relatively younger of the two species.

The origin of *A. villosicarpum* and *A. asperuloides* is more difficult to imagine. The establishment of the former in montane habitats, i.e., dry *Olea-Juniperus* forest, can perhaps be explained as follows: The habitats of *A. villosicarpum* are virtually "surrounded" by a dry type of woodland. It, therefore, could be theorized that the ancestral stock of the *A. ternatum* alliance was at one stage more widely distributed in the woodlands surrounding the rain forest region (i.e., in ± the Sudano-Zambezian region). In the NE part of this area, that part of the stock that gave rise to *A. villosicarpum* could have radiated to the mountainous terrain in which the species occurs today.

Even if an ancestral stock of the *A. ternatum* alliance is assumed to have been present, which was at one stage widely distributed in the woodland (± Sudano- Zambezian) region and extended into W Africa, the occurrence of *A. asperuloides* in the high mountains of Cameroun and Fernando Po cannot be explained by a mere radiation of this stock to these mountains, as the mountains were probably continuously separated from the woodlands by the lowland rainforest region. Only LDD offers a reasonable explanation for its present occurrence and distribution. In this context, it is also noteworthy and should be stressed that the alliances of *A. asperuloides* are fundamentally different from those found in numerous other afromontane species and species groups (of other and not necessarily rubiaceous genera) which occur in the W African mountains. Many of such afromontane groups have their main distribution in the E and S African mountains, and the species occurring in the W are either the same as those in the E or S(E) or are endemic but then, nevertheless, show a clear and close affinity to E African montane species. *A. asperuloides,* however, clearly does not belong to such a species group. It certainly is not a member of the afromontane, very widely and disjunctly distributed *A. usambarense* Group (1.2.). Habit, sex distributions, corolla shapes and fruit morphological features provide concrete evidence against such an alliance (also see comments, p. 297).

1.6. The *Anthospermum herbaceum* Group
(Species 25.–28.)

is comprised of perennial herbs, ± subshrubs or ± (presumably secondarily woody) dwarf shrubs. The plants of the group have non-dioecious sex distributions ($\female\male$, $\female\male + \female$, $\female$, $\male + \female\male + \female$ ...) except for *A.*

pachyrrhizum which is either dioecious or ± dioecious. The leaves are typically decussately arranged, only in *A. pachyrrhizum* is the leaf arrangement variable (decussate and, rarely, also in whorls of three); the blades are glabrous or nearly so (exception: *A. thymoides* subsp. *antsirabense*). The stipules, in *A. herbaceum,* normally bear several setae; in the other taxa there is mostly only one median seta. The ♀ (♂) flowers generally have rather long, cylindrical to (narrowly) funnel-shaped corolla tubes (relatively shorter in *A. pachyrrhizum*). The fruit characteristics appear to be rather heterogenous – i.e., indumentum: covered with rather long hairs (*A. pachyrrhizum*), glabrous or papillate or shortly hairy (the other taxa); persistent calyx lobes: long, ± unequal (*A. pachyrrhizum*), small but well discernible (*A. herbaceum*), ± absent (*A. herbaceum, A. thymoides, A. palustre*). The main discrepancies, however, lie in the shape of the fruits and the arrangement and anatomy of the mericarps: In *A. herbaceum,* there is a ± broad vertical groove (made up of sterile exocarp tissue) between the two mericarps (cf. Fig. 35 j) which is absent in all of the other species of this group; the mericarps, furthermore, are often rather broader in side view than in the remaining species. In *A. thymoides* and *A. palustre,* the mericarps of a fruit only touch each other in the lowermost part (commissure extremely short! Cf. Fig. 37 c and 84), and the mericarps contain two air-filled pockets (cf. Fig. 40 b); both of these features are absent in the other species of this group – and of the genus as a whole.

The group has an African – Madagascan ± afromontane distribution range (Map, Fig. 119). *A. herbaceum* typically occurs in moist to wet habitats (e.g., at the edge of streams, in dambos, etc.) (but also see the comments below!), a feature which it shares at least with *A. palustre.* In the second Madagascan species, *A. thymoides,* a trend towards the occupation of drier (± rocky?) habitats is recognizable. Also *A. pachyrrhizum* typically occurs in rocky areas (cracks of rocks).

As can be concluded from the above, this group is by far not as uniform as, for example, the *A. usambarense* (1.2.) or the *A. pumilum* Group (1.7., below).

The two Madagascan species are doubtlessly very closely related to each other (see also PUFF 1986 for details!), but their alliance to the other species of the group, notably *A. herbaceum,* is rather remote. Although largely agreeing in habit, sex distributions, corolla tube shapes and, at least in part, in their ecology, the Madagascan species differ fundamentally in the deviating fruit structure described above. This appears to indicate that the stock that gave rise to them must have reached Madagscar (presumably by LDD – for the same reasons as discussed in 1.2.!) at a relatively early period of time.

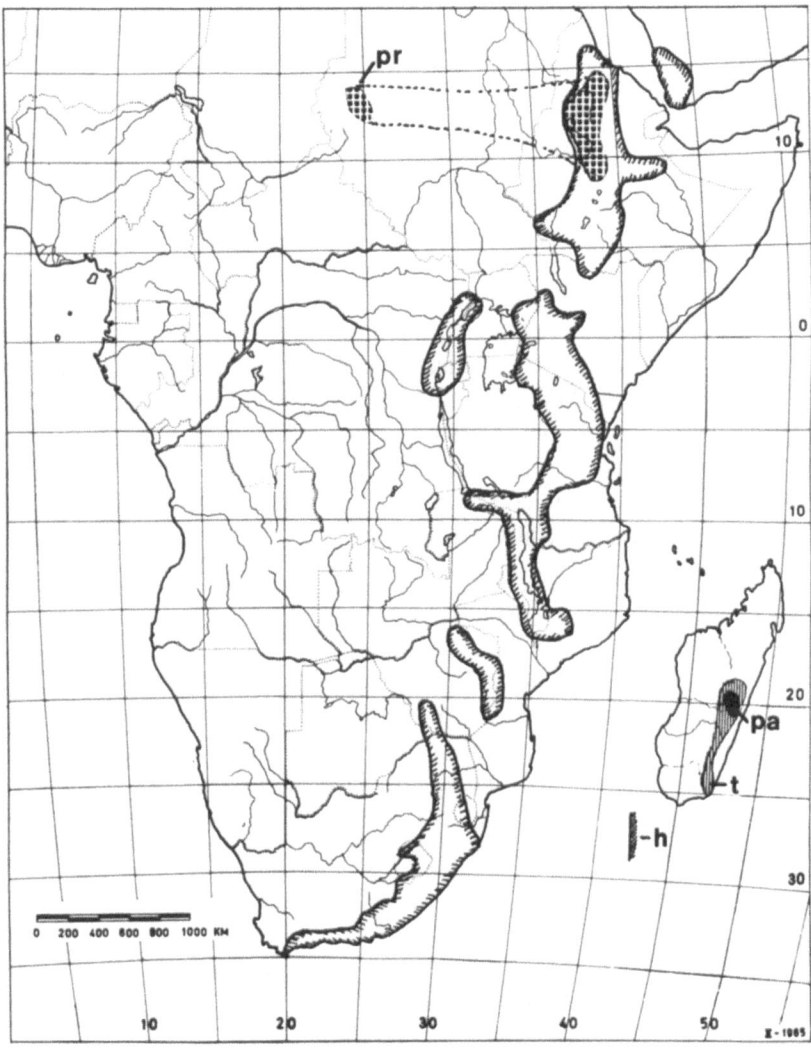

Fig. 119. Distribution ranges of *Anthospermum herbaceum* (h) and presumably
allied species. – pr *A. pachyrrhizum*, pa *A. palustre*, t *A. thymoides*

Also the relationship between *A. herbaceum* and *A. pachyrrhizum* is a
rather remote one. The more dwarf-shrubby habit of *A. pachyrrhizum*, its
(although not perfect) dioecy and the different fruits with their conspicuous calyx lobes make a very close affinity and a relatively recent
"separation" of *A. herbaceum* and *A. pachyrrhizum* seem unlikely. I,

nevertheless, am convinced that there must be a direct genetic link between *A. herbaceum* and *A. pachyrrhizum* (which has no obvious affinities to any other species of the genus) – not least because of phytogeographical considerations. In my opinion, it seems to be quite likely that the NE-afromontane *A. pachyrrhizum*, at a relatively early stage, evolved from the same stock as *A. herbaceum* in ± the same area where both species occur together today – i.e., in the Ethiopian highlands. The peculiar Ethiopia – Sudan (Jebel Marra and satellites) disjunction of *A. pachyrrhizum* may be attributed to subsequent LDD. Of course, it stands to reason that the presumed relatively "early" split of *A. herbaceum* and *A. pachyrrhizum* must, in geological terms, have actually happened in a quite recent epoch for it is well documented that the Ethiopian "flood basalt province" only came into existence some 30–15 million ago, i.e., in the mid Tertiary (Oligo- to Miocene and Miocene; see MOHR 1983 for details). Subsequently, the presumed "segregation" and origin of *A. pachyrrhizum* from a common ancestral stock that gave rise to both *A. herbaceum* and *A. pachyrrhizum* could not have occurred much earlier than the Miocene, but possibly even later (see also 1.10.!).

Assuming a tropical African (montane?) center of origin for *A. herbaceum* (see, in this context also 1.10.), the species must have gradually spread to both the NE (and eventually to the mountains of the Yemen Arab Republic and nearby Saudi Arabia) and the S, essentially following the major mountain ranges. Especially in SE Africa, it experienced a considerable ecological radiation (away from moist to wet habitats to dune sand localities; advancing to higher altitudes – i.e., to austro-afroalpine areas, etc.). For a description of the various "Forms" of *A. herbaceum* the reader is referred to the Systematic Part (and Figs. 80–82).

1.7. The *Anthospermum pumilum* Group
(Species 29.–34.)

is comprised of subshrubs or (presumably secondarily woody) dwarf shrubs with decussate, glabrous or subglabrous leaves whose blades are often rather narrow and variously rolled. The sex distributions range from non-dioecious (⚥, ⚥ + ♀ or ♀ individuals are most common) to dioecious. The ⚥ and ♂ corolla tubes are moderately long to ± short and ± funnel-shaped. The fruits have rather broad, roundish mericarps (e.g., *A. streyi*, Fig. 37 a) and are usually crowned by small to minute persistent calyx lobes.

The taxa of this group are either endemic to the SW Cape Floristic

Region or occur in the remainder of S Africa (Map, Fig. 120; but also see
A. pumilum subsp. *pumilum,* below!). They may represent a group which
has successfully "conquered" the SW Cape and experienced a "species
explosion" in that region (details below).

Within the group, *A. pumilum* subsp. *pumilum* is the most widely
distributed taxon which extends N to Angola, Zambia, Mozambique and

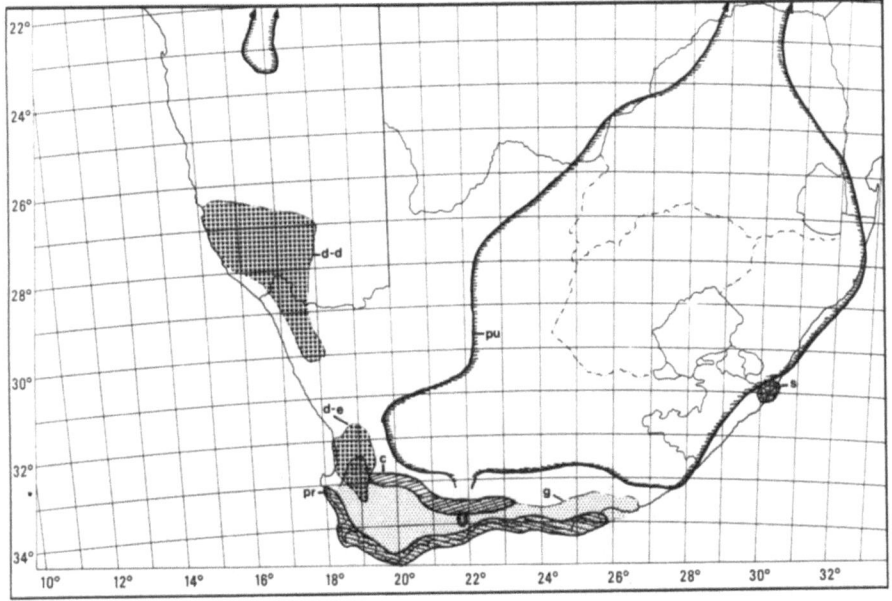

Fig. 120. Distribution ranges of *Anthospermum pumilum* (pu; extending to S
Tanzania in the E and to Angola in the W) and presumably allied species. – c *A.
comptonii,* d *A. dregei* (d–d subsp. *dregei*; d–e subsp. *ecklonis*); g *A. galioides*; pr *A.
prostratum*; s *A. streyi*

even S Tanzania (not shown in Map, Fig. 120); its range outside S Africa
appears to be rather discontinuous but, especially in the Flora Zambesiaca
area and Angola, this is, in my opinion, (at least in part) due to
undercollecting. It is difficult to make any sensible speculations as to the
possible center of origin of the taxon but it is, in my mind, clear that it was
somewhere outside the SW Cape Region (possibly in SE Africa, which has
a rather high concentration of *Anthospermum* species – cf. Fig. 121).
Subsp. *rigidum,* which differs primarily in its ecology, growth form and in
its clear trend to dioecy (some populations have become strictly dioecious!), is centered in the drier, karroid parts of S Africa (mainly interior of

32*

the Cape Prov. and the O.F.S.), but there is some overlap in the geographical range of the two subspecies. Subsp. *rigidum* may have established itself in periods of aridity (during the climatic fluctuations in S Africa in the Pleistocene?) in its present distribution area.

Morphologically very close to *A. pumilum* is *A. streyi,* which is confined to a small area on the Natal S Coast. The geological situation (sandstone!) and the resultant poor, acid soils of that region (and perhaps also climatic conditions?) may have favoured the evolution of this species.

The SW Cape species of this alliance form a very close-knit group which, in turn, is closely allied to *A. pumilum.* The ancestral stock of the *A. pumilum* Group apparently experienced an extensive radiation in the SW Cape Floristic Region:

Within the SW Cape subgroup, *A. prostratum* is ecologically well distinguished in that it has "conquered" coastal dune sand habitats to which it now is confined (next to the SE African *A. littoreum* – see 1.2.! – the second species in the genus to occur in such habitats)[1]. The distinct prostrate habit cf. Figs. 3 b – c) is most likely linked to and an adaptation to such habitats. The species has, furthermore, progressed to dioecy[1].

A. dregei subsp. *ecklonis* is confined to the W part of the SW Cape region and appears to be one of the few SW Cape taxa of *Anthospermum* which is strictly confined to a specific substrate – i.e., sandstone. Its distribution range contains small disjunctions which find their explanation in the, in part, scattered distribution of sandstone outcrops in the W Cape (i.e., W of the Cape folded mountain ranges). There are sometimes distances ranging from ca 50 to 100 km from one sandstone outcrop to the next. It is believed that its present distribution range could well have come about by dispersal of the mericarps by strong winds from one sandstone area to another. *A. dregei* subsp. *ecklonis* is morphologically very close to *A. galioides.*

The two subspecies of *A. galioides* (subsp. *galioides* in particular) are, in contrast to the former, extremely variably morphologically, ecologi-

---

[1] Considering the repeated modification of the shoreline due to changes in sea level in the more recent past, i.e., in later quaternary times, and since all lower-lying ground in the Cape and Malmesbury Distr. was under the sea then, as well as much of the areas between Bredasdorp and Mossel Bay (cf. LEVYNS 1962), its habitats, consequently, are very "young". ADAMSON (1959), in an analysis of the phytogeography of the Flora of the Cape Peninsula, writes that "the [Cape] Flats flora is more recent. Much of it is derived directly from that of the mountains...". Both the "young" habitats and the derived character states support my ideas of the "youthfulness" of *A. prostratum.* ADAMSON's (l.c.) general conclusions of a recent origin of many of the Cape Flats constituents may thus also hold true for *A. prostratum,* and also his notion of an origin from a Cape Mountain ancestor.

cally eurytopic and also widely distributed in the SW Cape. The "Forms" (± in the sense of ecotypes or races) of *A. galioides* (discussed in detail in the Systematic Part) seem to indicate that the process of active evolution is going on at a rapid rate in this species. At least the "Papillatum Form" of subsp. *galioides* is morphologically very close to *A. pumilum* subsp. *pumilum*. Whether this is an indication of a direct, close relationship between the two taxa or a superficial resemblance (convergence!) remains uncertain although I personally favour the second possibility.

A. comptonii, morphologically very close to both *A. galioides* and *A. dregei,* is clearly only marginally associated with the typical SW Cape fynbos vegetation. It is largely confined to the drier inland margins of the SW Cape and, within the SW Cape Region, to hot and dry karroid areas. There are possibly two ways to interpret its distribution and origin: (1) After an inital spreading of the ancestral stock of the *A. pumilum* Group to the SW Cape, there was, at a later period, a "secondary" radiation away from the SW Cape fynbos areas of a portion of the stock towards the N and NE (i.e., to the interior of the Cape). The climatic conditions in combination with geographical isolation provided the basis for the evolution of the species. (2) In the (well documented) climatic fluctuations of the Pleistocene, *A. comptonii* evolved in its present area during a period of aridity from the *common* ancestral stock of the *A. pumilum* Group. During subsequent wetter periods, the (remainder of the) stock continued to extend SW-ward, i.e., spread to the SW Cape Region s.str., where it radiated extensively, while *A. comptonii* remained ("was left behind") in its relatively drier surroundings. Various, obviously derived morphological features (many-stemmed, presumably *secondarily woody* dwarf shrubs!) and its dioecy seem to indicate that *A. comptonii* is quite a young taxon. It thus seems that of the two interpretations presented above, the first one may be the more likely one.

A. dregei subsp. *dregei* is confined to Namaqualand and S Namibia. Its origin and present distribution range can perhaps be explained in a similar way as that of *A. comptonii* [interpretation (1), above]: Part of the stock of *A. dregei* established itself in the sandstone areas of the W Cape to give rise to subsp. *ecklonis;* another portion of the stock radiated to the drier areas of the N, established itself there and gave rise to subsp. *dregei.* The situation with subsp. *dregei* is, however, not as straightforward. The reason is that the taxon seems to be closely linked morphologically not only to *A. dregei* subsp. *ecklonis* but also to *A. pumilum* subsp. *rigidum.* Purely on morphological grounds it does not appear to be 100% certain whether subsp. *dregei* should really be put into one species with subsp. *ecklonis.* Both the more distinctly dwarf-shrubby habit and the clear trend towards dioecy are characters which are shared by both *A. dregei* subsp.

dregei and *A. pumilum* subsp. *rigidum*; in addition, leaf sizes and shapes are similar. All this could be interpreted as points in favour of an association between subsp. *dregei* and subsp. *rigidum*. I, however, favour the idea that these similarities find their explanation in the convergent evolution of characters under similar ecological conditions (and I, therefore, stand to my proposed subdivision of *A. dregei* into subsp. *dregei* and subsp. *ecklonis*).

To sum up my ideas in short, I believe that in the *A. pumilum* Group, after an initial establishment of the stock in the SW Cape (mountains?), there was a subsequent radiation away from the SW Cape s.str. into the drier surrounding areas, in which different portions of the stock were involved. This resulted in the "formation" of *A. dregei* subsp. *dregei* and *A. comptonii*. *A. pumilum* subsp. *rigidum,* on the other hand, "separated" from the extra-SW Cape *A. pumilum* subsp. *pumilum* and, in turn, established itself in drier areas.

1.8. Isolated SW Cape Taxa (Species 35.–39.; 15.?), including general comments on taxa occurring in the SW Cape Floristic Region

The genus *Anthospermum* − and the *Anthosperminae* − have their highest concentration of taxa in the SW Cape Floristic Region (cf. Fig. 121; see also the Appendix to this subchapter!).

The species occurring in this region, however, show diverse affinities. Some species belong to or appear to be associated with essentially afromontane groups (e.g., *A. herbaceum,* see 1.6.; the SW Cape endemic *A. aethiopicum,* see 1.2.). Others belong to non-afromontane species groups which are (widely) distributed both in- and outside the SW Cape Region [e.g., the *A. pumilum* (1.7., above) and the *A. spathulatum* Group (1.3.)]. Thirdly, there are a number of (endemic) SW Cape taxa whose alliances are not fully understood. These species − *A. paniculatum* (± an exception as it also occurs in the E Cape), *A. esterhuysenianum, A. hirtum, A. bergianum, A. ericifolium* and *A. bicorne* − all occupy ± isolated positions within the genus and do not appear to have close or more remote allies in either the SW Cape Floristic Region or outside. Most of them, furthermore, do not seem to be closely related to each other either, so that this group of "Isolated SW Cape Taxa" is probably an artificial one. Except for *A. paniculatum,* all species of the group are characterized by flowers with 5-merous (rather than the common 4-merous) corollas, but this is about the only character that holds them together. Each of them is distinguished by a specific set or a combination of both "primitive" and very "derived" features (i.e., rather long and narrow-tubed corollas in

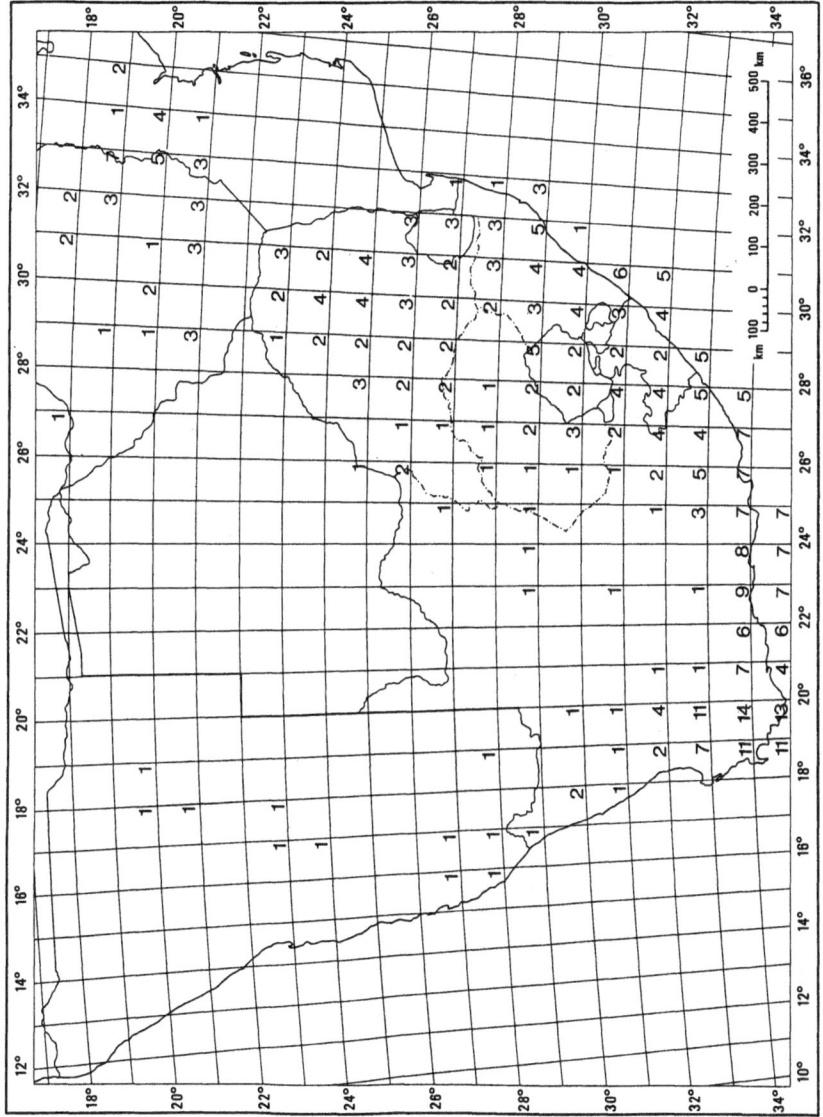

Fig. 121. *Anthospermum.* — Number of taxa (species and subspecies) per one degree square

combination with either very reduced inflorescences or with fruits with only one fertile carpel, or dioecy in combination with a "typical" thyrsus, etc.).

A. paniculatum is isolated in that it has "typical" thyrsic inflorescences with discrete peduncles and pedicels (as opposed to "axillary flower clusters" like in all other species of the genus; cf. Fig. 29 b) and in having fruits anatomically similar to those of certain *Nenax* species (presumably a convergence!).

Both *A. hirtum* and *A. bergianum* have similar, rather elongated fruits without persistent calyx lobes; no such fruits occur in any of the other SW Cape taxa of *Anthospermum*. The two species differ from each other in habit, leaves (including arrangement), and in sex distributions (mostly dioecious in *A. bergianum,* non-dioecious in *A. hirtum*) so that not more than a remote alliance between the two can be presumed. Neither of them has obviously allied species either in- or outside the SW Cape.

The high Cape mountain endemic *A. esterhuysenianum* differs markedly from other species in its habit and inflorescences (solitary flowers! Cf. Fig. 29 a) and, in that respect, is unique in the genus. Not even a guess as to its alliances can be attempted.

Both *A. ericifolium* and *A. bicorne* are unique in the genus in that one of the two carpels of the gynoecium is reduced, modified and sterile. In both, the sterile carpel is terminated by two calyx lobes which, however, are enlarged in *A. bicorne* only. Another difference between the two is that in *A. ericifolium* there are still two stigmas[1], while in *A. bicorne* there is only one. Moreover, leaf morphologically and anatomically the two species show considerable differences so that, in my opinion, the possibility cannot be ruled out completely that the reduction of one of the two carpels has occurred independently in each of the two species.

Since most of the above species are so different from one another, it is tempting to assume that they are not derived from a common ancestral SW Cape stock of *Anthospermum*. Thus the species of Group 1.8., together with the SW Cape species discussed in 1.2., 1.3. and 1.7., seem to represent a rather heterogenous agglomeration. It, however, seems clear that the ancestors of the species 35.–39. must have reached the SW Cape at a

---

[1] The situation that there is only one fertile carpel/ovule but still *two* stigmas is certainly unusual but by no means unique. In numerous *Rubiaceae-Gardenieae* with unisexual flowers, for example, the ♂ flowers have greatly reduced ovaries without any trace of ovules but style and stigma are still present and serve as pollen presenter ("receptaculum pollinis"; "ixoroid pollination mechanism"; ROBRECHT & PUFF 1986).

quite early stage – i.e., earlier than, for example, that portion of the ancestral stock which gave rise to the *A. pumilum* Group (1.7., above). Only with sufficient time could these taxa have diversified and acquired their specific set of features that makes it impossible now to determine their affinities.

The habit of the above species (never tall, *markedly* woody or many-stemmed, distinctly woody dwarf shrubs), the sex distributions (mostly non-dioecious; prevalence of ⚥ + ♀ or ♀) and the mostly quite long and narrow corolla tubes may, nevertheless, be indicative of some very remote alliance to *A. herbaceum*-like, perennially-herbaceous ancestors [rather than to woody, shrubby and dioecious ancestors – i.e., plants remotely similar to those grouped together in the *A. usambarense* Group (1.2.)].

To sum up, the species of *Anthospermum* confined to or centered in the SW Cape, thus – very much the same way as the Madagascan *Anthospermum* species (cf. PUFF 1986) – belong to several, rather diverse groups of species with quite different histories, i.e., the SW Cape species represent an agglomeration of very remotely related (or at least not closely allied) species or species groups which arrived there at different times.

If one, finally, also looks at the areas of distribution of all species of *Anthospermum* centered in or endemic to SW Cape, it, again, transpires that there is no uniform trend or pattern but rather diverse situations which repeat themselves in various (p.p. unrelated) species and species groups. A survey of the recognizable trends and patterns is given below:

1) a trend "away" from the SW Cape Floristic Region into (a) the drier interior of the Cape and to the NE (very pronounced, for example, in *A. spathulatum* subsp. *spathulatum*; cf. map, Fig. 69 b) and (b) to the E Cape (and Transkei) (e.g., *A. spathulatum* subsp. *uitenhagense*; map, Fig. 69 d; possibly also *A. paniculatum*, although there is no concrete evidence as to whether the species originated in the SW Cape Region or whether it is an E Cape species extending into the SW Cape).

2) ± wide, essentially (dry) mountain fynbos distributions (e.g., *A. bergianum*, *A. hirtum*, *A. bicorne*; maps, Figs. 95 e, g–h).

3) as above, but confined to the highest parts of the mountains or to the mountain tops (in the W part of the SW Cape Region: *A. esterhuysenianum*; wider distribution: *A. spathulatum* subsp. *saxatile*; maps, Figs. 95 c–d and 71 a, respectively).

4) rather restricted distribution ranges centered in the Cape Flats or ± confined to sandy to gravelly lower-lying areas of the SW Cape s.str. (e.g., *A. ericifolium*; map, Fig. 95 f). These are mostly relatively "young" areas and habitats (cf. ADAMSON 1959, LEVYNS 1952, and also see comments, p. 490 and 503).

Appendix. Notes on the *Rubiaceae* (as a whole) occurring in the SW Cape Floristic Region

It has been stressed repeatedly that the genus *Anthospermum* and the *Anthosperminae* as a whole have their highest concentration of taxa in the SW Cape Region. But how does this compare with the occurrence of other *Rubiaceae* in the SW Cape?

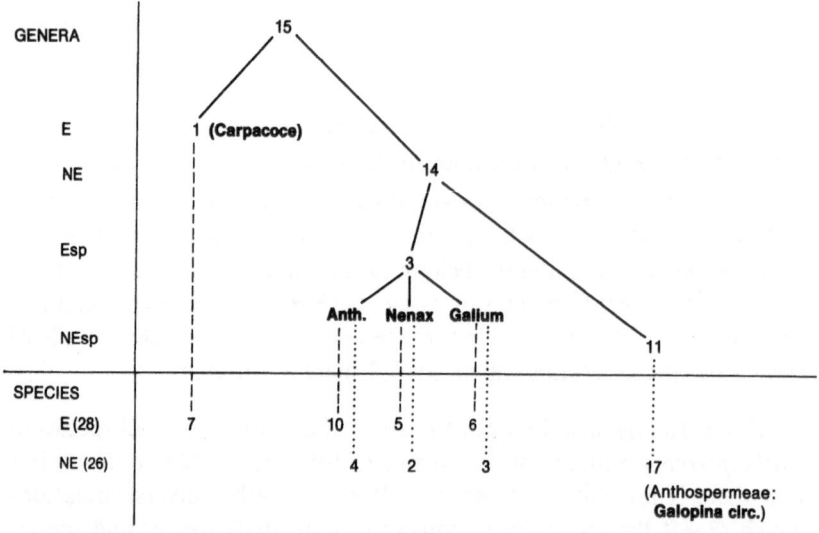

Fig. 122. *Rubiaceae* occurring in the SW Cape Floristic Region [based on the data given in BOND & GOLDBLATT (1984)]. – E endemic; NE non-endemic; Esp genera with some species endemic to the SW Cape; NEsp genera without SW Cape endemic species

Fig. 122 shows that
1) 54% of all species of the *Rubiaceae* occurring in the SW Cape belong to the *Anthospermeae* (29 : 54 species),
2) 41% of all species of the *Rubiaceae* occurring in the SW Cape are *endemic Anthospermeae* (22 : 54 species),
3) 78.5% of the species of the *Rubiaceae endemic* to the SW Cape belong to the *Anthospermeae* (22 : 28 species) (the remaining species belong to the genus *Galium*),
4) 76% of the species of the *Anthospermeae* occurring in the SW Cape are endemic (or nearly so) (22 : 29 species), and that
5) the only genus of the *Rubiaceae* endemic to the SW Cape belongs to the *Anthospermeae* (*Carpacoce*).

Thus, the majority of the genera of the *Rubiaceae* occurring in the SW Cape (11 : 15 genera = ca 73%) has no endemic species in the SW Cape. All of these genera are either SE African (amongst the *Anthospermeae*: *Galopina*) or are more widely distributed. Only a few species of these genera extend (in several cases just barely so) into the SW Cape. Amongst these genera, there are both herbaceous and woody representatives (the latter category, however, is the larger one: i.e., 6 : 11 genera = 54.5%, or 12 : 17 species = 70%). Most of the woody genera without

SW Cape endemic species are associated with (afromontane) forest and as such "follow" the patches of kloof forest, etc. into the SW Cape.

The three genera which also evolved SW Cape endemic species belong to the *Anthospermeae* (*Anthospermum* and *Nenax*) and the *Rubieae* (*Galium*). Of these, *Galium* has the widest general distribution range (± world-wide) and *Anthospermum* is widely distributed in Africa S of the Sahara and in Madagascar, but *Nenax* is confined to S Africa.

Anthospermum has the highest concentration of species in the Flora of Southern Africa area (21:39 species = 54%; 18:39 species = 46% endemic) [8:39 species = 20.5% occur in and are endemic to Madagascar, the remainder occurs in tropical Africa], and in the comparatively very small area of the SW Cape Region (14:39 species = 36%; 10:39 species = 25.5% endemic).

What are the possible reasons for the predominance of members of the *Anthospermeae* in the SW Cape, and how can this situation, which is so entirely different from that of the other *Rubiaceae* occurring in the region, be explained? *Anthospermum* may, from its "genetic make-up", have more "potential" for radiation. This can perhaps explain its more "aggressive" behaviour (as far as radiation in the SW Cape is concerned). In this, it seems to differ considerably from other ± afromontane rubiaceous shrubs or trees whose distribution ranges just "taper out" toward the SW Cape. Also the habitats of *Anthospermum* (the ancestral stock of *Anthospermum* is presumed to have occurred in the forest edge vegetation of tropical African montane areas; see elsewhere for details) may have been more favourable for such a radiation – at least more so than those of forest (undergrowth) taxa, although both of these categories (both presumably having originated in the Tropics) must have reached the SW Cape by S-ward migration. Furthermore, the relative youth of the *Anthosperminae* (for which there is a considerable amount of evidence; details elsewhere) may have played some role: As a relatively young "newcomer" the ancestral stock may have been still "flexible" enough to spread and radiate and conquer new environments. This very same argument may, by the way, also hold true for *Galium*, the only other rubiaceous genus outside the *Anthospermeae* that has evolved SW Cape endemics.

1.9. Isolated Madagascan Taxa
(Species 23. and 24.)

Analogous to the SW Cape taxa of *Anthospermum* which stand ± isolated and do not show clear affinities to any other taxa (see 1.8., above), Madagascar has two endemic species which do not seem to have any apparent close or more distant allies in either Madagascar or on the African mainland. The two species, *A. ibityense* and *A. longisepalum,* however, are doubtlessly very closely related to each other and appear to form a vicarious species pair (see PUFF 1986 for details).

It seems to be particularly noteworthy that these two species, similar to the isolated SW Cape species, have flowers with rather long, cylindrical tubes and a ± dwarf-shrubby habit (several- to many-stemmed plants innovating from the base, rather than tall, single-stemmed shrubs). *A. ibityense* is also similar in its non-dioecious sex distribution (⚥ and ♀, but perhaps also ♀̵ + ♀ and ♀̵ + ♂ plants occur), but *A. longisepalum* may, judging from the little flowering material available, have progressed to dioecy.

Thus, as it was speculated for the isolated SW Cape taxa in the previous subchapter, the two Madagascan species may also have some very remote alliance to essentially herbaceous, perennial, possibly *A. herbaceum*-like ancestors rather than to a "primarily" woody, large shrubby, dioecious (?) ancestral form with perhaps rather short ♂ corolla tubes.

1.10. Concluding Remarks

If one accepts the presumed "primitive" features of the ancestral form of *Anthospermum* as presented in 1.1., it becomes apparent that there seem to be two major lines of evolution (also see Figs. 109–114):

(1) a "shrubby (primarily woody) line" which is primitive in its growth form (mostly single-stemmed, large shrubs) but derived in many other features (dioecy; short and broad corolla tubes; leaves often in whorls of three, etc.). This "shrubby line", represented by the *A. usambarense* Group (1.2.), is confined to ± afromontane areas.

The *A. spathulatum* Group (1.3.) may be loosely connected with this "shrubby line" but its affinities remain somewhat obscure. Firstly, the species of this group are not afromontane; secondly, there are considerable growth form (and also wood anatomical) differences.

(2) a "herbaceous line" which is derived in its growth form but remarkably primitive in many of its other features (non-dioecious sex distributions; relatively long, ± cylindrical corolla tubes; rather large and broad, not distinctly "ericoid" foliage, mostly arranged decussately, etc.). Representatives of this line are, at least in part, also ± afromontane species and often grow in ± moist to wet habitats. Species centered around *A. herbaceum* belong here.

From this "herbaceous line", evolution could have progressed to

(a) rather short-lived shrubs ("woody herbs") or subshrubs which in several of their characters are still rather similar to the "herbaceous line" (e.g., often quite long and narrow corolla tubes; predominance of non-dioecious sex distributions) but somewhat more derived in some other features (e.g., leaves often in whorls of three or even four rather than strictly decussate, etc.). The species are mostly no longer ± afromontane. The *A. ternatum* Group (1.5.) should be mentioned here; it still shows some remote morphological affinities to the proper "herbaceous line" (i.e., the *A. herbaceum* Group, 1.6.) but also appears to be loosely linked with the *A. whyteanum* Group (1.4.; see below).

(b) [or (3)?] several- to many-stemmed, presumably secondarily woody dwarf shrubs or (fire-adapted) subshrubs — i.e., a "dwarf-shrubby line" — which, apart from its presumably derived growth form, is also ± derived to very derived in other characters (e.g., more "ericoid", often narrow or revolutely rolled leaves; relatively shorter corolla tubes; repeated trends to dioecy and establishment of dioecy in certain species, etc.). This "dwarf-shrubby line" does not occur in afromontane areas. Both the *A. whyteanum* Group (1.4.) and the *A. pumilum* Group (1.7.) could be listed here. These two groups, however, are by no means very closely related to each other. It is, on the contrary, quite likely that, on account of sex distributions, habit, corolla tube shapes and general fruit morphology, the former shows some affinities to the *A. ternatum* Group (1.5.). The *A. pumilum* Group, on the other hand, appears to be more isolated — not least because of its fruit morphology (see 1.7.); it, furthermore, has a much more "S" distribution (with numerous SW Cape species), whereas the *A. whyteanum* Group is SE- Central African and Madagascan in distribution.

The isolated SW Cape and the isolated Madagascan taxa (1.8., 1.9.) may, as suggested, also link up with the "herbaceous line" rather than the "shrubby line".

From the above, the following conclusions as regards the time-space development of *Anthospermum* can be drawn:

There must have been an early split of the ancestral stock of *Anthospermum* into a (primarily) "woody" and a "herbaceous" line. Both of these, possibly originally occurring in ± montane tropical African areas [perhaps associated with montane rainforest — as forest edge dwellers?; according to Axelrod & Raven (1978), montane rainforest may have been in existence since the Paleogene], subsequently continued to develop independently.

The "woody" line retained its shrubby habit but, in the course of time, also acquired numerous derived features (see above). It also remained in its presumably "original" tropical montane habitats, gradually spreading to all afromontane areas ("Regional Mountain Systems" sensu White 1978) except W Africa and also extended its range to the Madagascan mountains. This process of expansion could have started in the Miocene when E Africa was elevated and may have been concurrent with the spreading of the montane rain forest (cf. Axelrod & Raven 1978). Some time after the formation of the E African volcanoes (from the Miocene onwards; cf. HEDBERG 1970), an extension of the range to afroalpine regions could have occurred (e.g., *A. usambarense*[1]). Part of this "woody" stock could – at a later stage? – have spread to SE Africa (into no longer strictly montane areas) to give rise to *A. galpinii* and *A. littoreum* (from the geological point of view there is no objection to such a theory – cf. KING 1982); a further extension to the SW could, theoretically, explain the presumed loose links between the former two species and *A. aethiopicum* (and the *A. spathulatum* Group?).

The "herbaceous" line may have, as discussed above, experienced a further diversification.

One part of the herbaceous stock seems to have retained its, except for growth form, largely primitive set of features (i.e., that part of the stock that gave rise to the *A. herbaceum* Group). It can be easily envisaged that it experienced a similar fate as the "woody" line described above and expanded its range together with the montane rain forest vegetation, possibly from the Miocene onwards. See, in this context, also p. 488 for comments on the possible origin of the Ethiopio-Sudanian *A. pachyrrhizum*.

Another part of this herbaceous stock (i.e., that which gave rise to the *A. whyteanum*, *A. ternatum* and *A. pumilum* Group, and possibly also to the species grouped together as "Isolated SW Cape" and "Isolated Madagascan Taxa") underwent, in the course of time, considerable further changes and subsequent splitting. When and where these partial stocks split off and how they reached their respective distribution areas is difficult to imagine.

Still easiest, perhaps, are speculations on the origin of the *A. ternatum* and the *A. whyteanum* Group. The stock of the *A. ternatum* Group could have separated from its presumably herbaceous, montane ancestors in the tropics and, at first, established itself in savanna and woodland areas which, with the trend to a drier climate in the Neogene, had attained a rather wide distribution (cf. AXELROD & RAVEN 1978). Parallel with its establishment in a "new environment", its deviating and more derived features could have evolved. The present-day disjunct distribution of the species of this group could have come about subsequently, i.e., the "fragmentation" of the stock and their "secondary" establishment in

---

[1] The type locality of this species, the Usambara Mts., are a geologically very old crystalline formation (cf. PÓCS 1982). Assuming that *A. usambarense* had established itself there at a relatively early stage, it follows that the expansion of its range to the afroalpine region of the relatively younger E African high mountains of volcanic origin must have been a subsequent – "secondary" – event.

afromontane areas (W Africa: *A. asperuloides;* N Kenya – S Ethiopia: *A. villosicarpum*) could have been more recent events.

Similarly, the even more derived, presumably secondarily woody, dwarf-shrubby *A. whyteanum* Group could have separated from tropical montane, herbaceous ancestors, established itself in its specialized (rocky) habitats, experienced an initial expansion of its range to S Africa and Madagascar and, subsequently (primarily due to geographical isolation?), gave rise to three species.

Judging from the present-day distribution of its species, the *A. pumilum* Group is likely to have had a more S (SE African?) origin. How the stock of the *A. pumilum* Group links up with an originally herbaceous and possibly tropical montane ancestor – i.e., the "herbaceous line" of *Anthospermum* – is not at all certain and I am at present not prepared to make any speculations. The rather clear relationships and affinities within the Group are discussed in 1.7.

The ancestors of both the "Isolated SW Cape" and "Isolated Madagascan Taxa" must have reached their respective areas in a relatively early epoch. Considering the presumably relatively young age of the Groups described above, it is, however, tempting to assume that their arrival in the SW Cape and Madagascar respectively does not date back further than to the late Paleogene. At least as far as the SW Cape taxa are concerned it could be argued that the unique situation of that area (varied topography, largely infertile soils, climate, etc.; cf. BOND & GOLDBLATT 1984) contributed considerably to the rapid rate of evolution of these species and, subsequently, to the development of their rather unusual set of characters. Thus a *much* greater age is not necessarily the reason for the unexplicable relationships and affinities of these species. Although mentioned repeatedly, it is stressed again here that the high concentration of *Anthospermum* species in the SW Cape, thus, is not indicative of a SW Cape origin of the genus (cf., for example, also LEVYNS 1962!).

2. Affinities Between *Nenax* and *Anthospermum*, and Relationships Within *Nenax*

There are numerous indications that *Nenax* is a (relatively) very young genus, which is very closely allied to and presumably a "segregate" of *Anthospermum*. The two genera share a similar flower structure and (in part) similar fruits; growth forms of certain species are virtually identical; also the leaf arrangement, morphology and anatomy of some species may be very similar (also see comments, p. 394). *Nenax*, however, is characterized by almost invariably "derived" characters states such as, for example,

the presumably secondarily woody, dwarf-shrubby habit, mostly distinctly ericoid leaves, rather reduced (few-flowered) inflorescences or the
consistent unisexuality of the plants (dioecy).

The genus is confined to the Flora of S Africa area. Its center of origin
is presumably the SW Cape Floristic Region (where *Nenax* has the highest

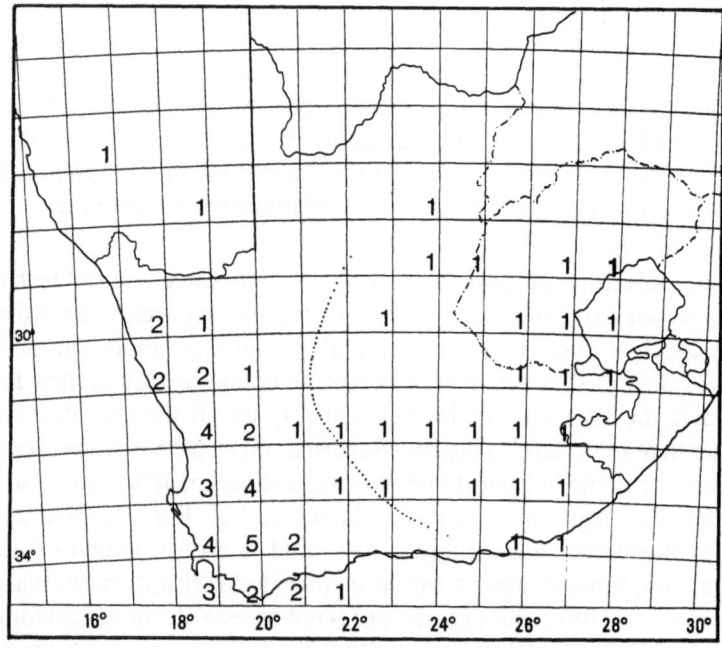

Fig. 123. *Nenax.* − Number of taxa (species and subspecies) per one degree
square. *N. microphylla* is the only species occurring E of the dotted line

concentration of taxa − cf. Fig. 123). I suspect that the ancestral stock of
Nenax may have separated from a portion of the stock of *Anthospermum*
which, at the time, had already established itself in the SW Cape Region.
When this split or segregation occurred is, of course, rather uncertain. In
my opinion and based on the thoughts expressed by both AXELROD &
RAVEN (1978) and BOND & GOLDBLATT (1984), it is, however, not likely to
have happened much earlier than in the Pliocene (-Pleistocene?) period.
There must have also been a subsequent marked radiation away from the
SW Cape s. str. into the drier (N and NE) parts of the Cape Province. *N.
microphylla*, *N. cinerea* and *N. namaquensis*, for example, do not occur in
the SW Cape s.str. or only just barely extend to that region; *N. elsieae*

occurs on the dry inland margin of the SW Cape Region[1]. Whether these taxa of arid regions established themselves during the climatic fluctuations in the Pleistocene – or, more precisely, during pluvial periods when the Cape vegetation was displaced to the N (into the present semidesert areas) – and, subsequently, remained there, or whether this radiation occurred even more recently, is not certain[2].

It is also noteworthy that even some of the taxa occurring in the SW Cape proper occur in "recent" habitats such as the broad basins which originated after the folding of the Cape mountains and which are covered by recent sands, or on the "young" limestone areas of the S Cape[3], or in the W Coast Strandveld vegetation (e.g., *N. hirta, N. acerosa, N. arenicola*). Again, this appears to support the "youthfulness" of the genus.

Within *Nenax*, there seem to be three ± distinct groups of species (also see Figs. 109–114):

1. Species with *Anthospermum*-like Fruits (Species 3.–6.)

The fruits of these species dehisce into two non-inflated mericarps (cf. *N. coronata*, Fig. 37 h). The leaves are ± needle-like or "ericoid"; the ♂ flowers have rather short and broad corolla tubes, the corollas are 4-, 4–5- or 5-merous (*N. divaricata* and *N. elsieae* → *N. coronata* → *N. namaquensis*); the fruits are glabrous, glabrous to very shortly hairy or ± long-hairy (*N. elsieae*) and are crowned by small to somewhat enlarged calyx lobes.

The group occurs from the SW and W part of the SW Cape Region N-wards (Map, Fig. 124). *N. divaricata* has its main distribution in the SW Cape Region (mostly in dry Mountain Fynbos), *N. elsieae* and *N. coronata* are only "marginally" associated with SW Cape vegetation, and *N. namaquensis* is confined to the NW Cape Prov. (mainly Namaqualand).

These species seem to be rather closely related with each other and appear to have closer affinities to *Anthospermum* than to the species of the following two groups.

---

[1] Similar distribution trends are found, for example, in *Homeria*, an iridaceous genus perhaps derived from the large African genus *Moraea* (cf. GOLDBLATT 1981).

[2] According to TANKARD & ROGERS (1978), the earliest evidence of the modern semi-arid environment and winter rainfall on the W Coast of the SW Cape Prov. dates to the Pliocene.

[3] The limestones as such are of mid-Tertiary age, but were under the sea; the shoreline retreated in the Pleistocene to expose much of the continental shelf, including the limestones. In later Quaternary times, however, the sea once more encroached on the land. The "final" emergence is thus a very recent event (cf. LEVYNS 1952).

2. Species with ± Inflated, Dehiscent Fruits (Species 1. and 2.)

The conspicuous fruits of *N. microphylla* and *N. cinerea* (cf. Figs. 37 g, 38 a–b and 41 f), although of rather different appearance than those of Group 1., show a similar basic anatomical structure (compare with *N. coronata,* Fig. 40 c); only the "free" (air-filled) space between the two

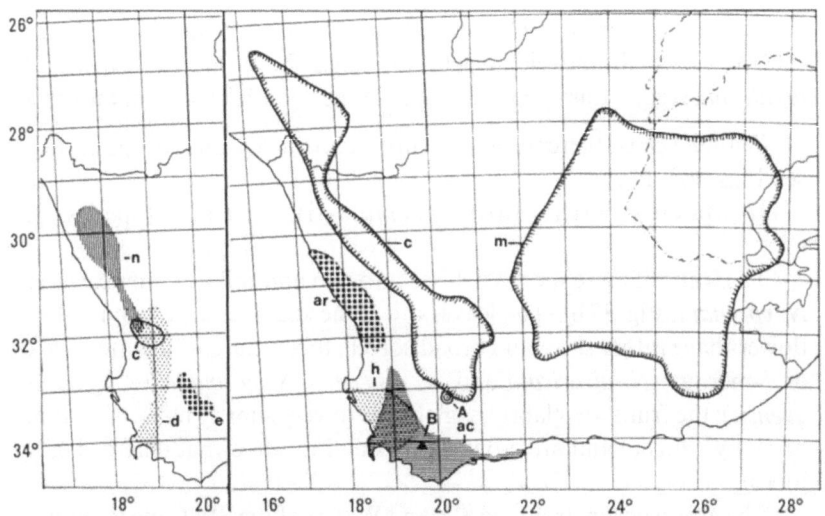

Fig. 124. Distribution ranges of *Nenax* species. – Left side: *Nenax* species with *Anthospermum*-like fruits (c *N. coronata;* d *N. divaricata;* e *N. elsieae;* n *N. namaquensis*). Right side: *Nenax* species with inflated fruits (c *N. cinerea;* m *N. microphylla*) and with indehiscent fruits (ac *N. acerosa;* ar *N. arenicola;* h *N. hirta;* A *Nenax* sp. A; B *Nenax* sp. B)

mericarps is much enlarged. The diaspores of these two species are undoubtedly better adapted to wind dispersal (increase in surface area!) and to their habitats (rather dry country, dominanted by low-growing vegetation) than those of other *Nenax* or *Anthospermum* species growing under similar situations. Also the rapid germination (one or two days after the onset of rains – cf. C.12.7.) is an apparent adaptation to the specific climatic and ecological conditions in which they are found.

None of the two species has a SW Cape distribution; *N. cinerea* is marginally associated with SW Cape vegetation only in the S-most part of its range; *N. microphylla* has a more E and NE distribution (extending to the O.F.S. and Lesotho) which is totally outside the SW Cape Region (Map, Fig. 124).

While the two species share the fruit characters discussed above, they differ in various morphological (habit and leaves!) and anatomical aspects. Especially because the leaf anatomy of the two species is so strikingly different (see C.4.5. for details on the unique leaf anatomical features of *N. cinerea*), I am not fully convinced that the two species are really very closely allied; there may have been a relatively early split in the ancestral stock of this group, which perhaps also finds some support in the present-day divergent distribution areas of the two species. It, furthermore, cannot be completely ruled out that the similarity in the fruit structure in the two species may be due to convergence (evolved under similar ecological conditions). It, nevertheless, seems clear that the two species of this group must somehow be linked to the species of Group 1.

3. Species with Indehiscent Fruits (Species 7. – 11.)

The fruits of the species of this group, in section, show a continuous ring of sclerenchymatic endocarp (e.g., *N. arenicola,* Fig. 41 c) thus preventing the fruits from dehiscing into two mericarps (doubtlessly a very "derived" feature which occurs in no other members of the *Anthosperminae*!). Except for *Nenax* sp. A., this sclerenchymatic endocarp is quite thick; consequently the fruits are very hard. Other than that, the general anatomy of the fruits, however, is not drastically different from that of the other *Nenax* species. The leaves are decussate, mostly glabrous and ± needle-like or "ericoid" (except for *N. hirta:* in whorls of three and hairy); the ♂ flowers have rather short and broad corolla tubes, the corollas are 4-, 4–5-, or 5-merous; the fruits are mostly glabrous except for *N. hirta*. Tetraploidy is documented for *N. acerosa* and *N. arenicola, N. hirta* is diploid (the other species of the group are karyologically unknown).

The species of this group occur in the SW Cape Floristic Region except for *N. arenicola* (W Coast Strandveld) (Map, Fig. 124). They are largely confined to lower-lying sandy to gravelly habitats; they can, in part, occur in Coastal Fynbos; none of them is a typical Mountain Fynbos dweller, although some do extend slightly into these areas. It thus seems that the species of this group are largely restricted to rather "recent" habitats.

The combination of indehiscent fruits and tetraploidy appears to indicate that these species are the most highly "derived" in the genus. If their habitats are taken into consideration as well, it may furthermore be concluded that the species of this group may also be the youngest in the genus. A further indication for this may be that in species of this group (notably *N. acerosa*), there are "Forms" (mostly in the sense of populations or races) which no longer morphologically fully match a given taxon. In my opinion, their occurrence suggests that, especially in this group, active evolution towards the formation of new taxa is presently

33*

going on at a rapid rate. *Nenax* sp. A, only known from one locality, is perhaps the best example for such a very "recent" new species.

Nenax hirta appears to be the "odd man out" in this group. It is (still) diploid but, instead, differs in its ternate leaves and indumentum. Its origin and its relationships to the remaining species seem somewhat obscure. There, nevertheless, can in my opinion be no doubt that this whole group must have initially separated from a (SW Cape?) portion of the ancestral stock of *Nenax* (with still dehiscent fruits) as an entity. Subsequently, it must have undergone a further morphological, geographical and ecological differentiation, during which also (?auto-)polyploidization took place.

3. Relationships Within *Galopina* and Its Affinities to Other Anthospermeae

Galopina is a well defined and "compact", essentially SE African genus of perennial herbs with rather large and broad decussately arranged foliage and terminal paniculate to thyrsic inflorescences; it cannot easily be confused with any other genus of the tribe.

The four species of *Galopina*, nevertheless, show some rather conspicuous character "progressions". The most prominent morphological progression is found in the inflorescence: extensive, ± spheroidal, paniculate → narrowly cylindrical, more reduced (fewer-flowered), thyrsic (*G. circaeoides* → *G. tomentosa* → *G. aspera* → *G. crocyllioides;* cf. Fig. 28). Other progressions concern (a) the fruit indumentum: "warty" or ± tuberculate (*G. circaeoides, G. tomentosa, G. aspera)* → long hairy (*G. crocyllioides*; compare Fig. 39 a–h), (b) the underground parts: long, branched, creeping rhizomes (e.g., *G. circaeoides*) → short, ± erect rhizomes (*G. crocyllioides*; compare Figs. 13 a–b and d–e) and (c) the leafy short shoots: present (*G. circaeoides, G. tomentosa, G. aspera*) → absent (only *G. crocyllioides*); short shoot leaves much smaller than long shoot leaves (*G. tomentosa, G. circaeoides*) → short shoot leaves enlarged, often almost as large as the long shoot leaves (*G. aspera*). Ecologically, there seems to be a clear radiation from ± shady, moist to wet habitats [streamsides, (montane) forest edge vegetation, etc.] to open, sunny grassland habitats (*G. circaeoides* → *G. tomentosa* → *G. aspera* → *G. crocyllioides*).

The degree of derivation of the species appears to be ± inversely proportional to their respective distribution areas. The species considered to be most primitive, *G. circaeoides,* has a wide and, in part, disjunct distribution range, whereas the presumably most derived species, *G. crocyllioides,* has a rather restricted distribution range, largely confined to S Natal except for a few "outposts" elsewhere.

The genus as a whole may well have had its center of origin in SE Africa (possibly in Natal?; originally occurring in forest edge vegetation?), from where each of the species spread both to the N(E) and SW to varying degrees: *G. circaeoides* spread in afromontane areas, also extended its range to many "outposts" of afromontane vegetation in the Transvaal and seems to have "followed" the patches of afromontane forest to the SW Cape Floristic Region (see also comments, p. 432); an extension of its range to the E highlands of Zimbabwe and neighbouring parts of Mozambique and to the Zomba Plateau in Malawi may have been achieved by LDD. The presumably most closely allied species, *G. tomentosa,* could have established itself in generally more open, sunnier habitats (forest/grassland border, open grassland), possibly in lower-lying areas relatively near the Indian Ocean coast, from where it spread both NE to Tongaland and SW to the Transkei Wild Coast. Likewise *G. aspera* which, however, in time attained a somewhat wider distribution stretching from the Transvaal Drakensberg escarpment to the E Cape. The (fire-prone) grassland species, *G. crocyllioides,* finally, retained a rather confined (largely S Natal) distribution range; extensions of its range to both Swaziland and the Transkei may be the result of LDD of its mericarps (in relatively recent times?).

While the herbaceous habit, the ± subcampanulate corollas and the unusual fruits characters (short commissure, anatomy − cf. C.7.2.) seem to indicate a rather derived state of the genus, the quite extensive paniculate to thyrsic inflorescences and the non-dioecious sex distributions are typically "primitive" characters. This combination of primitive on the one hand and derived features on the other (indicating differential "speeds" of character differentiation) make it difficult to assess the position and affinities of the genus within the *Anthosperminae* and to speculate on its time of origin. There are clearly no obvious, direct and close links between *Galopina* and species of *Anthospermum.* One, therefore, can certainly not argue that *Galopina* may be "derived" directly from the stock of *Anthospermum.* On the other hand, there are some conspicuous similarities to the Macaronesion genus *Phyllis* (inflorescence; leaves − including their anatomy), but these are, in my opinion, most certainly convergences[1]. The most reasonable assumption, perhaps, is that the

---

[1] SUNDING's (1979) groupment of *Phyllis* amongst [Macaronesian] "Endemics of South African affinity" and statement that "its closest relative" is *Galopina* is, therefore, not accepted. Also see p. 518.

ancestral stock of *Galopina* separated from the common (tropical) stock of
the *Anthosperminae* at a relatively early point of time[1]. Considering the
"afromontane affinities" of the *Galopina* species, it is tempting to
speculate that the modern *Galopina* evolved from an originally montane
tropical ancestral stock. Such a stock could have only reached SE Africa,
which is believed to be the center of origin of the modern *Galopina*, from
the Miocene onwards, i.e., when E Africa was uplifted, thus allowing the
ancestral stock to migrate S-wards[2]. This theory could also help to
account for the presence of derived character states in the genus: During
its migration S, the ancestral stock of *Galopina* could have, in the course of
time, acquired some of its more derived characters. Thus, when the
modern *Galopina* started to evolve in SE Africa, it already *had* its
characteristic set of, or its combination of both primitive and derived
features.

4. The Isolated Position of *Carpacoce* and Relationships Within the Genus

The SW Cape genus *Carpacoce* doubtlessly occupies a somewhat
isolated position within the *Anthosperminae*. Reasons for this are
primarily the markedly different fruits (dehiscence and anatomy; cf.
C.7.3.), their large, ± leaf-like calyx lobes, the remarkable modification of
one of the two carpels in all species but one (anatomically entirely different
from the situation in the two *Anthospermum* species in which one of the
two carpels is modified and sterile; compare Figs. 42 and 44), the
differences in the corolla morphology ("hooded" corollas – cf. Figs. 34 a–
d, f), and deviating leaf anatomical features such as the complete absence
of cells containing tannins and the arrangement of the parenchyma
sheaths. The chromosomes of *Carpacoce*, moreover, seem to be much
bigger than those of the remaining *Anthosperminae*; also the average
pollen diameters are considerably larger, although presumably all taxa are
diploid.

---

[1] See footnote on previous page.

[2] In this context, the Malawi (Zomba Plateau) and E Zimbabwean localities of
the presumably relatively "primitive" *G. circaeoides* require some comments: It
has been argued above that the species experienced an expansion of its essentially
SE African range to the mentioned areas by LDD, but could it not be that these
localities were the first to be occupied in the course of a general S-ward movement?
Theoretically yes – but this would mean that *G. circaeoides,* at the time, had
already diverged from the common ancestral stock and evolved into a discrete
species. If that had been the case, it would be, in my opinion, more difficult to
understand the origin of the remaining SE African species, notably that of the very
closely allied *G. tomentosa.*

As opposed to some SW Cape species of *Nenax* and certain SW Cape species groups of *Anthospermum*, *Carpacoce* contains mostly very well defined species which are well distinguished by good separation characters. One thus gets the impression that *Carpacoce* is a quite *old* genus (i.e., a genus at least *relatively* older than *Nenax* and numerous SW Cape species of *Anthospermum*) which may have established itself in the SW Cape vegetation a long time ago.

Its roots remain obscure. It may have separated from the stock of the *Anthosperminae* at a very early stage but, in my opinion, it cannot be ruled out completely that it may have segregated from the original *Anthospermeae* stock even earlier on. Provided that latter assumption is correct, it follows that it must have reached the SW Cape and established itself there independently from and earlier than the remaining *Anthosperminae* (*Anthospermum*; *Nenax*). Consequently, this would also lead to the conclusion that its placement in the *Anthosperminae* is not correct, and that the genus should perhaps be placed its own subtribe (as it also differs very markedly from all extra-African *Anthospermeae*). At present, however, I will refrain from doing so because there is neither sufficient nor concrete evidence for either of the above speculations, and I leave the genus, last but not least for practical reasons, in the *Anthosperminae*.

In context with the presumably great age of *Carpacoce*, the preference of wet to moist habitats of certain species (notably *C. spermacocea*) may have some significance: To explain the origin and isolated position of SW Cape endemic families such as the *Roridulaceae* (cf. CARLQUIST 1976) etc., the theory has been brought forward that they may have been derived from an ancient S African stock representing mesophytic relics surviving in wet montane habitats in a subarid region; perhaps, a similar theory could also be applied to *Carpacoce*.

The following "character progressions" are recognizable within *Carpacoce*:

Leaves and stipules: blades thin, ± coriaceous (e.g., *C. spermacocea*) → leathery, thick and broad (e.g., *C. curvifolia*) or ± round in section ("ericoid") (e.g., *C. vaginellata*); stipular sheaths with numerous setae (*C. spermacocea*) → seta 1 or ± 0 (other species); leaf arrangement: decussate → whorls of three (occasionally in *C. curvifolia*); inflorescences: relatively many-flowered and extensive (e.g., *C. spermacocea* subsp. *orientalis*) → very few-flowered, or flowers solitary (e.g., *C. gigantea*).

The number of the leaf-like calyx lobes and the corolla lobes differs greatly, and the number of the calyx lobes often does not correspond to the number of the corolla lobes [e.g., *C. gigantea:* calyx 3-merous, corolla 5(6)-merous, or *C. heteromorpha,* calyx 4-merous, corolla (5)6–7-

merous]. 5-merous calyces and corollas appear to be the basic situation. The presence of only 4 corolla and 4 calyx lobes (e.g., *C. curvifolia*) as well as (5)6–7 corolla and 5 calyx lobes (e.g., *C. heteromorpha*) appear to be derived.

As regards the number of fertile carpels in *Carpacoce* (*one* in all species but *C. scabra*), I am not quite certain what the "primitive" or "derived" state might be. Within the *Rubiaceae* – and even the *Anthosperminae* – the presence of only a single fertile carpel is certainly a derived feature, but does this also hold true for *Carpacoce*? Two theories, offering contradicting explanations can be brought forward:

(1) *C. scabra*, with two fertile carpels, represents the "primitive" state. The other species are derived to varying degrees: a fertile and a second, still quite large but sterile, modified carpel (e.g., *C. vaginellata;* cf. Fig. 44 j) → a fertile and a small to very small sterile carpel (a ± round "plug" of spongy tissue – cf. Figs. 44 a–f).

(2) The one-fertile-carpel condition represents the "basic" situation in the genus (i.e., the original stock itself was derived in this character). The two fertile carpels represent a "secondary" state, i.e., ± a "reversal" to an ancient two-carpel situation.

For the following reasons, I consider the theory (2) to be the more likely: In the species with a still relatively large, sterile, modified carpel, i.e., *C. vaginellata,* such "reversals" to a two-fertile-carpels state were actually observed in some populations which were carefully studied in the field. Next to plants with flowers which had only the "typical" single fertile carpel, there were plants with (i) flowers with both one *and* two fertile carpels (some), and (ii) plants with only two fertile carpels (very few plants of a population). Also the presence of other derived character states in *C. scabra* [inflorescences; growth form: presumably secondarily woody dwarf shrubs; possibly also the ericoid foliage (but also see the relevant comments below!)] can be used to argue against theory (1). These features make it seem unlikely that *C. scabra* can be considered a "primitive" species ± resembling the ancestral stock from which *Carpacoce* was derived.

It does not seem unlikely that the ± needle-like ("ericoid") leaves evolved independently in different species of *Carpacoce* (e.g., in hardly closely related species like *C. scabra* and *C. vaginellata*) and that, subsequently, the ericoid habit of several species is due to convergence, which is likely to have come about as an adaptation to the similar environmental conditions in which the respective species grow. Under these circumstances it is, of course, difficult to draw any definite conclusions as to the relationships and affinities between the taxa. It, nevertheless, seems possible to distinguish three species groups:

1. *Carpacoce spermacocea* Group (Species 1.–3.)

All three species seem to show a (in part very clear) preference for moist to wet habitats. Except for *C. curvifolia*, they have rather large, not distinctly "ericoid" foliage and stipular sheaths with several setae. The distribution patterns are diverse: *C. spermacocea,* subdivided into an "E" and a "SW" subspecies, is widely distributed; *C. gigantea* is only known from the Langeberg range above Swellendam; *C. curvifolia* shows the most interesting range: it only occurs on the tops of the (± drier?) E inland mountain ranges of the SW Cape Region (a "Karro-Mountain" and "South-Eastern Endem" sensu Weimarck 1941).

2. *Carpacoce scabra* and *C. burchellii*

The two species are very similar in their habit and general morphology, but are easily distinguished by having flowers with one and two fertile carpels respectively. Both primarily occur in Mountain Fynbos and are confined to the SW Cape s.str.

3. *Carpacoce vaginellata* and *C. heteromorpha*

The relationships between these two species are not entirely certain. They may be quite similar in habit but differ quite markedly in fruit characters, corollas and leaves (base of leaf blades, stipules). *C. vaginellata* is very widely distributed in the SW Cape Region, occurring in a wide range of altitudes and in rather diverse habitats. *C. heteromorpha,* on the other hand, is confined to a quite small area in the S part of the SW Cape s.str.

Finally, a note on the species concentrations of *Carpacoe* (Fig. 125) and on its possible center of origin:

The highest concentration of the taxa of *Carpacoce* is found in the one-degree-square 3319, followed by 3320, 3418 and 3419, i.e., in the SW Cape s.str. All taxa except for *C. gigantea, C. curvifolia* and *C. spermacocea* subsp. *orientalis* occur in 3319[1]. Does the high concentration in 3319 also allow the conclusion that this area is actually the center of origin of the genus? The question is difficult, if not impossible to answer. The area of square 3319 is characterized by a varied topography (high mountains and low, hot basins, etc.) and a diversity of soils, and possibly also by rather diverse local climatic conditions. The interpretation of these facts can be twofold: (i) the diverse conditions of the area provided an ideal "base" for

---

[1] Such a pattern is, by the way, characteristic of many "typical" families and genera of the SW Cape Flora (cf. Taylor 1980 and literature cited therein).

the "diversification" and the evolution of (new) taxa, or (ii) the area has merely become a "reservoir" for species of different phytogeographic affinities and/or diverse ecological preferences. On the basis of the assumption that *Carpacoce* originated from an originally montane, rather mesophytic stock (see above), the first interpretation is perhaps to be favoured over the second in that the stock could have started to spread from originally wet montane habitats to the surrounding (including the lower and drier) regions.

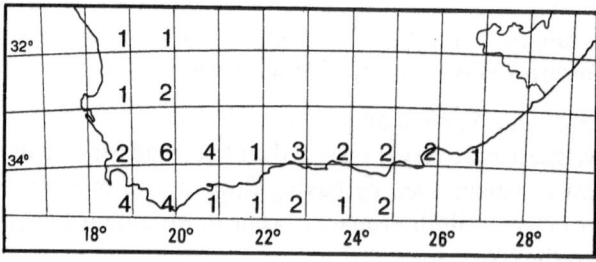

Fig. 125. *Carpacoce.* – Number of taxa (species and subspecies) per one degree square

5. Relationships Between *Anthosperminae* and Extra-African *Anthospermeae*

The *Anthosperminae* are a well circumscribed subtribe comprised of clearly allied genera, amongst which only *Carpacoce* (see 4., above) occupies a somewhat isolated position. Morphologically, the subtribe is primarily distinguished by its dry fruits dehiscing into two mericarps (except for some *Nenax* species with indehiscent, but basically similar fruits, and *Carpacoce*) and by its often rather "ericoid" leaves (if larger, ± broad leaves occur, they are rather thin and short-lived). With its African (incl. Macaronesian) – Madagascan distribution, it is also geographically well separated from the remaining two subtribes recognized in the *Anthospermeae*, the *Coprosminae* and the *Operculariinae* (Map, Fig. 126).

There appear to be closer affinities between the *Anthosperminae* and the *Coprosminae* than between the *Anthosperminae* and the *Operculariinae*, although the relationships between the genera of the *Anthosperminae* and those of the *Coprosminae* can hardly be considered as being *very* close. One of the main distinguishing characters are the fleshy fruits of the *Coprosminae* (although in some taxa – especially in *Normandia* and in

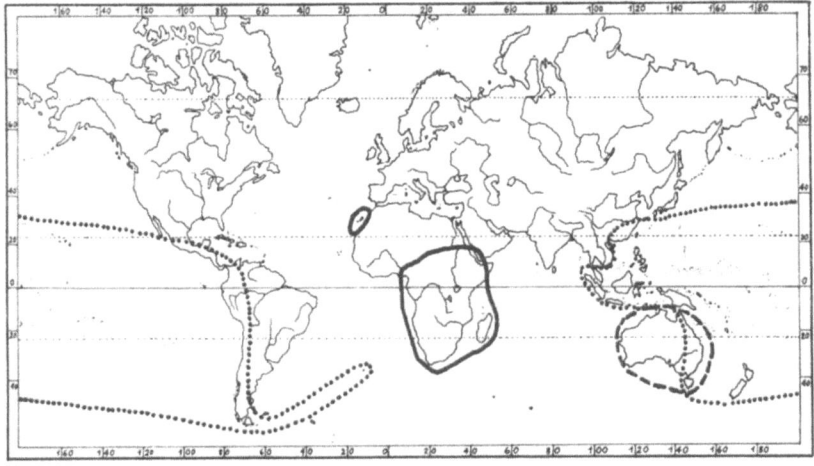

Fig. 126. Distribution of the tribe *Anthospermeae.* – Subtribe *Anthosperminae* (———), subtribe *Coprosminae* (·····), subtribe *Operculariinae* (– – – –)

species of *Leptostigma* – there is a trend to somewhat drier fruits with less distinctly fleshly exocarps); the leaves, furthermore, are often rather thick and leathery (notably in *Coprosma* species). There is, apart from the anemophilous flowers, agreement in numerous other features – e.g., in the diversity of growth forms (i.e., in the large genus *Coprosma*), or in the presence of taxa with ternate leaves next to those with decussately arranged foliage, but these are most likely convergent developments.

The *Operculariinae*, primarily on account of the fusion of the flowers of a partial inflorescence and the resultant peculiar infructescences which open by means of an operculum, occupy a rather isolated position within the tribe. The "ericoid" leaves of various taxa of that subtribe can most certainly not be used to favour a closer alliance to the *Anthosperminae* but are rather more likely the result of parallel evolution. See PUFF (1982 a) for further details.

Speculations on the origin and the time-space development of the *Anthospermeae* and the three subtribes are discussed below.

6. Thoughts on the Evolution of the *Anthospermeae*

There can be no doubt that the *Anthospermeae* must have originally been derived from a tropical, zoophilous rubiaceous stock which eventually evolved ("progressed to") secondary anemophily. The shift to anemophily required complex changes of various character states (cf. C.12.1.!), a process which is likely to have occurred gradually over a rather long period of time. Since, within the *Anthospermeae*, there are no "progressions" from animal (insect) pollinated flowers to wind pollinated ones via a series of intermediate forms as they are, for example, exhibited in other families *within* genera (e.g., *Acer!*) or in groups of related genera (*Rosaceae-Sanguisorbeae:* e.g., the genus pair *Leucosidea* and *Hagenia!*), i.e., since *all* present-day *Anthospermeae* are strictly and definitely anemophilous, it seems reasonable to assume that this switch to anemophily must have come about a rather long time ago (also see comments, p. 133).

A hypothetical ancestral form may thus have been characterized by the following set of features:

    – anemophily, but flowers perhaps not yet showing *all* typical features of the anemophilous pollination syndrome – i.e., flowers with still rather long and narrow (cylindrical) corolla tubes (rather than short tubes and rolled back corolla lobes for better exposure of the anthers and/or stigmas); receptive surface of the stigmas not yet as extremely increased as in the present-day *Anthospermeae*; flowers perhaps still mostly ♂ or (?) plants with ♀ and unisexual flowers

    – (primarily) woody plants – i.e., trees or shrubs with

    – ± broad, evergreen, ± coriaceous leaves with

    – conspicuous, perhaps ± collar-like stipular sheaths bearing a median seta

    – terminal, ± extensive, paniculate to thyrsic inflorescences

    – bicarpellate, fleshy(?) fruits with a sclerenchymatic endocarp.

On the basis of the above set of features, it transpires that the genera of the *Coprosminae*, and notably certain *Coprosma* species and *Normandia*, appear to be characterized by the, in general, most "primitive" character set, followed by the *Anthosperminae* and the *Operculariinae*.

Within the *Coprosminae*, there is a "woody line" represented by *Coprosma* and *Normandia*, and a "herbaceous line", represented by the genera *Nertera* and *Leptostigma*[1]. Of these two, the "woody line" appears

---

[1] FOSBERG (1982) considers the genus *Corynula* to be congeneric with *Leptostigma*; he also includes some species in *Leptostigma* which were previously placed in the genus *Nertera*. In my paper on the delimitation of the tribes *Anthospermeae* and *Paederieae*, which coincidentally was published in the same year (PUFF 1982a), I upheld *Corynula* as a distinct genus and considered *Leptostigma* to be a synonym of *Nertera*. I am, however, now convinced that FOSBERG's treatment of this group of genera should be given preference over mine.

to be the more "primitive" one. The large genus *Coprosma* [90 species according to OLIVER (1935) but possibly more] experienced considerable diversification and radiation and now has a wide, ± (S-temperate) amphi-transpacific distribution (but it does not reach S America); in the course of time some elements of the genus have become much more "derived" than others, so that the genus – comparable to the situation in the large genus *Anthospermum* – appears to be "heterogenous" with regard to its character states. The majority of the species of *Coprosma* are dioecious and thus appear to be "derived" in this character. Interestingly, both *Coprosma* and *Nertera* (see below) have a chromosome base number of $2n = 44$ (as opposed to $2n = 22$ in the remaining *Anthospermeae*[1]). The two genera, or at least *Coprosma*, may be palaeopoly(tetra)ploid. While in FEDOROV (1969) some species of *Coprosma* are listed as being diploid, recent counts of the same species by BEUZENBERG (1983) have shown that they are, in fact, tetraploid. The only aberrant published count for a *Coprosma* species ($2n = 32$ for *C. montana* HILLEBR.; SKOTTSBERG 1955) is probably also erroneous. Following the publication of BEUZENBERG's (1983) counts, roughly two thirds of the *Coprosma* species are known karyologically. Of these more than half have ploidy levels higher than $2n = 44$ (chromosome numbers of up to $2n = $ ca 220). The polyploidy in *Coprosma* is particularly remarkable in view of the dioeciousness of most taxa. Unless polyploidy was established before dioecy, or unless "leaky dioecism" (see comments, p. 517) played some role in the establishment of polyploidy, the process of polyploidization must have been incomparably more "complicated" than in plants with ⚥ flowers (taking place separately and independently and at ± the same time in both ♂ and ♀ flowers or plants, the chance of which seems to be very slim). It is very difficult to imagine how highly polyploid species like, for example, *C. pumila* ($2n = 132$) or *C. ernodeoides* ($2n = 220$) could have originated[2].

The monotypic genus *Normandia* appears to be as "primitive" in its character states as the more "primitive" elements of *Coprosma* (shrubs or small trees; evergreen, leathery leaves; terminal, ± extensive inflorescences; non-dioecious sex distributions, etc.) and appears to be "advanced" only in its less fleshy fruits in which the exocarp at maturity separates from the endocarp (cf. PUFF 1982a, Fig. 7). Unfortunately, no

---

[1] Except for some tetraploid *Nenax* species and *Anthospermum spathulatum* with di-, tetra- and hexaploids.

[2] In his monograph of *Coprosma*, OLIVER (1935) considers the two mentioned species to be "the living species which most closely resemble the stock from which *Coprosma* is derived". Whether these highly polyploid, small, creeping woody plants are really more "primitive" than shrubs or small trees like *C. baueri* or *C. robusta* (both tetraploid) appears very doubtful.

chromosomal data are available for the genus, although the small pollen grains, agreeing in size with or smaller than those of known tetraploid *Coprosma* species, could indicate diploidy. *Normandia* may well be derived from the same stock as *Coprosma*; it may, at a relatively early stage, have "separated" and established itself on New Caledonia, where it is endemic.

The herbaceous, (exclusively?) tetraploid genus *Nertera* and its ally *Leptostigma* appear to be much more derived than the above two genera (often very reduced inflorescences – i.e., solitary, axillary flowers, etc.). They are possibly much younger than the previous two genera; both eventually attained (most likely by LDD) a wide distribution range [*Leptostigma*: trans-S Pacific – cf. FOSBERG (1982, Fig. 1); *Nertera*: distribution range basically similar to that of *Coprosma*, but even wider in that it also extends further N and S – cf. VAN STEENIS (1962, Fig. 14)].

The character states of (certain members of) *Coprosma* and of *Normandia*, more "primitive" than those of the remaining *Anthospermae*, seem to indicate that amongst the two genera there are perhaps those present-day species which most closely resemble the ancestral form of the *Anthospermeae*. Their distribution patterns also, to some extent, agree with my ideas that the (zoophilous) ancestral stock of the *Anthospermeae* could have been SE Asiatic in origin.

Presuming such a SE Asiatic origin of *Coprosma* or rather of the ancestral stock of the *Coprosminae* (and, subsequently, of the entire *Anthospermeae*?) and considering that today only relatively few species of *Coprosma* are represented in SE Asia (Thailand, Borneo, New Guinea), it follows that the ancestral stock must have experienced a considerable radiation away from that region.

It is well documented that the area that included SE Asia and N Australia to New Caledonia and Fiji is geologically composite and only came into existence with the Miocene arrival of the Australian plate in the vicinity of Asia (see RAVEN & AXELROD 1974 and literature cited therein). As the origin of the stock of the entire *Anthospermeae* is likely to date back to a much earlier period (perhaps as far back as the later Cretaceous?), the (strictly SE Asiatic) stock of the *Coprosminae* could only have reached Australia and New Caledonia (to give rise to *Normandia,* see above) by LDD at an early stage. A further expansion of its range to New Zealand could be explained similarly (also see THORNE 1978). *Coprosma* has an unusually high concentration of species in New Zealand; in my opinion, this phenomenon should not be interpreted as an indication for an origin of the genus in that region. I rather believe that the high species concentration of the genus in that region may be attributed to a "secondary" species explosion there – analogous to the situation of

Anthospermum in the SW Cape and, to a lesser extent, in Madagascar. The "final" expansion of *Coprosma* to its present-day fragmented distribution range can, again, only be explained by LDD. In this context, the phenomenon of "leaky dioecism", discussed recently by BAKER & COX (1984), is particularly noteworthy as it, in the essentially dioecious *Coprosma* taxa (and, similarly, in *Nertera*), could have facilitated the radiation of these taxa to islands, etc. and made their establishment there much easier. Indirect reference to the occurrence of "leaky dioecism" in *Coprosma* species (i.e., odd ♀ flowers on ♂ plants, etc.) had already been made by SKOTTSBERG (1922), FOSBERG (1937) and GODLEY (1979).

The thoughts expressed above imply that a common ancestral stock of the *Anthospermeae* must have experienced a "split" into three segments (i.e., those that gave rise to the genera of the *Coprosminae, Anthosperminae* and *Operculariinae*) at a relatively early stage. The fate of that portion of the stock that gave rise to the genera of the *Coprosminae* is outlined above, but how can the origin of the genera of the *Operculariinae* be envisaged? After the presumably relatively early arrival of the stock of the *Operculariinae* in Australia (incl. Tasmania), the stock must have experienced a further diversification and differentiation which eventually led to the separation of the extant genera *Pomax* and *Opercularia* [*Eleutheranthus*, upheld as a distinct genus in PUFF (1982 a), has meanwhile been shown to be congeneric with *Opercularia*; see ROBBRECHT (1982)]. It seems clear, though, that the stock of the *Operculariinae* must have acquired some of its derived features – notably the fused flowers of a partial inflorescence and the resultant infructescences opening by means of an operculum – *before* the split into the two genera occurred as it seems to be highly unlikely that these rather complex structures evolved independently in each of the two genera. Further differentiation, however, is likely to have occurred at a later stage (i.e., the development of the umbel-like inflorescences of *Pomax* and the axillary, clustered inflorescences of *Opercularia* respectively).

Finally, it needs to be explained how the stock of the *Anthosperminae* could have reached the African continent and what its subsequent fate might have been. It cannot be ascertained whether this portion of the *Anthospermeae* stock reached Africa via India and the adjoining (and now largely submersed) Mascarene plateau (Malagasy – Mascarene subcontinent; cf. BRIDEN & al. 1974) or perhaps via a more N route (fossil leaves from the Paleocene of rainforest taxa are known, for example, from the Red Sea coast; see AXELROD & RAVEN 1978, and literature cited therein).

Had a migration taken place via the first mentioned route, one could perhaps expect that some "primitive" taxa, possibly faintly resembling the ancestral form, may have survived in Madagascar. There are, however definitely no such species in the modern Madagascan flora. As such primitive forms may well have disappeared in the course of time, conclusive arguments either in favour of or against such a route cannot be brought forward. Perhaps the time factor speaks against an "Indian-Mascarene Plateau route": Such a direct migration route was only possible until the late Cretaceous (thereafter, the land masses in question started to drift away from each other), but this may have been too early a date for the arrival of the stock of the *Anthosperminae* in Africa. Not only for this reason do I favour the second mentioned "N" route: A "N" route seems to allow for a more reasonable explanation of the distribution patterns of the genera of the *Anthosperminae*. In my arguments, the Macaronesian range of the genus *Phyllis* plays an important role. I envisage that the stock of the *Anthosperminae* — perhaps some time in the Paleogene (?) — reached tropical Africa and established itself in montane (rainforest) areas. A portion of that stock, possibly rather soon after the arrival of the entire stock, could have, via the E-W corridor provided by the highlands in the central Saharan region (Hoggar, Air, Tibesti massifs), started to migrate W-ward to the Canary Islands and eventually gave rise to the genus *Phyllis*. Vegetation allied to montane rainforest occurs today in the Canary Islands, a volcanic archipelago which is not older than mid Miocene. The adjacent coast of NW and N Africa is likely to have supported a similar vegetation in the middle Tertiary (see AXELROD & RAVEN 1978). As there is also considerable evidence for numerous (and continuous) links between the Canarian laurel forest and the montane forest of E Africa until during the Miocene (AXELROD & RAVEN 1978), such a migration route — or, in other words, such a link between the stock of *Phyllis* and the remainder of the stock of the *Anthosperminae* — seems quite plausible. Noteworthy, in this context, is also that *Phyllis nobla* is typically associated with the present Canarian laurel forest and occurs mostly in the forest edge vegetation [the other species of the genus, *P. viscosa,* is obviously more derived (dioecy, more congested inflorescences, hairy fruits, etc.) and occurs in drier habitats such as light succulent (*Euphorbia*) bush, etc. (cf. MENDOZA-HEUER 1972, 1977)]. The deviating wood anatomical features of *Phyllis* (cf. KOEK-NOORMAN & PUFF 1983) are perhaps also an argument in favour of the assumption voiced above that the stock of *Phyllis* may have "separated" from the main stock of the continental African *Anthosperminae* at a relatively early stage.

The relationships between and the possible centers of origin and the time-space development of the remaining *Anthosperminae* have already been discussed in the previous chapters (1.10.–4.) and require no further comments.

In a discussion of the evolution, origin and diversification of the anemophilous *Anthospermeae* mention should also be made of the genus *Theligonum*. On account of the, in several respects, rather deviating character set of the genus, it had been associated with various plant families (notably with those of the *Caryophyllales*) until WUNDERLICH (1971), in a detailed study of the genus, pointed out its affinities to the *Rubiaceae* and suggested a placement of the genus in a tribe *Theligoneae*. Although various authors subsequently expressed doubts about its placement in the *Rubiaceae* on the basis of the study of individual character states (e.g., pollen: PRAGLOWSKI 1973, NOWICKE & SKVARLA 1979), I am fully convinced that *Theligonum* belongs here. Not least because of its anemophilous flowers (making *Theligonum* the only other rubiaceous genus outside the *Anthospermeae* to show this phenomenon)[1], I am quite certain that the genus/tribe should be placed near the *Anthospermeae* (see PUFF 1982a for further details). It is, therefore, attempted below to elucidate the evolutionary relationships of these two tribes.

As it is unlikely that the complex structural changes associated with and leading to the evolution of anemophilous flowers occurred entirely separately and independently within the *Rubiaceae,* it may perhaps be assumed that both the *Theligoneae* and *Anthospermeae* have "common roots". There must have been, however, a very early split of a common ancestral stock, so that both tribes had sufficient time to evolve their deviating character set. *Theligonum* "advanced" to a herbaceous habit (both perennial and annual species occur) and evolved a rather extreme type of anisophylly (lowermost leaves decussate, gradually becoming unequal in size; upper leaves seemingly spirally arranged and solitary at a node due to suppression of the other leaf; cf. ULBRICH 1934, Figs. 151 and 154) which, however, should not be overemphasized as at least "trends" to anisophylly occur in numerous other *Rubiaceae* (and in members of the *Anthospermeae*). Also its sexually dimorphic, unisexual flowers acquired some rather unusual features (increased number of stamens in ♂; ♀ with one fertile carpel and almost gynobasic attachment of the style, etc.; cf. ULBRICH, l.c.) in which it differs quite markedly from the *Anthospermeae.*

---

[1] The Australian genus *Durringtonia,* placed into its own tribe by HENDERSON & GUYMER (1985), is also anemophilous. It is very closely allied to (and perhaps to be included in?) the *Anthospermeae.* See also footnote, p. 3.

While the *Anthospermeae* are an essentially S-hemispherical tribe (cf. Map, Fig. 126), *Theligonum* has a strictly N-hemispherical range (details below). These diverse distribution patterns, however, do not contradict a presumed SE Asiatic center of origin of a common ancestral stock of both *Anthospermeae* and *Theligoneae*. From tropical SE Asia s.str., *Theligonum,* or its stock, could have radiated N-, NE- and NW-wards and, in the course of time, attained its present-day fragmented distribution [one species in Japan, another in C Asia (E Tibet), the third in the Mediterranean region (from Syria to Portugal and Morocco) and in the Canary Islands; data from ULBRICH, l.c.]. Interesting is also the further differentiation and diversification of *Theligonum* that obviously went along with the fragmentation of its range: the Japanese species is perennial, the others annual; the C Asiatic species is supposedly dioecious, the other two species are monoecious.

As regards the relationships and affinities of both the *Anthospermeae* and the *Theligoneae* to the tribe *Paederieae* see PUFF (1982a) for details.

Acknowledgements

I wish to thank all the botanists—too numerous to mention most of them—who have over the years generously supported this project in one way or another. Particularly helpful have been Prof. F. EHRENDORFER (Vienna; continued overall support), Dr. W. GUTERMANN (Vienna; nomenclatural advice), Mr. A. IGERSHEIM and Mr. M. KIEHN (Vienna; technical assistance), Dr. D. E. MANTELL (Vienna; linguistic revision of the manuscript), Dr. E. ROBBRECHT (Meise; palynological contributions; continued interest and stimulating discussions), Prof. ST. VOGEL (Mainz; discussion of pollination biological matters) and numerous students of the Dept. of Botany, University of the Witwatersrand (Johannesburg; technical assistance). I, furthermore, am indebted to the Directors and Curators of the institutions and herbaria listed on p. 6 for providing facilities to study at their institution or for sending specimens on loan, to the staff of the EM Unit of the University of the Witwatersrand and of the University of Vienna for assistance at the SEM, and to numerous European and African colleagues who provided details on localities and additional material. Last but not least, my sincere thanks go to the authorities of the countries visited (p. 4) for their kind cooperation (research clearances, permission to do field work in "difficult" areas, logistic support, etc.) and to my African and Madagascan friends and colleagues, without whose help much of the field work could not have been carried out.

This project was supported by the "Fonds zur Förderung der wissenschaftlichen Forschung" (project 3681) and by grants and scholarships of the Austrian Ministry of Science and Research, the "Hochschuljubiläumsstiftung der Stadt Wien", the S African Dept. of Agriculture and Water Supply, and the S Africa Council for Scientific and Industrial Research (Grant CSR/P75).

References[1]

ACOCKS, J. P. H., 1975: Veld types of South Africa. 2nd Ed. − Mem. Bot. Surv. S. Afr. **40**, 1−128.

ADAMSON, R. S., 1935: The plant communities of Table Mountain. III. A six years' study of regeneration after fire. − J. Ecol. **23**, 44−55.

− 1938: Notes on the vegetation of the Kamiesberg. − Mem. Bot. Surv. S. Afr. **18**, 1−25.

− 1959: Notes on the phytogeography of the flora of the Cape Peninsula. − Trans. Roy. Soc. S. Afr. **35**, 443−462.

− SALTER, T. M., 1950: Flora of the Cape Peninsula. − Cape Town, Johannesburg, Juta.

ALLAN, H. H., 1961: Flora of New Zealand, vol. 1. − Wellington, Government Printer.

ANDERSON, W. R., 1973: A morphological hypothesis for the origin of heterostyly in the *Rubiaceae*. − Taxon **22**, 537−542.

ANDRÉ, R., DELAVEAU, P., JACQUEMIN, H., 1976: Recherches phytochimiques sur quelques Rubiacées malgaches. − Pl. méd. phytothér. **10**, 233−242.

AXELROD, D. I., RAVEN, P. H., 1978: Late cretaceous and tertiary vegetation history of Africa. − In: WERGER, M. J. A. (ed.), Biogeography and ecology of southern Africa, 77−130. − The Hague, W. Junk.

BABCOCK, E. B., 1947: The genus *Crepis*. − Univ. Calif. Publ. Bot. **21** and **22**, 1−1030. Berkeley, Los Angeles, University of California Press.

BAILLON, H., 1880: Histoire des Plantes, vol. 7. − Paris, Hachette.

BAKER, H. G., 1948: Corolla-size in gynodioecious and gynomonoecius species of flowering plants. − Proc. Leeds Phil. & Lit. Soc., Scient. Sect. **5** (2), 136−139.

− 1955: Self-compatibility and establishment after "long-distance" dispersal. − Evolution **9**, 347−349.

− Cox, P. A., 1984: Further thoughts on dioecism and islands. − Ann. Missouri Bot. Gard. **71**, 244−253.

BAMPS, P., 1976, 1981: Catalogue of the phanerogamic families dealt with in the main floras of tropical Africa. − Boissiera **24**, 667−686; updated, mimeographed list: Jardin Botanique National de Belgique, Meise. 24 pp.

BARRETT, S. C., HELENURM, K., 1981: Floral sex ratios and life history in *Aralia nudicaulis* (*Araliaceae*). − Evolution **35**, 752−762.

BATE-SMITH, E. C., 1962: Phenolic constituents of plants and their taxonomic significance. 1. Dicotyledons. − J. Linn. Soc. (Bot.) **58**, 95−173.

BAWA, K. S., 1980: Evolution of dioecy in flowering plants. − Ann. Rev. Ecol. Syst. **11**, 15−39.

− 1984: The evolution of dioecy−concluding remarks. − Ann. Missouri Bot. Gard. **71**, 294−296.

− BEACH, J. H., 1981: Evolution of sexual systems in flowering plants. − Ann. Missouri Bot. Gard. **68**, 254−274.

− KEEGAN, C. R., & VOSS, R. H., 1982: Sexual dimorphism in *Aralia nudicaulis*. − Evolution **36**, 371−378.

---

[1] General deadline: mid 1984; see p. 7.

34*

BEACH, J. H., BAWA, K. S., 1980: Role of pollinators in the evolution of dioecy from distyly. − Evolution **34**, 1138−1142.

BEUZENBERG, E. J., 1983: Contributions to a chromosome atlas of the New Zealand flora − 24. *Coprosma (Rubiaceae).* − New Zeal. J. Bot. **21**, 9−12.

BIR BAHADUR, 1968: Heterostyly in *Rubiaceae*: A review. − Osmania Univ. J. Sc. **4** (1/2) (Golden Jubilee Special Volume), 207−238.

BOLUS, L., 1917: *Rubiaceae.* − In: PEARSON, H. H. W., List of the plants collected in the Percy Sladen Memorial Expeditions. − Ann. S. Afr. Mus. **9**, 214−217.

BOND, P., GOLDBLATT, P., 1984: Plants of the Cape flora. A descriptive catalogue. − J. S. Afr. Bot., Suppl. **13**, 1−455.

BOUCHER, C., JARMAN, M. L., 1977: The vegetation of the Langebaan area, South Africa. − Trans. Roy. Soc. S. Afr. **42**, 241−272.

BOUGHEY, A. S., 1955: The vegetation of the mountains of Biàfra. − Proc. Linn. Soc. **165**, 144−150.

BRANFORD, J., 1978: A dictionary of South African English. − Cape Town, Oxford University Press.

BRENAN, J. P. M., 1965: Map of the extent of floristic exploration in Africa South of the Sahara [translation]. − Webbia **19**, 911−914, Map.

BRIDEN, J. C., DREWRY, G. E., SMITH, A. G., 1974: Phanerozoic equal-area world maps. − J. Geol. **82**, 555−574.

BRIGGS, B. G., JOHNSON, L. A. S., 1979: Evolution in the *Myrtaceae* − evidence from inflorescence structure. − Proc. Linn. Soc. New South Wales **102**, 157−272.

BROCKMANN, I., BOCQUET, G., 1978: Ökologische Einflüsse auf die Geschlechtsverteilung bei *Silene vulgaris* (MOENCH) GARCKE (*Caryophyllaceae*). − Ber. Deutsch. Bot. Ges. **91**, 217−230.

BULLOCK, S. H., BAWA, K. S., 1981: Sexual dimorphism and the annual flowering pattern in *Jacartia dolichaula* (D. SMITH) WOODSON (*Caricaceae*) in a Costa Rican rain forest. − Ecology **62**, 1494−1504.

CARLQUIST, S., 1962: A theory of paedomorphosis in dicotyledonous woods. − Phytomorphology **12**, 30−45.

− 1976: Wood anatomy of *Roridulaceae*: ecological and phylogenetic implications. − Amer. J. Bot. **63**, 1003−1008.

− 1978: Vegetative anatomy and systematics of *Grubbiaceae*. − Bot. Not. **131**, 117−126.

− 1980: Further concepts in ecological wood anatomy, with comments on recent work in wood anatomy and evolution. − Aliso **9**, 499−553.

CHAPMAN, J. D., 1962: The vegetation of the Mlanje Mountains, Nyasaland. − Zomba, Government Printer.

COMPTON, R. H., 1931: The flora of the Whitehill District. − Trans. Roy. Soc. S. Afr. **19**, 269−326.

CORBINEAU, F., CÔME, D., 1980: Some particularities of the germination of *Oldenlandia corymbosa* L. seeds (tropical *Rubiaceae*). − Israel J. Bot. **29**, 157−167.

− − 1982: Effect of the intensity and duration of light at various temperatures on the germination of *Oldenlandia corymbosa* L. seeds. − Plant Physiol. **70**, 1518−1520.

CORNER, E. J. H., 1976: The seeds of Dicotyledons. Vol. 1. − Cambridge, University Press.

COX, P. A., 1981: Niche partitioning between sexes of dioecious plants. − Am. Nat. **117**, 295−307.

CRUSE, G., 1825: De *Rubiaceis* capensibus praecipue de genere *Anthospermo*. − Berlin.

- 1831: De *Anthospermeis Rubiacearum* sectione. – Linnaea **6**, 1–21.
- 1832: Plantae Ecklonianae. *Rubiaceae*. – Linnaea **7**, 132–135.
DARWIN, C., 1877: The different forms of flowers on plants of the same species. – London, John Murray.
DE VOGEL, E. F., 1980: Seedlings of dicotyledons. – Wageningen, Centre for Agricultural Publishing and Documentation.
DOMMÉE, B., ASSOUAD, M. W., VALDEYRON, G., 1978: Natural selection and gynodioecy in *Thymus vulgaris* L. – Bot. J. Linn. Soc. **77**, 17–28.
DRÈGE, J. F., 1843: Zwei pflanzengeographische Documente. – Flora **26**(2) (= „Neue Serie, Jg. 1, Bd. 2"), Besondere Beigabe: 1–230.
- 1847a: Standörter-Verzeichniss der von C. L. ZEYHER in Südafrika gesammelten Pflanzen. – Linnaea **19**, 583–598.
- 1847b: Vergleichungen der von ECKLON und ZEYHER und von DRÈGE gesammelten südafrikanischen Pflanzen. – Linnaea **19**, 599–680.
- 1848: Nachtrag zum Standörter-Verzeichniss. – Linnaea **20**, 258.
DUKE, J. A., 1969: On tropical tree seedlings. 1. Seeds, seedlings, systems, and systematics. – Ann. Missouri Bot. Gard. **56**, 125–161.
DULBERGER, R., HOROWITZ, A., 1984: Gender polymorphism in flowers of *Silene vulgaris* (MOENCH) GARCKE (*Caryophyllaceae*). – Bot. J. Linn. Soc. **89**, 101–117.
ECKLON, C. F., ZEYHER, C., 1836: Enumeratio plantarum Africae australis extratropicae, pars 2. – Hamburg, Perthes & Besser.
EDWARDS, D., LEISTNER, O. A., 1971: A degree reference system for citing biological records in southern Africa. – Mitt. Bot. Staatssamml. München **10**, 424–437.
EHRENDORFER, F., 1983: Evolution und Systematik: Allgemeine Grundlagen. – In: STRASBURGER, E., Lehrbuch der Botanik für Hochschulen, 32. Aufl., 484–545. – Stuttgart, New York, Fischer.
ERDTMAN, G., 1952: Pollen morphology and plant taxonomy: Angiosperms. – Stockholm, Almqvist & Wiksell.
ESAU, K., 1953, 1965: Plant anatomy. 1st and 2nd Ed. – New York, J. Wiley & Sons.
ESTABROOK, G. F., 1977: Does common equal primitive? – Syst. Bot. **2**, 36–42.
FAEGRI, K., VAN DER PIJL, L., 1971: The principles of pollination ecology. Second revised ed. – Oxford, Pergamon Press.
FAGERLIND, F., 1936: Embryologische Beobachtungen über die Gattung *Phyllis*. – Bot. Not. **1936**, 577–584.
- 1937: Embryologische, zytologische und bestäubungsexperimentelle Studien in der Familie *Rubiaceae* nebst Bemerkungen über einige Polyploiditätsprobleme. – Acta Horti Berg. **11**, 195–470.
FEDOROV, A. (Ed.), 1969: Chromosome numbers of flowering plants. – Leningrad, Academic Press.
FERNANDES, A. (Ed.), 1962: Histoire de l'exploration botanique de l'Afrique au Sud du Sahara. In: Comptes rendus de la IVe réunion plénière de l'association pour l'étude taxonomique de la flore d'afrique tropicale, 45–248. – Lisboa: Junta de Invest. Ultramar.
FOSBERG, F. R., 1937: Some *Rubiaceae* of south-eastern Polynesia. – Occas. Pap. Bernice P. Bishop Mus. **13**, 245–293.
- 1982: A preliminary conspectus of the genus *Leptostigma* (*Rubiaceae*). – Acta Phytotax. Geobot. **33**, 73–83.
FOWERAKER, C. E., 1916: Mat-plants and cushion-plants of Cass River bed. – Trans. New Zeal. Inst. **49**, 1–45.

FREEMAN, D. C., KLIKOFF, L. G., HARPER, K. T., 1976: Differential resource utilization by the sexes of dioecious plants. − Science 193, 597−599.
− MCARTHUR, E. D., HARPER, K. T., 1984: The adaptive significance of sexual liability in plants using *Atriplex canescens* as a principal example. − Ann. Missouri Bot. Gard. 71, 265−277.

FUGGLE, R. F., ASHTON, E. R., 1979: Climate. − In: DAY, J., SIEGFRIED, W. R., LOUW, G. N., JARMAN, M. L. (eds.), Fynbos ecology: a preliminary synthesis. South African National Scientific Programmes Report 40, 7−15. − Pretoria.

GANDOGER, M., 1913: L'herbier africain de SONDER. − Bull. Soc. Bot. Fr. 60, 414−422.

GILLETT, J. B., 1972: W. G. SCHIMPER's botanical collecting localities in Ethiopia. − Kew Bull. 27, 115−128.

GODLEY, E. J., 1964: Breeding systems in New Zealand plants. 3. Sex ratios in some natural populations. − New Zeal. J. Bot. 2, 205−212.
− 1979: Flower biology in New Zealand. − New Zeal. J. Bot. 17, 441−446.

GOLDBLATT, P., 1978: An analysis of the flora of southern Africa: its characteristics, relationships and origins. − Ann. Missouri Bot. Gard. 65, 369−436.
− 1981: Systematics and biology of *Homeria* (*Iridaceae*). − Ann. Missouri Bot. Gard. 68, 413−503.

GOODIER, R., PHIPPS, J. B., 1962: A vegetation map of the Chimanimani National Park. − Kirkia 3, 2−7.

GORDON, H. D., 1959: Sex ratio in *Coprosma repens* (*Rubiaceae*). − Wellington Bot. Soc. Bull. 31, 11.

GRANT, M. C., MITTON, J. B., 1979: Elevational gradients in adult sex ratios and sexual differentiation in vegetative growth rates of *Populus tremuloides* MICHX. − Evolution 33, 914−918.

GRONEMEYER, 1967: Pollenallergie. − In: HANSEN, K., WERNER, M. (eds.), Handbuch der klinischen Allergie, 167−178. − Stuttgart, Thieme.

GUNN, M., CODD, L. E., 1981: Botanical exploration of southern Africa. − Cape Town, Balkema.

GUSTAFSON, F. G., 1942: Pathenocarpy: natural and artificial. − Bot. Rev. 8, 599−654.

HANCOCK, J. F., BRINGHURST, R. S., 1980: Sexual dimorphism in the strawberry *Fragaria chiloensis*. − Evolution 34, 762−768.

HASHMI, S., SIDDIQUI, S. A., 1974: Development of endosperm, embryo and seed in *Paederia scandens* LOUR. − Geobios (Jodhpur) 1, 183−184.

HASLAM, E., 1981: Vegetable tannins. − In: CONN, E. E., (ed.), The biochemistry of plants, Vol. 7. Secondary plant products, 527−556. − New York, Academic Press.

HEDBERG, I., HEDBERG, O., 1977: Chromosome numbers of afroalpine and afromontane angiosperms. − Bot. Not. 130, 1−24.

HEDBERG, O., 1951: Vegetation belts of the East African mountains. − Svensk bot. Tidskr. 45, 140−202.
− 1957: Afroalpine vascular plants. A taxonomic revision. − Symb. Bot. Ups. 15, 1, 1–411.
− 1964: Features of afroalpine plant ecology. − Acta Phytogeogr. Suecica 49, 1−144.
− 1970: Evolution of the afroalpine flora. − Biotropica 2, 16−23.

HEGNAUER, R., 1973: Chemotaxonomie der Pflanzen. Vol. 6. − Basel & Stuttgart, Birkhäuser.

HENDERSON, R. J. F., GUYMER, G. P., 1985: *Durringtonia* (*Durringtonieae*), a new genus and tribe of *Rubiaceae* from Australia. − Kew Bull. 40, 97−106.

HEPPER, F. N., 1963: *Anthospermum*. – In: HEPPER, F. N. (ed.), Flora of West Tropical Africa, 2nd Ed., Vol. 2, 222–223. – London, Crown Agents.

– BAMPS, P., 1971: A list of geographical and botanical gazetteers of Africa South of the Sahara. – AETFAT Bull. **21**, 22–26.

HESLOP-HARRISON, Y., SHIVANNA, K. R., 1977: The receptive surface of the angiosperm stigmas. – Ann. Bot. (London), ser. 2, **41**, 1233–1258.

HESSE, M., 1979: Ultrastruktur und Verteilung des Pollenkitts in der insekten- und windblütigen Gattung *Acer* (*Aceraceae*). – Pl. Syst. Evol. **131**, 277–289.

HILLIARD, O. M., 1978: The geographical distribution of *Compositae* native to Natal. – Notes Roy. Bot. Gard. Edinburgh **36**: 407–425.

– BURTT, B. L., 1985: Notes on some plants of southern Africa chiefly from Natal: XI. – Notes Roy. Bot. Gard. Edinburgh **42**, 227–260.

HOBSON, N. K., JESSOP, J. P., 1975: Veld plants of southern Africa. – Johannesburg, Macmillan.

HOLM, T., 1907: *Rubiaceae*: Anatomical studies of North American representatives of *Cephalanthus*, *Oldenlandia*, *Houstonia*, *Mitchella*, *Diodia* and *Galium*. – Bot. Gaz. **43**, 153–184.

HOLMGREN, P. K., KEUKEN, W., SCHOFIELD, E. K., 1981: Index Herbariorum. Part 1, The herbaria of the world, ed. 7. – Regnum Vegetabile **106**. Utrecht, Bohn, Scheltema & Holkema.

HOOKER, J. D., 1862: On the vegetation of Clearance Peak, Fernando Po; with descriptions of the plants collected by Mr. GUSTAV MANN on the higher parts of that mountain. – J. Linn. Soc. (Bot.) **6**, 1–23.

– 1864: On the plants of the temperate regions of the Cameroons Mountains and islands in the bight of Benin; collected by Mr. GUSTAV MANN, Government Botanist. – J. Linn. Soc. (Bot.) **7**, 171–240.

HUMBERT, H., 1955: Les territoires phytogéographiques de Madagascar. – In Colloques Internationaux du C.N.R.S. **59**: Les divisions écologiques du Monde. – Année Biol. (Paris), sér. 3, **31**, 439–448.

– 1965: Description des types de végétation. – In: HUMBERT, H., COURS-DARNE, G., Notice de la Carte Madagascar. – Trav. Sect. Sci. Tech. Inst. Franç. Pondichery, H. S., **6**, 46–78.

HUTCHINSON, J., 1946: A botanist in southern Africa. – London, Gawthorn.

JACKSON, J. K., 1956: Vegetation of the Imatong Mountains, Sudan. – J. Ecol. **44**, 341–374.

JESSOP, J. P., 1964: Itinerary of RUDOLF SCHLECHTER's collecting trips in southern Africa. – J. S. Afr. Bot. **30**, 129–146.

KAPIL, R. N., RAO, P. R. M., 1966: Embryology and systematic position of *Theligonum* LINN. – Proc. Nat. Inst. Sci. India **32**, B: 218–232.

KASSNER, T., 1911: My journey from Rhodesia to Egypt including an ascent of Ruwenzori and a short account of the route from Cape Town to Broken Hill and Lado to Alexandria. – London, E. Hutchinson & Co.

KAY, Q. O. N., LACK, A. J., BAMBER, F. C., DAVIES, C. R., 1984: Differences between sexes in floral in floral morpholgy, nectar production and insect visits in a dioecious species, *Silene dioica*. – New Phytol. **98**, 515–529.

KIEHN, M., 1986: Karyologische Untersuchungen und DNA-Messungen an Rubiaceen und ihre Bedeutung für die Systematik dieser Familie. – Diss. Form. Naturwiss. Fak. Univ. Wien (in prep.).

KILLICK, D. J. B., 1959: An account of the plant ecology of the Table Mountain area of Pietermaritzburg, Natal. – Mem. Bot. Surv. S. Afr. **32**, 1–133.

– 1963: An account of the plant ecology of the Cathedral Peak area of the Natal Drakensberg. – Mem. Bot. Surv. S. Afr. **34**, 1–178.

KING, L., 1982: The Natal monocline. Second revised Ed. — Pietermaritzburg, University of Natal Press.

KNOBLAUCH, E., 1896: Ökologische Anatomie der Holzpflanzen der südafrikanischen immergrünen Buschregion. — Habilitationsschrift Phil. Fak. Univ. Giessen. — Tübingen.

KOECHLIN, J., 1972: Flora and vegetation of Madagascar. — In: BATTISTINI, R., RICHARD-VINDGARD, G. (eds.), Biogeography and ecology in Madagascar, 145–190. — The Hague, W. Junk.

KOEK-NOORMAN, J., PUFF, C., 1983: The wood anatomy of Rubiaceae tribes Anthospermeae and Paederieae. — Pl. Syst. Evol. 143, 17–45.

KOOIMAN, P., 1969: The occurrence of asperulosidic glycosides in the Rubiaceae. — Acta Bot. Neerl. 18, 124–137.

KRAUSE, K., 1909: Über harzsecernierende Drüsen an den Nebenblättern von Rubiaceen. — Ber. Deutsch. Bot. Ges. 27, 446–452.

KRUGER, F. J., 1979: Plant ecology. — In: DAY, J., SIEGFRIED, W. R., LOUW, G. N., JARMAN, M. L. (eds.), Fynbos ecology: a preliminary synthesis. South African National Scientific Programmes Report 40, 88–126. — Pretoria.

KUMMEROW, J., 1973: Comparative anatomy of sclerophylls of mediterranean climatic areas. — In: DI CASTRI, F., MOONEY, H. A. (eds.), Mediterranean type ecosystems, 157–167. — Heidelberg, New York.

KUNTZE, O., 1898: Revisio generum plantarum. Pars III, 2. — Leipzig, A. Felix.

LAMBRECHTS, J. J. N., 1979: Geology, geomorphology and soils. — In: DAY, J., SIEGFRIED, W. R., LOUW, G. N., JARMAN, M. L. (eds.), Fynbos ecology: a preliminary synthesis. South African National Scientific Programmes Report 40, 16–26. — Pretoria.

LAUNERT, E., ROESSLER, H., 1966: Rubiaceae. — In: MERXMÜLLER, H. (ed.), Prodromus einer Flora von Südwestafrika, 115. — Lehre.

LEISTNER, O. A., 1967: The plant ecology of the southern Kalahari. — Mem. Bot. Surv. S. Afr. 38, 1–172.

— MORRIS, J. W., 1976: Southern African place names. — Ann. Cape Prov. Mus. 12, 1–565.

LÉONARD, J., 1984: Contributions à la connaissance de la flore de l'Iran. — VI. Le "complexe Gaillonia A. RICH. ex DC." (Rubiaceae). — Bull. Jard. Bot. Nat. Belg. 54, 493–497.

LEVYNS, M. R., 1935: Germination in some South African seeds. — J. S. Afr. Bot. 1, 89–103.

— 1937: The ovary and fruit of Carpacoce. — J. S. Afr. Bot. 3, 117–123.

— 1952: Clues to the past in the Cape flora of today. — S. Afr. J. Sc. 49, 155–164.

— 1962: Plant migrations in South Africa. — Ann. Cape Prov. Mus. 3, 7–10.

LEWIS, W. H., 1966: Chromosome numbers of phanerogams. I. — Ann. Missouri Bot. Gard. 53, 100–103.

LLOYD, D. G., BAWA, K. S., 1984: Modification of the gender of seed plants in varying conditions. — Evol. biol. 17, 255–338.

— HORNING, D. S., 1979: Distribution of sex in Coprosma pumila on Macquarie Island, Australia. — New Zeal. J. Bot. 17, 5–7.

— WEBB, C. J., 1977: Secondary sex characters in plants. — Bot. Rev. 43, 177–216.

LOVETT DOUST, J., 1980: Floral sex ratios in andromonoecious Umbelliferae. — New Phytol. 85, 265–273.

MABBERLEY, D. J., 1974: Pachycauly, vessel-elements, islands and the evolution of arborescence in "herbaceous" families. — New Phytol. 73, 977–984.

- 1982: On Dr. CARLQUIST's defence of paedomorphosis. - New Phytol. **90**, 751-755.

MAHESHWARI, P., 1950: An introduction to the embryology of angiosperms. - New York, Toronto, London, McGraw-Hill.

MANTELL, D. E., 1985: The Afro-South-west Asiatic genus *Kohautia* CHAM. & SCHLECHTD. (*Rubiaceae-Rubioideae-Hedyotideae*): Morphology, anatomy, taxonomy, phytogeography and evolution. - Diss. Form. Naturwiss. Fak. Univ. Wien (unpubl.).

MARTIN, A. R. H., 1966: Plant ecology of the Grahamstown Nature Reserve. II. Some effects of burning. - J. S. Afr. Bot. **32**, 1-40.

MCKAY, H. M., 1943: Sketch map of BURCHELL's trek. - J. S. Afr. Bot. **9**, 27-78.

MEAGHER, T. R., 1980: Population biology of *Chamaelirium luteum*, a dioecious lily. I. Spatial distributions of males and females. - Evolution **34**, 1127-1137.

MELAMPY, M. N., 1981: Sex-linked niche differentiation in two species of *Thalictrum*. - Am. Midl. Nat. **106**, 325-334.

- HOWE, H. F., 1977: Sex ratio in the tropical tree *Triplaris americana* (*Polygonaceae*). - Evolution **31**, 867-872.

MENDOZA-HEUER, I., 1972: Datos para la determinación de especes en el género *Phyllis* (*Rubiaceae*). - Cuad. Bot. Canar. **14-15**, 5-9.

- 1977: Die Rubiaceen der Kanarischen Inseln. - Ber. Deutsch. Bot. Ges. **90**, 211-217.

METCALFE, C. R., CHALK, L., 1950: Anatomy of the Dicotyledons. - Oxford, University Press.

MOHR, P., 1983: Ethiopian flood basalt province. - Nature **303** (5918), 577-584.

MOLL, E. J., NICHOLSON, H. B., FOSTER, R. T., STANNARD, J., 1982: Notes on the phenology of "fynbos" species from the Umtamvuna Nature Reserve, Natal. - J. Dendr. **2**, 25-28.

MOORE, S., 1902: Alabastra diversa. Part IX. Dr. RAND's Rhodesian *Rubiaceae*. - J. Bot. **40**, 250-254.

MORTON, J. K., 1972: Phytogeography of the West African mountains. - In: VALENTINE, D. H. (ed.), Taxonomy, phytogeography and evolution, 221-236. - London, New York, Academic Press.

MUIR, J., 1929: The vegetation of the Riversdale area, Cape Prov. - Mem. Bot. Surv. S. Afr. **13**, 1-82.

MUKERJI, S. K., 1936: Contributions to the autecology of *Mercurialis perennis* L. - J. Ecol. **24**, 38-81, 317-339.

MULLAN, D. P., 1933: Observations on the biology and physiological anatomy of some Indian halophytes. - J. Indian Bot. Soc. **12**, 165-187.

NAPP-ZINN, K., 1973-1974: Anatomie des Blattes. II. Blattanatomie der Angiospermen. A. Entwicklungsgeschichtliche und topographische Anatomie des Angiospermenblattes. - In: LINSBAUER, K. (ed.), Handbuch der Pflanzenanatomie. Spezieller Teil. Ed. 2, **8** (2 A). - Berlin, Stuttgart, Gebr. Borntraeger.

NETOLITZKY, F., 1905: Bestimmungsschlüssel und mikroskopische Beschreibung der einheimischen Dicotyledonenblätter. Kennzeichen der Gruppe: Raphidenkristalle. - Wien, Perles.

- 1926: Anatomie der Angiospermen-Samen. - In: LINSBAUER, K. (ed.), Handbuch der Pflanzenanatomie. **10**, 2 (2). - Berlin, Gebr. Borntraeger.

NEUBAUER, H. F., 1981: Der Knotenbau einiger *Rubiaceae*. - Pl. Syst. Evol. **139**, 103-111.

NIGTEVECHT, G., VAN, 1966: Genetic studies in dioecious *Melandrium*. I. Sex-linked and sex-influenced inheritance in *Melandrium album* and *Melandrium dioicum*. — Genetica **37**, 281 – 306.

NONTCHEFF, P., 1909: Recherches sur l'anatomie des feuilles du genre *Cliffortia*. — Bull. Inst. Bot. Univ. Genève **8**, 2, 1 – 96.

NORDBORG, G., 1967: The genus *Sanguisorba* section *Poterium*. Experimental studies and taxonomy. — Opera Bot. **16**, 1 – 166.

NORDENSTAM, B., 1980: The herbaria of LEHMANN and SONDER in Stockholm, with special reference to the ECKLON & ZEYHER collection. — Taxon **29**, 279 – 291.

NOWICKE, J. W., SKVARLA, J. J., 1979: Pollen morphology: the potential influence in higher order systematics. — Ann. Missouri Bot. Gard. **66**, 633 – 700.

OLIVER, W. R. B., 1935: The genus *Coprosma*. — Bernice P. Bishop Museum Bull. **132**, 1 – 207.

OPLER, P. A., BAWA, K. S., 1978: Sex ratios in tropical forest trees. — Evolution **32**, 812 – 821.

ORTH, R., 1940: Vegetationsbilder aus Zentralafrika und Angola. — In: WALTER, H. (ed.), Vegetationsbilder **25**, 8, plates 43 – 48. — Jena, Fischer.

PARIS, R. R., JACQUEMIN, H., 1975: Sur quelques Rubiacées malgaches. — Pl. méd. phytothér. **9**, 118 – 124.

PATEL, J. D., 1979: A new morphological classification of stomatal complexes. — Phytomorph. **29**, 218 – 229.

PERRIER DE LA BÂTHIE, H., 1936: Biogéographie des plantes de Madagascar. — Paris, Soc. d'Edit. Géogr., Maritimes et Coloniales.

PETIT, E., 1964: *Rubiaceae* africanae XIII. Le mode de ramification chez certaines Rubiacées et sa signification pour la systématique. — Bull. Jard. Bot. État **34**, 527 – 535.

PHIPPS, J. B., GOODIER, R., 1962: A preliminary account of the plant ecology of the Chimanimani Mountains. — J. Ecol. **50**, 291 – 319.

PIJL, L., VAN DER, 1982: Principles of dispersal in higher plants. 3rd Ed. — Berlin, Heidelberg, New York, Springer.

PLACK, A., 1957: Sexual dimorphism in *Labiatae*. — Nature **180**, 1218 – 1219.

— 1958: Effect of gibberellic acid on corolla size. — Nature **182**, 610.

PÓCS, T., 1982: The forest flora and vegetation of the old crystalline mountains of Tanzania and their importance for soil and water conservation. — Seminar/Workshop of Forest Conservation in Tanzania, Tanga; mimeogr., 5 pp.

PRAGLOWSKI, J., 1973: The pollen morphology of the *Theligonaceae* with reference to taxonomy. — Pollen et Spores **15**, 385 – 396.

PUFF, C., 1979a: The distribution of *Galium* L. (*Rubiaceae*) in Africa South of the Sahara. — Bull Jard. Bot. Nat. Belg. **49**, 361 – 382.

— 1979b: *Rubiaceae*. — Distr. Pl. Afr. **17**, 563 – 588.

— 1981: Studies in *Otiophora* ZUCC. (*Rubiaceae*): 2. A new classification of the southern African taxa — J. S. Afr. Bot. **47**, 297 – 329.

— 1982a: The delimitation of the tribe *Anthospermeae* and its affinities to the *Paederieae* (*Rubiaceae*). — Bot. J. Linn. Soc. **84**, 355 – 377.

— 1982b: Proposal to reject the name *Anthospermum ciliare* L. (1763) (*Rubiaceae*). — Taxon **31**, 759 – 760.

— 1986: The relationships and affinities of the Madagascan *Anthospermum* species (*Rubiaceae*). — Ann. Missouri Bot. Gard. (in prep.; Proc. 11th AETFAT Congress, June 1985).

— MANTELL, D. E., 1982a: The tribal position and affinities of *Crocyllis* (*Rubiaceae*). — Bot. Jahrb. Syst. **103**, 89 – 106.

– – 1982b: Revision and affinities of *Galium* (*Rubiaceae*) in Madagascar. – Pl. Syst. Evol. **140**, 57 – 73.

– ROBBRECHT, E., RANDRIANASOLO, V., 1984: Observations on the SE African – Madagascan genus *Alberta* and its ally *Nematostylis* (*Rubiaceae, Alberteae*), with a survey of the species and a discussion of the taxonomic position. – Bull. Jard. Bot. Nat. Belg. **54**, 293 – 366.

PUTWAIN, P. D., HARPER, J. L., 1972: Studies in the dynamics of plant populations. V. Mechanisms governing the sex ratio in *Rumex acetosa* and *R. acetosella*. – J. Ecol. **60**, 113 – 129.

QARER, M., 1973: Die Gattung *Gaillonia* A. RICH. (*Rubiaceae-Anthospermeae*) und ihre Differenzierung im Saharo-Sindischen und Irano-Turanischen Raum. – Diss. phil. Fak. Univ. Wien (unpubl.).

QUÉZEL, P., 1969: Flore et végétation des plateaux du Darfur Nord-occidental et du jebel Gourgeil (Rép. du Soudan). – CNRS. Dossier 5 de la Recherche Coopérative sur Programme 45, Populations anciennes et actuelles des confins Tchado-Soudanais.

RAUH, W., 1973: Über die Zonierung und Differenzierung der Vegetation Madagaskars. – Trop. und subtrop. Pflanzenwelt **1**, 1 – 146.

RAVEN, P. H., AXELROD, D. I., 1974: Angiosperm biogeography and past continental movements. – Ann. Missouri Bot. Gard. **61**, 539 – 673.

RICHARDSON, G. R., LUBKE, R. A., JACOT GUILLARMOD, A., 1984: Regeneration of grassy fynbos near Grahamstown (eastern Cape) after fire. – S. Afr. J. Bot. **3**, 153 – 162.

ROBBRECHT, E., 1979: The African genus *Tricalysia* A. RICH. (*Rubiaceae-Coffeeae*). 1. revision of the species of subgenus *Empogona*. – Bull. Jard. Bot. Nat. Belg. **49**, 239 – 360.

– 1982: Pollen morphology of the tribes *Anthospermeae* and *Paederieae* (*Rubiaceae*) in relation to taxonomy. – Bull. Jard. Bot. Nat. Belg. **52**, 349 – 366.

– 1985: Further observations on the pollen morphology of the South African genus *Carpacoce* (*Rubiaceae-Anthospermeae*). – Rev. Palaeobot. Palynol. **45**, 361 – 371.

– PUFF, C., 1981: *Mericocalyx* Bamps – synonymous with *Otiophora* ZUCC. (*Rubiaceae*). – Bull. Jard. Bot. Nat. Belg. **51**, 143 – 151.

– – 1986: A survey of the *Gardenieae* and related tribes (*Rubiaceae*). – Bot. Jahrb. Syst. (in print).

ROBERTS, B. R., 1966: Observations on the temperate affinities of the vegetation of Hangklip Mountain near Queenstown, Cape Province. – J. S. Afr. Bot. **32**, 243 – 260.

ROURKE, J. P., 1972: Taxonomic studies on *Leucospermum* R. BR. – J. S. Afr. Bot., Suppl. **8**, 1 – 194.

ROUSSEAU, J., 1953: Localisation des cellules sécrétices chez quelques Rubiacées. – Bull. Soc. Bot. Fr. **100**, 36 – 39.

SALTER, T. M., 1937: Notes on some species in the family *Rubiaceae* in the Cape Peninsula. – J. S. Afr. Bot. **3**, 109 – 116.

– 1950: *Rubiaceae*. – In: ADAMSON, R.S., SALTER, T.M. (eds.), Flora of the Cape Peninsula. – Cape Town, Johannesburg, Juta.

SCHAFFNER, J. H., 1935: Observations and experiments on sex in plants. – Bull. Torrey Bot. Cl. **62**, 387 – 401.

SCHLECHTENDAL, D. F. L. VON, 1826: Plantarum capensium descriptiones ex schedis derelictis BERGIANIS. Linnaea **1**, 250 – 258.

SCHUMANN, K., 1902: *Rubiaceae.* − In: ENGLER, A. (ed.), Beiträge zur Flora von Afrika XXII. Berichte über die botanischen Ergebnisse der Nyassa-See- und Kinga-Gebirge-Expedition der Hermann-und-Elise-(geb. Heckmann-) Wentzel-Stiftung. IV. Die von W. GOETZE am Rukwa-See und Nyassa-See sowie zwischen beiden Seen gelegenen Gebirgsländern, insbesondere dem Kinga-Gebirge gesammelten Pflanzen, nebst einigen Nachträgen (durch * gekennzeichnet) zu Bericht III. − Engl. Bot. Jahrb. **30**, 239−445.

SKEAD, C. J., 1973: Zoo-historical gazetteer. − Ann. Cape Prov. Mus. **10**, 1−259.

SKOTTSBERG, C., 1922: The phanerogams of the Juan Fernandez Islands. − Natural History of Juan Fernandez & Easter Island **2**, 95−240. Uppsala, Almqvist & Wiksell.

− 1944: On the flower dimorphism in Hawaiian *Rubiaceae.* − Arkiv f. Bot. **31 A**, 4, 1−28.

− 1955: Chromosome numbers of Hawaiian flowering plants. − Arkiv f. Bot. **3**, 63−70.

SOLEREDER, H., 1893: Ein Beitrag zur anatomischen Charakteristik und zur Systematik der Rubiaceen. − Bull. Herb. Boissier **1**, 167−183, 269−286, 309−326.

− 1899: Systematische Anatomie der Dicotyledonen. Ein Handbuch für Laboratorien der wissenschaftlichen und angewandten Botanik. − Stuttgart, Enke.

SONDER, O. W., 1865: *Rubiaceae.* − In: HARVEY, W. H., SONDER, O. W. (ed.), Flora Capensis **3**, 1−39. − Dublin.

STRAKA, H., 1960: Über Moore und Torf auf Madagaskar und den Maskarenen. − Erdkunde (Bonn) **14**, 2, 81−98.

SUESSENGUTH, K., MERXMÜLLER, H., 1951: A contribution to the flora of the Marandellas district, Southern Rhodesia. − Trans. Rod. Sc. Ass. **43**, 1−86.

SUNDING, P., 1979: Origins of the Macaronesian flora. − In: BRAMWELL, D. (ed.), Plants and islands, 13−40. − London, New York, Academic Press.

TANKARD, A. J., ROGERS, J., 1978: Late Cenozoic palaeoenvironments on the west coast of Southern Africa. − J. Biogeogr. **5**, 319−337.

TAYLOR, A. O., 1964: A biochemical approach to some taxonomic problems in the genus *Coprosma.* − New Phyt. **63**, 135−139.

TAYLOR, B. W., 1954: An example of long distance dispersal. − Ecology **35**, 569−572.

TAYLOR, H. C., 1972: Notes on the vegetation of the Cape Flats. − Bothalia **10**, 637−646.

− 1978: Capensis. − In: WERGER, M. J. A. (ed.), Biogeography and ecology of southern Africa, 171−229. − The Hague, Junk.

− 1979: Observations on the flora and phytogeography of Rooiberg, a dry fynbos mountain in the southern Cape Province, South Africa. − Phytocoenologia **6**, 524−531.

− 1980: Phytogeography of fynbos. − Bothalia **13**, 231−235.

THORNE, R. F., 1978: Plate tectonics and angiosperm distribution. − Notes Roy. Bot. Gard. Edinburgh **36**, 297−315.

TROLL, W., 1964: Die Infloreszenzen, Bd. 1. − Jena, Fischer.

− 1969: Die Infloreszenzen, Bd. 2 (1). − Stuttgart, Fischer.

ULBRICH, E., 1934: *Thelygonaceae.* − In: ENGLER, A., PRANTL, K. (ed.), Die natürlichen Pflanzenfamilien, 2. Aufl.. **16 c**, 368−378. − Leipzig, Engelmann.

VAN DAMME, J. M. M., VAN DELDEN, W., 1982: Gynodioecy in *Plantago lanceolata* L. 1. Polymorphism for plasmon type. − Heredity **49**, 303−318.

VAN DER MERWE, P., 1966: Die flora van Swartboskloof, Stellenbosch en die herstel von die soorte na 'n brand. − Ann. Univ. Stellenbosch, Vol. **41**, Ser. A, **14**.

VAN DER PIJL: see PIJL, ...

VAN NIGTEVECHT: see NIGTEVECHT, ...

VAN STEENIS, C. G. G. J., 1962: The land-bridge theory in botany. — Blumea 11, 235–372.

VENTER, H. J. T., BEUKES, G. J., 1982: A new species of *Aloe* (*Liliaceae*) from South Africa. — Kew Bull. 36, 675–678.

VERDCOURT, B., 1958: Remarks on the classification of the *Rubiaceae*. — Bull. Jard. Bot. Brux. 28, 209–290.

— 1975: Studies in the *Rubiaceae-Rubioideae* for the "Flora of Tropical East Africa": I. — Kew Bull. 30, 247–326.

— 1976: *Rubiaceae*, Part 1. — In: POLHILL, R. M. (ed.), Flora of Tropical East Africa, 1–414. — London, Crown Agents.

— 1983: The identity of *Anthospermum mazzocchii-alemannii* CHIOV. (*Rubiaceae*). — Kew Bull. 38, 196.

VOLLESEN, K., 1981: *Catunaregam pygmaea* (*Rubiaceae*) — a new geoxylic suffrutex from the woodlands of SE Tanzania. — Nordic J. Bot. 1, 735–740.

VON SCHLECHTENDAL: see SCHLECHTENDAL, ...

WAGENITZ, G., 1964: *Rubiaceae*. — In: MELCHIOR, H. (ed.), A. ENGLERS Syllabus der Pflanzenfamilien, Bd. 2, ed. 12, 417–422. — Berlin, Borntraeger.

WALLACE, C. S., RUNDEL, P. W., 1979: Sexual dimorphism and resource allocation in male and female shrubs of *Simmondsia chinensis*. — Oecologia (Berlin) 44, 34–39.

WEBERLING, F., 1977: Beiträge zur Morphologie der Rubiaceen-Infloreszenzen. — Ber. Deutsch. Bot. Ges. 90, 191–209.

WEIMARCK, H., 1934: Monograph of the genus *Cliffortia*. — Lund, Ohlsson.

— 1941: Phytogeographical groups, centres and intervals within the Cape flora. — Lunds Univ. Arsskr. N. F., Avd. 2, 37 (5), 1–143.

WERGER, M. J. A., 1978a: Biogeographical divisions of southern Africa. — In: WERGER, M. J. A. (ed.), Biogeography and ecology of southern Africa, 145–170. — The Hague, Junk.

— 1978b: The Karoo-Namib Region. — In: WERGER, M. J. A. (ed.), Biogeography and ecology of southern Africa, 231–299. — The Hague, Junk.

— COETZEE, B. J., 1978: The Sudano-Zambezian Region. — In: WERGER, M. J. A. (ed.), Biogeography and ecology of southern Africa, 301–462. — The Hague, Junk.

WHITE, F., 1965: The savanna woodlands of the Zambezian and Sudanian Domains: an ecological and phytogeographical comparison. — Webbia 19, 651–681.

— 1971: The taxonomic and ecological basis of chorology. — Mitt. Bot. Staatssamml. München 10, 91–112.

— 1976: The underground forests of Africa: a preliminary review. — Garden's Bull. (Singapore) 29, 57–71.

— 1978: The Afromontane Region. — In: WERGER, M. J. A. (ed.), Biogeography and ecology of southern Africa, 463–513. — The Hague, Junk.

— 1983: The vegetation of Africa. A descriptive memoir to accompany the UNESCO/AETFAT/UNSO vegetation map. UNESCO Natural Resources Research 20, 1–356. — Paris.

WICKENS, G. E., 1976: The flora of Jebel Marra (Sudan Republic) and its geographical affinities. — Kew Bull. Add. Ser. 5, 1–368.

WILD, G. V., ZOTOV, V. D., 1930: Notes on the sexual expression in certain species of New Zealand *Coprosmas*. — Trans. New Zeal. Inst. 60, 547–556.

532 References

WILD, H., 1964: The endemic species of the Chimanimani Mountains and their significance. − Kirkia **4**, 125−157.

WILSON, R. D., 1979: Chemotaxonomic studies in the *Rubiaceae*. 1. Methods for the identification of hybridization in the genus *Coprosma* J. R. et G. FORST. using flavonoids. − New Zeal. J. Bot. **17**, 113−116.

− 1984: Chemotaxonomic studies in the *Rubiaceae*. 2. Leaf flavonoids of New Zealand Coprosmas. − New Zeal. J. Bot. **22**, 195−200.

WOODHOUSE, R. P., 1971: Hayfever plants. 2nd Ed. − New York, Hafner.

WUNDERLICH, R., 1971: Die systematische Stellung von *Theligonum*. − Österr. Bot. Z. **119**, 329−394.

Index of Taxa[1]

Ambraria CRUSE 394
Ambraria HEISTER ex FABRICIUS 174
Ambraria acerosa SONDER 410, 417
Ambraria glabra CRUSE 417
 − var. *papillata* SONDER 417
 − var. *tulbaghica* SONDER 417
Ambraria hirta CRUSE 414
 − var. *macrocarpa* ECKLON &
  ZEYHER 421
Ambraria microphylla SONDER 398
Anthospermum L. 174
Anthospermum aberdaricum
  K. KRAUSE 196
Anthospermum aethiopicum L. 227
 − var. *alpinum* ECKLON &
  ZEYHER 234
 − var. *ciliare* (L.) O. KUNTZE 385
 − var. *ecklonianum* CRUSE 251
 − var. *montanum* SONDER 234
 − var. *oppositifolium* CRUSE 234
 − var. *papillatum* (SONDER)
  O. KUNTZE 354
 − var. *reflexifolium* O. KUNTZE 360
 − var. *ternifolium* CRUSE 227
 − var. *tulbaghense* ECKLON &
  ZEYHER 255
 − var. *uitenhagense* ECKLON &
  ZEYHER 245
Anthospermum albohirtum MILDBR.
  268
Anthospermum ambiguum GREVES
  225
Anthospermum ambrosiacum MOENCH
  227
Anthospermum ammannioides
  S. Moore 209
Anthospermum ammannioides sensu
  auctt. Afr. austr., non
  S. MOORE 190
Anthospermum andringitrense
  HOMOLLE 217

Anthospermum antsirabense
  HOMOLLE 328
Anthospermum arenicola GREVES 275
Anthospermum aromaticum SALISB.
  227
Anthospermum asperuloides HOOK. f.
  294
Anthospermum basuticum PUFF 262
Anthospermum bergianum CRUSE 384
Anthospermum bicorne PUFF 392
Anthospermum burkei SONDER 274
(*Anthospermum calycophyllum*
  SONDER) 465
Anthospermum cameroonense HUTCH.
  & DALZ. 294
Anthospermum ciliare L. 385
 − var. *angustifolium* ECKLON &
  ZEYHER 353
 − var. *glabrifolium* SONDER 353
 − var. *latifolium* ECKLON &
  ZEYHER 353
 − var. *papillatum* SONDER 353
 − var. *scabrum* ECKLON & ZEYHER 353,
  360
Anthospermum ciliare sensu auctt., non
  L. 343, 349
Anthospermum cliffortioides
  K. SCHUM. 190
Anthospermum comptonii PUFF 373
Anthospermum confertum CRUSE 265
(*Anthospermum crocyllis* SONDER)` 465
Anthospermum dregei SONDER 368
 − **subsp. dregei** 369
 − **subsp. ecklonis** (SONDER) PUFF 370
Anthospermum ecklonis SONDER 370
Anthospermum emirnense BAKER 206
Anthospermum erectum SUESSENG.
  285
Anthospermum ericifolium (LICHTEN-
  STEIN ex ROEMER & SCHULTES) O.
  KUNTZE 389

[1] In Part D. (p. 173–466). **Bold**: recognized taxa; in brackets: taxa excluded from *Anthosperminae*.

Anthospermum ericoideum K. KRAUSE 329

Anthospermum esterhuysenianum PUFF 377
- **var. esterhuysenianum** 378
- **var. hirsutum** PUFF 379

Anthospermum ferrugineum ECKLON & ZEYHER 301

Anthospermum foetidum ECKLON 449

Anthospermum frutescens DINTER 329

Anthospermum galioides RCHB. 349
- subsp. **galioides** 353
- subsp. **reflexifolium** (O. KUNTZE) PUFF 360

Anthospermum galopina THUNB. 430

Anthospermum galpinii SCHLECHTER 220

Anthospermum hedyotideum SONDER 301

Anthospermum herbaceum L.f. 300
- var. *villosicarpum* VERDC. 292

Anthospermum hirsutum DC. 380

Anthospermum hirsutum sensu A. RICH., non DC. 321

Anthospermum hirtum CRUSE 380

Anthospermum hispidulum E. MEYER ex SONDER 274

(*Anthospermum holtzii* K. SCHUM.) 465

Anthospermum humile N. E. BR. 329

Anthospermum ibityense PUFF 298

Anthospermum isaloense HOMOLLE ex PUFF 218

Anthospermum keilii K. KRAUSE 196

Anthospermum lanceolatum THUNB. 301
- var. *hedyotideum* (SONDER) O. KUNTZE 301
- var. *latifolium* SONDER 301

Anthospermum latifolium E. MEYER 301

Anthospermum leuconeuron K. SCHUM. 196

Anthospermum lichtensteinii CRUSE 389

Anthospermum littoreum L. BOLUS 224

Anthospermum longisepalum HOMOLLE ex PUFF 299

Anthospermum madagascariense HOMOLLE ex PUFF 215
- var. *australe* HOMOLLE 215

- var. *humile* HOMOLLE 215
- var. *laxum* HOMOLLE 215
- var. *macrocarpum* HOMOLLE 215

(*Anthospermum mazzocchii-alemannii* CHIOV.) 465

Anthospermum mildbraedii K. KRAUSE 302

Anthospermum monticola PUFF 257

Anthospermum muriculatum HOCHST. ex A. RICH. 301

Anthospermum nodosum E. MEYER 302

Anthospermum pachyrrhizum HIERN 321

Anthospermum palustre HOMOLLE ex PUFF 324

Anthospermum paniculatum CRUSE 264
- var. *confertum* ECKLON & ZEYHER 264
- var. *elongatum* ECKLON & ZEYHER 265

Anthospermum perrieri HOMOLLE ex PUFF 280

(*Anthospermum plicatum* HILSENB. & BOJER ex BAKER) 465

(*Anthospermum polyacanthum* BAKER) 465

Anthospermum prittwitzii K. SCHUM. & K. KRAUSE 196

Anthospermum prostratum SONDER 364
- var. *glabrum* SONDER 364
- var. *velutinum* SONDER 365

Anthospermum pumilum SONDER 329
- subsp. **pumilum** 330
- subsp. **rigidum** (ECKLON & ZEYHER) PUFF 342
- var. *pilosum* PHILLIPS 329

Anthospermum randii S. MOORE 285

Anthospermum rigidum ECKLON & ZEYHER 342

Anthospermum rigidum sensu auctt. Afr. austr., non ECKLON & ZEYHER 329

Anthospermum rosmarinus K. SCHUM. 289

Anthospermum rubiaceum RCHB. 381

Anthospermum rubricaule K. SCHUM. 275

Anthospermum scabrum THUNB. 456

Anthospermum spathulatum Sprengel 232
 − subsp. *ecklonianum* (Cruse) Puff 251
 − subsp. *saxatile* Puff   248
 − subsp. *spathulatum*   234
 − subsp. *tulbaghense* Puff   255
 − subsp. *uitenhagense* Puff   245
 − var. *ecklonianum* (Cruse) Cruse 251
Anthospermum spermacoceum Rchb. 449
Anthospermum spicatum Suesseng. 329
Anthospermum streyi Puff   347
Anthospermum ternatum Hiern   281
 − subsp. *randii* (S. Moore) Puff   285
 − subsp. *ternatum*   283
Anthospermum thymifolium Dinter & K. Krause   369
Anthospermum thymoides Baker   326
 − subsp. *antsirabense* Puff   328
 − subsp. *thymoides*   327
Anthospermum tricostatum Sonder 235
Anthospermum unisetum Hochst.   321
Anthospermum usambarense K. Schum.   196
Anthospermum uwembae Gilli   190
Anthospermum vallicola S. Moore 212
Anthospermum villosicarpum (Verdc.) Puff   292
Anthospermum viscosum Webb   466
Anthospermum welwitschii Hiern   190
Anthospermum whyteanum Britten 268
Anthospermum zimbabwense Puff 201
Carpacoce Sonder   446
Carpacoce burchellii Puff   459
Carpacoce curvifolia Puff   454
Carpacoce gigantea Puff   453
Carpacoce heteromorpha (Buek) L. Bolus   463
Carpacoce scabra (Thunb.) Sonder 456
 − subsp. *rupestris* Puff   458
 − subsp. *scabra*   456
Carpacoce spermacocea (Rchb.) Sonder   449

 − subsp. *orientalis* Puff   451
 − subsp. *spermacocea*   449
Carpacoce vaginellata Salter   460
Cliffortia acerosa MS.   417
Cliffortia cinerea Thunb.   404
Cliffortia spicata Rchb.   385
Galopina Thunb.   427
Galopina aspera (Ecklon & Zeyher) Walp.   441
Galopina circaeoides Thunb.   430
 − var. *glabra* O. Kuntze   430
 − var. *pubescens* O. Kuntze   430
Galopina crocyllioides Bär ex Schinz   444
Galopina hirsuta E. Meyer   437
Galopina oxyspermum Steud.   441
Galopina tomentosa Hochst.   437
Lagotis E. Meyer   446
Lagotis spermacocea (Rchb.) E. Meyer   449
Merciera heteromorpha Buek   463
Nenax Gaertn.   393
Nenax acerosa Gaertn.   416
 − subsp. *acerosa*   418
 − subsp. *macrocarpa* (Ecklon & Zeyher) Puff   421
Nenax acerosa sensu Ecklon & Zeyher, non Gaertn.   410
Nenax arenicola Puff   425
Nenax cinera (Thunb.) Puff   404
Nenax coronata Puff   408
Nenax divaricata Salter   410
Nenax dregei L. Bolus   404
Nenax elsieae Puff   412
Nenax glabra (Cruse) O. Kuntze   417
Nenax hantamensis Schlechter   404
Nenax hirta (Cruse) Salter   413
 − subsp. *calciphila* Puff   415
 − subsp. *hirta*   414
Nenax microphylla (Sonder) Salter 398
Nenax namaquensis Puff   406
Nenax sp. A   426
Nenax sp. B   427
Oxyspermum Ecklon & Zeyher   427
Oxyspermum asperum Ecklon & Zeyher   441
Phyllis sensu Cruse, non L.   427
Phyllis galopina Cruse   430
Spermacoce ericaefolia Lichtenstein ex Roemer & Schultes   389